Electronic Devices and Circuit Theory

THIRD EDITION

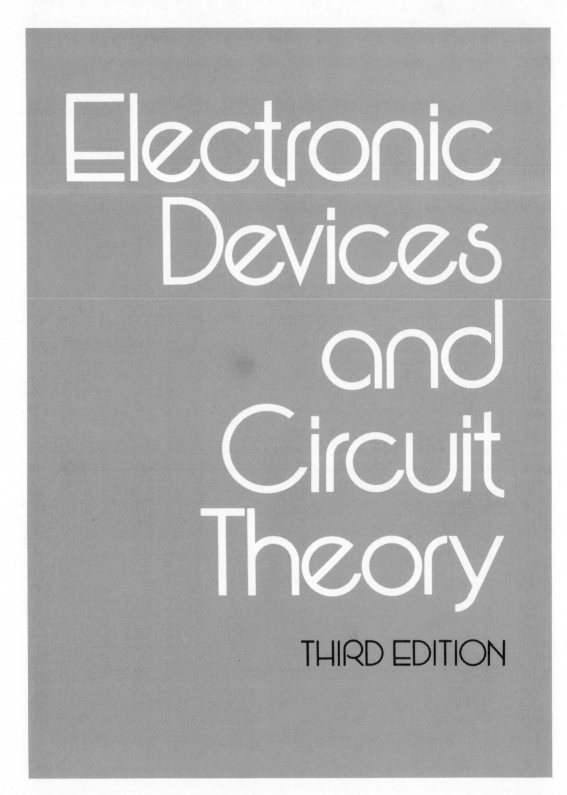

Electronic Devices and Circuit Theory

THIRD EDITION

ROBERT BOYLESTAD

Thayer School, Dartmouth College

LOUIS NASHELSKY

Queensborough Community College

PRENTICE-HALL, INC., *Englewood Cliffs, New Jersey 07632*

Library of Congress Cataloging in Publication Data

BOYLESTAD, ROBERT L. (date)
  Electronic devices and circuit theory.

  Includes index.
  1. Electronic circuits. 2. Electronic apparatus and
appliances. I. Nashelsky, Louis. II. Title.
TK7867.B66 1982   621.3815   81–15779
ISBN 0–13–250324–7   AACR2

ELECTRONIC DEVICES AND CIRCUIT THEORY, Third Edition
Robert Boylestad and Louis Nashelsky

Editorial/production supervision and
  interior design by *Ros Herion*
Cover design by *Sue Behnke*
Manufacturing buyer: *Gordon Osbourne*

dedicated to:
ELSE MARIE, ERIC, ALISON, and STACEY
and to
KATRIN, KIRA, AND LARREN

Printed in the United States of America

10 9 8 7 6 5 4 3 2 1

Prentice-Hall International, Inc., *London*
Prentice-Hall of Australia Pty. Limited, *Sydney*
Prentice-Hall of Canada, Ltd., *Toronto*
Prentice-Hall of India Private Limited, *New Delhi*
Prentice-Hall of Japan, Inc., *Tokyo*
Prentice-Hall of Southeast Asia Pte. Ltd., *Singapore*
Whitehall Books Limited, *Wellington, New Zealand*

Contents

Preface

This text is designed primarily for use in a two-semester or three-trimester sequence in the basic electronics area. It is expected that the student has taken a course in dc circuit analysis and has either taken or is taking a course in ac circuit analysis. This text requires only a mathematical background similar to that required for the ac circuit analysis course.

In an effort to aid the student, the text contains extensive examples that stress the main points of each chapter. There are also numerous illustrations to guide the student through the new concepts and techniques. Important conclusions are emphasized by boxed equations or boldface answers to make the student aware of the essential points covered.

The text is a result of a two-semester electronics course sequence that both authors were actively involved in teaching over a period of years. However, the 18 chapters actually contain more material than can be covered in two 15-week semesters (or three 10-week trimesters). This preface will suggest how the authors feel the material can be organized best.

This third edition was necessary for a number of reasons. Quite obviously, it is necessary to introduce the advances made in industry since 1978, when the second edition was published. Furthermore, the increased use of integrated circuits requires that these areas receive broader coverage in an introductory text of this nature. The third edition is also an opportunity to react to suggestions and criticisms received during the last few years. The problem set has been carefully reviewed and modified to exhibit a more practical orientation. The text and solutions manual have been reviewed very carefully to ensure that the third edition is as accurate as physically possible. Computer solutions to various problems have been added to match current trends and to reinforce some of the conclusions appearing in the text. In total, the third edition reflects the authors' efforts to ensure that the content is sensitive to current trends and advances.

Essentially, the first eight chapters provide the basic background to electronic devices—including construction, biasing, and operation as single stages. The material in these chapters can be included in the first semester with the option left to the teacher of stressing some areas more than others, or some not at all. The course should begin with the theory and operation of the two-terminal semiconductor diode. Since the theory course is usually taught in conjunction with a laboratory course, the material has been organized with regard to providing practical circuit examples for examination in the laboratory.

Chapter 2 includes a detailed description of a number of two-terminal devices such as the Zener diode, LED, LCD, solar cell, Schottky diode, etc., that have become increasingly important in recent years. There was a very serious attempt to expand on the coverage to ensure that some practical considerations associated with their use are fully understood and appreciated.

The construction and theory of operation of the BJT transistor is covered in detail in Chapter 3. The operation of the transistor is presented both mathematically and graphically; the amplifying action of the transistor is defined and demonstrated. Actual current directions are used in this introductory area, since both authors feel that the material is better absorbed using this approach. The investigation of the data sheet was expanded to ensure its readability and impact on the application of the device.

It is the authors' experience that the student can better comprehend the operation of the BJT transistor device if, initially, the dc bias and ac operation are treated separately. Thus, Chapter 4 deals only with the dc bias of the BJT transistor. This is done for common-emitter, common-base, and common-collector (emitter follower) configurations for a variety of bias current types. Numerous examples help to demonstrate the theory presented. Also, some design problems are included to provide a well-rounded treatment.

If possible, the material on the field-effect transistor (FET) should also be covered in the first semester of electronics. After the presentation and development of the concepts of dc bias of the BJT, Chapter 6 then covers a number of practical FET circuits. We had considered including the FET dc biasing in Chapters 4 and 5 and ac analysis in Chapter 6. It was our feeling from classroom experience that this would require spending too much time on each topic, and the FET would appear to be a minor device to the student. By covering the FET in a separate chapter, its significance is stressed, and its operation can be properly presented. The chapter has been extensively revised to include graphical techniques that permit the student to obtain directly dc bias levels for any FET device.

Chapter 7 is one of the most important in the basic coverage area and should be given sufficient time in any course. The development of the BJT transistor ac equivalent circuit model is covered in detail, followed by analysis of the ac operation of the full small-signal circuit. The treatment in this chapter (as in Chapter 4) is essentially mathematical. However, the mathematics are kept short and direct, with a generous number of examples provided so that the student will be able to follow the ideas presented. The hybrid equivalent circuit of the transistor is presented, and the usual engineering simplifications are included in ac analysis to provide a more practical, meaningful treatment. This is followed by an introduction to a simplified model that has received increased interest in the analysis of BJT circuits. Computer printouts have also been added to the chapter.

Chapter 8 then presents the ac small-signal analysis of the FET with numerous examples and some computer solutions.

Chapter 9, which covers the operation of multistage BJT and FET transistor circuits, would be the first topic in the second semester. Stage loading, overall gain calculations, and use of decibels are all covered in this important chapter. A number of examples help to emphasize the main points of the chapter. Increased emphasis has been placed on use of the approximate analysis techniques for multistage amplifiers. The material on frequency has been totally revised for increased clarity. In addition, computer analysis of some configurations has been added to confirm the results obtained.

Chapter 10 covers the operation of power transistors in a few basic power amplifier circuits. Most important is the operation of the push-pull circuit. Transistor push-pull circuits containing a transformer as well as transformerless circuits are covered. Additional material on quasi-complementary push-pull amplifiers and on class-B power and efficiency is provided.

Chapter 11 is a "catch-all" of a number of *pnpn* devices—covering their construction, operation, and circuit applications. It can be covered quickly or even passed over, if desired, without loss of continuity. This edition also includes an introduction to the UJT and PUT.

The fabrication and construction of integrated circuits (ICs) is examined in Chapter 12. The content has been substantially updated as the result of a recent visit to the Phoenix branch of the Motorola Corporation and includes many of the advances made in this area. If desired, the contents of this chapter can be assigned mainly as student reading.

Chapters 13–16 deal with a range of popular linear integrated circuits—their basic fabrication, operation, and, most important, their practical application. The topics covered in these chapters are representative of the newest concepts and developments taking place in the electronics field. This material is essentially new and is a major factor for the third edition.

Chapter 17 includes much of the previous edition's material on feedback amplifiers and oscillators. Chapter 18, on CROs, is a reduced treatment of the material presented in the previous edition.

To improve on the use of this text by both student and instructor there are numerous practical examples in most chapters. Problems at the end of these chapters are keyed to the particular section in which the problems are covered. The text has also been thoroughly revised to adhere to accepted industry standards for units and drawings.

We wish to thank Professors Aidala and Katz of the Electrical Technology Department at Queensborough Community College for their continued help and encouragement over the years. We also thank Mrs. Doris Topel and Mrs. Helene Rosenberg for their past help in typing manuscripts, and Mrs. Susan Kennedy for typing parts of the third edition manuscript. We thank Mr. Marshall G. Rothen and Mr. Lothar Stern of Motorola for their generous time and help in obtaining updated material on IC fabrication. Thanks are extended to Ros Herion of Prentice-Hall for her diligent efforts to produce the best possible text and Dave Boelio of Prentice-Hall for his encouragement and support in getting this third edition out so quickly and painlessly. Special thanks go to Bob Carter for proofreading the entire manuscript and for his many excellent suggestions, which should make the text not only more readable but technically and mathematically correct. Finally, we wish to thank each other again for a remarkably pleasant and rewarding collaboration.

ROBERT BOYLESTAD
Hanover, N. H.

LOUIS NASHELSKY
Great Neck, N.Y.

Semiconductor Diodes

1.1 INTRODUCTION

The few decades following the introduction of the semiconductor transistor in the late 1940s have seen a very dramatic change in the electronics industry. The miniaturization that has resulted leaves us to wonder about its limits. Complete systems now appear on a wafer thousands of times smaller than the single element of earlier networks.

The advantages associated with semiconductor systems as compared to the tube networks of prior years are, for the most part, immediately obvious: smaller and lightweight, no heater requirement or heater loss (as required for tubes), more rugged construction, more efficient, and not requiring a warm-up period. The tube still has a few isolated areas of application (very high power and high frequencies) but the use of tubes is becoming increasingly smaller in scope.

The miniaturization of recent years has resulted in semiconductor systems so small that the primary purpose of the container is simply to provide some means of handling the device and ensuring that the leads remain properly fixed to the semiconductor wafer. The limits of miniaturization appear to be limited by three factors: the quality of the semiconductor material itself, the network design technique, and the limits of the manufacturing and processing equipment.

1.2 GENERAL CHARACTERISTICS

The label *semiconductor* itself provides a hint as to its characteristics. The prefix *semi* is normally applied to anything midway between two limits. The term *conductor*

1

is applied to any material that will permit a generous flow of charge due to the application of a limited amount of external pressure. A semiconductor, therefore, is a material that has a conductivity level somewhere between the extremes of an insulator (very low conductivity) and a conductor, such as copper, which has a high level of conductivity. Inversely related to the conductivity of a material is its resistance to the flow of charge, or current. That is, the higher the conductivity level, the lower the resistance level. In tables, the term *resistivity* (ρ, Greek letter rho) is often used when comparing the resistance levels of materials. The resistivity of a material can be examined by noting the resistance of a sample having a length of 1 cm and a cross-sectional area of 1 cm² , as shown in Fig. 1.1. Recall that the equation for the

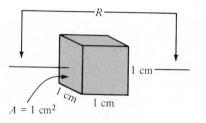

resistance of a material (at a particular temperature) is determined by $R = \rho l / A$, where R is measured in ohms, l is the length of the sample, A is its incident surface area, and ρ is the resistivity. If $l = 1$ cm and $A = 1$ cm², then $R = \rho$, as indicated above. The magnitude of the resistance of a 1-cm³ sample of the material is therefore determined by the resistivity. Or, in other words, the higher the resistivity, the more the magnitude of the resistance for such a sample. The units for ρ as defined by the equation

Figure 1.1

are as follows:

$$\rho = \frac{RA}{l} \Rightarrow \frac{\Omega \cdot \text{cm}^2}{\text{cm}} = \boxed{\Omega \cdot \text{cm}} \qquad (1.1)$$

Please be assured that this book is not heavily bent toward mathematical developments and complex algebraic techniques. The authors feel that a clear understanding of dimensions is an absolute necessity for proper development of an engineering background. We would assume that the somewhat abstract unit of measure for resistivity now has a measure of understanding and clarity. Incidentally, the actual resistance of a semiconductor material as measured above is called its *bulk* resistance. The resistance introduced by connecting the leads to the bulk material is called the *ohmic contact* resistance. These terms will appear in the description of devices to be introduced throughout.

In Table 1.1, typical resistivity values are provided for three broad categories of materials.

TABLE 1.1 Typical Resistivity Values
(At 300K—Room Temperature)

Conductor	Semiconductor	Insulator
$\rho \cong 10^{-6} \ \Omega \cdot \text{cm}$ (copper)	$\rho \cong 50 \ \Omega \cdot \text{cm}$ (germanium) $\rho \cong 50 \times 10^3 \ \Omega \cdot \text{cm}$ (silicon)	$\rho \cong 10^{12} \ \Omega \cdot \text{cm}$ (mica)

Although you may be familiar with the electrical properties of copper and mica from your past studies, the characteristics of the semiconductor materials of germanium (Ge) and silicon (Si) may be relatively new. As you will find in the chapters to follow, they are certainly not the only two semiconductor materials. They are, however, the two materials that have received the broadest range of interest in the development of semiconductor devices. In recent years the shift has been steadily toward silicon and away from germanium, but germanium is still in modest production.

Note in Table 1.1 the extreme range between the conductor and insulating materials for the 1-cm length of the material. Eighteen places separate the placement of the decimal point for one number from the other. Ge and Si have received the attention they have for a number of reasons. One very important consideration is the fact that they can be manufactured to a very high purity level. In fact, recent advances have reduced impurity levels in the pure material to 1 part in 10 billion (1:10,000,000,000). One might ask if these low impurity levels are really necessary. They certainly are if you consider that the addition of one part impurity (of the proper type) per million in a wafer of silicon material can change that material from a relatively poor conductor to a good conductor of electricity. We are obviously dealing with a whole new spectrum of comparison levels when we deal with the semiconductor medium. The ability to change the characteristics of the material significantly through this process, which is known as "doping," is yet another reason why Ge and Si have received such wide attention. Further reasons include the fact that their characteristics can be altered significantly through the application of heat or light—an important consideration in the development of heat- and light-sensitive devices.

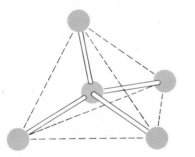

Figure 1.2 Ge and Si single crystal structure.

Some of the unique qualities of Ge and Si noted above are due to their atomic structure. The atoms of both materials form a very definite pattern that is periodic in nature (i.e., continually repeats itself). One complete pattern is called a *crystal* and the periodic arrangement of the atoms a *lattice*. For Ge and Si the crystal has the three-dimensional diamond structure of Fig. 1.2. Any material composed solely of repeating crystal structures of the same kind is called a *single-crystal* structure. For semiconductor materials of practical application in the electronics field, this single-crystal feature exists, and, in addition, the periodicity of the structure does not change significantly with the addition of impurities in the doping process.

Let us now examine the structure of the atom itself and note how it might affect the electrical characteristics of the material. As you are aware, the atom is composed of three basic particles: the *electron,* the *proton,* and the *neutron.* In the atomic lattice, the neutrons and protons form the *nucleus,* while the electrons revolve around the nucleus in a fixed *orbit.* The Bohr models of the two most commonly used semiconductors, *germanium* and *silicon,* are shown in Fig. 1.3.

As indicated by Fig. 1.3a, the germanium atom has 32 orbiting electrons, while silicon has 14 orbiting electrons. In each case, there are 4 electrons in the outermost *(valence)* shell. The potential *(ionization potential)* required to remove any one of these 4 valence electrons is lower than that required for any other electron in the structure. In a pure germanium or silicon crystal these 4 valence electrons are bonded to 4 adjoining atoms, as shown in Fig. 1.4 for silicon. Both Ge and Si are referred to as *tetravalent* atoms because they each have four valence electrons.

This type of bonding, formed by *sharing* electrons, is called *covalent bonding.*

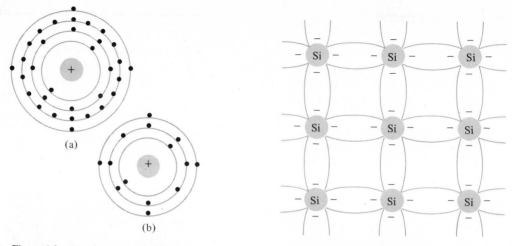

Figure 1.3 Atomic structure: (a) germanium; (b) silicon.

Figure 1.4 Covalent bonding of the silicon atom.

Although the covalent bond will result in a stronger bond between the valence electrons and their parent atom, it is still possible for the valence electrons to absorb sufficient kinetic energy from natural causes to break the covalent bond and assume the "free" state. These natural causes include effects such as light energy in the form of photons and thermal energy from the surrounding medium. At room temperature there are approximately 1.5×10^{10} free carriers in a cubic centimeter of intrinsic silicon material. *Intrinsic* materials are those semiconductors that have been carefully refined to reduce the impurities to a very low level—essentially as pure as can be made available through modern technology. The free electrons in the material due only to natural causes are referred to as *intrinsic carriers*. At the same temperature, intrinsic germanium material will have approximately 2.5×10^{13} free carriers per cubic centimeter. The ratio of the number of carriers in germanium to that of silicon is greater than 10^3 and would indicate that germanium is a much better conductor at room temperature. This may be true, but both are still considered poor conductors in the intrinsic state. Note in Table 1.1 that the resistivity also differs by a ratio of about $1000:1$, with silicon having the larger value. This should be the case, of course, since resistivity and conductivity are inversely related.

A change in the temperature of a semiconductor material can increase the number of free electrons quite substantially. As the temperature rises from absolute zero (0K), an increasing number of valence electrons absorb sufficient thermal energy to break the covalent bond and contribute to the number of free carriers as described above. This increased number of carriers will increase the conductivity index and result in a lower resistance level. Semiconductor materials such as Ge and Si that show a reduction in resistance with increase in temperature are said to have a *negative temperature coefficient*. You will probably recall that the resistance of most conductors will increase with temperature. This is due to the fact that the numbers of carriers in a conductor will not increase significantly with temperature, but their vibration pattern above a relatively fixed location will make it increasingly difficult for electrons to pass through. An increase in temperature therefore results in an increased resistance level and a *positive temperature coefficient*.

4

1.3 ENERGY LEVELS

In the isolated atomic structure there are discrete (individual) energy levels associated with each orbiting electron as shown in Fig. 1.5a. Each material will, in fact, have its own set of permissible energy levels for the electrons in its atomic structure. The more distant the electron from the nucleus, the higher the energy state, and any electron that has left its parent atom has a higher energy state than any electron in the atomic structure. Between the discrete energy levels are gaps in which no electrons in the isolated atomic structure can appear. As the atoms of a material are brought closer together to form the crystal lattice structure, there is an interaction between atoms that will result in the electrons in a particular orbit of one atom having slightly different energy levels from electrons in the same orbit of an adjoining atom. The net result is an expansion of the discrete levels of possible energy states for the valence electrons to that of bands as shown in Fig. 1.5b. Note that there are still boundary levels and maximum energy states in which any electron in the atomic lattice can find itself, and there remains a *forbidden* region between the valence band and the ionization level. Recall that ionization is the mechanism whereby an

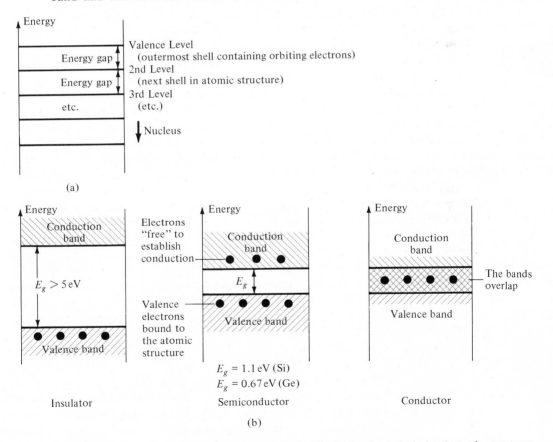

Figure 1.5 Energy levels: (a) discrete levels in isolated atomic structures; (b) conduction and valence bands of an insulator, semiconductor, and conductor.

electron can absorb sufficient energy to break away from the atomic structure and join the "free" carriers in the conduction band. You will note that energy is measured in *electron volts* (eV). The unit of measure is appropriate, since

$$W(\text{energy}) = P(\text{power}) \cdot t(\text{time})$$

but

$$P = VI$$

resulting in

$$W = VIt$$

but

$$I = \frac{Q}{t} \quad \text{or} \quad Q = It$$

and

$$\boxed{W = QV} \quad \text{joules} \tag{1.2}$$

Substituting the charge of an electron and a potential difference of 1 volt into Eq. (1.2) will result in an energy level referred to as one *electron volt*. Since energy is also measured in joules and the charge of one electron $= 1.6 \times 10^{-19}$ coulomb,

$$W = QV = (1.6 \times 10^{-19} \text{ C})(1 \text{ V})$$

and

$$\boxed{1 \text{ eV} = 1.6 \times 10^{-19} \text{ J}} \tag{1.3}$$

The small unit of measure is required to avoid reference to very small numbers in the discussion to follow.

At 0K or absolute zero, all the valence electrons of the semiconductor materials are in the valence bands. However, at room temperature (300K) a large number of electrons have acquired sufficient energy to enter the conduction band, that is, to bridge the 1.1-eV gap for silicon and 0.67 eV for germanium. The obviously lower E_g for germanium accounts for the increased number of carriers in that material as compared to silicon at room temperature. Note for the insulator that the energy gap is typically 5 eV or more. Very few electrons can acquire the required energy at room temperature, with the result that the material remains an insulator. The conductor has electrons in the conduction band even at 0K. Quite obviously, therefore, at room temperature there are more than enough free carriers to sustain a heavy flow of charge, or current.

We see in Section 1.4 that if certain impurities are added to the intrinsic semiconductor materials, the result will be permissible energy states in the forbidden band and a net reduction in E_g for both semiconductor materials—consequently increased carrier density in the conduction band at room temperature!

1.4 EXTRINSIC MATERIALS—*n*- AND *p*-TYPE

The characteristics of semiconductor materials can be altered significantly by the addition of certain impurity atoms into the relatively pure semiconductor material. These impurities, although only added to perhaps 1 part in 10 million, can alter the band structure sufficiently to totally change the electrical properties of the material. A semiconductor material that has been subjected to this *doping* process is called an *extrinsic* material. There are two extrinsic materials of immeasurable importance to semiconductor device fabrication: *n*-type and *p*-type. Each will be described in some detail in the following paragraphs.

n-Type Material

Both the *n*- and *p*-type materials are formed by adding a predetermined number of impurity atoms into a germanium or silicon base. The *n*-type is created by adding those impurity elements that have *five* valence electrons *(pentavalent)*, such as *antimony, arsenic,* and *phosphorus*. The effect of such impurity elements is indicated in Fig. 1.6 (using antimony as the impurity in a silicon base). Note that the four covalent bonds are still present. There is, however, an additional fifth electron due to the impurity atom, which is *unassociated* with any particular covalent bond. This remaining electron, loosely bound to its parent (antimony) atom, is relatively free to move within the newly formed *n*-type material. Since the inserted impurity atom has donated

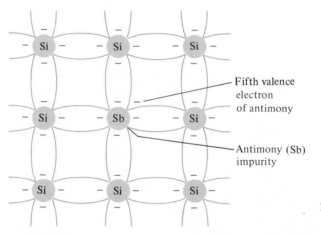

Fifth valence electron of antimony

Antimony (Sb) impurity

Figure 1.6 Antimony impurity in *n*-type material.

a relatively "free" electron to the structure, impurities with five valence electrons are called *donor* atoms. It is important to realize that even though a large number of "free" carriers have been established in the *n*-type material, it is still electrically *neutral* since ideally the number of positively charged protons in the nuclei is still equal to the number of "free" and orbiting negatively charged electrons in the structure.

The effect of this doping process on the relative conductivity can best be described through the use of the energy-band diagram of Fig. 1.7. Note that a discrete energy level (called the *donor* level) appears in the forbidden band with an E_g significantly less than that of the intrinsic material. Those "free" electrons due to the added impurity sit at this energy level and have absolutely no difficulty absorbing a sufficient measure of thermal energy to move into the conduction band at room temperature. The result is that at room temperature, there are a large number of carriers (electrons) in the conduction level and the conductivity of the material increases significantly. At room temperature in an intrinsic Si material there is about one free electron for every 10^{12} atoms (1 to 10^9 for Ge). If our dosage level were 1 in 10 million (10^7), the ratio ($10^{12}/10^7 = 10^5$) would indicate that the carrier concentration has increased by a ratio of 100,000 : 1.

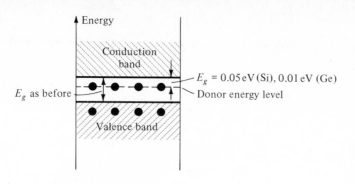

E_g as before

$E_g = 0.05\,\text{eV (Si)}, 0.01\,\text{eV (Ge)}$

Donor energy level

Figure 1.7 Effect of donor impurities on the energy band structure.

p-Type Material

The *p*-type material is formed by doping a pure germanium or silicon crystal with impurity atoms having *three* valence electrons. The elements most frequently used for this purpose are *boron, gallium,* and *indium.* The effect of one of these elements, boron, on a base of silicon is indicated in Fig. 1.8.

Note that there is now an insufficient number of electrons to complete the covalent bonds of the newly formed lattice. The resulting vacancy is called a *hole* and is represented by a small circle or positive sign due to the absence of a negative charge. Since the resulting vacancy will readily *accept* a "free" electron, the impurities added are called *acceptor* atoms. The resulting *p*-type material is electrically neutral, for the same reasons as for the *n*-type material.

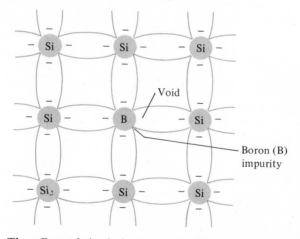

Void

Boron (B) impurity

Figure 1.8 Boron impurity in *p*-type material.

The effect of the hole on conduction is shown in Fig. 1.9. If a valence electron acquires sufficient kinetic energy to break its covalent bond and fills the void created by a hole, then a vacancy, or hole, will be created in the covalent bond that released the electron. There is therefore a transfer of holes to the left and electrons to the right, as shown in Fig. 1.9. The direction to be used in this text is that of *conventional* flow, which is indicated by the direction of hole flow.

In the intrinsic state, the number of free electrons in Ge or Si is due only to those few electrons in the valence band that have acquired sufficient energy from thermal or light sources to break the covalent bond or to the few impurities that could not be removed. The vacancies left behind in the covalent bonding structure represent our very limited supply of holes. In an *n*-type material, the number of

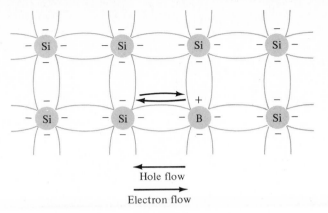

Hole flow

Electron flow

Figure 1.9 Electron vs. hole flow.

holes has not changed significantly from this intrinsic level. The net result, therefore, is that the number of electrons far outweighs the number of holes. For this reason the electron is called the *majority carrier* and the hole the *minority carrier,* as shown in Fig. 1.10a. Note in Fig. 1.10b that the reverse is true for the *p*-type material. When the fifth electron of a donor atom leaves the parent atom, the atom remaining acquires a net positive charge: hence the positive sign in the donor-ion representation. For similar reasons, the negative sign appears in the acceptor ion.

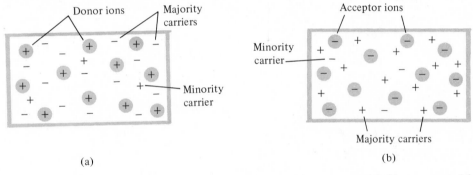

(a)

(b)

Figure 1.10 (a) *n*-type material; (b) *p*-type material.

The *n*- and *p*-type materials represent the basic building blocks of semiconductor devices. We will find later in this chapter that the joining of a single *n*-type material with a *p*-type material will result in a semiconductor element of considerable importance in electronic systems.

1.5 DRIFT AND DIFFUSION CURRENTS

The flow of charge, or current, through a semiconductor material is normally referred to as one of two types: drift and diffusion. *Drift current* relates directly to the mechanism encountered in the flow of charge in a conductor. When a voltage is applied across the material as shown in Fig. 1.11, the electrons are naturally drawn to the positive end of the sample. However, collisions with the other atoms, ions, and carriers encountered in their movement may result in an erratic path, as shown in the figure. The net result, however, is a drift of carriers to the positive end.

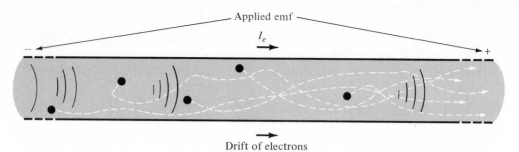

Drift of electrons

Figure 1.11 Drift current.

The concept of *diffusion current* is best described by considering the effect of placing a drop of dye into a clear pool of water. The heavy concentration of dye will eventually diffuse through the clear water. The darker color of the heavy concentration of dye will give way to a much lighter shade as it spreads out through the liquid. The same effect will take place in a semiconductor material if a heavy concentration of carriers is introduced to a region as shown in Fig. 1.12a. In time they will distribute themselves evenly through the material, as shown in Fig. 1.12b. This movement is due only to interaction between neighboring atoms; there is no applied source of energy as is required for drift current. Diffusion current is an important consideration in the examination of minority-carrier flow in *n*- and *p*-type materials. The diffusion phenomenon is also important as a technology for the doping process and must be carefully investigated when models are constructed for semiconductor devices (diffusion capacitance, etc.).

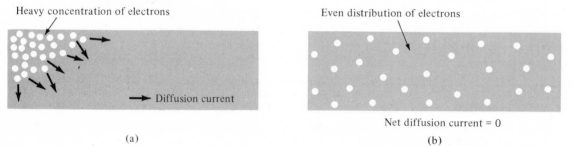

Figure 1.12 Diffusion current: (a) heavy introduction of carriers in a region of the semiconductor material; (b) the steady-state condition.

1.6 MANUFACTURING TECHNIQUES

The first step in the manufacture of any semiconductor device is to obtain semiconductor materials, such as germanium or silicon, of the desired purity level. Impurity levels of *less* than *one* part in *one billion* (1 in 1,000,000,000) are required for most semiconductor fabrications today.

The raw materials are first subjected to a series of chemical reactions and a zone-refining process to form a polycrystalline crystal of the desired purity level. The atoms of a polycrystalline crystal are haphazardly arranged, while in the single crystal desired the atoms are arranged in a symmetrical, uniform, geometrical lattice structure.

Zone-refining apparatus is shown in Fig. 1.13. It consists of a graphite or quartz boat for minimum contamination, a quartz container, and a set of RF (radio-frequency) induction coils. Either the coils or boat must be movable along the length of the quartz container. The same result will be obtained in either case, although moving coils are discussed here since it appears to be the more popular method. The interior of the quartz container is filled with either an inert (little or no chemical reaction) gas, or vacuum, to reduce further the chance of contamination. In the zone refining process, a bar of germanium is placed in the boat with the coils at one end of the bar as shown in Fig. 1.13. The radio-frequency signal is then applied to the coil, which will induce a flow of charge (eddy currents) in the germanium ingot. The magnitude of these currents is increased until sufficient heat is developed to melt that region of the semiconductor material. The impurities in the ingot will enter a more liquid state than the surrounding semiconductor material. If the induction coils of Fig. 1.13 are now slowly moved to the right to induce melting in the neighboring

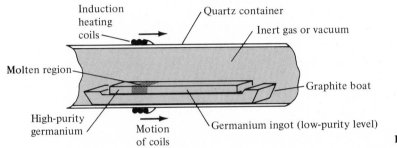

Figure 1.13   Zone refining process.

region, the "more fluidic" impurities will "follow" the molten region. The net result is that a large percentage of the impurities will appear at the right end of the ingot when the induction coils have reached this end. This end piece of impurities can then be cut off and the entire process repeated until the desired purity level is reached.

The final operation before semiconductor fabrication can take place is the formation of a single crystal of germanium or silicon. This can be accomplished using either the *Czochralski* or the *floating zone* technique, the latter being the more recently devised. The apparatus employed in the Czochralski technique is shown in Fig. 1.14a. The polycrystalline material is first transformed to the molten state by the RF induction coils. A single-crystal "seed" of the desired impurity level is then immersed in the molten germanium and gradually withdrawn while the shaft holding the seed is slowly turning. As the "seed" is withdrawn, a single-crystal germanium lattice structure will grow on the "seed" as shown in Fig. 1.14a. The resulting single-crystal ingots are typically 6 to 36 in. in length and 1 to 5 in. in diameter (Fig. 1.14b). Ingots having a length of 48 in. and a diameter of 3 in. have been grown. The weight of such a structure is about 28.5 lb.

The floating-zone technique eliminates the need for having both a zone refining and single-crystal forming process. Both can be accomplished at the same time using this technique. A second advantage of this method is the absence of the graphite or quartz boat, which often introduces impurities into the germanium or silicon ingot. Two clamps hold the bar of germanium or silicon in the vertical position within a set of movable RF induction coils as shown in Fig. 1.15. A small single-crystal "seed" of the desired impurity level is deposited at the lower end of the bar and heated

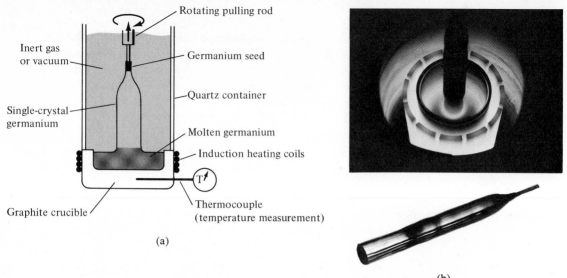

Figure 1.14 Czochralski technique and ingots. [(b) *top*, courtesy Texas Instruments Incorporated; *bottom*, courtesy Motorola Incorporated.)]

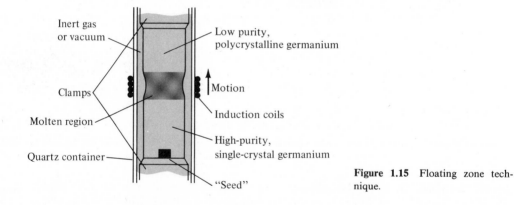

Figure 1.15 Floating zone technique.

with the germanium bar until the molten state is reached. The induction coils are then slowly moved up the germanium or silicon ingot while the bar is slowly rotating. As before, the impurities follow the molten state, resulting in an improved impurity level single-crystal germanium lattice below the molten zone. Through proper control of the process, there will always be sufficient surface tension present in the semiconductor material to ensure that the ingot does not rupture in the molten zone.

The single-crystal structure produced can then be cut into wafers sometimes as thin as $\frac{1}{1000}$ (or 0.001) of an inch ($\cong \frac{1}{5}$ the thickness of this paper). This cutting process can be accomplished using the setup of Fig. 1.16a or b. In Fig. 1.16a, tungsten wires (0.001 in. in diameter) with abrasive deposited surfaces are connected to supporting blocks at the proper spacing and then the entire system is moved back and forth as a saw. The system of Fig. 1.16b is self-explanatory.

In the next few chapters, as we introduce the various devices, we will pick up with the semiconductor wafer produced above and describe the full construction technique.

Other semiconductor materials will be introduced as their area of application is considered.

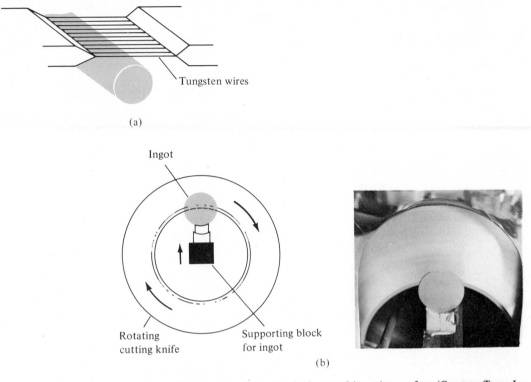

(a)

Ingot

Rotating cutting knife

Supporting block for ingot

(b)

Figure 1.16 Slicing the single-crystal ingot into wafers. (Courtesy Texas Instruments Incorporated.)

1.7 IDEAL DIODE

The semiconductor diode is one of the basic building blocks of the wide variety of electronic systems in use today. It will appear in a range of applications, extending from the simple to the very complex. In addition to the details of its construction and characteristics, we examine a few practical applications of the device. The very important data and graphs to be found on specification sheets will also be covered to ensure an understanding of the terminology employed and demonstrate the wealth of information typically available from manufacturers.

Before examining the construction and characteristics of an actual device, we first consider the ideal device, to provide a basis for comparison. The *ideal diode* is a *two-terminal* device having the symbol and characteristics shown in Fig. 1.17a and b, respectively.

In the description of the elements to follow, it is critical that the various *letter symbols, voltage polarities,* and *current directions* be defined. If the polarity of the applied voltage is consistent with that shown in Fig. 1.17a, the portion of the characteristics to be considered in Fig. 1.17b, is to the right of the vertical axis. If a reverse

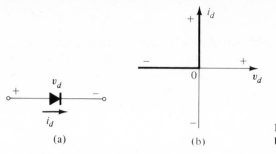

Figure 1.17 Ideal diode: (a) symbol; (b) characteristics.

voltage is applied, the characteristics to the left are pertinent. If the current through the diode has the direction indicated in Fig. 1.17a, the portion of the characteristics to be considered is above the horizontal axis, while a reversal in direction would require the use of the characteristics below the axis. For the majority of the device characteristics to appear in this text the *ordinate* will be the *current axis,* while the *abscissa* will be the *voltage axis.*

One of the important parameters for the diode is the resistance at the point or region of operation. If we consider the region defined by the direction of i_d and polarity of v_d in Fig. 1.17a (upper-right quadrant of Fig. 1.17b), we shall find that the value of the forward resistance, R_f, as defined by Ohm's law is

$$R_f = \frac{V_f}{I_f} = \frac{0}{2, 3 \text{ mA}, \ldots, \text{ or any positive value}} = 0\ \Omega$$

where V_f is the forward voltage across the diode and I_f is the forward current through the diode. *The ideal diode, therefore, is a short circuit for the forward region of conduction* ($i_d \neq 0$).

If we now consider the region of negatively applied potential (third quadrant) of Fig. 1.17b,

$$R_r = \frac{V_r}{I_r} = \frac{-5, -20, \text{ or any reverse-bias potential}}{0}$$

= very large number, which for our purposes
we shall consider to be infinite (∞)

where V_r is the reverse voltage across the diode and I_r is the reverse current in the diode. *The ideal diode, therefore, is an open circuit in the region of nonconduction* ($i_d = 0$).

In review, the conditions depicted in Fig. 1.18 are true.

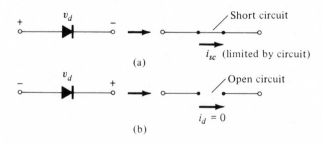

Figure 1.18 (a) Conduction and (b) nonconduction states of the ideal diode as determined by the applied bias.

In general, it is relatively simple to determine whether a diode is in the region of conduction or nonconduction by simply noting the direction of the current i_d to be established by an applied voltage. For conventional flow (opposite to that of electron flow), if the resultant diode current has the same direction as the arrowhead of the diode symbol, the diode is operating in the conducting region. This is depicted in Fig. 1.19.

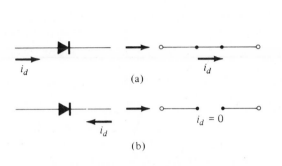

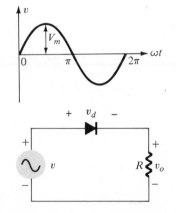

Figure 1.19 (a) Conduction and (b) non-conduction states of the ideal diode as determined by current direction of applied network.

Figure 1.20 Basic rectifying circuit.

As an introductory example of one practical application of the diode let us consider the process of rectification, by which an alternating voltage having zero average value is converted to one having a dc or average value greater than zero. The circuit required is shown in Fig. 1.20 with an ideal diode.

For the region defined by $0 \rightarrow \pi$ of the sinusoidal input voltage v, the polarity of the voltage drop across the diode would be such that the short-circuit representation would result and the circuit would appear as shown in Fig. 1.21a. For the region $\pi \rightarrow 2\pi$, the open-circuit representation would be applicable and the circuit would appear as shown in Fig. 1.21b.

For future reference, note the polarities of the input v for each circuit. For sinusoidal inputs the polarity indicated will be for the positive portion of the sinusoidal waveform, as shown in Fig. 1.20. For the situation shown in Fig. 1.21a, the output voltage v_o, will appear exactly the same as the input voltage, v, as long as the diode is forward-biased. In Fig. 1.21b, because of the open-circuit representation of the ideal diode, the output voltage v_o equals zero from π to 2π of the impressed voltage v. The complete resultant output waveform is shown in Fig. 1.21c for the entire sinusoidal input. For each cycle of the input voltage v, the waveform of v_o will repeat itself so that each waveform has the same frequency. A closer examination of the various figures will also reveal that the impressed voltage v and v_o are *in phase;* that is, the positive pulse of each appears during the same time period. Phase relationships will become increasingly important when we consider semiconductor amplifiers. The rectification process will be examined in greater detail in the concluding section of this chapter.

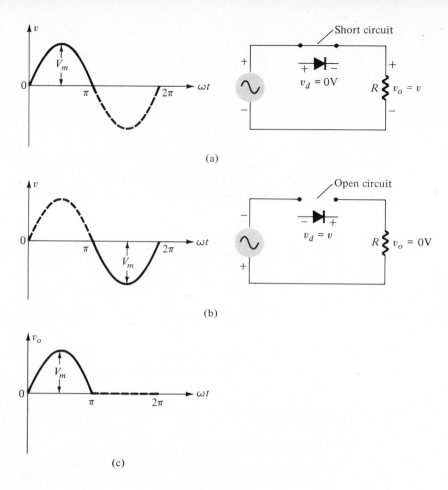

(a)

(b)

(c)

Figure 1.21 Rectifying action of the circuit of Fig. 1.20.

1.8 BASIC CONSTRUCTION AND CHARACTERISTICS

Earlier in this chapter both n- and p-type materials were introduced. The semiconductor diode is formed by simply bringing these materials together (constructed from the same base—Ge or Si), as shown in Fig. 1.22, using techniques to be described later in the chapter. At the instant the two materials are "joined" the electrons and holes in the region of the junction will combine resulting in a lack of carriers in the region near the junction. This region of uncovered positive and negative ions is called the *depletion* region due to the depletion of carriers in this region.

No Applied Bias

The minority carriers in the n-type material that find themselves within the depletion region will pass directly into the p-type material. The closer the minority carrier is to the junction, the greater the attraction for the layer of negative ions and the less the opposition of the positive ions in the depletion region of the n-type material. For the purposes of future discussions we shall assume that all the minority carriers

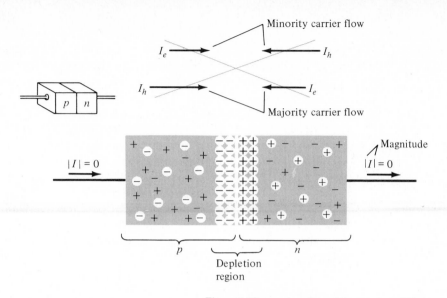

Figure 1.22 *p-n* junction with no external bias.

of the *n*-type material that find themselves in the depletion region due to their random motion will pass directly into the *p*-type material. Similar discussion can be applied to the minority carriers (electrons) of the *p*-type material. This carrier flow has been indicated in Fig. 1.22 for the minority carriers of each material.

The majority carriers in the *n*-type material must overcome the attractive forces of the layer of positive ions in the *n*-type material and the shield of negative ions in the *p*-type material in order to migrate into the neutral region of the *p*-type material. The number of majority carriers is so large in the *n*-type material, however, that there will be invariably a small number of majority carriers with sufficient kinetic energy to pass through the depletion region into the *p*-type material. Again, the same type of discussion can be applied to the majority carriers of the *p*-type material. The resulting flow due to the majority carriers is also shown in Fig. 1.22.

A close examination of Fig. 1.22 will reveal that the relative magnitudes of the flow vectors are such that the net flow in either direction is zero. This cancellation of vectors has been indicated by crossed lines. The length of the vector representing hole flow has been drawn longer than that for electron flow to demonstrate that the magnitude of each need not be the same for cancellation and that the doping levels for each material may result in an unequal carrier flow of holes and electrons. In summary, *the net flow of charge in any one direction with no applied voltage is zero.*

Reverse-Bias Condition

If an external potential of V volts is applied across the *p-n* junction such that the positive terminal is connected to the *n*-type material and the negative terminal is connected to the *p*-type material as shown in Fig. 1.23, the number of uncovered positive ions in the depletion region of the *n*-type material will increase due to the large number of "free" electrons drawn to the positive potential of the applied voltage. For similar reasons, the number of uncovered negative ions will increase in the *p*-

type material. The net effect, therefore, is a widening of the depletion region. This widening of the depletion region will establish too great a barrier for the majority carriers to overcome, effectively reducing the majority carrier flow to zero (Fig. 1.23).

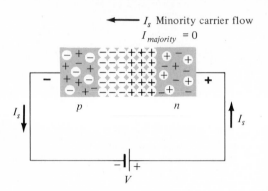

Figure 1.23 Reverse-biased *p-n* junction.

The number of minority carriers, however, that find themselves entering the depletion region will not change, resulting in minority-carrier flow vectors of the same magnitude indicated in Fig. 1.22 with no applied voltage. The current that exists under these conditions is called the *reverse saturation current* and is represented by the subscript *s*. It is seldom more than a few microamperes in magnitude except for high-power devices. The term "saturation" comes from the fact that it reaches its maximum value quickly and does not change significantly with increase in the reverse-bias potential. The situation depicted in Fig. 1.23 is referred to as a *reverse-bias* condition.

Forward-Bias Condition

A *forward-bias* condition is established by applying the positive potential to the *p*-type material and the negative potential to the *n*-type material (for future reference, note that the forward-bias condition is defined by the corresponding first letter in *p*-type and *p*ositive or in *n*-type and *n*egative), as shown in Fig. 1.24. Note that the minority-carrier flow has not changed in magnitude, but the reduction in the width of the depletion region has resulted in a heavy majority flow across the junction.

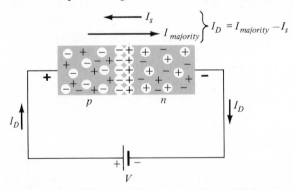

Figure 1.24 Forward-biased *p-n* junction.

The magnitude of the majority-carrier flow will increase exponentially with increasing forward bias, as indicated in Figs. 1.25 and 1.26. Note in Fig. 1.25 the similarities with the ideal diode except at a very large negative voltage. The offset in the first quadrant and the sharp drop in the third quadrant will be examined in this chapter. To reiterate, the first quadrant represents the forward-bias region, and the third quadrant the reverse-bias region. Note the extreme change in scales for both the voltage and current in Fig. 1.26. For the current it is a 5000:1 change. The vertical scale for the majority of smaller units is the milliampere, as shown in Fig. 1.26. However, semiconductor diodes are available today with ampere in the vertical scale, even though the diameter of the casing may not be greater than $\frac{1}{4}$ in.

CH. 1 SEMICONDUCTOR DIODES

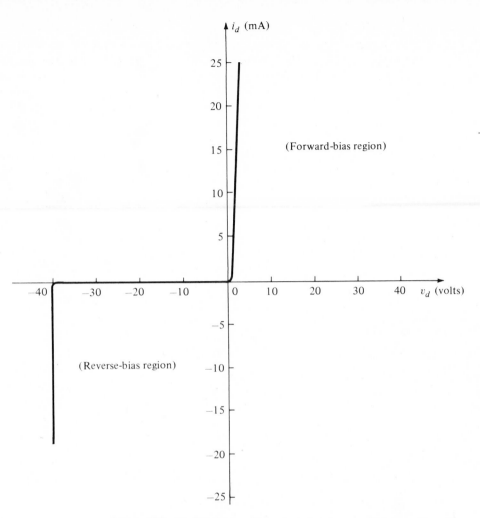

Figure 1.25 Semiconductor diode characteristics—continuous scale on the vertical and horizontal axes.

It can be demonstrated through the use of solid-state physics that this diode current can be mathematically related to temperature (T_K) and applied bias (V) in the following manner:

$$I = I_s(e^{kV/T_K} - 1) \qquad (1.4)$$

where

I_s = reverse saturation current

$k = 11{,}600/\eta$ with $\eta = 1$ for Ge and 2 for Si

$T_K = T_C + 273°$ (T_K—degrees Kelvin, T_C—degrees Celsius)

Note the exponential factor that will result in a very sharp increase in I with increasing levels of V. The characteristics of a commercially available silicon (Si) diode will differ slightly from the characteristics of Fig. 1.26 because of the *body* or *bulk* resistance of the semiconductor material and the *contact* resistance between the semiconductor material and the external metallic conductor. They will cause the curve to shift slightly in the forward-bias region, as indicated by the dashed line in Fig. 1.26. As construction techniques improve and these undesired resistance levels are reduced, the commercially available unit will approach the characteristic defined by Eq. (1.4).

In an effort to demonstrate that Eq. (1.4) does in fact represent the curves of Fig. 1.26, let us determine the current I for the forward-bias voltage of 0.5 V at room temperature (25°C).

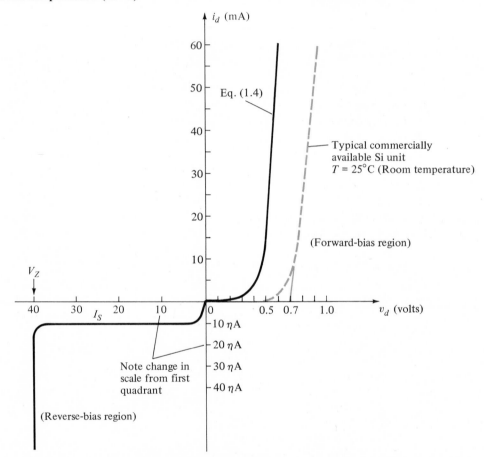

Figure 1.26 Semiconductor diode (Si) characteristics.

$$T_{\mathrm{K}} = T_{\mathrm{C}} + 273° = 25° + 273° = 298°$$

$$k(\mathrm{Si}) = \frac{11,600}{2} = 5800$$

$$\frac{kV}{T_{\mathrm{K}}} = \frac{(5800)(0.5)}{298} = 9.732$$

and $I = I_s(e^{9.732} - 1) = (1 \times 10^{-6})(16814 - 1) = 16.814 \times 10^{-3}$

so that $I \cong \textbf{16.8 mA}$

as verified by Fig. 1.26.

Temperature can have a marked effect on the diode current. This is clearly demonstrated by the factor T_K in Eq. (1.4). The effect of varying T_K will be determined for the forward-bias condition in the exercises appearing at the end of the chapter. In the reverse-bias region it has been found experimentally that the *reverse saturation current I_s will almost double in magnitude for every 10° C change in temperature.* It is not uncommon for a germanium diode with an I_s in the order of 1 or 2 μA at 25°C to have a leakage current of 100 μA = 0.1 mA at a temperature of 100°C. Current levels of this magnitude in the reverse-bias region would certainly question our desired open-circuit condition in the reverse-bias region. Fortunately, typical values of I_s for silicon at room temperature range from 1/100 to 1/1000 that of a similar application germanium diode so that even at higher temperature levels I_s does not usually reach levels of serious concern. For example, if $I_s = 1$ μA for a germanium diode and if I_s were, at the most, 1/100 of 1 μA = 0.01 μA for a silicon diode, then at 100°C it would be only (1/100)(100 μA) = 1 μA. The stability of an electronic system is highly dependent on the temperature sensitivity of its components. The fact that I_s will only approach 1 μA at 100°C for a silicon diode while it approaches 0.1 mA (100 μA) for a germanium diode is one very important reason that silicon devices enjoy a significantly higher level of attention. Fundamentally, the open-circuit equivalent in the reverse-bias region is better realized at any temperature with silicon than with germanium.

Zener Region

Note the sharp change in the characteristics of Fig. 1.27 at the reverse-bias potential V_Z (the subscript Z refers to the name Zener). This constant-voltage effect is induced by a high reverse-bias voltage across the diode. When the applied reverse potential becomes more and more negative, a point is eventually reached where the few free minority carriers have developed sufficient velocity to liberate additional carriers through ionization. That is, they collide with the valence electrons and impart sufficient energy to them to permit them to leave the parent atom. These additional carriers can then aid the ionization process to the point where a high *avalanche* current is established and the *avalanche breakdown* region determined.

The avalanche region (V_Z) can be brought closer to the vertical axis by increasing the doping levels in the p- and n-type materials. However, as V_Z decreases to very low levels, such as -5 V, another mechanism, called *Zener breakdown,* will contribute to the sharp change in the characteristic. It occurs because there is a strong electric field in the region of the junction that can disrupt the bonding forces within the atom and "generate" carriers. Although the Zener breakdown mechanism is only a significant contributor at lower levels of V_Z, this sharp change in the characteristic at any level is called the *Zener region* and diodes employing this unique portion of the characteristic of a *p-n* junction are called *Zener diodes.* They will be described in detail in Chapter 2.

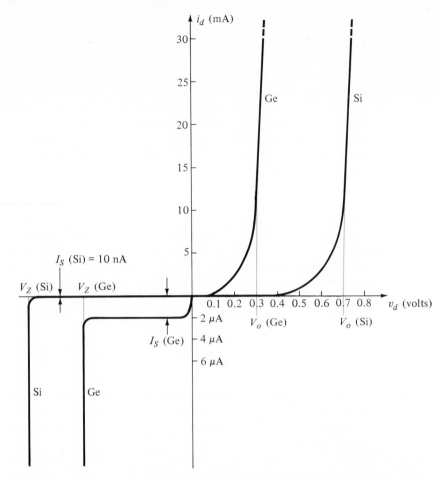

Figure 1.27 Comparison of Si and Ge semiconductor diodes.

The Zener region of the semiconductor diode described must be avoided if the response of a system is not to be completely altered by the sharp change in characteristics in this reverse-voltage region. The maximum reverse-bias potential that can be applied before entering this region is called the *peak inverse voltage* (referred to simply as the PIV rating), or the *peak reverse voltage* (denoted by PRV rating). If an application requires a PIV rating greater than that of a single unit, a number of diodes of the same characteristics can be connected in series. Diodes are also connected in parallel to increase the current-carrying capacity.

Silicon Versus Germanium

Silicon diodes have, in general, higher PIV and current ratings and wider temperature ranges than germanium diodes. PIV ratings for silicon can be in the neighborhood of 1000 V, whereas the maximum value for germanium is closer to 400 V. Silicon can be used for applications in which the temperature may rise to about 200°C (400°F), whereas germanium has a much lower maximum rating (100°C). The disadvantage of silicon, however, as compared to germanium, as indicated in Fig. 1.26, is the higher forward-bias voltage required to reach the region of upward swing. It is typically of the order of magnitude of 0.7 V for *commercially* available silicon diodes and 0.3 V for germanium diodes. The increased offshoot for silicon is due primarily to the factor η in Eq. (1.4). This factor only plays a part in determining the shape of the curve at very low current levels. Once the curve starts its vertical rise, the factor η drops to 1 (the continuous value for germanium). This is evidenced by the similarities in the curves once the offshoot potential is reached. The potential at which this rise occurs is very important in the circuit analysis to follow and therefore requires the specific notation V_o, as indicated on the figure. In review:

$$
V_o = 0.7 \text{ (Si)} \\
V_o = 0.3 \text{ (Ge)}
$$

Obviously, the closer the upward swing is to the vertical axis, the more "ideal" the device. However, the other characteristics of silicon as compared to germanium still make it the choice in the majority of commercially available units.

1.9 TRANSITION AND DIFFUSION CAPACITANCE

Electronic devices are inherently sensitive to very high frequencies. Most shunt capacitive effects that can be ignored at lower frequencies because the reactance $X_C = 1/2\pi f C$ is very large (open-circuit equivalent) cannot be ignored at very high frequencies. X_C will become sufficiently small due to the high value of f to introduce a low-reactance "shorting" path. In the *p-n* semiconductor diode, there are two capacitive effects to be considered. Both types of capacitance are present in the forward- and reverse-bias regions, but one so outweighs the other in each region that we consider the effects of only one in each region. In the reverse-bias region we have the *transition*- or *depletion*-region capacitance (C_T), while in the forward-bias region we have the *diffusion* (C_D) or *storage* capacitance.

Recall that the basic equation for the capacitance of a parallel-plate capacitor is defined by $C = \epsilon A/d$, where ϵ is the permittivity of the dielectric (insulator) between the plates of area A separated by a distance d. In the reverse-bias region there is a depletion region (free of carriers) that behaves essentially like an insulator between the layers of opposite charge. Since the depletion region will increase with increased reverse-bias potential, the resulting transition capacitance will decrease, as shown in Fig. 1.28. The fact that the capacitance is dependent on the applied reverse-bias

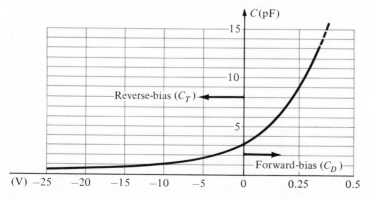

Figure 1.28 Transition and diffusion capacitance vs. applied bias for a silicon diode.

23

potential has application in a number of electronic systems. In fact, in Chapter 2 a diode will be introduced whose existence is wholly dependent on this phenomenon.

Although the effect described above will also be present in the forward-bias region, it is overshadowed by a capacitance effect directly dependent on the rate at which charge is injected into the regions just outside the depletion region. In other words, directly dependent on the resulting current of the diode. Increased levels of current will result in increased levels of diffusion capacitance. However, increased levels of current result in reduced levels of associated resistance (to be demonstrated shortly),

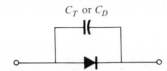

C_T or C_D

Figure 1.29 Including the effect of the transition or diffusion capacitance on the semiconductor diode.

and the resulting time constant $(\tau = RC)$, which is very important in high-speed applications, does not become excessive.

The capacitive effects described above are represented by a capacitor in parallel with the ideal diode, as shown in Fig. 1.29. For low- or mid-frequency applications (except in the power area), however, the capacitor is normally not included in the diode symbol.

1.10   REVERSE RECOVERY TIME

There are certain pieces of data that are normally provided on diode specification sheets provided by manufacturers. One such quantity that has not been considered yet is the reverse recovery time denoted by t_{rr}. In the forward-bias state it has been shown in an earlier section that there are a large number of electrons from the *n*-type material progressing through the *p*-type material and a large number of holes in the *n*-type—a requirement for conduction. The electrons in the *p*-type and of holes progressing through the *n*-type material establish a large number of minority carriers in each material. If the applied voltage should be reversed to establish a reverse-bias situation, we would ideally like to see the diode change instantaneously from the conduction state to the nonconduction state. However, because of the large number of minority carriers in each material, the diode will simply reverse as shown in Fig. 1.30 and stay at this measurable level for the period of time t_s (storage time) required for the minority carriers to return to their majority-carrier state in the opposite material. In essence, the diode will remain in the short-circuit state with a current I_{reverse} determined by the network parameters. Eventually, when this storage phase has passed, the current will reduce in level to that associated with the nonconduction state. This second period of time is denoted by t_t (transition interval). The reverse recovery time is the sum of these two intervals: $t_{rr} = t_s + t_t$. Naturally, it is an important consideration in high-speed switching applications. Most commercially available switching diodes have a t_{rr} in the range of a few nanoseconds to 1 μs. Units are available, however, with a t_{rr} of only a few hundred picoseconds (10^{-12}).

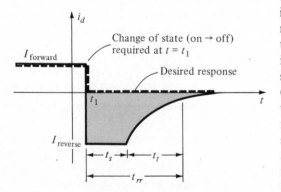

Figure 1.30 Defining the reverse recovery time.

1.11 TEMPERATURE EFFECTS

Temperature is a very important consideration in the design or analysis of electronic systems. It will affect virtually all of the characteristics of any semiconductor device. The change in characteristics of a semiconductor diode due to temperature variations above and below room temperature (25°C) is shown in Fig. 1.31. Note the reduced levels of forward voltage drop but increased levels of saturation current at 100°C. The Zener potential is also experiencing a pronounced change in level.

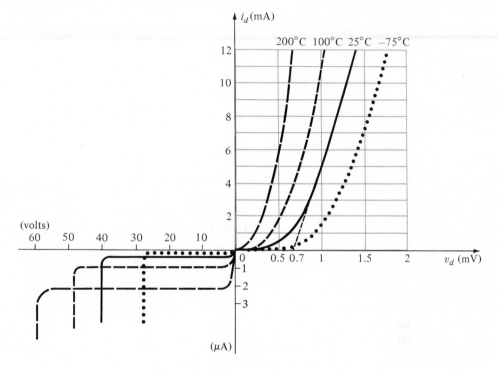

Figure 1.31 Variation in diode characteristics with temperature change.

Heat Sinks

Increased levels of current through any semiconductor device will result in increased junction temperatures. Although silicon materials can handle currents approaching 1000 A/in.², the changes in characteristics with temperature may result in an unstable system. Germanium materials have maximum allowable operating junction temperatures ranging from 85 to 100°C, while the range for silicon is 150 to 200°C (just one more case for the increased level of interest in silicon devices). Quite frequently, the maximum allowable junction temperature is exceeded before the maximum power dissipation level is established. In fact, we will find in this section that the maximum dissipation level is very dependent on temperature. With no applied bias and consequently zero current through the device, the temperature of the junction of the semiconductor diode is essentially the same as that of the surrounding air or medium. This temperature, called the *ambient* temperature, has the symbol T_A and is measured (as are all temperature levels in this discussion) in degrees Celsius.

As charge begins to flow through the device, there is an I^2R loss at the junction that will increase the temperature of the junction. The resulting heat is then transferred to the case and eventually to the surrounding air, as shown in Fig. 1.32a. If a *heat sink* is applied as shown in Fig. 1.32b, the resulting heat will pass from the case to the sink before being emitted to the surrounding medium. The sole purpose of the heat sink is to provide a large surface area through which the heat can be quickly removed from the device—thus permitting the device to work at higher dissipation levels. The insulator appearing in Fig. 1.32b is often necessary to insulate the semiconductor device from the heat sink. Its effect on the flow of heat will be considered shortly.

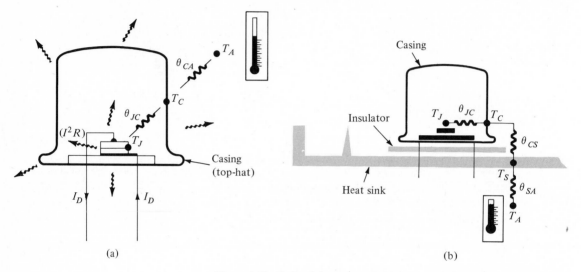

(a)   (b)

Figure 1.32   Path of thermal conductivity: (a) without a heat sink; (b) with a heat sink.

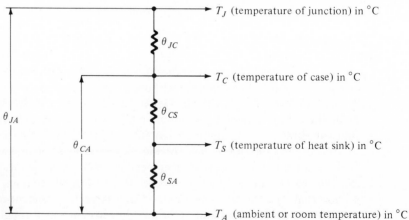

Figure 1.33   Thermal resistance path between the junction and ambient temperatures.

CH. 1   SEMICONDUCTOR DIODES

The ease with which heat can be transferred from one element to another is a measure of the *thermal conductivity* between the two. Inversely related to the conductivity is the *thermal resistance*—a measure of how much the medium will oppose the flow of heat. In our present design considerations, therefore, we are trying to establish the lowest possible thermal resistance path from the junction to the surrounding air. Thermal resistance has the symbol θ, is measured in °C/W, and has the resistance symbol as appearing in Fig. 1.32. In both parts of Fig. 1.32 the path of heat flow is indicated and the temperature of importance noted. A schematic representation for both systems appears in Fig. 1.33. For the system of Fig. 1.32a, θ_{CA} is simply the thermal resistance path from case to the surrounding medium—without consideration of a heat sink.

The series path between elements will result in a total thermal resistance from junction to surrounding air of

$$\boxed{\theta_{JA} = \theta_{JC} + \theta_{CS} + \theta_{SA}} \quad \text{with heat sink} \quad (°C/W) \qquad (1.5a)$$

$$\boxed{\theta_{JA} = \theta_{JC} + \theta_{CA}} \quad \text{without heat sink} \quad (°C/W) \qquad (1.5b)$$

The temperature of the junction is related to θ_{JA}, the power dissipated, and the ambient temperature T_A by

$$\boxed{T_J = P_D\theta_{JA} + T_A} \quad (°C) \qquad (1.6)$$

In words, it simply states that the junction temperature is equal to the temperature of the surrounding air plus the increased temperature due to the heat conversion at the junction. The better we are at removing the heat from the junction through lower values of θ_{JA}, the higher the power rating can be for the same junction temperature.

In total,

$$\boxed{T_J = P_D(\underbrace{\theta_{JC} + \theta_{CS} + \theta_{SA}}_{\theta_{CA}}) + T_A} \quad (°C) \qquad (1.7)$$

In Eq. (1.7), the maximum T_J is usually specified on a data sheet. It is always less than the maximum permissible value for that material. As indicated, T_A is simply the temperature of the surrounding medium measured in degrees Celsius. θ_{JA} is established by the device chosen—its power level, case construction, and so on. P_D is the level of power dissipation determined by $P = V_D I_D$ for a diode. θ_{CA} or θ_{JA} is normally provided on the data sheet for a specific semiconductor device. θ_{CS} is the interface thermal resistance that depends on the casing design and how it is mounted to the heat sink. The use of an insulator will increase the thermal resistance above that obtained with direct mounting. Keep in mind that the term "insulator" here is referring to charge flow and not heat flow. Our goal is to design an insulator that will block the flow of charge but not severely affect heat-flow levels. For a TO-3 case (top-hat appearance), an average value of θ_{CS} is 0.1°C/W for direct mounting and 0.5°C/W with an insulator. θ_{SA} is provided by the heat-sink design, as shown in Fig. 1.34.

(a)

(b)

Figure 1.34 Tran-tec semiconductor coolers: (a) Model 19, $\theta_{SA} = 2.5°C/W$; (b) Model 1128, $\theta_{SA} = 0.8°C/W$. (Courtesy Tran-tec Corporation.)

EXAMPLE 1.1

(a) Determine whether the junction temperature of a semiconductor diode is being exceeded under the following operating conditions:

$$P_D = 20 \text{ W}$$
$$T_A = 25°C$$
$$T_J \text{ (max)(specified)} = 150°C$$
$$\theta_{JC} = 2.0°C/W \qquad \theta_{CS} = 0.5°C/W \qquad \theta_{SA} = 2.5°C/W$$

(b) If $\theta_{JA} = 20°C/W$ for the device without a heat sink, has the junction temperature been exceeded?

Solution:

(a) From Eq. (1.7): $T_J = P_D(\theta_{JC} + \theta_{CS} + \theta_{SA}) + T_A$
$$= 20(2.0 + 0.5 + 2.5) + 25$$
$$= 100 + 25 = \mathbf{125°C}$$
$$T_J \text{ (specified)} = 150°C > 125°C \quad \text{(safe operation)}$$

(b) From Eq. (1.6): $T_J = P_D\theta_{JA} + T_A$
$$= 20(20) + 25$$
$$= 400 + 25 = \mathbf{425°C} \gg 150°C \quad \text{(permanent damage to the}$$
device can be expected)

In Example 1.1 the maximum permissible dissipation without a heat sink could be determined from Eq. (1.8):

$$\boxed{P_D = \frac{T_J - T_A}{\theta_{JA}}} \qquad \text{(W)} \qquad (1.8)$$

That is, $\qquad P_{D_{\text{max}}} = \dfrac{150 - 25}{20} = \dfrac{125}{20} = \mathbf{6.25 \text{ W}}$

The heat-sink requirement could be determined by

$$\theta_{SA} = \frac{T_J - T_A}{P_D} - (\theta_{JC} + \theta_{CS}) \qquad (^\circ C/W) \qquad\qquad (1.9)$$

Power-Derating Curve

It should be clear from the discussion above that the maximum power rating is closely related to the case temperature of the device. Once a certain case temperature is reached, the maximum power rating as provided in the data sheet will start to drop off linearly (straight line), as shown in Fig. 1.35. The resulting plot is called a *power-derating curve* for the device. On many specification sheets for semiconductor diodes, this will show a plot of forward current versus case temperature, but the effect on the quantity of interval is the same. The *derating factor* or measure of how quickly the curve will drop off is measured in W/°C and has the symbol D_F. In actuality, it is the inverse of the junction to case thermal resistance. That is, $D_F = 1/\theta_{JC}$.

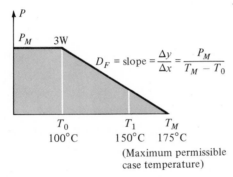

Figure 1.35 Maximum power rating vs. case temperature—the power derating curve.

In the linear sloped region, the power rating at any temperature T_1 compared to the maximum power (P_M) at temperature T_0 is determined by

$$P_{T_1} = P_M - (T_1 - T_0)(D_F) \qquad (W) \qquad\qquad (1.10)$$

D_F can be determined by

$$D_F = \frac{P_M}{T_M - T_0} \qquad (W/^\circ C) \qquad\qquad (1.11)$$

although it is often provided simply as a change of so many watts per degree.

EXAMPLE 1.2
(a) Determine the maximum power rating for a device having the derating curve of Fig. 1.35 at a temperature of 150°C.
(b) Calculate θ_{JC}.

Solution:

(a) From Eq. (1.11): $D_F = \dfrac{P_M}{T_M - T_0} = \dfrac{3}{175 - 100} = \dfrac{3}{75} = \mathbf{0.04 \ W/^\circ C}$

From Eq. (1.10): $P_{T_1} = P_M - (T_1 - T_0)(D_F)$

$$= 3 - (150 - 100)(0.04)$$

$$= 3 - (50)(0.04)$$

$$= 3 - 2 = \mathbf{1\ W} \quad \text{(a significant drop)}$$

(b) $\theta_{JC} = \dfrac{1}{D_F} = \dfrac{1}{0.04} = \mathbf{25°C/W}$

1.12   DIODE SPECIFICATION SHEETS

Data on specific semiconductor devices are normally provided by the manufacturer in two forms. One is a very brief description of a device that will permit a quick review of all devices available within a few pages. The other is a thorough examination of a device, including graphs, applications, and so on, which usually appears as a separate entity. The latter is normally only provided when specifically requested.

There are certain pieces of data, however, that normally appear on either one. They are included below:

1. The maximum forward voltage $V_{F(max)}$ (at a specified current and temperature).
2. The maximum forward current $I_{F(max)}$ (at a specified temperature).
3. The maximum reverse current $I_{R(max)}$ (at a specified temperature).
4. The reverse-voltage rating (PIV) or PRV or V(BR), where **BR** comes from the term "breakdown" (at a specified temperature).
5. Maximum capacitance.
6. Maximum t_{rr}.
7. The maximum operating (or case) temperature.

Depending on the type of diode being considered, additional data may also be provided, such as frequency range, noise level, switching time, thermal resistance levels, and peak repetitive values. For the application in mind, the significance of the data will usually be self-apparent. If the maximum power or dissipation rating is also provided, it is understood to be equal to the following product:

$$\boxed{P_{D_{max}} = V_D I_D} \tag{1.12}$$

where I_D and V_D are the diode current and voltage at a particular point of operation, each variable not to exceed its maximum value. The information in Table 1.2 was taken directly from a Texas Instruments, Inc., data book. Note that the forward voltage drop does not exceed 1 V, but the current has maximum values of 1 to 200 mA.

For the 1N463, if we establish maximum forward voltage and current conditions:

$$P_D = V_D I_D = (1)(1 \times 10^{-3}) = 1\ \text{mW} \quad \text{(a low-power device)}$$

TABLE 1.2 General-Purpose Diodes

Device Type	Forward Current		V_{BR} (V)	Maximum I_R			
				25° C		150° C	
	I_F (mA)	V_F (V)		V	μA	V	μA
1N463	1.0	1.0	200	175	0.5	175	30
1N462	5.0	1.0	70	60	0.5	60	30
1N459A	100.0	1.0	200	175	0.025	175	5
T151	200.0	1.0	20	10	1	—	—

Of course, a device may have a maximum dissipation less than that established by the maximum values. That is, if the voltage is a maximum, the current may have to be less than rated maximum value.

Note the increase in I_R for each device with temperature. For the 1N463, it is $30/0.5 = 60$ times larger.

An exact copy of the data provided by Fairchild Camera and Instrument Corporation for their BAY 73 and BA 129 high-voltage/low-leakage diodes appears in Figs. 1.36 and 1.37. This example would represent the expanded list of data and characteristics. Note that all but the average rectified current, peak repetitive forward current, and peak forward surge current have been defined in this chapter. The significance of these three quantities is as follows:

1. *Average Rectified Current:* The half-wave-rectified signal of Fig. 1.21 has an average value defined by $I_{av} = 0.318I_{peak}$. The average current rating is lower than the continuous forward current because a half-wave current waveform will have instantaneous values much higher than the average value.

2. *Peak Repetitive Forward Current:* This is the maximum instantaneous value of repetitive forward current. Note that since it is at this level for a brief period of time, its level can be higher than the continuous level.

3. *Peak Forward Surge Current:* On occasion during turn-on, malfunctions, and so on, there will be very high currents through the device for very brief intervals of time (that are not repetitive). This rating defines the maximum value and the time interval for such surges in current level.

Note the logarithmic scale appearing on some of the curves of Fig. 1.37. Each region is bisected such that the value of each horizontal line should be fairly obvious. For I_F versus V_F the horizontal lines between 1.0 and 10.0 mA are 2 mA, 4 mA, 6 mA, and 8 mA. Again, most of the axis variables on the graphs provided have been introduced, resulting in a set of curves that have some recognizable meaning. The temperature coefficient defines the change in voltage with temperature at different current levels. A range of values for the temperature coefficient is provided at each current level. The dynamic impedance (actually, simply the resistance of the device at that forward current) will be discussed in a later section. Note the effect of temperature in the power rating and current ratings of the device in the bottom-right figure.

DIFFUSED SILICON PLANAR

- **BV** . . . **125 V (MIN) @ 100 μA (BAY73)**
- **BV** . . . **200 V (MIN) @ 100 μA (BA129)**

ABSOLUTE MAXIMUM RATINGS (Note 1)

Temperatures
Storage Temperature Range	$-65°C$ to $+200°C$
Maximum Junction Operating Temperature	$+175°C$
Lead Temperature	$+260°C$

Power Dissipation (Note 2)
Maximum Total Power Dissipation at 25°C Ambient	500 mW
Linear Power Derating Factor (from 25°C)	3.33 mW / °C

Maximum Voltage and Currents
WIV	Working Inverse Voltage	BAY73	100 V
		BA129	180 V
I_O	Average Rectified Current		200 mA
I_F	Continuous Forward Current		500 mA
if	Peak Repetitive Forward Current		600 mA
if(surge)	Peak Forward Surge Current		
	Pulse Width = 1 s		1.0 A
	Pulse Width = 1 μs		4.0 A

DO-35 OUTLINE

1.0 (25.40) MIN

0.180 (4.57) / 0.140 (3.56)

0.021 (0.533) / 0.019 (0.483) DIA

0.075 (1.91) / 0.060 (1.52) DIA

NOTES:
Copper clad steel leads, tin plated
Gold plated leads available
Hermetically sealed glass package
Package weight is 0.14 gram

ELECTRICAL CHARACTERISTICS (25°C Ambient Temperature unless otherwise noted)

SYMBOL	CHARACTERISTIC	BAY73		BA129		UNITS	TEST CONDITIONS
		MIN	MAX	MIN	MAX		
V_F	Forward Voltage	0.85	1.00			V	I_F = 200 mA
		0.81	0.94			V	I_F = 100 mA
		0.78	0.88	0.78	1.00	V	I_F = 50 mA
		0.69	0.80	0.69	0.83	V	I_F = 10 mA
		0.67	0.75			V	I_F = 5.0 mA
		0.60	0.68	0.60	0.71	V	I_F = 1.0 mA
				0.51	0.60	V	I_F = 0.1 mA
I_R	Reverse Current		500			nA	V_R = 20 V, T_A = 125°C
			5.0			nA	V_R = 100 V
			1.0			μA	V_R = 100 V, T_A = 125°C
					10	nA	V_R = 180 V
					5.0	μA	V_R = 180 V, T_A = 100°C
BV	Breakdown Voltage	125		200		V	I_R = 100 μA
C	Capacitance		8.0		6.0	pF	V_R = 0, f = 1.0 MHz
t_{rr}	Reverse Recovery Time		3.0			μs	I_f = 10 mA, V_f = 35 V R_L = 1.0 to 100 KΩ C_L = 10 pf, JAN 256

NOTES:
1. These ratings are limiting values above which the serviceability of the diode may be impaired.
2. These are steady state limits. The factory should be consulted on applications involving pulses or low duty-cycle operation.

Figure 1.36 Electrical characteristics of the Bay 73 · BA 129 high-voltage, low-leakage diodes. (Courtesy Fairchild Camera and Instrument Corporation.)

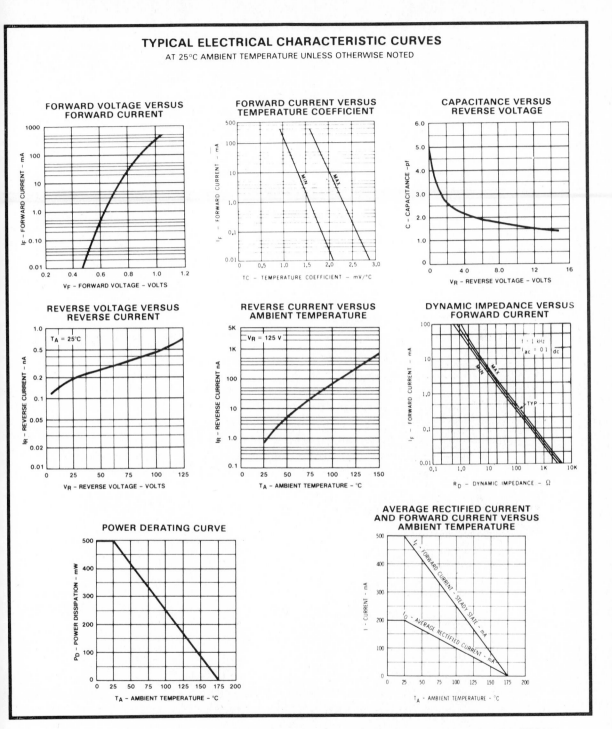

Figure 1.37 Terminal characteristics of the Fairchild Bay 73 · BA 129 high-voltage diodes. (Courtesy Fairchild Camera and Instrument Corporation.)

1.13 SEMICONDUCTOR DIODE NOTATION

The notation most frequently used for semiconductor diodes is provided in Fig. 1.38. For most diodes any marking such as a dot or band, as shown in Fig. 1.38, appears at the cathode end. The terminology anode and cathode is a carryover from vacuum-tube notation. The anode refers to the higher or positive potential and the cathode refers to the lower or negative terminal. This combination of bias levels will result in a forward-bias or "on" condition for the diode. In general, the maximum current-carrying capacity of the diodes of Fig. 1.38 increases from the left to the right. For each the size will increase with the current rating to ensure that it can handle the additional power dissipation. All but the stud type are limited to a few amperes.

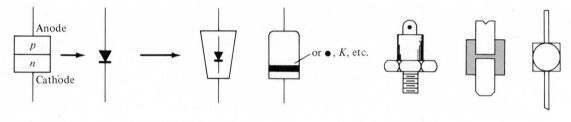

Figure 1.38 Semiconductor diode notation.

1.14 DIODE OHMMETER CHECK

The condition of a semiconductor diode can be quickly determined by using an ohmmeter such as is found on the standard **VOM.** The internal battery (often 1.5 V) of the ohmmeter section will either forward- or reverse-bias the diode when applied. If the positive (normally the red) lead is connected to the anode and the negative (normally the black) lead to the cathode, the diode is forward-biased and the meter should indicate a low resistance. The R $\times$ 1000 or R $\times$ 10,000 setting should be suitable for this measurement. With the reverse polarity the internal battery will back bias the diode and the resistance should be very large. A small reverse-bias resistance reading indicates a "short" condition while a large forward-bias resistance indicates an "open" situation. The basic connections for the tests appear in Fig. 1.39.

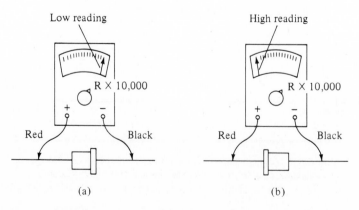

(a) (b)

Figure 1.39 Ohmmeter testing of a semiconductor diode: (a) forward-bias; (b) reverse-bias.

1.15 SEMICONDUCTOR DIODE FABRICATION

Semiconductor diodes are normally one of the following types: grown junction, alloy, diffusion, epitaxial growth, or point contact. Each will be described in some detail in this section. A rereading of Section 1.6 is suggested before proceeding with the following description.

Grown Junction

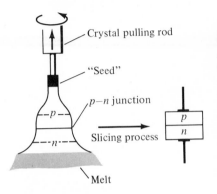

Figure 1.40 Grown junction diode.

Diodes of this type are formed during the Czochralski *crystal pulling* process. Impurities of *p*-and *n*-type can be alternately added to the molten semiconductor material in the crucible, resulting in a *p-n* junction, as indicated in Fig. 1.40, when the crystal is pulled. After slicing, the large-area device can then be cut into a large number (sometimes thousands) of smaller-area semiconductor diodes. The area of grown-junction diodes is sufficiently large to handle high currents (and therefore have high power ratings). The large area, however, will introduce undesired junction capacitive effects.

Alloy

The alloy process will result in a junction-type semiconductor diode that will also have a high current rating and large **PIV** rating. The junction capacitance is also large, however, due to the large junction area.

The *p-n* junction is formed by first placing a *p*-type impurity on an *n*-type substrate and heating the two until liquefaction occurs where the two materials meet (Fig. 1.41). An alloy will result that, when cooled, will produce a *p-n* junction at the boundary of the alloy and substrate. The roles played by the *n*- and *p*-type materials can be interchanged.

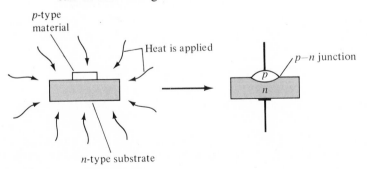

Figure 1.41 Alloy process diode.

Diffusion

The diffusion process of forming semiconductor junction diodes can employ either solid or gaseous diffusion. This process requires more time than the alloy process but it is relatively inexpensive and can be very accurately controlled. Diffusion is a process by which a heavy concentration of particles will "diffuse" into a surrounding region of lesser concentration. The primary difference between the diffusion and alloy process is the fact that liquefaction is not reached in the diffusion process. Heat is applied in the diffusion process only to increase the activity of the elements involved.

The process of solid diffusion commences with the "painting" of an acceptor impurity on an *n*-type substrate and heating the two until the impurity diffuses into the substrate to form the *p*-type layer (Fig. 1.42a).

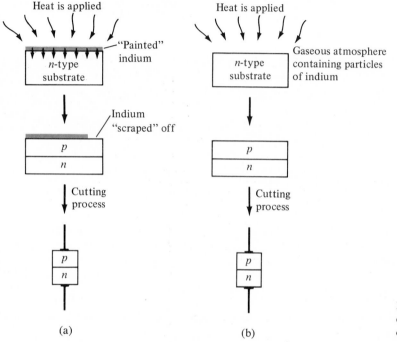

Figure 1.42 Diffusion process diodes: (a) solid diffusion; (b) gaseous diffusion.

In the process of gaseous diffusion, an *n*-type material is submerged in a gaseous atmosphere of acceptor impurities and then heated (Fig. 1.42b). The impurity diffuses into the substrate to form the *p*-type layer of the semiconductor diode. The roles of the *p*- and *n*-type materials can also be interchanged in each case. The diffusion process is the most frequently used today in the manufacture of semiconductor diodes.

Epitaxial Growth

The term "epitaxial" has its derivation from the Greek terms *epi* meaning "upon" and *taxis* meaning "arrangement." A base wafer of n^+ material is connected to a metallic conductor as shown in Fig. 1.43. The n^+ indicates a very high doping level for a reduced resistance characteristic. Its purpose is to act as a semiconductor extension of the conductor and not the *n*-type material of the *p-n* junction. The *n*-type layer is to be deposited on this layer as shown in Fig. 1.43 using a diffusion process. This technique of using an n^+ base gives the manufacturer definite design advantages. The *p*-type silicon is then applied by using a diffusion technique and the anode metallic connector added as indicated in Fig. 1.43.

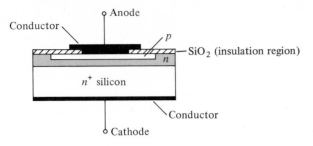

Figure 1.43 Epitaxial growth semi-conductor diode.

Point Contact

The point-contact semiconductor diode is constructed by pressing a phosphor-bronze spring (called a cat whisker) against an *n*-type substrate (Fig. 1.44). A high current is then passed through the whisker and substrate for a short period of time, resulting in a number of atoms passing from the wire into the *n*-type material to create a *p*-region in the wafer. The small area of the *p-n* junction results in a very small junction capacitance (typically 1 pF or less). For this reason, the point-contact diode is frequently used in applications where very high frequencies are encountered, such as in microwave mixers and detectors. The disadvantage of the small contact area is the resulting low current ratings and characteristics less ideal than those obtained from junction-type semiconductor diodes. The basic construction and photographs of point-contact diodes appear in Fig. 1.45. Various types of junction diodes appear in Fig. 1.46.

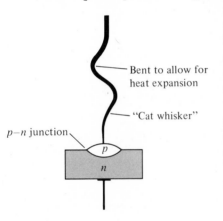

Figure 1.44 Point-contact diode.

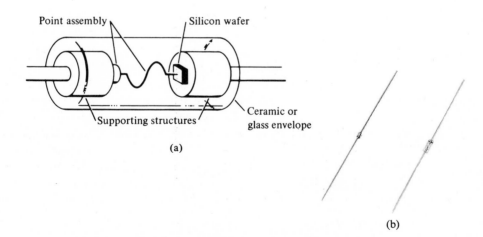

(a)

(b)

Figure 1.45 Point-contact diodes: (a) basic construction; (b) various types. (Courtesy General Electric Company.)

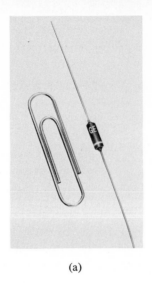

(a)

(b)

(c)

Figure 1.46 Various types of junction diodes. [(a) and (b) courtesy General Electric Company; (c) courtesy International Rectifier Corporation.]

1.16 DIODE ARRAYS—INTEGRATED CIRCUITS

The unique characteristics of integrated circuits will be introduced in Chapter 12. However, we have reached a plateau in our introduction to electronic circuits that permits at least a surface examination of diode arrays in the integrated-circuit package. You will find that the integrated circuit is not a unique device with characteristics totally different from those we will examine in these introductory chapters. It is simply a packaging technique that permits a significant reduction in the size of electronic systems. In other words, internal to the integrated circuit are systems and discrete devices that were available long before the integrated circuit as we know it today became a reality.

One possible array appears in Fig. 1.47. Note that eight diodes are internal to the Fairchild FSA 1410M diode array. That is, in the container shown in Fig. 1.48 there are diodes set in a single silicon wafer that have all the anodes connected to pin 1 and the cathodes of each to pins 2 through 9. Note in the same figure that pin 1 can be determined as being to the left of the small projection in the case if

FSA1410M
PLANAR AIR-ISOLATED MONOLITHIC DIODE ARRAY

- C...5.0 pF (MAX)
- ΔV_F...15 mV (MAX) @ 10 mA

ABSOLUTE MAXIMUM RATINGS (Note 1)

Temperatures

Storage Temperature Range	$-55°C$ to $+200°C$
Maximum Junction Operating Temperature	$+150°C$
Lead Temperature	$+260°C$

CONNECTION DIAGRAM

FSA1410M

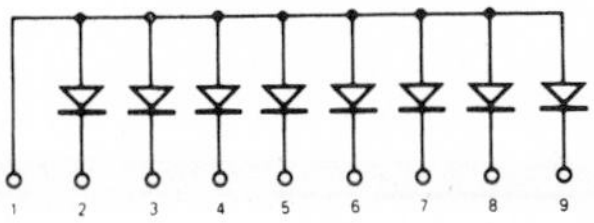

See Package Outline TO-96

Power Dissipation (Note 2)

Maximum Dissipation per Junction at 25°C Ambient	400 mW
per Package at 25°C Ambient	600 mW
Linear Derating Factor (from 25°C) Junction	3.2 mW/°C
Package	4.8 mW/°C

Maximum Voltage and Currents

WIV	Working Inverse Voltage	55 V
I_F	Continuous Forward Current	350 mA
$i_{f(surge)}$	Peak Forward Surge Current	
	Pulse Width = 1.0 s	1.0 A
	Pulse Width = 1.0 μs	2.0 A

ELECTRICAL CHARACTERISTICS (25°C Ambient Temperature unless otherwise noted)

SYMBOL	CHARACTERISTIC	MIN	MAX	UNITS	TEST CONDITIONS
B_V	Breakdown Voltage	60		V	$I_R = 10\ \mu A$
V_F	Forward Voltage (Note 3)		1.5	V	$I_F = 500$ mA
			1.1	V	$I_F = 200$ mA
			1.0	V	$I_F = 100$ mA
I_R	Reverse Current		100	nA	$V_R = 40$ V
	Reverse Current ($T_A = 150°C$)		100	μA	$V_R = 40$ V
C	Capacitance		5.0	pF	$V_R = 0$, f = 1 MHz
V_{FM}	Peak Forward Voltage		4.0	V	$I_f = 500$ mA, $t_r < 10$ ns
t_{fr}	Forward Recovery Time		40	ns	$I_f = 500$ mA, $t_r < 10$ ns
t_{rr}	Reverse Recovery Time		10	ns	$I_f = I_r = 10-200$ mA $R_L = 100\ \Omega$, Rec. to 0.1 I_r
			50	ns	$I_f = 500$ mA, $I_r = 50$ mA $R_L = 100\ \Omega$, Rec. to 5 mA
ΔV_F	Forward Voltage Match		15	mV	$I_F = 10$ mA

NOTES:
1. These ratings are limiting values above which life or satisfactory performance may be impaired.
2. These are steady state limits. The factory should be consulted on applications involving pulsed or low duty cycle operation.
3. V_F is measured using an 8 ms pulse.

Figure 1.47 Monolithic diode array. (Courtesy Fairchild Camera and Instrument Corporation.)

we look from the bottom toward the case. The other numbers then follow in sequence. If only one diode is to be used, then only pins 1 and 2 (or any number from 3 to 9) would be used. The remaining diodes would be left hanging and not affect the network that pins 1 and 2 are connected to.

Another diode array appears in Fig. 1.49. In this case the package is different but the numbering sequence appears in the outline. Pin 1 is the pin directly above the small indentation as you look down on the device.

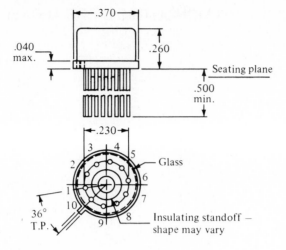

Notes:
Kovar leads, gold plated
Hermetically sealed package
Package weight is 1.32 grams

Figure 1.48 Package outline TO-96 for the FSA 1410M diode array. All dimensions are in inches. (Courtesy Fairchild Camera and Instrument Corporation.)

TO-116-2 Outline

Connection Diagrams
FSA2500M

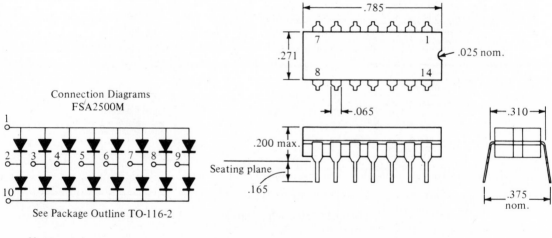

See Package Outline TO-116-2

Notes:
Alloy 42 pins, tin plated
Gold plated pins available
Hermetically sealed ceramic package

Figure 1.49 Monolithic diode array. All dimensions are in inches. (Courtesy Fairchild Camera and Instrument Corporation.)

1.17 DC CONDITIONS

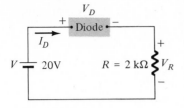

Figure 1.50 Fundamental diode circuit.

The analysis of earlier sections employed the ideal diode. We must now examine the effect of introducing a device that is less than perfect. Our first interest will be in the *dc* or *quiescent* conditions—the latter term meaning still, quiet, or inactive. The circuit of Fig. 1.50 is the simplest possible if the effects of the nonideal characteristics are to be examined. The block symbol was inserted to represent any diode that we may choose to use in this particular circuit.

Applying Kirchhoff's voltage law around the indicated loop will result in the following equation:

$$V = V_D + V_R \qquad (1.13)$$

Solving for V_D and substituting $V_R = I_D R$, we have

$$V_D = V - I_D R$$

and

$$V_D = V - I_D R \qquad (1.14)$$

Equation (1.14) has two dependent variables (V_D and I_D) and two fixed values (V and R). Since a minimum of two equations is required to solve for two unknown dependent variables, Eq. (1.14) is not sufficient for a complete solution. The second equation necessary to determine the value of V_D and I_D determined by V and R is provided by the characteristics of the diode element in the enclosed container; that is, for the diode employed, we know that the current is a *function of* the voltage across the diode, or mathematically,

$$I_D = f(V_D) \qquad (1.15)$$

It is necessary, therefore, to find the common solution of the equation determined by the load circuit [Eq. (1.14)] and the characteristics of the diode. One method of finding this solution is the graphical method, which will now be outlined. It is extremely important that the procedure described in the next few paragraphs be fully understood since similar operations apply to other devices, such as the transistor, and FET.

Rewriting Eq. (1.14) in a slightly different form, we have

$$I_D = -\frac{1}{R} V_D + \frac{V}{R} \qquad (1.16)$$

$$y = \quad m \quad x + b \qquad \text{(straight-line equation)}$$

Below this newly formed equation, the general equation for a straight line has been included. Note that the slope of the line is negative (I_D decreases in magnitude with increase in V_D) with a magnitude $1/R$, while the *y*-intercept is V/R and I_D

and V_D are the y- and x-variables, respectively. The intercepts of this straight line with the axes of the graph of Fig. 1.51 can be found rather quickly by applying the following conditions. If we consider first that if $I_D = 0$ mA, we must be somewhere along the horizontal axis of Fig. 1.51 and if we apply this condition to Eq. (1.16), then

$$I_D = 0 = -\frac{V_D}{R} + \frac{V}{R}$$

and solving for V_D yields

$$\boxed{V_D = V\big|_{I_D=0}} \tag{1.17}$$

The intersection of the straight line with the horizontal axis is the applied voltage V. If we then consider that if $V_D = 0$, we must be somewhere along the vertical axis, and we must apply this condition to Eq. (1.16), then

$$I_D = -\frac{V_D}{R} + \frac{V}{R} = 0 + \frac{V}{R}$$

and

$$\boxed{I_D = \frac{V}{R}\bigg|_{V_D=0}} \tag{1.18}$$

The intersection, therefore, of the straight line with the vertical axis is determined by the ratio of the applied voltage and load. Both intersections have been indicated in Fig. 1.51. All that remains is to connect these two points by a straight line to obtain a graphical representation of Eq. (1.14). This resulting line is called the *load line* since it represents the properties of the applied voltage and load and tells us nothing about the diode's characteristics.

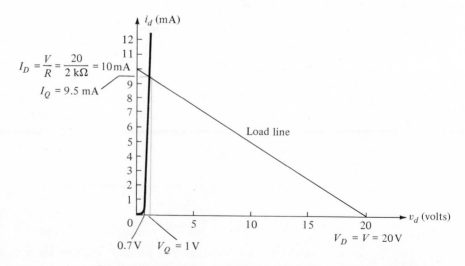

Figure 1.51 Sketching the load line and determining the Q-point for the network of Fig. 1.50.

In Fig. 1.51 the characteristics of a semiconductor diode have also been included. The intersection of the load line and the diode's characteristic curve will determine the point of operation for that diode. This point, due only to the dc input, is called the *quiescent* point. The voltage across and current through the diode can now be found by simply drawing a vertical and a horizontal line, respectively, to the voltage and current axis as indicated in Fig. 1.51. The subscript Q is used to denote quiescent values of current and voltage as shown in Fig. 1.51. The results are (to the accuracy possible)

$$V_Q = 1 \text{ V} \quad \text{and} \quad I_Q = 9.5 \text{ mA}$$

Substituting into Eq. (1.13):

$$V_R = V - V_D = V - V_Q = 20 - 1 = 19.0 \text{ V}$$

or

$$V_R = I_Q R = (9.5 \times 10^{-3})(2 \times 10^3) = 19.0 \text{ V}$$

The power delivered to the load is

$$P_L = I_Q^2 R = (9.5 \times 10^{-3})^2 \, (2 \times 10^3) = 180.5 \text{ mW}$$

or

$$P_L = P_S - P_D$$

where P_S is the power supplied by the source and P_D is the power dissipation of the diode, so that

$$P_L = VI_Q - V_Q I_Q = I_Q(V - V_Q)$$
$$= (9.5 \times 10^{-3})(20 - 1) = 180.5 \text{ mW}$$

Take special note of how closely the semiconductor diode approaches that of the ideal diode for the magnitudes of current and voltage indicated.

1.18   STATIC RESISTANCE

A second glance at Fig. 1.51 will reveal that the diode has a fixed voltage and current associated with the point of operation. Applying Ohm's law to this value will result in the *static* or *dc* resistance of the diode at the quiescent point.

$$R_{dc} = \frac{V_D}{I_D} \tag{1.19}$$

In this case,

$$R_{dc} = \frac{V_D}{I_D} = \frac{1}{9.5 \times 10^{-3}} = 105.3 \text{ } \Omega$$

For the reverse-bias region of a semiconductor diode with $V_D = -20$ V (for example) and $I_S = 1 \text{ } \mu A$:

$$R_{dc} = \frac{V_D}{I_D} = \frac{20}{1 \text{ } \mu A} = 20 \text{ M}\Omega \gg 105.2 \text{ } \Omega \quad \text{(obtained above)}$$

Once the dc resistance has been determined, the diode can be replaced by a resistor of this value. Any change in the applied voltage or load resistor, however, will result in a different Q-point and therefore different dc resistance.

1.19 DYNAMIC RESISTANCE

It is obvious from Fig. 1.51 that the dc resistance of a diode is independent of the shape of the characteristic in the region surrounding the point of interest. If a sinusoidal rather than dc input is applied to the circuit of Fig. 1.51, the situation will change completely. Consider the circuit of Fig. 1.52a, which has as its input a sinusoidal signal on a dc level. Since the magnitude of the dc level is much greater than that of the sinusoidal signal at any instant of time, the diode will always be forward-biased and current will exist continuously in the circuit in the direction shown.

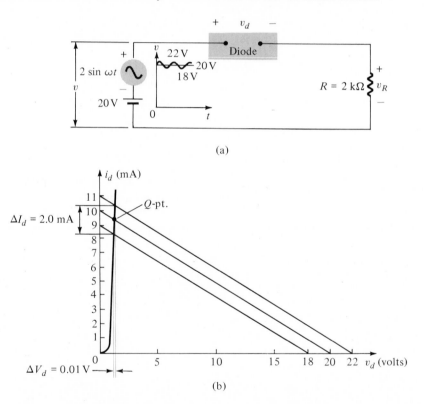

Figure 1.52   ac resistance: (a) circuit; (b) resulting region of operation.

The dc load line resulting from the dc input of 20 V is shown in Fig. 1.52b. The effect of the ac signal is also demonstrated pictorially in the same figure. Note that two additional load lines have been drawn at the positive and negative peaks of the input signal. At the instant the sinusoidal signal is at its positive peak value the input could be replaced by a dc battery with a magnitude of 22 V and the resultant load line drawn as shown. For the negative peak, $V_{dc} = 18$ V. A moment

CH. 1   SEMICONDUCTOR DIODES

of thought, however, should reveal the relative simplicity of superimposing the sinusoidal signal on the dc load line and drawing the load lines coinciding with the positive and negative peaks of the sinusoidal signal.

Note that we are now interested in a region of the diode characteristics as determined by the sinusoidal signal rather than a single point, as was the case for purely dc inputs. Since the resistance will vary from point to point along this region of interest, which value should we use to represent this portion of the characteristic curve? The value chosen is determined by drawing a straight line tangent to the curve at the quiescent point as shown in Fig. 1.53 for the semiconductor diode.

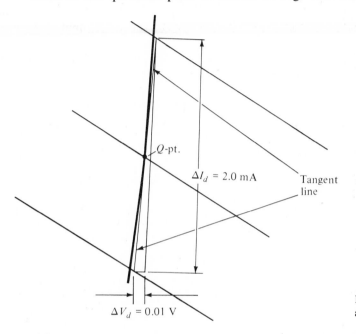

Figure 1.53 Semiconductor diode ac resistance.

The tangent line should "best fit" the characteristics in the region of interest as shown. The resultant resistance, called the *dynamic* or *ac* resistance, is then calculated on an approximate basis, in the following manner:

$$r_d = \frac{\Delta V_d}{\Delta I_d}\bigg|_{\text{tangent line}}$$

(1.20)

and

$$r_d = \frac{\Delta V_d}{\Delta I_d} \cong \frac{0.01}{2 \times 10^{-3}} = 5\ \Omega$$

There is a basic definition in differential calculus that states that *the derivative of a function at a point is equal to the slope of a tangent line drawn at that point.* Equation (1.20), as defined by Fig. 1.53 is, therefore, essentially finding the derivative of the function at the Q-point of operation. If we find the derivative of the general equation [Eq. (1.4)] for the semiconductor diode with respect to the applied forward bias and then invert the result, we will have an equation for the dynamic or ac resistance in that region. That is, taking the derivative of Eq. (1.4) with respect to the applied bias will result in

$$\frac{dI}{dV} = \frac{k}{T_K}(I + I_s)$$

For values of $I \gg I_s$, $I + I_s \cong I$ and, as indicated earlier, $\eta = 1$ for Ge *and* Si in the vertical rise section of the characteristics. Therefore,

$$k = \frac{11,600}{\eta} = \frac{11,600}{1} = 11,600$$

with (at room temperature)

$$T_K = T_C + 273° = 25 + 273° = 298°$$

and

$$\frac{dI}{dV} = \frac{11,600}{298} I \cong 38.93 I$$

or

$$\frac{dV}{dI} = \frac{1}{38.93 I} \cong \frac{0.026}{I}$$

and

$$\boxed{r_d' = \frac{dV}{dI} = \frac{0.026 \text{ V}}{I_D} = \frac{26 \text{ mV}}{I_D \text{ (mA)}}}_{\text{Ge, Si}} \qquad (1.21)$$

The significance of Eq. (1.21) must be understood. It implies that the dynamic resistance can be found by simply substituting the quiescent value of the diode current into the equation. There is no need to have the characteristics available or to worry about sketching tangent lines as defined by Eq. (1.20). Its use will be demonstrated below.

We already realize from Eq. (1.20) that the shape of the curve will have an effect on the dynamic resistance. The fact that the silicon and germanium curves in Fig. 1.27 are almost identical after they begin their vertical rise would suggest that the equation for the dynamic resistance of each might be the same as indicated by Eq. (1.21).

It was already noted on Fig. 1.26 that the characteristics of the commercial unit are slightly different from those determined by Eq. (1.4) because of the bulk and contact resistance of the semiconductor device. This additional resistance level must be included in Eq. (1.21) by adding a factor denoted r_B as appearing in Eq. (1.22).

$$\boxed{r_d = \frac{26 \text{ mV}}{I_D \text{ (mA)}} + r_B} \qquad \text{(ohms)} \qquad (1.22)$$

The factor r_B (measured in ohms) can range from typically 0.1 for high-power devices to 2 for some low-power, general-purpose diodes. As construction techniques improve, this additional factor will continue to decrease in importance until it can be dropped and Eq. (1.21) applied. For values of I_D in mA, the units of the first term are like those of r_B: ohms. For low levels of current, the first factor of Eq. (1.22) will certainly predominate. Consider

$$I_D = 1 \text{ mA}$$

with $$r_B = 2 \; \Omega$$

Then $$r_d = \frac{26}{1} + 2 = \mathbf{28 \; \Omega}$$

At higher levels of current the second factor may predominate. Consider

$$I_D = 52 \; \text{mA}$$

with $$r_B = 2 \; \Omega$$

Then $$r_d = \frac{26}{52} + 2 = 0.5 + 2 = \mathbf{2.5 \; \Omega}$$

For the example provided earlier where r_d was graphically determined to be 5 Ω, if we choose $r_B = 2 \; \Omega$, then

$$r_d = \frac{26}{9.8} + 2 = 2.65 + 2 = 4.65 \; \Omega$$

which is very close to the graphically determined value.

The question of how one is to determine which value to choose for r_B will probably arise. For some devices 2 Ω will be an excellent choice, while for others the approximate average of 1 Ω will perhaps be more appropriate. Certainly, 2 Ω could always be used as a worst-case design approach. However, it would appear that technology is reaching the point where an average value of 1 Ω would, in general, be more appropriate. Of course, the problem of choosing a correct value only arises in the intermediate range of current levels. At low levels of current either choice of r_B would be an insignificant factor. At higher levels the resistance level is so low in comparison to the other series elements that it can probably be ignored. For the purposes of this text the value of r_B chosen for an example will be directly related to the current level; it will extend from a minimum value at high currents of 0.1 Ω to a maximum value of 2 Ω at low levels. Experience will develop a sense for what value to choose, and, indeed, whether it is a factor of significance at all.

In summary, keep in mind that the static or dc resistance of a diode is determined solely by the point of operation, while the dynamic resistance is determined by the shape of the curve in the region of interest.

1.20 AVERAGE AC RESISTANCE

If the input signal is sufficiently large to produce the type of swing indicated in Fig. 1.54, the resistance associated with the device for this region is called the *average ac resistance*. The average ac resistance is, by definition, the resistance determined by a straight line drawn between the two intersections determined by the maximum and minimum values of input voltage. In equation form (note Fig. 1.54)

$$r_{\text{av}} = \frac{\Delta V_d}{\Delta I_d} \bigg|_{\text{pt. to pt.}} \qquad (1.23)$$

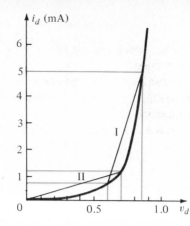

Figure 1.54 Average ac resistance.

For the situation indicated by Fig. 1.54 in region I,

$$r_{av} = \frac{\Delta V_d}{\Delta I_d}\bigg|_{\text{pt. to pt.}} = \frac{0.85 - 0.6}{(5 - 0.75) \times 10^{-3}} = \frac{0.25}{4.25 \times 10^{-3}} = 58.8 \ \Omega$$

For region II:

$$r_{av} = \frac{\Delta V_d}{\Delta I_d}\bigg|_{\text{pt. to pt.}} = \frac{0.7 - 0}{(1.2 - 0) \times 10^{-3}} = 583.3 \ \Omega$$

Note the significant increase in resistance as you progress down the curve. For curves in which the current is the vertical axis and the voltage the horizontal, it is useful to remember that the more horizontal the region, the higher the resistance. Or, if preferred, the more vertical the region, the less the resistance.

It is important to note in this discussion of average ac resistance that the resistance to be associated with the element is determined *only* by the region of interest, *not* by the entire characteristic.

1.21 EQUIVALENT CIRCUITS

An equivalent circuit is a combination of elements properly chosen to best represent the actual terminal characteristics of a device, system, and so on. That is, once the equivalent circuit is determined, the device symbol can be removed from a schematic and the equivalent circuit inserted in its place without severely affecting the behavior of the overall system.

One technique for obtaining an equivalent circuit for a diode is to approximate the characteristics of the device by straight-line segments, as shown in Fig. 1.55. This type of equivalent circuit is called a *piecewise-linear equivalent circuit*. It should be obvious from each curve that the straight-line segments do not result in an exact equivalence between the characteristics and the equivalent circuit. It will, however, at least provide a *first approximation* to its terminal behavior. In each case the resistance chosen is the average ac resistance as defined by Eq. (1.23). The equivalent circuit appears below the curve in Fig. 1.55. The ideal diode was included to indicate that there is only one direction of conduction through the device and that the reverse-bias state is an open-circuit state.

Since a silicon semiconductor diode does not reach the conduction state until approximately 0.7 V, an opposing battery V_o of this value must appear in the equivalent circuit. This indicates that the total forward voltage V_D across the diode must be greater than V_o before the ideal diode in the equivalent circuit will be forward-biased.

Keep in mind, however, that V_o is not an independent source of energy in a system. You will not measure a voltage $V_o = 0.7$ V across an isolated silicon diode using simply a voltmeter. It is simply a useful way of representing the horizontal offset of the semiconductor diode.

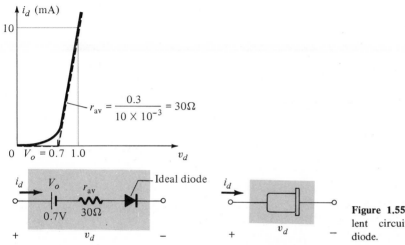

Figure 1.55 Piecewise linear equivalent circuits for a semiconductor diode.

The value of r_{av} can usually be determined purely from a few numerical values given on a specification sheet. The complete characteristics, therefore, are usually unnecessary for this calculation. For instance, for a semiconductor diode, if $I_F = 10$ mA, at 1 V, we know that for silicon a shift of 0.7 V is required before the characteristics rise and

$$r_{av} = \frac{1 - 0.7}{10\,\text{mA}} = \frac{0.3}{10\,\text{mA}} = 30\,\Omega$$

For a germanium diode it would be

$$\frac{1 - 0.3}{10\,\text{mA}} = \frac{0.7}{10\,\text{mA}} = 70\,\Omega$$

The use of the derived equivalent circuit can best be demonstrated by a few examples.

EXAMPLE 1.3 For the network of Fig. 1.50, determine the voltage across R, the total diode drop V_D, and the equivalent dc resistance of the diode.

Solution: The complete equivalent is substituted in Fig. 1.56.

Since the applied voltage of 20 V is much greater than 0.7 V, the ideal diode is forward-biased and the short-circuit equivalent can be substituted. Then, using the voltage-divider rule, we get

$$V_R = \frac{(2\,\text{k}\Omega)(20 - 0.7)}{2\,\text{k}\Omega + 30} = \frac{(2\,\text{k}\Omega)(19.3)}{2030} = \textbf{19.0 V}$$

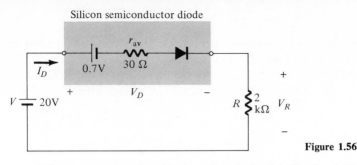

Silicon semiconductor diode

Figure 1.56

This compares exactly with the value obtained in Section 1.17.

The dc current through the circuit is

$$I_D = \frac{20 - 0.7}{2030} = \textbf{9.51 mA}$$

as compared to 9.5 mA and

$$V_D = 0.7 + I_D(r_{av}) = 0.7 + (9.85 \times 10^{-3})(30) = 0.7 + 0.296 \cong \textbf{1 V}$$

as determined earlier.

Finally,

$$R_{dc} = \frac{V_D}{I_D} = \frac{1}{9.51 \text{ mA}} = \textbf{105.15 } \Omega \text{ versus } 102 \text{ } \Omega$$

The results in this case were excellent. It must be realized, however, that when such an equivalent is used, this type of accuracy cannot always be expected, although a good first approximation is usually provided.

Examining the results above, we see that it should be obvious that the 30-Ω forward resistance of the diode is swamped by the 2-kΩ resistor and could be effectively eliminated from the equivalent circuit and still obtain a good first approximate solution to the circuit. That is,

$$V_R = E - V_0 = 20 - 0.7 = \textbf{19.3 V}$$

$$I_D = \frac{20}{2 \text{ k}\Omega} = \textbf{10 mA} \text{ versus } 9.51 \text{ mA} \quad \text{and} \quad V_D = \textbf{0.7 V}$$

This removal of r_{av} from the equivalent circuit is the same as implying that the characteristics of the diode appear as shown in Fig. 1.57. Indeed, this approximation is frequently employed in semiconductor circuit analysis. The reduced equivalent circuit appears in the same figure. It states that a forward-biased silicon diode in an electronic system under dc conditions has a drop of 0.7 V across it in the conduction state no matter what the diode current (within rated values, of course).

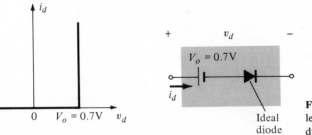

Figure 1.57 Approximate equivalent circuit for the silicon semiconductor diode.

In fact, we can now go a step further and say that the 0.7 V in comparison to the applied 20 V can be ignored leaving only the ideal diode as an equivalent for the semiconductor device. It is for this very reason that many of the applications to follow in later sections use ideal diodes rather than the complete equivalent. Except for small applied voltages or series resistances, it is never too far from the actual response and it does not cloud the application with a great deal of mathematical exercises.

EXAMPLE 1.4 For the input shown in Fig. 1.58, determine the output voltage by using the semiconductor diode of Fig. 1.55. Use the complete equivalent circuit.

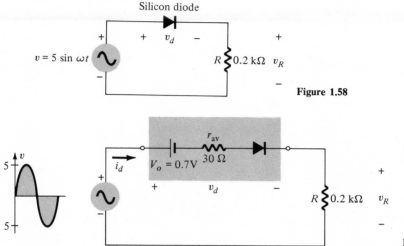

Figure 1.58

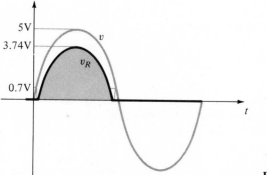

Figure 1.59

Solution: The equivalent circuit is inserted in Fig. 1.59.

The diode will not enter its conduction state (i clockwise through the circuit corresponding with the arrow in the diode symbol) until the applied voltage is greater than 0.7 V. This is shown in the output solution of Fig. 1.60. With the input at its maximum value of 5 V, the output is determined by the voltage divider rule.

$$V_R = \frac{(200)(5 - 0.7)}{200 + 30} = \frac{200}{230}(4.3) = 3.74 \text{ V}$$

Figure 1.60

For an intermediate value such as 3 V:

$$V_R = \frac{(200)(3 - 0.7)}{230} = \frac{200}{230}(2.3) = 2 \text{ V}$$

For v less than 0 V, the ideal diode is certainly reverse-biased, and $v_R = 0$ V, as shown in Fig. 1.60. In this type of application in which the input swing extends throughout a wide range of the characteristics, the piecewise equivalent circuit that employs r_{av} will give excellent results as a first approximation. However, let us now consider the small-signal situation in which the region of operation is very limited.

At high currents, such as 10 mA, if we use Eq. (1.22), we find the resistance to be

$$r_d = \frac{26}{I_D} + 2 = \frac{26}{10} + 2 = 4.65 \text{ } \Omega$$

while at 0.5 mA it is

$$r_d = \frac{26}{I_D} + 2 = \frac{26}{0.5} + 2 = 52 + 2 = 54 \text{ } \Omega$$

Depending on the region of operation, therefore, the average value of 30 Ω may be far from accurate. In this case, since the dc diode current will probably be discernible by first using the approximation of Fig. 1.57, it would be best to substitute this value into Eq. (1.22) and use this level of resistance. For small-signal applications in which the signal rides on a dc level such as in Fig. 1.52, the diode will always be forward-biased and the ideal diode can be replaced by its short-circuit equivalent. The dc level of V_o can be removed from the equivalent for the ac response since it will only determine the "riding" dc level and not affect the peak-to-peak (p–p) ac response. We will find in the analysis of electronic systems that the dc and ac response can normally be determined separately. That is, the theorem of superposition can usually be applied. This will be demonstrated in Example 1.5.

EXAMPLE 1.5 Determine the voltage v_d across the diode for the input shown in Fig. 1.61.

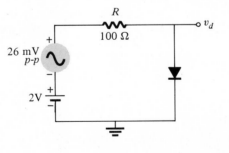

Figure 1.61

Solution: The 2-V dc level will ensure that the diode is always forward-biased. The "dc" equivalent appears in Fig. 1.62 using the approximation of Fig. 1.57. The dc diode current is found to be

$$I_D \cong \frac{2 - 0.7}{100} = \frac{1.3}{100} = 13 \text{ mA}$$

Substituting into Eq. (1.22),

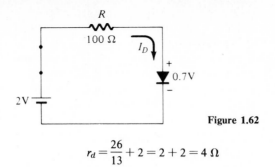

Figure 1.62

$$r_d = \frac{26}{13} + 2 = 2 + 2 = 4\ \Omega$$

The "ac" equivalent is then drawn in Fig. 1.63, and the ac voltage across the diode is determined.

$$v_{d_{ac}} = \frac{4(26\ \text{mV}_{p-p})}{4 + 100} = 1\ \text{mV}_{(p-p)}$$

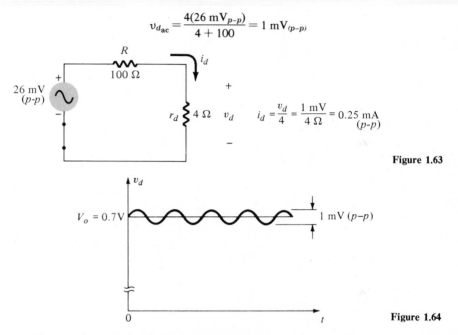

$$i_d = \frac{v_d}{4} = \frac{1\ \text{mV}}{4\ \Omega} = 0.25\ \text{mA}_{(p-p)}$$

Figure 1.63

Figure 1.64

The complete solution (applying the superposition theorem) appears in Fig. 1.64. The technique demonstrated above will be used frequently in the analysis in later chapters.

1.22 CLIPPERS AND CLAMPERS

Clippers and clampers are diode waveshaping circuits. Each performs the function indicated by its name. The output of clipping circuits appears as if a portion of the input signal were clipped off. Clamping circuits simply clamp the waveform to a different dc level.

Clippers

A clipping circuit requires at least two fundamental components, a diode and a resistor. A dc battery, however, is also frequently used. The output waveform can be clipped at different levels simply by interchanging the position of the various elements and changing the magnitude of the dc battery. Only ideal diodes appear in the examples to follow. As indicated in the preceding section, however, the response would not be severely altered if semiconductor devices were used.

For networks of this type it is often helpful to consider particular instants of the time-varying input signal to determine the state of the diode. Keep in mind that *at any instant of time* a varying signal can simply be replaced by a dc source of the same value. This is clearly shown in Example 1.6.

EXAMPLE 1.6
Clipper: Find the output voltage waveshape (v_o) for the inputs shown in Fig. 1.65.

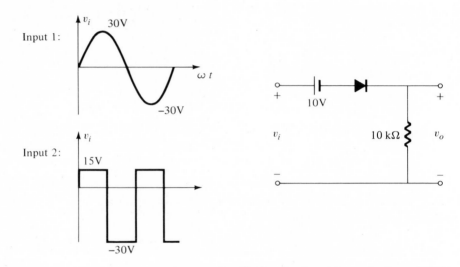

Figure 1.65 Clipping circuit and inputs for Example 1.6.

Solution:
Input 1: For any value of $v_i > 10$ V the ideal diode is forward-biased and $v_o = v_i - 10$. For example, at $v_i = 15$ V (Fig. 1.66), the result is $v_o = 15 - 10 = 5$ V.

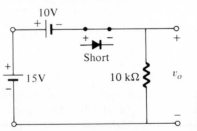

Figure 1.66 Clipping circuit of Fig. 1.65 at instant $v_i = 15$ V of input 1.

For any value of $v_i < 10$ V the ideal diode is reverse-biased and $v_o = 0$ since the current in the circuit is zero. For example, with $v_i = 5$ V (Fig. 1.67), $v_o = 0$, and $v_d = 5$ V (with polarity shown).

The output waveform v_o appears as if the entire input were clipped off except the positive peak (Fig. 1.68).

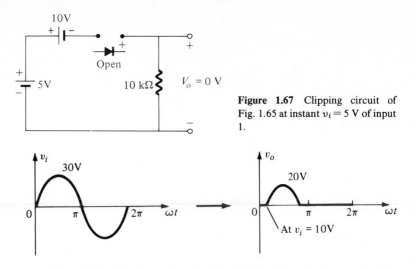

Figure 1.67 Clipping circuit of Fig. 1.65 at instant $v_i = 5$ V of input 1.

Figure 1.68 Output waveform (v_o) for input 1 to the clipping circuit of Fig. 1.65.

Input 2: The diode will change state at the same levels indicated for the first input. The output waveform appears as shown in Fig. 1.69.

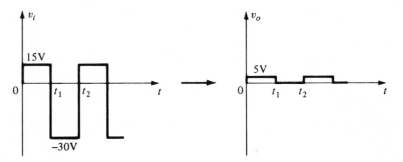

Figure 1.69 Output waveform (v_o) for input 2 to the clipping circuit of Fig. 1.65.

Another technique, other than treating instantaneous values of the input as dc levels, is to redraw the applied voltage as shown in Fig. 1.70. Note that the 10-V dc level has only shifted the sinusoidal input down 10 V (Fig. 1.70). In order for

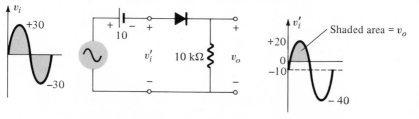

Figure 1.70

the diode to be forward-biased, the input v_i' must be positive. This region as clearly shown in the same figure represents the only region that will pass through to the load. A number of clipping circuits and their effect on the applied signal appear in Fig. 1.71.

SIMPLE SERIES CLIPPERS

POSITIVE

NEGATIVE

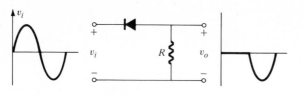

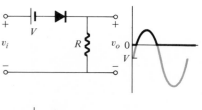

BIASED SERIES CLIPPERS

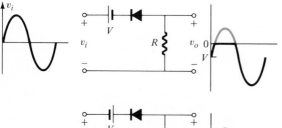

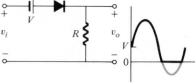

SIMPLE PARALLEL CLIPPERS

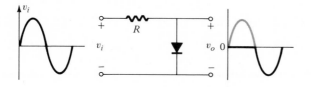

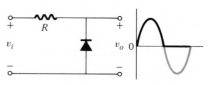

BIASED PARALLEL CLIPPERS

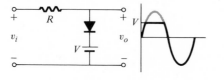

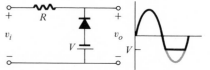

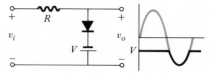

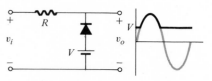

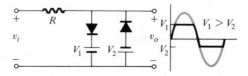

Figure 1.71 Clipping circuits.

56

Clampers

The clamping circuit has a minimum requirement of three elements: a diode, a capacitor, and a resistor. The clamping circuit may also be augmented by a dc battery. The magnitudes of R and C must be chosen such that the time constant $\tau = RC$ is large enough to ensure that the voltage across the capacitor does not change significantly during the interval of time, determined by the input, that both R and C affect the output waveform. The need for this condition will be demonstrated in Example 1.7. Throughout the discussion, we shall assume that for all practical purposes a capacitor will charge to its final value in five time constants.

It is usually advantageous when examining clamping circuits to first consider the conditions that exist when the input is such that the diode is forward-biased.

EXAMPLE 1.7

Clamper: Draw the output voltage waveform (v_o) for the input shown (Fig. 1.72).

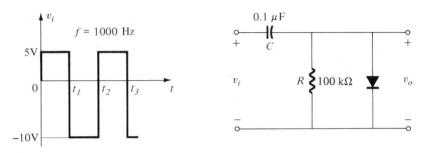

Figure 1.72 Clamping circuit and input for Example 1.7.

Solution: At the instant the input switches to the +5-V state the circuit will appear as shown in Fig. 1.73. The input will remain the +5-V state for an interval of time equal to one-half the period of the waveform since the time interval $0 \rightarrow t_1$ is equal to the interval $t_1 \rightarrow t_2$.

The period of v_i is $T = 1/f = 1/1000 = 1$ ms and the time interval of the +5-V state is $T/2 = 0.5$ ms.

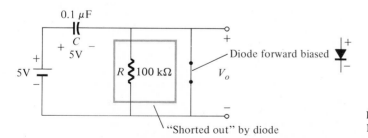

Figure 1.73 Clamping circuit of Fig. 1.72, when $v_i = 5$ V $(0 \rightarrow t_1)$.

Since the output is taken from directly across the diode it is 0 V for this interval of time. The capacitor, however, will rapidly charge to 5 V, since the time constant of the network is now $\tau = RC \cong 0C = 0$.

When the input switches to -10 V, the circuit of Fig. 1.74 will result.

The time constant for the circuit of Fig. 1.74 is

$$\tau = RC = (100 \times 10^3)(0.1 \times 10^{-6}) = 10 \text{ ms}$$

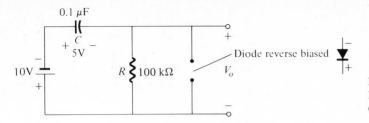

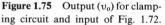

Figure 1.74 Clamping circuit of Fig. 1.72, when $v_i = -10$ V ($t_1 \rightarrow t_2$).

Since it takes approximately five time constants or 50 ms for the capacitor to discharge, and the input is only in this state for 0.5 ms, to assume the voltage across the capacitor does not change appreciably during this interval of time is certainly a reasonable approximation. The output is therefore

$$V_o = -10 \underset{\text{supply}\rule{0pt}{1.2em}}{-} \overset{\text{capacitor}}{5} = -15 \text{ V}$$

The resulting output waveform (v_o) is provided in Fig. 1.75. As indicated, the output is clamped to zero and will repeat itself at the same frequency as the input signal. Note that the swing of the input and output voltages is the same: 15 V. *For all clamping circuits the voltage swing of the input and output waveforms will be the same.* This is certainly not the case for clipping circuits.

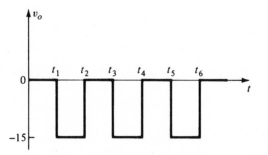

Figure 1.75 Output (v_o) for clamping circuit and input of Fig. 1.72.

If, for the sake of discussion, the 100-kΩ resistor were replaced by a 1-kΩ resistor, the time constant

$$\tau = RC = (10^3)(0.1 \times 10^{-6}) = 0.1 \text{ ms}$$

and $\quad\quad 5\tau = 0.5$ ms   (approximate total discharge time)

The capacitor, therefore, would discharge during the interval in which the voltage is 10 V since the time intervals match. The output waveform would then appear as shown in Fig. 1.76. Clamping networks must have time constants determined by the product RC, which will result in $5RC$ being significantly greater than the time interval in which the diode is reverse-biased or the waveform will be severely distorted.

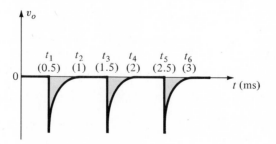

Figure 1.76 $5\tau = T/2$ in Example 1.7.

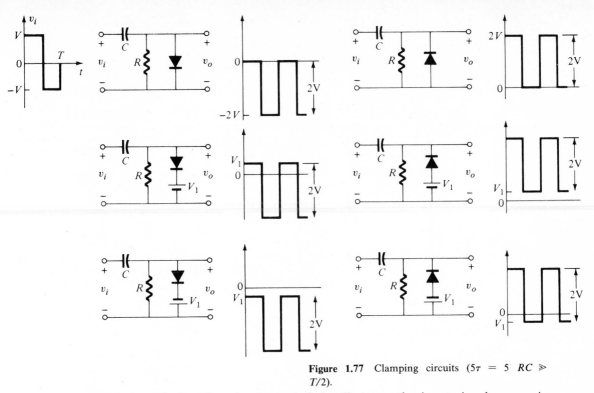

Figure 1.77 Clamping circuits ($5\tau = 5\ RC \gg T/2$).

A number of clamping circuits and their effects on the input signal appear in Fig. 1.77.

1.23 HALF-WAVE RECTIFICATION

The half-wave rectification process was introduced in Section 1.7 as an example in the use of the ideal diode. The results were repeated in Fig. 1.78a for comparison

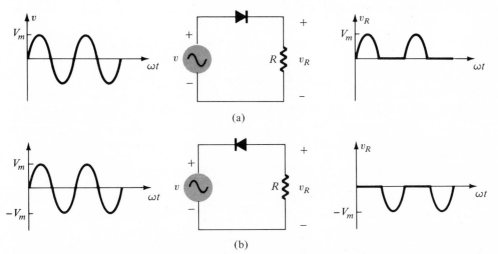

Figure 1.78 Half-wave rectifier circuits: (a) positive half-cycles; (b) negative half-cycles.

with the effect of reversing the diode as appearing in Fig. 1.78b. The analysis of the system of Fig. 1.78b is for obvious reasons very similar to the discussion provided in Section 1.7.

Rectifying networks of this type are most commonly employed in the conversion of sinusoidal ac signals (with zero average value) to pulsating waveforms having positive or negative average (dc) levels. Through filtering action, a very steady dc level can be obtained to meet the internal dc requirements of a system without having to introduce an external dc supply.

To determine the average value of the rectified signal we can calculate the area under the curve of Fig. 1.79 and divide this value by the period of the rectified waveform. To calculate the area under the half-cycle curve of the rectified signal we must integrate the rectified signal.[1] Doing this integration procedure (and dividing by the period) results in

$$\boxed{V_{dc} = 0.318\,V_m} \qquad \text{(half-wave)} \qquad (1.24)$$

where V_m = maximum (peak) value of ac voltage, and
V_{dc} = average value of rectified voltage.

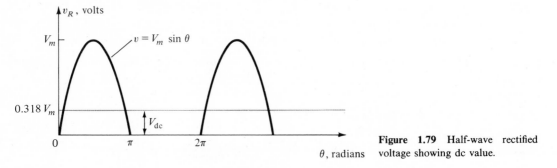

Figure 1.79 Half-wave rectified voltage showing dc value.

EXAMPLE 1.8   Calculate the average voltage of the rectified signal obtained from the circuit of Fig. 1.80.

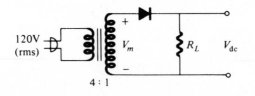

Figure 1.80 Half-wave rectifier circuit for Example 1.8.

[1] The dc signal can be expressed as $v = V_m \sin \theta$ for $0 \le \theta \le \pi$ radians (see Fig. 1.79). For θ from 0 to 2π radians the average value is calculated to be

$$V_{dc} = V_{av} = \frac{1}{T} \int v\, dt = \frac{1}{2\pi} \int_0^\pi (V_m \sin \theta)\, d\theta$$

$$V_{dc} = \frac{V_m}{2\pi}[-\cos \theta]_0^\pi = \frac{V_m}{2\pi}[-1(-1)-(-1)] = \frac{V_m}{\pi} = 0.318\,V_m$$

CH. 1   SEMICONDUCTOR DIODES

Solution:

$$V_m = \tfrac{1}{4}(1.414 \times 120) = 42.42 \text{ V}$$

$$V_{dc} = 0.318\,V_m = (0.318)(42.42) = \mathbf{13.49} \text{ V}$$

An important diode rating is the peak inverse voltage, PIV of the diode (the maximum reverse-bias voltage). For the half-wave rectifier circuit of Fig. 1.81 the peak voltage across the diode when the diode is reverse-biased is shown to be V_m in value.

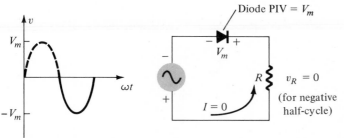

Figure 1.81 Half-wave rectifier circuit showing peak inverse voltage (PIV) across diode.

In Example 1.8 the peak inverse voltage was $V_m = 42.42$ V. The dc voltage obtained, however, was only $V_{dc} = 13.49$ V. This clearly points out one poor feature of the half-wave circuit, namely, the diode PIV rating must be considerably larger than the dc voltage obtained using the circuit. The forward current rating of the diode must equal, at least, the average current through it: V_{dc}/R. The peak current through the diode is V_m/R and must be less than the peak current rating for the diode.

1.24 FULL-WAVE RECTIFICATION

Center-Tapped Transformer Full-Wave Rectifier

It would be preferable to obtain a larger dc voltage compared to the maximum voltage than that of $0.318\,V_m$ for a half-wave rectified signal. In addition, we note that although an average voltage is obtained using a half-wave rectifier, no voltage is developed for half of the cycle. Using two diodes, as shown in Fig. 1.82, it is possible to rectify a sinusoidal signal to obtain one having the same polarity half-cycle for *each* of the half-cycles of input signal. This *full-wave* rectified signal provides a signal that has twice the dc value of the comparable half-wave rectified signal.

The full-wave rectifier circuit of Fig. 1.82 requires a center-tapped transformer and two diodes to develop a full-wave rectified output voltage. To understand how the output waveform is developed, we shall consider the detailed circuit operation for each half-cycle of secondary voltage. Figure 1.83a, shows the circuit operation for the positive half-cycle of secondary voltage. The transformer is center tapped and a peak voltage, V_m, is developed across each half of the transformer during the positive cycle.

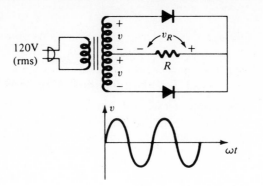

Figure 1.82 Full-wave rectifier circuit.

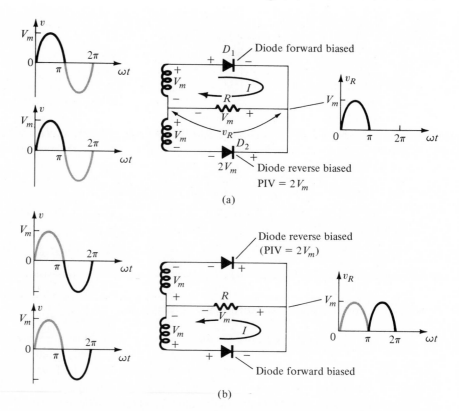

Figure 1.83 Full-wave rectifier, detail circuit operation: (a) positive half-cycle; (b) negative half-cycle.

During the entire positive half-cycle the polarity of the signal across the upper half of the transformer is in a direction to forward-bias diode D_1 causing it to conduct. With diode D_1 conducting, a positive half-cycle of voltage is developed across resistor R as shown in Fig. 1.83a. The figure shows the voltages in the circuit at the time of the peak positive voltage and as shown, there is a voltage V_m across the resistor at this time.

The current in the upper transformer half flows through the transformer, diode D_1, and the load resistor. For a perfect diode ($V_D = 0$, when conducting) the voltage

CH. 1 SEMICONDUCTOR DIODES

across the resistor will equal that of the transformer. At the time the transformer voltage is V_m the voltage across the resistor is also V_m in magnitude as shown in Fig. 1.83a. The voltage developed across the resistor is thus a half-cycle of signal.

The polarity of the voltage developed across the lower half of the transformer results in diode D_2 being back-biased. In addition, the reverse-bias voltage across the diode, which is maximum at the time the maximum voltage V_m is present, is $2V_m$. This is due to the fact that the voltage across reverse-biased diode D_2 is equal to the sum of the voltages across the lower half of the transformer and the load resistor since they are of the same polarity. A diode in this circuit must therefore be capable of handling a reverse-bias voltage equal to twice the value of the peak voltage developed across the output.

During the negative half-cycle diode D_2 in Fig. 1.83b is forward-biased, and diode D_1 is reverse-biased. Current flows through the lower half of the transformer but in the same direction through resistor R as shown in Fig. 1.83b. The output voltage developed across the resistor for the negative half-cycle of input signal is, then, of the same polarity as for the positive half-cycle of input signal. The peak inverse voltage across diode D_1 is $2V_m$, so that each diode must be capable of withstanding a reverse-bias voltage of $2V_m$ sometime during a cycle of operation. The resulting output voltage for a full cycle of input voltage is two positive-going half-cycles.

The average voltage for a full-wave rectified signal is twice that for the half-wave rectified, so that

$$\boxed{V_{\text{dc}} = 2(0.318\,V_m) = 0.636\,V_m} \qquad \text{(full-wave)} \qquad (1.25)$$

The full-wave rectifier circuit of Fig. 1.82 has the advantage of developing a larger dc voltage for the same peak voltage rating. It has, however, the disadvantage of requiring a diode rating of twice the peak inverse voltage, and a center-tapped transformer having twice the overall voltage rating.

EXAMPLE 1.9   Calculate the dc voltage obtained from a center-tapped full-wave rectifier for which the peak rectified voltage is 100 V, and the peak inverse voltage developed across the diode.

Solution:

$$V_{\text{dc}} = 0.636\,V_m = 0.636(100) = \textbf{63.6 V}$$

$$\text{diode PIV} = 2V_m = 2(100) = \textbf{200 V}$$

Bridge Rectifier Circuit

Another circuit variation of a full-wave rectifier is the bridge circuit of Fig. 1.84. This circuit requires four diodes for full-wave rectification but the transformer used is not center tapped and develops a maximum voltage of only V_m. In addition, the diode PIV rating will be shown to be only V_m, rather than $2V_m$.

In considering how the circuit operates we must understand how the conduction and nonconduction paths are formed during each half of the ac cycle. During the positive half-cycle the voltage across the transformer (measured from top to bottom)

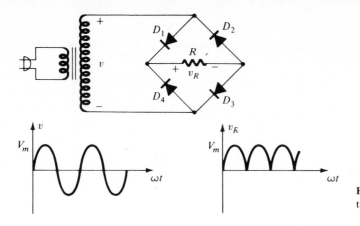

Figure 1.84 Full-wave bridge rectifier circuit.

is positive and the conduction path is shown in Fig. 1.85. Figure 1.85a shows the voltages at the time of the peak positive voltage, V_m. Since the diodes shown are forward-biased, the voltage drop across each is 0 V and the peak voltage from the transformer appears across resistor, R, at this time.

At the same time the voltage polarity is such as to reverse bias diodes D_2 and D_4, as shown in Fig. 1.85b. This represents the nonconduction path during the positive half-cycle of the input ac signal. Resistor R has a voltage developed across it by the current in the conducting path of diodes D_1 and D_3. If the voltage drops around the nonconducting loop are summed, then the transformer voltage and resistor voltage at the time of the peak voltage add up to $2V_m$. Since there are two diodes in the path, the voltage across each reverse-biased diode is V_m. This is half the developed peak inverse voltage across the diodes in the previous full-wave rectifier circuit (Fig. 1.82).

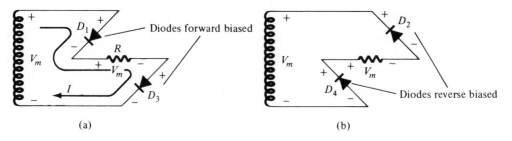

(a)   (b)

Figure 1.85 Bridge circuit, positive half-cycle operation: (a) conduction path; (b) nonconduction path.

During the negative half-cycle the conduction and nonconduction paths are shown in Fig. 1.86. Figure 1.86a shows that diodes D_4 and D_2 are forward-biased. Note carefully that the current, I, goes through resistor R in the same direction as did the current on the previous half-cycle. The voltage across resistor R is thus of the same polarity during each half-cycle of the input signal. During the negative-polarity half-cycle the path of diodes D_1 and D_3 is nonconducting as shown in Fig. 1.86b and the peak inverse voltage developed across each of the diodes is V_m.

To summarize, the addition of two diodes above the number in the center-tapped full-wave circuit provides improvement of two main factors. One, the transformer used need not be center tapped, requiring a maximum voltage across the transformer of V_m. Two, the peak inverse voltage (PIV) required of each diode is half that for the center-tapped full-wave circuit, only V_m. For low values of secondary maximum voltage the center-tapped full-wave circuit will be acceptable, whereas for high values of maximum secondary voltage the use of the bridge to reduce the maximum transformer rating and diode PIV rating is usually necessary.

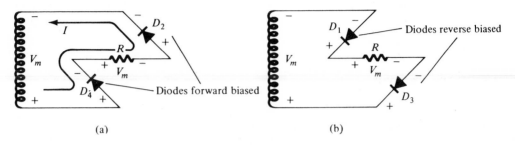

(a)   (b)

Figure 1.86   Bridge circuit, negative half-cycle operation: (a) conduction path; (b) nonconduction path.

EXAMPLE 1.10   The dc voltage developed by a full-wave bridge rectifier circuit is 325 V. Calculate the diode peak inverse voltage rating required for the diodes selected for this circuit.

Solution:

$$V_m = \frac{V_{dc}}{0.636} = \frac{32.5}{0.636} = 51.1 \text{ V}$$

For the bridge rectifier the value of diode PIV is V_m so that diode PIV $= V_m$ = **51.10 V.**

PROBLEMS

§ 1.2

1. In your own words, define semiconductor, resistivity, bulk resistance, and ohmic contact resistance.

2. (a)   Using Table 1.1, determine the resistance of a silicon sample having an area of 1 cm² and a length of 3 cm.
   (b)   Repeat part (a) if the length is 1 cm and the area 4 cm².
   (c)   Repeat part (a) if the length is 8 cm and the area 0.5 cm².
   (d)   Repeat part (a) for copper and compare the results.

3. Sketch the atomic structure of copper and discuss why it is a good conductor and how its structure is different from germanium and silicon.

4. Define, in your own words, an intrinsic material, a negative temperature coefficient, and covalent bonding.

5. Consult your reference library and list three materials that have a negative temperature coefficient and three that have a positive temperature coefficient.

6. How much energy in joules is required to move a charge of 6 C through a difference in potential of 3 V?

7. If 48 eV of energy is required to move a charge through a potential difference of 12 V, determine the charge involved.

8. Consult your reference library and determine the level of E_g for GaP and ZnS, two semiconductor materials of practical value. In addition, determine the written name for each material.

9. Describe the difference between n-type and p-type semiconductor materials.

10. Describe the difference between donor and acceptor impurities.

11. Describe the difference between majority and minority carriers.

12. Sketch the atomic structure of silicon and insert an impurity of arsenic as demonstrated for germanium in Fig. 1.6.

13. Repeat Problem 12 but insert an impurity of indium.

14. Consult your reference library and find another explanation of hole versus electron flow. Using both descriptions, describe in your own words the process of hole conduction.

§ 1.5

15. Describe another example of the diffusion process.

16. How is drift current different from diffusion current?

§ 1.6

17. Describe the Czochralski method for fabricating a single crystal of germanium or silicon.

18. How is the floating-zone technique different from the Czochralski method?

19. What is induction heating? Describe in your own words.

§ 1.7

20. Describe, in your own words, the characteristics of the *ideal* diode and how they determine the on and off state of the device. That is, describe why the short-circuit and open-circuit equivalents are appropriate.

21. (a) For the network of Fig. 1.20, sketch the waveform across the resistor R if its value is 6 kΩ and the input signal has a peak value of 120 V.
(b) Repeat part (a) for the voltage across the diode.
(c) Sketch the waveform for the current in the network for one full cycle of the input voltage.

22. Repeat Problem 21 if the diode is reversed.

23. If a second resistor of 2 kΩ is added in series with the 6-kΩ resistor of Problem 21(a), sketch the waveform of the voltage across the 2-kΩ resistor for the same 120-V input.

24. For the network of Fig. 1.87, determine the voltage across the resistor R_L for the input signals shown in the same figure. Assume an ideal diode.

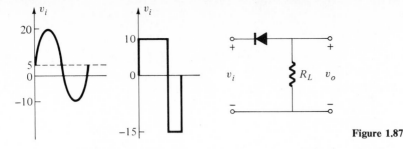

Figure 1.87

25. Repeat Problem 24 for the voltages across the diode.

26. Determine the voltage across the diode in Fig. 1.88 for the inputs appearing in Fig. 1.88. Assume an ideal diode.

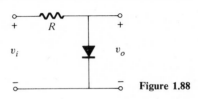

Figure 1.88

§ 1.8

27. Describe in your own words the conditions established by a forward- and reverse-bias condition on a *p-n* junction diode and how it effects the resulting current.

28. Describe how you will remember the forward- and reverse-bias states of the *p-n* junction diode. That is, how will you remember which potential (positive or negative) is applied to which terminal?

29. Referring to Fig. 1.26, determine the average difference in voltage between the typical commercially available Si unit and the characteristic determined by Eq. (1.4) for the range $i_d = 10$ mA to 50 mA.

30. Using Eq. (1.4), determine the diode current at 20°C for a silicon diode with $I_s = 50$ μA and an applied forward bias of 0.6 V.

31. Repeat Problem 30 for $T = 100$°C (boiling point of water). Assume I_s has increased to 50.0 μA.

32. In the reverse-bias region the saturation current of a silicon diode is about 0.1 μA ($T = 20$°C). Determine its approximate value if the temperature is increased 40°C.

33. Compare the characteristics of a silicon and germanium diode and determine which you would prefer to use for most practical applications. Give some detail. Refer to a manufacturer's listing and compare the characteristic of a germanium and silicon diode of similar maximum ratings.

§ 1.9

34. (a) Referring to Fig. 1.28, determine the transition capacitance at reverse-bias potentials of −25 V and −10 V. What is the ratio of the change in capacitance to the change in voltage?
    (b) Repeat part (a) for reverse-bias potentials of −10 V and −1 V. Determine the ratio of the change in capacitance to the change in voltage.

(c) How do the ratios determined in parts (a) and (b) compare? What does it tell you about which range may have more areas of practical application?

35. (a) Referring to Fig. 1.28, determine the diffusion capacitance at 0 V and 0.25 V.
     (b) What is the ratio of the change in level of capacitance to the change in voltage level?

36. Describe in your own words how diffusion and transition capacitances differ.

37. Determine the reactance offered by a diode described by the characteristics of Fig. 1.28 at a forward potential of 0.2 V and a reverse potential of −20 V if the applied frequency is 6 MHz.

§ 1.10

38. Sketch the waveform for i of the network of Fig. 1.89 if $t_t = 2t_s$ and the total reverse recovery time is 9 ns.

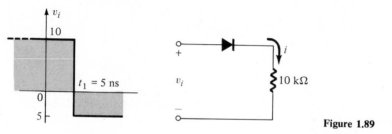

Figure 1.89

§ 1.11

39. Determine the forward voltage drop across the diode whose characteristics appear in Fig. 1.31 at temperatures of −75°C, 25°C, 100°C, and 200°C and a current of 10 mA. For each temperature, determine the level of saturation current. Compare the extremes of each and comment on the ratio of the two.

40. Determine the maximum power dissipation for a semiconductor diode under the following operating conditions: $T_J = 125°C$, $T_A = 25°C$, $\theta_{JC} = 2.0°C/W$, $\theta_{CS} = 0.5°C/W$, and $\theta_{SA} = 2.0°C/W$.

41. Repeat Problem 40 if the heat sink was not employed and $\theta_{JA} = 20°C/W$.

42. (a) If the maximum permissible power dissipation of a diode with a heat sink is 15 W and $T_A = 25°C$, $T_J = 150°C$, and $\theta_{CA} = 3.0°C/W$, determine θ_{JC}.
     (b) What is θ_{JA}?

43. Determine the heat-sink requirement θ_{SA} for the device of Problem 40 if the power demand had to be increased by 25%.

44. (a) If P_M were increased to 5 W in Fig. 1.35 at $T_0 = 100°C$ and T_M remained the same, determine the maximum power rating of the device at 150°C. How does it compare to the results of Example 1.2?
     (b) What is the new value of D_F and θ_{JC}?
     (c) At what temperature T_1 will the power rating drop to 4 W?
     (d) If $\theta_{SA} = 4\theta_{CS}$ and $\theta_{JA} = 25°C/W$, determine θ_{CS}, θ_{SA}, and θ_{CA}. Use the level of θ_{JC} obtained in part (b).

45. Determine the maximum power dissipation for the T151 diode in the forward-bias region. What is the maximum reverse-bias dissipation at $V = -10$ V ($T = 25°C$)?

46. Repeat Problem 45 for the 1N459A diode. Assume I_R does not change significantly with reverse bias.

47. Using the data of Fig. 1.36, sketch the power derating curve for the BAY73 diode and find its rated power level at 50°C.

48. Plot a curve of I_F (ordinate) versus V_F (max) (abscissa) for the BAY73 diode from the data of Fig. 1.36 and make any noteworthy comments.

49. Repeat Problem 48 for the reverse current ($T_A = 25°C$).

50. What is the capacitive reactance of the BA129 diode at a frequency of 1 MHz?

51. If $t_t = t_s$, sketch a curve of the reverse recovery period for the BAY73 diode.

52. Determine the peak current associated with a rated average rectified level of $I_0 = 200$ mA (half-wave rectified signal).

53. How does the curve of V_F versus I_F in Fig. 1.37 compare with the results of Problem 48?

54. (a) Referring to Fig. 1.37, determine the temperature coefficient at a forward current of 1.0 mA under maximum conditions.
(b) Using the results of part (a), determine the change in forward voltage if the temperature should increase 20°C.

55. Compare the power derating curve of Fig. 1.37 with the results of Problem 47.

56. What is the change in dynamic impedance if the current should be reduced from 10 mA to 0.1 mA under maximum conditions? Refer to Fig. 1.37.

§ **1.15**

57. List the types of diode fabrication introduced in Section 1.15 and discuss the disadvantages and advantages when applied to various practical applications.

§ **1.16**

58. What is the maximum power dissipation at 75°C for each diode in the FSA 1410M array? Sketch the power derating curve.

59. Referring to Fig. 1.47, list the detrimental effects of increasing the forward current (per diode) beyond 100 mA and the temperature above room temperature (25°C).

60. If each diode of the FSA 1410M has a current of 40 mA, what is the current through terminal 1? What is the forward voltage drop from pin 1 to 9 if diodes 1, 5, and 9 are active as forward currents (I_F) of 100 mA?

61. If each diode in Fig. 1.49 has a V_F of 0.7 V and an I_F equal to 30 mA, determine the current through terminals 1 and 10 and the voltage from across pins 1 and 10.

§ 1.17

62. Sketch the load line on the characteristics of Fig. 1.51 if the supply voltage is 18 V and $R = 3$ kΩ. Determine the Q-point for the diode and the corresponding levels of current and voltage. What is the dc dissipation of the diode at this quiescent point of operation? Determine the power delivered by the supply and to the load.

63. Repeat Problem 62 by cutting the voltage down to 9 V and the resistor down to 1.5 kΩ. Are the results one-half of those obtained in Problem 62? If not, why?

64. How would you expect the results obtained in Problem 62 to change if a germanium diode were inserted in place of the silicon device?

§ 1.18

65. Determine the static or dc resistance of the diode of Fig. 1.51 at a forward current of 5 mA.

66. Repeat Problem 65 at a forward current of 10 mA. Has the dc resistance been cut in half? If not, why?

67. Determine the dc resistance of the diode of Problem 62 at the Q-point.

§ 1.19

68. Repeat Problem 65 for the dynamic resistance. How do the static and dynamic resistances compare? (Make the necessary approximations, due to the size of the characteristic, for the semiconductor diode.)

69. Determine the dynamic resistance for the conditions of Problem 68 using Eq. (1.22) if r_B is 2 Ω.

70. Using Eq. (1.22), determine the dynamic resistance at 10 mA for the diode of Fig. 1.52 and compare with the results you obtain using Eq. (1.20). Use $r_B = 2$ Ω, and assume the curve changes at a rate of 0.005 V/mA in this region.

§ 1.20

71. Determine the average ac resistance for the diode of Fig. 1.51 for the region between 0.4 V and 0.8 V.

§ 1.21

72. Find the piecewise-linear equivalent circuit for the diode of Fig. 1.51. Assume that the straight-line segment for the semiconductor diode intersects the horizontal axis at 0.7 V (Si).

73. Determine V_R, I_D, V_D, and R_{dc} for the network of Fig. 1.56 using the equivalent circuit obtained in Problem 72 and $V = 10$ V with $R = 0.5$ kΩ.

74. If V in Problem 73 was increased to 50 V, would it be a reasonable approximation to assume that we are working with an ideal diode as in Fig. 1.17? Why?

75. Using the approximate characteristics of Fig. 1.57 for a silicon semiconductor diode, determine the level and appearance of the output voltage for the network of Fig. 1.58 and compare with the results of Example 1.4. Sketch the waveform of v_d using the approximate and piecewise-linear equivalent models for the diode.

CH. 1   SEMICONDUCTOR DIODES

76. If the supply voltage and resistance R of Fig. 1.59 were increased to 40 sin ωt and 2 kΩ, respectively, sketch the appearance of v_R and v_d using the approximate and piecewise-linear equivalent models for the diode. Is it reasonable to assume that we are simply working with an ideal diode as in Fig. 1.17? Why?

77. Determine the peak-to-peak value of v_d in Fig. 1.61 if the dc level is increased to 4 V. Sketch the waveform of the ac sinusoidal voltage across the 100-Ω resistor.

78. Repeat Problem 77 for a dc level of 4 V and the 100-Ω resistor replaced by one of 1 kΩ.

§ 1.22

79. Assuming an ideal diode in the circuit of Fig. 1.90, determine the output waveform for each of the input signals of Fig. 1.91.

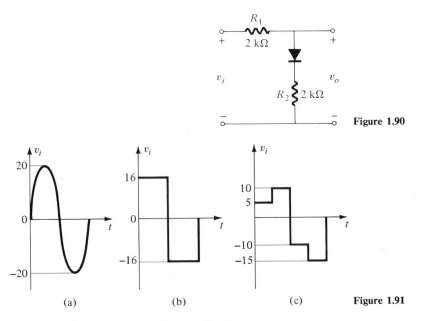

Figure 1.90

Figure 1.91

80. Repeat Problem 79 with the diode reversed.

81. Repeat Problem 79 with the diode and resistor R_1 interchanged.

82. Draw the output waveform for the circuit of Fig. 1.92a for each of the input signals of Fig. 1.91.

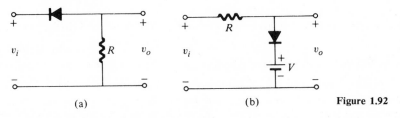

Figure 1.92

83. Draw the output waveform for the circuit of Fig. 1.92b for each of the input signals of Fig. 1.91. Use $V = 5$ V.

84. Repeat Problem 83 for $V = -10$ V.

85. Sketch the output waveform for the network of Fig. 1.93a for the input of Fig. 1.91b.

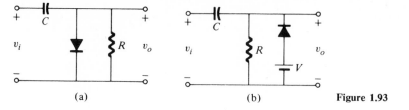

(a)   (b)   **Figure 1.93**

86. Sketch the output waveform for the network of Fig. 1.93b for the input of Fig. 1.91b. Use $V = 5$ V.

87. Repeat Problem 86 for $V = -10$ V.

88. Design a network that will only permit the +5- to +10-V swing of the input of Fig. 81.91c to pass through to v_o.

89. Design a network that will shift the input of Fig. 1.91b to a -10- to -42-V swing.

§ 1.23

90. Calculate the average (dc) voltage of a 90-V, peak rectified half-wave signal.

91. A half-wave rectifier operates off the 120-V (rms) line voltage through a $3:1$ step-down transformer. Calculate the dc voltage of the rectified signal.

92. A half-wave rectifier circuit develops a dc voltage of 180 V. What is the minimum diode peak inverse voltage rating for this circuit?

93. Draw the circuit diagram of a transformer-fed half-wave rectifier providing an output having negative half-cycles.

94. A half-wave rectifier circuit provides 40 mA (dc) to a 2-kΩ load resistance. Calculate the output dc voltage and diode PIV rating for this circuit.

95. A half-wave diode rectifier circuit is transformer-fed from the 120-V line. Calculate the turns ratio, and diode PIV rating, if the circuit provides an output of 12 V dc.

§ 1.24

96. Calculate the dc voltage obtained from a full-wave rectifier having a peak rectified voltage of 90 V.

97. Calculate the peak voltage rating of each half of a center-tapped transformer used in a full-wave rectifier circuit whose output dc voltage is 120 V.

98. Calculate the PIV rating for the diodes of a center-tapped full-wave rectifier circuit having an output dc voltage of 80 V.

99. Draw the circuit diagram of a center-tapped full-wave rectifier circuit developing an output rectified voltage of negative-going voltage cycles.

100. Draw the circuit diagram of a bridge full-wave rectifier circuit developing an output rectified voltage of negative-going voltage cycles. Show the conduction path through the circuit for each half-cycle of the input sinusoidal ac voltage.

101. Calculate the diode PIV rating for a bridge rectifier developing 50 V dc.

102. Design rectifier circuits to provide an output of 100 V dc using (a) half-wave; (b) center-tapped full-wave; and (c) bridge full-wave circuits. For each circuit calculate the transformer peak voltage rating, the diode PIV ratings, and the transformer turns ratio, if power is taken from the 120-V line ac supply.

CHAPTER

2

Zeners and Other
Two-Terminal Devices

2.1 INTRODUCTION

There are a number of two-terminal devices having a single *p-n* junction like the semiconductor diode but with different modes of operation, terminal characteristics, and areas of application. A number, including the Zener, Schottky, tunnel, and varicap diodes, photodiodes, LEDs, and solar cells will be introduced in this chapter. In addition, two-terminal devices of a different construction, such as the photoconductive cell, LCD (liquid-crystal display), and thermistor will be examined.

2.2 ZENER DIODES

The Zener and avalanche region of the semiconductor diode were discussed in detail in Section 1.8. It occurs at a reverse-bias potential of V_Z for the diode of Fig. 2.1a. The Zener diode is a device that is designed to make full use of this Zener region. If we present the characteristics as shown in Fig. 2.1b (the mirror image of Fig. 2.1a) to emphasize the region of interest by placing it in the first quadrant, a similarity appears between the characteristics and those of the ideal diode introduced in Section 1.7. The vertical rise approaches the ideal, although it is offset by a voltage V_Z. Any voltage from 0 to V_Z will result in an open-circuit equivalent as occurred below zero for the ideal diode. In the third quadrant the ideal diode remains an open-circuit while the Zener diode approaches the short-circuit state once again. In total, the almost vertical rise at V_Z has found numerous areas of application that make it a valuable device in electronic design. The first quadrant of Fig. 2.1b is defined by

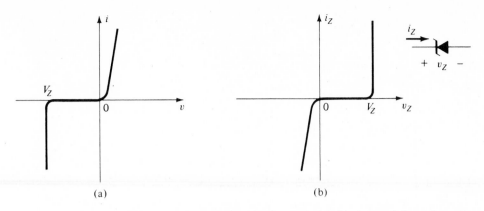

Figure 2.1 Zener diodes: (a) Zener potential; (b) characteristics and notation.

the polarities and current direction appearing next to the Zener diode symbol in the same figure.

The location of the Zener region can be controlled by varying the doping levels. An increase in doping, producing an increase in the number of added impurities, will decrease the Zener potential. Zener diodes are available having Zener potentials of 2.4 to 200 V with power ratings from $\frac{1}{4}$ to 50 W. Because of its higher temperature and current capability, silicon is usually preferred in the manufacture of Zener diodes.

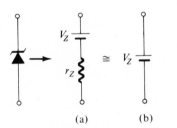

Figure 2.2 Zener equivalent circuit: (a) complete; (b) approximate.

The complete equivalent circuit of the Zener diode in the Zener region includes a small dynamic resistance and dc battery equal to the Zener potential, as shown in Fig. 2.2. For all applications to follow, however, we shall assume as a first approximation that the external resistors are much larger in magnitude than the Zener-equivalent resistor and that the equivalent circuit is simply that indicated in Fig. 2.2b.

A larger drawing of the Zener region is provided in Fig. 2.3 to permit a description of the Zener nameplate data appearing in Table 2.1 for a 1N961, Fairchild, 500 mW, 20% diode.

The term "nominal" associated with V_Z indicates that it is a typical average value. Since this is a 20% diode, the Zener potential can be expected to vary as 10 V $\pm$ 20% or from 8 to 12 V in its range of application. Also available are 10% and 5% diodes with the same specifications. The test current I_{ZT} is a typical operating level and Z_{ZT} is the dynamic impedance at this current level. The maximum knee impedance occurs at the knee current of I_{ZK}. The reverse saturation current is provided at a particular potential level and I_{ZM} is the maximum current for the 20% unit.

The temperature coefficient reflects the percent change in V_Z with temperature. It is defined by the equation

$$T_C = \frac{\Delta V_Z}{V_Z(T_1 - T_0)} \times 100\% \qquad (\%/°C) \qquad (2.1)$$

where ΔV_Z is the resulting change in Zener potential due to the temperature variation.

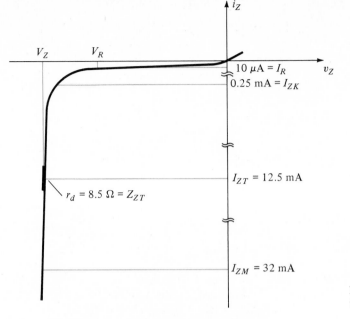

$10 \ \mu A = I_R$

$0.25 \ mA = I_{ZK}$

$I_{ZT} = 12.5 \ mA$

$r_d = 8.5 \ \Omega = Z_{ZT}$

$I_{ZM} = 32 \ mA$

Figure 2.3 Zener test characteristics (Fairchild 1N961).

TABLE 2.1 Electrical Characteristics (25°C Ambient Temperature Unless Otherwise Noted)

Jedec Type	Zener Voltage Nominal, V_Z (V)	Test Current, I_{ZT} (mA)	Max Dynamic Impedance, $Z_{ZT}@I_{ZT}$ (Ω)	Maximum Knee Impedance, $Z_{ZK}@I_{ZK}$ (Ω) (mA)	Maximum Reverse Current, $I_R@V_R$ (μA)	Test Voltage, V_R (V)	Maximum Regulator Current, I_{ZM} (mA)	Typical Temperature Coefficient (%/°C)
1N961	10	12.5	8.5	700 0.25	10	7.2	32	+0.072

Note in Fig. 2.4a that the temperature coefficient can be positive, negative, or even zero for different Zener levels. A positive value would reflect an increase in V_Z with an increase in temperature, while a negative value would result in a decrease in value with increase in temperature. The 24-V, 6.8-V, and 3.6-V levels refer to three Zener diodes having these nominal values within the same family of Zeners as the 1N961. The curve for the 10-V 1N961 Zener would naturally lie between the curves of the 6.8-V and 24-V devices. Note that it has a positive temperature coefficient for the entire region. Returning to Eq. (2.1), T_0 is the temperature at which V_Z is provided (normally room temperature — 25°C) and T_1 is the new level. Example 2.1 will demonstrate the use of Eq. (3.1).

> EXAMPLE 2.1 Determine the nominal voltage for a 1N961 Fairchild Zener diode at a temperature of 100°C.
>
> **Solution:**
> From Eq. (2.1): $\Delta V_Z = \dfrac{T_C V_Z}{100}(T_1 - T_0)$
>
> Substituting yields: $\Delta V_Z = \dfrac{(0.072)(10)}{100}(100 - 25)$

$$= (0.0072)(75)$$

$$= 0.54 \text{ V}$$

and because of the positive temperature coefficient, the new Zener potential, defined by V_Z', is

$$V_z = V_z + 0.54$$

$$= \mathbf{10.54 \ V}$$

The variation in dynamic impedance (fundamentally, its series resistance) with current appears in Fig. 2.4b. Again, the 10-V Zener appears between the 6.8-V and 24-V Zeners. Note that the heavier the current (or the farther up the vertical rise you are in Fig. 2.1b), the less the resistance value. And also note that as you approach the knee of the curve and beyond, the resistance increases to significant levels.

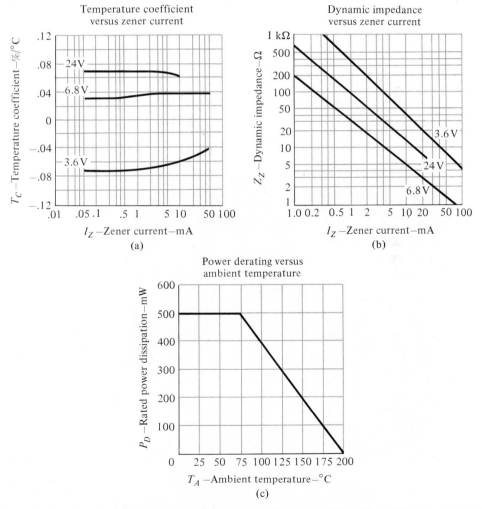

Figure 2.4 Electrical characteristics for a 500-mW Fairchild Zener diode. (Courtesy Fairchild Camera and Instrument Corporation.)

The power derating curve of Fig. 2.4c is very similar to that described for the semiconductor diode. The linear power derating factor for Zener diodes in this family is 3.33 mW/°C. The power rating at the 100°C employed in Example 2.1 can be determined from the derating factor in the following manner:

$$P_{100°C} = P_{75°C} - (D_F)(\Delta T)$$
$$= 500 \times 10^{-3} - (3.33 \times 10^{-3})(25)$$
$$= 500 \times 10^{-3} - 83.25 \times 10^{-3}$$
$$= \mathbf{416.75 \ mW}$$

while the graph would provide a value of approximately 400 mW.

The terminal identification and the casing for a variety of Zener diodes appear in Fig. 2.5

Figure 2.6 is an actual photograph of a variety of Zener devices. Note that their appearance is very similar to the semiconductor diode. A few areas of application for the Zener diode will now be examined.

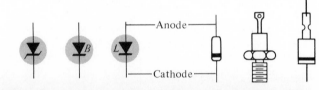

Figure 2.5 Zener terminal identification and symbols.

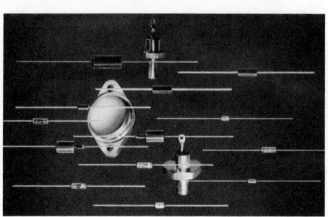

Figure 2.6 Zener diodes (Courtesy Siemens Corporation.)

It is often necessary to have a fixed *reference* voltage in a network for biasing and comparison purposes. This can be accomplished using a Zener diode as shown in Fig. 2.7. The variation in dc supply voltage due to any number of reasons has been included as a small sinusoidal signal.

Since v_i is always greater than 10 V, the Zener diode will always be in the "on" state, a condition defined by the vertical-rise region of Fig. 2.1b. Our analysis here will employ only the reduced equivalent circuit of Fig. 2.2. The output voltage V_o, therefore, will remain fixed at the Zener potential of 10 V, our *reference* potential.

The voltage appearing across the 5-kΩ resistor is then the difference between the two as defined by Kirchhoff's voltage law:

$$v_R = v_i - v_o$$

$$= (12 + 1 \sin \omega t) - 10$$

$$= 2 + 1 \sin \omega t$$

as shown in Fig. 2.7.

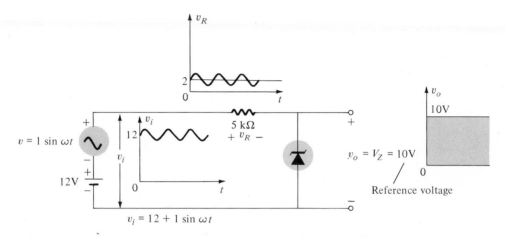

Figure 2.7 Reference voltage.

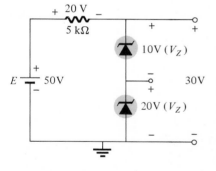

Figure 2.8 Two reference voltages.

Two reference levels can be established by the network of Fig. 2.8. Two back-to-back Zeners can also be used as an ac regulator as shown in Fig. 2.9. For the sinusoidal signal v_i the circuit will appear as shown in Fig. 2.9b at the instant $v_i = 10$ V. The region of operation for each diode is indicated in the adjoining figure. Note that the impedance associated with Z_1 is very small, or essentially a short, since it is in series with 5 kΩ, while the impedance of Z_2 is very large corresponding to the open-circuit representation. Since Z_2 is an open circuit, $v_o = v_i = 10$ V. This will continue to be the case until v_i is slightly greater than 20 V. Then Z_2 will enter the low-resistance region (Zener region) and Z_1 will for all practical purposes be a short circuit and Z_2 will be replaced by $V_Z = 20$ V. The resultant output waveform is indicated in the same figure. Note that the waveform is not purely sinusoidal, but its rms value is closer to the desired 20-V peak sinusoidal waveform than the sinusoidal input having a peak value of 22 V (the rms value of a square wave is its peak value, while the rms value of a sinusoidal function is 0.707 times the peak value).

The circuit of Fig. 2.9a can be extended to that of a simple square-wave generator

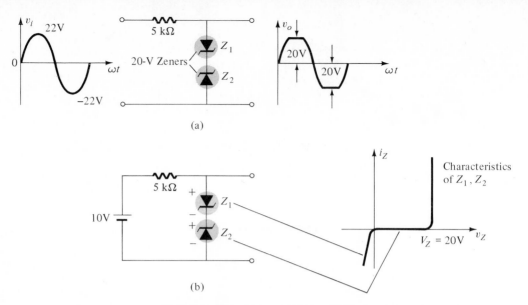

(a)

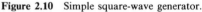

(b)

Figure 2.9 Sinusoidal ac regulation: (a) 40-V peak-to-peak sinusoidal ac regulator; (b) circuit operation at $V_i = 10$ V.

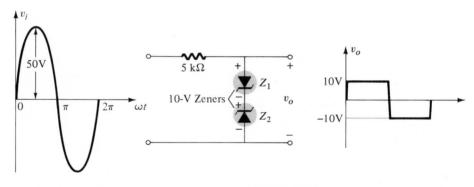

Figure 2.10 Simple square-wave generator.

(due to its clipping action) if the signal v_i is increased to perhaps a 50-V peak with 10-V Zeners. The resulting waveform appears in Fig. 2.10.

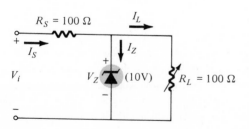

Figure 2.11 Zener regulator.

Finally, the Zener diode will be used to maintain a fixed voltage across a load for a variation in the load impedance (a voltage regulator). The basic Zener regulator appears in Fig. 2.11. In order to maintain V_Z at 10 V (the Zener potential), the diode must first be in the "on" state. Certainly, for values of v_i less than 10 V the Zener device has not reached its Zener potential. Will it "fire" at 12 V? To determine its state, first remove the Zener diode from the network and find the voltage across its open-circuit terminals using the voltage-divider rule (Fig. 2.12).

The resulting voltage is still less than the needed 10 V and the diode remains

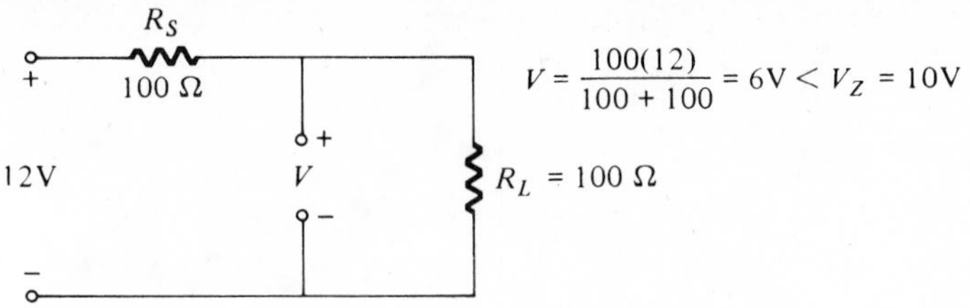

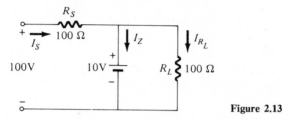

$$V = \frac{100(12)}{100 + 100} = 6V < V_Z = 10V$$

Figure 2.12

open. By setting V_Z to 10 V, we can calculate the required applied voltage in the following manner:

$$V_Z = 10 = \frac{100(V_i)}{100 + 100} \quad \text{and} \quad V_i = 20 \text{ V}$$

Let us now examine the behavior of the network with $V_i = 100$ V (Fig. 2.13).

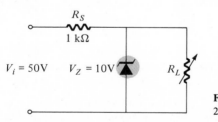

Figure 2.13

The voltage

$$V_{R_S} = 100 - 10 = 90 \text{ V}$$

$$I_S = \frac{90}{100} = 900 \text{ mA}$$

with

$$I_{R_L} = \frac{10}{100} = 100 \text{ mA}$$

and

$$I_Z = I_S - I_L = (900 - 100) \times 10^{-3} = 800 \text{ mA}$$

EXAMPLE 2.2 For the network of Fig. 2.14, determine the range of I_L that will result in V_{R_L} being maintained at 10 V.

R_S
1 kΩ

$V_i = 50V$ $V_Z = 10V$ R_L

Figure 2.14   Network for Example 2.2.

Solution: The minimum value of R_L that will ensure that the diode is in the on state can be determined through Fig. 2.15.

$$V_{R_L} = V_Z = 10 = \frac{R_{l(min)}(50 \text{ V})}{R_{L(min)} + 1000}$$

and

$$10{,}000 + 10R_{L(min)} = 50R_{L(min)}$$

or
$$40R_{L(\text{min})} = 10{,}000$$

and
$$R_{L(\text{min})} = 250 \ \Omega$$

The minimum load resistance corresponds with the maximum I_L and

$$I_{L(\text{max})} = \frac{V_Z}{R_{L(\text{min})}} = \frac{10}{250} = \mathbf{40 \ mA}$$

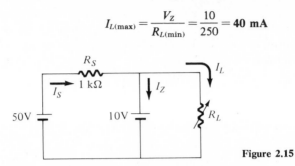

Figure 2.15

$I_{L(\text{min})}$ is determined by examining the network with maximum I_Z (I_{ZM}). When the diode is in the "on" state, $V_S = 50 - 10 = 40$ V and I_S is maintained at $I_S = 40/1000 = 40$ mA. Since $I_L = I_S - I_Z$, I_L will be a minimum when I_Z is a maximum, or I_{ZM} (= 32 mA, Table 2.1). Therefore,

$$I_{L(\text{min})} = I_S - I_{ZM} = (40 - 32) \times 10^{-3} = \mathbf{8 \ mA}$$

In addition:

$$R_{L(\text{max})} = \frac{V_Z}{I_{L(\text{min})}} = \frac{10}{8 \ \text{mA}} = 1.25 \ \text{k}\Omega$$

A plot of V_L versus I_L appears in Fig. 2.16.

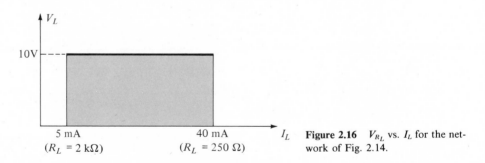

Figure 2.16 V_{R_L} vs. I_L for the network of Fig. 2.14.

2.3 SCHOTTKY BARRIER (HOT-CARRIER) DIODE

In recent years there has been increasing interest in a two-terminal device referred to as a *Schottky-barrier, surface-barrier,* or *hot-carrier* diode. Its areas of application were first limited to the very high frequency range as a competitor for the point-contact diode. It succeeded in this test because it was significantly more rugged and had a quicker response time (especially important at high frequencies) and a lower noise figure (a quantity of real importance in high-frequency applications). In recent years, however, it is appearing more and more in low-voltage/high-current power supplies and ac-to-dc converters. Other areas of application of the device include

CH. 2 ZENERS AND OTHER TWO-TERMINAL DEVICES

radar systems, Schottky TTL logic for computers, mixers and detectors in communication equipment, instrumentation, and analog-to-digital converters.

Its construction is very different from the conventional *p-n* junction in that a metal-semiconductor junction is created such as shown in Fig. 2.17. The semiconductor is normally *n*-type silicon (although *p*-type silicon is sometimes used), while a host of different metals, such as molybdenum, platinum, chrome, or tungsten, are used. Different construction techniques will result in a different set of characteristics for the device, such as increased frequency range, lower forward bias, and so on. Priorities do not permit an examination of each technique here, but information will usually be provided by the manufacturer. In general, however, Schottky diode construction results in a more uniform junction region and increased ruggedness compared to the point-contact diode.

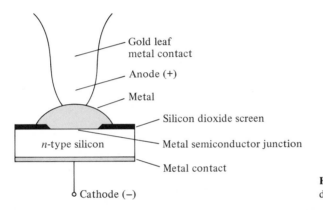

Figure 2.17 Passivated hot-carrier diode.

In both materials, the electron is the majority carrier. In the metal, the level of minority carriers (holes) is insignificant. When the materials are joined the electrons in the *n*-type silicon semiconductor material immediately flow into the adjoining metal, establishing a heavy flow of majority carriers. Since the injected carriers have a very high kinetic energy level compared to the electrons of the metal, they are commonly called "hot carriers." In the conventional *p-n* junction there was the injection of minority carriers into the adjoining region. Here the electrons are injected into a region of the same electron plurality. Schottky diodes are therefore unique, in that conduction is entirely by majority carriers. The heavy flow of electrons into the metal creates a region near the junction surface depleted of carriers in the silicon material—much like the depletion region in the *p-n* junction diode. The additional carriers in the metal establish a negative wall in the metal at the boundary between the two materials. The net result is a "surface barrier" between the two materials, preventing any further current. That is, any electrons (negatively charged) in the silicon material face a carrier-free region and a negative wall at the surface of the metal.

The application of a forward bias as shown in Fig. 2.17 will reduce the strength of the negative barrier through the attraction of the applied positive potential for electrons from this region. The result is a return to the heavy flow of electrons across the boundary, the magnitude of which is controlled by the level of the applied bias potential. The barrier at the junction for a Schottky diode is less than that of the *p-n* junction device in both the forward- and reverse-bias regions. The result is therefore a higher current at the same applied bias in the forward- and reverse-bias regions. This is a desirable effect in the forward-bias region but highly undesirable in the reverse-bias region.

The exponential rise in current with forward bias is described by Eq. (1.4) but with η dependent on the construction technique (1.05 for the metal whisker type of construction, which is somewhat similar to the germanium diode). In the reverse-bias region the current I_s is due primarily to those electrons in the metal passing into the semiconductor material. One of the areas of continuing research on the Schottky diode centers on reducing the high leakage currents that result with temperatures over 100°C. Through design, improvement units are now becoming available that have a temperature range from −65°C to +150°C. At room temperature, I_s is typically in the microampere range for low-power units and milliampere range for high-power devices, although it is typically larger than that encountered using conventional *p-n* junction devices with the same current limits. In addition, even though Schottky diodes exhibit better characteristics than the point-contact diodes in the reverse-bias region as shown in Fig. 2.18, the PIV of these diodes is usually significantly less than that of a comparable *p-n* junction unit. Typically, for a 50-A unit, the PIV of the Schottky diode would be about 50 V as compared to 150 V for the *p-n* junction variety. Recent advances, however, have resulted in Schottky diodes with PIVs greater than 100 V at this current level. It is obvious from the characteristics of Fig. 2.18 that the Schottky diode is closer to the ideal set of characteristics than the point contact and has levels of V_o less than the typical silicon semiconductor *p-n* junction. The level of V_o for the "hot-carrier" diode is controlled to a large measure by the metal employed. There exists a required trade-off between temperature range and level of V_o. An increase in one appears to correspond to a resulting increase in the other. In addition, the lower the range of allowable current levels, the lower

Figure 2.18 Comparison of characteristics of hot-carrier, point-contact, and *p-n* junction diodes.

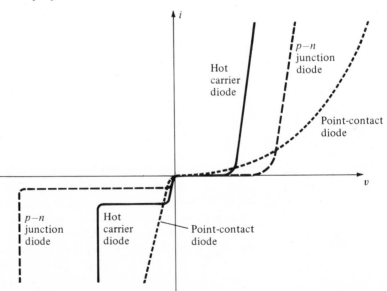

the value of V_0. For some low-level units, the value of V_0 can be assumed to be essentially zero on an approximate basis. For the middle and high range, however, a value of 0.2 V would appear to be a good representative value.

The maximum current rating of the device is presently limited to about 75 A, although 100-A units appear to be on the horizon. One of the primary areas of application of this diode is in *switching power supplies* that operate in the frequency range of 20 kHz or more. A typical unit at 25°C may be rated at 50 A at a forward voltage of 0.6 V with a recovery time of 10 ns for use in one of these supplies. A *p-n* junction device with the same current limit of 50 A may have a forward voltage drop of 1.1 V and a recovery time of 30 to 50 ns. The difference in forward voltage may not appear significant, but consider the power dissipation difference: $P_{\text{hot carrier}}$ = (0.6)(50) = 30 W compared to $P_{p\text{-}n}$ = (1.1)(50) = 55 W, which is a measurable difference when efficiency criteria must be met. There will, of course, be a higher dissipation in the reverse-bias region for the Schottky diode due to the higher leakage current, but the total power loss in the forward- and reverse-bias regions is still significantly improved as compared to the *p-n* junction device.

Recall from our discussion of reverse recovery time for the semiconductor diode that the injected minority carriers accounted for the high level of t_{rr} (the reverse recovery time). The absence of minority carriers at any appreciable level in the Schottky diode results in a reverse recovery time of significantly lower levels, as indicated above. This is the primary reason Schottky diodes are so effective at frequencies approaching 20 GHz, where the device must switch states at a very high rate. For higher frequencies the point-contact diode, with its very small junction area, is still employed.

The equivalent circuit for the device (with typical values) and a commonly used symbol appear in Fig. 2.19. A number of manufacturers prefer to use the standard

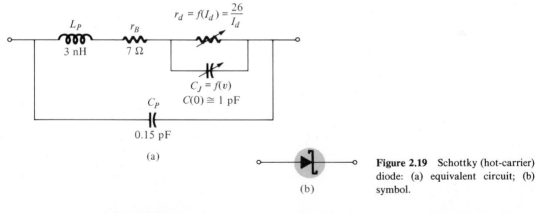

(a)

(b)

Figure 2.19 Schottky (hot-carrier) diode: (a) equivalent circuit; (b) symbol.

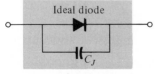

Figure 2.20 Approximate equivalent circuit for the Schottky diode.

diode symbol for the device, since its function is essentially the same. The inductance L_P and capacitance C_P are package values, and r_B is the series resistance, which includes the contact and bulk resistance. The resistance r_d and capacitance C_J are values defined by equations introduced in earlier sections. For many applications, an excellent ap-

I_O, Average rectified forward current (amperes)

	0.5 51-02 (DO-7) Glass	1.0 59-04 Plastic	3.0 267 Plastic	3.0 60 Metal	5.0 60 Metal	15 245 (DO-4) Metal	25 245 (DO-4) Metal	40 257 (DO-5) Metal	40 430-2 (DO-21) Metal
V_{RRM} (Volts)	Case / Anode: / Cathode:								
20	MBR020	1N5817 MBR120P	1N5820 MBR320P	MBR320M	1N5823	1N5826 MBR1520	1N5829 MBR2520	1N5832 MBR4020	MBR4020PF
30	MBR030	1N5818 MBR130P	1N5821 MBR330P	MBR330M	1N5824	1N5827 MBR1530	1N5830 MBR2530	1N5833 MBR4030	MBR4030PF
35		MBR135P	MBR335P	MBR335M		MBR1535	MBR2535	MBR4035	MBR4035PF
40		1N5819 MBR140P	1N5822 MBR340P	MBR340M	1N5825	1N5828 MBR1540	1N5831 MBR2540	1N5834 MBR4040	
I_{FSM} (Amps)	5.0	100	250	200	500	500	500	800	800
T_C @ Rated I_O (°C)		50			85	80	80	75	50
T_J Max (°C)	125	125	125	125	125	125	125	125	125
Max V_F @ $I_{FM}=I_O$	*0.60	0.65	*0.525	0.60	*0.38	*0.50	0.55	*0.59	0.63

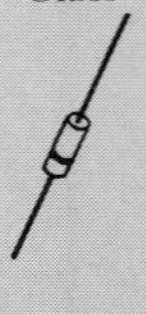

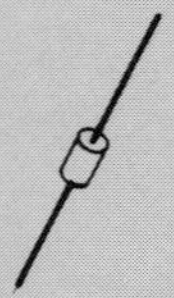

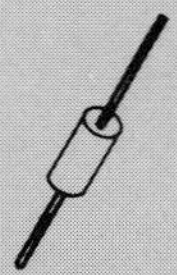

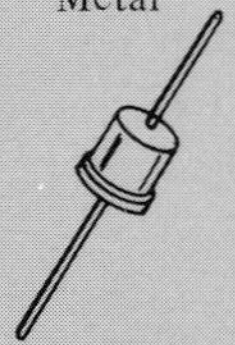

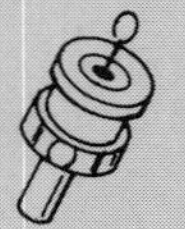

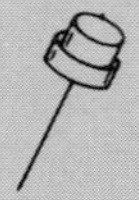

... Schottky barrier devices, ideal for use in low-voltage, high-frequency power supplies and as free-wheeling diodes. These units feature very low forward voltages and switching times estimated at less than 10 ns. They are offered in current ranges of 0.5 to 5.0 amperes and in voltages to 40.

V_{RRM}—respective peak reverse voltage
I_{FSM}—forward current, surge peak
I_{FM}—forward current, maximum

Figure 2.21 Motorola Schottky barrier devices. (Courtesy Motorola Semiconductor Products, Incorporated.)

proximate equivalent circuit simply includes an ideal diode in parallel with the junction capacitance as shown in Fig. 2.20.

A number of hot-carrier rectifiers manufactured by Motorola Semiconductor Products, Inc., appear in Fig. 2.21 with their specifications and terminal identification. Note that the maximum forward voltage drop V_F does not exceed 0.65 V for any of the devices, while this was essentially V_o for a silicon diode.

Three sets of curves for the Hewlett-Packard 5082-2300 series of general-purpose Schottky barrier diodes are provided in Fig. 2.22. Note at $T = 100°C$ in Fig. 2.22a that V_F is only 0.1 V at a current of 0.01 mA. Note also that the reverse current has been limited to nanoamperes in Fig. 2.22b and the capacitance to 1 pF in Fig. 2.22c to ensure a high switching rate.

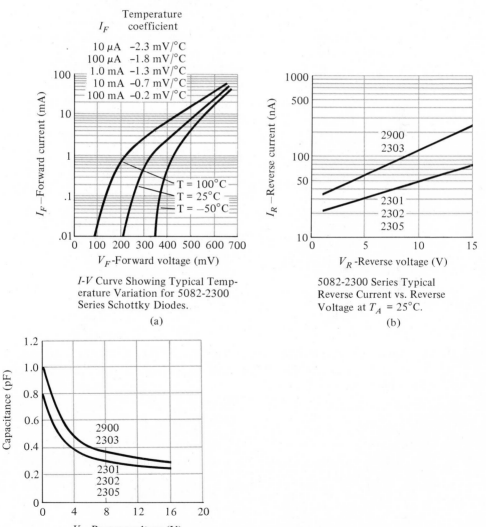

I-V Curve Showing Typical Temperature Variation for 5082-2300 Series Schottky Diodes.

(a)

5082-2300 Series Typical Reverse Current vs. Reverse Voltage at $T_A = 25°C$.

(b)

5082-2300 Series Typical Capacitance vs. Reverse Voltage at $T_A = 25°C$.

(c)

Figure 2.22 Characteristic curves for Hewlett-Packard 5082–2300 series of general-purpose Schottky barrier diodes. (Courtesy Hewlett-Packard Corporation.)

87

2.4 VARACTOR (VARICAP) DIODES

Varactor [also called varicap, VVC (voltage-variable capacitance), or tuning] diodes are semiconductor, voltage-dependent, variable capacitors. Their mode of operation depends on the capacitance that exists at the *p-n* junction when the element is reverse-biased. Under reverse-bias conditions, it was established in Section 1.8 that there is a region of uncovered charge on either side of the junction that together make up the depletion region and define the depletion width W_d. The transition capacitance (C_T) established by the isolated uncovered charges is determined by

$$C_T = \frac{\epsilon A}{W_d} \qquad (2.2)$$

where ϵ is the permittivity of the semiconductor materials, A is the *p-n* junction area, and W_d is the depletion width.

As the reverse-bias potential increases, the width of the depletion region increases, which in turn reduces the transition capacitance. The characteristics of a typical commercially available varicap diode appear in Fig. 2.23. Note the initial sharp decline

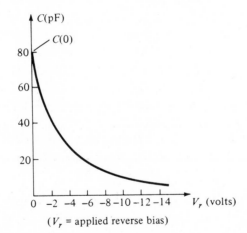

$(V_r$ = applied reverse bias)

Figure 2.23 Varicap characteristics: C(pF) vs. V_r.

in C_T with increase in reverse bias. The normal range of V_r for VVC diodes is limited to about 20 V. In terms of the applied reverse bias, the transition capacitance is given by

$$C_T = \frac{K}{(V_o + V_r)^n} \qquad (2.3)$$

where K = constant determined by the semiconductor material and construction technique

V_o = knee potential as defined in Section 1.8
V_r = applied reverse-bias potential
$n = \frac{1}{2}$ for alloy junctions and $\frac{1}{3}$ for diffused junctions

In terms of the capacitance at the zero-bias condition $C(0)$, the capacitance as a function of V_r is given by

$$C_T(V_r) = \frac{C(0)}{\left(1 + \dfrac{V_r}{V_o}\right)^n} \qquad (2.4)$$

The symbols most commonly used for the varicap diode and a first approximation for its equivalent circuit in the reverse-bias region are shown in Fig. 2.24. Since we are in the reverse-bias region, the resistance in the equivalent circuit is very large in magnitude—typically 1 MΩ or larger—while R_S, the geometric resistance of the diode, is, as indicated in Fig. 2.24, very small. The magnitude of C will vary from about 2 to 100 pF depending on the varicap considered. To ensure that R_r is as large (for minimum leakage current) as possible, silicon is normally used in varicap

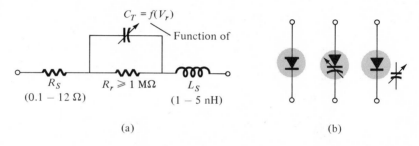

(a)   (b)

Figure 2.24 Varicap diode: (a) equivalent circuit in the reverse-bias region; (b) symbols.

diodes. The fact that the device will be employed at very high frequencies requires that we include the inductance L_s even though it is measured in nanohenries. Recall that $X_L = 2\pi f L$ and a frequency of 10 GHz with $L_s = 1$ nH will result in an $X_{Ls} = 2\pi f L = (6.28)(10^{12})(10^{-9}) = 62.8$ Ω. There is obviously, therefore, a frequency limit associated with the use of each varicap diode.

Assuming the proper frequency range, a low value of R_S, and X_{L_s} compared to the other series elements, then the equivalent circuit for the varicap of Fig. 2.24a can be replaced by the variable capacitor alone. The complete data sheet and its characteristic curves appear in Figs. 2.25 and 2.26, respectively. The C_3/C_{25} ratio in Fig. 2.25 is the ratio of capacitance levels at reverse-bias potentials of 3 and 25 V. It provides a quick estimate of how much the capacitance will change with reverse-bias potential. The figure of merit is a quantity of consideration in the application of the device and is a measure of the ratio of energy stored by the capacitive device per cycle to the energy dissipated (or lost) per cycle. Since energy loss is seldom considered a positive attribute, the higher its relative value the better. The resonant frequency of the device is determined by $f_0 = 1/2\pi\sqrt{LC}$ and affects the range of application of the device.

In Fig. 2.26, most quantities are self-explanatory. However, the *capacitance temperature coefficient* is defined by

$$TC_C = \frac{\Delta C}{C_0(T_1 - T_0)} \times 100\% \qquad (\%/°C) \qquad (2.5)$$

BB139

VHF / FM VARACTOR DIODE
DIFFUSED SILICON PLANAR

- C_3/C_{25}... 5.0–6.5
- MATCHED SETS (Note 2)

ABSOLUTE MAXIMUM RATINGS (Note 1)

Temperatures

Storage Temperature Range	−55°C to +150°C
Maximum Junction Operating Temperature	+150°C
Lead Temperature	+260°C

Maximum Voltage

WIV   Working Inverse Voltage   30 V

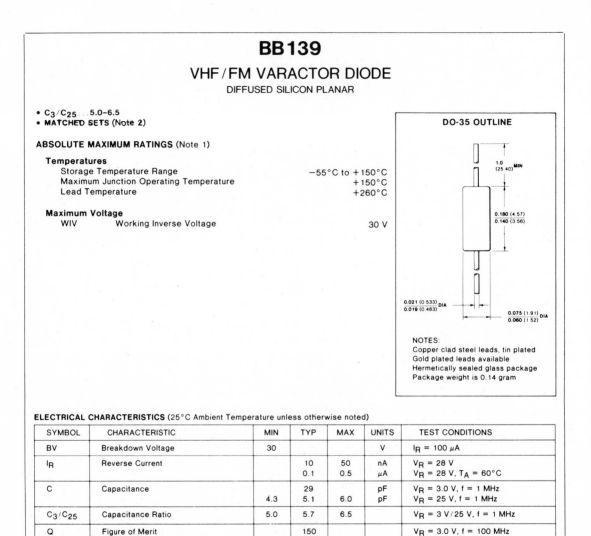

DO-35 OUTLINE

1.0 (25.40) MIN
0.180 (4.57)
0.140 (3.56)
0.021 (0.533) DIA
0.019 (0.483)
0.075 (1.91) DIA
0.060 (1.52)

NOTES:
Copper clad steel leads, tin plated
Gold plated leads available
Hermetically sealed glass package
Package weight is 0.14 gram

ELECTRICAL CHARACTERISTICS (25°C Ambient Temperature unless otherwise noted)

SYMBOL	CHARACTERISTIC	MIN	TYP	MAX	UNITS	TEST CONDITIONS
BV	Breakdown Voltage	30			V	$I_R = 100\ \mu A$
I_R	Reverse Current		10	50	nA	$V_R = 28$ V
			0.1	0.5	μA	$V_R = 28$ V, $T_A = 60°C$
C	Capacitance		29		pF	$V_R = 3.0$ V, f = 1 MHz
		4.3	5.1	6.0	pF	$V_R = 25$ V, f = 1 MHz
C_3/C_{25}	Capacitance Ratio	5.0	5.7	6.5		$V_R = 3$ V/25 V, f = 1 MHz
Q	Figure of Merit		150			$V_R = 3.0$ V, f = 100 MHz
R_S	Series Resistance		0.35		Ω	C = 10 pF, f = 600 MHz
L_S	Series Inductance		2.5		nH	1.5 mm from case
f_0	Series Resonant Frequency		1.4		GHz	$V_R = 25$ V

NOTES:
1. These ratings are limiting values above which the serviceability of the diode may be impaired.
2. The capacitance difference between any two diodes in one set is less than 3% over the reverse voltage range of 0.5 V to 28 V

Figure 2.25   Electrical characteristics for a VHF/FM Fairchild Varactor Diode. (Courtesy Fairchild Camera and Instrument Corporation.)

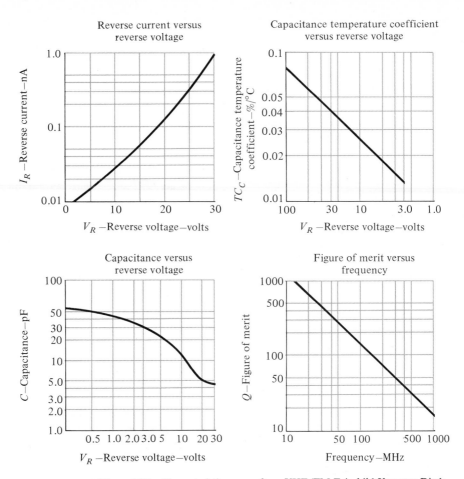

Figure 2.26 Characteristic curves for a VHF/FM Fairchild Varactor Diode. (Courtesy Fairchild Camera and Instrument Corporation.)

where ΔC is the change in capacitance due to the temperature change $T_1 - T_0$ and C_0 is the capacitance at T_0 for a particular reverse-bias potential. For example, Fig. 2.25 indicates that $C_0 = 29$ pF with $V_R = 3$ V and $T_0 = 25°C$. A change in capacitance ΔC could then be determined using Eq. (2.5) by simply substituting the new temperature T_1 and the TC_C as determined from the graph (= 0.013). At a new V_R the value of TC_C would change accordingly. Returning to Fig. 2.25, note that the maximum frequency appearing is 600 MHz. At this frequency,

$$X_L = 2\pi fL = (6.28)(600 \times 10^6)(2.5 \times 10^{-9}) = 9.42 \ \Omega$$

normally a quantity of sufficiently small magnitude to be ignored.

Some of the high-frequency (as defined by the small capacitance levels) areas of application include FM modulators, automatic-frequency-control devices, adjustable band-pass filters, and parametric amplifiers.

2.5 POWER DIODES

There are a number of diodes designed specifically to handle the high-power and high-temperature demands of some applications. The most frequent use of power diodes occurs in the rectification process, in which ac signals (having zero average value) are converted to ones having an average or dc level. When used in this capacity, diodes are normally referred to as *rectifiers*.

The majority of the power diodes are constructed using silicon because of its higher current, temperature, and PIV ratings. The higher current demands require that the junction area be larger, to ensure that there is a low forward diode resistance. If the forward resistance were too large, the I^2R losses would be excessive. The current capability of power diodes can be increased by placing two or more in parallel and the PIV rating can be increased by stacking the diodes in series.

Various types of power diodes and their current rating have been provided in Fig. 2.27a. The high temperatures resulting from the heavy current require, in many cases, that heat sinks be used to draw the heat away from the element. A few of the various types of heat sinks available are shown in Fig. 2.27b. If heat sinks are not employed, stud diodes are designed to be attached directly to the chassis, which in turn will act as the heat sink. The analysis associated with the temperature limitations is exactly the same as presented for semiconductor diodes in Section 1.11.

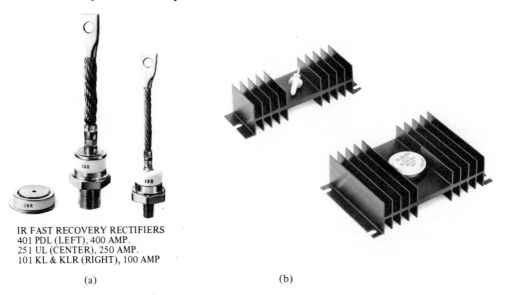

IR FAST RECOVERY RECTIFIERS
401 PDL (LEFT), 400 AMP.
251 UL (CENTER), 250 AMP.
101 KL & KLR (RIGHT), 100 AMP

(a)   (b)

Figure 2.27 Power diodes and heat sinks. (Courtesy International Rectifier Corporation.)

2.6 TUNNEL DIODES

The tunnel diode was first introduced by Leo Esaki in 1958. Its characteristics, shown in Fig. 2.28, are different from any diode discussed thus far in that it has a negative-resistance region. In this region, an increase in terminal voltage results in a reduction in diode current.

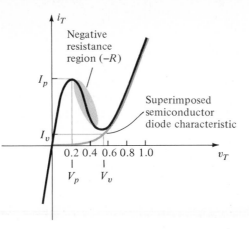

Figure 2.28 Tunnel diode characteristic.

The tunnel diode is fabricated by doping the semiconductor materials that will form the *p-n* junction at a level one hundred to several thousand times that of a typical semiconductor diode. This will result in a greatly reduced depletion region, of the order of magnitude of 10^{-6} cm, or typically about $\frac{1}{100}$ the width of this region for a typical semiconductor diode. It is this thin depletion region that many carriers can "tunnel" through, rather than attempt to surmount, at low forward-bias potentials that accounts for the peak in the curve of Fig. 2.28. For comparison purposes, a typical semiconductor diode characteristic has been superimposed on the tunnel-diode characteristic of Fig. 2.28.

This reduced depletion region results in carriers "punching through" at velocities that far exceed those available with conventional diodes. The tunnel diode can therefore be used in high-speed applications such as in computers, where switching times in the order of nanoseconds or picoseconds are desirable.

You will recall from a previous section that an increase in the doping level will drop the Zener potential. Note the effect of a very high doping level on this region in Fig. 2.28. The semiconductor materials most frequently used in the manufacture of tunnel diodes are germanium and gallium arsenide. The ratio I_p/I_v is very important for computer applications. For germanium it is typically 10:1, while for gallium arsenide it is closer to 20:1.

The peak current, I_p, of a tunnel diode can vary from a few microamperes to several hundred amperes. The peak voltage, however, is limited to about 600 mV. For this reason, a simple VOM with an internal dc battery potential of 1.5 V can severely damage a tunnel diode if used improperly.

The tunnel diode equivalent circuit in the negative-resistance region is provided

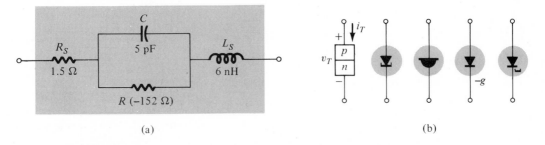

Figure 2.29 Tunnel diode: (a) equivalent circuit; (b) symbols.

in Fig. 2.29, with the symbols most frequently employed for tunnel diodes. The values for each parameter are for the 1N2939 GE tunnel diode whose specifications appear in Table 2.2. The inductor L_S is due mainly to the terminal leads. The resistor R_S is due to the leads, ohmic contact at the lead-semiconductor junction, and the semiconductor materials themselves. The capacitance C is the junction diffusion capacitance and the R is the negative resistance of the region. The negative resistance finds application in oscillators to be described later.

TABLE 2.2 Specifications: Ge 1N2939

	Min.	Typ.	Max.	
Absolute maximum ratings (25°C)				
Forward current (−55 to + 100°C)		5 mA		
Reverse current (−55 to + 100°C)		10 mA		
Electrical characteristics (25°C)($\frac{1}{8}$ in. leads)				
I_P	0.9	1.0	1.1	mA
I_V		0.1	0.14	mA
V_P	50	60	65	mV
V_V		350		mV
Reverse voltage ($I_R = 1.0$ mA)			30	mV
Forward peak point current voltage, V_{fp}	450	500	600	mV
I_p/I_v		10		
$-R$		−152		Ω
C		5	15	pF
L_S		6		nH
R_S		1.5	4.0	Ω

Note the lead length of $\frac{1}{8}$ in. included in the specifications. An increase in this length will cause L_S to increase. In fact, it was given for this device that L_S will vary 1 to 12 nH, depending on lead length. At high frequencies ($X_{L_S} = 2\pi f L_S$) this factor can take its toll.

The fact that $V_{fp} = 500$ mV (typ.) and $I_{forward}$ (max.) = 5 mA indicates that tunnel diodes are low-power devices [$P_D = (0.5)(5 \times 10^{-3}) = 2.5$ mW], which is also excellent for computer applications. A rendering of the device appears in Fig. 2.30.

Figure 2.30 A Ge IN2939 tunnel diode. (Courtesy General Electric Corporation.)

Although the use of tunnel diodes in present-day high-frequency systems has been dramatically stalled by manufacturing techniques that suggest alternatives to the tunnel diode, its simplicity, linearity, low power drain, and reliability ensure its continued life and application. The basic construction of an advance design tunnel diode appears in Fig. 2.31 with a photograph of the actual junction.

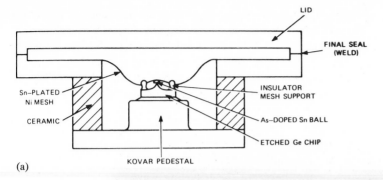

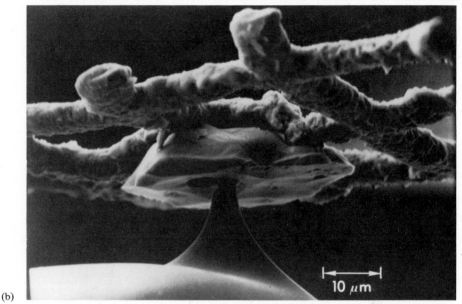

Figure 2.31 Tunnel diode: (a) construction; (b) photograph. (Courtesy *COMSAT Technical Review*, Vol. 3, No. 1, Spring 1973, P. F. Varadi and T. D. Kirkendall.)

2.7 PHOTODIODE

The interest in light-sensitive devices has been increasing at an almost exponential rate in recent years. The resulting field of *optoelectronics* will be receiving a great deal of research interest as efforts are made to improve efficiency levels.

Through the advertising media, the layperson has become quite aware that light sources offer a unique source of energy. This energy, transmitted as discrete packages called *photons*, has a level directly related to the frequency of the traveling light wave as determined by the following equation:

$$W = hf \qquad \text{(joules)} \qquad (2.6)$$

where h is called Planck's constant and is equal to 6.624×10^{-34} joule-seconds. It clearly states that since K is a constant, the energy associated with incident light waves is directly related to the frequency of the traveling wave.

The frequency is, in turn, directly related to the wavelength (distance between successive peaks) of the traveling wave by the following equation:

$$\lambda = \frac{v}{f} \tag{2.7}$$

where
λ = wavelength, meters
v = velocity of light, 3×10^8 m/s
f = frequency, hertz, of the traveling wave

The wavelength is usually measured in angstrom units ($\mathring{A}$) or microns (μ), where

$$1\ \mathring{A} = 10^{-10}\ m \quad \text{and} \quad 1\ \mu = 10^{-6}\ m$$

The wavelength is important because it will determine the material to be used in the optoelectronic device. The relative spectral response for Ge, Si, and selenium is provided in Fig. 2.32. The visible light spectrum has also been indicated with an indication of the wavelength associated with the various colors.

The number of free electrons generated in each material is proportional to the

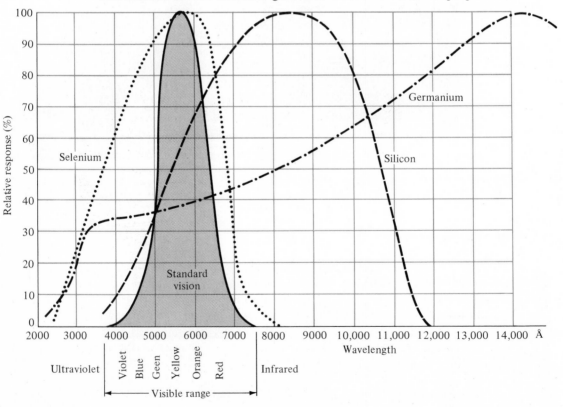
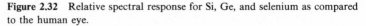

Figure 2.32 Relative spectral response for Si, Ge, and selenium as compared to the human eye.

CH. 2 ZENERS AND OTHER TWO-TERMINAL DEVICES

intensity of the incident light. Light intensity is a measure of the amount of *luminous flux* falling in a particular surface area. Luminous flux is normally measured in *lumens* (lm) or watts. The two units are related by

$$1 \text{ lm} = 1.496 \times 10^{-10} \text{ W}$$

The light intensity is normally measured in lm/ft^2, foot-candles (fc), or W/m^2, where

$$1 \text{ lm/ft}^2 = 1 \text{ fc} = 1.609 \times 10^{-12} \text{ W/m}^2$$

The photodiode is a semiconductor *p-n* junction device whose region of operation is limited to the reverse-bias region. The basic biasing arrangement, construction, and symbol for the device appear in Fig. 2.33.

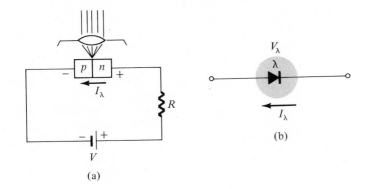

(a)

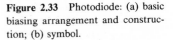

(b)

Figure 2.33 Photodiode: (a) basic biasing arrangement and construction; (b) symbol.

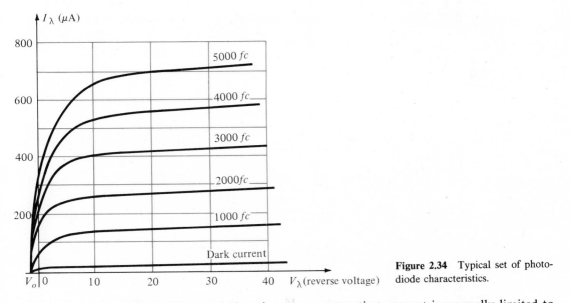

Figure 2.34 Typical set of photodiode characteristics.

Recall from Chapter 1 that the reverse saturation current is normally limited to a few microamperes. It is due solely to the thermally generated minority carriers in the *n*- and *p*-type materials. The application of light to the junction will result in

Figure 2.35 Photograph of Hewlett-Packard 5082–4200 S pin photo-diodes. (Courtesy Hewlett-Packard Corporation.)

a transfer of energy from the incident traveling light waves (in the form of photons) to the atomic structure, resulting in an increased number of minority carriers and an increased level of reverse current. This is clearly shown in Fig. 2.34 for different intensity levels. The *dark* current is that current that will exist with no applied illumination. Note that the current will only return to zero with a positive applied bias equal to V_o. In addition, Fig. 2.33 demonstrates the use of a lens to concentrate the light on the junction region. An actual device showing the lens in the cap appears in Fig. 2.35.

The almost equal spacing between the curves for the same increment in luminous flux reveals that the reverse current and luminous flux are almost linerally related. In other words, an increase in light intensity will result in a similar increase in reverse current. A plot of the two to show this linear relationship appears in Fig. 2.36 for a fixed voltage V_λ of 20 V. On the relative basis we can assume that the reverse current is essentially zero in the absence of incident light. Since the rise and fall times (change-of-state parameters) are very small for this device (in the nanosecond range), the device can be used for high-speed counting or switching applications. Returning to Fig. 2.32, we note that Ge encompasses a wider spectrum of wavelengths than Si. This would make it suitable for incident light in the infrared region as provided by lasers and IR (infrared) light sources to be described shortly. Of course, Ge has a higher dark current than silicon, but it also has a higher level of reverse current. The level of current generated by the incident light on a photodiode is not such that it could be used as a direct control, but it can be amplified for this purpose.

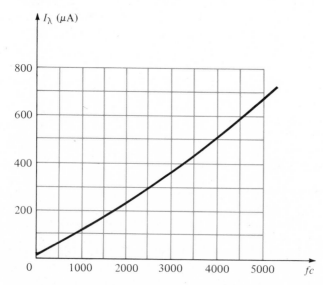

Figure 2.36 I_λ vs. *fc* (at $V_A = 20$ V) for the photodiode of Fig. 2.34.

2.8 PHOTOCONDUCTIVE CELL

The photoconductive cell is a two-terminal semiconductor device whose terminal resistance will vary (linearly) with the intensity of the incident light. For obvious reasons, it is frequently called a photoresistive device. A typical photoconductive cell and the most widely used graphic symbol for the device appear in Fig. 2.37.

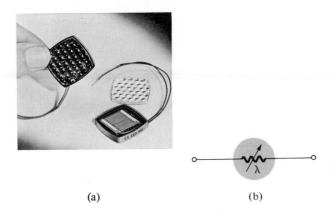

(a)

(b)

Figure 2.37 Photoconductive cell: (a) appearance; (b) symbol. [(a) Courtesy International Rectifier Corporation.]

The photoconductive materials most frequently used include cadmium sulfide (CdS) and cadmium selenide (CdSe). The peak spectral response of CdS occurs at approximately 5100 Å and for CdSe at 6150 Å, as shown in Fig. 2.32. The response time of CdS units is about 100 ms and 10 ms for CdSe cells.

The photoconductive cell does not have a junction like the photodiode. A thin layer of the material connected between terminals is simply exposed to the incident light energy.

As the illumination on the device increases in intensity, the energy state of a larger number of electrons in the structure will also increase, because of the increased availability of the photon packages of energy. The result is an increasing number of relatively "free" electrons in the structure and a decrease in the terminal resistance. The sensitivity curve for a typical photoconductive device appears in Fig. 2.38. Note

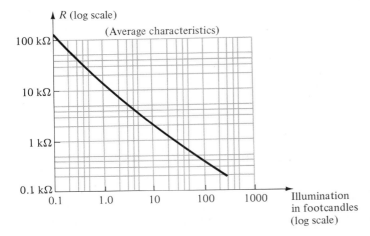

Figure 2.38 Photoconductive cell— terminal characteristics (GE type B425).

the linearity (when plotted using a log-log scale) of the resulting curve and the large change in resistance (100 kΩ → 100 Ω) for the indicated change in illumination.

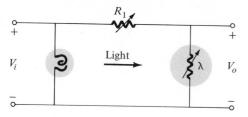

Figure 2.39 Voltage regulator employing a photoconductive cell.

One rather simple, but interesting, application of the device appears in Fig. 2.39. The purpose of the system is to maintain V_o at a fixed level even through V_i may fluctuate from its rated value. As indicated in the figure, the photoconductive cell, bulb, and resistor, all form part of this voltage-regulator system. If V_i should drop in magnitude for any number of reasons, the brightness of the bulb would also decrease. The decrease in illumination would result in an increase in the resistance (R_λ) of the photoconductive cell to maintain V_o at its rated level as determined by the voltage-divider rule; that is,

$$V_o = \frac{R_\lambda V_i}{R_\lambda + R_i} \tag{2.8}$$

In an effort to demonstrate the wealth of material available on each device from manufacturers, consider the CdS (cadmium sulfide) photoconductive cell described in Fig. 2.40. Note again the concern with temperature and response time.

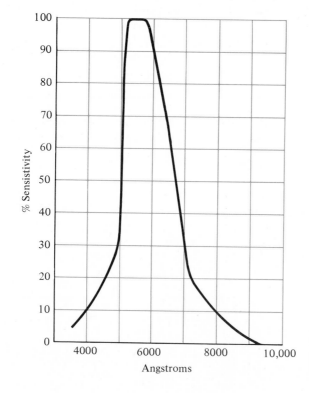

Figure 2.40 Characteristics of a Clairex CdS photoconductive cell. (Courtesy Clairex Electronics.)

**Variation of Conductance
With Temperature and Light**

Foot Candles	.01	0.1	1.0	10	100
Temperature	% Conductance				
−25°C	103	104	104	102	106
0	98	102	102	100	103
25°C	100	100	100	100	100
50°C	98	102	103	104	99
75°C	90	106	108	109	104

Response Time Versus Light

Foot Candles	.01	0.1	1.0	10	100
Rise (seconds)*	0.5	.095	.022	.005	.002
Decay (seconds)**	.125	.021	.005	.002	.001

* Time to $(1 - 1/e)$ of final reading after 5 seconds Dark adaptation.
** Time to $1/e$ of initial reading.

Figure 2.40 (Continued)

2.9 IR EMITTERS

Infrared-emitting diodes are solid-state gallium arsenide devices that emit a beam of radiant flux when forward-biased. The basic construction of the device is shown in Fig. 2.41. When the junction is forward-biased, electrons from the *n*-region will

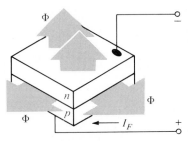

Figure 2.41 General structure of a semiconductor IR-emitting diode. (Courtesy RCA Solid State Division.)

recombine with excess holes of the *p*-material in a specially designed recombination region sandwiched between the *p*- and *n*-type materials. During this recombination process, energy is radiated away from the device in the form of photons. The generated photons will either be reabsorbed in the structure or leave the surface of the device as radiant energy, as shown in Fig. 2.41.

The radiant flux in mW versus the dc forward current for a typical device appears in Fig. 2.42. Note the almost linear relationship between the two. An interesting pattern for such devices is provided in Fig. 2.43. Note the very narrow pattern for devices with an internal collimating system. One such device appears in Fig. 2.44, with its internal construction and graphic symbol. A few areas of application for such devices include card and paper-tape readers, shaft encoders, data-transmission systems, and intrusion alarms.

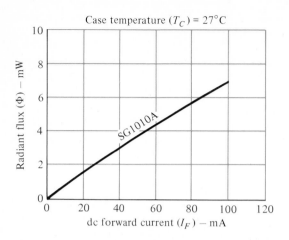

Figure 2.42 Typical radiant flux vs. dc forward current for an IR-emitting diode. (Courtesy RCA Solid State Division.)

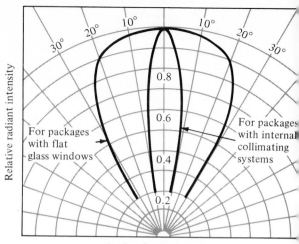

Angle of radiation–degrees

Figure 2.43 Typical radiant intensity patterns of RCA IR-emitting diodes. (Courtesy RCA Solid State Division.)

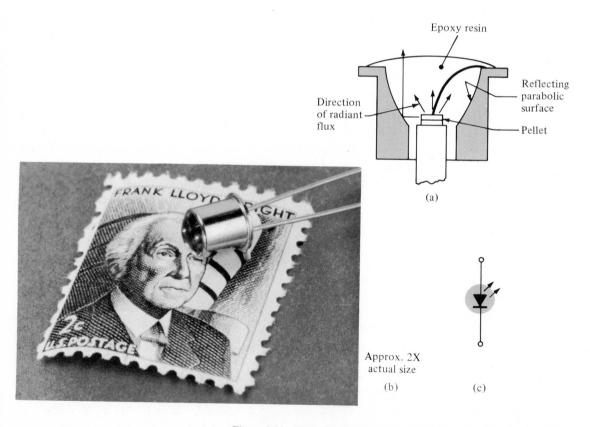

Figure 2.44 RCA IR-emitting diode: (a) construction; (b) photo; (c) symbol. (Courtesy RCA Solid State Division.)

The light-emitting diode (LED) is, as the name implies, a diode that will give off visible light when it is energized. In any forward-biased *p-n* junction there is, within the structure and primarily close to the junction, a recombination of holes and electrons. This recombination requires that the energy possessed by the unbound free electron be transferred to another state. In all semiconductor *p-n* junctions some of this energy will be given off as heat and some in the form of photons. In silicon and germanium the greater percentage is given up in the form of heat and the emitted light is insignificant. In other materials, such as gallium arsenide phosphide (GaAsP) or gallium phosphide (GaP), the number of photons of light energy emitted is sufficient to create a very visible light source. The process of giving off light by applying an electrical source of energy is called *electroluminescence*. As shown in Fig. 2.45, the conducting surface connected to the *p*-material is much smaller, to permit the emergence of the maximum number of photons of light energy. Note in the figure that the recombination of the injected carriers due to the forward-biased junction is resulting in emitted light at the site of recombination. There may, of course, be some absorption of the packages of photon energy in the structure itself, but a very large percentage are able to leave, as shown in the figure.

The appearance and characteristics of a subminiature high-efficiency solid-state lamp manufactured by Hewlett-Packard appears in Fig. 2.46. Two quantities yet undefined appear under the heading electrical/optical characteristics at $T_A = 25°C$. They are the *axial luminous intensity (I_V)* and the *luminous efficacy (η_v)*. Light intensity is measured in *candela*. One candela emits a light flux of 4π lumens and establishes an illumination of 1 foot-candle on a 1-ft² area 1 ft from the light source. Even though this description may not provide a clear understanding of the candela as a unit of measurement, its level can certainly be compared between similar devices. The term "efficacy" is, by definition, a measure of the ability of a device to produce a desired effect. For the LED this is the ratio of the number of lumens generated per applied watt of electrical energy. Note the high efficacy for the high-efficiency red LED. The relative efficiency is defined by the luminous intensity per unit current, as shown in Fig. 2.46h. Note also how close the peak wavelength of the light waves produced by each LED (red, yellow, green) approaches the defined wavelength (λ_d) for each color. The relative intensity of each color versus wavelength appears in Fig. 2.46e.

Since the LED is a *p-n* junction device, it will have a forward-biased characteristic (Fig. 2.46f) similar to the diode response curves introduced in Chapter 1. Note, however, that a change from Ge to a Si base has resulted in a much larger V_o. Note the almost linear increase in relative luminous intensity with forward current (Fig. 2.46g). Figure 2.46i reveals that the longer the pulse duration at a particular

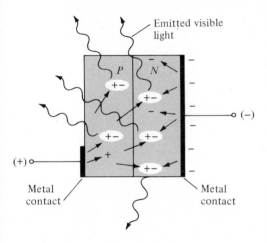

Figure 2.45 The process of electroluminescence in the LED.

(a)

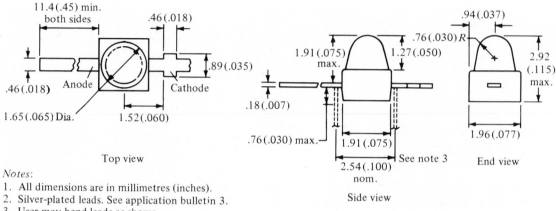

11.4(.45) min.
both sides

.46(.018)

.89(.035)

.46(.018) Anode   Cathode

1.65(.065) Dia.   1.52(.060)

Top view

.94(.037)

.76(.030)R

1.91(.075)
max.

1.27(.050)

2.92
(.115)
max.

.18(.007)

.76(.030) max.

1.91(.075)

2.54(.100)
nom.

See note 3

1.96(.077)

End view

Side view

Notes:
1. All dimensions are in millimetres (inches).
2. Silver-plated leads. See application bulletin 3.
3. User may bend leads as shown.

(b)

Absolute Maximum Ratings at $T_A = 25°C$					
Parameter	Red 4100/4101	High Eff. Red 4160	Yellow 4150	Green 4190	Units
Power dissipation	100	120	120	120	mW
Average forward current	50[1]	20[1]	20[1]	30[2]	mA
Peak forward current	1000	60	60	60	mA
Operating and storage temperature range	−55°C to 100°C				
Lead soldering temperature [1.6mm (0.063 in.) from body]	230°C for 3 seconds				

[1]. Derate from 50°C at 0.2mA/°C
[2]. Derate from 50°C at 0.4mA/°C

(c)

Figure 2.46 Hewlett-Packard subminiature high-efficiency red solid-state lamp: (a) appearance; (b) package dimensions; and (c) absolute maximum ratings. (Continued on pp. 105 and 106.) (Courtesy Hewlett-Packard Corporation.)

Electrical/Optical Characteristics at $T_A = 25°C$

Symbol	Description	5082-4100/4101			5082-4160			5082-4150			5082-4190			Units	Test Conditions
		Min.	Typ.	Max.	Min.	Typ.	Max.	Min.	Typ.	Max.	Min.	Typ.	Max.		
I_V	Axial luminous intensity	−/0.5	.7/1.0	2.0	1.0	3.0		1.0	2.0		0.8	1.5		mcd	$I_F = 10mA$, At $I_F = 20mA$
$2\theta_{1/2}$	Included angle between half luminous intensity points		45			80			90			70		deg.	Note 1
λ_{peak}	Peak wavelength		655			635			583			565		nm	Measurement at peak
λ_d	Dominant wavelength		640			628			585			572		nm	Note 2
τ_S	Speed of response		15			90			90			200		ns	
C	Capacitance		100			11			15			13		pF	$V_F = 0; f = 1$ MHz
θ_{JC}	Thermal resistance		125			120			100			100		°C/W	Junction to cathode lead at 0.79mm (.031 in) from body
V_F	Forward voltage		1.6	2.0		2.2	3.0		2.2	3.0		2.4	3.0	V	$I_F = 10mA$, At $I_F = 20mA$
BV_R	Reverse breakdown voltage	3.0	10		5.0			5.0			5.0			V	$I_R = 100\mu A$
η_v	Luminous efficacy		55			147			570			665		lm/W	Note 3

NOTES:
1. $\theta_{1/2}$ is the off-axis angle at which the luminous intensity is half the axial luminous intensity.
2. The dominant wavelength, λ_d, is derived from the CIE chromaticity diagram and represents the single wavelength which defines the color of the device.
3. Radiant intensity, I_e, in watts/steradian, may be found from the equation $I_e = I_v/\eta_v$, where I_v is the luminous intensity in candelas and η_v is the luminous efficacy in lumens/watt.

(d)

Figure 2.46 (continued)   (d) Electrical/optical characteristics.

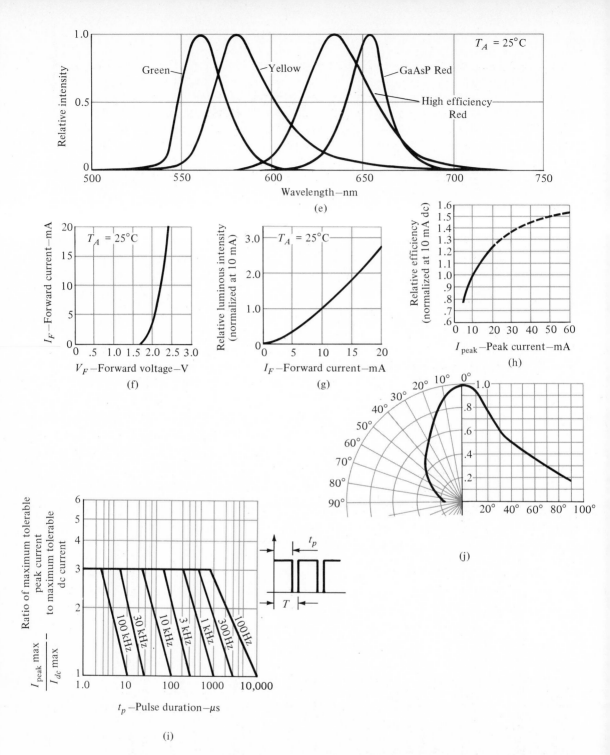

Figure 2.46 (continued) (e) Relative intensity vs. wavelength; (f) forward current vs. forward voltage; (g) relative luminous intensity vs. forward current; (h) relative efficiency vs. peak current; (i) maximum peak current vs. pulse duration; and (j) relative luminous intensity vs. angular displacement.

frequency, the lower the permitted peak current (after you pass the break value of t_p). Figure 2.46j simply reveals that the intensity is greater at 0° (or head on) and the least at 90° (when you view the device from the side).

LED displays are available today in many different sizes and shapes. The light-emitting region is available in lengths from 0.1 to 1 in. Numbers can be created by segments such as shown in Fig. 2.47. By applying a forward bias to the proper p-type material segment, any number from 0 to 9 can be displayed.

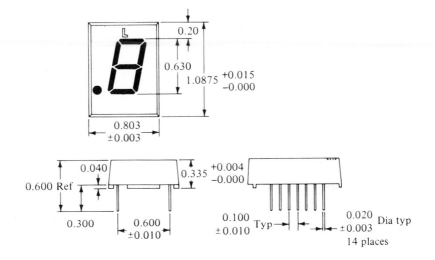

Figure 2.47 Litronix segment display.

The display of Fig. 2.48 is used in calculators and will provide eight digits. There are also two-lead LED lamps that contain two LEDs, so that a reversal in biasing will change the color from green to red, or vice versa. LEDs are presently available in red, green, yellow, orange, and white. It would appear that the introduction of the color blue is a possibility in the very near future. In general, LEDs operate at voltage levels from 1.7 to 3.3 V, which makes them completely compatible with solid-state circuits. They have a fast response time (nanoseconds) and offer good contrast ratios for visibility. The power requirement is typically from 10 to 150 mW with a lifetime of 100,000+ hours. Their semiconductor construction adds a significant ruggedness factor.

Figure 2.48 Eight-digit and a sign calculator display. (Courtesy Hewlett-Packard Corporation.)

2.11 LIQUID-CRYSTAL DISPLAYS

The liquid-crystal display (LCD) has the distinct advantage of having a lower power requirement than the LED. It is typically in the order of microwatts for the display, as compared to the same order of milliwatts for LEDs. It does, however, require an external or internal light source, is limited to a temperature range of about 0° to 60°C, and lifetime is an area of concern because LCDs can chemically degrade. The types receiving the major interest today are the field-effect and dynamic-scattering units. Each will be covered in some detail in this section.

A liquid crystal is a material (normally organic for LCDs) that will flow like a liquid but whose molecular structure has some properties normally associated with solids. For the light-scattering units, the greatest interest is in the *nematic liquid crystal,* having the crystal structure shown in Fig. 2.49. The individual molecules have a rodlike appearance as shown in the figure. The indium oxide conducting

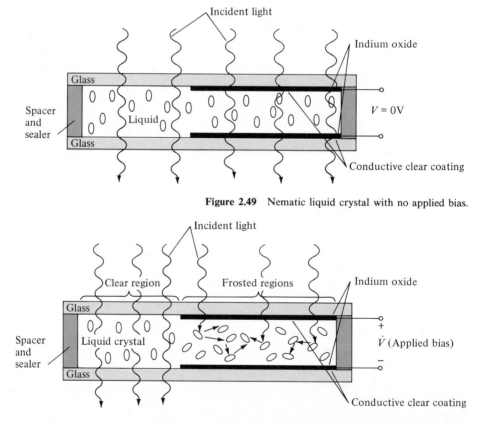

Figure 2.49 Nematic liquid crystal with no applied bias.

Figure 2.50 Nematic liquid crystal with applied bias.

surface is transparent and, under the condition shown in the figure, the incident light will simply pass through and the liquid-crystal structure will appear clear. If a voltage (for commercial units the threshold level is usually between 6 and 20 V) is applied across the conducting surfaces, as shown in Fig. 2.50, the molecular arrange-

CH. 2 ZENERS AND OTHER TWO-TERMINAL DEVICES

ment is disturbed, with the result that regions will be established with different indices of refraction. The incident light is, therefore, reflected in different directions at the interface between regions of different indices of refraction (referred to as *dynamic scattering*—first studied by RCA in 1968) with the result that the scattered light has a frosted-glass appearance. Note in Fig. 2.50, however, that the frosted look occurs only where the conducting surfaces are opposite each other and that the remaining areas remain translucent.

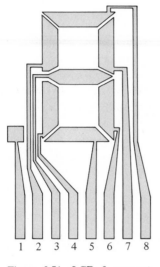

Figure 2.51 LCD 8 segment digit display.

A digit on an LCD display may have the segment appearance shown in Fig. 2.51. The black area is actually a clear conducting surface connected to the terminals below for external control. Two similar masks are placed on opposite sides of a sealed thick layer of liquid-crystal material. If the number 2 were required, the terminals 8, 7, 3, 4, and 5 would be energized and only those regions would be frosted while the other areas would remain clear.

As indicated earlier, the LCD does not generate its own light but depends on an external or internal source. Under dark conditions it would be necessary for the unit to have its own internal light source either behind or to the side of the LCD. During the day, or in lighted areas, a reflector can be put behind the LCD to reflect the light back through the display for maximum intensity. For optimum operation, current watch manufacturers are using a combination of the transmissive (own light source) and reflective modes called *transflective*.

The *field-effect* or *twisted nematic* LCD has the same segment appearance and thin layer of encapsulated liquid crystal, but its mode of operation is very different. Similar to the dynamic-scattering LCD, the field effect can be operated in the reflective or transmissive mode with an internal source. The transmissive display appears in Fig. 2.52. The internal light source is on the right and the viewer is on the left.

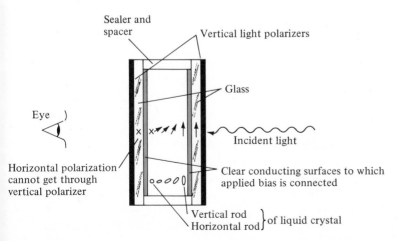

Figure 2.52 Transmissive field-effect LCD with no applied bias.

This figure is most noticeably different from Fig. 2.49 in that there is an addition of a *light polarizer*. Only the vertical component of the entering light on the right can pass through the vertical-light polarizer on the right. In the field-effect LCD, either the clear conducting surface to the right is chemically etched or an organic film is applied to orient the molecules in the liquid crystal in the vertical plane, parallel to the cell wall. Note the rods to the far right in the liquid crystal. The opposite conducting surface is also treated to ensure that the molecules are 90° out of phase in the direction shown (horizontal) but still parallel to the cell wall. In between the two walls of the liquid crystal there is a general drift from one polarization to the other, as shown in the figure. The left-hand light polarizer is also such that it permits the passage of only the vertically polarized incident light. If there is no applied voltage to the conducting surfaces, the vertically polarized light enters the liquid-crystal region and follows the 90° bending of the molecular structure. Its horizontal polarization at the left-hand vertical light polarizer does not allow it to pass through, and the viewer sees a uniformly dark pattern across the entire display. When a threshold voltage is applied (for commercial units from 2 to 8 V), the rodlike molecules align themselves with the field (perpendicular to the wall) and the light passes directly through without the 90° shift. The vertically incident light can then pass directly through the second vertically polarized screen and a light area is seen by the viewer. Through proper excitation of the segments of each digit the pattern will appear as shown in Fig. 2.53. The reflective type field effect is shown in Fig. 2.54. In this case the horizontally polarized light at the far left encounters a horizontally polarized filter and passes through to the reflector, where it is reflected back into the liquid crystal, bent back to the other vertical polari-

Figure 2.53 Reflective-type LCD. (Courtesy RCA Solid State Division.)

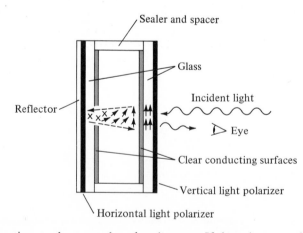

Figure 2.54 Reflective field-effect LCD with no applied bias.

zation, and returned to the observer. If there is no applied voltage, there is a uniformly lit display. The application of a voltage results in a vertically incident light encountering a horizontally polarized filter at the left which will not be able to pass through and be reflected. A dark area results on the crystal, and the pattern as shown in Fig. 2.55 appears.

Field-effect LCDs are normally used when a source of energy is a prime factor

Figure 2.55 Transmissive-type LCD. (Courtesy RCA Solid State Division.)

(e.g., in watches, portable instrumentation, etc.) since they absorb considerably less power than the light-scattering types—the microwatt range compared to the low-milliwatt range. The cost is typically higher for field-effect units, and their height is limited to about 2 in., while light-scattering units are available up to 8 in. in height.

A further consideration in displays is turn-on and turn-off time. LCDs are characteristically much slower than LEDs. LCDs typically have response times in the range 100 to 300 ms, while LEDs are available with response times below 100 ns. However, there are numerous applications, such as in a watch, where the difference between 100 ns and 100 ms ($\frac{1}{10}$ of a second) is of little consequence. For such applications the lower power demand of LCDs is a very attractive characteristic. The lifetime of LCD units is steadily increasing beyond the 10,000+ hours limit. Since the color generated by LCD units is dependent on the source of illumination, there is a greater range of color choice.

2.12 SOLAR CELLS

In recent years there has been increasing interest in the solar cell as an alternative source of energy. When we consider that the power density received from the sun at sea level is about 100 mW/cm² (1 kW/m²), it is certainly an energy source that requires further research and development to maximize the conversion efficiency from solar to electrical energy.

The basic construction of a silicon *p-n* junction solar cell appears in Fig. 2.56. As shown in the top view, every effort is made to ensure that the surface area perpendicular to the sun is a maximum. Also, note that the metallic conductor connected to the *p*-type material and the thickness of the *p*-type material are such that they ensure that a maximum number of photons of light energy will reach the junction. A photon of light energy in this region may collide with a valence electron and impart to it sufficient energy to leave the parent atom. The result is a generation of free electrons and holes. This phenomenon will occur on each side of the junction. In the *p*-type material the newly generated electrons are minority carriers and will move rather freely across the junction as explained for the basic *p-n* junction with

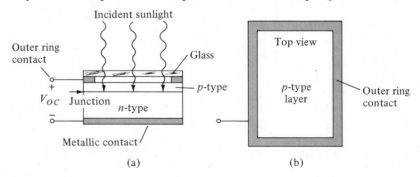

Figure 2.56 Solar cell: (a) cross section; (b) top view.

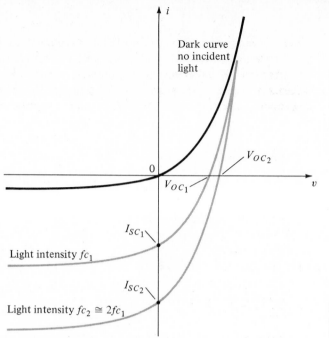

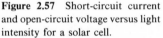

Figure 2.57 Short-circuit current and open-circuit voltage versus light intensity for a solar cell.

no applied bias. A similar discussion is true for the holes generated in the *n*-type material. The result is an increase in the minority carrier flow which is opposite in direction to the conventional forward current of a *p-n* junction. This increase in reverse current is shown in Fig. 2.57. Since $V = 0$ anywhere on the vertical axis and represents a short-circuit condition, the current at this intersection is called the *short-circuit current* and is represented by the notation I_{sc}. Under open-circuit conditions ($i_d = 0$) the *photovoltaic* voltage V_{oc} will result. This is a logarithmic function of the illumination, as shown in Fig. 2.58. V_{oc} is the terminal voltage of a battery under no-load (open-circuit) conditions. Note, however, in the same figure that the short-circuit current is a linear function of the illumination. That is, it will double for the same increase in illumination (f_{c_1} and $2f_{c_1}$ in Fig. 2.58) while the change in V_{oc} is less for this region. The major increase in V_{oc} occurs for lower-level increases in illumination. Eventually, a further increase in illumination will have very little effect on V_{oc}, although I_{sc} will increase, causing the power capabilities to increase.

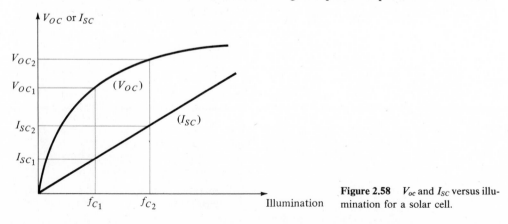

Figure 2.58 V_{oc} and I_{SC} versus illumination for a solar cell.

Selenium and silicon are the most widely used materials for solar cells, although gallium arsenide, indium arsenide, and cadmium sulfide, among others, are also used. The wavelength of the incident light will affect the response of the *p-n* junction to the incident photons. Note in Fig. 2.59 how closely the selenium cell response curve matches that of the eye. This fact has widespread application in photographic equip-

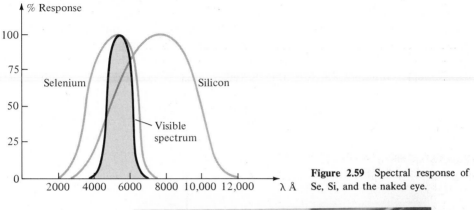

Figure 2.59 Spectral response of Se, Si, and the naked eye.

Electrical Characteristics*

IR Number	Load Voltage (volts) (min.)	Load Current (milliamps) (min.)	Power (milliwatts) (min.)
SP2A40B	1.6	36	58
SP2B48B	1.6	40	64
SP4C40B	3.2	36	115
SP2C80B	1.6	72	115
SP4D48B	3.2	40	128
SP2D96B	1.6	80	129
S2900E5M	.4	60	24
S2900E7M	.4	90	36
S2900E9.5M	.4	120	48

*Current Voltage characteristics are based on an illumination level of 100 mW/cm² (bright average sunlight).

Figure 2.60 Typical solar cell and its electrical characteristics. (Courtesy International Rectifier Corporation.)

ment such as exposure meters and automatic exposure diaphragms. Silicon also overlaps the visible spectrum but has its peak at the 0.8 μ (8000 Å) wavelength, which is in the infrared region. In general, silicon has a higher conversion efficiency and greater stability and is less subject to fatigue. Both materials have excellent temperature characteristics. That is, they can withstand extreme high or low temperatures without a significant drop-off in efficiency. A typical solar cell, with its electrical characteristics, appears in Fig. 2.60.

A very recent innovation in the use of solar cells appears in Fig. 2.61. The series arrangement of solar cells permits a voltage beyond that of a single element. The performance of a typical four-cell array appears in the same figure. At a current of approximately 2.6 mA the output voltage is about 1.6 V, resulting in an output power of 4.16 mW. The Schottky barrier diode is included to prevent battery current drain through the power converter. That is, the resistance of the Schottky diode is

(a)

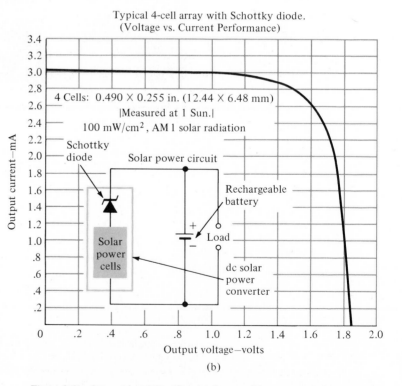

(b)

Figure 2.61 International Rectifier 4-cell array: (a) appearance; (b) characteristics. (Courtesy International Rectifier Corporation.)

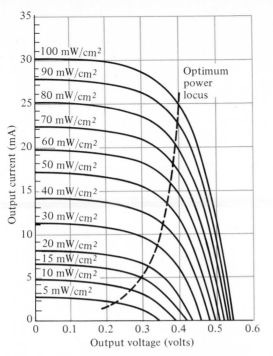

Figure 2.62 Typical output characteristics for silicon solar cells of 10% efficiency having an active area of 1 cm². Cell temperature is 30°C.

so high to charge flowing down through (+ to −) the power converter that it will appear as an open circuit to the rechargeable battery and not draw current from it.

It might be of interest to note that the Lockheed Missiles and Space Company has been awarded a grant from the National Aeronautics and Space Administration to develop a massive solar-array wing for the space shuttle. The wing will measure 13.5 ft by 105 ft when extended and will contain 41 panels, each carrying 3060 silicon solar cells. The wing can generate a total of 12.5 kW of electrical power.

The efficiency of operation of a solar cell is determined by the electrical power output divided by the power provided by the light source. That is,

$$\eta = \frac{P_{o(\text{electrical})}}{P_{i(\text{light energy})}} \times 100\% = \frac{P_{\text{max(device)}}}{(\text{area in cm}^2)(100 \text{ mW/cm}^2)} \times 100\% \qquad (2.9)$$

Typical levels of efficiency range from 10 to 14%—a level that should improve measurably if the present interest continues. A typical set of output characteristics for silicon solar cells of 10% efficiency with an active area of 1 cm² appears in Fig. 2.62. Note the optimum power locus and the almost linear increase in output current with luminous flux for a fixed voltage.

2.13 THERMISTORS

The thermistor is, as the name implies, a temperature-sensitive resistor; that is, its terminal resistance is related to its body temperature. It has a negative temperature coefficient, indicating that its resistance will decrease with an increase in its body temperature. It is not a junction device and is constructed of Ge, Si, or a mixture of oxides of cobalt, nickel, strontium, or manganese.

Specific resistance — (ohm—cm, the resistance between faces of 1 cm^3 of the material) (log scale)

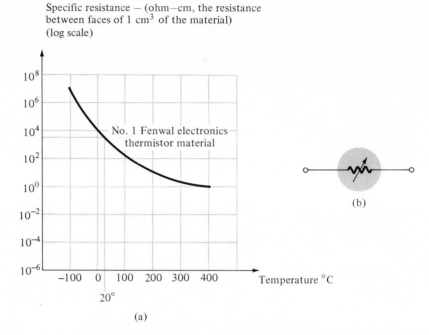

(a)

(b)

Figure 2.63 Thermistor: (a) typical set of characteristics; (b) symbol.

The characteristics of a representative thermistor are provided in Fig. 2.63, with the commonly used symbol for the device. Note, in particular, that at room temperature (20°C) the resistance of the thermistor is approximately 5000 Ω, while at 100°C (212°F) the resistance has decreased to 100 Ω. A temperature span of 80°C has therefore resulted in a 50:1 change in resistance. It is typically 3% to 5% per degree change in temperature. There are, fundamentally, two ways to change the temperature of the device: internally and externally. A simple change in current through the device will result in an internal change in temperature. A small applied voltage will result in a current too small to raise the body temperature above that of the surroundings. In this region, as shown in Fig. 2.64, the thermistor will act like a resistor

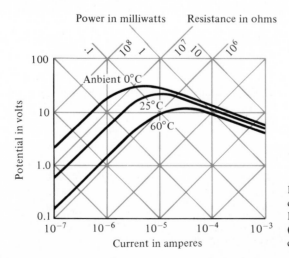

Figure 2.64 Steady-state voltage-current characteristics of Fenwal Electronics BK65V1 Thermistor. (Courtesy Fenwal Electronics, Incorporated.)

CH. 2 ZENERS AND OTHER TWO-TERMINAL DEVICES

and have a positive temperature coefficient. However, as the current increases, the temperature will rise to the point where the negative temperature coefficient will appear as shown in Fig. 2.64. The fact that the rate of internal flow can have such an effect on the resistance of the device introduces a wide vista of applications in control, measuring techniques, and so on. An external change would require changing the temperature of the surrounding medium or immersing the device in a hot or cold solution.

A photograph of a number of commercially available thermistors is provided in Fig. 2.65.

A simple temperature-indicating circuit appears in Fig. 2.66. Any increase in the temperature of the surrounding medium will result in a decrease in the resistance of the thermistor and an increase in the current I_T. An increase in I_T will produce an increased movement deflection, which when properly calibrated will accurately indicate the higher temperature. The variable resistance was added for calibration purposes.

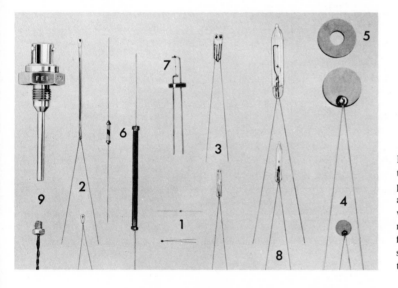

Figure 2.65 Various types of thermistors: (1) beads; (2) glass probes; (3) iso-curve interchangeable probes and beads; (4) discs; (5) washers; (6) rods; (7) specially mounted beads; (8) vacuum and gas-filled probes; (9) special probe assemblies. (Courtesy Fenwal Electronics, Incorporated.)

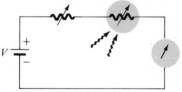

Sensitive movement — temperature calibrated

Figure 2.66 Temperature-indicating circuit.

PROBLEMS

§ 2.2

1. The following characteristics are specified for a particular Zener diode: $V_Z = 20$ V, $V_R = 16.8$ V, $I_{ZT} = 10$ mA, $I_R = 20$ μA, and $I_{ZM} = 40$ mA. Sketch the characteristic curve in the manner displayed in Fig. 2.3.

2. At what temperature will the IN961 10-V Fairchild Zener diode have a nominal voltage of 10.75 V? (*Hint:* Note data in Table 2.1.)

3. Determine the temperature coefficient of a 5-V Zener diode (rated 25°C value) if the nominal voltage drops to 4.8 V at a temperature of 100°C.

4. Using the curves of Fig. 2.4a, what level of temperature coefficient would you expect for a 20-V diode? Repeat for a 5-V diode. Assume a linear scale between nominal voltage levels and a current level of 0.1 mA.

5. Determine the dynamic impedance for the 24-V diode at $I_Z = 10$ mA from Fig. 2.4b. Note that it is a log scale.

6. Compare the levels of dynamic impedance for the 24-V diode of Fig. 2.4b at current levels of 0.2 mA, 1 mA, and 10 mA. How do the results relate to the shape of the characteristics in this region?

7. Using the curve of Fig. 2.4c, determine the temperature at which the rated power drops to 100 mW. How does this value compare with the results obtained using $D_F = 3.33$ mW/°C? Does the graph suggest another value for D_F?

8. Sketch the output (v_o) of Fig. 2.7 if the 12-V battery is reduced to 10 V. Sketch v_R also.

9. Sketch the output of Fig. 2.9a if v_i is reduced to peak values of 10 V. Assume straightline segments for the diode with zero dynamic impedance in the vertical-rise region of Fig. 2.1. Repeat for peak values of 20 V.

10. For the Zener diode network of Fig. 2.11, if $V_i = 40$ V, $R_S = 200$ Ω, and $V_Z = 22$ V:
    (a) Find the minimum value of R_L to ensure that the Zener diode has fired.
    (b) From part (a) determine the maximum load current under Zener regulation.
    (c) Find the minimum I_L if the maximum $R_L = 5$ kΩ.
    (d) Sketch the terminal voltage V_L versus I_L for the range of I_L.
    (e) At $R_L = 1$ kΩ determine I_Z and the current through the 40-V source.
    (f) From the results of part (e), determine the power dissipated by the Zener diode.

11. Using the 10-V Zener, R_L with a range from 50 Ω to 2 kΩ, $V_i = 30$ V, find a suitable value of R_S to ensure that the Zener diode maintains an "on" condition (use the circuit configuration in Fig. 2.11).

§ 2.3

12. (a) Describe in your own words how the construction of the hot-carrier diode is significantly different from the conventional semiconductor diode.
    (b) In addition, describe its mode of operation.

13. (a) Consult Fig. 2.18. How would you compare the dynamic resistances of the diodes in the forward-bias regions?
    (b) How do they compare at any level of reverse current more negative than I_S?

14. Referring to Fig. 2.21, how does the maximum surge current I_{FSM} relate to the average rectified forward current? Is it typically greater than 20:1? Why is it possible to have such high levels of current? What noticeable difference is there in construction as the current rating increases?

15. Referring to Fig. 2.22a, at what temperature is the forward voltage drop 300 mV at a

current of 1 mA? Which current levels have the highest levels of temperature coefficients? Assume a linear progression between temperature levels.

16. For the curve of Fig. 2.22b denoted 2900/2303, determine the percent change in I_R for a change in reverse voltage from 5 to 10 V. At what reverse voltage would you expect to reach a reverse current of 1 μA? Note the log scale for I_R.

17. Determine the percent change in capacitance between 0 and 2 V for the 2900/2303 curve of Fig. 2.22c. How does this compare to the change between 8 and 10 V?

§ 2.4

18. (a) Determine the transition capacitance of a diffused junction varicap diode at a reverse potential of 4.2 V if $C(0) = 80$ pF and $V_o = 0.7$ V.
    (b) From the information of part (a), determine the constant K in Eq. (2.3).

19. (a) For a varicap diode having the characteristics of Fig. 2.23, determine the difference in capacitance between reverse-bias potentials of -3 and -12 V.
    (b) Determine the incremental rate of change ($\Delta C/\Delta V_r$) at $V = -8$ V. How does this value compare with the incremental change determined at -2 V?

20. (a) The resonant frequency of a series RLC network is determined by $f_0 = 1/(2\pi\sqrt{LC})$. Using the value of f_0 and L_S provided in Fig. 2.25, determine the value of C.
    (b) How does the value calculated in part (a) compare with that determined by the curve in Fig. 2.26 at $V_R = 25$ V?

21. Referring to Fig. 2.26, determine the ratio of capacitance at $V_R = 3$ V to $V_R = 25$ V and compare to the value of C_3/C_{25} given in Fig. 2.25 (maximum = 6.5).

22. Determine T_1 for a varactor diode if $C_0 = 22$ pF, $TC_C = 0.02\%/°$C, and $\Delta C = 0.11$ pF due to an increase in temperature above $T_0 = 25°$C.

23. What region of V_R would appear to have the greatest change in capacitance per change in reverse voltage for the BB139 varactor diode? Be aware that the scales are nonlinear.

24. If $Q = X_L/R = 2\pi f L/R$, determine the figure of merit (Q) at 600 MHz using the fact that $R_S = 0.35$ Ω and $L_S = 2.5$ mH. Comment on the change in Q with frequency and the support or nonsupport of the curve in Fig. 3.26.

§ 2.5

25. Consult a manufacturer's data book and compare the general characteristics of a high-power device (> 10 A) to a low-power unit (<100 mA). Is there a significant change in the data and characteristics provided? Why?

§ 2.6

26. What are the essential differences between a semiconductor junction diode and a tunnel diode?

27. Note in the equivalent circuit of Fig. 2.29 that the capacitor appears in parallel with the negative resistance. Determine the reactance of the capacitor at 1 MHz and 100 MHz if $C = 5$ pF, and determine the total impedance of the parallel combination (with $R = -152$ Ω) at each frequency. Is the magnitude of the inductive reactance anything to be overly concerned about at either of these frequencies if $L_S = 6$ nH?

28. Why do you believe the maximum reverse current rating for the tunnel diode can be greater than the forward current rating? (*Hint:* Note the characteristics and consider the power rating.)

§ 2.7

29. Determine the energy associated with the photons of green light if the wavelength is 5000 Å. Give your answer in joules and electron volts.

30. (a)   Referring to Fig. 2.32, what would appear to be the frequencies associated with the upper and lower limits of the visible spectrum?
(b)   What is the wavelength in microns associated with the peak relative response of silicon?
(c)   If we define the bandwidth of the spectral response of each material to occur at 70% of its peak level, what is the bandwidth of silicon?

31. Referring to Fig. 2.34, determine I_λ if $V_\lambda = 30$ V and the light intensity is 4×10^{-9} W/m².

32. (a)   Which material of Fig. 2.32 would appear to provide the best response to yellow, red, green, and infrared (less than 11,000 Å) light sources?
(b)   At a frequency of 0.5×10^{15} Hz, which color has the maximum spectral response?

33. Determine the voltage drop across the resistor of Fig. 2.33 if the incident flux is 3000 fc, $V_\lambda = 25$ V, and $R = 100$ kΩ. Use the characteristics of Fig. 2.34.

§ 2.8

34. What is the approximate rate of change of resistance with illumination for a photoconductive cell with the characteristics of Fig. 2.38 for the ranges (a) $0.1 \rightarrow 1$ kΩ; (b) $1 \rightarrow 10$ kΩ; (c) $10 \rightarrow 100$ kΩ? (Note that this is a log scale.) Which region has the greatest rate of change in resistance with illumination?

35. What is the "dark current" of a photodiode?

36. If the illumination on the photoconductive diode in Fig. 2.39 is 10 fc, determine the magnitude of V_i to establish 6 V across the cell if R_1 is equal to 5 kΩ. Use the characteristics of Fig. 2.38.

37. Using the data provided in Fig. 2.40, sketch a curve of percent conductance versus temperature for 0.01, 1.0, and 100 fc. Are there any noticeable effects?

38. (a)   Sketch a curve of rise time versus illumination using the data from Fig. 2.40.
(b)   Repeat part (a) for the decay time.
(c)   Discuss any noticeable effects of illumination in parts (a) and (b).

39. Which colors is the CdS unit of Fig. 2.40 most sensitive to?

§ 2.9

40. (a)   Determine the radiant flux at a dc forward current of 70 mA for the device of Fig. 2.42.
(b)   Determine the radiant flux in lumens at a dc forward current of 45 mA.

41. (a)   Through the use of Fig. 2.43, determine the relative radiant intensity at an angle of 25° for a package with a flat glass window.
(b)   Plot a curve of relative radiant intensity versus degrees for the flat package.

42. If 60 mA of dc forward current is applied to an SG1010A IR emitter, what will be the incident radiant flux in lumens 5° off the center if the package has an internal collimating system? Refer to Figs. 2.42 and 2.43.

§ 2.10

43. (a) Convert the horizontal scale of Fig. 2.46e to angstrom units.
(b) How do the peak values of the relative intensity fall into the color bands provided in Fig. 2.32?

44. Referring to Fig. 2.46f, what would appear to be an appropriate value of V_0 for this device? How does it compare to the value of V_0 for silicon and germanium?

45. Using the information provided in Fig. 2.46, determine the forward voltage across the diode if the relative luminous intensity is 1.5.

46. (a) What is the percent increase in relative efficiency of the device of Fig. 2.46 if the peak current is increased from 5 to 10 mA?
(b) Repeat part (a) for 30 to 35 mA (the same increase in current).
(c) Compare the percent increase from parts (a) and (b). At what point on the curve would you say there is little gained by further increasing the peak current?

47. (a) Referring to Fig. 2.46i, determine the maximum tolerable peak current if the period of the pulse duration is 1 ms, the frequency is 300 Hz, and the maximum tolerable dc current is 20 mA.
(b) Repeat part (a) for a frequency of 100 Hz.

48. (a) If the luminous intensity at 0° angular displacement is 3.0 mcd for the device of Fig. 2.46, at what angle will it be 0.75 mcd?
(b) At what angle does the loss of luminous intensity drop below the 50% level?

49. Sketch the current derating curve for the average forward current of the High Eff. Red LED of Fig. 2.46 as determined by temperature. (Note the absolute maximum ratings.)

§ 2.11

50. Referring to Fig. 2.51, which terminals must be energized to display number 7?

51. In your own words, describe the basic operation of an LCD.

52. Discuss the relative differences in mode of operation between an LED and an LCD display.

53. What are the relative advantages and disadvantages of an LCD display as compared to an LED display?

§ 2.12

54. A 1 cm by 2 cm solar cell has a conversion efficiency of 9%. Determine the maximum power rating of the device.

55. If the power rating of a solar cell is determined on a very rough scale by the product $V_{OC}I_{SC}$, is the greatest rate of increase obtained at lower or higher levels of illumination? Explain your reasoning.

56. (a) Referring to Fig. 2.62, what power density is required to establish a current of 24 mA at an output voltage of 0.25 V?

(b) Why is 100 mW/cm² the maximum power density in Fig. 2.62?

(c) Determine the output current if the power is 40 mW/cm² and the output voltage is 0.3 V.

57. (a) Sketch a curve of output current versus power density at an output voltage of 0.15 V using the characteristics of Fig. 2.62.

(b) Sketch a curve of output voltage versus power density at a current of 19 mA.

(c) Is either of the curves from (a) and (b) linear within the limits of the maximum power limitation?

§ 2.13

58. For the thermistor of Fig. 2.63, determine the dynamic rate of change in specific resistance with temperature at $T = 20°C$. How does this compare to the value determined at $T = 300°C$? From the results, determine whether the greatest change in resistance per unit change in temperature occurs at lower or higher levels of temperature. Note the vertical log scale.

59. Using the information provided in Fig. 2.63, determine the total resistance of a 2-cm length of the material having a perpendicular surface area of 1 cm² at a temperature of 0°C. Note the vertical log scale.

60. (a) Referring to Fig. 2.64, determine the current at which a 25°C sample of the material changes from a positive to negative temperature coefficient. (Figure 2.64 is a log scale.)

(b) Determine the power and resistance levels of the device (Fig. 2.64) at the peak of the 0°C curve.

(c) At a temperature of 25°C, determine the power rating if the resistance level is 1 M.

61. In Fig. 2.66, $V = 0.2$ V, and $R_{variable} = 10 \ \Omega$. If the current through the sensitive movement is 2 mA and the voltage drop across the movement is 0 V, what is the resistance of the thermistor?

CHAPTER 3

Bipolar Junction Transistors (BJTs)

3.1 INTRODUCTION

During the period 1904–1947 the vacuum tube was undoubtedly the electronic device of interest and development. In 1904 the vacuum-tube diode was introduced by J. A. Fleming. Shortly thereafter, in 1906, Lee De Forest added a third element, called the *control grid,* to the vacuum diode, resulting in the first amplifier, the *triode.* In the following years, radio and television provided great stimulation to the tube industry. Production rose from about 1 million tubes in 1922 to about 100 million in 1937. In the early 1930s the four-element tetrode and five-element pentode gained prominence in the electron-tube industry. In the years to follow, the industry became one of primary importance and rapid advances were made in design, manufacturing techniques, high-power and high-frequency applications, and miniaturization.

On December 23, 1947, however, the electronics industry was to experience the advent of a completely new direction of interest and development. It was on the afternoon of this day that Walter H. Brattain and John Bardeen demonstrated the amplifying action of the first transistor at the Bell Telephone Laboratories. The original transistor (a point-contact transistor) is shown in Fig. 3.1. The advantages of this three-terminal solid-state device over the tube were immediately obvious: it was smaller and lightweight; had no heater requirement or heater loss; had rugged construction; and was more efficient since less power was absorbed by the device itself; it was instantly available for use, requiring no warm-up period; and lower operating voltages were possible. Note in the discussion above that this chapter is our first discussion of devices with three or more terminals. You will find that most amplifiers (devices

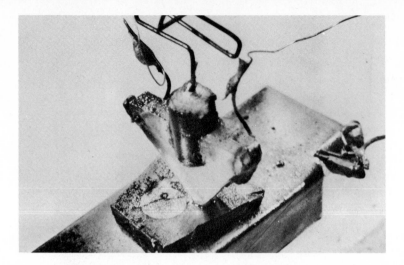

Figure 3.1 The first transistor (Courtesy Bell Telephone Laboratories).

that increase the voltage, current, or power level) will have three terminals—one for the control of the flow between the other two.

3.2 TRANSISTOR CONSTRUCTION

The transistor is a three-layer semiconductor device consisting of either two *n*- and one *p*-type layers of material or two *p*- and one *n*-type layers of material. The former is called an *npn transistor*, while the latter is called a *pnp transistor*. Both are shown in Fig. 3.2 with the proper dc biasing. We will find in Chapter 4 that the dc biasing is necessary to establish the proper region of operation for ac amplification. The outer layers of the transistor are heavily doped semiconductor materials having widths much greater than those of the sandwiched *p*- or *n*-type material. For the transistors shown in Fig. 3.2 the ratio of the total width to that of the center layer is 0.150/ 0.001 = 150:1. The doping of the sandwiched layer is also considerably less than that of the outer layers (typically 10:1 or less). This lower doping level decreases the conductivity (increases the resistance) of this material by limiting the number of "free" carriers.

For the biasing shown in Fig. 3.2 the terminals have been indicated by the capital letters E for *emitter*, C for *collector*, and B for *base*. An appreciation for this choice of notation will develop when we discuss the basic operation of the transistor.

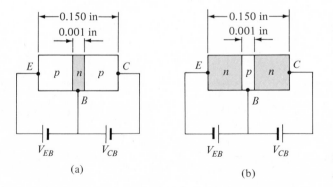

Figure 3.2 Types of transistors: (a) *pnp;* (b) *npn.*

CH. 3   BIPOLAR JUNCTION TRANSISTORS (BJTs)

The abbreviation BJT, from *bipolar junction transistor,* is often applied to this three-terminal device. The term *bipolar* reflects the fact that holes *and* electrons participate in the injection process into the oppositely polarized material. If only one carrier is employed (electron or hole), it is considered a *unipolar* device. Recall that the Schottky diode was such a device.

3.3 TRANSISTOR OPERATION

The basic operation of the transistor will now be described using the *pnp* transistor of Fig. 3.2a. The operation of the *npn* transistor is exactly the same if the roles played by the electron and hole are interchanged.

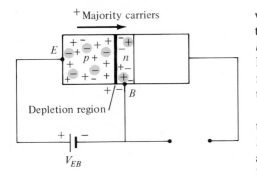

Figure 3.3 Forward-biased junction of a *pnp* transistor.

In Fig. 3.3 the *pnp* transistor has been redrawn without the base-to-collector bias. Note the similarities between this situation and that of the *forward-biased* diode in Chapter 1. The depletion region has been reduced in width due to the applied bias, resulting in a heavy flow of majority carriers from the *p*- to the *n*-type material.

Let us now remove the base-to-emitter bias of the *pnp* transistor of Fig. 3.2a as shown in Fig. 3.4. Consider the similarities between this situation and that of the *reverse-biased* diode of Section 1.8. Recall that the flow of majority carriers is zero, resulting in only a minority-carrier flow, as indicated in Fig. 3.4. *In summary, therefore, one* p-n *junction of a transistor is reverse-biased, while the other is forward-biased.*

In Fig. 3.5 both biasing potentials have been applied to a *pnp* transistor, with the resulting majority- and minority-carrier flow indicated. Note in Fig. 3.5 the widths of the depletion regions, indicating clearly which junction is forward-biased and which is reverse-biased. As indicated in Fig. 3.5, a large number of majority carriers will diffuse across the forward-biased *p-n* junction into the *n*-type material. The question then is whether these carriers will contribute directly to the base current I_B or pass

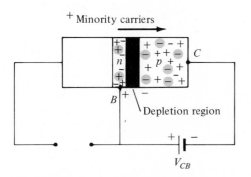

Figure 3.4 Reverse-biased junction of a *pnp* transistor.

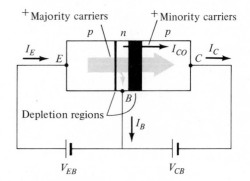

Figure 3.5 Majority and minority carrier flow of a *pnp* transistor.

directly into the p-type material. Since the sandwiched n-type material is very thin and has a low conductivity, a very small number of these carriers will take this path of high resistance to the base terminal. The magnitude of the base current is typically on the order of microamperes as compared to milliamperes for the emitter and collector currents. The larger number of these majority carriers will diffuse across the reverse-biased junction into the p-type material connected to the collector terminal as indicated in Fig. 3.5. The reason for the relative ease with which the majority carriers can cross the reverse-biased junction is easily understood if we consider that for the reverse-biased diode the injected majority carriers will appear as minority carriers in the n-type material. In other words, there has been an *injection* of minority carriers into the n-type base region material. Combining this with the fact that all the minority carriers in the depletion region will cross the reverse-biased junction of a diode accounts for the flow indicated in Fig. 3.5.

Applying Kirchhoff's current law to the transistor of Fig. 3.5 as if it were a single node, we obtain

$$I_E = I_C + I_B \tag{3.1}$$

and find that the emitter current is the sum of the collector and base currents. The collector current, however, is comprised of two components—the majority and minority carriers as indicated in Fig. 3.5. The minority current component is called the *leakage current* and is given the symbol I_{CO} (I_C current with emitter terminal Open). The collector current, therefore, is determined in total by Eq. (3.2).

$$I_C = I_{C_{\text{majority}}} + I_{CO_{\text{minority}}} \tag{3.2}$$

For general-purpose transistors, I_C is measured in milliamperes, while I_{CO} is measured in microamperes or nanoamperes. I_{CO}, like I_S for a reverse-biased diode, is temperature-sensitive and must be examined carefully when applications of wide temperature ranges are considered. It can severely affect the stability of a system at high temperatures if not considered properly.

Improvements in construction techniques have resulted in significantly lower levels of I_{CO}, to the point where its effect can often be ignored. However, higher-power devices still typically have values of I_{CO} in the microampere range.

The configuration shown in Fig. 3.2 for the *pnp* and *npn* transistors is called the *common-base* configuration since the base is common to both the emitter and collector terminals. For fixed values of V_{CB} in the common-base configuration the ratio of a small change in I_C to a small change in I_E is commonly called the *common-base, short-circuit amplification factor* and is given the symbol α (alpha).

In equation form, the magnitude of α is given by

$$\alpha = \frac{\Delta I_C}{\Delta I_E} \bigg|_{V_{CB} = \text{constant}} \tag{3.3}$$

The term "short circuit" indicates that the load is short-circuited when α is determined. More will be said about the necessity for shorting the load and the operations involved with using equations of the type indicated by Eq. (3.3) when we consider equivalent circuits in Chapter 7. Typical values of α vary from 0.90 to 0.998. For most practical applications, a first approximation for the magnitude of α, usually correct to within a few percent, can be obtained using the following equation:

$$\alpha \cong \frac{I_C}{I_E} \tag{3.4}$$

where I_C and I_E are the magnitude of the collector and emitter currents, respectively, at a particular point on the transistor characteristics.

Equations (3.3) and (3.4) are employed to determine α from the device characteristics or network conditions. However, in the strictest sense, α is only a measure of the percentage of holes (majority carriers) originating in the emitter p-material of Fig. 3.5 that reach the collector terminal. As defined by Eq. (3.2), therefore,

$$I_C = \alpha I_{E_{\text{majority}}} + I_{CO_{\text{minority}}} \tag{3.5}$$

3.4 TRANSISTOR AMPLIFYING ACTION

The basic voltage-amplifying action of the common-base configuration can now be described using the circuit of Fig. 3.6. The dc biasing does not appear in the figure since our interest will be limited to the ac response. For the common-base configuration,

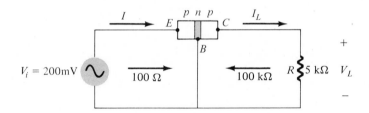

Figure 3.6 Basic voltage amplification action of the common-base configuration.

the input resistance between the emitter and the base of a transistor will typically vary from 20 to 200 Ω, while the output resistance may vary from 100 kΩ to 1 MΩ. The difference in resistance is due to the forward-biased junction at the input (base to emitter) and the reverse-biased junction at the output (base to collector). Using effective values and an average value of 100 Ω for the input resistance, we find

$$I = \frac{200 \times 10^{-3}}{100} = 2 \text{ mA}$$

If we assume for the moment that $\alpha = 1$ $(I_C = I_E)$,

$$I_L = I = 2 \text{ mA}$$

and

$$V_L = I_L R$$

$$= (2 \times 10^{-3})(5 \times 10^{+3})$$

$$V_L = 10 \text{ V}$$

The voltage amplification is

$$A_v = \frac{V_L}{V_i} = \frac{10}{200 \times 10^{-3}} = \mathbf{50}$$

Typical values of voltage amplification for the common-base configuration vary from 20 to 100. The current amplification (I_C/I_E) is always less than 1 for the common-base configuration. This latter characteristic should be obvious since $I_C = \alpha I_E$ and α is always less than 1.

The basic amplifying action was produced by *transferring* a current I from a low- to a high-*resistance* circuit. The combination of the two terms in italics results in the label *transistor;* that is,

*tran*sfer $+$ re*sistor* $\rightarrow$ *transistor*

3.5 COMMON-BASE CONFIGURATION

The notation and symbols used in conjunction with the transistor in the majority of texts and manuals published today are indicated in Fig. 3.7 for the common-base configuration with *pnp* and *npn* transistors. Throughout this text all current directions will refer to the conventional (hole flow) rather than the electron flow. This choice was based primarily on the fact that the vast majority of past and present publications in electrical engineering use conventional current.

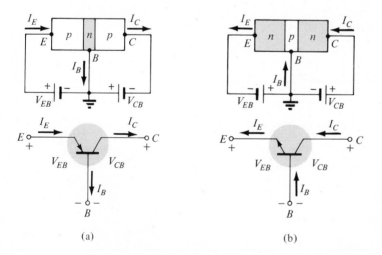

(a)   (b)

Figure 3.7 Notation and symbols used with the common-base configuration: (a) *pnp* transistor; (b) *npn* transistor.

Some texts prefer to show all the currents entering in Fig. 3.7 when they describe the basic operation of the transistor and simply include negative signs when appropriate. In other words, if the actual conventional flow direction is the opposite direction, a negative sign is included along with the magnitude. For clarity, all currents, as indicated in Fig. 3.7, will indicate the actual flow direction for the active region. *Note that the arrow in the symbol is the same as the direction of I_E (only true for conventional flow).* On specification sheets negative signs indicate that all currents are entering.

For the common-base configuration the applied potentials are written with respect to the base potential resulting in V_{EB} and V_{CB}. In other words, the second subscript will always indicate the transistor configuration. In all cases the first subscript is defined to be the point of higher potential, as shown in Fig. 3.7. For the *pnp* transistor, therefore, V_{EB} is positive and V_{CB} is negative (since the battery V_{CB} sets the collector at the lower potential), as indicated on the characteristics of Fig. 3.8. For the *npn* transistor V_{EB} is negative and V_{CB} is positive.

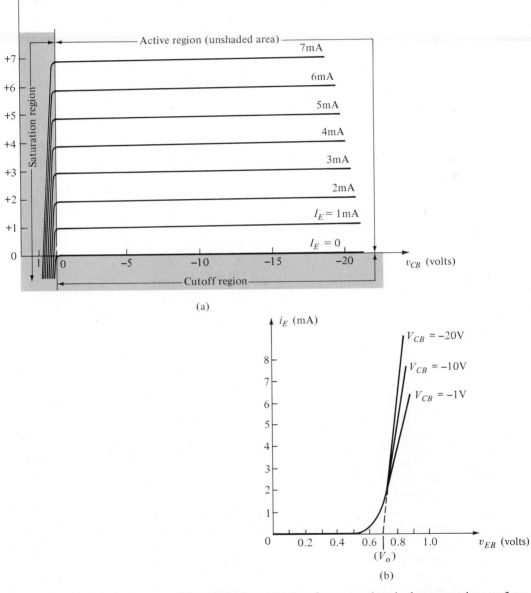

(a)

(b)

Figure 3.8 Characteristics of a *pnp* transistor in the common-base configuration: (a) collector characteristics; (b) emitter characteristics.

In addition, note that two sets of characteristics are necessary to represent the behavior of the *pnp* common-base transistor of Fig. 3.7: the *driving point* (or input) and the *output* set.

The output or collector characteristics of Fig. 3.8a relate the collector current to the collector-to-base voltage and emitter current. The collector characteristics have three basic regions of interest, as indicated in Fig. 3.8a: the *active, cutoff,* and *saturation* regions.

In the active region the collector junction is reverse-biased, while the emitter junction is forward-biased. These conditions refer to the situation of Fig. 3.5. The active region is the only region employed for the amplification of signals with minimum distortion. When the emitter current (I_E) is zero, the collector current is simply that due to the reverse saturation current I_{CO}, as indicated in Fig. 3.8a. The current I_{CO} is so small (microamperes) in magnitude compared to the vertical scale of I_C (milliamperes) that it appears on virtually the same horizontal line as $I_C = 0$. The circuit conditions that exist when $I_E = 0$ for the common-base configuration are shown in Fig. 3.9.

The notation most frequently used for I_{CO} on data and specification sheets is, as indicated in Fig. 3.9, I_{CBO}. Because of improved construction techniques, the level of I_{CBO} for general-purpose transistors (especially silicon) in the low- and mid-power ranges is usually so low that its effect can be ignored. However, for higher power units I_{CBO} will still appear in the microampere range. In addition, keep in mind that I_{CBO} like I_S for the diode (both reverse leakage currents) is temperature-sensitive. At higher temperatures the effect of I_{CBO} for any power level unit may become an important factor since it increases so rapidly with temperature.

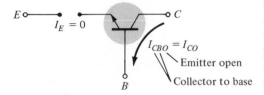
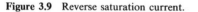

Figure 3.9 Reverse saturation current.

Note in Fig. 3.8a that as the emitter current increases above zero, the collector current increases to a magnitude slightly less ($\alpha < 1$) than that of the emitter current as determined by the basic transistor-current relations. Note also the almost negligible effect of V_{CB} on the collector current for the active region. The curves clearly indicate that *a first approximation to the relationship between I_E and I_C in the active region is given by $I_E = I_C$.*

In the cutoff region the collector and emitter junctions are both reverse-biased, resulting in negligible collector current as demonstrated in Fig. 3.8a.

The horizontal scale for V_{CB} has been expanded to the left of 0 V to represent clearly the characteristics in this region. *In the region called the* saturation region, *the collector and emitter junctions are forward-biased,* resulting in the exponential change in collector current with small changes in collector-to-base potential.

The input or emitter characteristics have only one region of interest, as illustrated by Fig. 3.8b. For fixed values of collector voltage *(V_{CB})*, as the emitter-to-base potential increases, the emitter current increases, as shown. Increasing levels of V_{CB} result in a reduced level of V_{EB} to establish the same current.

Note the tight grouping of the curves for the wide range of values for V_{CB}. In addition, consider how closely the average value of the curves appears to begin its rise or about $V_0 = 0.7$ V for the silicon transistor. As with the semiconductor silicon diode, *a first approximation for the forward-biased base-emitter junction in the dc mode would be that $V_{EB} \cong 0.7$ V for all levels of V_{CB}.*

EXAMPLE 3.1   Using the characteristics of Fig. 3.8:
(a)   Find the resulting collector current if $I_E = 3$ mA and $V_{CB} = -10$ V.
(b)   On the input characteristics $I_E = 3.5$ mA at the intersection of $V_{EB} = 750$ mV, $V_{CB} = -10$ V, and $I_C \cong I_E = \textbf{3.5 mA}$.
(c)   Find V_{EB} for the conditions $I_C = 5$ mA and $V_{CB} = -1$ V.

Solution:
(a) $I_C \cong I_E = \textbf{3 mA}$
(b)   On the input characteristics $I_E = 3.5$ mA at the intersection of $V_{EB} = 750$ mV, $V_{CB} = -10$ V, and $I_C \cong I_E = \textbf{3.5 mA}$.
(c) $I_E \cong I_C = 5$ mA
         On the input characteristics the intersection of $I_E = 5$ mA and $V_{CB} = -1$ V results in $V_{EB} \cong 800$ mV $= \textbf{0.8 V}$.

3.6 COMMON-EMITTER CONFIGURATION

The most frequently encountered transistor configuration is shown in Fig. 3.10 for the *pnp* and *npn* transistors. It is called the *common-emitter configuration* since the emitter is common to both the base and collector terminals. Two sets of characteristics are again necessary to describe fully the behavior of the common-emitter configuration: one for the input or base circuit and one for the output or collector circuit. Both are shown in Fig. 3.11.

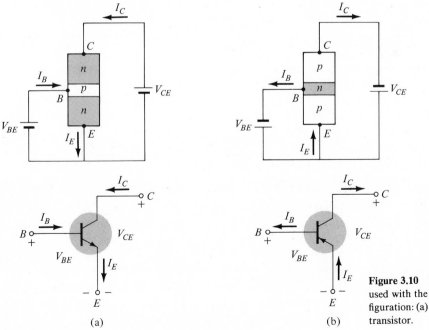

Figure 3.10 Notation and symbols used with the common-emitter configuration: (a) *npn* transistor; (b) *pnp* transistor.

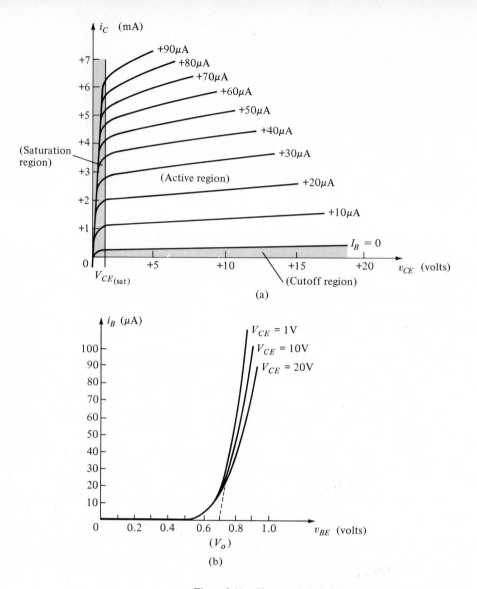

Figure 3.11 Characteristics of a *pnp* transistor in the common-emitter configuration: (a) collector characteristics; (b) base characteristics.

The emitter, collector, and base currents are shown in their actual conventional current direction, while the potentials have the capital letter E as the second subscript to indicate the configuration. Even though the transistor configuration has changed, the current relations developed earlier for the common-base configuration are still applicable.

For the common-emitter configuration the output characteristics will be a plot of the output voltage (V_{CE}) versus output current (I_C) for a range of values of input current (I_B). The input characteristics are a plot of the input current (I_B) versus the input voltage (V_{BE}) for a range of values of output voltage (V_{CE}).

Note that on the characteristics of Fig. 3.11 the magnitude of I_B is in microamperes as compared to milliamperes for I_C. Consider also that the curves of I_B are not as horizontal as those obtained for I_E in the common-base configuration, indicating that the collector-to-emitter voltage will influence the magnitude of the collector current.

The active region for the common-emitter configuration is that portion of the upper-right quadrant that has the greatest linearity, that is, that region in which the curves for I_B are nearly straight and equally spaced. In Fig. 3.11a this region exists to the right of the vertical dashed line at $V_{CE_{sat}}$ and above the curve for I_B equal to zero. The region to the left of $V_{CE_{sat}}$ is called the saturation region. *In the active region the collector junction is reverse-biased, while the emitter junction is forward-biased.* You will recall that these were the same conditions that existed in the active region of the common-base configuration. The active region of the common-emitter configuration can be employed for voltage, current, or power amplification.

The cutoff region for the common-emitter configuration is not as well defined as for the common-base configuration. Note on the collector characteristics of Fig. 3.11 that I_C is not equal to zero when I_B is zero. For the common-base configuration, when the input current I_E was equal to zero, the collector current was equal only to the reverse saturation current I_{CO}, so that the curve $I_E = 0$ and the voltage axis were, for all practical purposes, one.

The reason for this difference in collector characteristics can be derived through the proper manipulation of Eqs. (3.1) and (3.5). That is,

$$I_C = \alpha I_E + I_{CO} \qquad \text{[Eq. (3.5)]}$$

but
$$I_E = I_C + I_B \qquad \text{[Eq. (3.1)]}$$

Therefore,
$$I_C = \alpha (I_C + I_B) + I_{CO} = \alpha I_C + \alpha I_B + I_{CO}$$

and
$$I_C(1 - \alpha) = \alpha I_B + I_{CO}$$

with
$$I_C = \frac{\alpha I_B}{1 - \alpha} + \frac{I_{CO}}{1 - \alpha} \qquad (3.6)$$

If we consider the case discussed earlier, where $I_B = 0$, and substitute this value into Eq. (3.6), then

$$I_C = \frac{I_{CO}}{1 - \alpha} \bigg|_{I_B=0} \qquad (3.7)$$

For $\alpha = 0.996$,

$$I_C = \frac{I_{CO}}{1 - 0.996} = \frac{I_{CO}}{0.004}$$

and
$$I_C = 250 I_{CO} \big|_{I_B=0}$$

which accounts for the vertical shift in the $I_B = 0$ curve from the horizontal voltage axis.

For future reference, the collector current defined by Eq. (3.7) will be assigned the notation indicated by Eq. (3.8).

$$I_{CEO} = \frac{I_{CO}}{1 - \alpha}\bigg|_{I_B = 0} \tag{3.8}$$

In Fig. 3.12 the conditions surrounding this newly defined current are demonstrated with its assigned reference direction.

The magnitude of I_{CEO} is typically much smaller for silicon materials than for germanium materials. For transistors with similar ratings I_{CEO} would typically be a few microamperes for silicon but perhaps a few hundred microamperes for germanium.

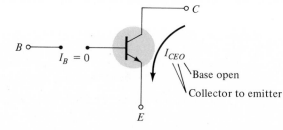

Figure 3.12 Circuit conditions related to I_{CEO}.

For linear (least distortion) amplification purposes, cutoff for the common-emitter configuration will be (for this text) determined by $I_C = I_{CEO}$. In other words, the region below $I_B = 0$ is to be avoided if an undistorted output signal is required.

When employed as a switch in the logic circuitry of a computer, a transistor will have two points of operation of interest: one in the cutoff and one in the saturation region. The cutoff condition should ideally be $I_C = 0$ for the chosen V_{CE} voltage. Since I_{CEO} is typically low in magnitude for silicon materials, *cutoff will exist for switching purposes when $I_B = 0$ or $I_C = I_{CEO}$ for silicon transistors only. For germanium transistors, however, cutoff for switching purposes will be defined as those conditions that exist when $I_C = I_{CBO} = I_{CO}$.* This condition can normally be obtained for germanium transistors by reverse-biasing the normally forward-biased base-to-emitter junction a few tenths of a volt.

EXAMPLE 3.2 Using the characteristics of Fig. 3.11:
    (a)   Find the value of I_C corresponding to $V_{BE} = +800$ mV and $V_{CE} = +10$ V.
    (b)   Find the value of V_{CE} and V_{BE} corresponding to $I_C = +4$ mA and $I_B = +40$ μA.

Solution:
(a)   On the input characteristics the intersection of $V_{BE} = +800$ mV and $V_{CE} = +10$ V results in

$$I_B \cong 50 \ \mu A$$

On the output characteristics the intersection of $I_B = 50$ μA and $V_{CE} = 10$ V results in

$$I_C \cong \mathbf{5.1 \ mA}$$

(b)   On the output characteristics the intersection of $I_C = +4$ mA and $I_B = +40$ μA results in

$$V_{CE} = \mathbf{+6.2 \ V}$$

On the input characteristics the intersection of $I_B = +40$ μA and $V_{CE} = +6.2$ V results in

$$V_{BE} \cong \mathbf{770\ mV}$$

In Section 3.3 the symbol alpha (α) was assigned to the forward current transfer ratio of the common-base configuration. For the common-emitter configuration, the ratio of a small change in collector current to the corresponding change in base current at a fixed collector-to-emitter voltage (V_{CE}) is assigned the Greek letter beta (β) and is commonly called the *common-emitter forward-current amplification factor*. In equation form, the magnitude of β is given by

$$\beta = \frac{\Delta I_C}{\Delta I_B} \bigg|_{V_{CE}=\text{constant}} \qquad (3.9)$$

As a first, but close, approximation, the magnitude of beta (β) can be determined by the following equation:

$$\beta \cong \frac{I_C}{I_B} \qquad (3.10)$$

where I_C and I_B are collector and base currents of a particular operating point in the linear region (i.e., where the horizontal base current lines of the common-emitter characteristics are closest to being parallel and equally spaced). Since I_C and I_B in Eq. (3.10) are fixed or dc values, the value obtained for β from Eq. (3.10) is frequently called the *dc beta,* while that obtained by Eq. (3.9) is called the *ac* or *dynamic* value. Typical values of β vary from 20 to 600. Through the following manipulations of Eqs. (3.1), (3.4), and (3.10):

[Eq. (3.10)] $\qquad \beta = \dfrac{I_C}{I_B}$ resulting in $I_B = \dfrac{I_C}{\beta}$

[Eq. (3.4)] $\qquad \alpha = \dfrac{I_C}{I_E}$ resulting in $I_E = \dfrac{I_C}{\alpha}$

[Eq. (3.1)] $\qquad I_E = I_C + I_B$

substituting: $\qquad \dfrac{I_C}{\alpha} = I_C + \dfrac{I_C}{\beta}$

and dividing by I_C: $\qquad \dfrac{1}{\alpha} = 1 + \dfrac{1}{\beta}$

and $\qquad \beta = \alpha\beta + \alpha$

or $\qquad \beta(1 - \alpha) = \alpha$

we obtain

$$\boxed{\beta = \frac{\alpha}{1 - \alpha}} \tag{3.11}$$

or

$$\boxed{\alpha = \frac{\beta}{\beta + 1}} \tag{3.12}$$

In addition, since

$$I_{CEO} = \frac{I_{CO}}{1 - \alpha} = \frac{I_{CBO}}{1 - \alpha}$$

then

$$\boxed{I_{CEO} = (\beta + 1) I_{CBO} \cong \beta I_{CBO}} \tag{3.13}$$

EXAMPLE 3.3

(a) Find the dc beta at an operating point of $V_{CE} = +10$ V and $I_C = +3$ mA on the characteristics of Fig. 3.11.
(b) Find the value of α corresponding with this operating point.
(c) At $V_{CE} = +10$ V find the corresponding value of I_{CEO}.
(d) Calculate the approximate value of I_{CBO} using the β_{dc} obtained in part (a).

Solution:

(a) At the intersection of $V_{CE} = +10$ V and $I_C = +3$ mA, $I_B = +25$ μA,

so that
$$\beta_{dc} = \frac{I_C}{I_B} = \frac{3 \times 10^{-3}}{25 \times 10^{-6}} = 120$$

(b) $\alpha = \dfrac{\beta}{\beta + 1} = \dfrac{120}{121} \cong 0.992$

(c) $I_{CEO} = 300 \ \mu A$

(d) $I_{CBO} \cong \dfrac{I_{CEO}}{\beta} = \dfrac{300 \ \mu A}{120} = 2.5 \ \mu A$

The input characteristics for the common-emitter configuration are very similar to those obtained for the common-base configuration (Fig. 3.11). In both cases, the increase in input current is due to an increase in majority carriers crossing the base-to-emitter junction with increasing forward-bias potential. Note also that the variation in output voltage (V_{CE} for the CE configuration and V_{CB} for the CB configuration) does not result in a large relocation of the characteristics. In fact, for the dc voltage levels commonly encountered the variation in base-to-emitter voltage with change in output terminal voltage can, as a first approximation, be ignored. On this basis, if we use an average value, the curve of Fig. 3.13 for the CE configuration will result. Note the similarities with the silicon-

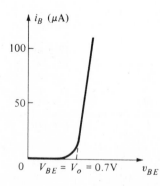

Figure 3.13 Reproduction of Fig. 3.11b ignoring the effects of V_{CE}.

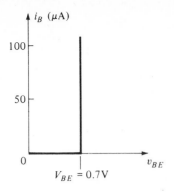

Figure 3.14 Approximate reproduction of Fig. 3.13 for dc analysis.

diode characteristics. Recall also from the description of the semiconductor diode that for dc analysis we approximated the curve of Fig. 3.13 with that indicated in Fig. 3.14. Essentially, therefore, for dc analysis a first approximation to the base-to-emitter voltage of a transistor configuration is to assume that $V_{BE} \cong 0.7$ V for silicon and 0.3 V for germanium. If insufficient voltage is present to provide the 0.7 V bias (for silicon transistors) with the proper polarity, the transistor cannot be in the active region. A number of applications of this approximation will appear in Chapter 7. Since the CB characteristics had a similar set of input characteristics [also true for the common-collector (CC) configuration to be discussed], we can conclude *as a first approximation for dc analysis that the base-to-emitter voltage of a BJT is assumed to be V_0 when biased in the active region of the characteristics.*

Further, we found for the output characteristics of the CB configuration that $I_C \cong I_E$. For the CE configuration $I_C = \beta I_B$, where β is determined by the operating conditions.

In manuals, data sheets, and other transistor publications the common-emitter characteristics are the most frequently presented. The common-base characteristics can be obtained directly from the common-emitter characteristics using the basic current relations derived in the past few sections. In other words, for each point on the characteristics of the common-emitter configuration a sufficient number of variables can be obtained to substitute into the equations derived to come up with a point on the common-base characteristics. This process is, of course, time consuming, but it will result in the desired characteristics.

3.7 COMMON-COLLECTOR CONFIGURATION

The third and final transistor configuration is the *common-collector configuration,* shown in Fig. 3.15 with the proper current directions and voltage notation.

The common-collector configuration is used primarily for impedance-matching purposes since it has a high input impedance and low output impedance, opposite to that which is true of the common-base and common-emitter configurations.

The common-collector circuit configuration is generally as shown in Fig. 3.16 with the load resistor from emitter to ground. Note that the collector is tied to ground even though the transistor is connected in a manner similar to the common-emitter configuration. From a design viewpoint, there is no need for a set of common-collector characteristics to choose the parameters of the circuit of Fig. 3.16. It can be designed using the common-emitter characteristics of Section 3.6. For all practical purposes, the output characteristics of the common-collector configuration are the same as for the common-emitter configuration. For the common-collector configuration the output characteristics are a plot of I_E versus V_{EC} for a range of values of I_B. The input current, therefore, is the same for both the common-emitter and common-collector characteristics. The horizontal voltage axis for the common-collector

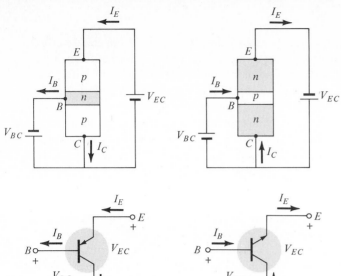

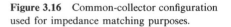

Figure 3.15 Notation and symbols used with the common-collector configuration: (a) *pnp* transistor; (b) *npn* transistor.

Figure 3.16 Common-collector configuration used for impedance matching purposes.

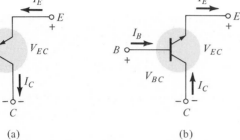

(a)   (b)

configuration is obtained by simply changing the sign of the collector-to-emitter voltage of the common-emitter characteristics since $V_{EC} = -V_{CE}$. Finally, there is an almost unnoticeable change in the vertical scale of I_C of the common-emitter characteristics if I_C is replaced by I_E for the common-collector characteristics (since $\alpha \cong 1$). For the input circuit of the common-collector configuration the common-emitter base characteristics are sufficient for obtaining any required information by simply writing Kirchhoff's voltage law around the loop indicated in Fig. 3.16 and performing the proper mathematical manipulations.

3.8 TRANSISTOR MAXIMUM RATINGS

The standard transistor data sheet will include at least three maximum ratings: *collector dissipation, collector voltage, and collector current.*

For the transistor whose characteristics were presented in Fig. 3.11, the following maximum ratings were indicated:

$$P_{C_{max}} = 30 \text{ mW}$$
$$I_{C_{max}} = 6 \text{ mA}$$
$$V_{CE_{max}} = 20 \text{ V}$$

The power or dissipation rating is the product of the collector voltage and current. For the common-emitter configuration,

$$\boxed{P_{C_{max}} = V_{CE}I_C} \tag{3.14}$$

The nonlinear curve determined by this equation is indicated in Fig. 3.17. The curve was obtained by choosing various values of V_{CE} or I_C and finding the other variable using Eq. (3.14). For example, at $V_{CE} = 10$ V,

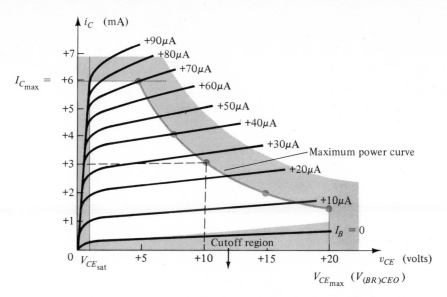

Figure 3.17 Region of operation for amplification purposes.

$$I_C = \frac{P_{C_{\max}}}{V_{CE}} = \frac{30 \times 10^{-3}}{10} = 3 \text{ mA}$$

as indicated in Fig. 3.17. The region above this curve must be avoided in the design of systems using this particular transistor if the maximum power rating is not to be exceeded. The maximum collector voltage, in this case V_{CE}, is indicated as a vertical line in Fig. 3.17. The maximum collector current is also indicated as a horizontal line.

For the common-base configuration the collector dissipation is determined by the following equation. The maximum collector voltage would refer to V_{CB}.

$$\boxed{P_{C_{\max}} = V_{CB} I_C} \tag{3.15}$$

For amplification purposes the nonlinear characteristics of the saturation and the cutoff regions are also avoided. The saturation region has been indicated by the vertical line at $V_{CE_{\mathrm{sat}}}$ and the cutoff region by $I_B = 0$ in Fig. 3.17. The unshaded region remaining is the region employed for amplification purposes. Although it appears as though the area of operation has been drastically reduced, we must keep in mind that many signals are in the microvolt or millivolt range, while the horizontal axis of the characteristics is measured in volts. In addition to maximum ratings, data and specification sheets on transistors also include other important information about their operation. The discussion of these additional data will not be considered until each parameter is fully defined.

3.9 TRANSISTOR SPECIFICATION SHEET

The information provided in the RCA power transistors data book for the 2N1711 transistor appears in Figs. 3.18 through 3.25. As noted, this is a general-purpose small-signal/medium-power device.

Electrical Characteristics

Characteristic	Symbol	Case Tempera-ture °C	Fre-quency kHz	dc Collector-to-Base Voltage V V_{CB}	dc Collector-to-Emitter Voltage V V_{CE}	dc Emitter-to-Base Voltage V V_{EB}	dc Collector Current mA I_C	dc Emitter Current mA I_E	dc Base Current mA I_B	Limits Min.	Limits Max.	Units
Collector-cutoff current	I_{CBO}	25		60			0	0		—	0.01	μA
		150		60			0	0		—	10	μA
Emitter-cutoff current	I_{EBO}	25				5	0			—	0.005	μA
dc-pulse forward-current transfer ratio[a]	h_{FE}	25			10		10			75	—	
		25			10		150			100	300	
		25			10		500			40	—	
dc forward-current transfer ratio	h_{FE}	25			10		0.01			20	—	
		25			10		0.1			35	—	
		−55			10		10			35	—	
Collector-to-base breakdown voltage	$V_{(BR)CBO}$	25					0.1	0		75	—	V
Emitter-to-base breakdown voltage	$V_{(BR)EBO}$	25					0	0.1		7	—	V
Collector-to-emitter reach-through voltage	V_{RT}	25				1.5[b]	0.1			75	—	V
Collector-to-emitter sustaining voltage with external base-to-emitter resistance = 10 ohms	$V_{CER}(sus)$	25					100 (pulsed)			50	—	V

Parameter	Symbol						Min	Max	Units
Collector-to-emitter saturation voltage	V_{CE}(sat)	25		150		15	—	1.5	V
Base-to-emitter saturation voltage	V_{BE}(sat)	25		150		15	—	1.3	V
Small-signal forward-current transfer ratio	h_{fe}	25	1	1	5		50	200	
		25	1	5	10		70	300	
		25	20 MHz	50	10		5	—	
Noise figure: Generator resistance $(R_G) = 510$ ohms, circuit bandwidth (BW) = 1 cycle	NF	25	1	0.3	10		—	8	dB
Output capacitance	C_{ob}	25	1	0	10	0	—	15	pF
Input capacitance	C_{ib}	25	1	1		0.5	—	80	pF
Input resistance	h_{ib}	25	1	5	5		24	34	Ω
		25	1	5	10		4	8	
Voltage-feedback ratio	h_{rb}	25	1	1	5		—	5×10^{-4}	
		25	1	5	10		—	5×10^{-4}	
Output conductance	h_{ob}	25	1	1	5		0.1	0.5	μS
		25	1	5	10		0.1	1	
Thermal resistance: Junction-to-case	$R_{\theta JC}$	—					—	58.3	°C/W
Junction-to-free air	$R_{\theta JA}$	—					—	219	

a Pulse duration = 300 μs; duty factor ≤2%.
b V_{EBF} = Emitter-to-base floating potential.

Figure 3.18 Electrical characteristics of the RCA 2N1711 power transistor. (Courtesy RCA Solid State Division.)

The letter o at the end of a parameter indicates that the terminal not listed is left open. Note in Fig. 3.18 that I_{CBO} is only 0.01 μA at 25°C but 10 μA at 150°C. The quantity h_{FE}, which is synonymous with $\beta_{dc} = I_C/I_B$, has a minimum value of 20. Priorities do not permit an introduction to all the quantities listed in Fig. 3.18. However, most companies carefully define these quantities in the introductory pages of their catalogs. A number of the remaining quantities in the figure will be introduced

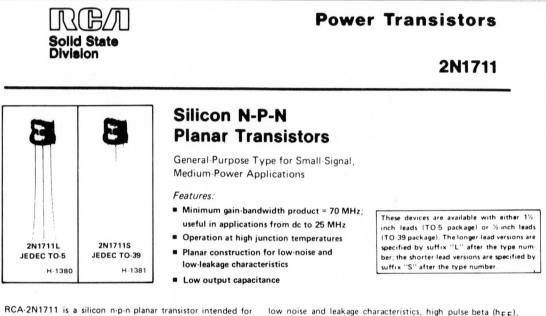

Power Transistors

Solid State Division

2N1711

Silicon N-P-N Planar Transistors

General-Purpose Type for Small-Signal, Medium-Power Applications

Features:

- Minimum gain-bandwidth product = 70 MHz; useful in applications from dc to 25 MHz
- Operation at high junction temperatures
- Planar construction for low-noise and low-leakage characteristics
- Low output capacitance

2N1711L
JEDEC TO-5
H-1380

2N1711S
JEDEC TO-39
H-1381

These devices are available with either 1½-inch leads (TO-5 package) or ½-inch leads (TO-39 package). The longer-lead versions are specified by suffix "L" after the type number; the shorter-lead versions are specified by suffix "S" after the type number.

RCA-2N1711 is a silicon n-p-n planar transistor intended for a wide variety of small-signal and medium-power applications in military and industrial equipment. It features exceptionally low noise and leakage characteristics, high pulse beta (h_{FE}), high breakdown-voltage ratings, low saturation voltages, high sustaining voltages, and low output capacitance.

MAXIMUM RATINGS, *Absolute-Maximum Values:*

COLLECTOR-TO-BASE VOLTAGE	V_{CBO}	75	V
COLLECTOR-TO-EMITTER VOLTAGE:			
With external base-to-emitter resistance (R_{BE}) $\leqslant$ 10 Ω	V_{CER}	50	V
EMITTER-TO-BASE VOLTAGE	V_{EBO}	7	V
COLLECTOR CURRENT	I_C	1	A
TRANSISTOR DISSIPATION:	P_T		
At case temperatures up to 25°C		3	W
At case temperatures above 25°C		See Fig. 3.22	
At free-air temperatures up to 25°C		0.8	W
At free-air temperatures above 25°C		See Fig. 3.22	
TEMPERATURE RANGE:			
Storage and Operating (Junction)		−65 to +200	°C
LEAD TEMPERATURE (During soldering):			
At distance $\geqslant$ 1/16 in. (1.58 mm) from seating plane for 10 s max.		230	°C

Figure 3.19 RCA 2N1711 power transistor. (Courtesy RCA Solid State Division.)

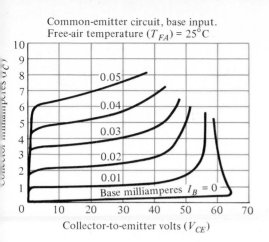

Common-emitter circuit, base input.
Free-air temperature $(T_{FA}) = 25°C$

Collector milliamperes (I_C)

0.05
0.04
0.03
0.02
0.01
Base milliamperes $I_B = 0$

Collector-to-emitter volts (V_{CE})

Figure 3.20 RCA 2N1711 output characteristics. (Courtesy RCA Solid State Division.)

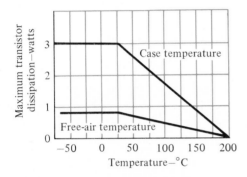

Maximum transistor dissipation—watts

Case temperature

Free-air temperature

Temperature—°C

Figure 3.22 RCA 2N1711 derating curves. (Courtesy RCA Solid State Division.)

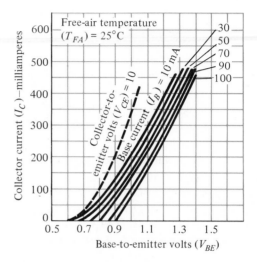

Collector current (I_C)—milliamperes

Free-air temperature $(T_{FA}) = 25°C$

30
50
70
90
100

Collector-to-emitter volts $(V_{CE}) = 10$
Base current $(I_B) = 10\ mA$

Base-to-emitter volts (V_{BE})

Figure 3.21 RCA 2N1711 transfer characteristics. (Courtesy RCA Solid State Division.)

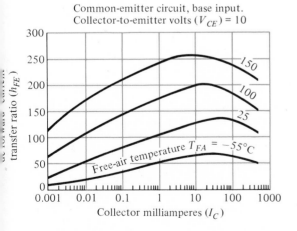

Common-emitter circuit, base input.
Collector-to-emitter volts $(V_{CE}) = 10$

dc forward current transfer ratio (h_{FE})

150
100
25
Free-air temperature $T_{FA} = -55°C$

Collector milliamperes (I_C)

Figure 3.23 RCA 2N1711 dc beta characteristics. (Courtesy RCA Solid State Division.)

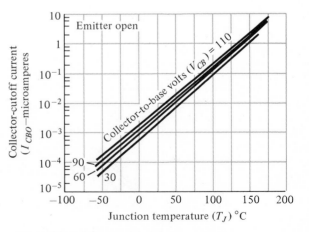

Collector-cutoff current (I_{CBO})—microamperes

Emitter open

Collector-to-base volts $(V_{CB}) = 110$

90
60
30

Junction temperature (T_J) °C

Figure 3.24 RCA 2N1711 collector-cutoff-current characteristics. (Courtesy RCA Solid State Division.)

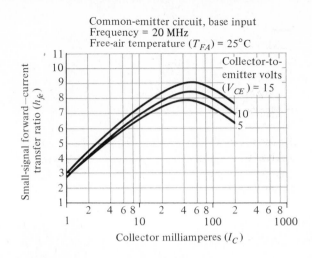

Common-emitter circuit, base input
Frequency = 20 MHz
Free-air temperature (T_{FA}) = 25°C

Figure 3.25 RCA 2N1711 small-signal beta characteristics. (Courtesy RCA Solid State Division.)

in a later chapter. Certainly, V_{CE}(sat.), the capacitance levels and thermal resistance quantities are familiar. The *hybrid* parameters h_{fe}, h_{ib}, h_{rb}, and h_{ob} will be introduced in Chapter 7.

The output characteristics for the device appear in Fig. 3.20. Note the resulting distortion at high levels of voltage and current—a region that must be avoided for linear operation. Linear operation suggests that the output waveform has the same appearance as the input (but amplified) and is not distorted by the amplifying device.

Note the shift in the V_{BE} versus I_C curve for increasing levels of base current. Consider also that I_B is measured in milliamperes since it is a power device. It would appear that our use of $V_0 = 0.7$ V is a good average value since base currents will probably not exceed 25 mA, as shown in Fig. 3.26.

The familiar power derating curve is provided in Fig. 3.22. Note that a curve has been provided for the case and free-air temperature.

The variation in h_{FE} is provided versus I_C for a range of temperatures. Note that the ratio is less at each temperature if the collector current is too high.

In Fig. 3.24 the level of I_{CBO} is provided versus junction temperature for different levels of collector-to-base voltage. It appears the I_{CBO} will not reach 1 μA until the

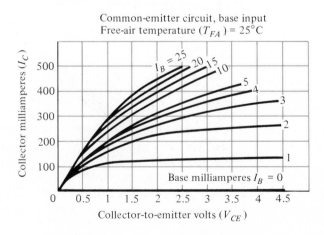

Figure 3.26 RCA 2N1711 output characteristics (high I_C). (Courtesy RCA Solid State Division.)

CH. 3 BIPOLAR JUNCTION TRANSISTORS (BJTs)

junction temperature approaches 135°C. Even with a β of 100, I_{CEO} is still limited to 100 $\mu A = 0.1$ mA at this temperature. For a medium-power device, this is a fairly low level for this undesirable effect.

The small-signal forward-current transfer ratio *(h_{fe})* will be defined in Chapter 7. Briefly, it is a measure of the small-signal ac gain of the device (the increase in the peak-to-peak level of a sinusoidal signal).

Figure 3.26 demonstrates the change in collector characteristics at the high levels of current. The nice even spacing between curves in Fig. 3.20 is no longer present, but the increasing density of the lines follows a fairly linear pattern.

We will refer to these figures as we introduce new quantities of importance in succeeding chapters.

3.10 TRANSISTOR FABRICATION

The majority of the methods used to fabricate transistors are simply extensions of the methods used to manufacture semiconductor diodes. The methods most frequently employed today include *point-contact, alloy junction, grown junction, and diffusion.* The following discussion of each method will be brief, but the fundamental steps included in each will be presented. A detailed discussion of each method would require a text in itself.

Point-Contact

The point-contact transistor is manufactured in a manner very similar to that used for point-contact semiconductor diodes. In this case two wires are placed next to an *n*-type wafer as shown in Fig. 3.27. Electrical pulses are then applied to each wire resulting in a *p-n* junction at the boundary of each wire and the semiconductor wafer. The result is a *pnp* transistor as shown in Fig. 3.27. This method of fabrication is today limited to high-frequency/low-power devices. It was the method used in the fabrication of the first transistor shown in Fig. 3.1.

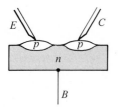

Figure 3.27 Point-contact transistor.

Alloy Junction

The alloy junction technique is also an extension of the alloy method of manufacturing semiconductor diodes. For a transistor, however, two dots of the same impurity are deposited on each side of a semiconductor wafer having the opposite impurity as shown in Fig. 3.28. The entire structure is then heated until melting occurs and each dot is alloyed to the base wafer resulting in the *p-n* junctions indicated in Fig. 3.28 as described for semiconductor diodes.

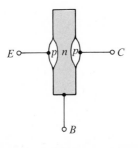

Figure 3.28 Alloy junction transistor.

The collector dot and resulting junction are larger, to withstand the heavy current and power dissipation at the collector-base junction. This method is not employed as much as the diffusion technique to be described shortly, but it is still used extensively in the manufacture of high-power diodes.

Grown Junction

The Czochralski technique (Section 1.6) is used to form the two *p-n* junctions of a grown-junction transistor. The process, as depicted in Fig. 3.29, requires that the impurity control and withdrawal rate be such as to ensure the proper base width and doping levels of the *n*- and *p*-type materials. Transistors of this type are, in general, limited to less than $\frac{1}{4}$-W rating.

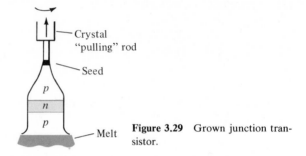

Figure 3.29 Grown junction transistor.

Diffusion

The most frequently employed method of manufacturing transistors today is the diffusion technique. The basic process was introduced in the discussion of semiconductor diode fabrication. The diffusion technique is employed in the production of *mesa* and *planar* transistors, each of which can be of the *diffused* or *epitaxial* type.

In the *pnp,* diffusion-type mesa transistor the first process is an *n*-type diffusion into a *p*-type wafer, as shown in Fig. 3.30, to form the base region. Next, the *p*-type emitter is diffused or alloyed to the *n*-type base as shown in the figure. Etching is done to reduce the capacitance of the collector junction. The term "mesa" is

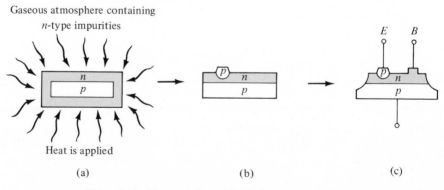

Figure 3.30 Mesa transistor: (a) diffusion process; (b) alloy process; (c) etching process.

derived from its similarities with the geographical formation. As mentioned earlier in the discussion of diode fabrication, the diffusion technique permits very tight control of the doping levels and thicknesses of the various regions.

The major difference between the epitaxial mesa transistor and the mesa transistor is the addition of an epitaxial layer on the original collector substrate. The term epitaxial is derived from the Greek word *epi*—upon, and *taxi*—arrange, which describe the process involved in forming this additional layer. The original *p*-type substrate (collector of Fig. 3.31) is placed in a closed container having a vapor of the same impurity. Through proper temperature control, the atoms of the vapor will *fall upon* and *arrange* themselves on the original *p*-type substrate resulting in the epitaxial layer indicated in Fig. 3.31. Once this layer is established, the process continues, as described above for the mesa transistor, to form the base and emitter regions. The original *p*-type substrate will have a higher doping level and correspondingly less resistance than the epitaxial layer. The result is a low-resistance connection to the collector lead that will reduce the dissipation losses of the transistor.

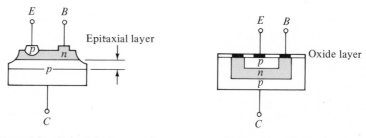

Figure 3.31 Epitaxial mesa transistor.   **Figure 3.32** Planar transistor.

The planar and epitaxial planar transistors are fabricated using two diffusion processes to form the base and emitter regions. The planar transistor, as shown in Fig. 3.32, has a flat surface, which accounts for the term "planar." An oxide layer is added as shown in Fig. 3.32 to eliminate exposed junctions, which will reduce substantially the surface leakage loss (leakage currents on the surface rather than through the junction).

3.11 TRANSISTOR CASING AND TERMINAL IDENTIFICATION

After the transistor has been manufactured using one of the techniques indicated in Section 3.10, leads of, typically, gold, aluminum, or nickel are then attached and the entire structure is encapsulated in a container such as that shown in Fig. 3.33. Those with the studs and heat sinks are high-power devices, while those with the small can (top hat) or plastic body are low- to medium-power devices.

Whenever possible, the transistor casing will have some marking to indicate which leads are connected to the emitter, collector, or base of a transistor. A few of the methods commonly used are indicated in Fig. 3.34.

The internal construction of a TO-92 package in the Fairchild line appears in Fig. 3.35. Note the very small size of the actual semiconductor device. There are gold bond wires, a copper frame, and an epoxy encapsulation.

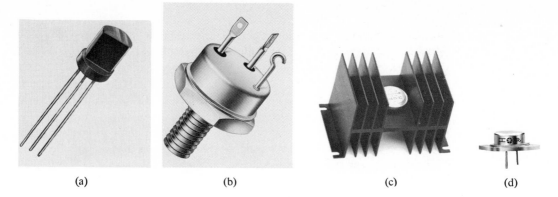

(a)   (b)   (c)   (d)

Figure 3.33 Various types of transistors. [(a) and (b), courtesy General Electric Company; (c) and (d), courtesy International Rectifier Corporation.]

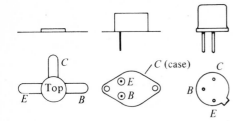
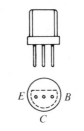
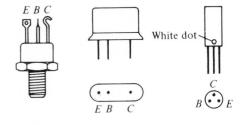

Figure 3.34 Transistor terminal identification.

Figure 3.35 Internal construction of a Fairchild transistor in a TO-92 package. (Courtesy Fairchild Camera and Instrument Corporation.)

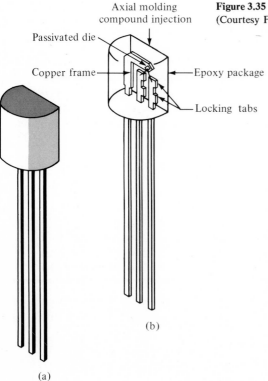

Axial molding compound injection

Passivated die

Copper frame

Epoxy package

Locking tabs

(a)

(b)

(c)

Four (quad) individual *pnp* silicon transistors can be housed in the 14-pin plastic dual-in-line package appearing in Fig. 3.36a. The internal pin connections appear in Fig. 3.36b. As with the diode IC package, the indentation in the top surface reveals the number 1 and 14 pins.

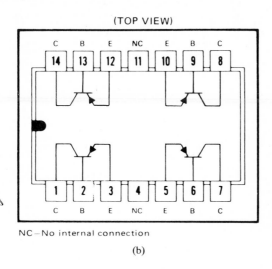

(TOP VIEW)

NC—No internal connection

(a)   (b)

Figure 3.36 Type Q2T2905 Texas Instruments quad *pnp* silicon transistors: (a) appearance; (b) pin connections. (Courtesy Texas Instruments Incorporated.)

PROBLEMS

§ 3.2

1. What names are applied to the two types of transistors? Sketch each transistor and indicate the type of majority and minority carrier in each layer. Is any of this information altered by changing from silicon to germanium material?

2. What is the major difference between a bipolar and a unipolar device?

§ 3.3

3. How must the two transistor junctions be biased for proper transistor operation?

4. What is the source of the leakage current in a transistor?

5. Sketch a figure similar to Fig. 3.3 for the forward-biased junction of an *npn* transistor. Describe the resulting carrier motion.

6. Sketch a figure similar to Fig. 3.4 for the reverse-biased junction of an *npn* transistor. Describe the resulting carrier motion.

7. Sketch a figure similar to Fig. 3.5 for the majority- and minority-carrier flow of an *npn* transistor. Describe the resulting carrier motion.

8. Determine the resulting change in emitter current for a change in collector current of 2 mA and an alpha of 0.98.

9. A transistor has an emitter current of 8 mA and an α of 0.99. How large is the collector current?

10. Calculate the voltage gain $(A_v = V_o/V_i)$ for the circuit of Fig. 3.6 if $V_i = 500$ mV and $R = 1$ kΩ. (The other circuit values remain the same.)

11. Calculate the voltage gain $(A_v = V_o/V_i)$ for the circuit of Fig. 3.6 if the source has an internal resistance of 100 Ω in series with V_i.

§ 3.5

12. From memory, and memory only, sketch the common-base transistor configuration (for *npn* and *pnp*) and indicate the polarity of the applied bias and resulting current directions.

13. Using the characteristics of Fig. 3.8:
    (a) Find the resulting collector current if $I_E = 5$ mA and $V_{CB} = -10$ V.
    (b) Find the resulting collector current if $V_{EB} = 750$ mV and $V_{CB} = -10$ V.
    (c) Find V_{EB} for the conditions $I_C = 4$ mA and $V_{CB} = -1$ V.

14. The characteristics of Fig. 3.8b are for a silicon transistor. How would you expect them to be different for a germanium transistor? What would be a first approximation for the base-to-emitter voltage of the forward-biased junction?

§ 3.6

15. Define I_{CO} and I_{CEO}. How are they different? How are they related? Are they typically close in magnitude?

16. Using the characteristics of Fig. 3.11:
    (a) Find the value of I_C corresponding to $V_{BE} = +750$ mV and $V_{CE} = +5$ V.
    (b) Find the value of V_{CE} and V_{BE} corresponding to $I_C = 3$ mA and $I_B = 30$ μA.

17. (a) For the common-emitter characteristics of Fig. 3.11, find the dc beta at an operating point of $V_{CE} = +8$ V and $I_C = 2$ mA.
    (b) Find the value of α corresponding to this operating point.
    (c) At $V_{CE} = +8$ V, find the corresponding value of I_{CEO}.
    (d) Calculate the approximate value of I_{CBO} using the dc beta value obtained in part (a).

§ 3.7

18. An input voltage of 2 V rms (measured from base to ground) is applied to the circuit of Fig. 3.16. Assuming that the emitter voltage follows the base voltage exactly and that V_{be} (rms) $= 0.1$ V, calculate the circuit voltage amplification $(A_v = V_o/V_i)$ and emitter current for $R_E = 1$ kΩ.

19. For a transistor having the characteristics of Fig. 3.11, sketch the input and output characteristics of the common-collector configuration.

§ 3.8

20. Determine the region of operation for a transistor having the characteristics of Fig. 3.11 if $I_{C_{max}} = 5$ mA, $V_{CE_{max}} = 15$ V, and $P_{C_{max}} = 40$ mW.

21. Determine the region of operation for a transistor having the characteristics of Fig. 3.8 if $I_{C_{max}} = 6$ mA, $V_{CB_{max}} = -15$ V, and $P_{C_{max}} = 30$ mW.

22. Referring to Fig. 3.19, determine the temperature range for the device in degrees Fahrenheit.

23. Determine the value of α when $I_C = 0.1$ mA from the provided value of dc forward-current transfer ratio (Fig. 3.18).

24. Using the results of Problem 23 and Eq. (3.7), calculate I_{CO} and note whether it falls within the limits of I_{CBO} in Fig. 3.18. Use $I_C = 0.1$ mA at $I_B = 0$ μA with $T_{\text{case}} = 150°$C.

25. Using the information provided in Figs. 3.18 and 3.19, sketch the boundaries of the maximum power dissipation region for the CE configuration ($T_{\text{case}} = 25°$C).

26. Sketch the maximum power curve in the characteristics of Fig. 3.26 if the free-air temperature is 100°C.

27. Referring to Fig. 3.20, determine the following:
    (a) I_C if $V_{CE} = 30$ V, $I_B = 25$ μA.
    (b) V_{CE} if $I_C = 4$ mA, $I_B = 30$ μA.
    (c) I_B if $I_C = 3$ mA, $V_{CE} = 20$ V.

28. (a) At an operating point of $V_{CE} = 30$ V and $I_C = 7$ mA, determine the level of V_{BE} using Figs. 3.20 and 3.21. For base currents significantly less than 10 mA, use the curve of Fig. 3.21 denoted $V_{CE} = 10$ V.
    (b) At an operating point of $V_{CE} = 2$ V and $I_C = 300$ mA, determine the level of V_{BE} using Figs. 3.21 and 3.26.
    (c) What would be an appropriate level of V_o for the approximate equivalent circuit for each of the operating conditions of parts (a) and (b)?

29. (a) Calculate the power derating factor for each of the curves of Fig. 3.22.
    (b) Using the value obtained for the case-temperature curve, find the power rating at a temperature of 100°C.
    (c) Compare the results of part (b) with the value obtained from the graph.

30. (a) Determine the value of β_{dc} at $I_C = 5$ mA and a temperature of 100°C, using Fig. 3.23.
    (b) What is the level of α at this point?
    (c) What is the average difference in β level between room temperature and 100°C for the range $I_C = 0.01$ mA to $I_C = 10$ mA? Is it something we should carefully consider in a design problem? Why?

31. (a) Determine I_{CBO} from Fig. 3.24 for $V_{CB} = 30$ V and a junction temperature of 50°C. Note the log scale.
    (b) If $\beta = 200$, what is the level of I_{CEO}?
    (c) What is the rate of change of I_{CBO} per degree change in temperature near 50°C for $V_{CB} = 30$ V?

32. (a) Using Fig. 3.25, determine the rate of change of h_{fe} with change in collector current for the range $I_C = 2$ to 4 mA ($V_{CE} = 10$ V).
    (b) Why would you assume that the magnitude of h_{fe} in this particular figure is so small on the vertical scale when typical values of h_{fe} approach 100 or more?

§ 3.10

33. (a) Describe the basic differences among the various techniques of transistor construction.
(b) Which would you categorize as acceptable for high-power applications?
(c) Define the *diffusion* process.

§ 3.11

34. (a) Find three transistors with different casings, identify the terminals, and sketch the device.
(b) Search through a manufacturer's data book for another IC structure limited totally to transistors. Sketch the internal schematic and identify the terminals.

4

Dc Biasing: BJTs

4.1 GENERAL

Transistors are used in a large variety of applications and in many different ways. It would be difficult if not impossible to learn each area and application. Instead, one studies the more fundamental circuit operation so that enough is known to carry over this knowledge to slightly different or even completely different applications. This chapter covers the basic concepts in the dc biasing of bipolar junction transistors (BJTs).

To use these devices for amplification of voltage or current, or as control (*on* or *off*) elements, it is necessary first to *bias* the device. The usual reason for this biasing is to turn the device on, and in particular, to place it in operation in the region of its characteristic where the device operates most linearly.

Although the purpose of the bias network or biasing circuit is to cause the device to operate in this desired *linear* region of operation (which is best defined by the manufacturer for each device type), the bias components are still part of the overall application circuit—amplifier, waveform shaper, logic circuit, etc. We could treat the overall circuit and consider all aspects of the operation at once, but that would be complex and more confusing. Each type of circuit application would have to be studied for all aspects of operation without a more basic understanding of those common features of operation. This chapter therefore provides basic concepts of dc biasing of the bipolar transistor, with the understood aim of getting the device operating in a desired region of the device characteristic. If these concepts are well understood, many different circuits, even new circuit applications, can be studied and analyzed more easily because a basic understanding of the circuit has been established. Amplifier gain and other factors affecting ac operation will be considered in Chapter 7. Basic dc bias concepts are presented in this chapter and then applied in Chapters 7, 9, and 10.

Dc biasing is a *static* operation since it deals with setting a fixed (steady) level of current (through the device) with a desired fixed voltage drop across the device. Necessary information about the device can be obtained from the device's static characteristics.

4.2 OPERATING POINT

Since the aim of biasing is to achieve a certain condition of current and voltage called the *operating point* (*quiescent* point or *Q*-point), some attention is given to the selection of this point in the device characteristic. Figure 4.1 shows a general device characteristic with four indicated operating points. The biasing circuit may be designed to set the device operation at any of these points or others within the *operating region*. The operating region is the area of current or voltage within the maximum limits for the particular device. These maximum ratings are indicated on the characteristic of Fig. 4.1 by a horizontal line for the maximum current, I_{max}, and a vertical line for the maximum voltage, V_{max}. An additional consideration of maximum power (product of voltage and current) must also be taken into consideration in defining the operating region of a particular device, as shown by the line marked P_{max} on Fig. 4.1.

The BJT device could be biased to operate outside these maximum limit points but the result of such operation would be either a considerable shortening of the lifetime of the device or destruction of the device. Confining ourselves to the safe operating region we may select many different operating areas or points. The exact

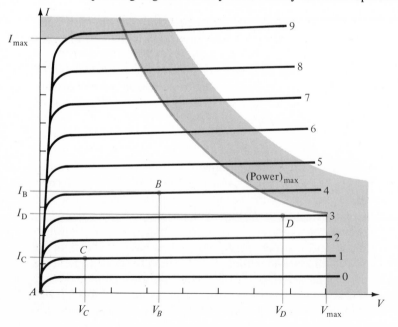

Figure 4.1 Various operating points on device static characteristics.

CH. 4   DC BIASING: BJTs

point or area often depends on the intended use of the circuit. Still, we can consider some differences between operation at the various points shown in Fig. 4.1 to present some basic ideas about the operating point and, thereby, the bias circuit.

If no bias were used, the device would initially be completely off, which would result in the current of point *A*—namely, zero current through the device (and zero voltage across it). It is necessary to bias the device so that it can respond or change in current and voltage for the entire range of an input signal. While point *A* would not be suitable, point *B* provides this desired operation. If a signal is applied to the circuit, *in addition to the bias level,* the device will vary in current and voltage from operating point *B*, allowing the device to react to (and possibly amplify) both the positive and negative part of the input signal. If, as could be the case, the input signal is small, the voltage and current of the device will vary but not enough to drive the device into *cutoff* or *saturation.* Cutoff is the condition in which the device no longer conducts. Saturation is the condition in which voltage across the device is as small as possible with the current in the device path reaching a limiting or saturating value depending on the external circuit. The usual amplifier action desired occurs within the operating region of the device, that is, between saturation and cutoff.

Point *C* would also allow some positive and negative variation with the device still operating, but the output voltage could not decrease much because bias point *C* is lower in voltage than point *B*. Point *C* is also a region of operation in which the current level in the device is smaller and the device gain is *not* linear, that is, the spacing in going from one curve to the next is unequal. This nonlinearity shows that the amount of gain of the device is smaller when biased lower on the characteristic and larger when biased higher up. It is preferable to operate where the gain of the device is most constant (or linear) so that the amount of amplification over the entire swing of input signal is the same. Point *B* is a region of more linear spacing and, therefore, more linear operation, as shown in Fig. 4.1.

Point *D* sets the device operating point near the maximum voltage level. The output voltage swing in the positive direction is thus limited if the maximum voltage is not to be exceeded. Point *B*, therefore, seems the best operating point in terms of linear gain or largest possible voltage and current swing. This is usually the desired condition for small-signal amplifiers (Chapter 7) but not necessarily for power amplifiers and logic circuits, which will be considered in Chapters 10 and 15, respectively. In this discussion, we will concentrate mainly on biasing the device for *small-signal* amplification operation.

One other very important biasing factor must be considered. Having selected and biased the BJT at a desired operating point, the effect of temperature must also be taken into account. Temperature causes device characteristics such as the transistor current gain and the transistor leakage current to change. Higher temperature results in more current in the device than at room temperature, thereby upsetting the operating condition set by the bias circuit. Because of this, the bias circuit must also provide a degree of *temperature stability* to the circuit so that temperature changes at the device produce minimum change in its operating point. This maintenance of operating point may be specified by a *stability factor, S,* indicating the amount of change in operating-point current due to temperature. A highly stable circuit is desirable and the stability of a few basic bias circuits will be compared.

Bipolar transistor operation may be specified sufficiently well by device parameters, and mathematical techniques can be used to determine its biasing. Nevertheless, the transistor characteristic still provides a convenient picture for understanding device operation and will be used on occasion.

4.3 COMMON-BASE (CB) BIAS CIRCUIT

The common-base (CB) configuration provides a relatively straightforward and simple starting point in our dc bias considerations. Figure 4.2a shows a common-base circuit configuration. By CB we mean that the base is the reference (common) point for measurement for both input (emitter, in this case) and output (collector).

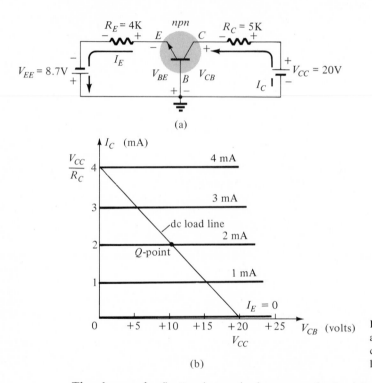

(a)

(b)

Figure 4.2 Common-base circuit and transistor characteristic: (a) common-base circuit; (b) dc load line on common-base characteristic.

The dc supply (battery) terminals are marked with a double-letter designation. V_{EE} is the dc voltage supply associated with the emitter section, and V_{CC} is the dc supply for the collector section of the circuit. Two separate voltage supplies may be required for the CB configuration. Resistor R_E is basically a current-limiting resistor for setting the emitter current I_E. Resistor R_C is the collector (output or load) resistor and the output ac signal is developed across it. It is also one of the components that is used to set a desired operating point.

Figure 4.2b shows the CB collector characteristic, which is the output characteristic for this circuit connection. The abscissa is the collector-to-base voltage V_{CB}, and the ordinate is the collector current, I_C. The family of curves for the characteristic is for various emitter currents.

The theory developed for either the *pnp* or *npn* transistor device can be equally applied to the other device by merely changing all current directions and all voltage polarities. We shall consider mostly *npn* transistors in developing concepts because it is the more popular unit at present.

It is possible to consider biasing the CB circuit by analyzing separately the input (base-emitter loop) and the output (base-collector loop) sections of the circuit. Although in reality there is some interaction between the operation of the base-collector section and that of the base-emitter section, this can be neglected with excellent practical results obtained.

Input Section

The input loop (see Fig. 4.3a) is composed of the battery, V_{EE}, the resistor, R_E, and the base-emitter junction of the transistor, V_{BE}. Writing the voltage loop equation (using Kirchhoff's voltage law) for the input loop

$$-V_{EE} + I_E R_E + V_{BE} = 0$$

from which we get

$$I_E = \frac{V_{EE} - V_{BE}}{R_E} \qquad (4.1a)$$

where all terms have been defined above.

When forward-biased, the base-emitter voltage V_{EB} is small—on the order of 0.3 V for germanium transistors and 0.7 V for silicon. Although the actual emitter-base voltage is slightly affected by the collector-base voltage, this effect can be neglected for practical considerations. In fact, if the supply voltage V_{EE} is, say, 10 V or more, the emitter-base voltage could be neglected, giving

$$I_E \cong \frac{V_{EE}}{R_E} \qquad (4.1b)$$

as a good approximation. Observe that the emitter current is set essentially by the emitter supply voltage and the emitter resistor. But the supply voltage is usually fixed since it is required to provide voltage to other parts of the electronic circuit. The emitter current, therefore, is specifically determined by the emitter resistor R_E, whose value is selected to give the desired emitter current.

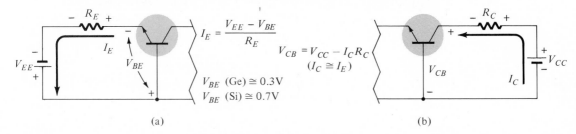

(a)   (b)

Figure 4.3 Input and output sections for common-base circuit: (a) input (emitter-base) section only; (b) output (collector-base) section only.

Output Section

The output loop (see Fig. 4.3b) consists of the battery, V_{CC}, the resistor R_C, and the voltage across the collector-base junction of the transistor, V_{CB}. For operation as an amplifier the *collector-base* junction must be *reverse-biased* in addition to the *base-emitter* being *forward-biased*. This *reverse bias* is provided by the V_{CC} battery voltage connected in polarity so that the *p*ositive battery terminal connects to the *n*-material and the *n*egative battery terminal to the *p*-material. (Note the letter opposites for *p* and *n* for battery and transistor type.)

Summing the voltage drops around the output or collector-base loop of the circuit of Fig. 4.3b, we get

$$+V_{CC} - I_C R_C - V_{CB} = 0$$

Solving for the collector-base voltage results in

$$\boxed{V_{CB} = V_{CC} - I_C R_C} \tag{4.2}$$

The collector current I_C is approximately the same magnitude as the emitter current I_E [obtained from Eqs. (4.1a) or (4.1b)]. This is a very good approximation for any type of transistor connection used. For the purposes of calculating circuit bias values we may write the relation as

$$\boxed{I_C \cong I_E} \tag{4.3}$$

Actually, $I_C = \alpha I_E$, where α (alpha) is typically 0.9–0.998 in value.

Complete Solution of Bias Conditions
for Common-Base Circuit

Having presented the essential circuit operation of the CB connection we can consider the complete solution of bias currents and voltages. The results obtained will apply to both *npn* and *pnp* CB transistor circuits. To help in the solution of bias conditions for a CB circuit as in Fig. 4.2a and to provide a structured procedure to be followed in later types of circuit connections we shall formalize the solution into a step-by-step calculation. To solve for the bias voltages and currents of a CB bias circuit as in Fig. 4.2a proceed as follows:

1. With emitter-base voltage polarity providing forward bias, assume approximate voltages of

    $$V_{BE} \cong 0.3 \text{ V} \quad \text{(germanium), or}$$
    $$V_{BE} \cong 0.7 \text{ V} \quad \text{(silicon)}$$

    (Forward bias is provided if the battery voltage connection results in *p*ositive voltage at *p*-type material and *n*egative voltage at *n*-type material of the transistor.)

2. Calculate the emitter current I_E using

    $$I_E = \frac{V_{EE} - V_{BE}}{R_E} \cong \frac{V_{EE}}{R_E}$$

3. Collector current is approximately the emitter current calculated in step (2) above:

$$I_C \cong I_E$$

4. Collector-base voltage is calculated from

$$V_{CB} = V_{CC} - I_C R_C$$

EXAMPLE 4.1   Calculate bias voltages V_{BE} and V_{CB} and currents I_E and I_C for the circuit of Fig. 4.4. The circuit contains an *npn* silicon transistor.

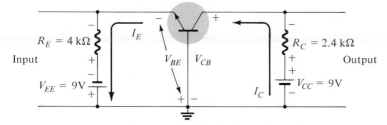

Figure 4.4   Common-base bias circuit for Example 4.1.

Solution: Using the step-by-step procedure outlined previously, we get:
(a) $V_{BE} = \textbf{0.7 V}$ (silicon)
(b) $I_E = (V_{EE} - V_{BE})/R_E = (9 \text{ V} - 0.7 \text{ V})/4 \text{ k}\Omega = \textbf{2.075 mA}$
(c) $I_C = I_E = \textbf{2.075 mA}$
(d) $V_{CB} = V_{CC} - I_C R_C = 9 \text{ V} - (2.075 \text{ mA})(2.4 \text{ k}\Omega) = \textbf{4.02 V}$

4.4   COMMON-EMITTER (CE)—GENERAL BIAS CONSIDERATIONS

A more popular amplifier connection applies the input signal to the base of the transistor with the emitter as common terminal. The CE circuit of Fig. 4.5 shows only one supply voltage. Recall that the circuit connection of Fig. 4.2 used two supply voltages, one to forward bias the base-emitter and the second to provide reverse bias for the base-collector. Both forward- and reverse-bias conditions are achieved in the CE connection using one voltage supply. Later we will show a number of other important advantages of the CE circuit relating to input and output impedances, current and voltage gain, which apply to the ac operation of the circuit (considered in Chapter 5).

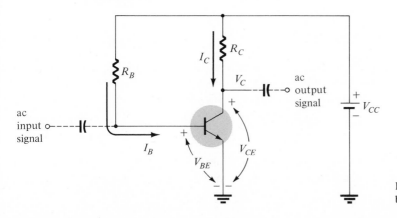

Figure 4.5   Common-emitter fixed-bias circuit.

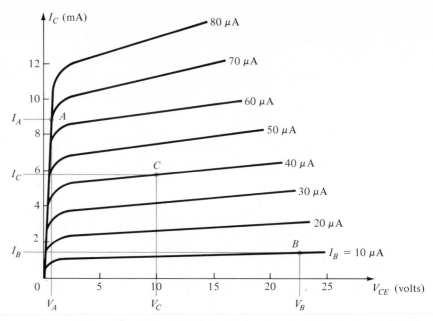

Figure 4.6 Common-emitter collector characteristics showing typical bias points.

Although a mathematical (rather than graphical) technique may be used to obtain the bias voltages and currents in the circuit, we shall refer to the CE collector characteristic of the transistor to provide a reason for choosing a particular operating point. Figure 4.6 shows the CE collector characteristic of a transistor indicating a few bias points. Bias at point *C* would provide the largest collector-emitter voltage variation around the bias point.

The remainder of this chapter will concentrate on (1) determining what bias point actually will result for a given circuit containing given circuit elements *(analysis)* and (2) how to obtain the circuit elements to provide a desired bias point of collector current and collector-emitter voltage *(synthesis)*. To present the basic circuit theory and provide some appreciation of the circuit operation, analysis of a given circuit will be considered first. After that, the design (synthesis) of a circuit to obtain a desired bias condition will be covered.

4.5 COMMON-EMITTER FIXED-BIAS CIRCUIT

Given the fixed-bias circuit of Fig. 4.5, how can we determine the dc bias currents and voltages for the base and collector of the transistor? This section will develop the procedure that will determine the answers to the above question.

Input Section

Consider first the base-emitter circuit loop shown in the partial circuit diagram of Fig. 4.7a. Writing the Kirchhoff voltage equation for the loop, we get

$$+V_{CC} - I_B R_B - V_{BE} = 0$$

We can solve the foregoing equation for the base current I_B

$$I_B = \frac{V_{CC} - V_{BE}}{R_B}$$ (4.4a)

Since the supply voltage V_{CC} and the base-emitter voltage V_{BE} are fixed values of voltage, the selection of a base bias resistor fixes the value of the base current. As a good approximation we may even neglect the few tenths volt drop across the forward-biased base-emitter V_{BE}, obtaining the simplified form for calculating base current,

$$I_B \cong \frac{V_{CC}}{R_B}$$ (4.4b)

Output Section

The output section of the circuit (Fig. 4.7b) consists of the supply battery, the collector (lead) resistor, and the transistor collector-emitter junctions. The currents in the collector and emitter are about the same since I_B is small in comparison to

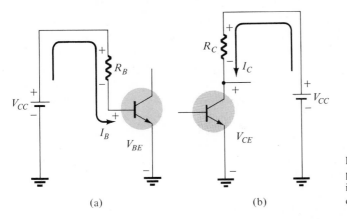

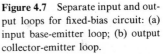

Figure 4.7 Separate input and output loops for fixed-bias circuit: (a) input base-emitter loop; (b) output collector-emitter loop.

(a)   (b)

either. For linear amplifier operation the collector current is related to the base current by the transistor current gain, beta (β) or h_{FE}. Expressed mathematically,

$$I_C = \beta I_B$$ (4.5)

The base current is determined from the operation of the base-emitter section of the circuit as provided by Eq. (4.4a) or (4.4b). The collector current as shown by Eq. (4.5) is β times greater than the base current *and* not at all dependent on the resistance in the collector circuit. From the previous consideration of the common-base circuit we know that the collector current is controlled in the base-emitter section of the circuit and not in the collector-base (or collector-emitter, in this case) section of the circuit.

Calculating voltage drops in the output loop, we get

$$V_{CC} - I_C R_C - V_{CE} = 0$$

$$V_{CE} = V_{CC} - I_C R_C$$ (4.6)

Transistor Saturation

One additional consideration must be included in the above solution steps. The relation between collector and base current, namely, that $I_C = \beta I_B$, is true *only* if the transistor is properly biased in the linear region of the transistor's operation. If the transistor, for example, is biased in the *saturation* region, Eqs. (4.5) and (4.6) lead to incorrect results.

For the transistor to be biased in a region of linear amplifier operation (as opposed to regions of cutoff or saturation) the base-emitter junction must be forward biased *and* the base-collector junction reverse biased. Our concern here is with the second bias condition—that the collector-base be properly reverse biased. This is true only as long as the collector-emitter voltage V_{CE} is larger in value than the base-emitter forward-bias voltage V_{BE}. Since the collector-emitter voltage V_{CE} given by Eq. (4.6) is the difference between the supply voltage V_{CC} and the voltage drop across the collector resistor $(I_C R_C)$, the latter must be less than V_{CC} or in terms of the collector current, I_C must be less than V_{CC}/R_C. Stated mathematically,

$$I_C < \frac{V_{CC}}{R_C} \tag{4.7}$$

for the transistor to be biased in the active (linear) region of operation. A quick check of Eq. (4.5) would therefore be in order using Eq. (4.7) when performing the calculations of collector-emitter voltage to make sure that the condition just stated is correct in the circuit under consideration. If so, the three solution steps outlined above can be carried out, as representing the operation of the circuit. If, however, the above relation of maximum I_C allowable for operation in the transistor linear region is exceeded, the transistor is operating in the saturation region. In this case the collector current will be the maximum value set by the circuit:

$$I_{C_{sat}} \cong \frac{V_{CC}}{R_C} \tag{4.8}$$

and $\qquad\qquad V_{CE_{sat}} \cong 0 \text{ V} \qquad$ (actually a few tenths volt) $\qquad\qquad$ (4.9)

The base current calculated from Eq. (4.4) is correct in any case.

If the circuit to be analyzed is used as an amplifier, we shall not expect it to be biased in the saturation region. If, however, some value used is incorrect, or some wiring error occurs, the resulting operation might possibly bias the transistor into saturation and we must be aware of this condition. (Keep in mind that the saturation condition is undesirable only for amplifier operation. For operation in computer switching circuits the saturation region of operation is important and we shall consider it fully under that topic in Chapter 15.)

EXAMPLE 4.2   Compute the dc bias voltages and currents for the *npn* CE circuit of Fig. 4.8.

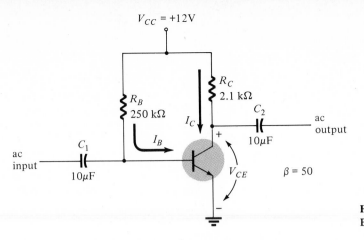

$V_{CC} = +12V$

R_C
$2.1\ k\Omega$

R_B
$250\ k\Omega$

I_C

C_2

ac output

$10\mu F$

I_B

ac input

C_1

$10\mu F$

V_{CE}

$\beta = 50$

Figure 4.8 Dc fixed-bias circuit for Example 4.2.

Solution:

(a) $I_B = (V_{CC} - V_{BE})/R_B \cong V_{CC}/R_B = 12\ V/250\ k\Omega = \mathbf{48\ \mu A}$

(b) $I_C = \beta I_B = 50(48\ \mu A) = \mathbf{2.4\ mA}$

(c) $V_{CE} = V_{CC} - I_C R_C = 12\ V - (2.4\ mA)(2.1\ k\Omega) = \mathbf{6.96\ V}$

4.6 BIAS STABILIZATION

While the fixed-bias circuit provides suitable gain as an amplifier it has difficulty maintaining bias stability. In any amplifier circuit the collector current, I_C, will vary with change in temperature because of the three following main factors:

1. Reverse saturation current (leakage current), I_{CO}, which doubles for every $10°$ increase in temperature.

2. Base-emitter voltage, V_{BE}, which decreases by 2.5 mV per °C.

3. Transistor current gain, β, which increases with temperature.

Any or all of these factors can cause the bias point to shift from the values originally set by the circuit because of a change in temperature. Table 4.1 lists typical parameter values for silicon transistors.

TABLE 4.1 Typical Silicon Transistor Parameters

T (°C)	I_{CO} (nA)	β	V_{BE} (V)
-65	0.2×10^{-3}	20	0.85
25	0.1	50	0.65
100	20	80	0.48
175	3.3×10^3	120	0.3

We first demonstrate the effect of leakage current and current gain change on the dc bias point initially set by the circuit. Consider the graphs of Figs. 4.9a and 4.9b, which show a transistor collector characteristic at room temperature (25°C) and the same transistor at some elevated temperature (100°C). Notice that the significant increase of leakage current not only causes the curves to rise but also that an increase in beta occurs as shown by the larger spacing between the curves at the higher temperature.

The operating point may be specified by drawing the circuit dc load line on the graph of the collector characteristic and noting the intersection of the load line and the dc base current set by the input circuit. An arbitrary point is marked as an example in Fig. 4.9a. Since the fixed-bias circuit provides a base current whose value depends approximately on the supply voltage and base resistor, neither of which is affected by temperature or the change in leakage current or beta, the same base current magnitude will exist at high temperatures as indicated on the graph of Fig. 4.9b. As the figure shows, this will result in the dc bias point's shifting to a higher collector current and a lower collector-emitter voltage operating point. In the extreme, the transistor could be driven into saturation. In any case the new operating point may not be at all satisfactory and considerable distortion may result because of the bias-point shift. A better bias circuit is needed, one that will stabilize or maintain the dc bias initially set, so that the amplifier can be used in a changing-temperature environment.

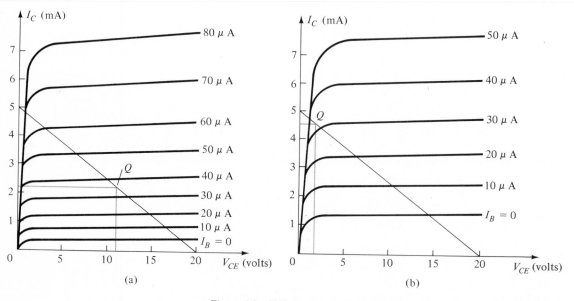

Figure 4.9 Shift in dc bias point (Q-point) due to change in temperature: (a) 25°C; (b) 100°C.

Stability Factor, *S*

A stability factor, *S*, can be defined for each of the parameters affecting bias stability. These are

$$S(I_{CO}) = \frac{\Delta I_C}{\Delta I_{CO}} \qquad S(V_{BE}) = \frac{\Delta I_C}{\Delta V_{BE}} \qquad S(\beta) = \frac{\Delta I_C}{\Delta \beta}$$

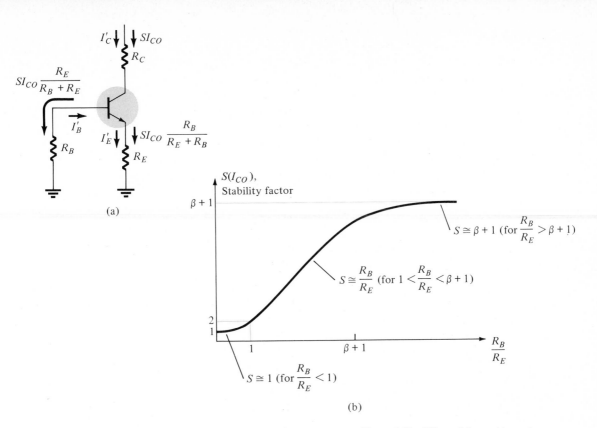

Figure 4.10 Effect of I_{CO} on bias point.

The stability factor is a numerical quantity representing the amount the collector current changes due to changes in each of the parameters because of temperature. The results of detailed mathematical analysis will be used to provide a comparison of how the device and circuit components affect bias stability to allow comparison of the effects on I_C by each of the transistor parameters.

$S(I_{CO})$: Figure 4.10a shows a basic transistor circuit and the effect of I_{CO}. Figure 4.10b uses the result of analyzing the stability based on change due only to I_{CO} (β and V_{BE} considered held constant). Referring to Fig. 4.10b, we see that the stability factor varies from the ideal case (best condition) of $S = 1$ up to a maximum value of $S = \beta + 1$, which occurs for the fixed-bias circuit, or when the ratio R_B/R_E is greater than $\beta + 1$. Essentially, the stability factor is smallest for larger values of R_E so that inclusion of an emitter resistor improves the bias stability (makes S smaller).

$$S(I_{CO}) = \frac{(\beta + 1)(1 + R_B/R_E)}{(\beta + 1) + (R_B/R_E)} \tag{4.10}$$

The value of I_{CO} in modern transistors is so low (refer to Table 4.1), however, that the change in bias point in a circuit with even a large value of $S(I_{CO})$ would not be considerable, as the following example illustrates.

EXAMPLE 4.3 In a circuit using a transistor typified by the parameters in Table 4.1 calculate the change in I_C from 25°C for: (a) fixed bias; (b) $R_B/R_E = 10$; and (c) $R_B/R_E = 1$.

Solution: From 25°C to 100°C the change in I_{CO} is

$$\Delta I_{CO} = (20 - 0.1) \text{ nA} \cong 20 \text{ nA}$$

(a)   For fixed bias: $S = \beta + 1 = 51$.
Using the definition of stability, we get

$$\Delta I_C = S(\Delta I_{CO}) = 51(20 \text{ nA}) \cong \textbf{1 μA}$$

(b)   For $S = R_B/R_E = 10$:

$$\Delta I_C = 10 \cdot \Delta I_{CO} = 10(20 \text{ nA}) = \textbf{0.2 μA}$$

(c)   For $S = 1$:

$$\Delta I_C = 1(20 \text{ nA}) = \textbf{20 nA}$$

While the change in I_C is considerably different in a circuit having ideal stability ($S = 1$) and one having the maximum stability factor ($S = 51$, in this example), the change in I_C is not significant. For example, the amount of change in I_C from a dc bias current set at, say, 2 mA, would be from 2 mA to 2.001 mA (only 0.05%) in the worst case, which is obviously small enough to be ignored. Some power transistors exhibit larger leakage current, but for most amplifier circuits the effect of I_{CO} change with temperature is slight.

$S(V_{BE})$:   Analysis of the stability factor due to change in V_{BE} will result in

$$S(V_{BE}) = \frac{\Delta I_C}{\Delta V_{BE}} = \frac{-\beta}{R_B + R_E(\beta + 1)} \tag{4.11}$$

$$= \frac{-1}{R_E}, \qquad \text{for } (\beta + 1) \gg \frac{R_B}{R_E} \text{ and } \beta \gg 1$$

Since smaller values of S indicate better stability, the larger the value of R_E the better the circuit stability due to changes in V_{BE} with temperature.

> **EXAMPLE 4.4**   Determine the change in I_C for a transistor having parameters as listed in Table 4.1 over a temperature range from 25°C to 100°C for a circuit having $R_E = 1$ kΩ (and $\beta + 1 \gg R_B/R_E$).
>
> **Solution:**   Using $S(V_{BE}) = -1/R_E = -1/1$ kΩ $= -10^{-3}$ and $V_{BE} = (0.65 - 0.48)$ $= 0.17$ V, from 25°C to 100°C
>
> $$I_C = S \cdot \Delta V_{BE} = -10^{-3}(0.17 \text{ V}) = -170 \text{ μA}$$
>
> which is seen to be a reasonably large current change compared to that resulting from a change in I_{CO}. Using a typical collector current of 2 mA, we see that the collector current would change from 2 mA at 25°C to 1.830 mA at 100°C: a Á **8.5% change**.

The effect of V_{BE} changing with temperature can be somewhat compensated by the use of diode compensation in which the voltage change across a diode compensates for the change in V_{BE} to maintain the bias value of I_C. Thermistor and transistor compensation techniques are also used when needed.

S(β): Analysis of the effect of β changing with temperature on the circuit bias stability results in

$$\frac{\Delta I_C}{I_C(T_1)} = \left(1 + \frac{R_B}{R_E}\right)\frac{\Delta\beta}{\beta(T_1)\beta(T_2)} = \left(1 + \frac{R_B}{R_E}\right)\frac{\dfrac{\beta(T_2)}{\beta(T_1)} - 1}{\beta(T_2)} \qquad (4.12)$$

where $\beta(T_1) =$ beta at temperature T_1
    $\beta(T_2) =$ beta at temperature T_2
    $I_C(T_1) =$ collector current at temperature T_1

    EXAMPLE 4.5 Calculate the change in collector current for the transistor having parameters as given in Table 4.1 from room temperature to 100°C. Assume that $R_B/R_E = 20$ for the circuit used and that I_C at room temperature is 2 mA.

    Solution: The change in I_C is calculated to be

$$\Delta I_C = I_C(T_1)\left[\left(1 + \frac{R_B}{R_E}\right)\cdot\frac{\dfrac{\beta(T_2)}{\beta(T_1)} - 1}{\beta(T_2)}\right]$$

$$= 2\text{ mA}\left[(1+20)\frac{\left(\dfrac{80}{50}-1\right)}{80}\right] = 0.315\text{ mA} = \mathbf{315\ \mu A}$$

    The collector current changing from 2 mA at room temperature to 2.315 mA at 100°C represents a change of about **16%.**

Comparison of the three examples shows that of the three parameters affecting bias stability the change due to β variation is probably greatest. These changes in parameter value need not be only due to temperature. Although the value of I_{CO} at room temperature varies negligibly between transistors, even of the same manufacture type, as does V_{BE}, the value of β varies considerably. For example, the same numbered transistor may have $\beta = 125$ for one device and $\beta = 300$ for another. In addition, the value of β for specific transistors will be different at different values of bias current. For all these reasons the design of a good bias-stabilized circuit usually concentrates most on stabilizing the effect of changes in transistor beta.

4.7 DC BIAS CIRCUIT WITH EMITTER RESISTOR

The dc bias circuit of Fig. 4.11 contains an emitter resistor to provide better bias stability than the fixed-bias circuit considered in Section 4.5. For the analysis of the circuit operation we shall deal separately with the base-emitter loop of the circuit and the collector-emitter loop of Fig. 4.11.

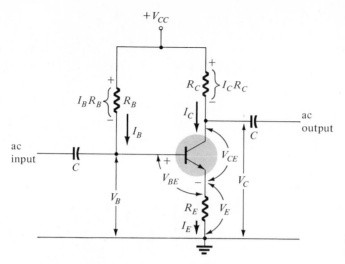

Figure 4.11 Dc bias circuit with emitter-stabilization resistor.

Input Section (Base-Emitter Loop)

A partial circuit diagram of the base-emitter loop is shown in Fig. 4.12a. Writing Kirchhoff's voltage equation for the loop, we get

$$V_{CC} - I_B R_B - V_{BE} - I_E R_E = 0$$

We can replace I_E with $(\beta + 1)I_B$ so that the above equation can be written as

$$V_{CC} - I_B R_B - V_{BE} - (\beta + 1)I_B R_E = 0$$

Solving for the base current, we get

$$I_B = \frac{V_{CC} - V_{BE}}{R_B + (\beta + 1)R_E} \cong \frac{V_{CC}}{R_B + \beta R_E} \tag{4.13}$$

Note that the difference between the fixed-bias current calculation [Eq. (4.4)] and Eq. (4.13) is the additional term of $(\beta + 1)R_E$ in the denominator.

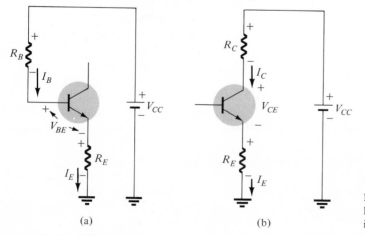

(a)

(b)

Figure 4.12 Input and output loops of the circuit of Fig. 4.11: (a) input loop; (b) output loop.

Output Section (Collector-Emitter Loop)

The collector-emitter loop is shown in Fig. 4.12b. Writing the Kirchhoff voltage equation for this loop, we get

$$V_{CC} - I_C R_C - V_{CE} - I_E R_E = 0$$

Using the relation

$$I_C \cong I_E$$

we can solve for the voltage across the collector-emitter:

$$\boxed{V_{CE} \cong V_{CC} - I_C(R_C + R_E)} \qquad (4.14)$$

The voltage measured from emitter to ground is

$$V_E = I_E R_E \cong I_C R_E$$

and the voltage measured from collector to ground is

$$V_C = V_{CC} - I_C R_C$$

The voltage at which the transistor is biased is measured from collector to emitter, V_{CE}, which is given by Eq. (4.14) and may also be calculated as

$$V_{CE} = V_C - V_E$$

EXAMPLE 4.6 Calculate all dc bias voltages and currents in the circuit of Fig. 4.13.

  Solution:
  (a) $I_B = (V_{CC} - V_{BE})/(R_B + \beta R_E) = (20 \text{ V} - 0.7 \text{ V})/[400 \text{ k}\Omega + 100(1 \text{ k}\Omega)]$
      $= 19.3 \text{ V}/500 \text{ k}\Omega = \textbf{38.6 } \boldsymbol{\mu}\textbf{A}$
  (b) $I_C = \beta I_B = 100(38.6 \text{ } \mu\text{A}) = \textbf{3.86 mA} \cong I_E$
  (c) $V_{CE} = V_{CC} - I_C R_C - I_E R_E = 20 \text{ V} - 3.86 \text{ mA}(2 \text{ k}\Omega) - 3.86 \text{ mA}(1 \text{ k}\Omega)$
      $= 20 \text{ V} - 7.72 \text{ V} - 3.86 \text{ V} = \textbf{8.42 V}$

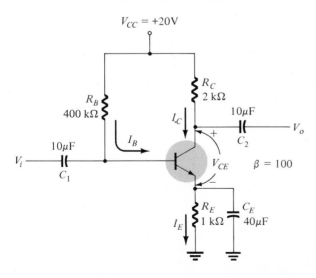

Figure 4.13 Emitter-stabilized bias circuit for Example 4.6.

4.8 DC BIAS CIRCUIT INDEPENDENT OF BETA

Approximate Analysis

In the previous dc bias circuits the values of the bias current and voltage of the collector depended on the current gain (β) of the transistor. But the value of beta is temperature-sensitive, especially for silicon transistors and since, also, the nominal value of beta is not well defined, it would be desirable for these as well as other reasons (transistor replacement and stability) to provide a dc bias circuit that is *independent* of the transistor beta. The circuit of Fig. 4.14 meets these conditions and is thus a very popular bias circuit.

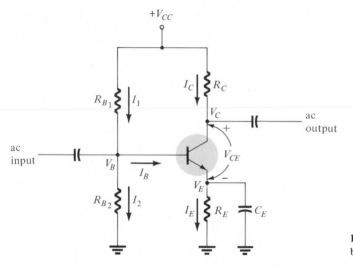

Figure 4.14 Beta-independent dc bias circuit.

Let us first analyze the base-emitter input circuit. If the resistance seen looking into the base (see Fig. 4.15) is much larger than that of resistor R_{B2}, then the base voltage is set by the voltage divider of R_{B1} and R_{B2}. If this is so, then the current through R_{B1} goes almost completely into R_{B2} and the two resistors may be considered effectively in series. The voltage at the junction of the resistors, which is also the voltage of the base of the transistor, is then determined simply by the voltage-divider network of R_{B1} and R_{B2} and the supply voltage. Calculating the voltage at the transistor

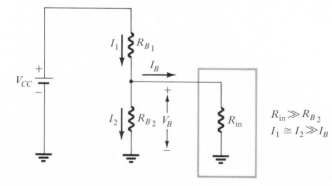

Figure 4.15 Partial bias circuit for calculating base voltage V_B.

base due to the voltage-divider network of resistors R_{B1} and R_{B2}, we get

$$V_B = \frac{R_{B2}}{R_{B1} + R_{B2}} \, V_{CC}$$ (4.15)

where V_B is the voltage measured from base to ground.

We can calculate the voltage at the emitter from

$$V_E = V_B - V_{BE}$$ (4.16)

The current in the emitter may then be calculated from

$$I_E = \frac{V_E}{R_E}$$ (4.17a)

and the collector current is then

$$I_C \cong I_E$$ (4.17b)

The voltage drop across the collector resistor is

$$V_{RC} = I_C R_C$$

The voltage at the collector (measured with respect to ground) can then be obtained

$$V_C = V_{CC} - V_{RC} = V_{CC} - I_C R_C$$ (4.18)

and, finally, the voltage from collector to emitter is calculated from

$$V_{CE} = V_C - V_E$$

$$V_{CE} = V_{CC} - I_C R_C - I_E R_E$$ (4.19)

Look back at the procedure just outlined and notice that the value of beta was never used in Eqs. (4.15)–(4.19). The base voltage is set by resistors R_{B1} and R_{B2} and the supply voltage. The emitter voltage is fixed at approximately the same voltage value as the base. Resistor R_E then determines emitter and collector currents. Finally, R_C determines the collector voltage and, thereby, the collector-emitter bias voltage.

The base voltage V_B is best adjusted using resistor R_{B2}, the collector current by resistor R_E, and the collector-emitter voltage by resistor R_C. Varying other components will have less effect on the dc bias adjustments. The capacitor components are part of the ac amplifier operation but have no effect on the dc bias and will not be discussed at this time.

EXAMPLE 4.7   Calculate the dc bias voltages and currents for the circuit of Fig. 4.16.

Solution:
(a) $V_B = [R_{B2}/(R_{B1} + R_{B2})](V_{CC}) = [4/(40 + 4)](22) = \mathbf{2 \ V}$
(b) $V_E = V_B - V_{BE} = 2 - 0.7 = \mathbf{1.3 \ V}$
(c) $I_E = V_E/R_E \cong I_C = 1.3 \ \text{V}/1.5 \ \text{k}\Omega = \mathbf{0.867 \ mA}$

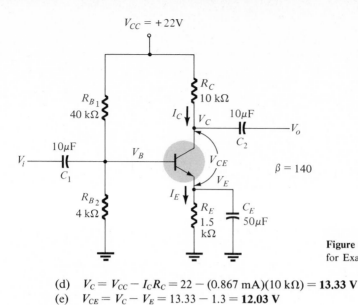

Figure 4.16 Beta-stabilized circuit for Example 4.7.

(d) $V_C = V_{CC} - I_C R_C = 22 - (0.867 \text{ mA})(10 \text{ k}\Omega) = \textbf{13.33 V}$

(e) $V_{CE} = V_C - V_E = 13.33 - 1.3 = \textbf{12.03 V}$

Exact Analysis

The circuit of Fig. 4.14 can be analyzed as covered above only if the voltage divider is not loaded by the dc impedance looking into the transistor. A more exact analysis can be obtained by using the Thévenin equivalent of the voltage divider as described by the following analysis: The Thévenin equivalent resistance of R_{B1} and R_{B2} is

$$R_{BB} = \frac{R_{B1} R_{B2}}{R_{B1} + R_{B2}} \qquad (4.20)$$

The Thévenin equivalent voltage is

$$V_{BB} = \frac{R_{B2}}{R_{B1} + R_{B2}} \cdot V_{CC} \qquad (4.21)$$

The dc circuit to be analyzed than can be redrawn as in Fig. 4.17.

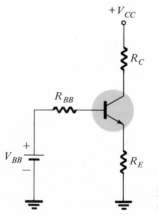

Figure 4.17 Dc circuit to analyze using Thévenin equivalent.

From Fig. 4.17, we can calculate I_B

$$I_B = \frac{V_{BB} - V_{BE}}{R_{BB} + (\beta + 1)R_E} \tag{4.22}$$

The collector current is then

$$I_C = \beta I_B$$

and the value of V_{CE} can be obtained using Eq. (4.14). The complete analysis is demonstrated in the following example.

EXAMPLE 4.8   Calculate the dc bias voltages and currents for the circuit of Fig. 4.16.

Solution:

(a) $V_{BB} = \dfrac{R_{B2}}{R_{B1} + R_{B2}}(V_{CC}) = \dfrac{4 \text{ k}\Omega}{40 \text{ k}\Omega + 4 \text{ k}\Omega}(22 \text{ V}) = \textbf{2 V}$

(b) $R_{BB} = R_{B1} \cdot R_{B2}/(R_{B1} + R_{B2}) = 4 \text{ k}\Omega \cdot 40 \text{ k}\Omega/(4 \text{ k}\Omega + 40 \text{ k}\Omega) = 3.6 \text{ k}\Omega$

(c) $I_B = \dfrac{V_{BB} - V_{BE}}{R_{BB} + (\beta + 1)R_E} = \dfrac{2 \text{ V} - 0.7 \text{ V}}{3.6 \text{ k}\Omega + 141(1.5 \text{ k}\Omega)} = \textbf{6.04 } \boldsymbol{\mu}\textbf{A}$

(d) $I_C = \beta I_B = 140(6.04 \text{ }\mu\text{A}) = \textbf{0.85 mA} \cong I_E$

(e) $V_{CE} = V_{CC} - I_C(R_C + R_E) = 22 \text{ V} - 0.85 \text{ mA}(10 \text{ k}\Omega + 1.5 \text{ k}\Omega)$

    $= 22 \text{ V} - 9.8 \text{ V} = \textbf{12.2 V}$

Comparing these values with the values calculated in Example 4.7, we note the difference is less than 10% for

$$\beta R_E > 10 R_{BB}$$

A computer program to calculate the dc bias currents and voltages can be run (see Appendix D, Fig. D.1) as Printout 4.1 shows. Compare the results obtained in Example 4.8 with those in Printout 4.1.

```
10 DIGITS 6,3
20 PRINT "THIS PROGRAM CALCULATES THE DC BIAS FOR "
30 PRINT "A VOLTAGE-DIVIDER BIAS CIRCUIT"
40 PRINT
50 PRINT "INPUT THE FOLLOWING CIRCUIT VALUES:"
60 INPUT "RB1=";R1:INPUT "RB2=";R2
70 INPUT "RE=";RE
80 INPUT "RC=";RC
90 INPUT "VCC=";SV
100 INPUT "AND TRANSISTOR BETA=";B
110 PRINT
120 PRINT "THE RESULTING DC BIAS CURRENTS ARE:"
130 VT=R2*SV/(R1+R2)
140 RT=(R2/(R1+R2))*R1
150 IB=(VT-0.7)/(RT+(B+1)*RE)
160 IC=B*IB
170 IE=(B+1)*IB
180 PRINT "IB=";IB,"IC=";IC;" IE =";IE
190 PRINT
200 PRINT "AND THE DC VOLTAGES ARE:"
210 VE=IE*RE
220 VB = VE+0.7
230 VC = SV - IC*RC
240 PRINT "VB= ";VB;" VE= ";VE;" VC= ";VC
250 PRINT "and VCE= ";VC-VE
260 END
```

Printout 4.1 (continued on p. 174)

READY

RUN

THIS PROGRAM CALCULATES THE DC BIAS FOR
A VOLTAGE-DIVIDER BIAS CIRCUIT

INPUT THE FOLLOWING CIRCUIT VALUES:
RB1=? 40E3
RB2=? 4E3
RE=? 1.5E3
RC=? 10E3
VCC=? 22
AND TRANSISTOR BETA=? 140

THE RESULTING DC BIAS CURRENTS ARE:
IB= 6.043E-06   IC= 8.46E-04   IE = 8.52E-04

AND THE DC VOLTAGES ARE:
VB= 1.978 VE= 1.278 VC= 13.54
and VCE= 12.262

data input

computed
results

Printout 4.1   (continued)

4.9 DC BIAS VOLTAGE FEEDBACK

Apart from the use of an emitter resistor to provide improved bias stability, voltage feedback also provides improved dc bias stability. Figure 4.18 shows a dc bias circuit with voltage feedback. This section shows how to calculate the dc currents and voltages of this circuit.

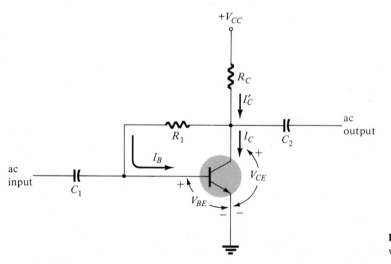

Figure 4.18 Dc bias circuit with voltage feedback.

Input Section

Figure 4.19a shows the input section (base-emitter) loop of the voltage feedback circuit. Writing the Kirchhoff voltage equation around the loop gives

$$+ V_{CC} - I'_C R_C - I_B R_1 - V_{BE} = 0$$

The current I'_C is the sum of I_C and I_B, but I_B is so much smaller than I_C that we can write the approximation:

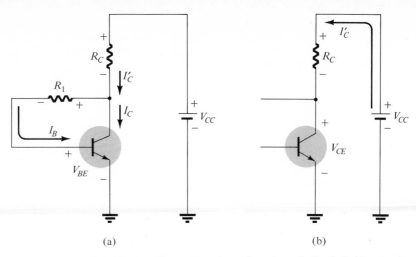

Figure 4.19 Input and output sections of a voltage feedback dc bias circuit: (a) input section; (b) output section.

$$I'_C \cong I_C \cong \beta I_B$$

which, substituted into the Kirchhoff voltage equation, yields

$$V_{CC} - \beta I_B R_C - I_B R_1 - V_{BE} = 0$$

Solving for the base current I_B, we get,

$$I_B = \frac{V_{CC} - V_{BE}}{R_1 + (\beta + 1)R_C} \cong \frac{V_{CC}}{R_1 + \beta R_C} \qquad (4.23)$$

Output Section

From the partial circuit diagram of the output section show in Fig. 4.19b the Kirchhoff voltage equation is

$$+ V_{CC} - I'_C R_C - V_{CE} = 0$$

and using $I'_C \cong I_C$, we can solve for V_{CE}

$$V_{CE} \cong V_{CC} - I_C R_C \qquad (4.24)$$

EXAMPLE 4.9 Calculate the dc bias currents and voltages for the practical dc bias circuit of Fig. 4.20 using voltage feedback.

Solution: The feedback resistor R_B is the sum of the resistors between collector and base (the capacitor in the feedback path provides for attenuating or blocking the ac feedback signal and has no effect on the dc bias calculation).
(a) $(V_{CC} - V_{BE})/(R_B + \beta R_C) \cong V_{CC}/(R_B + \beta R_C)$
$$= 10 \text{ V}/[250 \text{ k}\Omega + 50(3 \text{ k}\Omega)] = 10 \text{ V}/400 \text{ k}\Omega$$
$$= \textbf{25 } \mu\textbf{A}$$
(b) $I'_C \cong I_C = \beta I_B = 50(25 \ \mu\text{A}) = \textbf{1.25 mA}$
(c) $V_{CE} \cong V_{CC} - I_C R_C = 10 - (1.25 \text{ mA})(3 \text{ k}\Omega) = 10 - 3.75 = \textbf{6.25 V}$

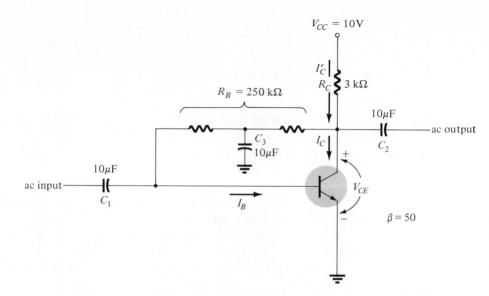

Figure 4.20 Voltage feedback stabilized bias circuit for Example 4.9.

The dc bias circuit may include *both* emitter-resistor stabilization and voltage-feedback stabilization as shown in the practical circuit of Fig. 4.21. An example will help demonstrate calculation of dc bias values.

EXAMPLE 4.10 Calculate the dc bias currents and voltages for the bias circuit of Fig. 4.21.

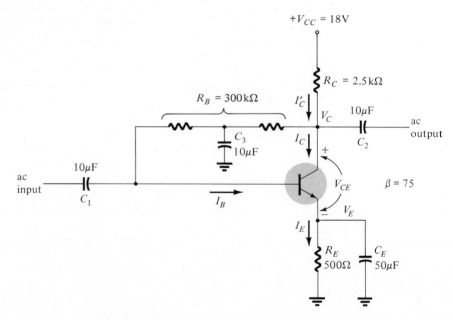

Figure 4.21 Dc bias circuit with emitter resistor and voltage feedback stabilization.

CH. 4 DC BIASING: BJTs

Solution:

(a) $I_B = \dfrac{V_{CC} - V_{BE}}{R_B + \beta R_C + (\beta + 1)R_E} = \dfrac{18\ \text{V} - 0.7\ \text{V}}{300\ \text{k}\Omega + 75(2.5\ \text{k}\Omega) + 76(0.5\ \text{k}\Omega)} = \mathbf{32.92\ \mu A}$

(b) $I'_C \cong I_C = \beta I_B = 75(32.92\ \mu\text{A}) = \mathbf{2.47\ mA}$

(c) $V_{CE} \cong V_{CC} - I_C(R_C + R_E) = 18\ \text{V} - (2.47\ \text{mA})(3\ \text{k}\Omega) = \mathbf{10.59\ V}$

4.10 COMMON-COLLECTOR (EMITTER-FOLLOWER) DC BIAS CIRCUIT

A third connection for the transistor provides input to the base circuit and output from the emitter circuit, with the collector common (CC) to the ac input and output signals. A simple CC circuit (usually referred to as emitter-follower) is shown in Fig. 4.22. The collector voltage is fixed at the positive supply voltage value. For V_{CE} to be approximately one-half the voltage of V_{CC}, allowing the widest voltage swing in the output before distortion occurs, the emitter voltage should be set at a voltage of about one-half V_{CC}.

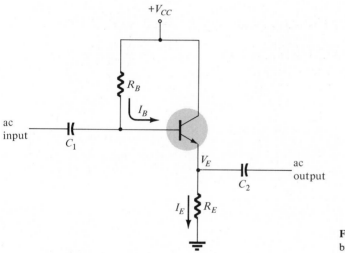

Figure 4.22 Emitter-follower dc bias circuit.

Input Section

For the input section of the circuit the voltages summed around the base-emitter loop give

$$+ V_{CC} - I_B R_B - V_{BE} - I_E R_E = 0$$

Using the current relation

$$I_E = (\beta + 1)I_B \cong \beta I_B$$

we can solve for the base current, obtaining

$$I_B = \dfrac{V_{CC} - V_{BE}}{R_B + (\beta + 1)R_E} \cong \dfrac{V_{CC}}{R_B + \beta R_E} \qquad (4.25)$$

Output Section

The voltage from emitter to ground is

$$V_E = I_E R_E \qquad (4.26)$$

and the collector-emitter voltage is

$$V_{CE} = V_{CC} - V_E = V_{CC} - I_E R_E \qquad (4.27)$$

EXAMPLE 4.11 Calculate all dc bias currents and voltages for the circuit of Fig. 4.23.

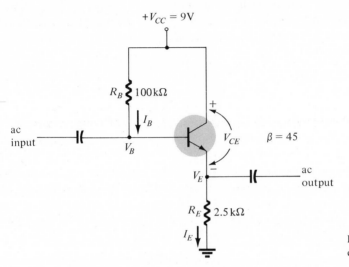

$+V_{CC} = 9\text{V}$

R_B $100\text{k}\Omega$

I_B

ac input

V_B

V_{CE} $\beta = 45$

V_E ac output

R_E $2.5\text{k}\Omega$

I_E

Figure 4.23 Emitter follower bias circuit for Example 4.11.

Solution:
(a) $I_B = (V_{CC} - V_{BE})/(R_B + \beta R_E) = (9\text{ V} - 0.7\text{ V})/(100\text{ k}\Omega + 45 \times 2.5\text{ k}\Omega) =$ **39.06 μA**
(b) $I_E = (\beta + 1)I_B = 46(39.06\text{ μA}) =$ **1.8 mA**
(c) $V_{CE} = V_{CC} - I_E R_E = 9\text{ V} - (1.8\text{ mA})(2.5\text{ k}\Omega) =$ **4.5 V**
(d) $V_E = I_E R_E = (1.8\text{ mA})(2.5\text{ k}\Omega) =$ **4.5 V**

A second emitter-follower dc bias circuit is shown in Fig. 4.24. Like the similar CE dc bias circuit of Fig. 4.14, this circuit provides a bias operating condition that depends not on the current gain (β) of the transistor, but only on the resistor components and the supply voltage.

EXAMPLE 4.12 Calculate all dc bias currents and voltages for the CC circuit of Fig. 4.24.

Solution:
(a) $V_B \cong (R_{B2}/[R_{B1} + R_{B2}])(V_{CC}) = (30/[30 + 30])(15) =$ **7.5 V**
(b) $V_E = V_B - V_{BE} = 7.5 - 0.7 =$ **6.8 V**
(c) $I_E = (V_E/R_E) = 6.8\text{ V}/3\text{ k}\Omega \cong$ **2.27 mA**
(d) $V_{CE} = V_{CC} - I_E R_E = 15 - (2.27\text{ mA})(3\text{ k}\Omega) =$ **8.2 V**

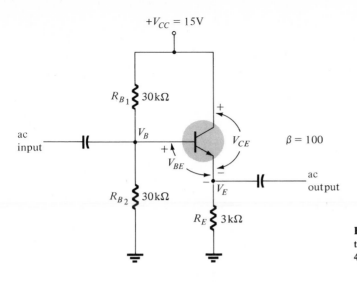

Figure 4.24 Beta-stabilized emitter follower bias circuit for Example 4.12.

4.11 GRAPHICAL DC BIAS ANALYSIS

The previous analysis of the dc bias currents and voltages was carried out mathematically for a number of transistor circuits. The only factors of interest used were the current gain (β) and base-emitter voltage (V_{BE}) when forward-biased. This section shows a graphical technique for finding the operating point of a transistor circuit. The graphical method demonstrated provides additional insight into the choice of operating point and leads into Section 4.12 on the design (or synthesis) of a dc bias circuit.

The typical CE collector characteristic, shown in Fig. 4.25, only defines the overall

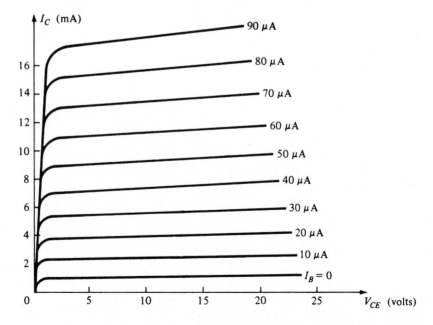

Figure 4.25 Transistor collector characteristics.

operation of the transistor device. The circuit constraints must also be taken into account in obtaining the actual operating point (called the *quiescent operating point* or *Q-point*). Eq. (4.28) for the output loop of the circuit is a straight line in a voltage-current plot (as in Fig. 4.26).

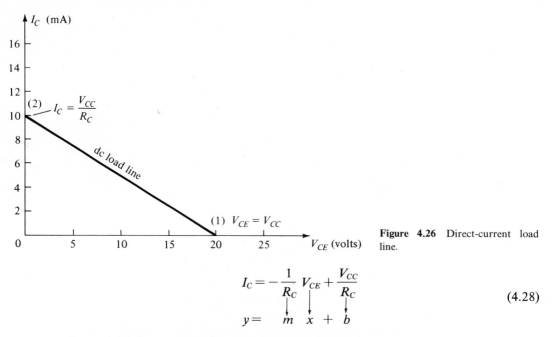

Figure 4.26 Direct-current load line.

$$I_C = -\frac{1}{R_C}\, V_{CE} + \frac{V_{CC}}{R_C}$$

$$y = \quad m \quad x \quad + \quad b$$

(4.28)

The straight line representing Eq. (4.28) can be drawn on the graph of Fig. 4.25 by obtaining the two extreme points of the straight line as follows:

1. For $I_C = 0$, $V_{CE} = V_{CC}$ in Eq. (4.28).
2. For $V_{CE} = 0$, $I_C = V_{CC}/R_C$ in Eq. (4.28).

These points are marked in Fig. 4.26 as (1) and (2), respectively, and the straight line connecting them is called the *dc load line*. Although the same voltage-current axis as that of the transistor collector characteristic is used, no characteristic is shown to reinforce the fact that the dc load line has nothing to do with the device itself. The load line drawn depends only on the supply voltage, V_{CC}, and the value of R_C, the collector resistor.

The slope of the load line depends only on the value of R_C. Figure 4.27a shows the load line slopes for R_C values smaller and larger than that of Fig. 4.26. Figure 4.27b shows that changing only the supply voltage will move the load line parallel to that of Fig. 4.26, and the slope remains the same since R_C has not changed.

Since circuit operation depends on both the transistor characteristic and the circuit elements, plotting *both* curves (transistor characteristic and dc load line) on *one* graph allows determination of the circuit *Q*-point. This is shown in Fig. 4.28. The typical dc bias point shown in Fig. 4.28 is somewhat in the center of the voltage range (0 to V_{CC}) and the center of the current range (0 to V_{CC}/R_C). A large-signal amplifier

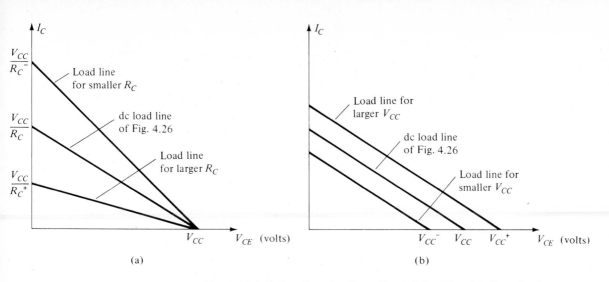

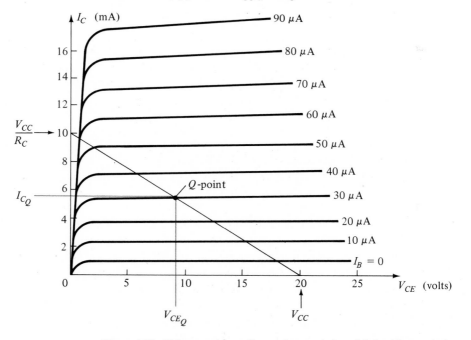

Figure 4.27 Effect of varying R_C or V_{CC} on dc load line: (a) effect of resistor on dc load line; (b) effect of supply voltage on dc load line.

Figure 4.28 Using transistor collector characteristic and dc load line to obtain quiescent operating point (Q-point).

with output voltage swing near the voltage range set by the voltage supply value would require a centered operating point.

For circuits other than amplifiers different bias points may be desired. The dc load line describes all the possible values of voltage and current in the output section of the circuit. Figure 4.28 shows a typical bias point set by the amount of base current and the dc load line. Adjusting the base current to higher values moves the operating point toward saturation along the load line, whereas reducing the base current moves the bias point toward transistor cutoff. For the characteristic and load line of Fig. 4.28 base currents in excess of 60 μA will drive the transistor into saturation. Note that for the load line and operating point indicated, an ac input, which adds to the dc base bias current, can go positive only by about 25 μA (from 30 to 55 μA) before the limiting condition (saturation) occurs. The variation of the ac base current, on the other hand, can go negative by 30 μA (30 to 0 μA) before cutoff is reached so that the particular bias point in Fig. 4.28 is not centered. For small-signal amplifiers with output voltage swings of less than 1 V, the exact centering of the Q-point is not essential—usually a region of largest transistor gain or most linear operation is sought.

4.12 DESIGN OF DC BIAS CIRCUITS

Up to now the discussion has been directed to the techniques of analyzing a given transistor circuit to determine the dc operating point. Although it is often necessary to determine the Q-point of a given circuit, it is also important to be able to design a circuit to operate at a desired or specified bias point. Often the manufacturer's specification (spec) sheets provide information stating a suitable operating point (or operating region) for the particular transistor. In addition, other circuit factors connected with the given amplifier stage may also dictate some conditions of current swing, voltage swing, value of common supply voltage, etc., which can be used in determining the Q-point in a design.

The techniques of synthesis (or design) readily follow from the previous discussions of circuit analysis. In almost all cases the calculation of the circuit elements proceeds in the reverse to those in the analysis consideration. Basically, the problem of concern in this section can be briefly stated as follows.

Given a desired point or region of operation for a particular transistor, design the bias circuit (resistor and supply voltage values) to obtain the specified operating point.

In actual practice many other factors may have to be considered and may go into the selection of the desired operating point. For the moment we shall concentrate, however, on determining the component values to obtain a specified operating point. Since the basic relations and operation of a number of bias circuits have already been considered, no new theory has to be developed.

Design of Bias Circuit with Emitter Feedback Resistor

Consider first the design of the dc bias components of an amplifier circuit having emitter-resistor bias stabilization. (See Fig. 4.29.) The supply voltage and operating point will be selected from the manufacturer's information on the transistor used in the amplifier.

The selection of collector and emitter resistors cannot proceed directly from the information just specified. In Eq. (4.14), which relates the voltages around the collector-emitter loop, there are two unknown quantities present—the values of the collector and emitter resistors, R_C and R_E, respectively. To make the solution to the problem easy (and meaningful), some engineering judgment may be used; that is, if some

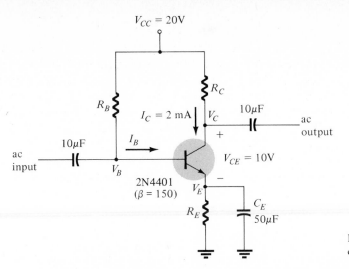

$V_{CC} = 20V$

R_C

R_B

$I_C = 2$ mA

V_C

$10\mu F$

ac output

$10\mu F$

ac input

I_B

V_B

$V_{CE} = 10V$

2N4401
($\beta = 150$)

V_E

R_E

C_E
$50\mu F$

Figure 4.29 Emitter-stabilized bias circuit for design consideration.

reasonable approximation of, say, the emitter voltage, can be made, the problem will then be straightforward.

Recall that the need for including a resistor from emitter to ground was to provide a means of dc bias stabilization so that the change of collector current due to leakage currents in the transistor and transistor β would not cause a large (if any) shift in the operating point. The emitter resistor cannot be unreasonably large because the voltage developed across it limits the range of voltage swing of the voltage from collector to emitter. The examples in Section 4.7 show that the voltage from emitter to ground, V_E, is typically around one-fifth to one-tenth of the supply voltage, V_{CC}. Selecting the emitter voltage in this way will permit calculating the emitter resistor, R_E, and then the collector resistor, R_C. Carrying out these calculations we get

$$V_{E_Q} \cong \frac{1}{10} V_{CC} = \frac{20 \text{ V}}{10} = 2 \text{ V}$$

$$R_E = \frac{V_{E_Q}}{I_{E_Q}} \cong \frac{V_{E_Q}}{I_{C_Q}} = \frac{2 \text{ V}}{2 \text{ mA}} = 1 \text{ k}\Omega$$

$$R_C = \frac{V_{CC} - V_{CE_Q} - V_{E_Q}}{I_{C_Q}} = \frac{(20 - 10 - 2) \text{ V}}{2 \text{ mA}} = 4 \text{ k}\Omega$$

$$I_{B_Q} = \frac{I_{C_Q}}{\beta} = \frac{2 \text{ mA}}{150} = 13.33 \ \mu A$$

$$R_B = \frac{V_{CC} - V_{BE} - V_{E_Q}}{I_{B_Q}} = \frac{(20 - 0.7 - 2) \text{ V}}{13.33 \ \mu A} \cong 1.3 \text{ M}\Omega$$

EXAMPLE 4.13 Calculate the resistor values R_E, R_C, and R_B for a transistor amplifier circuit having emitter-resistor stabilization (Fig. 4.29). The current gain of an *npn* 2N4401 transistor is typically 90 at a collector current of 5 mA. Use a supply voltage of 20 V.

Design Solution: The operating point selected from the information of supply voltage and transistor is $I_{C_Q} = 5$ mA and $V_{CE_Q} = 10$ V.

$$V_{E_Q} \cong \frac{1}{10}(V_{CC}) = \frac{1}{10}(20) = 2 \text{ V}$$

The emitter resistor is then

$$R_E \cong \frac{V_E}{I_{C_Q}} = \frac{2 \text{ V}}{5 \text{ mA}} = 400 \text{ }\Omega$$

The collector resistor is calculated to be

$$R_C = \frac{V_{CC} - V_{CE_Q} - V_{E_Q}}{I_{C_Q}} = \frac{(20 - 10 - 2) \text{ V}}{5 \text{ mA}} = \frac{8 \text{ V}}{5 \text{ mA}} = 1.6 \text{ k}\Omega$$

Calculating the base current using

$$I_{B_Q} = \frac{I_{C_Q}}{\beta} = \frac{5 \text{ mA}}{90} \cong 55.56 \text{ }\mu\text{A}$$

we see that the base resistor is calculated to be

$$R_B = \frac{V_{CC} - V_{BE} - V_{E_Q}}{I_{B_Q}} = \frac{(20 - 0.7 - 2) \text{ V}}{55.56 \text{ }\mu\text{A}} = \frac{17.3 \text{ V}}{55.56 \text{ }\mu\text{A}}$$

$$= 311 \text{ k}\Omega \quad (\text{use } R_B = 300 \text{ k}\Omega)$$

Design of Current Gain Stabilized Circuit

The circuit of Fig. 4.30 provides stabilization both for leakage current and current gain changes. The values of the four resistors shown must be obtained for a specified operating point. Engineering judgment in selecting a value of emitter voltage, V_E, as in the previous design consideration, leads to a simple straightforward solution for all the resistor values. The design steps are as follows:

The emitter voltage will be selected to be approximately one-tenth of the supply voltage (V_{CC}).

$$V_{E_Q} \cong \frac{1}{10} V_{CC} = \frac{1}{10}(20 \text{ V}) = 2 \text{ V}$$

Using this value of V_E, we see that the emitter-resistor value is calculated to be

$$R_E \cong \frac{V_{E_Q}}{I_{C_Q}} = \frac{2 \text{ V}}{10 \text{ mA}} = 200 \text{ }\Omega$$

The collector resistance is then obtained using

$$R_C = \frac{V_{CC} - V_{CE_Q} - V_{E_Q}}{I_{C_Q}} = \frac{(20 - 8 - 2) \text{ V}}{10 \text{ mA}} = 1 \text{ k}\Omega$$

The base voltage is approximately equal to the emitter voltage or, more exactly,

$$V_B = V_E + V_{BE} = 2 \text{ V} + 0.7 \text{ V} = 2.7 \text{ V}$$

The equation for calculation of the base resistors R_{B1} and R_{B2} will require a little thought. Using the values of base voltage calculated in step (4) and the value of the supply voltage will provide one equation—but there are two unknowns, R_{B1} and R_{B2}. An additional equation can be obtained from an understanding of the operation of these two resistors in providing the base voltage. For the circuit to operate properly the current through the two resistors should be approximately equal and therefore larger than the base current by an order of magnitude (at least 10 times larger). The two equations that will enable calculating the resistors R_{B1} and R_{B2} are

$$V_B = \frac{R_{B2}}{R_{B1} + R_{B2}} (V_{CC})$$

$$R_{B2} \leq \frac{1}{10}(\beta R_E)$$

Solving these equations results in

$$R_{B1} \cong 10 \text{ k}\Omega, \qquad R_{B2} \cong 1.6 \text{ k}\Omega$$

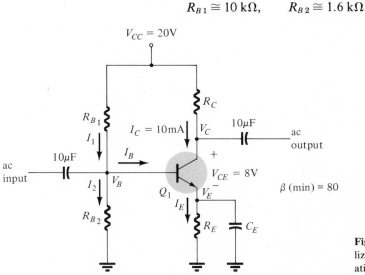

Figure 4.30 Current gain stabilized circuit for design considerations.

EXAMPLE 4.14 Design a dc bias circuit for an amplifier circuit as in Fig. 4.30. For this example, the manufacturer's spec states that the transistor has a current gain of 150, typical, at a collector current of 1 mA, and the supply voltage for the present circuit is 16 V.

Design Solution: The bias point to be designed is for $I_{C_Q} = 1$ mA, $V_{CE_Q} = 6$ V.
(a) Selecting $V_{E_Q} = \frac{1}{10}(V_{CC}) = \frac{1}{10}(16) = 1.6$ V
(b) Calculating R_E:

$$R_E \cong \frac{V_{E_Q}}{I_{C_Q}} = \frac{1.6 \text{ V}}{1 \text{ mA}} = \mathbf{1.6 \text{ k}\Omega}$$

(c) Calculating R_C:

$$R_C = \frac{V_{CC} - V_{CE_Q} - V_{E_Q}}{I_{C_Q}} = \frac{(16 - 6 - 1.6) \text{ V}}{1 \text{ mA}} = \mathbf{8.4 \text{ k}\Omega}$$

(d)   Calculating V_{B_Q}:

$$V_{B_Q} = V_{E_Q} + V_{BE} = 1.6 + 0.7 = 2.3 \text{ V}$$

(e)   Calculating R_{B1} and R_{B2}:

$$R_{B2} \leq \frac{1}{10}(\beta R_E) = \frac{150(1.6 \text{ k}\Omega)}{10} = 24 \text{ k}\Omega$$

and since

$$\frac{R_{B2}}{R_{B1} + R_{B2}} V_{CC} = V_{B_Q} = 2.3 \text{ V}$$

$$R_{B1} \cong 143 \text{ k}\Omega \quad (\text{use } 150 \text{ k}\Omega)$$

4.13   MISCELLANEOUS BIAS CIRCUITS

In practice, one finds that bias circuits do not always conform to the basic forms considered in this chapter. It should not be difficult to analyze the bias operation of a circuit slightly modified from those considered previously. To show various bias circuits a few examples are included in this section to demonstrate the use of the basic bias concepts discussed in this chapter.

EXAMPLE 4.15   Obtain the bias voltages and currents for the circuit of Fig. 4.31.

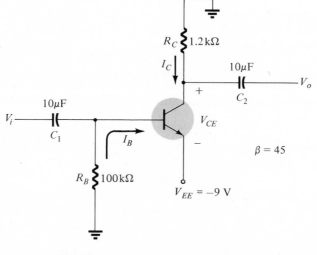

Figure 4.31   Bias circuit for Example 4.15.

Solution:

Input Loop:

$$-I_B R_B - V_{BE} + V_{EE} = 0$$

$$I_B = \frac{V_{EE} - V_{BE}}{R_B} = \frac{(9 - 0.7) \text{ V}}{100 \text{ k}\Omega} = 83 \text{ }\mu\text{A}$$

$$I_C = \beta I_B = 45(83 \text{ }\mu\text{A}) = \mathbf{3.735 \text{ mA}}$$

Output Loop:

$$-I_C R_C - V_{CE} + V_{EE} = 0$$

$$V_{CE} = V_{EE} - I_C R_C = 9 \text{ V} - (3.735 \text{ mA})(1.2 \text{ k}\Omega)$$
$$= 9 \text{ V} - 4.48 \text{ V} = \mathbf{4.52 \text{ V}}$$

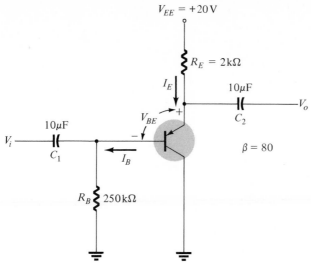

$V_{EE} = +20\,\text{V}$

$R_E = 2\,\text{k}\Omega$

I_E

$10\mu\text{F}$

C_2

V_o

V_{BE}

$10\mu\text{F}$

C_1

V_i

I_B

$\beta = 80$

$R_B \lessgtr 250\,\text{k}\Omega$

Figure 4.32   Bias circuit for Example 4.16.

EXAMPLE 4.16   Calculate the bias voltages and currents for the circuit of Fig. 4.32.

Solution: Writing the voltage loop equation

$$V_{EE} - I_E R_E - V_{EB} - I_B R_B = 0$$

$$(V_{EE} - V_{EB}) = (\beta + 1)I_B R_E + I_B R_B$$

$$I_B = \frac{V_{EE} - V_{EB}}{R_B + (\beta + 1)R_E} \cong \frac{V_{EE}}{R_B + \beta R_E} = \frac{20\,\text{V}}{250\,\text{k}\Omega + 80(2\,\text{k}\Omega)}$$

$$= \frac{20\,\text{V}}{410\,\text{k}\Omega} = 48.78\,\mu\text{A}$$

$$I_C = \beta I_B = 80(48.78\,\mu\text{A}) = \textbf{3.9 mA} \cong I_E$$

$$V_E = V_{EE} - I_E R_E = 20\,\text{V} - (3.9\,\text{mA})(2\,\text{k}\Omega) = (20 - 7.8)\,\text{V}$$

$$= \textbf{12.2 V}$$

EXAMPLE 4.17   Calculate the bias voltages and currents for the circuit of Fig. 4.33.

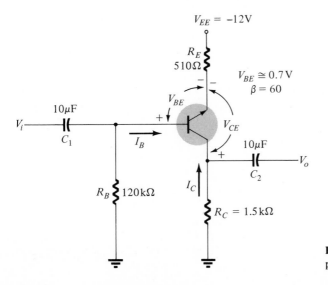

$V_{EE} = -12\text{V}$

R_E
510Ω

$V_{BE} \cong 0.7\,\text{V}$
$\beta = 60$

V_{BE}

$10\mu\text{F}$

C_1

V_i

I_B

V_{CE}

$10\mu\text{F}$

C_2

V_o

$R_B \lessgtr 120\,\text{k}\Omega$

I_C

$R_C = 1.5\,\text{k}\Omega$

Figure 4.33   Bias circuit for Example 4.17.

Solution:

Input Loop: $-I_B R_B - V_{BE} - I_E R_E + V_{EE} = 0$

$$I_B \cong \frac{V_{EE} - V_{BE}}{R_B + \beta R_E} = \frac{12\,V - 0.7\,V}{120\,k\Omega + 60(0.510\,k\Omega)} = 75.03\,\mu A$$

$$I_C = \beta I_B = 60(75.03\,\mu A) = \textbf{4.5 mA}$$

Output Loop: $-I_E R_E - I_C R_C - V_{CE} = 0$

$$V_{CE} \cong V_{EE} - I_C(R_C + R_E) = 12\,V - (4.5mA)(1.5\,k\Omega + 0.510\,k\Omega) = \textbf{2.96 V}$$

EXAMPLE 4.18 Calculate the bias currents and voltages for the circuit of Fig. 4.34. (Use the approximate voltage divider method.)

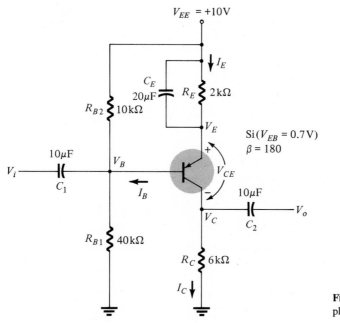

Figure 4.34 Bias circuit for Example 4.18.

Solution:
(a) $V_B \cong [R_{B1}/(R_{B1} + R_{B2})]V_{CC} = [40/(10 + 40)]\,10 = 8\,V$
(b) $V_E = V_B + V_{EB} = 8\,V + 0.7\,V = 8.7\,V$
(c) $I_E = (V_{EE} - V_E)/R_E = (10\,V - 8.7\,V)/2\,k\Omega = \textbf{0.65 mA} \cong I_C$
(d) $V_C = I_C R_C = (0.65\,mA)(6\,k\Omega) = 3.9\,V$
(e) $V_{CE} = V_C - V_E = 3.9\,V - 8.7\,V = \textbf{-4.8 V}$

PROBLEMS

§ 4.3

1. Calculate the dc bias voltage V_{CB} and current I_C for a *npn* common-base bias circuit. Assume transistor values of $\alpha = 0.985$ and $V_{BE} = +0.7$ V. Circuit components are $R_E = 720\ \Omega$ and $R_C = 3.9$ kΩ and supply voltages are $V_{CC} = 9$ V and $V_{EE} = 1.5$ V. (See Fig. 4.2 for circuit connection.)

2. Calculate the collector-base voltage for *npn* common-base bias circuit (as in Fig. 4.4) for the following circuit values: $R_E = 1.8$ kΩ, $R_C = 2.7$ kΩ, $V_{EE} = 9$ V, $V_{CC} = 22$ V, $\alpha = 0.995$, and $V_{BE} = +0.7$ V.

§ 4.5

3. For a fixed-bias common-emitter circuit, as in Fig. 4.5, calculate the bias current I_C and voltage V_{CE} for the following circuit values: $R_B = 150$ kΩ, $R_C = 2.1$ kΩ, $V_{CC} = 9$ V, $V_{BE} = +0.7$ V, and $\beta = 45$.

4. Using an *npn* fixed-bias circuit as in Fig. 4.8, calculate the collector-emitter bias voltage (V_{CE}) for the following circuit values: $R_B = 250$ kΩ, $R_C = 1.8$ kΩ, $V_{CC} = 12$ V, $V_{BE} = 0.7$ V, and $\beta = 70$.

§ 4.7

5. Calculate the dc bias voltage V_{CE} and current I_C for an emitter-stabilized bias circuit as in Fig. 4.11 for the following circuit values: $R_B = 47$ kΩ, $R_E = 750$ Ω, $R_C = 0.5$ kΩ, $V_{BE} = 0.7$ V, $\beta = 55$, and $V_{CC} = 18$ V.

6. Calculate the collector-emitter voltage (V_{CE}) for an *npn* emitter-stabilized bias circuit as in Fig. 4.11 for the following circuit values: $R_B = 75$ kΩ, $R_C = 0.5$ kΩ, $R_E = 470$ Ω, $V_{BE} = 0.7$ V, $V_{CC} = 10$ V, and $\beta = 80$.

§ 4.8

7. Calculate the bias voltage V_{CE} and current I_C for a circuit as in Fig. 4.14 for the following circuit values: $R_{B1} = 56$ kΩ, $R_{B2} = 4.7$ kΩ, $R_E = 750$ Ω, $R_C = 6.8$ kΩ, $V_{CC} = 24$ V, $V_{BE} = 0.7$ V, and $\beta = 55$.

8. Calculate the collector voltage V_C for a bias circuit as in Fig. 4.14 for the following values: $R_{B1} = 12$ kΩ, $R_{B2} = 1.5$ kΩ, $R_E = 1$ kΩ, $R_C = 4.7$ kΩ, $V_{CC} = 9$ V, $V_{BE} = +0.7$ V, and $\beta = 75$.

§ 4.9

9. Calculate the bias current I_C and voltage V_C for a circuit as in Fig. 4.20 using the following circuit values: $R_B = 100$ kΩ, $R_C = 5$ kΩ, $V_{CC} = 10$ V, $V_{BE} = 0.7$ V, and $\beta = 60$.

10. For a bias circuit as in Fig. 4.20 calculate the dc voltage measured from collector to ground (V_C) for the following circuit values: $R_B = 68$ kΩ, $R_C = 2.4$ kΩ, $V_{CC} = 15$ V, $V_{BE} = 0.7$ V, and $\beta = 48$.

11. Calculate the bias voltage V_{CE} and current I_C for a circuit as in Fig. 4.21 for the following circuit values: $R_B = 200$ kΩ, $R_C = 3.6$ kΩ, $R_E = 270$ Ω, $V_{BE} = 0.7$ V, $\beta = 40$, and $V_{CC} = 16$ V.

§ 4.10

12. For an emitter follower circuit as in Fig. 4.22 calculate the dc bias current I_E and voltage V_E for the following circuit values: $R_B = 240$ kΩ, $R_E = 1.8$ kΩ, $V_{CC} = 9$ V, $V_{BE} = 0.7$ V, and $\beta = 85$.

13. For an emitter-follower circuit as in Fig. 4.22 calculate the dc voltage across the emitter resistor for the following circuit values: $R_B = 91$ kΩ, $R_E = 1.2$ kΩ, $V_{CC} = 25$ V, $V_{BE} = 0.7$ V, and $\beta = 60$.

14. Calculate the emitter voltage (with respect to ground) for a bias circuit as in Fig. **4.24** for the following values: $R_{B1} = 270$ kΩ, $R_{B2} = 330$ kΩ, $R_E = 4.7$ kΩ, $V_{CC} = 15$ V, $V_{BE} = 0.7$ V, and $\beta = 150$.

§ 4.11

15. For a fixed-bias circuit as in Fig. 4.5 with circuit values of $R_B = 80$ kΩ, $R_C = 4$ kΩ, $V_{CC} = 20$ V, $V_{BE} = 0.7$ V, and transistor collector characteristic as shown in Fig. 4.35, do the following:
  (a) Draw the dc load line.
  (b) Obtain the quiescent operating point (Q-point).
  (c) Find the operating point if R_C is changed to 8 kΩ.
  (d) Find the operating point if V_{CC} is changed to 15 V (R_C still is 4 kΩ).

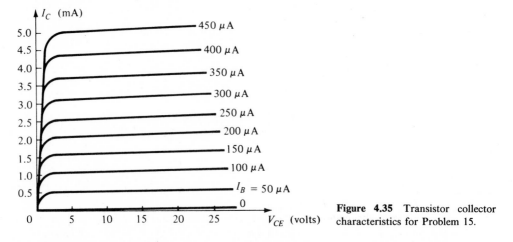

Figure 4.35 Transistor collector characteristics for Problem 15.

16. Determine graphically the operating point for a fixed-bias circuit using a *pnp* transistor having a collector characteristic as in Fig. 4.36. The circuit values are as follows: $R_B = 150$ kΩ, $R_C = 2$ kΩ, $V_{CC} = -20$ V, and $V_{BE} = -0.7$ V.

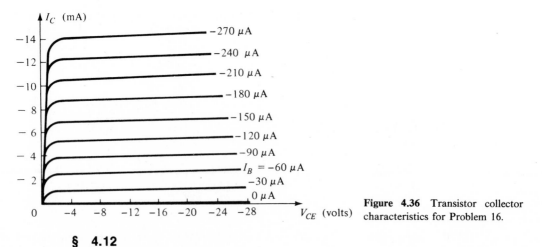

Figure 4.36 Transistor collector characteristics for Problem 16.

§ 4.12

17. Design a fixed-bias common-emitter circuit using an *npn* transistor as in Fig. 4.5. The

CH. 4   DC BIASING: BJTs

transistor has a current gain (β) of 80 and is to be operated at $I_{C_Q} = 2$ mA and $V_{CE_Q} = 10$ V. Use a collector supply of 22 V.

18. Design a fixed-bias common-emitter circuit as in Fig. 4.5 using an *npn* transistor rated to have a current gain of 250 at a collector current of 10 mA and a collector-emitter voltage of 10 V. Use a supply of 22 V.

19. Calculate the resistor values for an emitter-stabilized amplifier circuit as in Fig. 4.29. Use a silicon *npn* transistor having a current gain of 250 when biased at $I_{C_Q} = 10$ mA, $V_{CE_Q} = 15$ V. A 30-V supply is provided.

20. Design a dc bias circuit as in Fig. 4.29 using an *npn* transistor having a current gain of 400 at $I_{C_Q} = 10$ mA and $V_{CE_Q} = 10$ V. Use a supply voltage of 22 V.

21. Design a voltage divider bias circuit as in Fig. 4.30 for operation at $V_{CE_Q} = 8$ V, $I_{C_Q} = 5$ mA, using a transistor with current gain of $\beta = 130$ and a voltage supply, $V_{CC} = 18$ V.

§ 4.13

22. Obtain the value of dc voltage measured from collector to ground for the circuit of Fig. 4.31 for the following circuit values: $V_{EE} = -15$ V, $R_B = 47$ kΩ, $R_C = 1.2$ kΩ, and $\beta = 30$.

23. Calculate the base voltage (with respect to ground) for the circuit of Fig. 4.32 using the following values: $R_B = 210$ kΩ, $R_E = 4.3$ kΩ, $V_{EE} = +12$ V, $V_{BE} = -0.7$ V, and $\beta = 120$.

24. Calculate the collector current for the circuit of Fig. 4.33 using the following values: $R_B = 80$ kΩ, $R_C = 1.8$ kΩ, $R_E = 750$ Ω, $V_{EE} = -9$ V, $V_{BE} = 0.7$ V, and $\beta = 135$.

25. Calculate the collector-emitter voltage for a circuit as in Fig. 4.34 for the following circuit values: $R_{B1} = 120$ kΩ, $R_{B2} = 15$ kΩ, $R_E = 3.9$ kΩ, $R_C = 12$ kΩ, $V_{EE} = 18$ V, $V_{BE} = 0.7$ V, and $\beta = 200$.

COMPUTER PROBLEMS

Write BASIC programs to:

1. Compute the dc bias values, I_C and V_{CE} for a fixed-bias circuit.

2. Compute I_C for an emitter-stabilized bias circuit.

3. Compute V_B for a circuit using voltage divider biasing using the approximate method.

4. Compute V_B for a voltage divider biasing circuit using the exact method (Thévenin equivalent).

5. Compute I_C and V_{CE} for a voltage divider biasing circuit using the exact method.

6. Compute I_C and V_{CE} using the approximate method in a voltage divider biasing circuit.

7. Calculate the bias values I_C and V_{CE} for a collector feedback-stabilized bias circuit.

CHAPTER

5

Field-Effect
Transistors (FETs)

5.1 GENERAL DESCRIPTION OF FET

A bipolar junction transistor (BJT) made as *npn* or as *pnp* is a current-controlled device in which both electron current and hole current are involved. The field-effect transistor (FET) is a unipolar device. It operates as a voltage-controlled device with either electron current in an *n*-channel FET or hole current in a *p*-channel FET. Either BJT or FET devices can be used to operate in an amplifier circuit (or other similar electronic circuits), with different bias considerations.

A few general comparisons between FET and BJT devices and resulting circuits can be made.

1. The FET has an extremely high input resistance with about 100 MΩ typical.
2. The FET has no offset voltage when used as a switch (or chopper).
3. The FET is relatively immune to radiation, but the BJT is very sensitive (beta is particularly affected).
4. The FET is less "noisy" than a BJT and thus more suitable for input stages of low-level amplifiers (it is used extensively in hi-fi FM receivers).
5. The FET can be operated to provide greater thermal stability than a BJT.

Some disadvantages of the FET are the relatively small gain-bandwidth of the device compared to the BJT and the greater susceptibility to damage in handling the FET.

The FET is a three-terminal device containing one basic *p-n* junction and can be built as either a Junction FET (JFET) or a Metal-Oxide-Semiconductor FET (MOS-FET). Although the FET was one of the earliest solid-state devices proposed[1] for amplifier operation, until the mid-1960s the development of a commercially useful device lagged because of manufacturing construction limitations. Large and very large-scale integrated circuits are built primarily using MOSFET transistor devices.

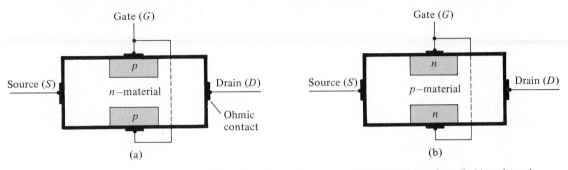

Figure 5.1 Physical structure of a JFET: (a) *n*-channel; (b) *p*-channel.

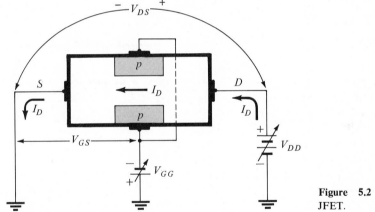

Figure 5.2 Basic operation of JFET.

JFET Structure

The physical structure of a JFET is shown in Fig. 5.1. The *n*-channel JFET shown in Fig. 5.1a is constructed using a bar of *n*-type material into which a pair of *p*-type regions are diffused. A *p*-channel JFET is made using a bar of *p*-type material with *n*-type diffused regions as shown in Fig. 5.1b.

To examine how the device is operated, consider the *n*-channel JFET of Fig. 5.2, shown with applied bias voltage to operate the device. The supply voltage, V_{DD}, provides a voltage across drain-source, V_{DS}, which results in a current, I_D, from

---

[1] W. Shockley, *Electrons and Holes in Semiconductors* (New York: Van Nostrand, 1953).

drain to source (electrons in an *n*-channel actually move from the source, hence name, to drain). This drain current passes through the *channel* surrounded by the *p*-type gate. A voltage between gate and source, V_{GS}, is shown here to be set by a voltage supply, V_{GG}. Since this gate-source voltage will reverse bias the gate-source junction, no gate current will result. The effect of the gate voltage will be to create a depletion region in the channel and thereby reduce the channel width to increase the drain-source resistance resulting in less drain current.

We shall first consider the device operation with $V_{GS} = 0$ V and then with the reverse-bias voltage, V_{GS}, increased (made more negative for an *n*-channel device). Figure 5.3a shows that drain current through the *n*-material of the drain-source provides a voltage drop along the channel, which is more positive at the drain-gate junction than at the source-gate junction. This reverse-bias potential across the *p-n* junction causes a depletion region to form as shown in Fig. 5.3a. As the voltage, V_{DD}, is increased, the current, I_D, increases, resulting in a larger depletion region with increased channel resistance from drain to source. As the voltage V_{DD} is increased, the depletion region forms fully across the channel as shown in Fig. 5.3b. At this point any further increase in V_{DD} will result in further increasing the voltage across only the depletion region, leaving the voltage drop from depletion point to ground

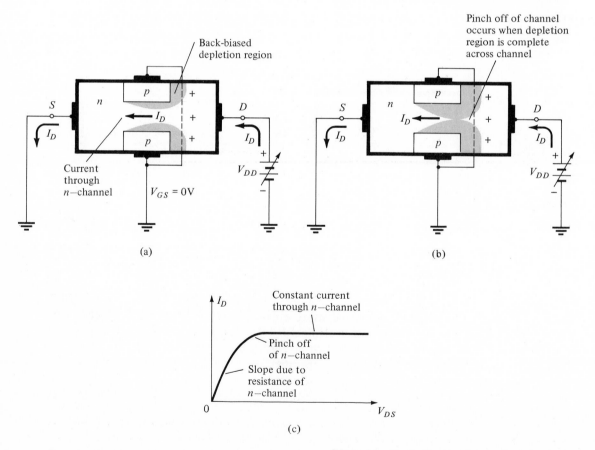

Figure 5.3 Pinch-off action due to channel current.

constant, the current I_D then remaining constant. This operation is depicted in the $V_{GS} = 0$ characteristic curve of Fig. 5.3c. As V_{DS} increases, the current I_D increases until the depletion region is fully formed across the channel after which, as shown, the drain current saturates and remains a constant value for increased voltage, V_{DS}. This value of drain current, for $V_{GS} = 0$ V is an important parameter used to specify the operation of the FJET and is referred to as I_{DSS}, the *d*rain to *s*ource current with gate-source *s*horted ($V_{GS} = 0$ V).

Drain-Source Characteristic

If V_{GS} is increased (more negative for an *n*-channel device), the channel will develop a depletion region so that the amount of current needed to close off the channel is less, as shown by the $V_{GS} = -1$ V curve of Fig.5.4a. We then see that the gate voltage acts as a control, reducing the amount of drain current (at a specific voltage V_{DS}). For a *p*-channel JFET the drain current is reduced from I_{DSS} as V_{GS} is made more positive (Fig. 5.4b).

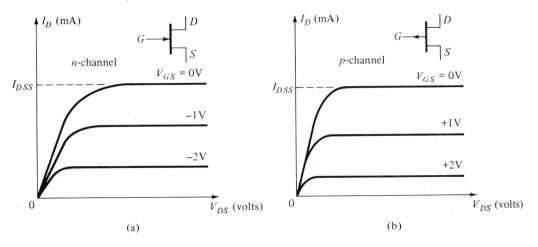

Figure 5.4   JFET drain-source characteristic.

As the value of V_{GS} is increased, with reduced drain current, a voltage is reached, after which no drain current will result, regardless of the voltage, V_{DS}. This gate-source pinch-off voltage, V_P [or $V_{GS(OFF)}$] is also an important parameter used to specify the operation of the JFET. The drain-source characteristics of Fig. 5.4 show that for an *n*-channel FET, V_P is a negative voltage and that for a *p*-channel FET, V_P is positive.

Transfer Characteristic

Another form of device characteristic is the transfer characteristic which is a plot of drain current, I_D as a function of gate-source voltage, V_{GS}, for a constant value of drain-source voltage, V_{DS}. The transfer characteristic can be directly viewed on a curve tracer unit, obtained directly by measurement of device operation, or drawn from the drain characteristic as shown in Fig. 5.5. Two important points of

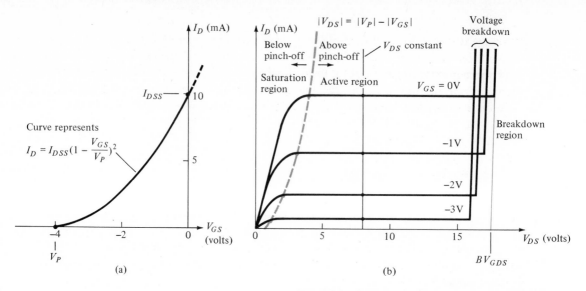

Figure 5.5 JFET drain-source and transfer characteristics.

the transfer curve shown are the values I_{DSS} and V_P. When these points are fixed, the rest of the curve can be seen on the transfer characteristic or obtained from theoretical consideration of the physical process occurring in the JFET, leading to the relation[2]

$$I_D = I_{DSS}\left(1 - \frac{V_{GS}}{V_P}\right)^2 \tag{5.1}$$

which represents the transfer characteristic curve of Fig. 5.5a. Notice that when $V_{GS} = 0$, $I_D = I_{DSS}$ and that when $I_D = 0$, $V_{GS} = V_P$ as seen on the transfer characteristic. The drain characteristic shows that the current saturates when the channel pinches off, the pinch-off occurring at lower values of V_{DS} for more negative values of V_{GS} (as shown by the dashed line on the drain characteristic in Fig. 5.5b). The JFET device is normally biased to operate above pinch-off in the region of current saturation. Within this region the device operation is more easily described by the transfer characteristic or by Shockley's equation (Eq. 5.1).

> **EXAMPLE 5.1** Determine the drain current of an n-channel JFET having pinch-off voltage $V_P = -4$ V and drain-source saturation current $I_{DSS} = 12$ mA at the following gate-source voltages:
> (a) $V_{GS} = 0$ V; (b) $V_{GS} = -1.2$ V; and (c) $V_{GS} = -2$ V.
>
> **Solution:** Using the JFET transfer characteristic described by Eq. (5.1):
> (a) $I_D = I_{DSS}\left(1 - \frac{V_{GS}}{V_P}\right)^2 = 12$ mA $\left(1 - \frac{0}{-4}\right)^2 =$ **12 mA**
>
> (b) $I_D = 12$ mA $\left(1 - \frac{-1.2}{-4}\right)^2 =$ **5.88 mA**
>
> (c) $I_D = 12$ mA $\left(1 - \frac{-2}{-4}\right)^2 =$ **3 mA**

---

[2] Shockley's equation applies above pinch-off region for JFET device.

CH. 5 FIELD-EFFECT TRANSISTORS (FETs)

5.3 JFET PARAMETERS

Manufacturers specify a number of parameters to describe the JFET device and to provide data necessary for selecting between various units. Some of the more useful parameters specified are:

1. I_{DSS}, the drain-source saturation current;
2. $V_P = V_{GS(OFF)}$, the pinch-off or gate-source off voltage;
3. BV_{GSS}, the device breakdown voltage with drain-source shorted;
4. $g_m = g_{fs}$, the device transconductance; and
5. $r_{ds(on)}$, the drain-source resistance when the device is turned on.

A number of other parameters relating to device capacitance, noise voltage, turn-on and turn-off times, and power handling are also usually provided in the manufacturer's spec sheets.

I_{DSS}, Drain-Source Saturation Current

The current at which the channel pinches off when gate-source is shorted ($V_{GS} = 0$) is a most important device parameter. The value of I_{DSS} can be easily measured using a circuit as in Fig. 5.6a and represents the largest drain current using the JFET reverse-biased. For a small-signal device, this current is typically milliamperes (mA).

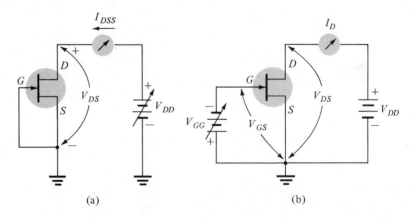

Figure 5.6 (a) Circuit to measure I_{DSS}; (b) circuit to measure $V_{GS(OFF)}$ ($=V_P$).

$V_{GS(OFF)} = V_P$, Gate-Source Cutoff (Pinch-off) Voltage

The gate-source voltage at which the drain-source channel is cut off or pinched off, resulting in essentially no drain current, is $V_{GS(OFF)}$ or V_P in the manufacturer's specifications sheets. Practical measurements require determination of the cutoff voltage at a specified low value of drain current (some microamperes) since zero current or little drain current is too ambiguous a value to measure. A circuit to provide measurement of $V_{GS(OFF)}$ is shown in Fig. 5.6b. For a specified value of V_{DS}, I_D and V_{GS} are measured as V_{GS} is varied until I_D is reduced to the sufficiently small current value indicating cutoff.

BV_{GSS}, Source-Gate Breakdown Voltage

The breakdown voltage of the source-gate junction, BV_{GSS}, is measured at a specified current with the drain-source shorted. The value of breakdown voltage indicates a limiting value of voltage across gate-source, above which the device current must be limited by the external circuit or else the device may be permanently damaged. The breakdown voltage provides a limiting voltage value for use in selecting the drain supply voltage.

$g_{fs} = g_m$, Common-Source Forward Transfer Transconductance

The measured device common-source forward transconductance, specified as g_{fs}, is also referred to in the text as g_m. It is measured with drain-source shorted; that is

$$g_{fs} = \frac{\Delta I_D}{\Delta V_{GS}}\bigg|_{V_{DS}=0}$$

and is an indication of the JFET ac amplification. The value of g_{fs} (or g_m) tells how much the ac current will change due to an applied ac gate-source voltage.

The value of g_m is measured in *siemens* (S)[3] with typical values from 1 mS to 10 mS or 1000 μS $-$ 10,000 μS. Mathematical derivation can be used to show that the relation given by Eq. (5.1) when mathematically differentiated results in

$$g_m = g_{mo}\left(1 - \frac{V_{GS}}{V_{GS(OFF)}}\right) \qquad (5.2)$$

where

$$g_{mo} = \frac{2I_{DSS}}{|V_{GS(OFF)}|} = \frac{2I_{DSS}}{|V_P|} \qquad (5.3)$$

The value of g_{mo} is the maximum ac gain parameter of the JFET occurring at a bias of $V_{GS} = 0$ V. For any other bias condition, the value of g_m is less, as calculated using Eqs. (5.2) and (5.3).

> **EXAMPLE 5.2** Calculate the transconductance, g_m, of a JFET having specified values of $I_{DSS} = 12$ mA and $V_{GS(OFF)} = -4$ V at bias points:
> (a) $V_{GS} = 0$ V; and (b) $V_{GS} = -1.5$ V.
>
> **Solution:** Using Eqs. (5.2) and (5.3)

---

[3] Formerly the unit was mho.

CH. 5   FIELD-EFFECT TRANSISTORS (FETs)

$$g_{m0} = \frac{2\,I_{DSS}}{|V_{GS(OFF)}|} = \frac{2(12\text{ mA})}{|-4\text{ V}|} = \frac{24 \times 10^{-3}}{|-4|} = 6 \times 10^{-3}\text{ S} = 6\text{ mS} = 6000\ \mu\text{S}$$

(a) $\quad g_m = g_{m0}\left(1 - \dfrac{V_{GS}}{V_P}\right) = 6\text{ mS}\left(1 - \dfrac{0}{-4}\right) = 6\text{ mS} = \mathbf{6000\ \mu S}$

(b) $\quad g_m = 6\text{ mS}\left(1 - \dfrac{-1.5}{-4}\right) = 3.75\text{ mS} = \mathbf{3750\ \mu S}$

$r_{ds(on)}$, Drain-Source on Resistance

The drain-source on resistance, measured at a specified gate-source voltage and drain current is important when using the JFET as a switch. When the JFET is biased in its saturation or ohmic region of operation, it exhibits a resistance between drain and source from ten to a few hundred ohms, as specified by the value of $r_{ds(on)}$.

5.4 MOSFET CONSTRUCTION AND CHARACTERISTICS

A field-effect transistor can be constructed with the gate terminal insulated from the channel. The popular Metal-Oxide-Semiconductor FET (MOSFET) is constructed as either a *depletion* MOSFET (Fig. 5.7a) or *enhancement* MOSFET (Fig. 5.7b). In the depletion-mode construction a channel is physically constructed and current between drain and source will result from a voltage connected across the drain-source. The enhancement MOSFET structure has *no* channel formed when the device is constructed. Voltage must be applied at the gate to develop a channel of charge carriers so that a current results when a voltage is applied across the drain-source terminals.

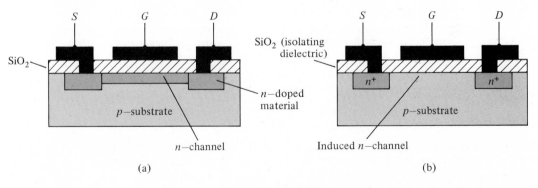

Figure 5.7 MOSFET construction: (a) depletion; (b) enhancement.

Depletion MOSFET

The *n*-channel depletion MOSFET device of Fig. 5.7a is formed on a *p*-substrate (*p*-doped silicon material used as the starting material onto which the FET structure is formed). The source and drain are connected by metal (aluminum) to *n*-doped source and drain regions which are connected internally by an *n*-doped channel region.

A metal layer is deposited above the *n*-channel on a layer of silicon dioxide (SiO₂) which is an insulating layer. This combination of a *metal* gate on an *oxide* layer over a *semi*conductor substrate forms the depletion MOSFET device. For the *n*-channel device of Fig. 5.7a negative gate-source voltages push electrons out of the channel region to deplete the channel and a large enough negative gate-source voltage will pinch off the channel. Positive gate-source voltage *on the other hand* will result in an increase in the channel size (pushing away *p*-type carriers), allowing more charge carriers and therefore greater channel current to result.

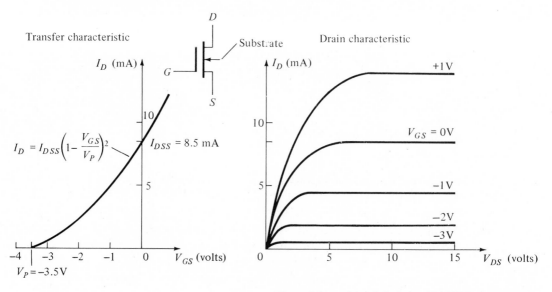

Figure 5.8 *n*-channel depletion MOSFET characteristic.

An *n*-channel depletion MOSFET device characteristic is shown in Fig. 5.8. The device is shown to operate with either positive or negative gate-source voltage, negative values of V_{GS} reducing the drain current until the pinch-off voltage, after which no drain current occurs. The transfer characteristic is the same for negative gate-source voltages, but it continues for positive values of V_{GS}. Since the gate is isolated from the channel for both negative and positive values of V_{GS}, the device can be operated with either polarity of V_{GS}—no gate current resulting in either case. The device schematic symbol in Fig. 5.8 shows the addition of a substrate terminal (in addition to gate, source, and drain leads) on which the device type is indicated, the arrow here indicating a *p*-substrate and thus *n*-channel device. A *p*-channel depletion MOSFET characteristic is shown in Fig. 5.9.

EXAMPLE 5.3 A depletion MOSFET has $I_{DSS} = 12$ mA and $V_{GS(OFF)} = -4.5$ V. Calculate the drain current at gate-source voltages of:
(a) 0 V; (b) -2 V; and (c) -3 V.

Solution: Using Eq. (5.1)

(a) $I_D = I_{DSS}\left(1 - \dfrac{V_{GS}}{V_P}\right)^2 = 12\text{ mA}\left(1 - \dfrac{0}{-4.5}\right)^2 = \mathbf{12\ mA}$

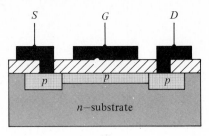

(a)

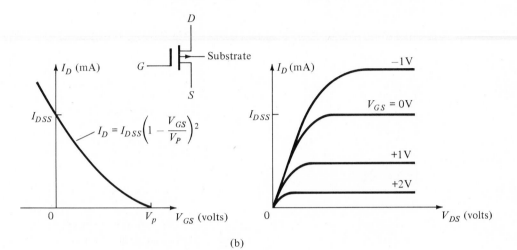

(b)

Figure 5.9 *p*-channel depletion MOSFET: (a) device structure; (b) device characteristic.

(b) $I_D = 12 \text{ mA} \left(1 - \dfrac{-2}{-4.5}\right)^2 = \textbf{3.7 mA}$

(c) $I_D = 12 \text{ mA} \left(1 - \dfrac{-3}{-4.5}\right)^2 = \textbf{1.33 mA}$

Enhancement MOSFET

The enhancement MOSFET of Fig. 5.10 has no channel between drain and source as part of the basic device construction. Application of a positive gate-source voltage will repel holes in the substrate region under the gate leaving a depletion region.

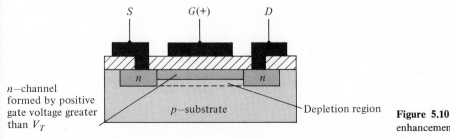

n–channel formed by positive gate voltage greater than V_T

Figure 5.10 *n*-channel formed in enhancement MOSFET.

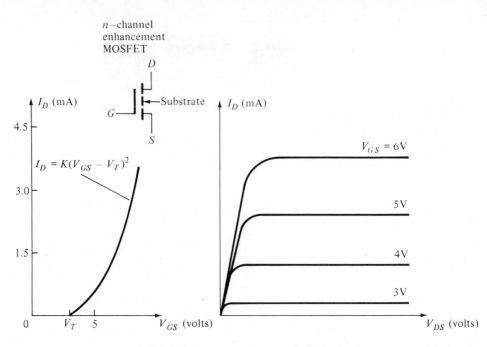

Figure 5.11 Device characteristic for *n*-channel enhancement MOSFET.

When the gate voltage is sufficiently positive, electrons are attracted into this depletion region making it then act as an *n*-channel between drain and source. The resulting *n*-channel enhancement MOSFET characteristic is shown in Fig. 5.11. There is no drain current until the gate-source voltage exceeds the threshold value, V_T. Positive voltages above this threshold value result in increased drain current, the transfer characteristic being described by[4]

$$I_D = K(V_{GS} - V_T)^2 \qquad (5.4)$$

where K, typically 0.3 mA/volt², is a property of the device construction. Note that no value I_{DSS} can be associated with an enhancement MOSFET because no drain current occurs with $V_{GS} = 0$ V. Although the enhancement MOSFET is more restricted in operating range than is the depletion device, the enhancement device is very useful in large-scale integrated circuits in which the simpler construction and smaller size make it a suitable device. The enhancement schematic symbol shows a broken line between drain and source indicating that there is no initial channel for the enhancement device. The substrate terminal arrow shows a *p*-substrate and an *n*-channel. *P*-channel enhancement MOSFETS can also be constructed, the device and characteristic being shown in Fig. 5.12.

> **EXAMPLE 5.4** For an *n*-channel enhancement MOSFET with threshold voltage of 2.5 V, determine the current at values of gate-source voltages:
> (a) $V_{GS} = 2.5$ V; (b) $V_{GS} = 4$ V; and (c) $V_{GS} = 6$ V.

---

[4] Equation (5.4) is only valid for $|V_{GS}| > |V_T|$.

202   CH. 5   FIELD-EFFECT TRANSISTORS (FETs)

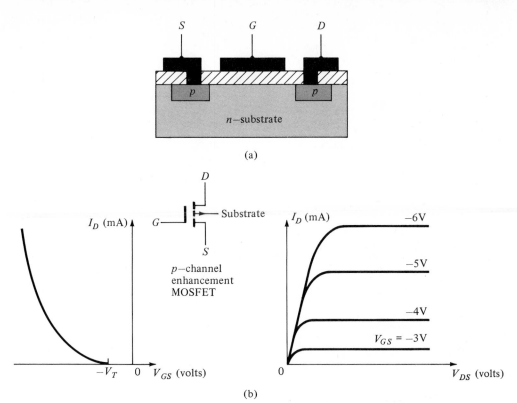

Figure 5.12 *p*-channel enhancement MOSFET: (a) device structure; (b) device characteristic.

Solution: Using $K = 0.3$ mA/V^2 in Eq. (5.4):
(a) $I_D = 0.3 \times 10^{-3}(V_{GS} - V_T)^2 = 0.3 \times 10^{-3}(2.5 - 2.5)^2 = \textbf{0 mA}$
(b) $I_D = 0.3 \times 10^{-3}(4 - 2.5)^2 = \textbf{0.675 mA}$
(c) $I_D = 0.3 \times 10^{-3}(6 - 2.5)^2 = \textbf{3.675 mA}$

As with a JFET or depletion MOSFET, a value of transconductance can also be obtained for an enhancement MOSFET, the relation in this case being[5]

$$g_m = 2\ K(V_{GS} - V_T) \tag{5.5}$$

EXAMPLE 5.5 Determine the value of transconductance for an *n*-channel enhancement MOSFET having threshold voltage $V_T = 3$ V at the following operating points:
(a) 6 V; and (b) 8 V.

Solution:
(a) $g_m = 2(0.3 \times 10^{-3})\ (6 - 3) = \textbf{1.8 mS}$
(b) $g_m = 2(0.3 \times 10^{-3})\ (8 - 3) = \textbf{3 mS}$

---

[5] Obtained from derivating Eq. (5.4)

$$g_m = \frac{\Delta I_D}{\Delta V_{GS}} = 2\ K(V_{GS} - V_T)$$

where $K = 0.3 \times 10^{-3}$ A/V^2(typical) $= 0.3$ mA/V^2.

In recent years there has been increasing interest in raising the power limits of the FET. One technique that appears to be taking hold in the commercial market is the V-FET construction appearing in Fig. 5.13a. The basic construction of the typical MOSFET appears in Fig. 5.13b for comparison purposes. Most noticeable are the four *diffused* layers in the V-FET as compared to the three regions of the MOSFET developed through *photolithographic* methods. The term V-FET is derived primarily

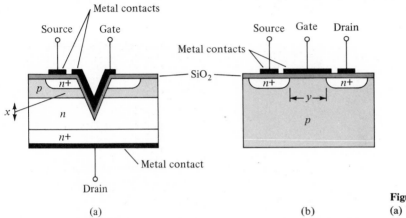

Figure 5.13   V-FET vs. MOSFET; (a) V-FET; (b) MOSFET.

(a)   (b)

from the fact that the drain-to-source current follows a "*vertical*" path rather than the horizontal path of the conventional MOSFET. Obviously, the V-type construction of the gate could also suggest this terminology. Increased currents are possible with the V-FET, due to the significantly reduced channel length (1:3) (x versus y in Fig. 5.13), the availability of two current paths from the lower drain to the separated source, and the fact that many other Vs of construction can be introduced resulting in a saw-edge pattern in the top layer and a significant increase in the number of paths from drain to source as shown in Fig. 5.14. In the past FETs have been limited to the milliampere range with relatively low power ratings (milliwatts to few watts). Some *n*-channel enhancement-mode V-FETs of rating 60-W, 2-A, are now available.

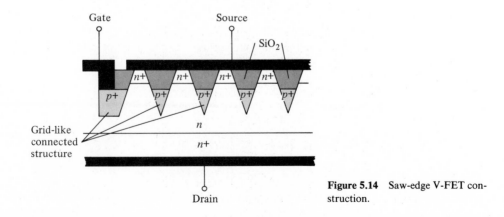

Figure 5.14   Saw-edge V-FET construction.

In Fig. 5.13a, the added *n*-type region will result in improved drain-to-source breakdown voltages and lower parasitic capacitance levels. For the V-FET the short channel results in a straight-line (linear) relationship between I_D and V_{GS}.

Other important characteristics of the V-FET as compared to the bipolar transistor include a negative temperature coefficient to remove the concern about thermal runaway, a very low leakage current (a few nanoamperes), and higher switching speeds (the VMP-1 produced by Siliconix can switch 2 A in 5 ns).

Further information on this commercial unit (VMP-1) includes gate threshold voltages from 0.8 to 1.8 V, a maximum gate voltage of 10 V, and a drain-to-source breakdown voltage of 60 V. Its minimum g_m is 200 mS.

5.6 CMOS

A popular connection used primarily in digital circuits connects enhancement *p*MOS and *n*MOS transistors into a complementary or CMOS device. Figure 5.15 shows the basic CMOS connection. The input is connected in common to the gate of both

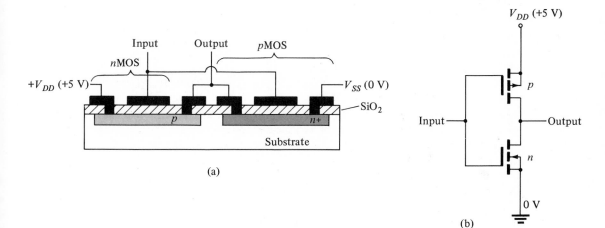

Figure 5.15 Basic CMOS connection.

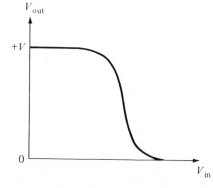

Figure 5.16 CMOS input–output relationship.

pMOS and nMOS transistors. A positive input drives the pMOS off, the nMOS on, with the output dropping to 0 V. A low-value input will correspondingly drive the pMOS device on, the nMOS device off, with the output voltage rising to $+V_{DD}$. A plot showing the relation between input and output voltages is provided in Fig. 5.16.

The CMOS device is used primarily in digital circuits, operating to provide output of either 0 V or +5 V while drawing very little power from the supply. Most low-power integrated circuits are built using CMOS transistors.

PROBLEMS

§ 5.2

1. An n-channel JFET has device parameters $V_{GS(OFF)} = -3$ V, and $I_{DSS} = 8$ mA. Calculate the drain current at gate-source voltages of: (a) -1 V; (b) 0 V; and (c) -2 V.

2. A p-channel JFET has device parameters of $I_{DSS} = 7.5$ mA and $V_{GS(OFF)} = +4$ V. Determine the drain current at gate-source voltages of: (a) $+2$ V; and (b) $+3$ V.

3. What gate-source voltage is required to obtain a drain current of 5 mA using an n-channel JFET having $V_{GS(OFF)} = -3$ V and $I_{DSS} = 8$ mA?

4. What gate-source voltage is required to obtain a drain current of 5 mA using a p-channel JFET having $V_{GS(OFF)} = +4$ V and $I_{DSS} = 7.5$ mA?

5. An n-channel JFET having value of $I_{DSS} = 8.5$ mA is operated with resulting measured values of $I_D = 2.125$ mA and $V_{GS} = -2.5$ V. What is the value of device $V_{GS(OFF)}$?

6. An n-channel JFET with $V_{GS(OFF)} = -6$ V is operated at $I_D = 6.75$ mA and $V_{GS} = -1.5$ V. Determine the value of I_{DSS} for the device.

7. An n-channel JFET with rated values of $I_{DSS} = 10$ mA and $V_{GS(OFF)} = -3.5$ V is operated with I_D measured to be 3.265 mA. What was the value of V_{GS}?

§ 5.3

8. Calculate the device transconductance, g_{m0}, for an n-channel JFET having $I_{DSS} = 8$ mA and $V_{GS(OFF)} = -4.5$ V.

9. What is the value of I_{DSS} for an n-channel JFET with $g_{m0} = 4.5$ mS and $V_{GS(OFF)} = -3$ V?

10. What is the value of $V_{GS(OFF)}$ of a p-channel JFET having $I_{DSS} = 12$ mA and $g_{m0} = 6500$ μS?

11. Determine the value of g_{m0} for a p-channel JFET having $V_{GS(OFF)} = -3.8$ V and $I_{DSS} = 6.8$ mA.

12. What value of transconductance, g_m, results when operating an n-channel JFET ($I_{DSS} = 8$ mA, $V_{GS(OFF)} = -4$ V) at $V_{GS} = -1.5$ V?

13. What value of transconductance results when operating a p-channel ($V_{GS(OFF)} = +3.5$ V, $I_{DSS} = 9$ mA) at $V_{GS} = +0.75$ V?

14. An n-channel JFET is biased at $V_{GS} = -1.5$ V and $I_D = 2.9$ mA. If $I_{DSS} = 7.5$ mA, what are the values of $V_{GS(OFF)}$ and g_m?

15. An n-channel JFET ($I_{DSS} = 7.8$ mA) is operated at $I_D = 3.82$ mA and $V_{GS} = -1.2$ V. What is the value of g_m at this operating point?

16. A p-channel JFET ($I_{DSS} = 13.5$ mA, $V_{GS(OFF)} = +5$ V) is operated at $I_D = 9.5$ mA. What is the value of g_m at this operating point?

17. An n-channel JFET ($V_{GS(OFF)} = -4.5$ V, $I_{DSS} = 8$ mA) is operated at $V_{GS} = -1.2$ V. Determine the value of device transconductance, g_{m0}, and transconductance at the operating point g_m.

18. A JFET having $g_{m0} = 4200$ μS is operated at $V_{GS} = -1$ V. What is the value of g_m at this operating point? ($V_{GS(OFF)} = -4$ V.)

19. A JFET ($I_{DSS} = 6$ mA, $V_{GS(OFF)} = -2.5$ V) is operated at $I_D = 5$ mA. What is the value of g_m at this operating point?

20. What is the maximum value of transconductance of a JFET ($V_{GS(OFF)} = -4$ V) if the transconductance is 4500 μS when operated at $V_{GS} = -1$ V.

§ 5.4

21. A depletion MOSFET ($I_{DSS} = 12$ mA, $V_{GS(OFF)} = -4$ V) is operated at $V_{GS} = -0.5$ V. What is the value of the transconductance at this operating point?

22. What is the value of transconductance of a depletion MOSFET ($I_{DSS} = 8$ mA, $V_{GS(OFF)} = -2$ V) when operated at $V_{GS} = 0$ V?

23. An enhancement MOSFET having threshold voltage of 3.5 V is operated at $V_{GS} = 5$ V. What current results (use $K = 0.3$ mA/V²)?

24. What is the value of threshold voltage for an n-channel enhancement MOSFET that operates at $I_D = 4.8$ mA when biased at 7 V?

25. Determine the value of circuit transconductance for an n-channel enhancement MOSFET having $V_T = 2.8$ V when operated a 6 V.

26. An enhancement MOSFET operated at $V_{GS} = 7.5$ V has transconductance of 2.5 mS. What is the value of device threshold voltage?

27. An n-channel enhancement MOSFET ($V_T = 2.5$ V) when operated at $I_D = 6$ mA has what value of transconductance?

CHAPTER 6

FET Biasing

6.1 FIXED BIAS

Dc bias of an FET device requires setting of the gate-source voltage, which results in a desired drain current. For JFET and depletion MOSFET devices operating in the depletion mode, the drain current is limited by the saturation current, I_{DSS}. An enhancement MOSFET requires biasing at a gate-source voltage greater than the threshold value to turn on the device. Since the FET has such a high impedance seen looking into the gate (either reverse-biased p-n junction or isolation by silicon-dioxide layer) the dc voltage of the gate set by a voltage divider or a fixed battery voltage is not affected or loaded by the FET. Fixed dc bias is obtained using a battery to set the gate-source reverse bias voltage as in Fig. 6.1a. Battery V_{GG} is used to set the reverse-bias voltage V_{GS} with no resulting current through R_G or the gate terminal. The resulting current I_D through resistor R_D then sets the value of drain voltage V_D by the amount of voltage drop from V_{DD} by the voltage across R_D; that is

$$V_{GS} = V_{GG}$$

and

$$V_D = V_{DD} - I_D R_D$$

The JFET drain characteristic (Fig. 6.1b) shows $I_{DSS} = 2$ mA and $V_P = V_{GS\,(OFF)} = -4$ V. Since the gate-source voltage is set by the battery voltage V_{GG}, we know that the curve, $V_{GS} = -2$ V, represents one of the conditions of dc bias. To determine the exact bias point graphically, we can solve by plotting a dc load line representing the Kirchhoff voltage equation around the drain-source loop

$$V_{DD} - I_{D_Q} R_D = V_{DS_Q}$$

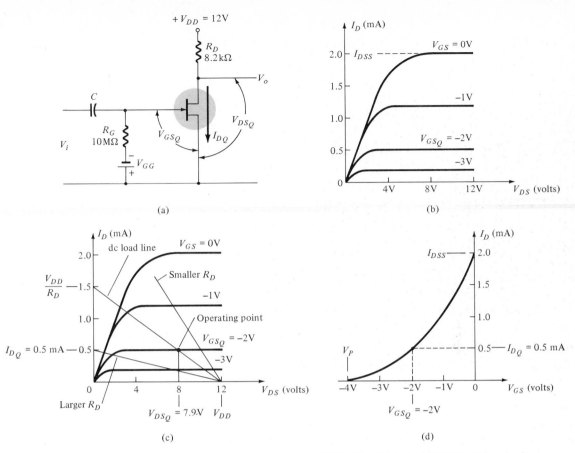

Figure 6.1 Dc bias in JFET amplifier circuit.

This condition of circuit operation is shown graphically as a dc load line in Fig. 6.1c, drawn by interconnecting two points of the straight line:

1. At $I_D = 0$, $V_{DS} = V_{DD}$.

2. At $V_{DS} = 0$, $I_D = \dfrac{V_{DD}}{R_D}$.

As seen in Fig. 6.1c, the intersection of the dc load line gate-source voltage curve occurs at approximately

$$V_{DS_Q} = 7.9 \text{ V} \qquad I_{D_Q} = 0.5 \text{ mA}$$

and the operating point is established in the circuit of Fig. 6.1c. Decreasing the magnitude of V_{GS} will result in the bias point moving to lower values of V_{DS} and larger I_D along the dc load line. Using a smaller value of R_D will provide a steeper dc load line, as shown in Fig. 6.1c, resulting in a higher value of V_{DS_Q} at the same value of I_{D_Q} (for V_{GS_Q} remaining at −2 V).

The dc bias point can also be determined using the transfer characteristic as shown in Fig. 6.1d. At $V_{GS_Q} = -2$ V, the transfer curve shows $I_{D_Q} = 0.5$ mA. The value of V_{DS_Q} can then be calculated.

$$V_{DS_Q} = V_{DD} - I_{D_Q}R_D = 12 - (0.5 \text{ mA})(8.2 \text{ k}\Omega) = 7.9 \text{ V}$$

The transfer characteristic shows more clearly the result of varying V_{GS} on the bias current I_{D_Q}. As V_{GS} is increased toward $V_P = -4$ V, the device drain current is reduced, while reducing the magnitude of V_{GS} toward 0 V increases I_{D_Q} toward I_{DSS}. The value of I_D at $V_{GS_Q} = -2$ V can also be calculated using Eq. (5.1).

$$I_{D_Q} = I_{DSS}\left(1 - \frac{V_{GS_Q}}{V_P}\right)^2 = 2 \text{ mA}\left(1 - \frac{-2}{-4}\right)^2 = 0.5 \text{ mA}$$

from which we calculate $V_{DS_Q} = 7.9$ V, as shown above.

EXAMPLE 6.1   Obtain the operating point for the fixed-bias circuit of Fig. 6.2a and n-channel JFET with drain characteristic shown in Fig. 6.2b.

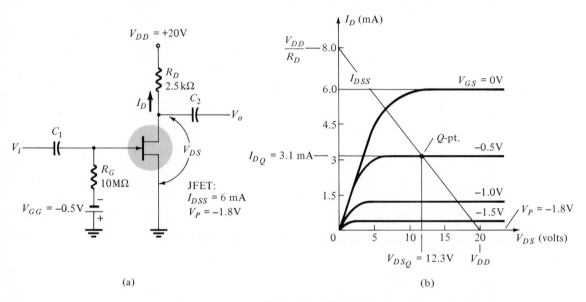

(a)   (b)

Figure 6.2   JFET circuit and drain characteristic for Example 6.1.

Solution: Drawing the dc load line between $V_{DD} = 20$ V and $V_{DD}/R_D = 20/2.5 \text{ k}\Omega = 8$ mA results in the Q-point (quiescent operating point) at

$$V_{DS_Q} = \textbf{12.3 V}, \qquad I_{D_Q} = \textbf{3.1 mA}$$

6.2   JFET AMPLIFIER WITH SELF-BIAS

A more practical version of a JFET amplifier has only a single voltage supply using a self-bias resistor, R_S, to obtain the gate-source bias voltage, as shown in Fig. 6.3a. The presence of resistor R_S results in a positive voltage V_S due to the voltage drop $I_D R_S$. Since the gate voltage, V_G, is 0 V (no dc current in gate or resistor R_G), the net voltage measured from gate (0 V) to source ($+ V_S$) is a negative voltage which is the gate-source bias voltage, V_{GS}. The gate circuit bias relation:

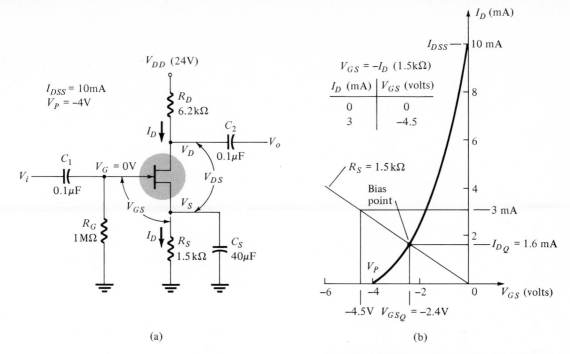

Figure 6.3 JFET amplifier circuit with self-bias.

$$V_{GS} = 0 - I_D R_S = -I_D R_S \qquad (6.1)$$

can be plotted on the I_D, V_{GS} characteristic as shown in Fig. 6.3b. To plot the straight-line equation of Eq. (6.1) we can use $V_{GS} = 0$ at $I_D = 0$ and any other corresponding value. At $I_D = 3$ mA, for example,

$$V_{GS} = -I_D R_S = -(3 \text{ mA})(1.5 \text{ k}\Omega) = -4.5 \text{ V}$$

The self-bias line, as shown, intersects the device characteristic at $I_{D_Q} = 1.6$ mA, $V_{GS_Q} = -2.4$ V, the bias point set by R_S.

We should note that increasing R_S would lower the R_S line, resulting in a lower value of I_{D_Q} and corresponding larger value V_{GS_Q}. Reducing R_S would result in a steeper R_S line with higher I_{D_Q} and less voltage V_{GS_Q}.

EXAMPLE 6.2 Determine the bias point of the circuit shown in Fig. 6.3a for $R_S = 1$ kΩ.

Solution: The self-bias line is drawn on the characteristic of Fig. 6.3b from the 0-axis ($V_{GS} = 0$, $I_D = 0$) passing through a point of selected value I_D, say $I_D = 4$ mA and corresponding voltage $V_{GS} = -I_D R_S = -(4 \text{ mA})(1 \text{ k}\Omega) = -4$ V. The intersection of this line with the device transfer characteristic is the bias point

$$I_{D_Q} = 2.2 \text{ mA}, \qquad V_{GS_Q} = -2.2 \text{ V}$$

We can then calculate

$$V_{DS_Q} = V_{DD} - I_{DQ}R_D - I_{DQ}R_S = 24 - 2.2 \text{ mA } (6.2 \text{ k}\Omega + 1.5 \text{ k}\Omega)$$
$$= 24 - 16.9 = 7.1 \text{ V}$$

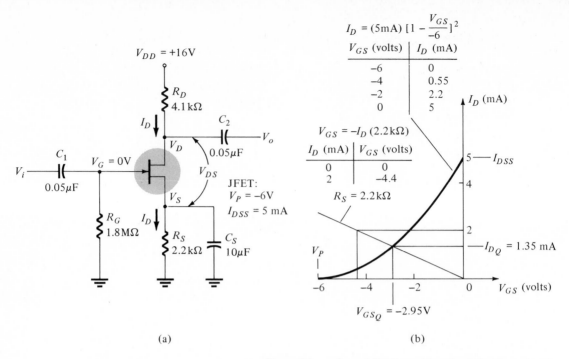

Figure 6.4 *n*-channel JFET amplifier circuit for Example 6.3.

EXAMPLE 6.3 Determine the operating point of the *n*-channel JFET amplifier shown in Fig. 6.4a. (See Appendix D, Fig. D.2.)

Solution: It is fairly easy to sketch the transfer characteristic using the values of V_P and I_{DSS} provided and the transfer curve formula [Eq. (5.1)].

$$I_D = I_{DSS}\left(1 - \frac{V_{GS}}{V_P}\right)^2$$

A sketch on a sheet of graph paper can be made for a few points of selected V_{GS} and corresponding calculated values of I_D. For example, the present JFET would result in a transfer curve sketched from $V_P(= -6 \text{ V})$ along the *x*-axis through points

$$V_{GS} = -4 \text{ V}: \quad I_D = 5 \text{ mA}\left(1 - \frac{-4}{-6}\right)^2 = 0.56 \text{ mA}$$

$$V_{GS} = -2 \text{ V}: \quad I_D = 5 \text{ mA}\left(1 - \frac{-2}{-6}\right)^2 = 2.22 \text{ mA}$$

to the point $I_D = I_{DSS} = 5 \text{ mA}$ along the *y*-axis.

Using this transfer characteristic (Fig. 6.4b), we can then plot the self-bias line for $R_S = 2.2 \text{ k}\Omega$. Choosing a current, say $I_D = 2 \text{ mA}$, we obtain

$$V_{GS} = -I_D R_S = -(2 \text{ mA})(2.2 \text{ k}\Omega) = -4.4 \text{ V}$$

The self-bias line then intersects the transfer curve at about

$$V_{GS_Q} = -2.95 \text{ V}, \qquad I_{D_Q} = 1.35 \text{ mA}$$

from which we calculate

$$V_{DS_Q} = V_{DD} - I_{D_Q}(R_S + R_D) = 16 - 1.35 \text{ mA}(2.2 \text{ k}\Omega + 4.1 \text{ }\Omega) = 7.495 \text{ V}$$

Voltage-Divider Biasing

Another form of dc bias circuit is that shown in Fig. 6.5a. Except for the gate voltage being set other than 0 V, the determination of bias voltage and current proceeds as discussed previously. The bias circuit of Fig. 6.5 provides a greater dc bias stability than the bias circuit of Fig. 6.3. The value of V_G obtained from the voltage-divider network is

$$V_G = V_{GG} = \frac{R_2}{R_1 + R_2} \cdot V_{DD} \tag{6.2}$$

and the bias voltage V_{GS_Q} is then

$$V_{GS_Q} = V_{GG} - I_D R_S \tag{6.3}$$

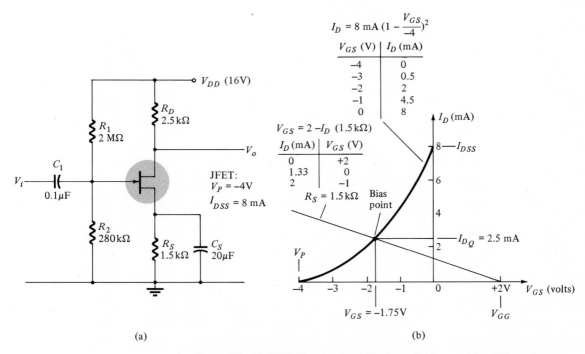

(a)   (b)

Figure 6.5 (a) JFET bias circuit with voltage divider gate bias; (b) transfer characteristic and self-bias line for Example 6.4.

EXAMPLE 6.4 Determine the dc bias of the JFET in the circuit of Fig. 6.5a.

Solution: The gate voltage is

$$V_G = V_{GG} = \frac{R_2}{R_1 + R_2} \cdot V_{DD} = \frac{280 \text{ k}\Omega}{2 \text{ M}\Omega + 280 \text{ k}\Omega} \cdot 16 \text{ V} = \mathbf{1.965 \text{ V}}$$

The result of voltage drop $I_D R_S$ is a gate-source voltage

$$V_{GS} = V_G - V_S = +1.965 - I_D R_S$$

An R_S-bias line (see Fig. 6.5b) can be drawn corresponding to the above circuit equation. For the JFET with $V_P = -4$ V and $I_{DSS} = 8$ mA we can also plot the transfer characteristic as in Fig. 6.5b, the intersection of self-bias line and transfer characteristic providing a dc bias at

$$I_{D_Q} = 2.5 \text{ mA}, \qquad V_{GS_Q} = -1.75 \text{ V}$$

We can then calculate

$$V_{D_Q} = V_{DD} - I_{D_Q}R_D = 16 \text{ V} - 2.5 \text{ mA}(2.5 \text{ k}\Omega) = \textbf{9.75 V}$$

$$V_{S_Q} = I_{D_Q}R_D = 2.5 \text{ mA}(1.5 \text{ k}\Omega) = \textbf{3.75 V}$$

and $\qquad V_{DS_Q} = V_{D_Q} - V_{S_Q} = 9.75 - 3.75 = \textbf{6 V}$

(Note that $V_{GS_Q} = V_{G_Q} = 1.965 - 3.75 = -1.785$ V.) (See Appendix D, Fig. D.2.)

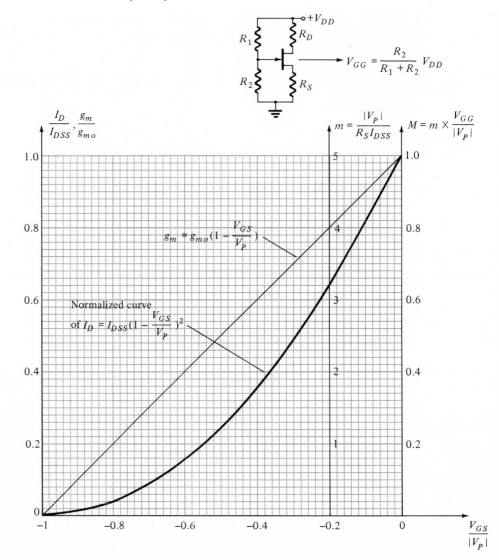

Figure 6.6 Universal JFET characteristic.

6.3 dc BIAS USING UNIVERSAL JFET BIAS CURVE

To reduce some of the effort in dc bias calculation with JFET (or depletion MOSFET) a normalized *n*-channel curve as shown in Fig. 6.6 may be used. The JFET transfer characteristic is plotted on normalized axes. To simplify the plotting of the R_S self-bias line an axis of normalized R_S values is plotted as the value *m*, where

$$m = \frac{|V_P|}{I_{DSS}R_S} \tag{6.4}$$

with V_P in volts, I_{DSS} in milliamperes and R_S in kilohms. For the voltage-divider bias stabilized circuit the values of *M* and V_{GG} are used, as will be demonstrated shortly.

$$V_{GG} = \frac{R_2}{R_1 + R_2} \cdot V_{DD} \tag{6.5}$$

$$M = m \times \frac{V_{GG}}{|V_P|} \tag{6.6}$$

An example for each type of bias circuit will help show how this universal characteristic is used.

EXAMPLE 6.5 Determine the dc bias voltages and currents for the circuit of Fig. 6.7a.

Solution: Calculating the value of *m*,

$$m = \frac{|V_P|}{I_{DSS}R_S} = \frac{|-3|}{6(1.6)} = 0.31$$

we plot the R_S bias line from the point through the point $m = 0.31$ along the *m*-axis. The resulting bias point is seen in Fig. 6.7b to be

$$\frac{I_D}{I_{DSS}} = 0.18, \qquad \frac{V_{GS}}{|V_P|} = 0.575$$

from which we calculate

$$I_{D_Q} = 0.18(6 \text{ mA}) = \mathbf{1.08 \text{ mA}}$$

$$V_{GS_Q} = 0.575(-3) = \mathbf{-1.73 \text{ V}}$$

Using the value of I_{D_Q} obtained, we then calculate

$$V_{DS_Q} = V_{DD} - I_{D_Q}(R_D + R_S) = 16 - 1.08 \text{ mA}(3.9 \text{ k}\Omega + 1.6 \text{ k}\Omega) = \mathbf{10.06 \text{ V}}$$

EXAMPLE 6.6 Determine the dc bias condition for the circuit in Fig. 6.7.

Solution: For the voltage-divider bias circuit of Fig. 6.7c we first calculate

$$m = \frac{|V_P|}{I_{DSS}R_S} = \frac{|-3|}{6(1.6)} = 0.31 \qquad \text{(same as in Fig. 6.7c)}$$

$$V_{GG} = \frac{R_{G_2}}{R_{G_1} + R_{G_2}} V_{DD} = \frac{500 \text{ k}\Omega}{4.5 \text{ M}\Omega + 500 \text{ k}\Omega} \cdot 16 = 1.6 \text{ V}$$

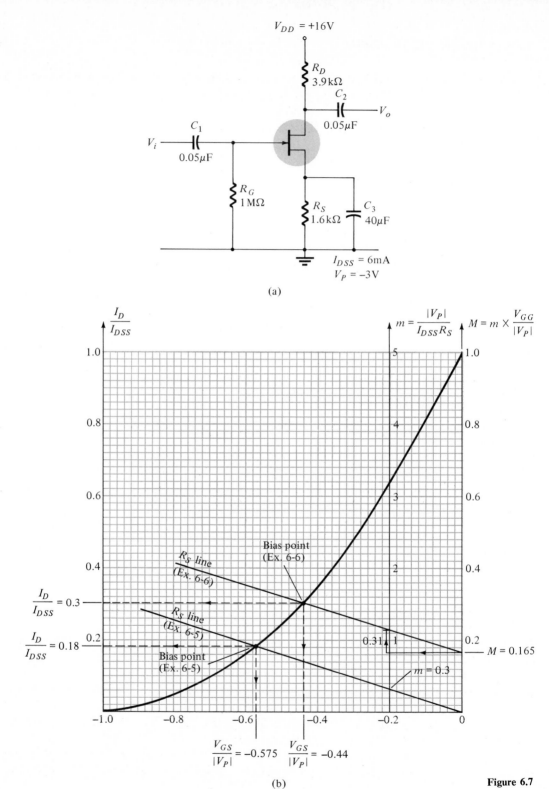

(a)

(b)

Figure 6.7

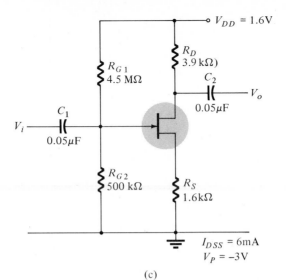

$V_{DD} = 1.6\text{V}$

R_{G1} 4.5 MΩ

R_D 3.9 kΩ)

C_2 0.05μF

V_o

C_1 0.05μF

V_i

R_{G2} 500 kΩ

R_S 1.6kΩ

$I_{DSS} = 6\text{mA}$
$V_P = -3\text{V}$

(c)

Figure 6.7 (continued)

$$M = m \cdot \frac{V_{GG}}{|V_P|} = 0.31 \frac{1.6}{|-3|} = 0.165$$

To plot the bias line we must connect a line having the same slope (same R_S and JFET values) as in the previous example passing through the $M = 0.165$ point. This is directly accomplished by connecting a line between the point $M = 0.165$ along the M-axis and a point along the m-axis which is 0.31 higher. From Fig. 6.7b we see that this line gives a bias point

$$\frac{V_{GS}}{|V_P|} = -0.44 \qquad \frac{I_D}{I_{DSS}} = 0.3$$

from which we calculate

$$V_{GS_Q} = 0.44(3 \text{ V}) = \mathbf{-1.32 \text{ V}}$$
$$I_{D_Q} = 0.3(6 \text{ mA}) = \mathbf{1.8 \text{ mA}}$$

We can then calculate

$$V_{DS_Q} = V_{DD} - I_{D_Q}(R_D + R_S) = 16 - 1.8 \text{ mA}(3.9 \text{ k}\Omega + 1.6 \text{ k}\Omega) = \mathbf{6.1 \text{ V}}$$

6.4 DEPLETION MOSFET DC BIAS CIRCUITS

The depletion MOSFET operation as described by its transfer characteristic shows that the depletion device operates the same as the JFET when reverse-biased. As an example, an n-channel depletion MOSFET operates in the depletion mode for negative gate-source voltages the same as a reverse-biased JFET (and operates in its enhancement mode for positive gate-source voltage). While the JFET should not be forward-biased when operated as an amplifier, the depletion MOSFET can be forward- or reverse-biased, since the oxide insulation of the gate prevents any gate current for either gate-source voltage polarity. A few examples based on the same bias circuit connection used with JFETs will demonstrate how the depletion MOSFET is operated.

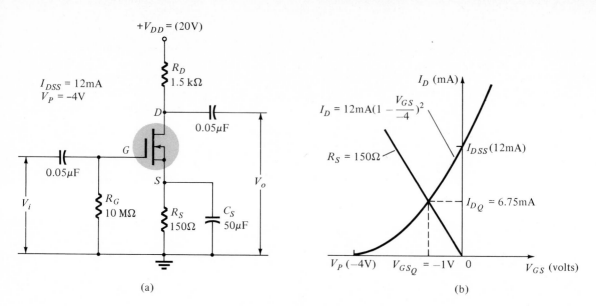

(a)

(b)

Figure 6.8 n-channel depletion MOSFET amplifier circuit and device characteristic for Example 6.7.

EXAMPLE 6.7 Determine the dc bias voltage V_{DS} and bias current I_D for the depletion MOSFET in the circuit of Fig. 6.8. The MOSFET specifications are $I_{DSS} = 12$ mA and $V_{GS(OFF)} = -4$ V.

Solution: Using the device transfer characteristic, a self-bias load line for the value of R_S is drawn as shown in Fig. 6.8. The bias point is seen to be at

$$V_{GS_Q} = -1 \text{ V} \quad \text{and} \quad I_{D_Q} = \textbf{6.75 mA}$$

The drain-source voltage is then

$$V_{DS_Q} = V_{DD} - I_{D_Q}R_D - I_{D_Q}R_S = 20 \text{ V} - 6.75 \text{ mA}(1.5 \text{ k}\Omega + 0.150 \text{ k}\Omega)$$
$$= \textbf{8.86 V}$$

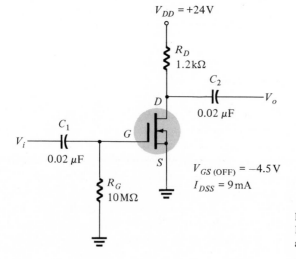

Figure 6.9 n-channel depletion MOSFET amplifier circuit for Example 6.8.

EXAMPLE 6.8 Determine the bias values I_{D_Q} and V_{DS_Q} for the circuit of Fig. 6.9 for a MOSFET with $V_{GS(OFF)} = -4.5$ V and $I_{DSS} = 9$ mA.

Solution: With both gate and source at 0 V, the gate-source is biased at $V_{GS_Q} = 0$ V for which $I_{D_Q} = I_{DSS} = 9$ mA. The resulting drain-source voltage is then

$$V_{DS_Q} = V_{DD} - I_D R_D = 24 \text{ V} - 9 \text{ mA}(1.2 \text{ k}\Omega) = \textbf{13.2 V}$$

It is possible to bias an *n*-channel depletion MOSFET at $V_{GS_Q} = 0$ V since an ac signal applied to the device can drive the gate-source positive without any gate current. In fact, good ac voltage gain is achieved when the device operates near $V_{GS_Q} = 0$ V. Bias with V_{GS} positive can also be used as demonstrated by the next example.

EXAMPLE 6.9 Determine the dc bias conditions for the circuit of Fig. 6.10 by using an *n*-channel depletion MOSFET with $I_{DSS} = 4$ mA and $V_{GS(OFF)} = -5$ V.

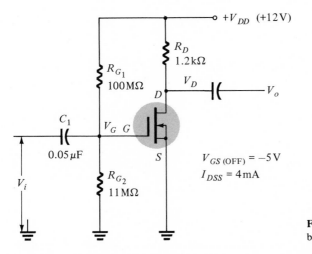

Figure 6.10 Depletion MOSFET bias for Example 6.9.

Solution: The gate bias voltage V_{GQ} is set by a voltage divider

$$V_{GQ} = \frac{R_{G2}}{R_{G1} + R_{G2}} (V_{DD}) = \frac{11 \text{ M}\Omega}{100 \text{ M}\Omega + 11 \text{ M}\Omega} (12 \text{ V}) = 1.19 \text{ V}$$

Since $\qquad\qquad V_{S_Q} = 0$ V,

$$V_{GS_Q} = V_{GQ} - V_{S_Q} = 1.19 - 0 = 1.19 \text{ V}$$

The resulting drain current is

$$I_{D_Q} = I_{DSS}\left(1 - \frac{V_{GS_Q}}{V_{GS(OFF)}}\right)^2 = 4 \text{ mA}\left(1 - \frac{1.19}{-5}\right)^2 = \textbf{6.13 mA}$$

The drain voltage is then

$$V_{D_Q} = V_{DD} - I_{D_Q} R_D = 12 \text{ V} - 6.13 \text{ mA}(1.2 \text{ k}\Omega) = 4.6 \text{ V}$$

and $\qquad V_{DS_Q} = V_{D_Q} - V_{S_Q} = 4.6 \text{ V} - 0 \text{ V} = \textbf{4.6 V}$

6.5 ENHANCEMENT MOSFET BIAS CIRCUIT

An enhancement MOSFET requires a gate-source voltage greater than the threshold voltage to drive the FET on. A popular circuit arrangement to bias the enhancement MOSFET is shown in Fig. 6.11a. The drain-source voltage is used to set the gate-source voltage. Since there is no gate current, there is no voltage drop across R_G and $V_{DS} = V_{GS}$. The resulting bias point can be determined by the intersection of the circuit dc load line and device transfer characteristic as demonstrated by the following example.

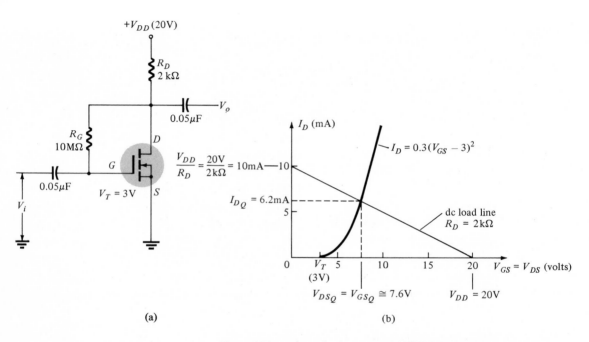

Figure 6.11 n-channel enhancement MOSFET dc bias circuit and characteristic for Example 6.10.

EXAMPLE 6.10 For the bias circuit of Fig. 6.11a, determine the dc bias drain-source voltage V_{DS_Q} and drain current I_{D_Q}. The MOSFET threshold voltage is $V_T = 3$ V.

Solution: The device transfer characteristic is drawn in Fig. 6.11b using

$$I_D = K(V_{GS} - V_T)^2 = 0.3 \text{ mA/V}^2 (V_{GS} - 3)^2$$

A dc load line is also drawn in Fig. 6.11b using

$$V_{GS} = V_{DS} = V_{DD} - I_D R_D = 20 \text{ V} - (2 \text{ k}\Omega)I_D$$

The intersection of load line and device characteristic result in a bias point of

$$V_{GS_Q} = V_{DS_Q} = \textbf{7.6 V}$$
$$I_{D_Q} = \textbf{6.2 mA}$$

PROBLEMS

§ 6.1

1. A fixed bias circuit as in Fig. 6.2a is operated with a JFET having $I_{DSS} = 8$ mA and $V_P = -4$ V. What is the resulting value of I_{D_Q}?

2. If the circuit of Fig. 6.2 is modified by $R_D = 2$ kΩ, what value of V_{DS_Q} results?

3. What is the value of V_{DS_Q} for the circuit of Fig. 6.2 when the JFET is changed to one having $I_{DSS} = 6$ mA and $V_P = -3$ V?

§ 6.2

4. Determine the dc bias voltage V_{D_Q} for the circuit of Fig. 6.3 when operated with JFET having $I_{DSS} = 8$ mA and $V_{GS(OFF)} = -6$ V.

5. Determine the dc bias voltage V_{S_Q} for the circuit of Fig. 6.3 with R_S changed to 1.2 kΩ.

6. What is the dc bias voltage V_{DS_Q} for the circuit of Fig. 6.3 with R_D changed to 5.1 kΩ?

7. What value of R_S is needed for the circuit of Fig. 6.3 to modify the bias point to $V_{GS_Q} = -2$ V?

8. Determine the source voltage V_{S_Q} for the circuit of Fig. 6.12.

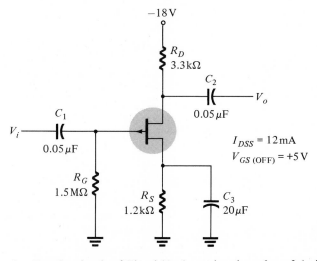

Figure 6.12 Circuit for Problems 8–10.

9. For the circuit of Fig. 6.12, determine the value of drain current.

10. What is the value of V_{DS_Q} for the circuit of Fig. 6.12?

11. What value of I_{D_Q} results when R_S is changed to 1.5 kΩ in the circuit of Fig. 6.4?

12. What value of V_{DS_Q} results when the JFET in the circuit of Fig. 6.4 is changed to one having $I_{DSS} = 7.5$ mA and $V_{GS(OFF)} = -4.5$ V?

13. What value of resistor R_D must be used in the circuit of Fig. 6.4 to alter the voltage V_{DS_Q} to 5.3 V?

14. What value of V_{S_Q} results when R_S is changed to 1.2 kΩ in the circuit of Fig. 6.5?

15. Determine the value of V_{DS_Q} in the circuit of Fig. 6.5 using a JFET having I_{DSS} = 6 mA and $V_{GS(OFF)}$ = −3.5 V.

16. What value of I_{D_Q} results in the circuit of Fig. 6.5 when R_2 is changed to 210 kΩ?

§ 6.3

17. Determine the dc bias voltage V_{GS_Q} for the circuit of Fig. 6.7a using a JFET with $V_{GS(OFF)}$ = −4.5 V and I_{DSS} = 8.5 mA using the universal JFET characteristic.

18. Using the universal JFET characteristic, determine the value of dc bias current I_{D_Q} for the circuit of Fig. 6.7a with JFET changed to one having I_{DSS} = 10 mA and $V_{GS(OFF)}$ = −5 V.

19. Using the universal JFET bias curves, determine the dc bias current for the circuit of Fig. 6.7a with R_S changed to 1.2 kΩ.

20. Using the universal JFET curve, determine the dc bias voltage V_{DS_Q} when R_{G2} is changed to 270 kΩ in Fig. 6.7c.

21. What value of V_{GS_Q} results in the circuit of Fig. 6.7c when the JFET used has I_{DSS} = 12 mA and $V_{GS(OFF)}$ = −4.5 V?

§ 6.4

22. Determine the dc bias voltage V_{DS_Q} for the circuit of Fig. 6.8 with depletion MOSFET having $V_{GS(OFF)}$ = −6 V and I_{DSS} = 8 mA.

23. What is the value of dc bias current in the circuit of Fig. 6.8 if R_S is changed to 330 Ω?

24. What value of V_{DS_Q} results in the circuit of Fig. 6.8 if R_D is changed to 1.8 kΩ?

25. What value of R_D is required in the circuit of Fig. 6.8 to obtain a bias voltage of V_{DS_Q} = 11.9 V?

26. What value of I_{D_Q} results if the JFET used in the circuit of Fig. 6.9 is changed to $V_{GS(OFF)}$ = −4 V and I_{DSS} = 12 mA?

27. What value of V_{DS_Q} results if the value of R_D is changed to 1.8 kΩ in the circuit of Fig. 6.9?

28. What value of I_{D_Q} results in the circuit of Fig. 6.10 if R_{G2} is changed to 20 MΩ?

§ 6.5

29. What value of I_{D_Q} results in the circuit of Fig. 6.11 if an enhancement MOSFET having V_T = 3.5 V is used?

30. What value of V_{DS_Q} results in the circuit of Fig. 6.11 if R_D is changed to 1.5 kΩ?

31. What value of V_{DS_Q} results in the circuit of Fig. 6.11 if the supply voltage is changed to +12 V?

32. What value of threshold voltage would result in a bias point of I_{D_Q} = 2.2 mA in the circuit of Fig. 6.11?

COMPUTER PROBLEMS

Write BASIC programs to:

1. Compute a table of points of I_D and V_{DS} for a JFET circuit for given values of I_{DSS} and $V_{GS(OFF)}$.

2. Compute I_D and V_{DS} for a JFET circuit using fixed battery biasing.

3. Compute I_D and V_{DS} for a self-biased JFET circuit.

4. Compute I_D and V_{DS} for a JFET circuit with voltage-divider biasing.

5. For a self-bias JFET circuit, tabulate the points to plot the transfer characteristic for a given device.

6. Compute and tabulate the values of I_D and V_{DS} for an enhancement MOSFET to plot the device drain characteristic.

CHAPTER
7

BJT Small-Signal Analysis

7.1 INTRODUCTION

The basic construction, appearance, and characteristics of the transistor were introduced in Chapter 3. The dc biasing of the device was then examined in Chapter 4. We will now consider the small-signal ac response of the BJT amplifier. Fortunately, the dc and ac analysis of a transistor configuration can be isolated and considered separately. The first effort will be to introduce an ac model for the transistor and examine its sinusoidal response.

It would be virtually impossible to examine all the possible configurations within a single chapter. Those of primary importance will be examined, however, and it would be assumed that an analysis of those systems not included could be undertaken based on the content of this chapter.

7.2 AMPLIFICATION IN THE AC DOMAIN

It was noted earlier that the transistor is an amplifying device. That is, the output sinusoidal signal is greater than the input signal or, stated another way, the output ac power is greater than the input ac power. The question often arises as to where this additional ac power has been generated. The sole purpose of the next few paragraphs will be to answer that question, since it is fundamental to the clear understanding of a number of important efficiency criteria to be defined later in the text.

Analogies are seldom perfect, but the following will be useful in describing the events leading to the foregoing conclusions. In Fig. 7.1a a steady heavy flow of a

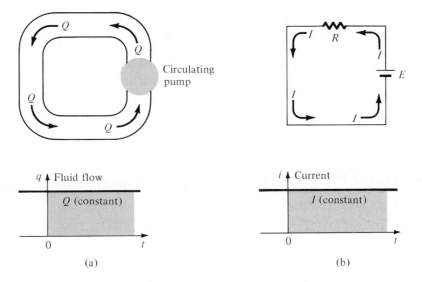

Figure 7.1 Fluid flow analogy of a series electrical circuit with a dc input: (a) fluidic system; (b) electrical system.

liquid has been established by the pump. The electrical analogy of this system appears in Fig. 7.1b. In each case there is some resistance to the flow, with the result that the magnitude of that flow is determined by an Ohm's law relationship. A graph of the flow versus time appears in each figure.

Let us now install a control mechanism in each system, as shown in Fig. 7.2. A small signal at the input to each of these control elements can have a marked effect on the established steady-state (dc) flow of each system. For the fluid system it could

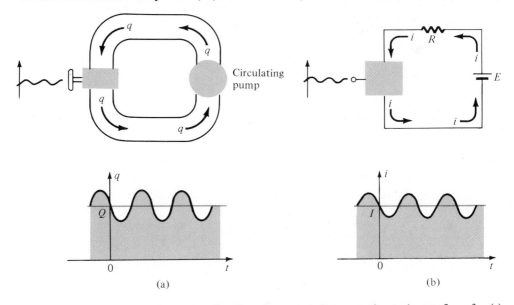

Figure 7.2 The effect of a control element on the steady-state flow of a: (a) fluid system; (b) electrical system.

be the oscillatory partial closing of the passage to limit the flow through the pipe. For the electrical system a mechanism is established for the control of the current i through the system. Recall that very small variations in I_B can have a pronounced effect on the collector current for the common-emitter transistor configuration.

In other words, a small input signal can have a pronounced effect on the steady-state flow of the system. Consider that the resulting output flow for the two systems may be as shown in Fig. 7.2. The sinusoidal swing of the output flow is certainly greater than the applied input—*amplification in the ac domain is therefore a reality!*

We can conclude, therefore, that most amplifiers are simply devices having a control point or terminal that can establish a heavy variation in flow between the other two terminals (normally part of the output circuit). The dc biasing circuits are necessary to establish the heavy flow of charge that will be very sensitive to the magnitude of the input signal. The increased ac power is only the result of the conversion of some of the dc power to the sinusoidal domain. The efficiency of an electronic amplifier is typically the ratio of the ac power out to the dc power in.

In ac (sinusoidal) analysis our first concern is the magnitude of the input signal. It will determine whether *small-signal* or *large-signal* techniques must be used. There is no set dividing line between the two, but the application, and the magnitude of the variables of interest *(i, υ)* relative to the scales of the device characteristics, will usually make it clear which is the case in point. The small-signal technique will be discussed in this chapter; large-signal applications will be considered in a later chapter.

7.3 TRANSISTOR MODELING

The key to the small-signal approach is the use of equivalent circuits (models) to be derived in this chapter. It is that combination of circuit elements, properly chosen, that will best approximate the actual semiconductor device in a particular operating region. Once the ac equivalent circuit has been determined, the graphic symbol of the device can be replaced in the schematic by this circuit and the basic methods of ac circuit analysis (branch-current analysis, mesh analysis, nodal analysis, and Thévenin's theorem) can be applied to determine the response of the circuit.

There are two schools of thought in prominence today regarding the equivalent circuit to be substituted for the transistor. For many years the industrial and educational institutions relied heavily on the *hybrid parameters* (to be introduced shortly). The hybrid-parameter equivalent continues to be very popular, although it must now share the spotlight with an equivalent circuit derived directly from the operating conditions of the transistor. Manufacturers continue to specify the hybrid parameters for a particular operating region on their specification sheets. The parameters (or components) of the other equivalent circuit can be derived directly from the hybrid parameters in this region. However, the hybrid equivalent circuit suffers from being limited to a particular set of operating conditions if it is to be considered accurate. The parameters of the other equivalent circuit can be determined for any region of operation within the active region and are not limited by the single set of parameters provided by the specification sheet. For the purposes of this text, if the operating region corresponds with that indicated on the specification sheet, then either equivalent will be used. If not specified, the equivalent circuit derived from the operating condi-

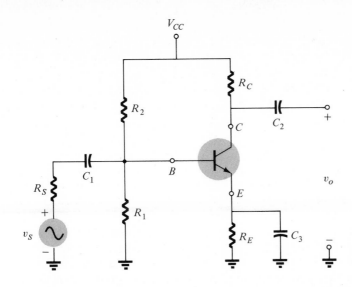

Figure 7.3 Transistor circuit under examination in this introductory discussion.

tions will be used. Be encouraged by the fact that both equivalent circuits are very similar in appearance and application. Developed skills with one will result in a measure of ability with the other.

In an effort to demonstrate the effect that the ac equivalent circuit will have on the analysis to follow, consider the circuit of Fig. 7.3. Let us assume for the moment that the small-signal ac equivalent circuit for the transistor has already been determined. Since we are interested only in the ac response of the circuit, all the dc supplies can be replaced by a zero-potential equivalent (short circuit) since they determine only the dc or quiescent level of the output voltage and not the magnitude of the swing of the ac output. This is clearly demonstrated by Fig. 7.4. The dc levels were simply important for determining the proper Q-point of operation. Once determined, the dc levels can be ignored for the ac analysis of the network. In addition, the coupling capacitors C_1 and C_2 and bypass capacitor C_3 were chosen to have a very small reactance at the frequency of application. Therefore, they too may for all practical purposes be replaced by a low-resistance path (short circuit). Note that this will result in the "shorting out" of the dc biasing resistor R_E. Connecting common grounds will result in a parallel combination for resistors R_1 and R_2, and R_C will appear from collector to emitter as shown in Fig. 7.5. Since the components of the transistor equivalent circuit inserted in Fig. 7.5 are those we are already familiar with (resistors, controlled sources, etc.), analysis techniques such as superposition, Thévenin's theorem, and so on, can be applied to determine the desired quantities.

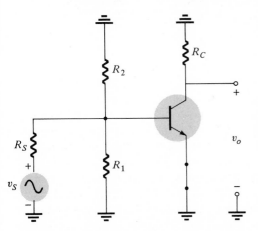

Figure 7.4 The network of Fig. 7.3 following the removal of the dc supply and inserting the short-circuit equivalent for the capacitors.

Let us further examine Fig. 7.5 and identify the important quantities to be determined for the system. Certainly, we would like to know the input and output impedance Z_i and Z_o as shown in Fig. 7.5. Since we know that the transistor is an amplifying

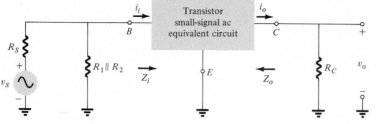

Figure 7.5 Circuit of Fig. 7.4 redrawn for small-signal ac analysis.

device, we would expect some indication of how the output current i_o is related to the input current—the *current gain*. Note in this case that $i_o = i_C$ and $I_i = i_B$. The ratio of these two quantities certainly relates directly to the β of the transistor. In Chapter 3 we found that the collector-to-emitter voltage did have some effect (if even slight) on the input relationship between i_B and v_{BE}. We might, therefore, expect some "feedback" from the output to input circuit in the equivalent circuit. The following section, through its brief introduction to *two-port theory*, will introduce the hybrid equivalent circuit, which will have parameters that will permit a determination of each of the quantities discussed above.

7.4 TRANSISTOR HYBRID EQUIVALENT CIRCUIT

The development that follows is an introduction to a subject called *two-port* theory. For the basic three-terminal device it is obvious, from Fig. 7.6, that there are two ports (pairs of terminals) of interest. For our purposes, the set at the left will represent the input terminals, and the set at the right, the output terminals. Note that, for each set of terminals, there are two variables of interest.

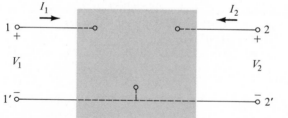

Figure 7.6 Two-port system.

The following set of equations (7.1) is only one of a number of ways in which the four variables can be related. It is the most frequently employed in transistor circuit analysis, however, and, therefore, will be discussed in detail in this chapter.

$$V_1 = h_{11}I_1 + h_{12}V_2 \tag{7.1a}$$

$$I_2 = h_{21}I_1 + h_{22}V_2 \tag{7.1b}$$

The parameters relating the four variables are called *h-parameters* from the word hybrid. The term hybrid was chosen because the mixture of variables (v & i) in each equation results in a "hybrid" set of units of measurement for the *h*-parameters.

A clearer understanding of what the various *h*-parameters represent and how we can expect to treat them later can be developed by isolating each and examining the resulting relationship.

If we arbitrarily set $V_2 = 0$ (short circuit the output terminals), and solve for h_{11} in Eq. (7.1a), the following will result:

$$h_{11} = \frac{V_1}{I_1}\bigg|_{V_2 = 0} \qquad \text{(ohms)} \qquad (7.2)$$

The ratio indicates that the parameter h_{11} is an impedance parameter to be measured in ohms. Since it is the ratio of the *input* voltage to the *input* current with the output terminals *shorted,* it is called the *short-circuit input-impedance parameter.* The subscript 11 of h_{11} defines the facts that the parameter is determined by a ratio of quantities measured at the input terminals.

If I_1 is set equal to zero by opening the input leads, the following will result for h_{12}:

$$h_{12} = \frac{V_1}{V_2}\bigg|_{I_1 = 0} \qquad \text{(unitless)} \qquad (7.3)$$

The parameter h_{12}, therefore, is the ratio of the input voltage to the output voltage with the input current equal to zero. It has no units since it is a ratio of voltage levels. It is called the *open-circuit reverse transfer voltage ratio parameter.* The subscript 12 of h_{12} reveals that the parameter is a transfer quantity determined by a ratio of input to output measurements. The first integer of the subscript defines the measured quantity to appear in the numerator; the second integer defines the source of the quantity to appear in the denominator. The term *reverse* is included to indicate that the voltage ratio is an input quantity over an output quantity rather than the reverse, which is usually the ratio of interest.

If in Eq. (7.1b), V_2 is equal to zero by again shorting the output terminals, the following will result for h_{21}:

$$h_{21} = \frac{I_2}{I_1}\bigg|_{V_2 = 0} \qquad \text{(unitless)} \qquad (7.4)$$

Note that we now have the ratio of an output quantity to an input quantity. The term *forward* will now be used rather than *reverse* as indicated for h_{12}. The parameter h_{21} is the ratio of the output current to the input current with the output terminals shorted. It is, for most applications, the parameter of greatest interest. This parameter, like h_{12}, has no units since it is the ratio of current levels. It is formally called the *short-circuit forward transfer current ratio parameter.* The subscript 21 again indicates that it is a transfer parameter with the output quantity in the numerator and the input quantity in the denominator.

The last parameter, h_{22}, can be found by again opening the input leads to set $I_1 = 0$ and solving for h_{22} in Eq. (7.1b).

$$h_{22} = \frac{I_2}{V_2}\bigg|_{I_1 = 0} \quad \text{(Siemens)} \tag{7.5}$$

Since it is the ratio of the output current to the output voltage, it is the output conductance parameter and is measured in siemens (S) (formerly mhos ℧). It is called the *open-circuit output conductance parameter*. The subscript 22 reveals that it is determined by a ratio of output quantities.

Since each term of Eq. (7.1a) has the unit volt, let us apply Kirchhoff's voltage law in reverse to find a circuit that "fits" the equation. Performing this operation will result in the circuit of Fig. 7.7. Since the parameter h_{11} has the unit ohm, it is represented as an impedance which for the transistor becomes a resistor in Fig. 7.7. The quantity h_{12} is dimensionless. Note that it is a "feedback" of the output voltage to the input circuit.

Each term of Eq. (7.1b) has the unit of current. Let us now apply Kirchhoff's current law in reverse to obtain the circuit of Fig. 7.8.

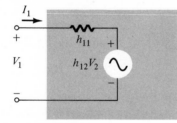 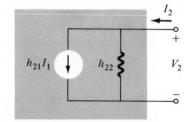

Figure 7.7 Hybrid input equivalent circuit.

Figure 7.8 Hybrid output equivalent circuit.

Since h_{22} has the unit of admittance, which for the transistor model is conductance, it is represented by the resistor symbol. Keep in mind, however, that the resistance in ohms of this resistor is equal to the reciprocal of conductance ($1/h_{22}$).

The complete "ac" equivalent circuit for the basic three-terminal linear device is indicated in Fig. 7.9 with a new set of subscripts for the *h*-parameters.

The notation of Fig. 7.9 is of a more practical nature since it relates the *h*-parameters to the resulting ratio obtained in the last few paragraphs. The choice of letters is obvious from the following listing:

$$h_{11} \rightarrow \textbf{\textit{i}}\text{nput resistance} \rightarrow h_i$$
$$h_{12} \rightarrow \textbf{\textit{r}}\text{everse transfer voltage ratio} \rightarrow h_r$$
$$h_{21} \rightarrow \textbf{\textit{f}}\text{orward transfer current ratio} \rightarrow h_f$$
$$h_{22} \rightarrow \textbf{\textit{o}}\text{utput conductance} \rightarrow h_o$$

The circuit of Fig. 7.9 is applicable to any linear three-terminal device with no internal independent sources. For the transistor, therefore, even though it has three basic configurations, *they are all three-terminal configurations,* so that the resulting equivalent circuit will have the same format as shown in Fig. 7.9. In each case the bottom of each configuration can be connected as shown in Fig. 7.10, since the potential level is the same. Essentially, therefore, the transistor model is a 3-terminal 2-port system. The *h*-parameters, however, will change with each configuration. To distin-

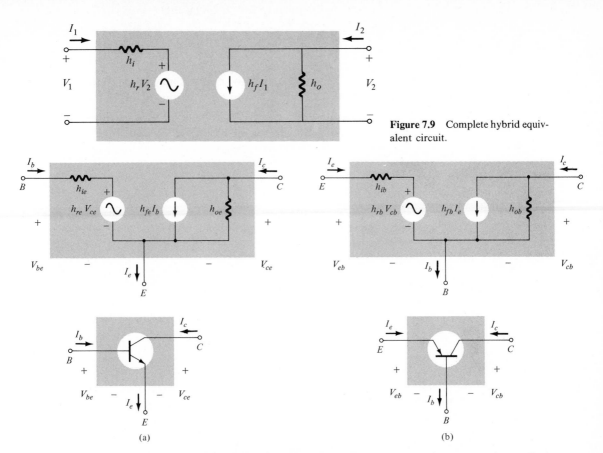

Figure 7.9 Complete hybrid equivalent circuit.

Figure 7.10 Complete hybrid equivalent circuits: (a) common-emitter configuration; (b) common-base configuration.

guish which parameter has been used or which is available, a second subscript has been added to the *h*-parameter notation. For the common-base configuration the lowercase letter *b* was added, while for the common-emitter and common-collector configurations the letters *e* and *c* were added, respectively. The hybrid equivalent circuit for the common-base and common-emitter configurations with the standard notation is presented in Fig. 7.10. The circuits of Fig. 7.10 are applicable for *pnp* or *npn* transistors.

The hybrid equivalent circuit of Fig. 7.9 is an extremely important one in the area of electronics today. It will appear over and over again in the analysis to follow. It would be time well spent, at this point, for the reader to memorize and draw from memory its basic construction and define the significance of the various parameters [see Eqs. (7.2)–(7.5)]. The fact that both a Thévenin and Norton circuit appear in the circuit of Fig. 7.9 was further impetus for calling the resultant circuit a *hybrid* equivalent circuit. Two additional transistor equivalent circuits, not to be discussed in this text, called the **z**-parameter and **y**-parameter equivalent circuits, use either the voltage source or the current source but not both in the same equivalent circuit. In Section 7.5 the magnitude of the various parameters will be found from the transistor characteristics in the region of operation resulting in the desired *small-signal equivalent circuit* for the transistor.

7.5 GRAPHICAL DETERMINATION OF THE h-PARAMETERS

Using partial derivatives (calculus), we can show that the magnitude of the h-parameters for the small-signal transistor equivalent circuit in the region of operation for the common-emitter configuration can be found using the following equations:

$$h_{ie} = \frac{\partial v_1}{\partial i_1} = \frac{\partial v_{BE}}{\partial i_B} \cong \frac{\Delta v_{BE}}{\Delta i_B}\bigg|_{v_{CE}=\text{constant}} \qquad (7.6)$$

$$h_{re} = \frac{\partial v_1}{\partial v_2} = \frac{\partial v_{BE}}{\partial v_{CE}} \cong \frac{\Delta v_{BE}}{\Delta v_{CE}}\bigg|_{i_B=\text{constant}} \qquad (7.7)$$

$$h_{fe} = \frac{\partial i_2}{\partial i_1} = \frac{\partial i_C}{\partial i_B} \cong \frac{\Delta i_C}{\Delta i_B}\bigg|_{v_{CE}=\text{constant}} \qquad (7.8)$$

$$h_{oe} = \frac{\partial i_2}{\partial v_2} = \frac{\partial i_C}{\partial v_{CE}} \cong \frac{\Delta i_C}{\Delta v_{CE}}\bigg|_{i_B=\text{constant}} \qquad (7.9)$$

In each case the symbol Δ refers to a small change in that quantity around the quiescent point of operation. In other words, the h-parameters are determined in the region of operation for the applied signal, so that the equivalent circuit will be the most accurate available. The constant values of V_{CE} and I_B in each case refer to a condition that must be met when the various parameters are determined from the characteristics of the transistor. For the common-base and common-collector configurations the proper equation can be obtained by simply substituting the proper values of v_1, v_2, i_1, and i_2. *In Appendix A a list has been provided that relates the hybrid parameters of the three basic transistor configurations.* In other words, if the h-parameters for the common-emitter configuration are known, the h-parameters for the common-base or common-collector configurations can be found using these tables.

The parameters h_{ie} and h_{re} are determined from the input or base characteristics, while the parameters h_{fe} and h_{oe} are obtained from the output or collector characteristics. Since h_{fe} is usually the parameter of greatest interest, we shall discuss the operations involved with equations, such as Eqs. (7.6)–(7.9), for this parameter first. The first step in determining any of the four hybrid parameters is to find the quiescent point of operation as indicated in Fig. 7.11. In Eq. (7.8) the condition V_{CE} = constant requires that the changes in base current and collector current be taken along a vertical straight line drawn through the Q-point representing a fixed collector-to-emitter voltage. Equation (7.8) then requires that a small change in collector current be divided by the corresponding change in base current. For the greatest accuracy these changes should be made as small as possible.

In Fig. 7.11 the change in i_B was chosen to extend from I_{B_1} to I_{B_2} along the perpendicular straight line at V_{CE}. The corresponding change in i_C is then found by drawing the horizontal lines from the intersections of I_{B_1} and I_{B_2} with V_{CE} = constant to the vertical axis. All that remains is to substitute the resultant changes of i_B and i_C into Eq. (7.8); that is,

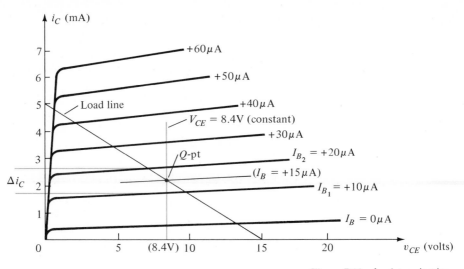

Figure 7.11 h_{fe} determination.

$$|h_{fe}| = \frac{\Delta i_C}{\Delta i_B}\bigg|_{v_{CE}=\text{constant}} = \frac{(2.7 - 1.7) \times 10^{-3}}{(20 - 10) \times 10^{-6}}\bigg|_{v_{CE}=8.4\text{ V}}$$

$$= \frac{10^{-3}}{10 \times 10^{-6}} = \mathbf{100}$$

In Fig. 7.12 a straight line is drawn tangent to the curve I_B through the Q-point to establish a line $I_B = \text{constant}$ as required by Eq. (7.9) for h_{oe}. A change in v_{CE} was then chosen and the corresponding change in i_C is determined by drawing the horizontal lines to the vertical axis at the intersections on the $I_B = \text{constant}$ line. Substituting into Eq. (7.9), we get

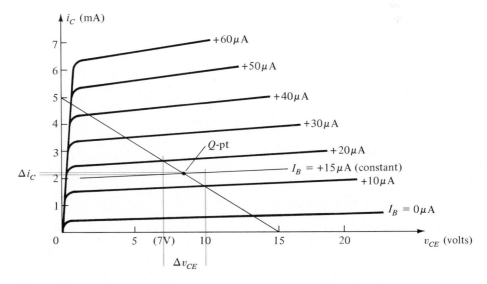

Figure 7.12 h_{oe} determination.

$$|h_{oe}| = \frac{\Delta i_C}{\Delta v_{CE}} \bigg|_{i_B = \text{constant}} = \frac{(2.2 - 2.1) \times 10^{-3}}{10 - 7} \bigg|_{I_B = +15\,\mu A}$$

$$= \frac{0.1 \times 10^{-3}}{3} = \textbf{33}\ \mu\textbf{A/V} = \textbf{33} \times \textbf{10}^{-6}\,\textbf{S} = \textbf{33}\ \mu\textbf{S}$$

To determine the parameters h_{ie} and h_{re} the Q-point must first be found on the input or base characteristics as indicated in Fig. 7.13. For h_{ie}, a line is drawn tangent to the curve $V_{CE} = 8.4$ V through the Q-point, to establish a line $V_{CE} = $ constant as required by Eq. (7.6). A small change in v_{BE} was then chosen, resulting in a

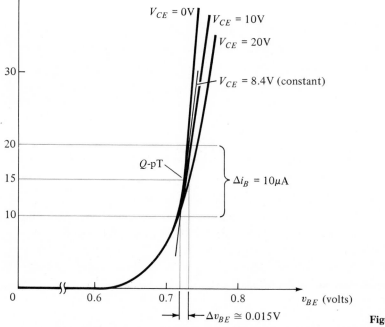

Figure 7.13 h_{ie} determination.

corresponding change in i_B. Substituting into Eq. (7.6), we get

$$|h_{ie}| = \frac{\Delta v_{BE}}{\Delta i_B} \bigg|_{v_{CE} = \text{constant}} = \frac{(733 - 718) \times 10^{-3}}{(20 - 10) \times 10^{-6}} \bigg|_{v_{CE} = 8.4 \text{V}}$$

$$= \frac{15 \times 10^{-3}}{10 \times 10^{-6}} = \textbf{1.5 k}\Omega$$

The last parameter, h_{re}, can be found by first drawing a horizontal line through the Q-point at $I_B = 15$ μA. The natural choice then is to pick a change in v_{CE} and find the resulting change in v_{BE} as shown in Fig. 7.14.

Substituting into Eq. (7.7), we get

$$|h_{re}| = \frac{\Delta v_{BE}}{\Delta v_{CE}} \bigg|_{i_B = \text{constant}} = \frac{(733 - 725) \times 10^{-3}}{20 - 0} = \frac{8 \times 10^{-3}}{20} = \textbf{4} \times \textbf{10}^{-4}$$

For the transistor whose characteristics have appeared in Figs. 7.11–7.14 the resulting hybrid small-signal equivalent circuit is shown in Fig. 7.15.

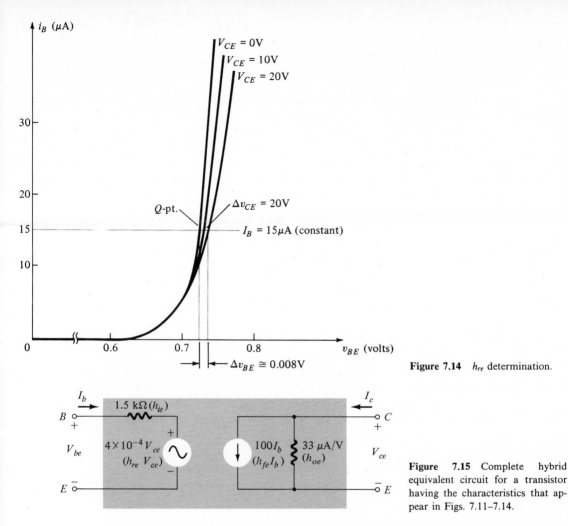

Figure 7.14 h_{re} determination.

Figure 7.15 Complete hybrid equivalent circuit for a transistor having the characteristics that appear in Figs. 7.11–7.14.

As mentioned earlier the hybrid parameters for the common-base and common-collector configurations can be found using the same basic equations with the proper variables and characteristics.

Typical values for each parameter for the broad range of transistors available today in each of its three configurations are provided in Table 7.1. The minus sign indicates that in Eq. (7.8) as one quantity increased in magnitude, within the change chosen, the other decreased in magnitude.

TABLE 7.1 Typical Parameter Values for the CE, CC, and CB Transistor Configurations

Parameter	CE	CC	CB
h_i	1 kΩ	1 kΩ	20 Ω
h_r	2.5×10^{-4}	$\cong 1$	3.0×10^{-4}
h_f	50	−50	−0.98
h_o	25 μA/V	25 μA/V	0.5 μA/V
$1/h_o$	40 kΩ	40 kΩ	2 MΩ

Note in retrospect (Section 3.4: Transistor Amplifying Action) that the input resistance of the common-base configuration is low, while the output resistance is high. Consider also that the short-circuit current gain is very close to 1. For the common-emitter and common-collector configurations note that the input resistance is much higher than that of the common-base configuration and that the ratio of output to input resistance is about 40:1. Consider also for the common-emitter and common-base configuration that h_r is very small in magnitude. Transistors are available today with values of h_{fe} that vary from 20 to 600. For any transistor the region of operation and conditions under which it is being used will have an effect on the various h-parameters. The effect of temperature and collector current and voltage on the h-parameters will be discussed in Section 7.6.

7.6 VARIATIONS OF TRANSISTOR PARAMETERS

There are a large number of curves that can be drawn to show the variations of the h-parameters with temperature, frequency, voltage, and current. The most interesting and useful at this stage of the development include the h-parameter variations with junction temperature and collector voltage and current.

In Fig. 7.16 the effect of the collector current on the h-parameter has been indicated. Take careful note of the logarithmic scale on the vertical and horizontal axes. Logarithmic scales will be examined in Chapter 9. The parameters have all been normalized to unity so that the relative change in magnitude with collector current can easily be determined. On every set of curves, such as in Fig 7.16, the operating point at which the parameters were found is always indicated. For this particular situation the quiescent point is at the intersection of $V_{CE} = 5.0$ V and $I_C = 1.0$ mA. Since the frequency and temperature of operation will also affect the h-parameters, these

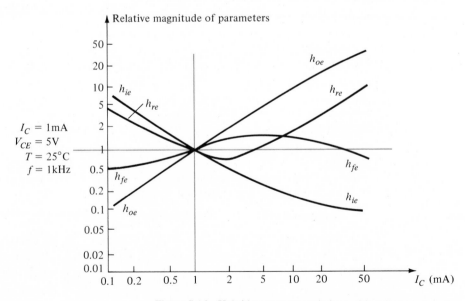

Figure 7.16 Hybrid parameter variations with collector current.

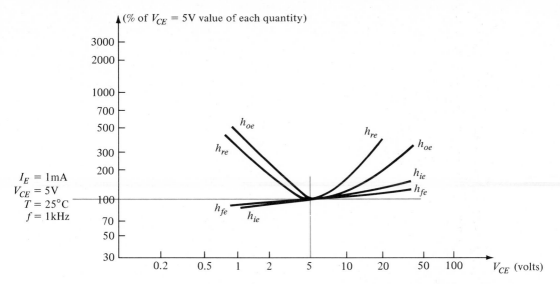

Figure 7.17 Hybrid parameter variations with collector-emitter potential.

quantities are also indicated on the curves. At 0.1 mA, h_{fe} is about 0.5 or 50% of its value at 1.0 mA, while at 3 mA, it is 1.5 or 150% of that value. In other words, h_{fe} has changed from a value of 0.5(50) = 25 to 1.5(50) = 75 with a change of I_C from 0.1 mA to 3 mA. In Section 7.8 we shall find that for the majority of applications it is a fairly good approximation to neglect the effects of h_{re} and h_{oe} in the equivalent circuit. Consider, however, the point of operation at $I_C = 50$ mA. The magnitude of h_{re} is now approximately 11 times that at the defined Q-point, a magnitude that may not permit eliminating this parameter from the equivalent circuit. The parameter h_{oe} is approximately 35 times the normalized value. This increase in h_{oe} will decrease the magnitude of the output resistance of the transistor to a point where it may approach the magnitude of the load resistor. There would then be no justification in eliminating h_{oe} from the equivalent circuit on an approximate basis.

In Fig. 7.17 the variation in magnitude of the *h*-parameters on a normalized basis has been indicated with change in collector voltage. This set of curves was normalized at the same operating point of the transistor discussed in Fig. 7.16 so that a comparison between the two sets of curves can be made. Note that h_{ie} and h_{fe} are relatively steady in magnitude, while h_{oe} and h_{re} are much larger to the left and right of the chosen operating point. In other words, h_{oe} and h_{re} are much more sensitive to changes in collector voltage than are h_{ie} and h_{fe}.

In Fig. 7.18 the variation in *h*-parameters has been plotted for changes in junction temperature. The normalization value is taken to be room temperature: $T = 25°C$. The horizontal scale is a linear scale rather than a logarithmic scale as was employed for Figs. 7.16 and 7.17. In general, all the parameters increase in magnitude with temperature. The parameter least affected, however, is h_{oe}, while the input impedance h_{ie} changes at the greatest rate. The fact that h_{fe} will change from 50% of its normalized value at −50°C to 150% of its normalized value at +150°C indicates clearly that the operating temperature must be carefully considered in the design of transistor circuits.

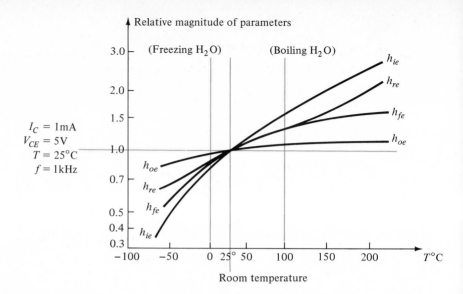

Figure 7.18 Hybrid parameter variations with temperature.

7.7 SMALL-SIGNAL ANALYSIS OF THE BASIC TRANSISTOR AMPLIFIER USING THE HYBRID EQUIVALENT CIRCUIT

In this section the basic transistor amplifier will be examined in detail using the hybrid equivalent circuit. It will not be specified whether the transistor is in the common-emitter, base, or collector configuration. The results, therefore, are applicable to each configuration, requiring only that the proper h-parameters be used for the configuration of interest. All amplifiers are basically two-port devices as indicated in Fig. 7.19; that is, there are a pair of input terminals and a pair of output terminals. For an amplifier there are six quantities of general interest: current gain, voltage gain, input impedance, output impedance, power gain, and phase relationships, each of which will be discussed in detail in this section. The load impedance Z_L can be any combination of resistive and reactive elements. In Fig. 7.19 all voltages and currents refer to the effective value of the sinusoidally varying quantities. The resistor R_s represents, in total, the internal resistance of the source and any resistance in series with the source V_s. Before continuing, keep in mind that the analysis to follow is for small-signal inputs. There is no discussion of dc levels and biasing arrangement.

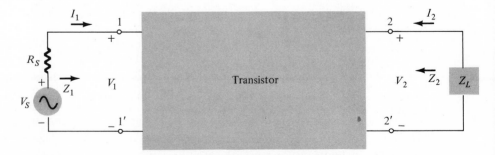

Figure 7.19 Basic transistor amplifier configuration.

For the resulting equations to be useful, however, the quiescent point must be established and the resulting h-parameters must be known.

The Current Gain $A_i = (I_2/I_1)$

Substituting the hybrid equivalent circuit for the transistor of Fig. 7.19 will result in the configuration of Fig. 7.20.

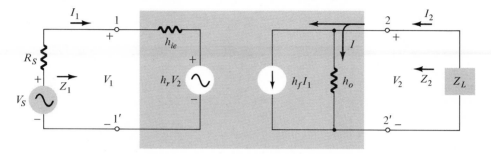

Figure 7.20 Transistor hybrid equivalent circuit substituted for the transistor of Fig. 7.19.

Applying Kirchhoff's current law to the output circuit:

$$I_2 = h_f I_1 + I = h_f I_1 + h_o V_2$$

Substituting $V_2 = -I_2 Z_L$:

$$I_2 = h_f I_1 - h_o Z_L I_2$$

The minus sign arises because the direction of I_2 as shown in Fig. 7.20 would result in a polarity across the load Z_L opposite to that shown in the same figure.

Rewriting the above equation:

$$I_2 + h_o Z_L I_2 = h_f I_1$$

and

$$I_2(1 + h_o Z_L) = h_f I_1$$

so that

$$\boxed{A_i = \frac{I_2}{I_1} = \frac{h_f}{1 + h_o Z_L}}$$
(7.10)

The Voltage Gain $A_v = (V_2/V_1)$

Applying Kirchhoff's voltage law to the input circuit results in

$$V_1 = I_1 h_i + h_r V_2$$

Substituting $I_1 = [(1 + h_o Z_L) I_2 / h_f]$ from Eq. (7.10) and $I_2 = -(V_2/Z_L)$ from above results in

$$V_1 = \frac{-(1 + h_o Z_L) h_i}{h_f Z_L} V_2 + h_r V_2$$

Solving for the ratio V_2/V_1:

$$A_v = \frac{V_2}{V_1} = \frac{h_f Z_L}{h_i + (h_i h_o - h_f h_r) Z_L} \qquad (7.11)$$

The Input Impedance $Z_1 = (V_1/I_1)$

For the input circuit

$$V_1 = h_i I_1 + h_r V_2$$

Substituting

$$V_2 = -I_2 Z_L$$

we have

$$V_1 = h_i I_1 - h_r Z_L I_2$$

Since

$$A_i = \frac{I_2}{I_1}$$

$$I_2 = A_i I_1$$

so that the above equation becomes

$$V_1 = h_i I_1 - h_r Z_L A_i I_1$$

Solving for the ratio V_1/I_1, we get

$$Z_1 = \frac{V_1}{I_1} = h_i - h_r Z_L A_i$$

and substituting

$$A_i = \frac{h_f}{1 + h_o Z_L}$$

yields

$$Z_1 = \frac{V_1}{I_1} = h_i - \frac{h_f h_r Z_L}{1 + h_o Z_L} \qquad (7.12)$$

The Output Impedance $Z_2 = (V_2/I_2)$

The output impedance of an amplifier is defined to be the ratio of the output voltage to the output current with the signal (V_s) set at zero.

For the input circuit with $V_s = 0$

$$I_1 = \frac{-h_r V_2}{R_s + h_i}$$

Substituting this relationship into the following equation obtained from the output circuit yields

$$I_2 = h_f I_1 + h_o V_2$$

$$I_2 = \frac{-h_f h_r V_2}{R_s + h_i} + h_o V_2$$

and the ratio

$$Z_2 = \frac{V_2}{I_2} \bigg|_{V_s=0} = \frac{1}{h_o - \left(\dfrac{h_f h_r}{h_i + R_s}\right)} \qquad (7.13)$$

so that the output admittance

$$Y_2 = \frac{I_2}{V_2}\bigg|_{V_s=0} = h_o - \frac{h_f h_r}{h_i + R_s} \tag{7.14}$$

The Power Gain $A_p = (P_L/P_i)$

The average power to the load is $V_L I_L \cos \theta,$[1] which, for the situation being discussed, is $-V_2 I_2 \cos \theta$. The minus sign arises again for the same reason discussed in the derivation of the equation for A_i. It indicates that the load is absorbing power and not supplying it to the circuit. If we limit our discussion to *purely resistive loads,* then $\cos \theta = 1$ and $P_L = P_2 = -V_2 I_2$. The input power is $V_1 I_1$ so that

$$A_p = \frac{P_L}{P_i} = \frac{-V_2 I_2}{V_1 I_1} \tag{7.15}$$

but

$$A_v = \frac{V_2}{V_1} \quad \text{and} \quad A_i = \frac{I_2}{I_1}$$

so that

$$A_p = \left[-\frac{V_2}{V_1}\right]\left[\frac{I_2}{I_1}\right]$$

is also

$$A_p = -A_v A_i \tag{7.16}$$

In terms of the *h*-parameters,

$$A_p = \frac{h_i^2 R_L}{(1 + h_o R_L)[h_i + (h_i h_o - h_f h_r)R_L]} \tag{7.17}$$

If we consider that

$$V_2 = -I_2 R_L$$

and substitute

$$I_2 = A_i I_1$$

then

$$V_2 = -A_i I_1 R_L$$

and

$$A_v = \frac{V_2}{V_1} = \frac{-A_i I_1 R_L}{V_1} = \frac{-A_i R_L}{V_1/I_1} = \frac{-A_i R_L}{Z_1}$$

so that

$$A_p = -A_v A_i = -\left(-\frac{A_i R_L}{R_1}\right)A_i$$

and

$$A_p = \frac{A_i^2 R_L}{R_1} \tag{7.18}$$

---

[1] In the following analysis, all voltages and currents are the magnitude of the rms value.

Phase Relationship

The phase relationship between the output current or voltage and the input current or voltage can be found by simply examining Eqs. (7.10) and (7.11), repeated below for convenience.

$$A_i = \frac{h_f}{1 + h_o Z_L}$$

$$A_v = \frac{-h_f Z_L}{h_i + (h_i h_o - h_f h_r) Z_L}$$

Recall from Section 7.5 that all the h-parameters are positive except h_f for the common-base and common-collector configurations. For the common-emitter configuration, therefore, it should be obvious from the above equations that for resistive loads the output current is in phase with the input current, while the output voltage, due to the negative sign, has a polarity opposite to that of the input voltage. That is, it will reach its negative maximum when the input signal reaches its positive maximum. For the common-collector and common-base configurations the opposite is true due to the difference in sign of h_f. Keep in mind, however, that the above discussion applies only to the current direction and voltage polarities as defined by Fig. 7.19.

EXAMPLE 7.1 Find the following for the fixed-bias transistor of Fig. 7.21.
  (a) Current gain $A_i = I_o/I_i$.
  (b) Voltage gain $A_v = V_o/V_i$.
  (c) Input impedance Z_i.
  (d) Output impedance Z_o.
  (e) Power gain A_p.

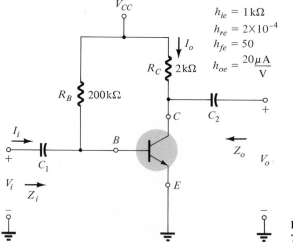

$h_{ie} = 1\,k\Omega$
$h_{re} = 2 \times 10^{-4}$
$h_{fe} = 50$
$h_{oe} = \dfrac{20\,\mu A}{V}$

Figure 7.21 Circuit for Example 7.1.

Solution: Replacing the dc supplies and capacitors by short circuits and substituting the transistor hybrid equivalent circuit will result in the configuration of Fig. 7.22.
  Redrawing the circuit we obtain Fig. 7.23.
  The basic hybrid equations will now be applied to the circuit of Fig. 7.23 to obtain the desired results. Note the similarities in the appearance of Fig. 7.23 as

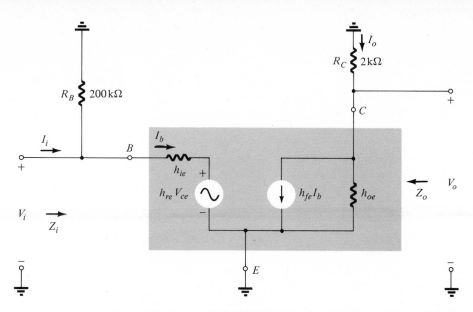

Figure 7.22 Circuit of Fig. 7.21 following the substitution of the small-signal hybrid equivalent circuit for transistor.

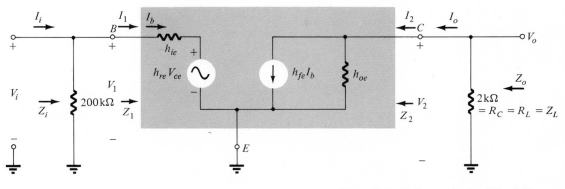

Figure 7.23 Redrawn circuit of Fig. 7.22.

compared to the fundamental configuration of Fig. 7.20. This is a basic requirement if the equations just derived from Fig. 7.20 are to be employed.

(a)   To determine A_i, Z_1 must first be found

$$Z_1 = h_{ie} - \frac{h_{fe}h_{re}Z_L}{1 + h_{oe}Z_L} = 1 \times 10^3 - \frac{(50)(2 \times 10^{-4})(2 \times 10^3)}{1 + (20 \times 10^{-6})(2 \times 10^3)}$$

$$= 1 \times 10^3 - \frac{20}{1.04} = 980.77 \ \Omega$$

Applying the current-divider rule:

$$I_1 = \frac{R_B(I_i)}{R_B + Z_1} = \frac{(200 \times 10^3)I_i}{200 \times 10^3 + 980.77}$$

and $I_1 = 0.995 I_i$

with $I_o = I_2$ as appearing in Fig. 7.23.

Then,

$$A_i = \frac{I_o}{I_i} = \frac{I_o}{I_1/0.995} = 0.995\left(\frac{I_o}{I_1}\right) = 0.995\left(\frac{I_2}{I_1}\right)$$

$$= 0.995\left[\frac{h_{fe}}{1 + h_{oe}Z_L}\right] = \frac{(0.995)(50)}{1 + (20 \times 10^{-6})(2 \times 10^3)}$$

$$= \frac{49.75}{1 + 40 \times 10^{-3}} = \frac{49.75}{1.040} = \mathbf{47.84}$$

(b) $$A_v = \frac{V_o}{V_i} = \frac{V_2}{V_1} = \frac{-h_{fe}Z_L}{h_{ie} + (h_{ie}h_{oe} - h_{fe}h_{re})Z_L}$$

$$= \frac{-50(2 \times 10^3)}{1 \times 10^3 + [(1 \times 10^3)(20 \times 10^{-6}) - (50)(2 \times 10^{-4})]2 \times 10^3}$$

$$= \frac{-100 \times 10^3}{1 \times 10^3 + 20} \cong \mathbf{-98.04}$$

(c) $Z_i = 200 \text{ k}\Omega \parallel Z_1 = 200 \text{ k}\Omega \parallel 980.77 \ \Omega = \mathbf{0.976 \text{ k}\Omega}$

(d) $Z_o = 2 \text{ k}\Omega \parallel Z_2$

where $$Z_2 = \frac{1}{h_{oe} - \dfrac{h_{fe}h_{re}}{h_{ie} + R_s}} = \frac{1}{20 \times 10^{-6} - \dfrac{(50)(2 \times 10^{-4})}{1 \times 10^3 + 0}}$$

$$= \frac{1}{(20 \times 10^{-6}) - (10 \times 10^{-6})} = \frac{1}{10 \times 10^{-6}} = 100 \text{ k}\Omega$$

and $Z_o = 2 \text{ k}\Omega \parallel 100 \text{ k}\Omega = \mathbf{1.961 \text{ k}\Omega}$

(e) $A_p = -A_v A_i = -(-98.04)(47.84) = \mathbf{4690.2}$

There appears in Appendix D the description of a computer program designed to solve networks of this type. The program is quite versatile in that a variety of networks can be examined by substituting (or not substituting) the integral components of each network. That is, the computer network is broader in scope and number of components than that appearing in Fig. 7.21. However, the network of Fig. 7.21 can be analyzed by simply substituting values such as R = 0 Ω or R = ∞ Ω (actually R = 1 × 10³⁰ Ω for this program since it requires a numerical value) for appropriate resistors in the system.

The following (Fig. 7.24) is the computer printout and the solution of Example 7.1. Keep in mind that the results were obtained almost instantaneously rather than going through the long arduous calculations appearing in Example 7.1. Furthermore, whole sets of new results could be obtained for different network values with a minimum of effort. The authors encourage readers to attempt the same program on the computer facilities available at their own institutions to initiate the development of this very important link with computer analysis of engineering systems.

```
THIS PROGRAM DOES AC CALCULATIONS USING THE COMPLETE
HYBRID EQUIVALENT CIRCUIT MODEL

ENTER THE FOLLOWING CIRCUIT COMPONENT VALUES:
RB1=? 200E3
RB2= (USE 1E30 IF INFINITE) ? 1E30
RC=? 2E3
```

Figure 7.24

```
RS= (UNBYPASSED VALUE ONLY) ? 0
LOAD RESISTANCE, RL= (USE 1E30 IF OUTPUT OPEN) ? 1E30

NOW ENTER VALUES FOR TRANSISTOR HYBRID EQUIVALENT CIRCUIT:
hie=? 1E3
hfe=? 50
hoe=? 20E-6
hre=? 2E-4

RESULTS OF AC CIRCUIT ANALYSIS ARE:
CURRENT GAIN, Ai= 47.842
VOLTAGE GAIN, Av= -98.039
INPUT IMPEDANCE, Zi= 975.983
OUTPUT IMPEDANCE, Zo= 1960.784
POWER GAIN, Ap= 4690.423
```

Figure 7.24   (continued)

EXAMPLE 7.2   Find the following for the network of Fig. 7.25:

(a) $A_v = V_o/V_i$.

(b) $A_i = I_o/I_i$.

(c) Z_i.

(d) Z_o.

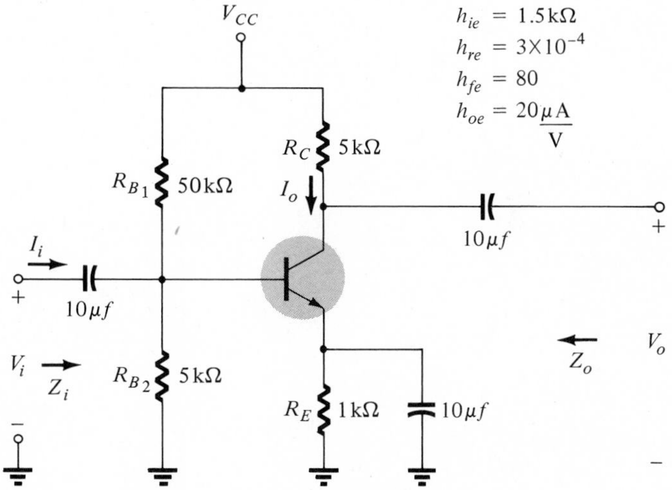

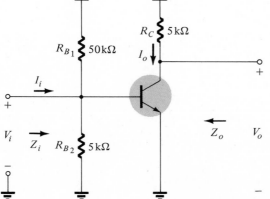

$h_{ie} = 1.5\,\text{k}\Omega$
$h_{re} = 3 \times 10^{-4}$
$h_{fe} = 80$
$h_{oe} = 20\,\mu\text{A}/\text{V}$

Figure 7.25   Circuit for Example 7.2.

Figure 7.26   Circuit of Fig. 7.25 re-drawn for small-signal ac analysis.

Solution: Replacing the dc supplies and capacitors by short circuits will result in the circuit of Fig. 7.26.

Substituting the hybrid equivalent circuit and redrawing the circuit will result in the following configuration (Fig. 7.27):

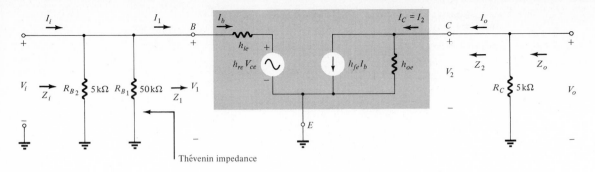

Thévenin impedance

Figure 7.27 Circuit of Fig. 7.26 following the substitution of the small-signal hybrid equivalent circuit for the transistor.

(a)

$$A_v = \frac{V_o}{V_i} = \frac{V_2}{V_1} = \frac{-h_{fe}Z_L}{h_{ie} + (h_{ie}h_{oe} - h_{fe}h_{re})Z_L}$$

$$= \frac{-80(5 \times 10^3)}{1.5 \times 10^3 + [(1.5 \times 10^3)(20 \times 10^{-6}) - (80)(3 \times 10^{-4})](5 \times 10^3)}$$

$$= \frac{-400 \times 10^3}{1.5 \times 10^3 + (6 \times 10^{-3})(5 \times 10^3)} = \frac{-400 \times 10^3}{1.5 \times 10^3 + 30}$$

$$= \frac{-400 \times 10^3}{1530} = \mathbf{-261.44}$$

(b) Z_1 is required to determine the relationship between I_i and I_1

$$Z_1 = h_{ie} - \frac{h_{fe}h_{re}Z_L}{1 + h_{oe}Z_L} = 1.5 \times 10^3 - \frac{(80)(3 \times 10^{-4})(5 \times 10^3)}{1 + (20 \times 10^{-6})(5 \times 10^3)}$$

$$= 1.5 \times 10^3 - \frac{120}{1 + 0.1} = 1.5 \times 10^3 - \frac{120}{1.1}$$

$$= 1.5 \times 10^3 - 109.09$$

$$Z_1 \cong 1391 \ \Omega$$

The current-divider rule can then be applied to the equivalent circuit of Fig. 7.28.

$$I_1 = \frac{4.55 \ \text{k}\Omega \ I_i}{4.55 \ \text{k}\Omega + 1.391 \ \text{k}\Omega} \cong 0.766 \ I_i$$

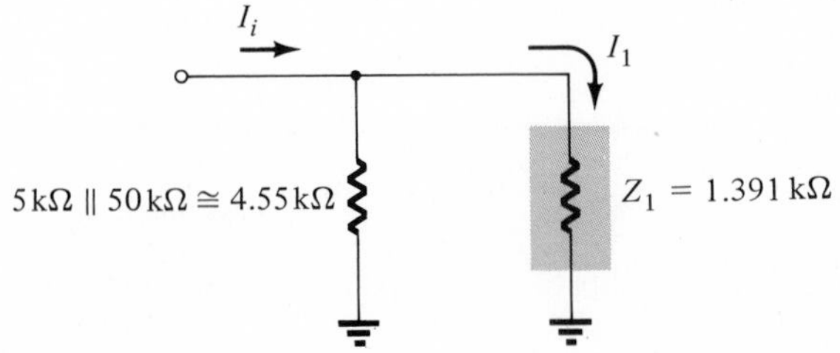

$5 \text{k}\Omega \parallel 50 \text{k}\Omega \cong 4.55 \text{k}\Omega$ $Z_1 = 1.391 \text{ k}\Omega$

Figure 7.28 Determining the relationship between I_1 and I_i for the circuit of Fig. 7.27.

CH. 7   BJT SMALL-SIGNAL ANALYSIS

The current gain

$$A_i = \frac{I_o}{I_i} = \frac{I_2}{I_i} = \left[\frac{I_1}{I_i}\right]\left[\frac{I_2}{I_1}\right] = [0.766]\left[\frac{h_{fe}}{1 + h_{oe}R_L}\right]$$

$$= [0.766]\left[\frac{80}{1 + (20 \times 10^{-6})(5 \times 10^3)}\right] = [0.766]\left[\frac{80}{1 + 0.1}\right]$$

$$= \frac{61.28}{1.1} \cong \mathbf{55.71}$$

(c) $Z_i = 5 \text{ k}\Omega \parallel 50 \text{ k}\Omega \parallel Z_1 = 4.55 \text{ k}\Omega \parallel 1.391 \text{ k}\Omega \cong 1.065 \text{ k}\Omega$

(d) R_s, as defined by Fig. 7.20, is required to determine Z_2. As in the previous example, it is simply the Thévenin impedance indicated in Fig. 7.27 and redrawn in Fig. 7.29.

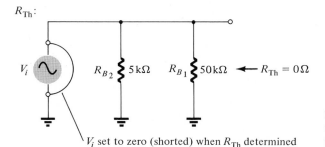

V_i set to zero (shorted) when R_{Th} determined

Figure 7.29 Determining R_{Th} for the portion of the circuit indicated in Fig. 7.27.

$$R_s = R_{Th} = 0$$

so that

$$Z_2 = \frac{1}{h_{oe} - \dfrac{h_{fe}h_{re}}{h_{ie} + R_s}} = \frac{1}{20 \times 10^{-6} - \dfrac{80(3 \times 10^{-4})}{1.5 \times 10^3}}$$

$$= \frac{1}{20 \times 10^{-6} - 16 \times 10^{-6}} = \frac{1}{4 \times 10^{-6}}$$

$$Z_2 = 250 \text{ k}\Omega$$

with

$$Z_o = Z_2 \parallel 5 \text{ k}\Omega = 250 \text{ k}\Omega \parallel 5 \text{ k}\Omega = \mathbf{4.90 \text{ k}\Omega}$$

The versatility of the program appearing in Appendix D and applied to the network of Example 7.1 is now demonstrated by applying it to the transistor network of Example 7.2. Note that it requires only that the elements to appear be defined and assigned the proper value. The results as appearing in Fig. 7.30 agree well with Example 7.2.

```
THIS PROGRAM DOES AC CALCULATIONS USING THE COMPLETE
HYBRID EQUIVALENT CIRCUIT MODEL

ENTER THE FOLLOWING CIRCUIT COMPONENT VALUES:
RB1=? 50E3
RB2= (USE 1E30 IF INFINITE) ? 5E3
RC=? 5E3
RS= (UNBYPASSED VALUE ONLY) ? 0
LOAD RESISTANCE, RL= (USE 1E30 IF OUTPUT OPEN) ? 1E30
```
Figure 7.30

NOW ENTER VALUES FOR TRANSISTOR HYBRID EQUIVALENT CIRCUIT:
hie=? 1.5E3
hfe=? 80
hoe=? 20E-6
hre=? 3E-4

RESULTS OF AC CIRCUIT ANALYSIS ARE:
CURRENT GAIN, Ai= 55.687
VOLTAGE GAIN, Av= -261.438
INPUT IMPEDANCE, Zi= 1065.015
OUTPUT IMPEDANCE, Zo= 4901.961
POWER GAIN, Ap= 14558.703

Figure 7.30 (continued)

EXAMPLE 7.3 Find the following quantities for the common-base configuration of Fig. 7.31:

(a) $A_i = I_o/I_i$.

(b) $A_{v1} = V_o/V_i$.

(c) Z_i.

(d) Z_o.

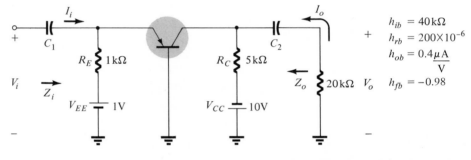

Figure 7.31 Network for Example 7.3.

Solution: Replacing the dc supplies and capacitors by short circuits will result in the following configuration (Fig. 7.32).

(a) The equivalent load $Z_L = R_L = 5 \text{ k}\Omega \parallel 20 \text{ k}\Omega = 4 \text{ k}\Omega$

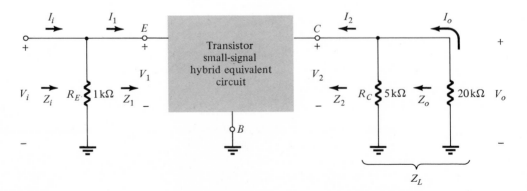

Figure 7.32 Circuit of Fig. 7.31 redrawn for small-signal ac analysis.

and
$$Z_1 = h_{ib} - \frac{h_{fb}h_{rb}\,Z_L}{1 + h_{ob}\,Z_L} = 40 - \frac{(-0.98)(200 \times 10^{-6})(4 \times 10^3)}{1 + (0.4 \times 10^{-6})(4 \times 10^3)}$$

$$= 40 - \frac{(-784 \times 10^{-3})}{1 + 1.6 \times 10^{-3}} = 40 + \frac{784 \times 10^{-3}}{1.0016}$$

$$Z_1 \cong 40.78 \ \Omega$$

The relationship between I_1 and I_i can then be determined by applying the current divider to the equivalent circuit of Fig. 7.33.

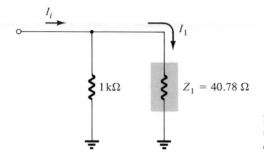

Figure 7.33 Determining the relationship between I_1 and I_i for the circuit of Fig. 7.32.

$$I_1 = \frac{1 \text{ k}\Omega \ (I_i)}{1 \text{ k}\Omega + 40.78 \ \Omega} = \frac{1 \text{ k}\Omega \ (I_i)}{1040.78} \cong 0.96 \ I_i$$

and the relationship between I_o and I_2 (Fig. 7.32).

Current-divider rule:

$$I_o = \frac{5 \text{ k}\Omega \ (I_2)}{5 \text{ k}\Omega + 20 \text{ k}\Omega} = 0.2 \ I_2$$

so that

$$A_i = \frac{I_o}{I_i} = \left[\frac{I_o}{I_2}\right]\left[\frac{I_2}{I_i}\right] = \left[\frac{I_o}{I_2}\right]\left[\frac{I_1}{I_i}\right]\left[\frac{I_2}{I_1}\right]$$

$$= [0.2]\,[0.96]\left[\frac{h_{fb}}{1 + h_{ob}\,Z_L}\right] = [0.192]\left[\frac{-0.98}{1 + (0.4 \times 10^{-6})(4 \times 10^3)}\right]$$

$$= \frac{-0.188}{1 + 1.6 \times 10^{-3}} = \frac{-0.188}{1.0016} \cong \mathbf{-0.188}$$

(b) $A_{v1} = \dfrac{V_o}{V_i} = \dfrac{V_2}{V_1} = \dfrac{-h_{fb}\,Z_L}{h_{ib} + (h_{ib}h_{ob} - h_{fb}h_{rb})Z_L}$

$$= \frac{-(-0.98)(4 \times 10^3)}{40 + [(40)(0.4 \times 10^{-6}) - (-0.98)(200 \times 10^{-6})]4 \times 10^3}$$

$$= \frac{3.92 \times 10^3}{40 + (16 \times 10^{-6} + 198 \times 10^{-6})(4 \times 10^3)}$$

$$= \frac{3.92 \times 10^3}{40 + (214 \times 10^{-6})(4 \times 10^3)} = \frac{3.92 \times 10^3}{40 + 0.856} = \mathbf{96}$$

(c) $Z_i = 1 \text{ k}\Omega \parallel Z_1 = 1 \text{ k}\Omega \parallel 40.78 \ \Omega \cong \mathbf{39.2 \ \Omega}$

(d) $\quad Z_2 = \dfrac{1}{h_{ob} - \dfrac{h_{fb}h_{rb}}{h_{ib} + R_s}} = \dfrac{1}{0.4 \times 10^{-6} - \dfrac{(-0.98)(200 \times 10^{-6})}{40 + 0}}$

$\quad\quad = \dfrac{1}{0.4 \times 10^{-6} - \dfrac{-196 \times 10^{-6}}{40}} = \dfrac{1}{0.4 \times 10^{-6} + 4.9 \times 10^{-6}}$

$\quad\quad = \dfrac{1}{5.3 \times 10^{-6}}$

$\quad Z_2 \cong 189 \text{ k}\Omega$

and $\quad\quad\quad\quad Z_o = 5 \text{ k}\Omega \| Z_2 = 5 \text{ k}\Omega \| 189 \text{ k}\Omega \cong \mathbf{4.87 \text{ k}\Omega}$

7.8 APPROXIMATIONS FREQUENTLY APPLIED WHEN USING THE HYBRID EQUIVALENT CIRCUIT AND ITS RELATED EQUATIONS

On many occasions an exact theoretical solution requiring the use of the complete hybrid equivalent circuit is not required. Rather, an approximate solution, requiring considerably less time, is often desirable. In fact, considering that the h-parameters of a particular type of transistor will vary, even if only slightly, from transistor to transistor of the same manufacturing series with the operating point and with temperature, an approximate solution will often be as "accurate" as one involving all the parameters. To derive and demonstrate the validity of some of the approximations to be made, the transistor parameters indicated in Table 7.1 will be used. The common-emitter parameters are repeated here for convenience.

$$h_{fe} = 50, \quad h_{ie} = 1000 \ \Omega, \quad h_{re} = 2.5 \times 10^{-4}, \quad h_{oe} = 25 \ \mu\text{A/V} = 25 \ \mu\text{S}$$

For the purposes of discussion we shall assume that the impedance of the source V_s is 1 kΩ and that the load Z_L is 2 kΩ (resistive).

The Current Gain A_i

$$A_i = \frac{h_{fe}}{(1 + h_{oe}Z_L)} \quad \text{[Eq. (7.10)]}$$

Substituting values:

$$(1 + h_{oe}Z_L) = [1 + (25 \times 10^{-6})(2 \times 10^3)] = (1 + 50 \times 10^{-3})$$
$$= (1 + 0.05) \cong 1$$

and $\quad\quad\quad\quad A_i = \dfrac{h_{fe}}{(1 + h_{oe}Z_L)} \cong \dfrac{50}{1} \cong 50$

resulting in $\quad\quad\quad \boxed{A_i \cong h_{fe}} \quad\quad\quad\quad\quad\quad (7.19)$

[Note that the term $h_{oe}Z_L$ in Eq. (7.10) is negligible because of the small value of h_{oe}.]

The Voltage Gain A_v

$$A_v = \frac{-h_{fe} Z_L}{h_{ie} + (h_{ie}h_{oe} - h_{fe}h_{re})Z_L} \qquad \text{[Eq. (7.11)]}$$

Substituting values:

$$(h_{ie}h_{oe} - h_{fe}h_{re}) = [(1 \times 10^3)(25 \times 10^{-6}) - (50)(2.5 \times 10^{-4})]$$
$$= (25 \times 10^{-3} - 125 \times 10^{-4})$$
$$= 125 \times 10^{-4}$$

and $\qquad h_{ie} + (h_{ie}h_{oe} - h_{fe}h_{re})Z_L = 1000 + (125 \times 10^{-4})(2 \times 10^3)$
$$= 1000 + 25 \cong 1000 = h_{ie}$$

and $$A_v \cong \frac{-50(2 \times 10^3)}{1 \times 10^3} = -100$$

so that

$$\boxed{A_v \cong \frac{-h_{fe}}{h_{ie}} Z_L} \qquad (7.20)$$

due to the fact that the term $Z_L(h_{ie}h_{oe} - h_{fe}h_{re})$ is negligible compared to h_{ie} in Eq. (7.11) because of the small values of h_{oe} and h_{re}.

The Input Impedance Z_1

$$Z_1 = h_{ie} - \frac{h_{fe}h_{re} Z_L}{1 + h_{oe} Z_L} \qquad \text{[Eq. (7.12)]}$$

Substituting values in Eq. (7.12):

$$h_{fe}h_{re} Z_L = (50)(2.5 \times 10^{-4})(2 \times 10^3) = 25$$

and $$\frac{h_{fe}h_{re} Z_L}{(1 + h_{oe} Z_L) \cong 1} \cong \frac{25}{1} = 25$$

and $$Z_1 = h_{ie} - \frac{h_{fe}h_{re} Z_L}{1 + h_{oe} Z_L} = 1000 - 25 \cong 1000 = h_{ie}$$

so that

$$\boxed{Z_1 \cong h_{ie}} \qquad (7.21)$$

The Output Impedance Z_2

$$Z_2 = \frac{1}{h_{oe} - \dfrac{h_{fe}h_{re}}{h_{ie} + R_s}} \qquad \text{[Eq. (7.13)]}$$

Equation (7.13) rewritten in a slightly different form:

$$Z_2 = \frac{h_{ie} + R_s}{h_{oe}(h_{ie} + R_s) - h_{fe}h_{re}}$$

Substituting values:

$$h_{oe}(h_{ie} + R_s) - h_{fe}h_{re} = 25 \times 10^{-6}(1000 + 1000) - 50(2.5 \times 10^{-4})$$
$$= 50 \times 10^{-3} - 12.5 \times 10^{-3}$$

For Z_2 the order of magnitude of $h_{oe}(h_{ie} + R_s)$ is too close to $h_{fe}h_{re}$ to warrant dropping the smaller quantity. Therefore, there is no clear-cut approximation that can be made for Z_2. Frequently, however, the relationship

$$\boxed{Z_2 > \frac{1}{h_{oe}}} \qquad (7.22)$$

is used to obtain some idea of the order of magnitude for the output impedance.

Power Gain A_p

$$A_p = \frac{A_i^2 R_L}{Z_i}$$

Through substitution in Eq. (7.18) of the complete expressions for A_i and Z_i and the typical values indicated for the parameters at the beginning of this section it can be shown that

$$A_p \cong \frac{h_{fe}^2 R_L}{h_{ie}} \qquad (7.23)$$

In summary, the exact equation and the frequently used approximate form for each quantity discussed above are presented in Table 7.2.

TABLE 7.2   Exact vs. Approximate Equations for the Quantities of Interest for the Basic Transistor Amplifier

Quantity	Exact	Approximate
A_i	$A_i = \dfrac{h_{fe}}{1 + h_{oe}Z_L}$	$A_i = h_{fe}$
A_v	$A_v = \dfrac{-h_{fe}Z_L}{h_{ie} + (h_{ie}h_{oe} - h_{fe}h_{re})Z_L}$	$A_v = \dfrac{-h_{fe}Z_L}{h_{ie}}$
Z_1	$Z_1 = h_{ie} - \dfrac{h_{fe}h_{re}Z_L}{1 + h_{oe}Z_L}$	$Z_1 = h_{ie}$
Z_2	$Z_2 = \dfrac{1}{h_{oe} - \dfrac{h_{fe}h_{re}}{h_{ie} + R_s}}$	$Z_2 > \dfrac{1}{h_{oe}}$
A_p	$A_p = \dfrac{A_i^2 R_L}{R_i}$	$A_p = \dfrac{h_{fe}^2 R_L}{h_{ie}}$

For comparison purposes the exact and approximate values for each quantity using the h-parameters employed in the above derivations have been listed in Table 7.3.

**TABLE 7.3 Exact vs. Approximate Results
for a Transistor Amplifier Having
the Parameter Values of Table 7.1**

Quantity	Exact	Approximate
A_i	47.62	50
A_v	-97.5	-100
Z_1	975 Ω	1000 Ω
Z_2	53.3 kΩ	$Z_2 > 40$ kΩ
A_p	4650	5000

Let us now turn our attention to the hybrid equivalent circuit and note the effect of these approximations on the circuit itself. In all cases, except for Z_2, the hybrid parameter h_{re} does not appear in the approximate equation (Table 7.2) for each quantity. On this basis, the voltage-controlled source $h_{re}V_2$ can be replaced by a short circuit whenever approximate values of A_i, A_v, and Z_i are desired. As far as Z_2 is concerned, we shall simply keep in mind that whenever $h_{re}V_2$ is eliminated, the resulting Z_2 will be somewhat smaller than the actual value. Removing $h_{re}V_2$ from the complete hybrid equivalent circuit will result in the approximate form of Fig. 7.34. For the circuit just discussed

$$\frac{1}{h_{oe}} = \frac{1}{25 \times 10^{-6}} = 40 \text{ k}\Omega$$

and $Z_L = 2$ kΩ. The parallel combination of the two:

$$40 \text{ k}\Omega \parallel 2 \text{ k}\Omega = \frac{40 \text{ k}\Omega \times 2 \text{ k}\Omega}{40 \text{ k}\Omega + 2 \text{ k}\Omega} = \frac{80 \text{ k}\Omega}{42 \text{ k}\Omega} = 1.91 \text{ k}\Omega \cong 2 \text{ k}\Omega = Z_L$$

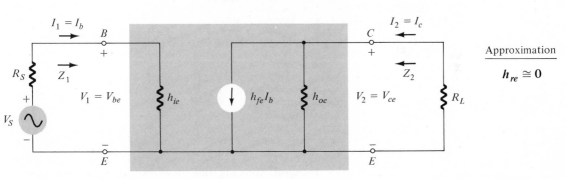

Figure 7.34 Circuit of Fig. 7.20 following the application of the approximation: $h_{re} \cong 0$.

For a number of applications the parallel combination of $1/h_{oe}$ and Z_L will result in an equivalent resistance close in magnitude to the load resistance. On this basis, the parameter h_{oe} can be eliminated as a second approximation from the circuit of Fig. 7.34, resulting in the reduced form of the hybrid equivalent circuit in Fig. 7.35.

The circuit of Fig. 7.35 would certainly require fewer calculations to obtain such quantities as the output voltage and current than the complete hybrid equivalent circuit. The conditions associated with the circuit of Fig. 7.35 must always be considered before it can be applied.

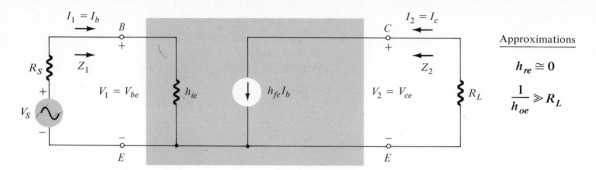

Figure 7.35 Circuit of Fig. 7.34 following the application of: $1/h_{oe} \gg R_L$.

In the examples that follow we shall find that since h_{re} and h_{oe} have been removed from the equivalent circuit, $Z_2 = \infty \ \Omega$ (open circuit). At first glance, this might appear to be a completely unacceptable approximation. We must keep in mind, however, that in this situation it is simply stating that compared to the other circuit parameters Z_2 is *sufficiently large* to be considered an open circuit.

There is one further approximation that can sometimes be made and that will certainly lead us to accept the fact that the transistor is a current-controlled device. In some cases, the resistance R_s, which was defined to be the "total" resistance of the source impedance and any resistance added in series with the source to limit the input current, is larger in magnitude than h_{ie}. On this basis, if we drop h_{ie} from the picture, the result is Fig. 7.36.

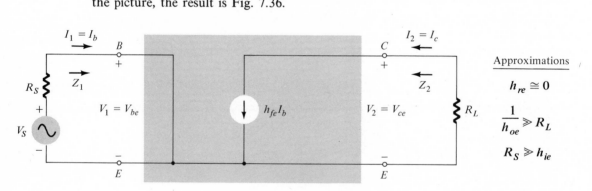

Figure 7.36 Circuit of Fig. 7.35 following the application of: $R_s \gg h_{ie}$.

It is obvious from Fig. 7.36 that the only link between the input and output circuits is a current source whose magnitude is controlled by the base current I_b. It would appear, however, that the number of conditions that must be satisfied before the circuit can be employed would substantially limit its use in transistor circuit analysis. It must be kept in mind, however, that in many manuals h_{fe} is the *only* parameter provided. This would require the use of the approximate equivalent circuit of Fig. 7.36 if some idea of the circuit's response is to be obtained.

EXAMPLE 7.4 Repeat Example 7.1 using the approximate equations and compare results.

Solution: ($h_{fe} = 50$, $h_{re} = 2 \times 10^{-4}$, $h_{oe} = 20 \times 10^{-6} \ \mu A/V$, $h_{ie} = 1 \ k\Omega$)

(a) $A_i \cong h_{fe} = \mathbf{50}$ vs. 47.84 obtained in Example 7.1.

(b) $A_v \cong -h_{fe}Z_L/h_{ie} = -50(2 \times 10^3)/1 \text{ k}\Omega = \mathbf{-100}$ vs. -98.04 obtained in Example 7.1.

(c) $Z_i = 200 \text{ k}\Omega \parallel Z_1 \cong 200 \text{ k}\Omega \parallel h_{ie} \cong h_{ie} = \mathbf{1} \text{ k}\Omega$ vs. 0.976 kΩ obtained in Example 7.1.

(d) $Z_2 > 1/h_{oe} = 1/20 \times 10^{-6} = \mathbf{50} \text{ k}\Omega$

and $Z_o = 2 \text{ k}\Omega \parallel > 50 \text{ k}\Omega \cong \mathbf{2} \text{ k}\Omega$ vs. 1.961 kΩ obtained in Example 7.1.

(e) $A_p = A_v A_i \cong (-100)(50) = \mathbf{5} \text{ Â } \mathbf{10^3}$ vs. 4690.2 obtained in Example 7.1.

In each case it is obvious that the approximate solution is in the "ball park" if a rough idea of a system's response or characteristics is required with a minimum of time and effort.

From Example 7.4 we can conclude for the fixed-bias circuit with typical values that

$$A \cong h_{fe}, \qquad A_v \cong \frac{-h_{fe}Z_L}{h_{ie}}, \qquad Z_i \cong h_{ie}, \qquad Z_2 > \frac{1}{h_{oe}}$$

and $Z_o \cong R_L$.

Applying the approximations above to the program of Appendix D (Fig. D.3) will result in the printout of Fig. 7.37. The slight differences in results are due to some of the parallel-resistor approximations that are not part of the computer program.

```
RUN

THIS PROGRAM DOES AC CALCULATIONS USING THE COMPLETE
HYBRID EQUIVALENT CIRCUIT MODEL

ENTER THE FOLLOWING CIRCUIT COMPONENT VALUES:
RB1=? 200E3
RB2= (USE 1E30 IF INFINITE) ? 1E30
RC=? 2E3
RS= (UNBYPASSED VALUE ONLY) ? 0
LOAD RESISTANCE, RL= (USE 1E30 IF OUTPUT OPEN) ? 1E30

NOW ENTER VALUES FOR TRANSISTOR HYBRID EQUIVALENT CIRCUIT:
hie=? 1E3
hfe=? 50
hoe=? 0
hre=? 0

RESULTS OF AC CIRCUIT ANALYSIS ARE:
CURRENT GAIN, Ai=  49.751
VOLTAGE GAIN, Av= -100
INPUT IMPEDANCE, Zi=  995.025
OUTPUT IMPEDANCE, Zo=  2000
POWER GAIN, Ap=  4975.124
```

Figure 7.37

EXAMPLE 7.5 Repeat Example 7.2 using the approximate equivalent circuit and compare results.

Solution: Figure 7.25 will appear as shown in Fig. 7.38 following the substitution of the approximate equivalent circuit.

h_{oe} does not appear since

$$\frac{1}{h_{oe}} = \frac{1}{20 \text{ } \mu\text{A/V}} = 50 \text{ k}\Omega \text{ and } 50 \text{ k}\Omega \parallel 5 \text{ k}\Omega \cong 5 \text{ k}\Omega \text{ (10:1 ratio)}$$

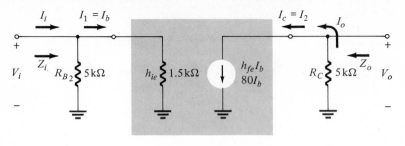

Figure 7.38 Approximate equivalent circuit for the network of Fig. 7.25.

Also R_{B_1} does not appear since $R_{B_1} \| R_{B_2} = 50 \text{ k}\Omega \| 5 \text{ k}\Omega \cong 5 \text{ k}\Omega$ (10:1 ratio)

(a) A_v:

$$V_o = -I_o R_c = -h_{fe} I_b R_c$$

and

$$I_b = \frac{V_i}{h_{ie}}$$

so that

$$V_o = -h_{fe}\left(\frac{V_i}{h_{ie}}\right) R_c$$

and

$$A_v = \frac{V_o}{V_i} = \frac{-h_{fe}}{h_{ie}} R_c$$

as obtained for the fixed-bias network in Example 7.4.

Substituting values:

$$A_v = \frac{-(80)}{1.5 \text{ k}\Omega}(5 \text{ k}\Omega) \cong -267$$

as compared to 261.44 obtained earlier using the complete model.

(b) A_i:

$$I_o = h_{fe} I_b$$

Current-divider rule:

$$I_b = \frac{5 \text{ k}\Omega (I_i)}{5 \text{ k}\Omega + 1.5 \text{ k}\Omega} = 0.769 I_i$$

and

$$I_o = h_{fe}(0.769 I_i) = (80)(0.769) I_i = 61.52 I_i$$

with

$$A_i = \frac{I_o}{I_i} \cong \mathbf{61.52}$$

as compared to 55.71 obtained earlier.

(c) Z_i:

$$Z_i \cong 5 \text{ k}\Omega \| 1.5 \text{ k}\Omega = \mathbf{1.15 \text{ k}\Omega}$$

as compared to 1.065 kΩ obtained earlier.

(d) Z_o: The output impedance is defined by the condition $V_i = 0$. Therefore,

$$I_b = \frac{V_i}{h_{ie}} = 0 \quad \text{and} \quad h_{fe} I_b = 0 \qquad \text{(an open-circuit equivalent)}$$

and

$$Z_o = \mathbf{5 \text{ k}\Omega}$$

as compared to 4.90 kΩ obtained earlier. Take a moment to compare the degree of effort to obtain the results above as compared to using the complete model.

```
RUN

THIS PROGRAM DOES AC CALCULATIONS USING THE COMPLETE
HYBRID EQUIVALENT CIRCUIT MODEL

ENTER THE FOLLOWING CIRCUIT COMPONENT VALUES:
RB1=? 50E3
RB2= (USE 1E30 IF INFINITE) ? 5E3
RC=? 5E3
RS= (UNBYPASSED VALUE ONLY) ? 0
LOAD RESISTANCE, RL= (USE 1E30 IF OUTPUT OPEN) ? 1E30

NOW ENTER VALUES FOR TRANSISTOR HYBRID EQUIVALENT CIRCUIT:
hie=? 1.5E3
hfe=? 80
hoe=? 0
hre=? 0

RESULTS OF AC CIRCUIT ANALYSIS ARE:
CURRENT GAIN, Ai=  60.15
VOLTAGE GAIN, Av= -266.667
INPUT IMPEDANCE, Zi=  1127.82
OUTPUT IMPEDANCE, Zo=  5000
POWER GAIN, Ap=  16040.1
```

Figure 7.39

Applying the approximations above to the program of Appendix D (Fig. D.3) will result in the printout of Fig. 7.39.

EXAMPLE 7.6   Determine A_v, A_i, Z_i, and Z_o for the configuration appearing in Fig. 7.40 using the approximate equivalent circuit. Conditions are such that the effect of h_{re} and h_{oe} can be ignored.

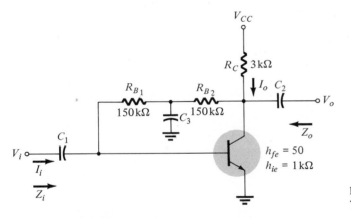

Figure 7.40   Network for Example 7.6.

Solution:
(a)   Connecting the dc supplies to ground potential and replacing all capacitors by a short-circuit equivalent will result in the network of Fig. 7.41.

Redrawing the circuit, we obtain Fig. 7.42 and, finally, Fig. 7.43.

From Fig. 7.43

$$V_o = -I_o R_L = -h_{fe} I_b R_L$$

but

$$I_b = \frac{V_i}{h_{ie}}$$

and

$$V_o = -h_{fe}\left(\frac{V_i}{h_{ie}}\right) R_L$$

with

$$A_v = \frac{V_o}{V_i} = \frac{-h_{fe}}{h_{ie}} R_L$$

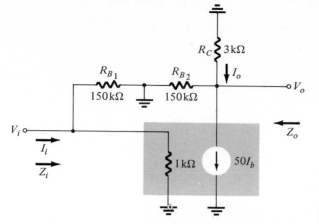

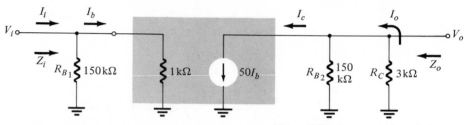

Figure 7.41 Network of Fig. 7.40 following the substitution of the approximate equivalent circuit and removing the dc levels.

Figure 7.42 Reconstruction of Fig. 7.41.

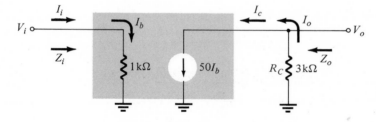

Figure 7.43 Network of Fig. 7.42 following the application of additional approximations.

as obtained for Examples 7.4 and 7.5.
Substituting numbers:

$$A_v = \frac{-50(3 \text{ k}\Omega)}{1 \text{ k}\Omega} = -150$$

(b) A_i: From Fig. 7.43

$$I_i = I_b$$

and
$$I_o = h_{fe}I_b$$
with result that

$$A_i = \frac{I_o}{I_i} = \frac{h_{fe}I_b}{I_b} = h_{fe} = 50$$

(c) Z_i: From Fig. 7.43

$$Z_i = h_{ie} = 1 \text{ k}\Omega$$

(d) Z_o: From Fig. 7.43

$$Z_o|_{v_i=0} = R_L = 3 \text{ k}\Omega$$

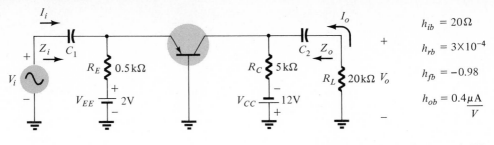

$h_{ib} = 20\,\Omega$

$h_{rb} = 3\times10^{-4}$

$h_{fb} = -0.98$

$h_{ob} = 0.4\,\mu\text{A}/V$

Figure 7.44 Circuit for Example 7.7.

EXAMPLE 7.7 For the common-base circuit of Fig. 7.44 with a load applied, find the following:

(a) $A_i = I_o/I_i$.
(b) $A_v = V_o/V_i$.
(c) Z_i.
(d) Z_o.

Solution: (a) Replacing the dc supplies and coupling capacitors by short circuits will result in the configuration of Fig. 7.45.

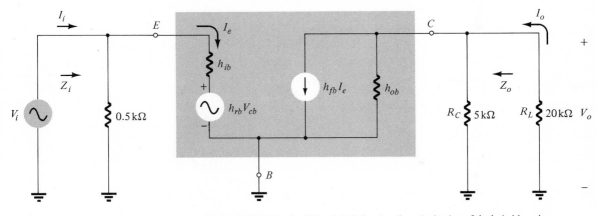

Figure 7.45 Circuit of Fig. 7.44 following the substitution of the hybrid equivalent circuit.

Note that the common-base equivalent circuit has exactly the same format as the common-emitter configuration but with the appropriate common-base parameters. Consider that now $h_{re}V_{ce}$ is $h_{rb}V_{cb}$ and $h_{fe}I_b$ is now $h_{fb}I_e$. Since the configuration is the same and $1/h_{ob}$ is greater than $1/h_{oe}$ with h_{re} usually close in magnitude to h_{rb}, the approximations ($1/h_{ob} \cong \infty\ \Omega$ and $h_{rb} \cong 0$) made for the common-emitter configuration are applicable here, also.

Applying $h_{rb} \cong 0$, we redraw the circuit (Fig. 7.46). Eliminating the 500-Ω and 2500-kΩ ($1/h_{ob}$) resistors due to the lower parallel resistor will result in the circuit of Fig. 7.47.

$$I_o = \frac{(5\ \text{k}\Omega)\,(I_2)}{5\ \text{k}\Omega + 20\ \text{k}\Omega} = 0.2I_2 \qquad \text{(current-divider rule)}$$

and $\qquad I_2 = h_{fb}I_e = h_{fb}I_1 = h_{fb}I_i$ with A_i (transistor) $= \dfrac{I_2}{I_i} = h_{fb}$

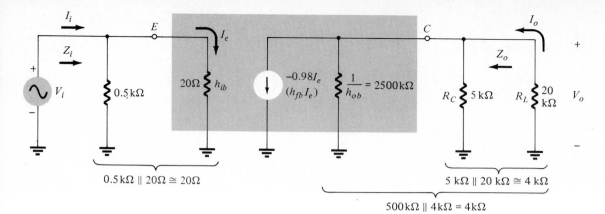

$$0.5\,\text{k}\Omega \parallel 20\Omega \cong 20\Omega$$

$$5\,\text{k}\Omega \parallel 20\,\text{k}\Omega \cong 4\,\text{k}\Omega$$

$$500\,\text{k}\Omega \parallel 4\,\text{k}\Omega = 4\,\text{k}\Omega$$

Figure 7.46 Redrawn circuit of Fig. 7.45 following the application of the approximation $h_{rb} \cong 0$.

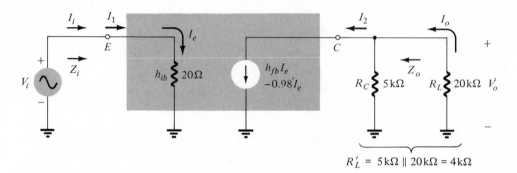

$$R_L' = 5\,\text{k}\Omega \parallel 20\,\text{k}\Omega = 4\,\text{k}\Omega$$

Figure 7.47 Circuit of Fig. 7.46 following the elimination (on an approximate basis) of certain parallel elements.

Therefore,

$$I_o = 0.2 I_2 = 0.2 h_{fb} I_i$$

and

$$A_i = \frac{I_o}{I_i} \cong 0.2\,(h_{fe}) = 0.2(-0.98) = \mathbf{-0.196}$$

indicating as before that the current gain of the common-base configuration is always less than one.

(b) $A_v = \dfrac{V_o}{V_i}$:

Using

$$R_L' = R_C \parallel R_L = 5\,\text{k}\Omega \parallel 20\,\text{k}\Omega$$

$$V_o = -I_2 R_L' = -h_{fb} I_e R_L'$$

and

$$I_e = I_i = \frac{V_i}{h_{ib}}$$

so that

$$V_o = -h_{fb}\left(\frac{V_i}{h_{ib}}\right) R_L'$$

with

$$A_v = \frac{V_o}{V_i} = \frac{-h_{fb}}{h_{ib}} R_L'$$

Note the similarities between this equation for voltage gain and that obtained

CH. 7 BJT SMALL-SIGNAL ANALYSIS

for the common-emitter configurations. Consider also the effect of the added load was only to change R_L to R'_L which is the parallel combination of R_C and the applied load R_L.

Substituting values:

$$A_v = -\frac{(-0.98)(4 \times 10^3)}{20} = \mathbf{196}$$

The voltage gain, therefore, can be significantly greater than one and results in an output that is *in phase* with the input (note the absence of the minus sign in the result).

(c) $Z_i \cong h_{ib} = \mathbf{20\ \Omega}$.

(d) $Z_o|_{v_{i=0}} \cong R_C = \mathbf{5\ k\Omega}$

EXAMPLE 7.8 Find the following for the circuit of Fig. 7.48:

(a) $A_i = I_o/I$.

(b) $A_v = V_o/V_i$.

(c) Z_i.

(d) Z_o.

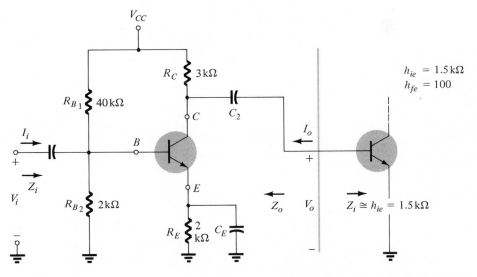

Figure 7.48 Circuit for Example 7.8.

Solution: Replacing the dc supplies and capacitors by short circuits and substituting the appropriate hybrid equivalent circuit will result in the configuration of Fig. 7.49. Note that the 2-kΩ emitter resistor has been "shorted out" by the capacitor C_E.

Considering parallel elements will result in the configuration of Fig. 7.50.

(a) $A_i = (I_o/I_i)$ From Fig. 7.50 $I_2 = 100I_b$. That is, A_i (transistor) $= I_c/I_b = h_{fe}$. Applying the current-divider rule to the input and output circuits:

$$I_b = \frac{2\ \text{k}\Omega\ I_i}{2\ \text{k}\Omega + 1.5\ \text{k}\Omega} = 0.571I_i$$

and

$$I_o = \frac{3\ \text{k}\Omega_2}{3\ \text{k}\Omega + 1.5\ \text{k}\Omega} = 0.667I_2$$

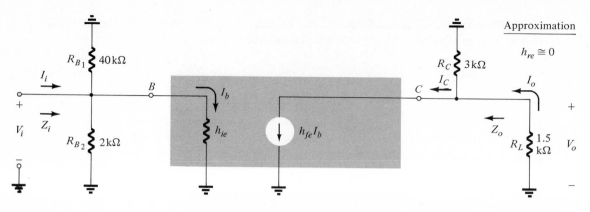

Figure 7.49 Circuit of Fig. 7.48 following the substitution of the approximate ($h_{re} \cong 0$) hybrid equivalent circuit.

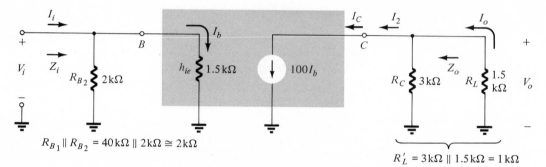

$R_{B_1} \parallel R_{B_2} = 40\,\text{k}\Omega \parallel 2\,\text{k}\Omega \cong 2\,\text{k}\Omega$

$R'_L = 3\,\text{k}\Omega \parallel 1.5\,\text{k}\Omega = 1\,\text{k}\Omega$

Figure 7.50 Circuit of Fig. 7.49 following the elimination (on an approximate basis) of certain parallel elements.

Substituting, we get

$$A_i = \frac{I_o}{I_i} = \left[\frac{I_o}{I_2}\right]\left[\frac{I_2}{I_i}\right] = \left[\frac{I_o}{I_2}\right]\left[\frac{I_2}{I_b}\right]\left[\frac{I_b}{I_i}\right]$$

$$= [0.667][100][0.571]$$
$$= \mathbf{38.1}$$

(b) $A_v = (V_o/V_i)$: The configuration is similar to that obtained for both the CE and CB configurations and

$$A_v = \frac{V_o}{V_i} = \frac{-h_{fe}R'_L}{h_{ie}}$$

$$= -\frac{(100)(1 \times 10^3)}{1.5 \times 10^3}$$

$$= \mathbf{-66.7}$$

(c) $Z_i \cong R_{B2} \parallel h_{ie} = 2\ \text{k}\Omega \parallel 1.5\ \text{k}\Omega = \mathbf{0.86\ k\Omega}$
(d) $Z_o|_{V_i=0} \cong R_C = \mathbf{3\ k\Omega}$

CH. 7 BJT SMALL-SIGNAL ANALYSIS

7.9 APPROXIMATE BASE, COLLECTOR, AND EMITTER EQUIVALENT CIRCUITS

In the following analysis it will prove very useful to know at a glance what the effect of loads and signal sources in another portion of the network will have on the base, collector, or emitter potential and current. In this section we shall find, on an approximate basis, the equivalent circuit "seen" looking into the base, collector, or emitter terminals of a transistor in the common-emitter configuration (Fig. 7.51).

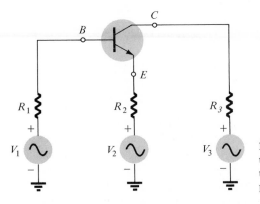

Figure 7.51 Representative circuit to be employed in the derivation of the base, collector, and emitter approximate equivalent circuits.

The approximations $h_{re} \cong 0$ and $(1/h_{oe}) \cong \infty \ \Omega$ (open circuit), employed continually in Section 7.8, will be used throughout this section (they are valid only if $1/h_{oe} > R_3, R_2$). To include the effects of a supply and a load connected to each terminal the circuit of Fig. 7.51 will be employed. The full benefit of the circuits to be derived will be more apparent when the equivalent circuits have been obtained.

Substituting the approximate equivalent circuit for the transistor of Fig. 7.51 will result in the circuit of Fig. 7.52.

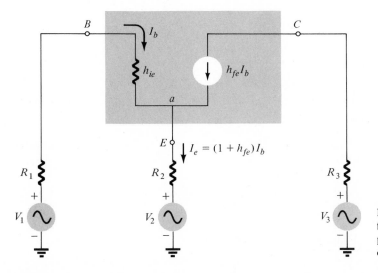

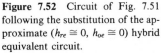

Figure 7.52 Circuit of Fig. 7.51 following the substitution of the approximate ($h_{re} \cong 0$, $h_{oe} \cong 0$) hybrid equivalent circuit.

Base Circuit

The current through the resistor R_2 can be found by applying Kirchhoff's current law at node a:

$$I_{R_2} = I_b + h_{fe}I_b = (1 + h_{fe})I_b$$

Applying Kirchhoff's voltage law in the loop indicated in Fig. 7.52, we get

$$V_1 - I_bR_1 - h_{ie}I_b - (1 + h_{fe})I_bR_2 - V_2 = 0$$

Rewriting, we get

$$I_bR_1 + h_{ie}I_b + (1 + h_{fe})I_bR_2 = V_1 - V_2$$

and solving for I_b, we get

$$I_b = \frac{V_1 - V_2}{R_1 + h_{ie} + (1 + h_{fe})R_2} \tag{7.24}$$

The circuit that "fits" Eq. (7.24) appears in Fig. 7.53.

We must now take a moment to fully appreciate the benefits of having the base equivalent circuit of Fig. 7.53. It "tells" us that any signal (V_3) or load resistor from collector to ground (R_3) is not reflected to the equivalent "base" circuit of a transistor on an approximate basis. It also clearly indicates what effect the source V_2 and load resistor R_2 in the emitter leg will have on the base current I_b. Note that for the common-emitter circuit, when R_2 and $V_2 = 0$, if we substitute these conditions into the circuit of Fig. 7.53, the base circuit of Fig. 7.54 will result, which should by now be familiar.

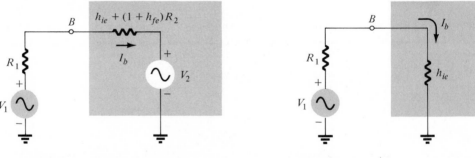

Figure 7.53 Approximate base equivalent circuit for the transistor.

Figure 7.54 Circuit of Fig. 7.53 with the conditions $R_2 = 0$, $V_2 = 0$.

A second glance at Fig. 7.53 will also reveal that any resistor in the emitter leg will appear as a much larger resistance in the base circuit due to the factor $(1 + h_{fe})$.

Typically $h_{fe} \gg 1$ and $(h_{fe} + 1) \cong h_{fe}$. In addition, it has been determined through practical experience that $(h_{fe} + 1)R_2 \cong h_{fe}R_2 \gg h_{ie}$, and h_{ie} can usually be dropped on an approximate basis. For the transistor configuration of Fig. 7.55, therefore, the input impedance Z_i is determined by

$$\boxed{Z_i \cong h_{fe}R_E} \tag{7.25}$$

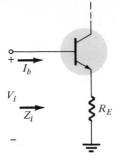

Figure 7.55 Effect of R_E on the input impedance.

with the result that

$$I_b \cong \frac{V_i}{Z_i} = \frac{V_i}{h_{fe}R_E}$$

Emitter Circuit

If Eq. (7.24) is multiplied by $(1 + h_{fe})$, the following equation will result:

$$(1 + h_{fe})\, I_b = (1 + h_{fe}) \left[\frac{V_1 - V_2}{R_1 + h_{ie} + (1 + h_{fe})\, R_2} \right]$$

Rewritten:

$$(1 + h_{fe})\, I_b = I_e = \frac{V_1 - V_2}{\dfrac{R_1 + h_{ie}}{1 + h_{fe}} + R_2} \qquad (7.26)$$

Constructing a circuit to "fit" Eq. (7.26), we obtain Fig. 7.56.

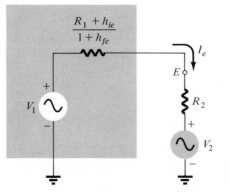

Figure 7.56 Approximate emitter equivalent circuit for the transistor.

The "emitter circuit" also clearly indicates that a source (V_3) and load resistor (R_3) in the collector circuit are not reflected to the "emitter circuit." Note also that V_1 will affect the current I_e but the reflected resistance is normally small in magnitude due to the division by the factor $(1 + h_{fe})$.

For the representative network of Fig. 7.57, if we use the approximation $(h_{fe} + 1) \cong h_{fe}$ and define $R_S' = R_S + h_{ie}$, then

$$Z_e = \frac{R_1 + h_{ie}}{1 + h_{fe}}$$

becomes

$$Z_e \cong \frac{R_S'}{h_{fe}} \bigg|_{R_S' = R_S + h_{ie}}$$ (7.27)

and for the network of Fig. 7.57

$$I_e \cong \frac{V_2}{R_2 + Z_e} = \frac{V_2}{R_2 + \dfrac{R_S'}{h_{fe}}}$$

which for the values provided is

$$I_e = \frac{2}{400 + \dfrac{(0.5 + 1.5) \times 10^3}{50}} = \frac{2}{400 + 40} = \frac{2}{440}$$

$$= 4.55 \text{ mA}$$

Note the small magnitude of the transferred resistance Z_e. This is typical due to the division by the factor h_{fe}.

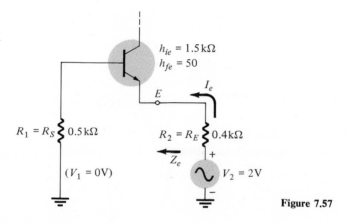

Figure 7.57

Collector Circuit

The current I_c is $h_{fe}I_b$ independent of the surrounding circuit. The resulting collec-

$$V_c = V_3 - I_c R_3 = V_3 - h_{fe}I_b R_3$$

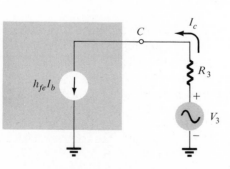

Figure 7.58 Approximate collector equivalent circuit for the transistor.

tor circuit is as shown in Fig. 7.58. The collector current or potential, therefore, can be found directly after I_b has been determined.

The beneficial aspects of the equivalent circuits just derived will become obvious in the examples to follow and in later chapters. It would be wise to memorize these equivalent circuits for future use. They can be very powerful tools in analysis of transistor networks.

EXAMPLE 7.9 The transistor configuration of Fig. 7.59, called the *emitter follower,* is frequently used for impedance-matching purposes; that is, it presents a high impedance at the input terminals (Z_i) and a low output impedance (Z_o), rather than the reverse, which is typical of the basic transistor amplifier. The effect of the emitter follower circuit is much like that obtained using a transformer to match a load to the source impedance for maximum power transfer. The following analysis will reflect that the voltage gain of the emitter follower circuit is always less than one.

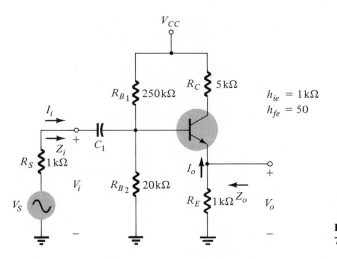

Figure 7.59 Circuit for Example 7.9.

For the circuit of Fig. 7.59 calculate the following:
(a) Z_i.
(b) Z_o.
(c) $A_{v_1} = V_o/V_s$ and $A_{v_2} = V_o/V_i$.
(d) $A_i = I_o/I_i$.

Solution: Eliminating the dc levels and replacing the capacitor by a short circuit will result in the circuit of Fig. 7.60.
(a) Using the base equivalent circuit (Fig. 7.61):

$$Z_1 = h_{ie} + (1 + h_{fe})R_2 = 52 \text{ k}\Omega$$

(certainly high compared to the typical input impedance $Z_i \cong h_{ie}$ for the basic transistor amplifier).

The resulting $Z_i = 20 \text{ k}\Omega \parallel 52 \text{ k}\Omega = \mathbf{14.44 \text{ k}\Omega}$

If we had used Eq. (7.25) and the conditions surrounding this approximation, the equivalent circuit would appear as shown in Fig. 7.62 and $Z_1 = h_{fe}R_E = 50 \text{ k}\Omega$

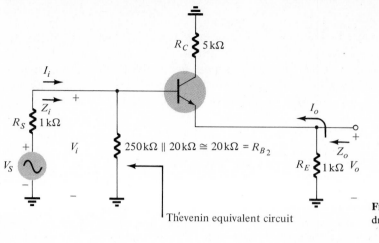

$R_C \gtrless 5\,k\Omega$

I_i

Z_i

$R_S \gtrless 1\,k\Omega$

V_i

$250\,k\Omega \parallel 20\,k\Omega \cong 20\,k\Omega = R_{B_2}$

V_S

I_o

Z_o

$R_E \gtrless 1\,k\Omega$ V_o

Thévenin equivalent circuit

Figure 7.60 Circuit of Fig. 7.59 re-drawn for small-signal ac analysis.

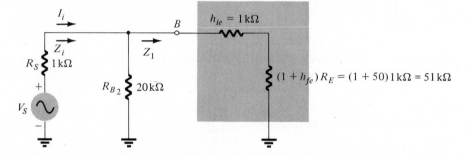

I_i

Z_i

Z_1

B

$h_{ie} = 1\,k\Omega$

$R_S \gtrless 1\,k\Omega$

$R_{B_2} \gtrless 20\,k\Omega$

V_S

$(1 + h_{fe})R_E = (1 + 50)1\,k\Omega = 51\,k\Omega$

Figure 7.61 Substitution of the approximate base equivalent circuit into the circuit of Fig. 7.60.

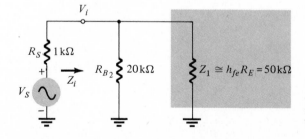

V_i

$R_S \gtrless 1\,k\Omega$

Z_i

$R_{B_2} \gtrless 20\,k\Omega$

$Z_1 \cong h_{fe}R_E = 50\,k\Omega$

V_S

Figure 7.62 Approximation frequently used to determine the input impedance of a network with an unbypassed emitter resistor.

and $Z_i = 20\ k\Omega \parallel 50\ k\Omega = 14.29\ k\Omega$. This result will usually be so close to the result obtained using the complete approximate equivalent that we will use Eq. (7.25) throughout the following analysis unless otherwise noted.

(b) The Thévenin equivalent circuit of the portion of the network indicated in Fig. 7.60 will now be found so that the input circuit will have the basic configuration of Fig. 7.51. That is, R_1 and V_1 will be found to ensure that the proper values are substituted into the emitter equivalent circuit to be employed in the determination of Z_o.

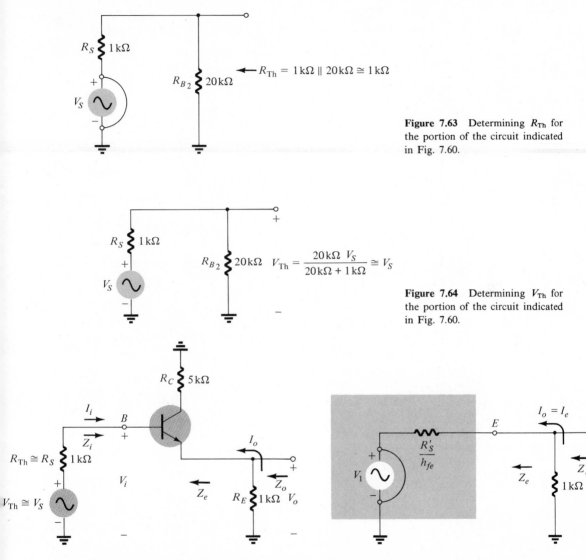

Figure 7.63 Determining R_{Th} for the portion of the circuit indicated in Fig. 7.60.

Figure 7.64 Determining V_{Th} for the portion of the circuit indicated in Fig. 7.60.

Figure 7.65 Circuit of Fig. 7.60 following the substitution of the Thévenin equivalent circuit.

Figure 7.66 Substitution of the emitter equivalent circuit with V_1 set to zero.

Substituting the Thévenin equivalent circuit into the circuit of Fig. 7.60 will result in the configuration of Fig. 7.65.

Using the emitter equivalent circuit with $V_s = 0$ (as required by definition) to determine Z_o (Fig. 7.66), we obtain

$$Z_e = \frac{R_S'}{h_{fe}} = \frac{1\ k\Omega\ +\ 1\ k\Omega}{50} \cong \mathbf{40}\ \mathbf{\Omega}$$

and

$$Z_o = Z_e \| R_L = 40 \| 1000 \cong \mathbf{40}\ \mathbf{\Omega}$$

(c)   A careful examination of Fig. 7.59 will reveal that the output voltage V_o is separated from the input voltage V_i by only the voltage drop V_{be} across the base-to-emitter junction. For ac operations and an unbypassed (no capacitor present) emitter resistor the approximation indicated by Eq. (7.28) is frequently employed.

$$\boxed{V_{be} \cong 0 \text{ V}} \quad \left(\substack{\text{ac operations and} \\ \text{unbypassed emitter resistor}}\right) \tag{7.28}$$

Using this approximation, we get

$$V_o \cong V_i \quad \text{and} \quad A_{v_2} = \frac{V_o}{V_i} \cong 1$$

For $A_{v_1} = V_o/V_s$ we can turn to the network of Fig. 7.62 where

$$V_i = \frac{(50 \text{ k}\Omega \parallel 20 \text{ k}\Omega) V_s}{(50 \text{ k}\Omega \parallel 20 \text{ k}\Omega) + 1 \text{ k}\Omega} \cong 0.93 \ V_s$$

and

$$A_{v_1} = \frac{V_o}{V_s} = \left[\frac{V_o}{V_i}\right]\left[\frac{V_i}{V_s}\right] = [\cong 1][0.93] = \mathbf{0.93}$$

The validity of Eq. (7.28) can be demonstrated using the emitter equivalent circuit of Fig. 7.67 where

$$V_o = \frac{1 \text{ k}\Omega \ (V_s)}{1 \text{ k}\Omega + 40} = 0.96 V_s$$

and

$$A_{v_1} = \frac{V_o}{V_s} = \mathbf{0.96} \cong 0.93 \quad \text{(as determined above)}$$

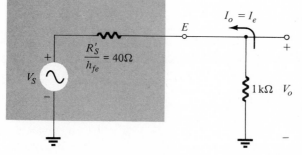

Figure 7.67 Circuit to be employed in determining A_{v1}.

(d)   From the emitter equivalent circuit:

$$I_o = I_e = \frac{V_s}{1.040 \text{ k}\Omega}$$

and from Fig. 7.62

$$I_i = \frac{V_s}{R_s + Z_i} = \frac{V_s}{1 \text{ k}\Omega + 14.29 \text{ k}\Omega} = \frac{V_s}{15.29 \text{ k}\Omega}$$

or $V_s = I_i \ 15.29 \text{ k}\Omega$

Substituting this result into the above equation:

$$I_o = -\frac{I_i \ 15.29 \text{ k}\Omega}{1.040 \text{ k}\Omega}$$

and
$$A_i = \frac{I_o}{I_i} = \frac{-15.29 \text{ k}\Omega}{1.040 \text{ k}\Omega} = -14.70$$

EXAMPLE 7.10 In Chapter 13 the *difference amplifier* will be discussed in detail. For the present, however, to demonstrate the usefulness of the base, collector, and emitter circuits, we shall consider a simple difference amplifier. In its basic form, a difference amplifier is simply a network that will produce a signal that is the difference of the two applied signals.

Figure 7.68 is such a circuit. Note that a signal has been applied to both the base and emitter leg of the transistor.

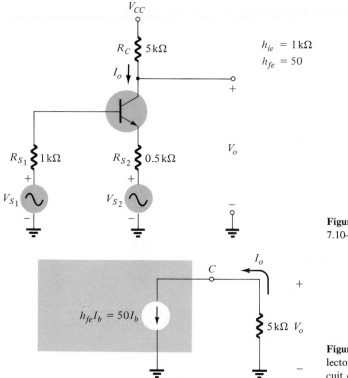

$h_{ie} = 1 \text{k}\Omega$
$h_{fe} = 50$

Figure 7.68 Circuit for Example 7.10—difference amplifier.

Figure 7.69 Application of the collector equivalent circuit to the circuit of Fig. 7.68.

Solution: Using the "collector equivalent circuit" after replacing the dc supply and capacitors by short circuits will result in the circuit of Fig. 7.69 where

$$V_o = -I_o 5 \text{ k}\Omega = -50 I_b 5 \text{ k}\Omega$$

Using the "base equivalent circuit" gives us Fig. 7.70 and

$$I_b \cong \frac{V_{s_1} - V_{s_2}}{R_{s_1} + h_{fe} R_{s_2}} = \frac{V_{s_1} - V_{s_2}}{1 \text{ k}\Omega + 25 \text{ k}\Omega} = \frac{V_{s_1} - V_{s_2}}{26}$$

Substituting into the equation for V_o

$$V_o = -50 I_b 5 \text{ k}\Omega = -50 \frac{(V_{s_1} - V_{s_2})}{26 \text{ k}\Omega} 5 \text{ k}\Omega = \frac{-250(V_{s_1} - V_{s_2})}{26}$$

and
$$V_o \cong -9.62(\mathbf{V}_{s_1} - \mathbf{V}_{s_2})$$

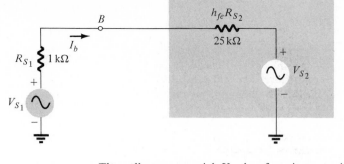

Figure 7.70 Application of the base equivalent circuit to the circuit of Fig. 7.68.

The collector potential V_o, therefore, is approximately 9.62 times the difference of the two applied signals. As mentioned earlier, difference amplifiers will be discussed in much greater detail in Chapter 13.

7.10 AN ALTERNATE APPROACH

In recent years there has been an increasing interest in an approximate equivalent circuit for the transistor in which one of the parameters is determined by the dc operating conditions. You will possibly recall from the transistor specification sheet provided in Chapter 3 that the hybrid parameter h_{ie} was specified at a particular operating point. Figure 7.16 revealed a significant variation in h_{ie} with I_C ($\cong I_E$). The question then arises of what one would do with the provided value of h_{ie} if the conditions of operation (level of $I_C \cong I_E$) were different from those indicated on the specification sheet. The equivalent circuit derived below will permit the determination of an equivalent h_{ie} using the dc operating conditions of the network, thereby not limiting itself to the data on the device as provided by the manufacturer.

The derivation of the alternate equivalent circuit begins with a close examination of the input and output characteristics of the CB transistor configuration, as redrawn in Fig. 7.71, on an approximate basis. Note that straight-line segments are used to represent the collector characteristics and a single diode characteristic for the emitter circuit (neglecting the variation in the input characteristics with the change in V_{CB}), resulting in an equivalent circuit such as shown in Fig. 7.72b. For ac conditions, therefore, the input impedance at the emitter of the CB transistor can be determined

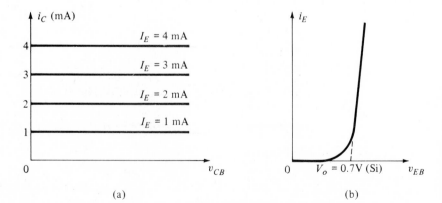

Figure 7.71 Approximate CB characteristics: (a) output; (b) input.

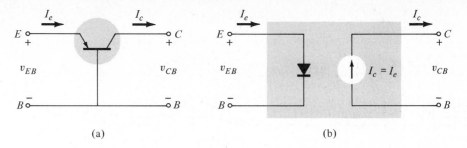

(a)   (b)

Figure 7.72 (a) CB configuration; (b) approximate CB equivalent circuit as defined by Fig. 7.71.

using Eq. (1.22) as introduced for the ac resistance of a diode. The factor r_B will be dropped to ensure that it does not affect the clarity of the introduction of this alternate technique. You will recall from Chapter 1 that in time it is conceivable that the factor r_B can be totally ignored with a negligible loss in accuracy if manufacturing techniques continue to improve. In time, when the alternate approach is developed, if you prefer to add a factor developed through experience for that transistor, then it can surely be introduced with little added confusion. For now we will define the input impedance for the CB configuration to be

$$r_e = \frac{26 \text{ mV}}{I_E \text{ (mA)}} \quad \text{(ohms)} \tag{7.29}$$

where I_E is the dc emitter current of the transistor. The emitter current is employed in Eq. (7.29) since it is the defined current of the diode in Fig. 7.71b. The result of the above discussion is the input equivalent circuit appearing in Fig. 7.73.

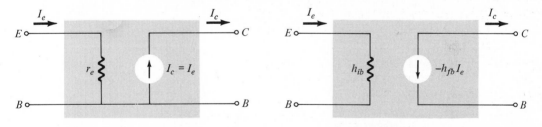

Figure 7.73 Approximate CB equivalent circuit.

Figure 7.74 Approximate CB hybrid equivalent circuit.

Figure 7.71a clearly indicates that the collector curves have been approximated to result in $i_C = i_E$ at any point on the characteristics. This would result in the output equivalent circuit appearing in Figs. 7.72 and 7.73. On an approximate basis, the alternate equivalent circuit is now defined. Note its similarities with the reduced hybrid equivalent circuit of Fig. 7.74. A comparison of the two clearly indicates that

$$h_{ib} = r_e \tag{7.30}$$
$$h_{fb} = 1$$

The following example will clarify the use of the alternate equivalent circuit.

EXAMPLE 7.11   For the network of Fig. 7.75, determine A_v, A_i, Z_i, and Z_o.

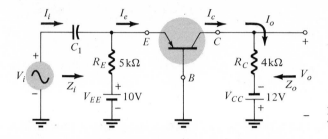

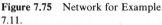

Figure 7.75   Network for Example 7.11.

Solution: *dc conditions:*

$$I_E = \frac{V_{EE} - V_{BE}}{R_E} = \frac{10 - 0.7}{5 \text{ k}\Omega} = \frac{9.3}{5 \text{ k}\Omega} = 1.86 \text{ mA}$$

and

$$r_e = \frac{26 \text{ mV}}{I_E} = \frac{26 \text{ mV}}{1.86 \text{ mA}} \cong 14 \text{ }\Omega$$

ac conditions: The network redrawn (Fig. 7.76):
(a) A_v:

$$V_o = I_c R_C = I_e R_C$$

and

$$I_e = \frac{V_i}{r_e}$$

so that

$$V_o = \left(\frac{V_i}{r_e}\right) R_C$$

and

$$\boxed{A_v = \frac{V_o}{V_i} = \frac{R_C}{r_e}}$$ (7.31)

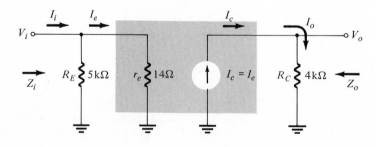

Figure 7.76   Ac equivalent for the network of Fig. 7.75.

Compare this result to that obtained in Example 7.7 where

$$|A_v| = \frac{h_{fb}}{h_{ib}} R_L'$$

but in this case $R_L' = R_C$, $h_{fb} = 1$, $h_{ib} = r_e$ which through substitution will result in Eq. (7.31). Substituting values, we get

$$A_v = \frac{R_C}{r_e} = \frac{4\ \text{k}\Omega}{14} = \mathbf{285.71}$$

(b)　A_i:　Since $R_E \parallel r_e \cong r_e$,

$$I_o = I_c = I_e = I_i$$

and

$$\boxed{A_i \cong \mathbf{1} \cong h_{fb}}$$

(c)　Z_i:

$$\boxed{Z_i \cong r_e = h_{ib}} = \mathbf{14\ \Omega}$$

(d)　Z_o:

$$\boxed{Z_o|_{v_i=0} = R_C} = \mathbf{4\ k\Omega}$$

For the common-emitter configuration appearing in Fig. 7.77a the input and output characteristics have been approximated by the set appearing in Fig. 7.77b and 7.77c,

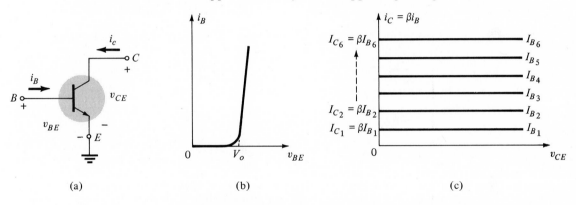

(a)　　　　　　　　　(b)　　　　　　　　　(c)

Figure 7.77　(a) CE configuration; (b) input characteristics; (c) output characteristics.

respectively. The base characteristics are again approximated to be those of a diode (the effect of V_{CE} on the characteristics is ignored) and

$$\boxed{r_{ac} = \frac{26\ \text{mV}}{I_B}} \qquad (7.32a)$$

But

$$I_E \cong I_C = \beta I_B \quad \text{and} \quad I_B \cong \frac{I_E}{\beta}$$

so that

$$r_{ac} = \frac{26\ \text{mV}}{I_B} = \frac{26\ \text{mV}}{I_E/\beta} = \beta\left(\frac{26\ \text{mV}}{I_E}\right)$$

or

$$\boxed{r_{ac} = \beta r_e} \qquad (7.32b)$$

Equation (7.32) has the same format as Eq. (7.25) $(Z_i \cong h_{fe}R_E)$ used to reflect an emitter resistor to the base circuit. In this case Eq. (7.32) can be directly derived using the same equation [Eq. (7.25)] and Fig. 7.78a where r_e appears as a resistor in the emitter leg.

For the situation of Fig. 7.78b,

$$r_{ac} = \beta\,(r_e + R_e\,) \cong \beta R_E \tag{7.33}$$

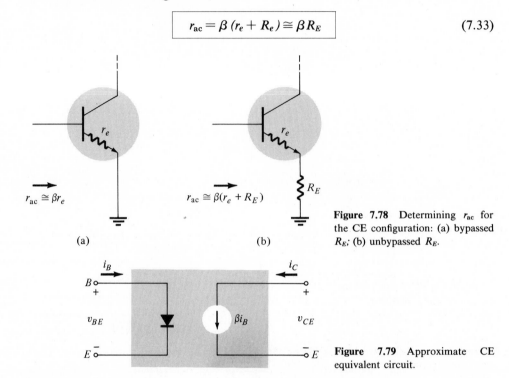

(a)   (b)

Figure 7.78 Determining r_{ac} for the CE configuration: (a) bypassed R_E; (b) unbypassed R_E.

Figure 7.79 Approximate CE equivalent circuit.

The input circuit for the CE configuration is approximated, for the reasons discussed above, by the diode circuit appearing in Fig. 7.79, but the input impedance appears as βr_e in Fig. 7.80a since r_e is determined by I_E and not I_B. In Fig. 7.77c the approximation was employed that β is the same throughout the device characteristics. This we know is absolutely untrue. However, its variation about the provided value for the typical application in the active region is assumed to be minimal and a fixed value a valid first approximation. For our analysis, we will consider it to be a constant at the provided value resulting in the output equivalent circuits of Figs. 7.79 and 7.80a. From the equivalent circuits of Fig. 7.80 we can readily note that

$$\begin{aligned} \beta &= h_{fe} \\ \beta r_e &= h_{ie} \end{aligned} \tag{7.34}$$

For the various configurations examined in detail earlier the resulting equations can be quickly converted to a set having only β and r_e in place of the hybrid parameters using the relationships of Eq. (7.34). It is also important to keep in mind that Eq. (7.34) permits a direct determination of h_{ie} at bias points different from the provided condition.

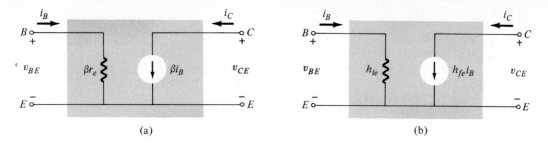

Figure 7.80 CE configuration: (a) alternate equivalent circuit; (b) approximate hybrid equivalent circuit.

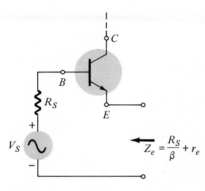

Figure 7.81 Emitter impedance Z_e.

For the case indicated in Fig. 7.81,

$$Z_e = \frac{R_S'}{h_{fe}} = \frac{R_S + h_{ie}}{h_{fe}} = \frac{R_S + \beta r_e}{\beta}$$

and

$$\boxed{Z_e = \frac{R_S}{\beta} + r_e}$$ (7.35)

A few examples will clarify the use of the alternate CE equivalent circuit.

EXAMPLE 7.12 Determine A_v, A_i, Z_i, and Z_o for the network of Fig. 7.82.

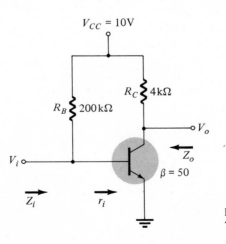

Figure 7.82 Network for Example 7.12.

Solution:

(a) A_v: For a network such as shown in Fig. 7.82 we have reached the point where it should be unnecessary to redraw the network for each calculation. Therefore, for:

dc conditions:

$$I_B = \frac{V_{CC} - V_{BE}}{R_B} = \frac{10 - 0.7}{200 \text{ k}\Omega} = \frac{9.3}{200 \text{ k}\Omega} \cong 46.5 \text{ }\mu\text{A}$$

and

$$I_E \cong I_C = \beta I_B = 50(46.5 \times 10^{-6}) = 2.325 \text{ mA}$$

so that

$$r_e = \frac{26 \text{ mV}}{I_E} = \frac{26}{2.325} \cong 11.2 \text{ }\Omega$$

ac conditions:

$$r_i = \beta r_e$$

and

$$I_b = \frac{V_i}{\beta r_e}$$

so that

$$V_o = -I_C R_C \cong \beta I_b R_C \cong \beta \left(\frac{V_i}{\beta r_e} \right) R_C = -\frac{R_C}{r_e} V_i$$

with

$$\boxed{A_v = \frac{V_o}{V_i} = -\frac{R_C}{r_e}} \tag{7.36}$$

Substituting numbers:

$$A_v = -\frac{R_C}{r_e} = -\frac{4 \text{ k}\Omega}{11.2} \cong \mathbf{-357.14}$$

(b) A_i:

$$R_B \| r_i = R_B \| \beta r_e \cong \beta r_e$$

therefore,

$$I_b \cong I_i$$

and

$$I_o = h_{fe} I_b = h_{fe} I_i$$

with

$$\boxed{A_i = \frac{I_o}{I_i} = h_{fe}} \tag{7.37}$$

and

$$A_i = \mathbf{50}$$

(c) Z_i:

$$\boxed{Z_i \cong \beta r_e} \tag{7.38}$$

$$= (50)(11.2) = \mathbf{560 \ \Omega}$$

(d) Z_o:

$$\boxed{Z_o \cong R_C} \tag{7.39}$$

$$= \mathbf{4 \text{ k}\Omega}$$

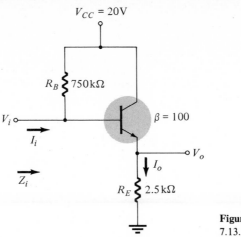

$V_{CC} = 20V$

R_B 750 kΩ

V_i

I_i

$\beta = 100$

V_o

I_o

Z_i

R_E 2.5 kΩ

Figure 7.83 Network for Example 7.13.

EXAMPLE 7.13 Determine A_v, A_i, Z_i, and Z_o for the network of Fig. 7.83.

Solution:

dc conditions:

From Chapter 4,

$$I_B = \frac{V_{CC} - V_{BE}}{R_B + \beta R_E} = \frac{20 - 0.7}{750 \text{ k}\Omega + (100)(2.5 \text{ k}\Omega)} = \frac{19.3}{750 \text{ k}\Omega + 250 \text{ k}\Omega} = \frac{19.3}{1 \times 10^6}$$

$$= 19.3 \text{ } \mu A$$

$$I_E \cong I_C = \beta I_B = (100)(19.3 \times 10^{-6}) = 1.93 \text{ mA}$$

and

$$r_e = \frac{26 \text{ mV}}{I_E} = \frac{26}{1.93} = 13.47 \text{ } \Omega$$

ac conditions:

A reduced ac equivalent circuit appears in Fig. 7.84.

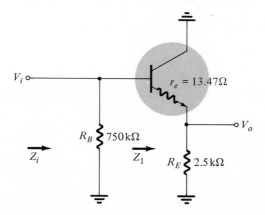

V_i

$r_e = 13.47\Omega$

V_o

R_B 750 kΩ

Z_i

Z_1

R_E 2.5 kΩ

Figure 7.84 Ac equivalent for the network of Fig. 7.83.

(a) A_v: In a previous section we introduced the approximation $V_{be} \cong 0$ V when a network had an unbypassed emitter resistor. With this in mind $V_o = V_i$

and

$$\boxed{A_v \cong 1}$$

(actually slightly less)

(7.40)

(b) A_i: The impedance Z_1:

$$Z_1 = \beta (r_e + R_E) = 100(13.47 + 2500)$$

Note here that r_e can realistically be ignored in comparison with R_E. Taking the approach, we get

$$Z_1 \cong \beta R_E = 100(2.5 \text{ k}\Omega) = 250 \text{ k}\Omega$$

and using Fig. 7.85, we get

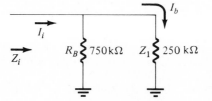

Figure 7.85 Determining the relationship between I_i and I_b.

$$I_b = \frac{750 \text{ k}\Omega I_i}{750 \text{ k}\Omega + 250 \text{ k}\Omega} = 0.75 I_i$$

and

$$A_i = \frac{I_o}{I_i} = \left[\frac{I_b}{I_i} \right] \left[\frac{I_o}{I_b} \right] = [0.75][\beta] = [0.75][100] = \mathbf{75}$$

(c) Z_i: From Fig. 7.85

$$\boxed{Z_i \cong R_B \| \beta R_E} \qquad (7.41)$$

$$= 750 \text{ k}\Omega \| 250 \text{ k}\Omega = 187.5 \text{ k}\Omega$$

(d) Z_o: With V_i set to zero, R_B is effectively "shorted out" and in Eq. (7.35) $R_S = 0$ Ω. Therefore,

$$Z_e = \frac{R_S}{\beta} + r_e = 0 + 13.47 = 13.47 \text{ } \Omega$$

and

$$\boxed{Z_o = R_E \| r_e \cong r_e} = \mathbf{13.47} \text{ } \Omega \qquad (7.42)$$

EXAMPLE 7.14 Determine A_v, A_i, Z_i, and Z_o for the network of Fig. 7.86.

Solution:
dc conditions:

From Chapter 4,

$$V_B \cong \frac{R_{B2} (V_{CC})}{R_{B2} + R_{B1}} = \frac{5 \text{ k}\Omega(20)}{5 \text{ k}\Omega + 50 \text{ k}\Omega} = \frac{5}{55}(20) = 1.818 \text{ V}$$

and

$$I_E = \frac{V_B - V_{BE}}{R_E} = \frac{1.818 - 0.7}{1 \text{ k}\Omega} = \frac{1.118}{1 \text{ k}\Omega} = 1.118 \text{ mA}$$

and

$$r_e = \frac{26 \text{ mV}}{I_E} = \frac{26}{1.118} \cong 23.26 \text{ } \Omega$$

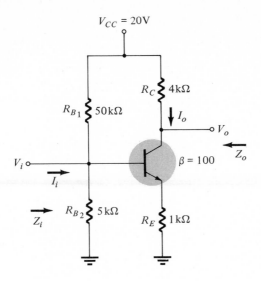

Figure 7.86 Network for Example 7.14.

ac conditions:

The network is redrawn as shown in Fig. 7.87.

(a) A_v:

$$V_o = -I_o R_C = -I_c R_C = -I_e R_C$$

but

$$I_e = \beta I_b$$

with

$$I_b \cong \frac{V_i}{\beta R_E}$$

so that

$$I_e = \beta \left(\frac{V_i}{\beta R_E} \right) = \frac{V_i}{R_E}$$

and

$$V_o = -\left(\frac{V_i}{R_E} \right) R_C$$

with

$$\boxed{A_v = \frac{V_o}{V_i} = -\frac{R_C}{R_E}}$$

(7.43)

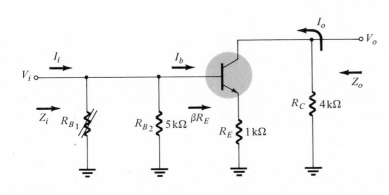

Figure 7.87 Network of Fig. 7.86 redrawn for ac conditions.

The distinct advantage of this configuration is now obvious; it is β independent. It is not concerned with the value of $\beta(=h_{fe})$ which will vary depending on the operating point and the particular transistor of a series used. As a consequence, however, there is a significant loss in gain with an unbypassed emitter resistor.

Substituting numbers, we get

$$A_v \cong \frac{-R_C}{R_E} = -\frac{4\ \text{k}\Omega}{1\ \text{k}\Omega} = -4$$

(b) A_i: Since $R_{B1} \| R_{B2} \cong R_{B2}$ as shown in Fig. 7.87,

$$I_b \cong \frac{R_{B2}I_i}{R_{B2} + \beta R_E}$$

and

$$I_o = I_c = \beta I_b = \beta \left(\frac{R_{B2}I_i}{R_{B2} + \beta R_E} \right)$$

with

$$A_i = \frac{I_o}{I_i} = \frac{\beta R_{B2}}{R_{B2} + \beta R_E} = \frac{R_{B2}}{\dfrac{R_{B2}}{\beta} + R_E}$$

$$\boxed{A_i \cong \frac{R_{B2}}{R_E}}\Bigg|_{R_{B1} \gg R_{B2}} \tag{7.44}$$

Substituting values, we get

$$A_i \cong \frac{5\ \text{k}\Omega}{1\ \text{k}\Omega} = 5$$

(c) Z_i: From Fig. 7.87

$$Z_i \cong R_{B2} \| \beta R_E = 5\ \text{k}\Omega \| 100\ \text{k}\Omega \cong 5\ \text{k}\Omega$$

and

$$\boxed{Z_i \cong R_{B2}} \tag{7.45}$$

(d) Z_o: From Fig. 7.87

$$Z_o = (Z_{\text{Xistor}} + R_E) \| R_C \cong (\infty\ \Omega + R_E) \| R_C \cong R_C$$

and

$$\boxed{Z_o \cong R_C} \tag{7.46}$$

By defining

$$\beta = h_{fe} = 100$$

and

$$\beta r_e = h_{ie} = 2326\ \Omega$$

we can apply the program appearing in Appendix D to obtain the desired solution.

For the following modifications (Fig. 7.88) of the network of Fig. 7.86, the results for A_v, A_i, Z_i, and Z_o are provided. The derivations will appear as exercises at the end of the chapter.

$$\boxed{A_v \cong -\frac{R_C}{R_E}} \tag{7.47}$$

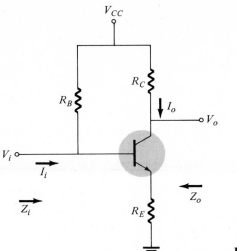

Figure 7.88

$$A_i \cong \frac{\beta R_B}{R_B + \beta R_E} \qquad (7.48)$$

$$Z_i \cong R_B \| \beta R_E \qquad (7.49)$$

$$Z_o \cong R_C \qquad (7.50)$$

For the configuration of Fig. 7.89 the following results are obtained.

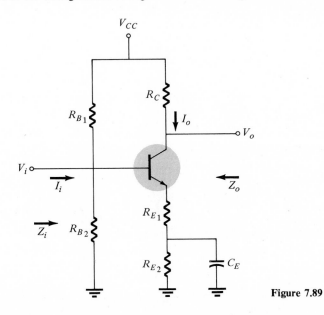

Figure 7.89

$$A_v = -\frac{R_C}{R_{E_1}} \qquad (7.51)$$

$$A_i \cong \frac{R_{B_2}}{R_{E_1}} \quad {}_{R_{B_1} \gg R_{B_2}} \qquad (7.52)$$

$$Z_i \cong R_{B_2} \| R_{E_1} \qquad (7.53)$$

$$Z_o \cong R_C \qquad (7.54)$$

7.11 COLLECTOR FEEDBACK

The last transistor configuration to be analyzed in detail in this chapter appears in Fig. 7.90. The analysis will be in terms of the hybrid parameters but the results will appear in each form using the direct conversion equations of the previous section. Note in this case that even though the collector is connected directly to the base through the feedback resistor R_F, it is not connected directly to V_{CC} so the output current I_o is not equal to the collector current of the transistor. This network provides a voltage feedback from the collector to the base to increase the stability of the system.

An approximate hybrid equivalent circuit has been substituted in Fig. 7.91.

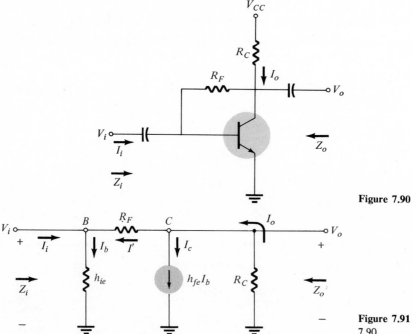

Figure 7.90

Figure 7.91 Ac equivalent of Fig. 7.90.

A_v: Applying Kirchhoff's current law at the collector terminal:

$$I_o = h_{fe}I_b + I'$$

and $$V_o = -I_oR_C = -(h_{fe}I_b + I') = -(h_{fe}I_bR_C + I'R_C)$$

but $$I_b = \frac{V_i}{h_{ie}} \quad \text{and} \quad I' = \frac{V_o - V_i}{R_F}$$

Substituting into the above equation for V_o results in

$$-V_o = h_{fe}\left(\frac{V_i}{h_{ie}}\right)R_C + \left(\frac{V_o - V_i}{R_F}\right)R_C$$

$$-V_o = \left(\frac{h_{fe}}{h_{ie}}R_C\right)V_i + \left(\frac{R_C}{R_F}\right)V_o - \left(\frac{R_C}{R_F}\right)V_i$$

or $$-V_o\left(1 - \frac{R_C}{R_F}\right) = V_i\left(\frac{h_{fe}}{h_{ie}}R_C - \frac{R_C}{R_F}\right)$$

and $$A_v = \frac{V_o}{V_i} = \frac{-\left(\dfrac{h_{fe}}{h_{ie}} - \dfrac{1}{R_F}\right)}{1 - \dfrac{R_L}{R_F}}R_C \cong \boxed{\frac{-h_{fe}}{h_{ie}}R_C} \qquad (7.55)$$

For the conversion:

$$A_v = -\frac{\beta}{\beta r_e}R_C = \boxed{\frac{-R_C}{r_e}} \qquad (7.56)$$

A_i: At node B, $I_i = I' = I_b$

or $$I' = I_b - I_i$$

At node C, $I_o = I' + I_c = I' + h_{fe}I_b$

or $$I_o = (I_b - I_i) + h_{fe}I_b$$

$$I_o = (h_{fe} - 1)I_b - I_i$$

As an approximation: $(h_{fe} + 1) \cong h_{fe}$

and $$I_b = \frac{I_o + I_i}{h_{fe}}$$

Applying Kirchhoff's voltage law around the outside network loop:

$$V_i + V_{R_F} - V_o = 0$$

or $$I_b h_{ie} + I'R_F + I_oR_C = 0$$

and $$I_b h_{ie} + (I_b - I_i)R_F + I_oR_C = 0$$

gathering terms:

$$I_b(h_{ie} + R_F) - I_iR_F + I_oR_C = 0$$

Substituting $I_b = (I_o + I_i)/h_{fe}$ into the above and arranging terms, we get

$$I_o(h_{ie} + R_F + h_{fe}R_C) + I_i(h_{ie} + R_F - h_{fe}R_F) = 0$$

with
$$A_i = \frac{I_o}{I_i} = \frac{-(h_{ie} + R_F - h_{fe}R_E)}{(h_{ie} + R_F + h_{fe}R_C)}$$

$$= \frac{-\left(\dfrac{h_{ie}}{h_{fe}} + \dfrac{R_F}{h_{fe}} - R_F\right)}{\left(\dfrac{h_{ie}}{h_{fe}} + \dfrac{R_F}{h_{fe}} + R_C\right)} \cong \frac{R_F}{\dfrac{R_F}{h_{fe}} + R_C} = \boxed{\frac{h_{fe}R_F}{R_F + h_{fe}R_C}} \tag{7.57}$$

For $h_{fe}R_L \gg R_F$

$$A_i \cong \boxed{\frac{R_F}{R_C}} \tag{7.58}$$

For the conversion

$$A_i = \frac{R_F}{R_C}\bigg|_{\beta R_C \gg R_F} \tag{7.59}$$

The above derivation could have been shortened significantly if we had assumed $I_o \cong h_{fe}I_b$. Then $I_b = (I_o/h_{fe})$ and not $(I_o + I_i)/h_{fe}$ and the mathematical manipulations would be substantially reduced.

Z_i:

$$V_i = I_b h_{ie}$$

and
$$I_b = I_i + I' = I_i + \frac{(V_o - V_i)}{R_F}$$

so that
$$V_i = \left(I_i + \frac{V_o - V_i}{R_F}\right)h_{ie}$$

with
$$V_i = I_i h_{ie} + \frac{h_{ie}}{R_F}V_o - \frac{h_{ie}}{R_F}V_i$$

but
$$V_o = A_v V_i$$

and
$$V_i = I_i h_{ie} + \frac{A_v h_{ie}}{R_F}V_i - \frac{h_{ie}}{R_F}V_i$$

with
$$V_i\left(1 - \frac{A_v h_{ie}}{R_F} + \frac{h_{ie}}{R_F}\right) = I_i h_{ie}$$

so that
$$Z_i = \frac{V_i}{I_i} = \frac{h_{ie}}{1 - \dfrac{A_v h_{ie}}{R_F}} = \frac{h_{ie}}{1 - \dfrac{h_{ie}}{\dfrac{R_F}{A_v}}}$$

Recall for parallel elements that

$$x \parallel y = \frac{xy}{x+y} = \frac{y}{1 + \dfrac{y}{x}}$$

In this case $y = h_{ie}$ and $x = (R_F/A_v)$

Therefore,

$$Z_i = \frac{V_i}{I_i} = \boxed{\frac{R_F}{A_v} \parallel h_{ie}} \qquad (7.60)$$

or

$$Z_i = \frac{V_i}{I_i} = \boxed{\frac{R_F}{A_v} \parallel \beta r_e} \qquad (7.61)$$

The resulting equations appear below for the configuration appearing in Fig. 7.92. The derivations will appear as exercises at the end of the chapter.

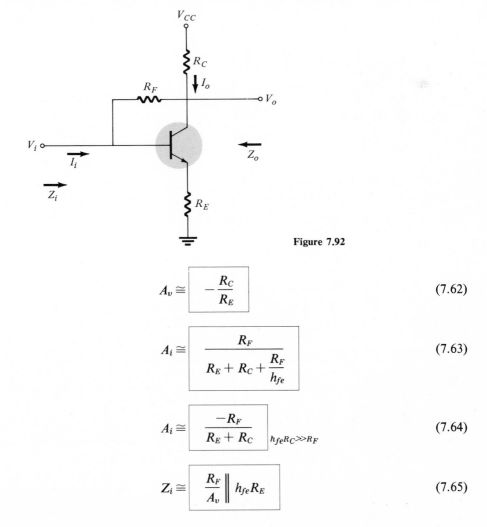

Figure 7.92

$$A_v \cong \boxed{-\frac{R_C}{R_E}} \qquad (7.62)$$

$$A_i \cong \boxed{\frac{R_F}{R_E + R_C + \dfrac{R_F}{h_{fe}}}} \qquad (7.63)$$

$$A_i \cong \boxed{\frac{-R_F}{R_E + R_C}}_{\,h_{fe}R_C \gg R_F} \qquad (7.64)$$

$$Z_i \cong \boxed{\frac{R_F}{A_v} \parallel h_{fe}R_E} \qquad (7.65)$$

TABLE 7.4 Summary Table of Transistor Configurations (A_v, Z_i, Z_o)
(In each case $R_L' = R_L \| R_C$ and $R_E' = R_L \| R_E$)

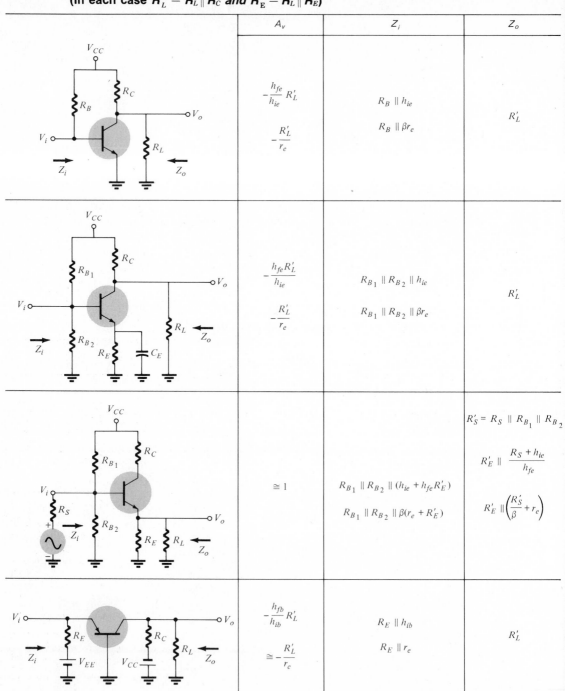

	A_v	Z_i	Z_o
	$-\dfrac{h_{fe}}{h_{ie}} R_L'$ $-\dfrac{R_L'}{r_e}$	$R_B \| h_{ie}$ $R_B \| \beta r_e$	R_L'
	$-\dfrac{h_{fe} R_L'}{h_{ie}}$ $-\dfrac{R_L'}{r_e}$	$R_{B_1} \| R_{B_2} \| h_{ie}$ $R_{B_1} \| R_{B_2} \| \beta r_e$	R_L'
	$\cong 1$	$R_{B_1} \| R_{B_2} \| (h_{ie} + h_{fe} R_E')$ $R_{B_1} \| R_{B_2} \| \beta(r_e + R_E')$	$R_S' = R_S \| R_{B_1} \| R_{B_2}$ $R_E' \| \dfrac{R_S + h_{ie}}{h_{fe}}$ $R_E' \| \left(\dfrac{R_S'}{\beta} + r_e\right)$
	$-\dfrac{h_{fb}}{h_{ib}} R_L'$ $\cong -\dfrac{R_L'}{r_e}$	$R_E \| h_{ib}$ $R_E \| r_e$	R_L'

288

TABLE 7.4 *(Continued)*

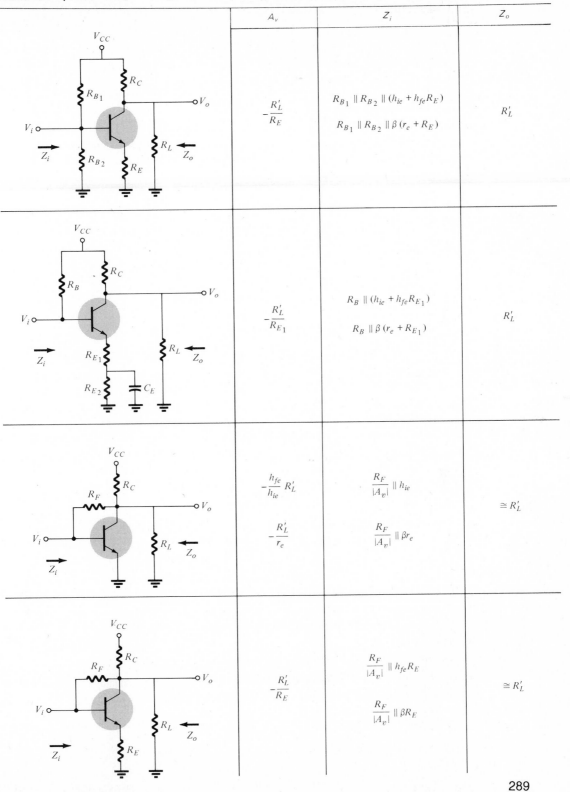

	A_v	Z_i	Z_o
	$-\dfrac{R_L'}{R_E}$	$R_{B_1} \parallel R_{B_2} \parallel (h_{ie} + h_{fe}R_E)$ $R_{B_1} \parallel R_{B_2} \parallel \beta(r_e + R_E)$	R_L'
	$-\dfrac{R_L'}{R_{E_1}}$	$R_B \parallel (h_{ie} + h_{fe}R_{E_1})$ $R_B \parallel \beta(r_e + R_{E_1})$	R_L'
	$-\dfrac{h_{fe}}{h_{ie}}R_L'$ $-\dfrac{R_L'}{r_e}$	$\dfrac{R_F}{|A_v|} \parallel h_{ie}$ $\dfrac{R_F}{|A_v|} \parallel \beta r_e$	$\cong R_L'$
	$-\dfrac{R_L'}{R_E}$	$\dfrac{R_F}{|A_v|} \parallel h_{fe}R_E$ $\dfrac{R_F}{|A_v|} \parallel \beta R_E$	$\cong R_L'$

289

7.12 SUMMARY TABLE

Table 7.4 summarizes the various configurations examined in this chapter. The approximation h_{re}, $h_{oe} \cong 0$ has been applied in each case. The latter case of $(1/h_{oe}) \cong \infty \ \Omega$ is one that should be checked for each application. If $(1/h_{oe})$ is close in magnitude to R_C, R_L, or their parallel combination $R_C \| R_L$, it cannot be dropped and its effects should be included with perhaps a hybrid equivalent circuit with just $h_{re} \cong 0$. If included for the reasons above, it will cut both the voltage and current gain.

In addition, if a situation is encountered where R_E and r_e are comparable in value, the r_e cannot be dropped in equations such as $\beta(R_E + r_e)$. It must be included and the equations modified.

PROBLEMS

§ 7.4

1. (a) Determine the hybrid parameters for the network of Fig. 7.93.

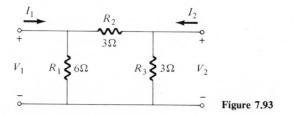

Figure 7.93

   (b) Sketch the hybrid equivalent circuit.

2. Sketch the complete hybrid equivalent circuit for the common-collector configuration and indicate the current directions as shown in Fig. 7.10a and b.

§ 7.5

3. Determine the hybrid parameters h_{fe} and h_{oe} from the collector characteristics of Fig. 7.11 at Q-pt of $V_{CE} = 5$ V and $I_C = 5$ mA and compare to those obtained in Section 7.5.

§ 7.6

4. For a change in I_C from 1 to 20 mA, which hybrid parameter exhibits the greatest change in Fig. 7.16? Which exhibits the least change?

5. For the range of V_{CE} from 1 to 50 V, which parameter in Fig. 7.17 exhibits the greatest change in value? Which exhibits the least change?

6. In Fig. 7.18, which parameter exhibits the least sensitivity to change in temperature? Which exhibits the most sensitivity?

§ 7.7

7. For the network of Fig. 7.94 determine the following:

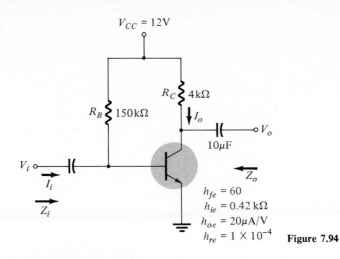

Figure 7.94

(a) Current gain $A_i = I_o/I_i$.
(b) Voltage gain $A_v = V_o/V_i$.
(c) Input impedance Z_i.
(d) Output impedance Z_o.
(e) Power gain A_p.

8. Repeat Problem 7 for the network of Fig. 7.95.

9. Repeat Problem 7 for the network of Fig. 7.96.

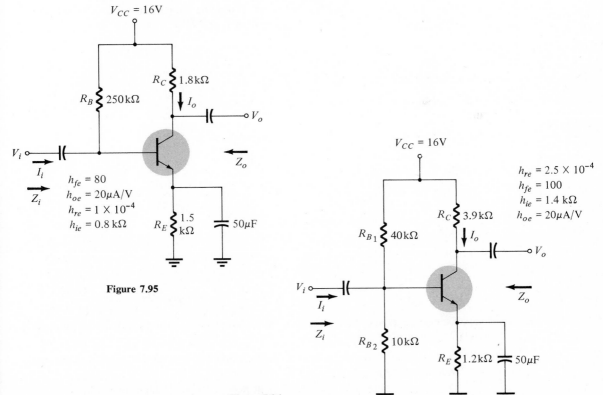

Figure 7.95

Figure 7.96

10. Repeat Problem 7 for the network of Fig. 7.97.

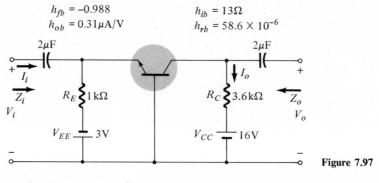

$$h_{fb} = -0.988$$
$$h_{ob} = 0.31 \mu A/V$$

$$h_{ib} = 13\Omega$$
$$h_{rb} = 58.6 \times 10^{-6}$$

Figure 7.97

§ 7.8

11. Using an appropriate approximate hybrid equivalent circuit, determine A_i, A_v, Z_i, Z_o, and A_p for the network of Fig. 7.94 (compare with the earlier results of Problem 7 if available).

12. Repeat Problem 11 for the network of Fig. 7.95.

13. Repeat Problem 11 for the network of Fig. 7.96.

14. Repeat Problem 11 for the network of Fig. 7.97.

15. (a) Using an approximate equivalent circuit, determine the current gain $A_i = I_o/I_i$ and voltage gain $A_v = V_o/V_i$ for the network of Fig. 7.98.

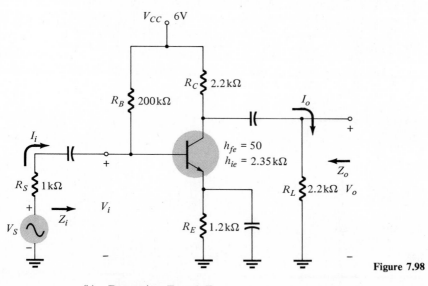

Figure 7.98

   (b) Determine Z_i and Z_o.
   (c) Determine $A_v = V_o/V_s$.

16. (a) Determine the new $A_v = V_o/V_i$ for Example 7.6 if a 5.6-kΩ load is connected from collector to ground.
   (b) Determine the new $A_i = I_o/I_i$ if I_o is now the current through the 5.6-kΩ load.

17. Determine the input impedance Z_i, the output impedance Z_o, and the voltage gain $A_v = V_o/V_i$ for the network of Fig. 7.99.

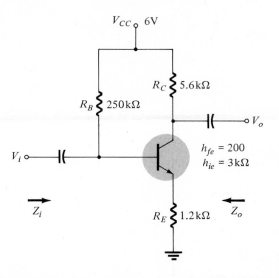

Figure 7.99

18. Determine Z_i, Z_o, A_v, A_i for the network of Fig. 7.100.

19. Determine Z_i, Z_o, A_v, A_i for the network of Fig. 7.101.

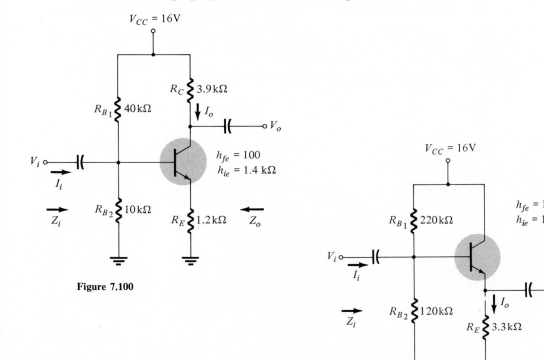

Figure 7.100

Figure 7.101

20. Determine V_o (in terms of V_1 and V_2) for the network of Fig. 7.102.

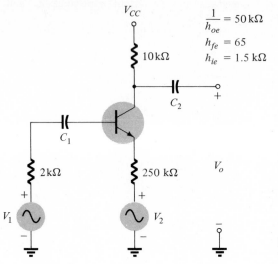

$$\frac{1}{h_{oe}} = 50 \,\text{k}\Omega$$
$$h_{fe} = 65$$
$$h_{ie} = 1.5 \,\text{k}\Omega$$

Figure 7.102 Circuit for Problem 7.20.

21. Using the approach introduced in Section 7.10 (use $r_B = 1.25 \,\Omega$), repeat Problem 7.

22. Using the approach introduced in Section 7.10 (use $r_B = 1.7 \,\Omega$), repeat Problem 10.

23. Using the approach introduced in Section 7.10 (use $r_B = 1.1 \,\Omega$), repeat Problem 17.

24. Using the approach introduced in Section 7.10 (use $r_B = 0.5 \,\Omega$), repeat Problem 18.

25. Using the approach introduced in Section 7.10 (use $r_B = 1.7 \,\Omega$), repeat Problem 19.

26. Determine A_v, A_i, Z_i, and Z_o for the network of Fig. 7.103.

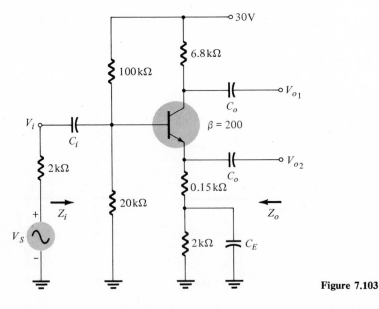

Figure 7.103

27. Determine A_v, A_i, Z_i, and Z_o for the network of Fig. 7.104.

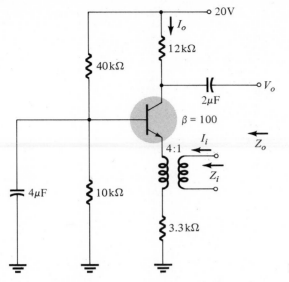

Figure 7.104

28. Derive the expressions for the network of Fig. 7.88.

29. Derive the expressions for the network of Fig. 7.89.

§ 7.11

30. Determine A_v, A_i, and Z_i for the network of Fig. 7.90 if $R_C = 5.6$ kΩ, $R_F = 120$ kΩ, $V_{CC} = 12$ V, and $\beta = 100$.

31. Determine A_v, A_i, and Z_i for the network of Fig. 7.92 if $R_C = 6.8$ kΩ, $R_F = 180$ kΩ, $R_E = 2.2$ kΩ, $V_{CC} = 16$ V, and $\beta = 150$.

32. Derive the expressions for the network of Fig. 7.92. That is, derive Eqs. (7.62)–(7.65).

COMPUTER PROBLEMS

Write BASIC programs to:

1. Compute the voltage gain of a BJT amplifier using the r_e equation formulation.

2. Compute the input and output impedances of a BJT amplifier circuit.

3. Compute the voltage gain for a range of load resistance values.

4. Compute the output voltage for a range of input source resistances.

5. Compute the voltage gain over a range of BJT current gains.

8

FET Small-Signal Analysis

8.1 GENERAL

FET devices can be used to build small-signal amplifier circuits providing voltage gain at very high input resistance. Both JFET and depletion MOSFET can operate with similar dc bias, providing the same magnitude of voltage gain with the MOSFET device and allowing much higher input resistance.

The common-source amplifier configuration provides the best voltage-gain operation. An input signal is applied to the gate, and the output signal is taken from the drain, the source terminal being the reference or common. A common-drain amplifier provides a noninverted output with near-unity gain. A common-gate amplifier connection is used less frequently, providing voltage gain with no polarity inversion.

The FET ac equivalent circuit is even simpler than that for a BJT, having only an output current source with value dependent on the device transconductance, g_m, the principal device factor. Values of FET transconductance vary from about 1 mS to 20 mS, providing larger voltage gain for larger values of g_m.

The FET can be used as a linear device or as a digital device. In linear circuits, JFET and depletion MOSFET devices are most often used. Digital FET circuits use the enhancement MOSFET especially in large-scale integrated (LSI) circuits and very-large-scale integrated (VLSI) circuits.

8.2 JFET/DEPLETION MOSFET SMALL-SIGNAL MODEL

To aid in an analysis of ac circuits that use FET devices, we first need to obtain an ac equivalent circuit for the device. Figure 8.1 shows a simple equivalent circuit of

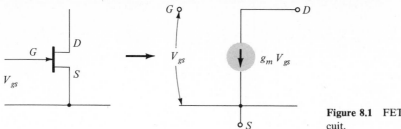

Figure 8.1 FET ac equivalent circuit.

the FET device. The ac voltage applied to the gate-source, V_{gs}, results in a drain current, I_d, of value $g_m V_{gs}$. The device transconductance, g_m, relates the amount of current resulting from an applied voltage across gate-source. The value of g_m can be obtained from the Shockley equation[1]

$$g_m = g_{mo}\left[1 - \frac{V_{GS}}{V_{GS(OFF)}}\right]$$ (8.1)

where

$$g_{mo} = \frac{2I_{DSS}}{|V_{GS(OFF)}|}$$ (8.2)

The value g_{mo} is the value of the transconductance at the $V_{GS} = 0$ V bias point and represents a fixed value of the maximum gain of the JFET device. The value of g_{mo} is a constant for a particular FET and is not affected by the choice of dc bias point. At any bias point in the reverse-bias operating region, a value of g_m smaller than the value of g_{mo} is obtained.

8.3 AC SMALL-SIGNAL OPERATION

The FET ac equivalent circuit introduced in Section 8.2 can now be used in analyzing various FET amplifier configurations for voltage gain and input and output resistances. To demonstrate use of the ac equivalent circuit, consider the FET amplifier circuit of Fig. 8.2a. The ac equivalent circuit is redrawn in Fig. 8.2b with capacitors replaced by a short for ac operation and with the FET device replaced by its equivalent circuit. The output ac voltage is seen to be

$$V_o = -(g_m V_{gs})R_D$$

---

[1] Since

$$i_D = I_{DSS}\left(1 - \frac{v_{GS}}{V_P}\right)^2$$

and

$$g_m = \frac{\partial i_D}{\partial v_{GS}}\bigg|_{v_{DS}=\text{constant}}$$

we obtain

$$g_m = -\frac{2I_{DSS}}{V_P}\left(1 - \frac{v_{GS}}{V_P}\right) = g_{mo}\left(1 - \frac{v_{GS}}{V_P}\right)$$

where

$$g_{mo} = A\frac{2I_{DSS}}{|V_P|}$$

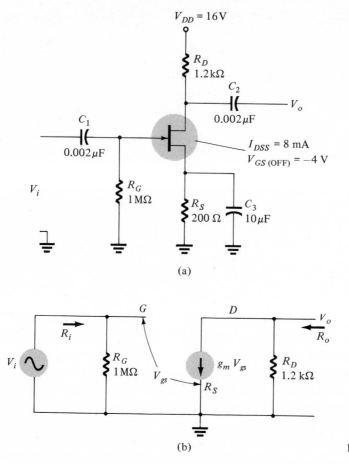

(a)

(b)

Figure 8.2 FET amplifier.

Since $V_i = V_{gs}$, the circuit voltage gain is

$$\boxed{A_v = \frac{V_o}{V_i} = -g_m R_D} \tag{8.3}$$

The ac impedance looking into the amplifier is

$$\boxed{R_i = R_G} \tag{8.4}$$

and the ac impedance seen looking from the load into the amplifier output terminal is

$$\boxed{R_o = R_D} \tag{8.4}$$

A circuit current gain can be determined using

$$\boxed{A_i = A_v \frac{R_i}{R_L} \quad \text{or} \quad = A_v \frac{R_i}{R_o}} \tag{8.6}$$

Example 8.1 Calculate the circuit factors A_v, R_i, and R_o for the JFET amplifier of Fig. 8.2a.

Solution: Dc bias as covered in Chapter 6 results in $V_{GS_Q} = -0.94$ V. At this bias point, the value of g_m, using Eqs. (8.1) and (8.2) is:

$$g_{mo} = \frac{2I_{DSS}}{V_{GS(OFF)}} = \frac{2(8 \text{ mA})}{|-4 \text{ V}|} = 4 \text{ mS}$$

and

$$g_m = g_{mo}\left[1 - \frac{V_{GS_Q}}{V_{GS(OFF)}}\right] = 4 \text{ mS}\left(1 - \frac{-0.94 \text{ V}}{-4}\right)$$

$$= 3.06 \text{ mS}$$

Using Eqs. (8.3) through (8.6):

$$A_v = -g_m R_D = -(3.06 \text{ mS})(1.2 \text{ k}\Omega) = -\mathbf{3.67}$$

$$R_i = R_G = \mathbf{1\ M\Omega}$$

$$R_o = R_D = \mathbf{1.2\ k\Omega}$$

$$A_i = A_v\frac{R_i}{R_o} = 3.67\ (1 \text{ M}\Omega/1.2 \text{ k}\Omega) = \mathbf{3.06 \times 10^3}$$

The voltage gain of magnitude 3.67 is typical of the lower values obtained using an FET circuit as opposed to a BJT circuit using discrete circuits.

Amplifier With Source Resistance

If the amplifier circuit is built with part of the source resistance unbypassed (see Fig. 8.3), the resulting voltage gain equation is

$$\mathbf{A}_v = \frac{-g_m R_D}{1 + g_m R_{S1}} \tag{8.7}$$

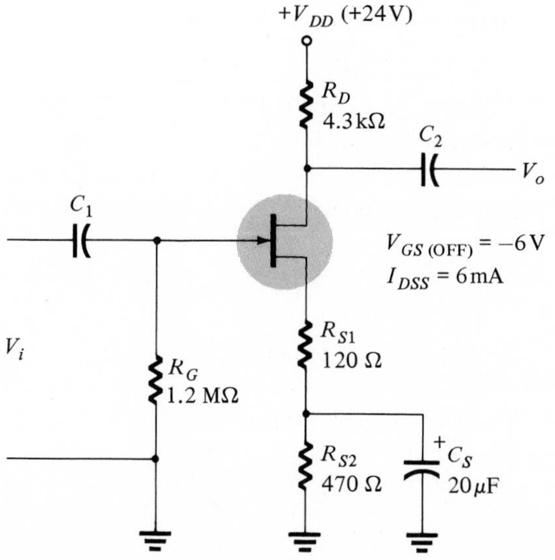

Figure 8.3 JFET amplifier with some source resistance.

The overall circuit voltage gain is reduced from that with the source resistance completely bypassed.

Example 8.2 Calculate the voltage gain of the circuit in Fig. 8.3.

Solution: Dc bias provides a gate-source voltage

$$V_{GS_Q} = -1.76 \text{ V}$$

The device transconductance is

$$g_{mo} = \frac{2I_{DSS}}{|V_{GS(OFF)}|} = \frac{2(6 \text{ mA})}{|-6 \text{ V}|} = 2 \text{ mS}$$

At the bias point established,

$$g_m = g_{mo}(1 - V_{GS_Q}/V_{GS(OFF)}) = 2 \text{ mS}\left(1 - \frac{-1.76 \text{ V}}{-6 \text{ V}}\right)$$

$$= 1.413 \text{ mS}$$

The circuit voltage gain is then

$$A_v = \frac{-g_m R_D}{1 + g_m R_S} = \frac{-1.413 \text{ mS}(4.3 \text{ k}\Omega)}{1 + (1.413 \text{ mS})(0.120 \text{ k}\Omega)}$$

$$= \frac{-6.08}{1.17} = -5.2$$

Alternate Equation for A_v

If we use the reciprocal of the value of g_m, we can express Eq. (8.7) in a form similar to that used in BJT circuits. Thus, we express the device factor as

$$r_s = 1/g_m \tag{8.8}$$

a resistance that is the numerical reciprocal of the device transconductance. Using Eq. (8.8) in Eq. (8.7), we get

$$A_v = \frac{-g_m R_D}{1 + g_m R_{S1}} = \frac{-(1/r_s)(R_D)}{1 + (1/r_s)(R_{S1})} = -\frac{R_D}{r_s + R_{S1}} \tag{8.9}$$

EXAMPLE 8.3 Use Eq. (8.9) to determine the voltage gain of the circuit in Example 8.2.

Solution: From Example 8.2

$$g_m = 1.413 \text{ mS}$$

so that

$$r_s = 1/g_m = 1/1.413 \text{ mS} = 707.7 \text{ }\Omega$$

Using Eq. (8.9)

$$A_v = -\frac{R_D}{r_s + R_{S1}} = -\frac{4.3 \text{ k}\Omega}{707.7 \text{ }\Omega + 120 \text{ }\Omega} = -5.2$$

Effect of Output Load

When the output of the circuit is connected to another circuit resulting in an effective load resistance, the overall voltage gain of the present amplifier stage will be reduced. This reduction in voltage gain may be accounted for by including the load resistance in parallel with the circuit output resistance as shown in the following example.

EXAMPLE 8.4 Determine the circuit factors A_v, R_i, and R_o and the output voltage V_o for the JFET circuit of Fig. 8.4.

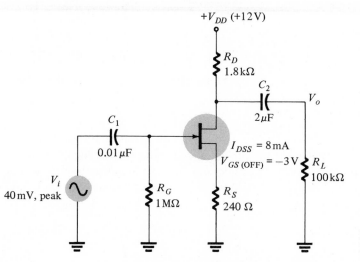

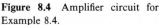

Figure 8.4 Amplifier circuit for Example 8.4.

Solution: The dc bias analysis determines that $V_{GS_Q} = -0.92$ V. The device transconductance at that bias point is then

$$g_m = \frac{2I_{DSS}}{V_{GS(OFF)}}\left[1 - \frac{V_{GS_Q}}{V_{GS(OFF)}}\right] = \frac{2(8 \text{ mA})}{-3 \text{ V}}\left(1 - \frac{-0.92 \text{ V}}{-3 \text{ V}}\right)$$

$$= \frac{16 \times 10^{-3}}{3}(0.693) = 3.696 \text{ mS}$$

and

$$r_s = 1/g_m = 1/3.696 \times 10^{-3} = 271 \text{ }\Omega$$

Using Eq. (8.9) with R_L included in parallel with R_D

$$A_v = -\frac{R_D \parallel R_L}{r_s + R_S} = -\frac{1.8 \text{ k}\Omega \parallel 100 \text{ k}\Omega}{271 \text{ }\Omega + 240 \text{ }\Omega} = -\frac{1.768 \text{ k}\Omega}{511 \text{ }\Omega} = -3.46$$

$$R_i = R_G = 1 \text{ M}\Omega$$

$$R_o = R_D = 1.8 \text{ k}\Omega$$

The resulting output voltage across the load is then

$$V_o = A_v V_i = -3.46(40 \text{ mV, peak}) = -138.4 \text{ mV, peak}$$

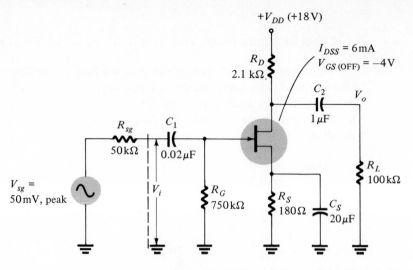

Figure 8.5 FET amplifier circuit for Example 8.5.

Effect of Input Source Resistance

When the input signal is provided by a source having resistance, the input signal to the amplifier, V_i, is reduced from the unloaded value of the signal generator, V_{sg}, by the loading between source resistance, R_{sg}, and the amplifier input resistance, R_i. Example 8.5 demonstrates how the input and output amplifier voltages can be determined using

$$V_i = \frac{R_i}{R_i + R_{sg}} V_{sg}$$

EXAMPLE 8.5 Calculate V_o for the circuit of Fig. 8.5.

Solution: The dc bias of Fig. 8.5 provides $V_{GS_Q} = -0.72$ V. At this bias, the device transconductance is

$$g_m = \frac{2(6 \times 10^{-3})}{|-4|}\left(1 - \frac{-0.72}{-4}\right) = 2.46 \text{ mS}$$

and the value of device ac resistance is

$$r_s = 1/g_m = 1/2.46 \times 10^{-3} = 406.5 \ \Omega$$

The circuit ac voltage gain is then

$$A_v = -\frac{R_D \parallel R_L}{r_s + R_{S1}} = -\frac{2.1 \text{ k}\Omega \parallel 100 \text{ k}\Omega}{406.5 \ \Omega + 0 \ \Omega} = -5.06$$

The circuit ac input resistance is

$$R_i = R_G = 750 \text{ k}\Omega$$

The resulting input voltage V_i is then

$$V_i = \frac{R_i}{R_i + R_{sg}} V_{sg} = \frac{750 \text{ k}\Omega}{50 \text{ k}\Omega + 750 \text{ k}\Omega} (50 \text{ mV, peak}) = 46.88 \text{ mV, peak}$$

The output voltage is then calculated to be

$$V_o = A_v V_i = -5.06 \,(46.88 \text{ mV, peak}) = -237.2 \text{ mV, peak}$$

Both dc bias calculations and ac circuit factors can be calculated using a general computer program such as that listed in Appendix D. Figure D.4 requests as input the information of the circuit and device and provides as output the dc bias and ac factors. Using this program for Examples 8.4 and 8.5 will demonstrate how the availability of a properly programmed computer can make calculations so easy.

```
(a) Example 8.4
PROGRAM TO CALCULATE DC BIAS AND THEN AC ANALYSIS
ENTER THE FOLLOWING CIRCUIT VALUES
RD = ? 1.8E3
RS1(UNBYPASSED SOURCE RESISTANCE)=? 240
RS2=? 0
GATE RESISTANCE TO GROUND, RG2=? 1E6
GATE RESISTANCE TO SUPPLY (USE 1E30 IF OPEN), RG1=? 1E30
SUPPLY VOLTAGE, VDD = ? 12
NOW FOR THE JFET DEVICE PARAMETERS
GATE-SOURCE OFF VOLTAGE, VGS(OFF) = ? -3
SATURATION CURRENT, IDSS = ? 8E-3

RESULTS OF BIAS CALCULATIONS ARE:
BIAS GATE-SOURCE VOLTAGE IS, VGSQ = -.922
BIAS DRAIN-SOURCE CURRENT,IDQ =  3.84E-03
DRAIN-SOURCE BIAS VOLTAGE IS  4.167
DRAIN VOLTAGE TO GROUND IS  5.088
AND SOURCE VOLTAGE TO GROUND IS  .922
SOURCE SIGNAL VOLTAGE =? 40E-3 = Vsg
AND SOURCE GENERATOR RESISTANCE =? 0 = Rsg
LOAD RESISTANCE (USE 1E30 IF OPEN) =? 100E3 = RL
JFET RESISTANCE IS  270.637 = rs
RESULTS OF AC ANALYSIS:
VOLTAGE GAIN,Av=-3.463
INPUT RESISTANCE, Ri= 1E+06
OUTPUT RESISTANCE, Ro= 1800
CURRENT GAIN, Ai= 34.627
AND THE OUTPUT VOLTAGE IS -.139



(b) Example 8.5
PROGRAM TO CALCULATE DC BIAS AND THEN AC ANALYSIS
ENTER THE FOLLOWING CIRCUIT VALUES
RD = ? 2.1E3
RS1(UNBYPASSED SOURCE RESISTANCE)=? 0
RS2=? 180
GATE RESISTANCE TO GROUND, RG2=? 750E3
GATE RESISTANCE TO SUPPLY (USE 1E30 IF OPEN), RG1=? 1E30
SUPPLY VOLTAGE, VDD = ? 18
NOW FOR THE JFET DEVICE PARAMETERS
GATE-SOURCE OFF VOLTAGE, VGS(OFF) = ? -4
SATURATION CURRENT, IDSS = ? 6E-3

RESULTS OF BIAS CALCULATIONS ARE:
BIAS GATE-SOURCE VOLTAGE IS, VGSQ = -.724
BIAS DRAIN-SOURCE CURRENT,IDQ =  4.024E-03
DRAIN-SOURCE BIAS VOLTAGE IS  8.826
DRAIN VOLTAGE TO GROUND IS  9.55
AND SOURCE VOLTAGE TO GROUND IS  .724
SOURCE SIGNAL VOLTAGE =? 50E-3 = Vsg
AND SOURCE GENERATOR RESISTANCE =? 50E3 = Rsg
LOAD RESISTANCE (USE 1E30 IF OPEN) =? 100E3 = RL
JFET RESISTANCE IS  407.037 = Vs
RESULTS OF AC ANALYSIS:
VOLTAGE GAIN,Av=-5.053
INPUT RESISTANCE, Ri= 750000
OUTPUT RESISTANCE, Ro= 2100
CURRENT GAIN, Ai= 37.898
AND THE OUTPUT VOLTAGE IS -.237
```

Figure 8.6 Computer programs for Example 8.6. (a) shows solution for Example 8.4. (b) shows solution for Example 8.5.

EXAMPLE 8.6 Use the computer program of Fig. D.4 to obtain solution to the circuits of Examples 8.4 and 8.5.

Solution: Run the program for the circuits of Figs. 8.4 and 8.5 as shown in Fig. 8.6. Comparing the results obtained in Examples 8.4 and 8.5 with those obtained using the computer program show very good agreement.

8.4 SOURCE FOLLOWER (COMMON-DRAIN) CIRCUIT

A second ac circuit configuration is the common-drain or source follower shown in Fig. 8.7. The circuit can be seen as a JFET version of a bipolar emitter-follower configuration. In fact, voltage gain is also less than unity with no polarity inversion, and the circuit provides high input resistance and lower output resistance than the common-source configuration.

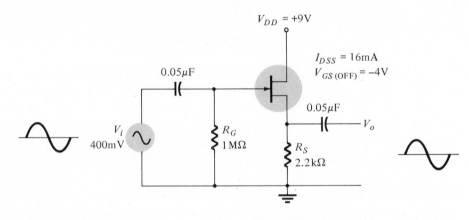

Figure 8.7 Source-follower amplifier circuit.

Source-Follower Amplifier (Output from Source Terminal)

VOLTAGE GAIN

If the output is taken from the source terminal (see Fig. 8.7), there is no polarity inversion between output and input, and the voltage amplitude is reduced from the input value, as given by[2]

---

[2] From ac equivalent circuit

$$V_o = g_m V_{gs} R_S, \qquad V_{gs} = V_i - V_o$$

$$V_{gs} = V_i - g_m V_{gs} R_S$$

$$(1 + g_m R_S) V_{gs} = V_i$$

so that

$$A_v = \frac{V_o}{V_i} = \frac{g_m R_S V_{gs}}{1 + g_m R_S V_{gs}} = \frac{g_m R_S}{1 + g_m R_S} = \frac{\frac{1}{r_s} R_S}{1 + \frac{1}{r_s} R_S} = \frac{R_S}{r_s + R_S}$$

CH. 8 FET SMALL-SIGNAL ANALYSIS

$$A_v = \frac{V_o}{V_i} = \frac{+g_m R_S}{1 + g_m R_S} = \frac{R_S}{r_s + R_S} \qquad (8.10)$$

EXAMPLE 8.7 Calculate the voltage gain of the amplifier circuit of Fig. 8.7.

Solution: From dc bias calculations we obtain $V_{GS_Q} = -2.85$ V and $I_{D_Q} = 1.3$ mA, at which point we calculate

$$g_m = g_{mo}\left(1 - \frac{V_{GS_Q}}{V_{GS(OFF)}}\right) = \frac{2(16 \times 10^{-3})}{|-4|}\left(1 - \frac{-2.85}{-4}\right) = 2.3 \text{ mS}$$

$$r_s = \frac{1}{g_m} = \frac{1}{2.3 \times 10^{-3}} = 435 \ \Omega$$

The voltage gain is then

$$A_v = \frac{V_o}{V_i} = \frac{2.2 \times 10^3}{435 + 2.2 \times 10^3} = \mathbf{0.835}$$

and $$V_o = A_v V_i = 0.835(400 \text{ mV}) = 334 \text{ mV}$$

EXAMPLE 8.8 Determine V_{o1}, V_{o2}, and R_i for the amplifier circuit of Fig. 8.8.

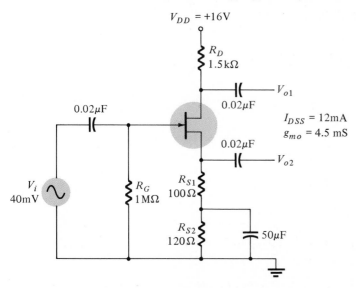

Figure 8.8 JFET amplifier circuit for Example 8.8.

Solution: Using the given JFET device parameters, we can determine $V_{GS(OFF)}$ to be

$$V_{GS(OFF)} = -\frac{2I_{DSS}}{g_{mo}} = -\frac{2(12 \times 10^{-3})}{4.5 \times 10^{-3}} = -5.33 \text{ V}$$

Using a plot of the transfer characteristic or the universal JFET curve, we find the value of the dc bias to be $V_{GS_Q} = -1.4$ V and $I_{D_Q} = 6.5$ mA, at which

$$g_m = g_{mo}\left[1 - \frac{V_{GS_Q}}{V_{GS(OFF)}}\right] = (4.5 \times 10^{-3})\left[1 - \frac{-1.4}{-5.33}\right] = 3.32 \text{ mS}$$

$$r_s = 1/g_m = 1/3.32 \times 10^{-3} = 301.2 \ \Omega$$

The voltage gains from input to output are

$$A_{v1} = \frac{V_{o1}}{V_i} = \frac{-R_D}{r_s + R_{S1}} = \frac{-(1.5 \times 10^3)}{301.2 + 100} = -3.74$$

$$A_{v2} = \frac{V_{o2}}{V_i} = \frac{R_{S1}}{r_s + R_{S1}} = \frac{100}{301.2 + 100} = 0.25$$

so that

$$V_{o1} = A_{v1}V_i = -3.74(40 \text{ mV}) = \textbf{149.6 mV}$$

$$V_{o2} = A_{v2}V_i = +0.25(40 \text{ mV}) = \textbf{10 mV}$$

The input impedance of the amplifier stage is

$$R_i = R_G = \textbf{1 M}\boldsymbol{\Omega}$$

OUTPUT RESISTANCE

The output resistance of a source follower is the parallel resistance of the source resistance, R_S, and the ac resistance seen looking into the JFET source terminal, r_s. The value of R_o can thus be determined using

$$R_o = r_s \,\|\, R_S \tag{8.11}$$

EXAMPLE 8.9   Determine the output resistance of the source-follower circuit of Fig. 8.7, and from output at V_{o2} in Fig. 8.8.

Solution: From Example 8.7, $g_m = 2.3$ mS and $r_s = 435 \ \Omega$. The output resistance of the amplifier is then

$$R_o = r_s \,\|\, R_S = 435 \ \Omega \,\|\, 2.2 \text{ k}\Omega = \textbf{363.2} \ \boldsymbol{\Omega}$$

The output resistance at V_{o2} in Fig. 8.8 is

$$R_o = r_s \,\|\, R_S = 301.2 \ \Omega \,\|\, 100 \ \Omega = \textbf{75.07} \ \boldsymbol{\Omega}$$

EFFECT OF LOAD AND SOURCE RESISTANCE

When a load is connected across the output (see Fig. 8.9), the equation for voltage gain is modified from Eq. (8.10) to that of Eq. (8.12):

$$A_v = V_o/V_i = \frac{R_S \,\|\, R_L}{r_s + R_S \,\|\, R_L} \tag{8.12}$$

If the source has resistance, the resulting input voltage to the amplifier can be obtained as

$$V_i = \frac{R_i}{R_{sg} + R_i} \cdot V_{sg} \tag{8.13}$$

EXAMPLE 8.10   Determine the output voltage of the circuit of Fig. 8.9 for an input of $V_{sg} = 100$ mV, peak with source resistance $R_{sg} = 100$ kΩ.

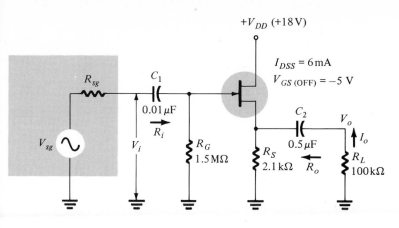

+V_{DD} (+18 V)

$I_{DSS} = 6$ mA
$V_{GS\,(OFF)} = -5$ V

C_1
0.01 μF

R_{sg}

R_G
1.5 MΩ

R_S
2.1 kΩ

C_2
0.5 μF

V_o

R_L
100 kΩ

V_{sg}

V_i

R_i

I_o

R_o

Figure 8.9 Source-follower circuit.

Solution: From the dc bias calculations, $V_{GS_Q} = -2.69$ V. The device transconductance is

$$g_m = \frac{2 I_{DSS}}{|V_{GS(OFF)}|}\left[1 - \frac{V_{GS_Q}}{V_{GS(OFF)}}\right] = \frac{2(6 \times 10^{-3})}{|-5|}\left(1 - \frac{-2.69}{-5}\right)$$

$$= 1.109 \text{ mS}$$

for which

$$r_s = 1/g_m = 1/1.109 \times 10^{-3} = 901.7\ \Omega$$

The circuit ac voltage gain is then

$$A_v = \frac{R_S \parallel R_L}{r_s + R_S \parallel R_L} = \frac{2.1\text{ k}\Omega \parallel 100\text{ k}\Omega}{901.7\ \Omega + 2.1\text{ k}\Omega \parallel 100\text{ k}\Omega} = 0.695$$

The input voltage is

$$V_i = \frac{R_i}{R_i + R_{sg}} \cdot V_{sg} = \frac{1.5\text{ M}\Omega}{(100\text{ k}\Omega + 1.5\text{ M}\Omega)}\qquad (100\text{ mV, peak})$$

$$= 93.75 \text{ mV, peak}$$

The output voltage is then

$$V_o = A_v V_i = 0.695\,(93.75\text{ mV, peak}) = \textbf{65.16 mV, peak}$$

8.5 COMMON-GATE CIRCUIT

A third circuit configuration (see Fig. 8.10) provides input to source, with output taken from drain, with the gate terminal common—this being the common-gate amplifier configuration. In this circuit connection, the input resistance is low, voltage gain is noninverting, and the output resistance is the same as in the common-source circuit.

VOLTAGE GAIN

Including a load across the output, the voltage gain of the circuit of Fig. 8.10 is

$$A_v = \frac{V_o}{V_i} = +g_m\,(R_D \parallel R_L) = \frac{R_D \parallel R_L}{r_s} \qquad (8.14)$$

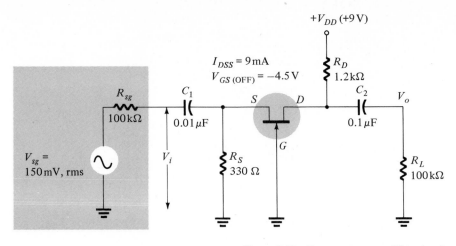

+V_DD (+9 V)

$I_{DSS} = 9$ mA
$V_{GS\,(OFF)} = -4.5$ V

R_{sg}
100 kΩ

C_1
0.01 μF

S   D

C_2
0.1 μF

V_o

R_D
1.2 kΩ

$V_{sg} =$
150 mV, rms

V_i

R_S
330 Ω

G

R_L
100 kΩ

Figure 8.10   Common-gate amplifier circuit.

INPUT AND OUTPUT RESISTANCE

The amplifier input resistance is

$$R_i = R_S \parallel r_s \tag{8.15}$$

and the output resistance is

$$R_o = R_D \tag{8.16}$$

EXAMPLE 8.11   Determine the voltage gain, input and output resistances of the amplifier circuit of Fig. 8.10, and the resulting output voltage, V_o.

Solution: Dc bias calculations result in $V_{GS_Q} = -1.41$ V. At this bias point, the device transconductance is

$$g_m = \frac{2I_{DSS}}{|V_{GS(OFF)}|}\left[1 - \frac{V_{GS_Q}}{V_{GS(OFF)}}\right] = \frac{2(9 \times 10^{-3})}{|-4.5|}\left(1 - \frac{-1.41}{-4.5}\right)$$

$$= 2.747 \text{ mS}$$

and $r_s = 1/g_m = 1/2.747 \text{ mS} = 364 \ \Omega$

The amplifier voltage gain is then

$$A_v = \frac{V_o}{V_i} = +\frac{R_D \parallel R_L}{r_s} = \frac{1.2 \text{ k}\Omega \parallel 100 \text{ k}\Omega}{364 \ \Omega} = +3.258$$

The circuit input resistance is

$$R_i = R_S \parallel r_s = 330 \ \Omega \parallel 364 \ \Omega = \mathbf{173.1 \ \Omega}$$

and the output resistance is

$$R_o = R_D = \mathbf{1.2 \text{ k}\Omega}$$

The value of the input voltage is

$$V_i = \frac{R_i}{R_{sg} + R_i}V_{sg} = \frac{173.1}{100 + 173.1}(150 \text{ mV, peak}) = 95.1 \text{ mV, peak}$$

CH. 8   FET SMALL-SIGNAL ANALYSIS

and the output voltage is then

$$V_o = A_v V_i = +3.258(95.1 \text{ mV, peak}) = \textbf{310 mV, peak}$$

8.6 DESIGN OF FET AMPLIFIER CIRCUITS

To complete our basic coverage of FET amplifier circuits we will consider the design of a few types of amplifiers. Starting with some description of FET and bias condition desired, the circuit resistor values are then determined.

EXAMPLE 8.12   Determine suitable resistor and capacitor values for the JFET amplifier of Fig. 8.11. Use a 2N4220 n-channel JFET having device parameters

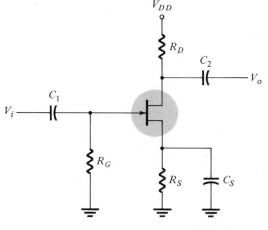

Figure 8.11   JFET amplifier circuit for Example 8.12.

$BV_{GSS(\min)} = 30 \text{ V}$ $g_{fs} = g_m$ is between 1000–4000 μS
$V_{GS(OFF)} = -4 \text{ V} = V_P$ $I_{DSS(\max)} = 3 \text{ mA}$

to achieve a mid-frequency gain of at least 10.

Solution: Our procedure will essentially work reverse from the analysis steps covered up to now.

For a voltage gain of 10 at the minimum g_m of 1000 μS, we have

$$A_v = -g_m R_D$$

$$R_D = \frac{A_v}{-g_m} = \frac{-10}{-1000 \times 10^{-6}} = 10 \text{ k}\Omega$$

The value of $BV_{GSS(\min)}$ is the minimum voltage at which the drain-source voltage will cause breakdown to occur and should not be exceeded. Using a voltage supply of $V_{DD} = 25$ V in this design will then be acceptable. To achieve a bias voltage V_{D_Q} about midway in the range from 0 V to 25 V we selected $V_{D_Q} = 10$ V as bias voltage, and with $R_D = 10$ kΩ

$$V_{D_Q} = V_{DD} - I_{D_Q} R_D$$

$$I_{D_Q} = \frac{V_{DD} - V_{D_Q}}{R_D} = \frac{25 - 10 \text{ V}}{10 \text{ k}\Omega} = 1.5 \text{ mA}$$

To achieve the bias current of 1.5 mA we now determine the value of R_S using the device transfer characteristic or universal JFET characteristic. From the universal characteristic the point

$$\frac{I_D}{I_{DSS}} = \frac{1.5 \text{ mA}}{3 \text{ mA}} = 0.5$$

corresponds to a point

$$\frac{V_{GS}}{|V_P|} = -0.3$$

so that
$$V_{GS_Q} = -0.3(4 \text{ V}) = -1.2 \text{ V}$$
We then calculate

$$V_{GS_Q} = -I_{D_Q}R_S$$

$$R_S = \frac{-V_{GS_Q}}{I_{D_Q}} = \frac{-(-1.2 \text{ V})}{1.5 \text{ mA}} = 800 \text{ } \Omega$$

The value of R_G can be selected $R_G = 1 \text{ M}\Omega$ since

$$I_{GSS(OFF)}R_G = (0.1 \text{ nA})(1 \times 10^6) = 10^{-4} \text{ V} = 0.1 \text{ mV}$$

where I_{GSS} the gate-source leakage current, will result in only a 0.1 mV dc voltage drop across a 1-MΩ gate resistor, a negligible voltage in this circuit. Values of C_1 and C_2 are selected from frequency considerations. *At mid-frequency* ($f = 1000$ Hz) values of $C_1 = C_2 = 0.1 \text{ } \mu\text{F}$ would provide an ac impedance magnitude of

$$X_C = \frac{1}{2\pi fC} = \frac{1}{2\pi(1000)(0.1 \times 10^{-6})} \cong 1.6 \text{ k}\Omega$$

which is small compared to $R_G = 1 \text{ M}\Omega$

A value of C_S for which X_S is at least 10 times smaller than R_S to provide good ac bypass would be

$$X_C \leq \frac{1}{10}R_S = \frac{800}{10} = 80 \text{ } \Omega$$

$$\frac{1}{2\pi fC_S} = 80 \text{ } \Omega$$

$$C_S = \frac{1}{2\pi(1000)(80)} = 1.9 \times 10^{-6} \cong 2 \text{ } \mu\text{F}$$

EXAMPLE 8.13 Design a FET amplifier as shown in Fig. 8.12 to operate from a 12-V supply with a gain of at least 5. Bias at $V_{D_Q} = 8$ V, $I_{D_Q} = 0.25$ mA.

Solution: To obtain the bias condition

$$V_{D_Q} = V_{DD} - I_{D_Q}R_D$$

$$R_D = \frac{V_{DD} - V_{D_Q}}{I_{D_Q}} = \frac{12 - 8 \text{ V}}{0.25 \text{ mA}} = 16 \text{ k}\Omega$$

Using universal characteristics, we can determine V_{GS} at bias using $I_D/I_{DSS} = 0.25$ mA/0.5 mA $= 0.5$ for which we get $(V_{GS_Q}/|V_P|) = -0.3$ so that

$$V_{GS_Q} = 0.3(-2) = -0.6 \text{ V}$$

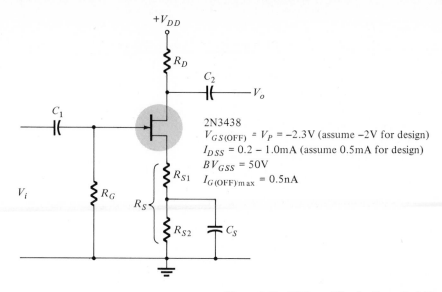

+V_{DD}

R_D

C_2

V_o

C_1

2N3438
$V_{GS(OFF)} = V_P = -2.3\text{V}$ (assume -2V for design)
$I_{DSS} = 0.2 - 1.0\text{mA}$ (assume 0.5mA for design)
$BV_{GSS} = 50\text{V}$
$I_{G(OFF)\max} = 0.5\text{nA}$

V_i

R_G

R_S

R_{S1}

R_{S2}

C_S

Figure 8.12 FET amplifier for Example 8.13.

The value of R_S needed is then

$$V_{GS_Q} = -I_{D_Q} R_S$$

$$R_S = -\frac{V_{GS_Q}}{I_{D_Q}} = \frac{-(-0.6 \text{ V})}{0.25 \text{ mA}} = 2.4 \text{ k}\Omega$$

At bias point we have

$$g_m = g_{mo}\left(1 - \frac{V_{GS_Q}}{V_P}\right) = \frac{2(0.5 \times 10^{-3})}{|-2|}\left(1 - \frac{-0.6}{-2}\right) = 0.35 \text{ mS}$$

$$r_s = \frac{1}{g_m} = 2.857 \text{ k}\Omega$$

For the voltage gain we can write

$$A_v = \frac{-R_D}{r_s + R_{S1}} = \frac{-(16 \text{ k}\Omega)}{2.857 \text{ k}\Omega + R_{S1}} \geq -5$$

which requires that $R_{S1} \leq 343 \ \Omega$.

Choosing $R_{S1} = 270 \ \Omega$ and $R_{S2} = 2.1 \text{ k}\Omega$ ($R_{S1} + R_{S2} = 2.4 \text{ k}\Omega$) would provide the bias values specified with gain of

$$A_v = \frac{16 \times 10^3}{2.857 \times 10^3 + 270} \cong -5.1$$

Choosing values of $R_G = 1 \text{ M}\Omega$ and $C_1 = C_2 = 0.1 \ \mu\text{F}$ as in Example 8.12 would be satisfactory.

The value of C_S is chosen so that at $f = 1000$ Hz

$$X_{C_s} \leq \frac{1}{10} R_{S2} = \frac{2.1 \text{ k}\Omega}{10} = 210 \ \Omega$$

$$C_s = \frac{1}{2\pi(1000)(210)} \cong 0.76 \times 10^{-6} \cong 1\mu\text{F}$$

8.7 THE FET AS A VOLTAGE-VARIABLE RESISTOR (VVR)

The drain-source resistance of a FET can be varied as a function of applied gate-source voltage. This control is fairly linear and applies to the device operating region shown in Fig. 8.13a. Note that this is only a limited part of the FET operating region and is not the linear region of operation as an amplifier. The current range shown is limited to only about 100 μA and a corresponding voltage range of only a few hundred millivolts. Within this limited operating region the FET can be used as a voltage-variable resistor (VVR).

An enlarged view of the low-level region in which the FET can be used as a VVR is shown in Fig. 8.13b. The slope representing the device resistance is seen to

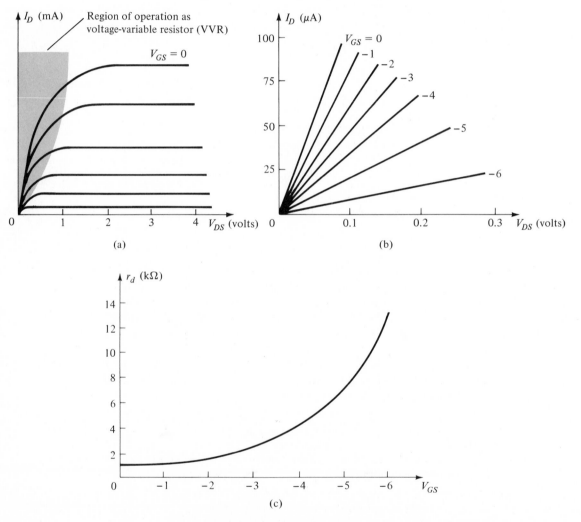

Figure 8.13 (a) and (b) operating action of FET as VVR; (c) resistance versus control voltage.

vary as a function of gate-source control voltage. For example, we see that the slope is steepest, and therefore resistance least, for $V_{GS} = 0$ V whereas the slope is least, and resistance greatest, for $V_{GS} = -6$ V. A graph of device resistance versus control voltage obtained from Fig. 8.13b can be made as shown in Fig. 8.13c. Here we also see that the resistance increases with larger control voltage, although not in a linear manner. The change in device resistance is greatest at larger values of gate-source voltage.

Applications of VVR

One common application of a VVR is to vary the gain of an amplifier chain in order to achieve gain control. If this gain control results from a control voltage derived from the output voltage, then an automatic gain control (AGC) action is obtained. A simplified circuit diagram for such operation is shown in Fig. 8.14. The

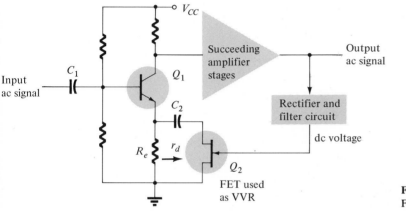

Figure 8.14 AGC amplifier using FET as VVR.

ac input signal is applied to an amplifier stage (Q_1) and then to succeeding amplifier stages to obtain an ac output signal. As a means of maintaining the output signal level constant the amplifier gain may be reduced as signal level increases. To achieve this gain control the output ac signal is rectified and filtered, thereby providing a dc voltage whose magnitude increases as the ac signal magnitude increases. This dc voltage is then applied to an FET used as a VVR. Notice that the FET is now used as a resistor, r_d, which effectively is in parallel (for ac operation) with resistor R_e. In this way the resistance of the FET from drain to source acts to vary the effective emitter degeneration resistance seen by amplifier Q_1. Capacitor C_2 serves only to prevent the operation of the VVR from affecting dc bias of stage Q_1.

The gain of transistor Q_1 is then decreased as the output signal level increases. This occurs since the resulting increased dc control voltage to the FET causes its resistance as a VVR to increase, thereby allowing a larger effective emitter degeneration resistance for stage Q_1 and less voltage gain by that stage.

Other applications of the FET as a VVR might include voltage-controlled bandwidth in an LC circuit, electronically tuned RC filter, expander, and compressor circuits in hi-fi, and so on.

8.8 FET APPLICATIONS

Bootstrap Source-Follower Circuit

Although the FET amplifier has a high input impedance due to the large value of R_G, there are applications in which a much higher input impedance is desired. The bootstrap source-follower circuit of Fig. 8.15 provides this increased input impedance. The "bootstrap" effect due to the 100-μF capacitor feeding a portion of the ac output voltage back to R_G makes that resistor have an effective Miller impedance of

$$(R_{G1})_{\text{effective}} = R'_{G1} = \frac{R_{G1}}{1 - A_v} = R_i \qquad (8.17)$$

where

$$A_v = \frac{V_o}{V_i} \cong \frac{R_S \| R_{G2}}{r_s + (R_S \| R_{G2})}$$

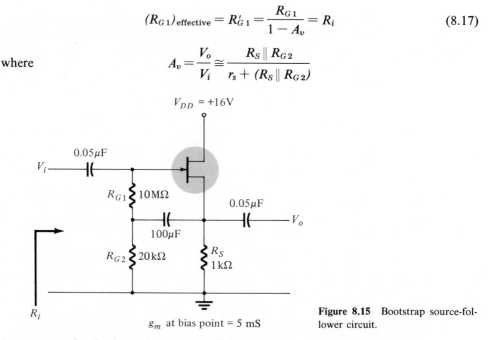

Figure 8.15 Bootstrap source-follower circuit.

g_m at bias point = 5 mS

As an example, the input impedance of the bootstrap circuit of Fig. 8.15 is

$$R_S \| R_{G2} = 1 \text{ k}\Omega \| 20 \text{ k}\Omega = 0.95 \text{ k}\Omega = 950 \ \Omega$$

$$r_s = \frac{1}{g_m} = \frac{1}{5 \times 10^{-3}} = 200 \ \Omega$$

$$A_v = \frac{950}{200 + 950} = 0.826$$

so that

$$R_i = R'_{G1} = \frac{10 \text{ M}\Omega}{1 - 0.826} = 57.47 \text{ M}\Omega$$

Ideal Buffer Amplifier

An ideal buffer amplifier would have infinite input impedance, zero output impedance, and a voltage gain of unity. One practical form of such a circuit can be obtained

CH. 8 FET SMALL-SIGNAL ANALYSIS

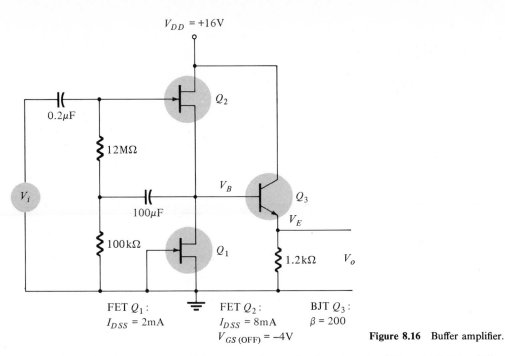

$V_{DD} = +16\text{V}$

$0.2\mu\text{F}$

$12\text{M}\Omega$

Q_2

V_B

$100\mu\text{F}$

Q_3

V_E

$100\text{k}\Omega$

Q_1

V_i

$1.2\text{k}\Omega$ V_o

FET Q_1:   FET Q_2:   BJT Q_3:
$I_{DSS} = 2\text{mA}$ $I_{DSS} = 8\text{mA}$ $\beta = 200$
                  $V_{GS\,(OFF)} = -4\text{V}$

Figure 8.16 Buffer amplifier.

using a bootstrap source-follower as shown in the circuit of Fig. 8.16. The following calculations show how the above circuit parameters can be calculated.

Since Q_1 has $V_{GS} = 0$ V, the drain current which is the current set for Q_2 is about 2 mA. From the transfer characteristic of Q_2 it can be determined that at $I_{DQ} = 2$ mA the value of V_{GSQ} is $V_{GSQ} = -2$ V. Since $V_Q = 0$ V, the value of V_S is

$$V_S = V_B = 2 \text{ V}$$

so that

$$V_E = 2 - 0.7 = 1.3 \text{ V}$$

and

$$I_E = \frac{V_E}{R_E} = \frac{1.3 \text{ V}}{1.2 \text{ k}\Omega} = 1.08 \text{ mA}$$

We can then calculate r_e

$$r_e = \frac{26}{1.08} + 2 \cong 26 \text{ }\Omega$$

The source impedance for Q_2 is then calculated from

$$R_S = 100 \text{ k}\Omega \parallel \beta(r_e + R_E) = 100 \text{ k}\Omega \parallel 200(1.2 \text{ k}\Omega + 26 \text{ }\Omega) \cong 71 \text{ k}\Omega$$

At bias point for Q_2 we calculate

$$g_m = g_{mo}\left(1 - \frac{V_{GSQ}}{V_P}\right) = \frac{2(8 \times 10^{-3})}{|-4|}\left(1 - \frac{-2}{-4}\right) = 2 \text{ mS}$$

$$r_s = \frac{1}{g_m} = \frac{1}{2 \text{ mS}} = 500 \text{ }\Omega$$

and

$$A_v = \frac{R_S}{r_s + R_S} = \frac{71 \times 10^3}{500 + 71 \times 10^3} = 0.993$$

The input impedance is then determined to be

$$R_i = R'_{G1} = \frac{R_{G1}}{1 - A_v} = \frac{12 \text{ M}\Omega}{1 - 0.993} = 1.714 \times 10^9 = 1714 \text{ M}\Omega$$

an extremely large value, as desired.

The voltage gain of the BJT emitter follower is

$$A_{v2} = \frac{R_E}{R_E + r_e} = \frac{1.2 \text{ k}\Omega}{1.2 \text{ k}\Omega + 26 \text{ }\Omega} = 0.9788$$

so that overall gain is

$$A_v = A_{v1}A_{v2} = (0.993)(0.9788) \cong 0.9719$$

which is very close to the ideal value of 1.

Finally, the output resistance of the circuit taken from the emitter-follower output is

$$R_o \cong r_e = 26 \text{ }\Omega$$

a low output resistance.

8.9 HIGH-FREQUENCY EFFECTS—MILLER CAPACITANCE

Ac analysis so far has been applied only to the mid-frequency operation of the circuit. At higher frequencies, device capacitances between terminals cause reduction in the amplifier gain due to decreased capacitive impedance with increased frequency. These circuit capacitances resulting from device construction (or stray wiring) are connected with dotted lines in the circuit of Fig. 8.17a to indicate they are not capacitances that are connected into the circuit but arise as a result of the circuit and device construction. While the device capacitance between each set of terminals affects the overall amplifier gain, the capacitance between input and output has the greatest effect because of the *Miller effect,* which results in the effective capacitance being multiplied by the gain of the amplifier as will now be described.

Miller Effect (Miller Capacitance)

The most pronounced effect due to these device capacitances is due to the capacitance between the input and output terminals, C_{gd}, in the circuit of Fig. 8.17a. While each individual capacitive impedance provides a loading at higher frequencies, the magnitude of the effective loading due to the input–output capacitance, C_{gd}, is magnified by the amplifier gain. Figure 8.17b shows an ac equivalent circuit of the JFET amplifier of Fig. 8.17a, emphasizing the effect of the interterminal capacitances. In Fig. 8.17b, the input current is

$$I_i = V_i Y_i = I_1 + I_2 = V_i Y_{gs} + (V_i - A_v V_i)Y_{gd}$$

from which we get

$$Y_i = Y_{gs} + (1 - A_v)Y_{gd}$$

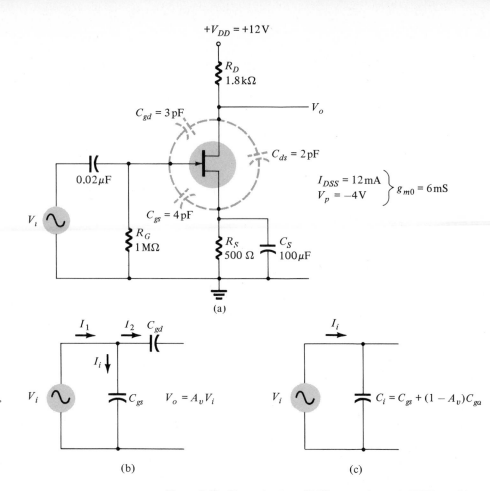

Figure 8.17 Determination of Miller capacitance in JFET amplifier.

which can be expressed in terms of the capacitance values

$$C_i = C_{gs} + (1 - A_v)C_{gd} \qquad (8.17a)$$

where

$$C_M = (1 - A_v)C_{gd} \qquad (8.17b)$$

C_M being the Miller capacitance.

As shown in Fig. 8.17b, the effective input capacitance due to both C_{gs} and C_{gd} is the effective capacitance given by Eq. (8.17), which is mostly that due to the Miller effect.

EXAMPLE 8.14 Calculate the value of Miller capacitance in the circuit of Fig. 8.17.

Solution: Calculating the dc bias to be $V_{GS_Q} = -1.8$ V, the device transconductance at the bias point is

$$g_m = (6 \times 10^{-3})\left(1 - \frac{-1.8}{-4}\right) = 3.3 \text{ mS}$$

The value of r_s is then

$$r_s = 1/g_m = 1/(3.3 \times 10^{-3}) = 303 \ \Omega$$

and the circuit mid-frequency gain is

$$A_v = \frac{-R_D}{r_s} = \frac{-1.8 \times 10^3}{303} = -5.94$$

The Miller capacitance is then

$$C_M = (1 - A_v)C_{gd} = [1 - (-5.94)] \ 3 \ \text{pF} = 20.82 \ \text{pF}$$

PROBLEMS

§ 8.3

1. Calculate the voltage gain for the circuit of Fig. 8.2 using a JFET having $I_{DSS} = 10$ mA and $V_{GS(OFF)} = -4$ V.

2. Calculate the input and output resistance of the amplifier circuit of Fig. 8.2 with JFET having $I_{DSS} = 10$ mA and $V_{GS(OFF)} = -4$ V.

3. Calculate the voltage gain of the amplifier of Fig. 8.2 with R_D replaced by $R_D = 1.8$ kΩ.

4. Calculate the voltage gain of the circuit of Fig. 8.2 with R_S replaced by 330 Ω.

5. Determine the voltage gain of the circuit of Fig. 8.3 using a JFET having $V_{GS(OFF)} = -5$ V and $I_{DSS} = 8$ mA.

6. What is the voltage gain of the circuit in Fig. 8.3 with R_D replaced by 5.6 kΩ?

7. Calculate the output voltage for the circuit of Fig. 8.3 with input of 100 mV, peak.

8. Calculate the output voltage of the circuit of Fig. 8.3 with $R_D = 3.6$ kΩ and $R_{S1} = 150$ Ω.

9. Determine the values of A_v, R_i, and R_o for the circuit of Fig. 8.4 with JFET having $I_{DSS} = 10$ mA and $V_{GS(OFF)} = -4$ V.

10. What output voltage results if the load resistance in Fig. 8.4 is changed to 10 kΩ?

11. Calculate the voltage gain of the amplifier circuit of Fig. 8.4 with $R_S = 180$ Ω.

12. Determine the output voltage for the circuit of Fig. 8.5 for input of $V_{sg} = 80$ mV, peak and load $R_L = 22$ kΩ.

13. Calculate the voltage gain of the circuit of Fig. 8.5 with $R_D = 3.3$ kΩ and $R_L = 50$ kΩ.

14. What is the output voltage of the circuit of Fig. 8.5 with $V_{sg} = 120$ mV, peak, $R_{sg} = 100$ kΩ, and $R_L = 20$ kΩ?

15. Determine the value of input voltage needed in the circuit of Fig. 8.5 to obtain an output of 300 mV, peak with load $R_L = 50$ kΩ.

16. Calculate the values of A_v, R_i, R_o, and V_o for the circuit of Fig. 8.5 with $R_D = 3.3$ kΩ, $R_L = 50$ kΩ, and JFET having $I_{DSS} = 8$ mA and $V_{GS(OFF)} = -5$ V.

§ 8.4

17. Calculate the voltage gain of the amplifier circuit of Fig. 8.7 with $R_S = 1.8$ kΩ.

18. Calculate the output voltage of the circuit of Fig. 8.7 with $V_i = 500$ mV, peak and $R_S = 1.5$ kΩ.

19. Determine the output voltage of the circuit of Fig. 8.7 with $R_S = 1.5$ kΩ.

20. What output voltage results in the circuit of Fig. 8.7 with JFET having $I_{DSS} = 12$ mA and $V_{GS(OFF)} = -5$ V?

21. What value of output voltage V_{o1} results in the circuit of Fig. 8.8 with $R_D = 1.8$ kΩ?

22. What output voltage V_{o1} results in the circuit of Fig. 8.8 with load of 10 kΩ across V_{o1}?

23. What output voltage V_{o2} results in the circuit of Fig. 8.8 with load of 10 kΩ across V_{o2}?

24. What is the voltage gain V_{o1}/V_i for the circuit of Fig. 8.8 with JFET having $I_{DSS} = 10$ mA and $V_{GS(OFF)} = -4$ V?

25. What is the voltage gain V_{o2}/V_i for the circuit of Fig. 8.8 with JFET having $I_{DSS} = 10$ mA and $V_{GS(OFF)} = -4$ V?

26. What is the output voltage in the circuit of Fig. 8.9 with $R_L = 10$ kΩ? ($V_{sg} = 100$ mV, peak and $R_{sg} = 100$ kΩ.)

27. What are the values of input and output resistance for the circuit of Fig. 8.9 with JFET having $I_{DSS} = 8$ mA and $V_{GS(OFF)} = -4$ V?

28. What value of input voltage, V_{sg}, is needed in the circuit of Fig. 8.9 to provide output $V_o = 150$ mV, peak if $R_{sg} = 50$ kΩ and $R_L = 10$ kΩ?

§ 8.5

29. Determine the voltage gain and input and output resistances of the circuit of Fig. 8.10 with $R_D = 1.8$ kΩ.

30. What is the output voltage V_o in the circuit of Fig. 8.10 with $R_L = 10$ kΩ?

§ 8.6

31. Design the amplifier circuit in Fig. 8.11 (select values of R_D and R_S) for a JFET having $V_{GS(OFF)} = -6$ V, $I_{DSS} = 12$ mA, and $g_m = 2000$ μS. The voltage gain required is -10. Use a supply of $V_{DD} = 20$ V.

32. Design a JFET amplifier as shown in Fig. 8.12 to operate from an 18-V supply with voltage gain of at least -45. Bias at $V_{DQ} = 9$ V, and $I_{DQ} = 0.4$ mA.

§ 8.8

33. Determine the input resistance of the circuit if Fig. 8.15 for a JFET having $g_m = 8$ mS at the bias point.

34. Determine the input and output resistances of the circuit of Fig. 8.16 using JFETs $Q1$ with $I_{DSS} = 3$ mA, $Q2$ with $I_{DSS} = 10$ mA, and bipolar transistor $Q3$ with $\beta = 150$.

35. Calculate the Miller capacitance of a circuit as in Fig. 8.17 with $R_D = 2.1$ kΩ and $C_{gd} = 4.5$ pF.

36. Calculate the input capacitance of a circuit as in Fig. 8.17 with $C_{gs} = 4.5$ pF and $R_D = 2.1$ kΩ.

COMPUTER PROBLEMS

Write BASIC programs to:

1. Calculate the voltage gain of a common-source circuit as in Fig. 8.4.

2. Calculate the input and output resistances of a JFET circuit as in Fig. 8.4.

3. Calculate the input and output resistances of a JFET source follower as in Fig. 8.7.

4. Calculate the voltage gain of a source follower as in Fig. 8.7.

5. Calculate the output voltage for a source follower as in Fig. 8.9.

6. Calculate the output voltage of a common-source amplifier as in Fig. 8.5.

9

Multistage Systems and Frequency Considerations

9.1 INTRODUCTION

This chapter will include, under the heading of multistage systems, both the *cascaded* and *compound* configurations. The *cascaded* system, for the purposes of this text, is one in which each stage and the connections between stages are very similar or identical. The *compound* system includes all other possible *multiple* active-device configurations, each stage of which can be completely different in appearance with a variety of interconnections.

The first few sections of this chapter examine multistage systems employing the technique of analysis developed in earlier chapters. This is followed by a detailed discussion of decibels (dB) and the effect of frequency on the response of single and multistage systems.

9.2 GENERAL CASCADED SYSTEMS

A discussion of cascaded systems is best initiated by considering the block-diagram representation of Fig. 9.1. The quantities of interest are indicated in the figure. The indicated A_v (voltage amplification) and A_i (current amplification) of each stage were determined with all stages connected as indicated in Fig. 9.1. In other words, A_v and A_i of each stage *do not* represent the gain of each stage on an independent basis. The loading effect of one stage on another was considered when these quantities were determined. All levels of gain, voltage, current, and impedance are magnitudes only and are not complex.

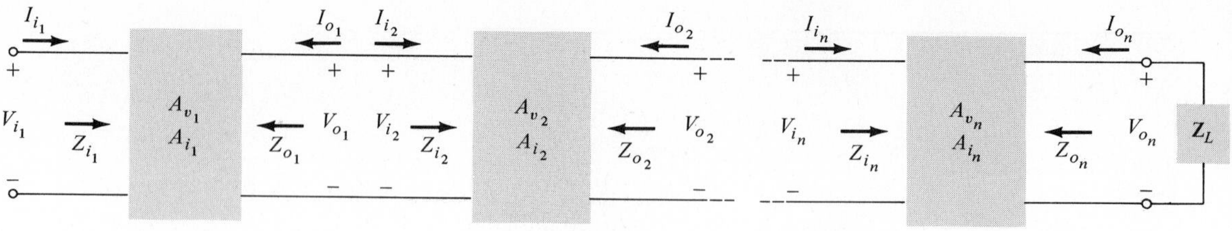

Figure 9.1 General cascaded system.

Rather than simply state the result for the overall gain of the system (voltage or current) a simple numerical example will clearly indicate the solution. If $A_{v_1} = -40$ and $A_{v_2} = -50$ with $V_{i_1} = 1$ mV, then $V_{o_1} = A_{v_1} \times V_{i_1} = -40(1 \text{ mV}) = -40$ mV. Since $V_{o_1} = V_{i_2}$,

$$V_{o_2} = A_{v_2} V_{i_2} = -50(-40 \text{ mV}) = 2000 \text{ mV} = 2 \text{ V}$$

The overall gain is $A_{v_T} = 2000 \text{ mV}/1 \text{ mV} = 2000$.

Obviously, the total gain of the two stages is simply the product of the individual gains A_{v_1} and A_{v_2}. In general, for n stages,

$$\boxed{A_{v_T} = \pm A_{v_1} A_{v_2} A_{v_3} \cdot \cdot \cdot A_{v_n}} \tag{9.1}$$

The same is true for the net current gain

$$\boxed{A_{i_T} = \pm A_{i_1} A_{i_2} A_{i_3} \cdot \cdot \cdot A_{i_n}} \tag{9.2}$$

The input and output impedance of each stage as indicated in Fig. 9.1 are also those values obtained by considering the effects of each and every stage of the system. There is no generally employed equation, such as Eq. (9.2), for the input or output impedances of the system in terms of the individual values. However, in a number of situations (transistor, FET, or tube) the input (or output) impedance can normally be determined to an acceptable degree of accuracy by considering only one, or perhaps two, stages of the system.

The magnitude of the overall voltage gain of the representative system of Fig. 9.1 can be written as

$$|A_{v_T}| = \left| \frac{V_{o_n}}{V_{i_1}} \right| = \left| \frac{-I_{o_n} Z_L}{I_{i_1} Z_{i_1}} \right|$$

so that

$$\boxed{|A_{v_T}| = |A_{i_T}| \cdot \left| \frac{Z_L}{Z_{i_1}} \right|} \tag{9.3}$$

Equation (9.3) will prove useful in the analysis to follow. To go a step further, if the product of the voltage and current gain is formed for resistive loads

$$|A_{v_T} A_{i_T}| = \left| \frac{I_{o_n} R_L}{I_{i_1} R_{i_1}} \right| \cdot \left| \frac{I_{o_n}}{I_{i_1}} \right| = \left| \frac{I_{o_n}^2 R_L}{I_{i_1}^2 R_{i_1}} \right| = \frac{P_o}{P_i}$$

and
$$\boxed{|A_{p_T}| = |A_{v_T}| \cdot |A_{i_T}|}$$
(magnitude only) (9.4)

this being the overall power gain of the system.

There are three types of coupling between stages of a system, such as in Fig. 9.1, that will be considered. The first to be described is the *RC-coupled* amplifier system, which is the most frequently applied of the three. This will be followed by the *transformer* and *direct-coupled* amplifier systems.

9.3 *RC*-COUPLED AMPLIFIERS

A cascaded *RC*-coupled transistor amplifier (two-stage) showing typical values and biasing techniques appears in Fig. 9.2. The terminology "*RC*-coupled" is derived from the biasing resistors and coupling capacitors employed between stages.

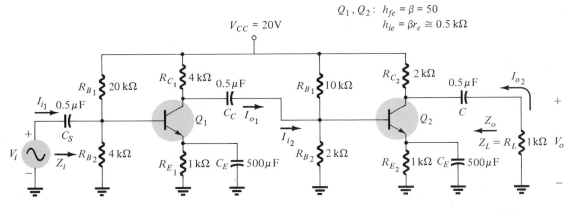

Figure 9.2 Two-stage *RC*-coupled amplifier.

The primary function of the approximate technique is to obtain a "ball-park" solution with a minimum of time and effort. A reduced time element obviously requires that the network be redrawn a minimum number of times. In fact, let us optimistically say that we can find the solutions for the network of Fig. 9.2 using only the original artwork.

Z_i

From the past experience with single-stage amplifiers and the analysis just completed, it should be clear that for the ac response, both the 4-kΩ and 20-kΩ resistors will appear in parallel if the network is redrawn. They are also in parallel with the input impedance of Q_1, which is approximately $h_{ie} = \beta r_e = 0.5$ kΩ since the emitter resistor is bypassed by C_E.

The parallel combination:

$$Z_{i_1} = 20 \text{ k}\Omega \| 4 \text{ k}\Omega \| 0.5 \text{ k}\Omega = \mathbf{0.435 \text{ k}\Omega}$$

Z_o

Recall that the approximate collector-to-emitter equivalent circuit of a transistor is simply a current source $h_{fe}I_b$. This being the case, when $V_i = 0$, then $I_{b_1} = 0$, and $I_{b_2} = 0$, resulting in $h_{fe}I_{b_2} = 0$, so that Z_o is simply R_{C_2} in parallel with the open-circuit representation of the controlled current source. That is,

$$Z_o|_{V_i=0} = R_{C_2} = 2\text{ k}\Omega$$

A_i

Applying the current-divider rule (Fig. 9.3):

$$I_{b_1} = \frac{R'I_{i_1}}{R' + h_{ie}}$$

$$= \frac{3.333\text{ k}\Omega\, I_{i_1}}{3.333\text{ k}\Omega + 0.5\text{ k}\Omega}$$

and
$$I_{b_1} \cong 0.87I_{i_1}$$

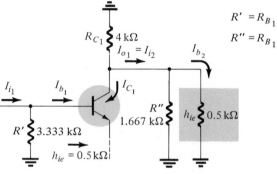

$R' = R_{B_1} \| R_{B_2} = 20\text{ k}\Omega \| 4\text{ k}\Omega$

$R'' = R_{B_1} \| R_{B_2} = 10\text{ k}\Omega \| 2\text{ k}\Omega$

Figure 9.3 Determining the relationship between I_{i_1} and I_{b_2}.

The collector current of the first stage $I_{C_1} \cong h_{fe}I_{b_1}$. However, I_{C_1} will divide between the 4-kΩ resistor and the *loading* of the second stage (Fig. 9.3). The parallel combination of the 10-kΩ, 2-kΩ, and 0.5-kΩ resistors as the loading of the next stage will result in an impedance of $= 0.385$ kΩ.

Applying the current-divider rule:

$$I_{o_1} = \frac{-R_{C_1}\,(I_{C_1})}{R_{C_1} + 0.385\text{ k}\Omega} = \frac{-4\text{ k}\Omega\,(I_{C_1})}{4\text{ k}\Omega + 0.385\text{ k}\Omega}$$

$$= \frac{-4\text{ k}\Omega\,(h_{fe}I_{b_1})}{4.385\text{ k}\Omega} = \frac{-4\text{ k}\Omega\,(50)(0.87\,I_{i_1})}{4.385\text{ k}\Omega}$$

and
$$A_{i_1} = \frac{I_{o_1}}{I_{i_1}} \cong -39.68$$

For the second stage:

$$I_{b_2} = \frac{R'' I_{i_2}}{R'' + h_{ie}} = \frac{1.667 \text{ k}\Omega \ (I_{i_2})}{1.667 \text{ k}\Omega + 0.5 \text{ k}\Omega} = 0.769 \ I_{i_2}$$

and
$$I_{C_2} = h_{fe}I_{b_2} = 50(0.769 I_{i_2}) = 38.45 I_{i_2}$$

Applying the current-divider rule to the output circuit:

$$I_{o_2} = \frac{R_{C_2}I_{C_2}}{R_{C_2} + R_L} = \frac{2 \text{ k}\Omega \ (h_{fe}I_{b_2})}{2 \text{ k}\Omega + 1 \text{ k}\Omega} = \frac{2 \text{ k}\Omega(38.45 I_{i_2})}{3 \text{ k}\Omega} = 25.63 I_{i_2}$$

and
$$A_{i_2} = \frac{I_{o_2}}{I_{i_2}} = 25.63$$

with
$$A_{i_T} = A_{i_1} \cdot A_{i_2} = (-39.68)(25.63) \cong -1017.0$$

A_v

The direct connection under ac conditions clearly indicates in Fig. 9.2 that V_i appears directly at the base of the transistor of the first stage. Since the transistor has a grounded emitter terminal, the ac voltage gain can be obtained (on an approximate basis) using the following equation:

$$A_v \cong \frac{-h_{fe}R_L}{h_{ie}} = \frac{-R_L}{r_e}$$

R_L, the loading on the first stage, is the parallel combination of R_{C_1}, R_{B_1}, R_{B_2}, and $h_{ie}(= \beta r_e)$ which is $= 0.3512 \text{ k}\Omega$. The voltage amplification, A_{v_1}, therefore, equals $[-(50)(0.3512 \text{ k}\Omega)]/0.5 \text{ k}\Omega = -35.12$. For the second stage,

$$A_{v2} = \frac{-(50) \ (R_{C_2} \| R_L \ )}{0.5 \text{ k}\Omega} = \frac{-(50)(2 \text{ k}\Omega \| 1 \text{ k}\Omega)}{0.5 \text{ k}\Omega} = \frac{-(50)(0.667 \text{ k}\Omega)}{0.5 \text{ k}\Omega} = -66.70$$

The net gain is, therefore,

$$A_{v_T} = A_{v_1}A_{v_2} = (-35.12)(-66.70)$$

$$A_{v_T} \cong \mathbf{2342.50}$$

Using Eq. (9.3)

$$|A_{v_T}| = |A_{i_T}| \left| \frac{Z_L}{Z_{i_1}} \right| = \frac{(1017.0)(1 \text{ k}\Omega)}{0.435 \text{ k}\Omega} = 2337.93$$

The very slight difference between the obtained values of A_{v_T} is due only to the decimal carry over in the individual solutions for A_{i_T} and A_{v_T}. In this case, the determination of A_{v_T} was markedly easier than that for A_{i_T}. In the future, therefore, there may be some savings in time if A_{v_T} is first determined, and A_{i_T} calculated from Eq. (9.3) in the following form:

$$|A_{i_T}| = |A_{v_T}| \cdot \left| \frac{Z_{i_1}}{Z_L} \right|$$

EXAMPLE 9.1 We shall calculate the input and output impedance, voltage gain, and current

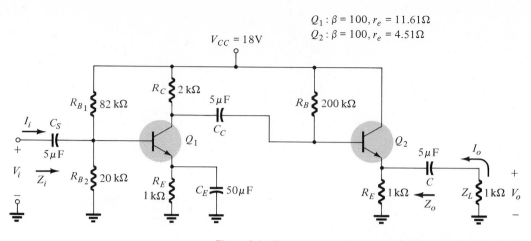

Figure 9.4 Two-stage transistor network to be examined in detail.

gain of the two-stage amplifier of Fig. 9.4. Note that the second stage is an emitter-follower configuration.

Solution:

(a) Z_i: For ac conditions, R_E of the first stage is bypassed by C_E and the input impedance to Q_1 is $\cong \beta r_e = (100)(11.61) = 1.161$ kΩ.
Then

$$Z_i = R_{B_1} \| R_{B_2} \| \beta r_e = 82 \text{ k}\Omega \| 20 \text{ k}\Omega \| 1.161 \text{ k}\Omega \cong \beta r_e = \mathbf{1.161 \text{ k}\Omega}$$

(b) Z_o: For ac conditions the network is redrawn as shown in Fig. 9.5. $Z_e = (R_s/\beta) + r_e$ where R_s is the source resistance connected to the base of the transistor.

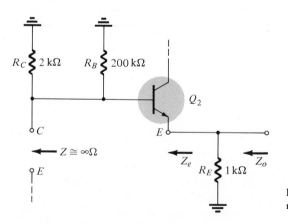

Figure 9.5 Determining Z_o for the network of Fig. 9.4.

In this case $R_s = 2$ k$\Omega \| 200$ k$\Omega \cong 2$ kΩ and

$$Z_e = \frac{2 \text{ k}\Omega}{100} + 4.51 = 20 + 4.51 = 24.51 \ \Omega$$

$$Z_o = Z_e \| R_E = 24.51 \| 1 \text{ k}\Omega \cong \mathbf{24.51 \ \Omega}$$

(c) A_v:
$$V_i = V_{b_1}$$

and

$$A_{v_1} \cong \frac{-R_L}{r_e} = \frac{-(R_C \| R_B \| \beta(R_E \| Z_L))}{r_e} = \frac{-(2 \text{ k}\Omega \| 200 \text{ k}\Omega \| 100(1 \text{ k}\Omega \| 1 \text{ k}\Omega))}{r_e}$$

$$= -\frac{(2 \text{ k}\Omega \| 200 \text{ k}\Omega \| 50 \text{ k}\Omega)}{11.61} \cong -\frac{2 \text{ k}\Omega}{11.61} = -172.27$$

$V_{be_2} \cong 0$ V and $V_{b_2} \cong V_o$ with $A_{v_2} = (V_{o_2}/V_{i_2}) = 1$. Then

$$A_{v_T} = A_{v_1} \cdot A_{v_2} = (-172.27)(1) = \mathbf{-172.27}$$

(d) A_i:
$$|A_{i_T}| = |A_{v_T}| \left| \frac{Z_{i_1}}{Z_L} \right|$$

$$= 172.27 \frac{(1.161 \text{ k}\Omega)}{1 \text{ k}\Omega}$$

$$\cong \mathbf{200}$$

Note above how rapidly the solutions to fairly complex configurations are developing using the approximations developed in Chapter 7. In the following example the values of r_e will have to be determined.

EXAMPLE 9.2   Determine Z_i, Z_o, A_v, A_i, and A_p for the network of Fig. 9.6.

Solution: The values of r_e must be determined. For Q_1:

$$V_{B_1} = \frac{[R_{B_2} \| \beta(R_{E_1} + R_{E_2})] V_{CC}}{[R_{B_2} \| \beta(R_{E_1} + R_{E_2})] + R_{B_1}}$$

but

$$[R_{B_2} \| \beta(R_{E_1} + R_{E_2})] = 22 \text{ k}\Omega \| 80(3 \text{ k}\Omega) = 22 \text{ k}\Omega \| 240 \text{ k}\Omega \cong 22 \text{ k}\Omega = R_{B_2}$$

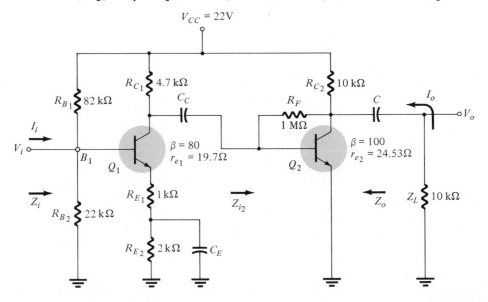

Figure 9.6

and
$$V_B \cong \frac{R_{B_2} V_{CC}}{R_{B_2} + R_{B_1}} = \frac{22\text{ k}\Omega\ (22)}{22\text{ k}\Omega + 82\text{ k}\Omega} = \frac{484}{104} = 4.65\text{ V}$$

and
$$V_E = V_B - V_{BE} = 4.65 - 0.7 = 3.95\text{ V}$$

with
$$I_E = \frac{V_E}{R_{E_1} + R_{E_2}} = \frac{3.95}{3\text{ k}\Omega} = 1.32\text{ mA}$$

and
$$r_{e_1} = \frac{26\text{ mV}}{I_E} = \frac{26}{1.32} = 19.70\ \Omega$$

For Q_2:

$$V_{CC} - (\beta + 1)I_B R_C - R_F I_B - V_{BE} = 0$$

$$22 - (101)I_B\ 10\text{ k}\Omega - 10^6 I_B - 0.7 = 0$$

$$21.3 = 2.01 \times 10^6 I_B$$

and
$$I_B = 10.6\ \mu\text{A}$$

with
$$I_E \cong I_C = \beta I_B = (100)(10.6\ \mu\text{A}) = 1.06\text{ mA}$$

and
$$r_{e_2} = \frac{26\text{ mV}}{I_E} = \frac{26}{1.06} = 24.53\ \Omega$$

(a) Z_i: $\qquad Z_i = R_{B_1} \| R_{B_2} \| \beta R_{E_1} = 82\text{ k}\Omega \| 22\text{ k}\Omega \| 80\text{ k}\Omega \cong \mathbf{14.26\text{ k}\Omega}$

(b) Z_o: $\qquad\qquad\qquad\qquad Z|_{V_i=0} \cong R_C = \mathbf{10\text{ k}\Omega}$

(c) A_v: $\quad V_{b_1} = V_i$ and $\quad A_{v_1} = \dfrac{-R_L}{(R_{E_1} + r_e)} = \dfrac{-(R_C \| Z_{i_2})}{R_{E_1} + r_e}$

From Chapter 7, $\qquad\qquad\qquad Z_{i_2} = \dfrac{R_F}{A_v} \| \beta r_e$

and
$$A_{v_2} = \frac{-R_L}{r_e} = \frac{-R_C \| Z_L}{r_e} = -\frac{5\text{ k}\Omega}{24.53} = -203.83$$

$$Z_{i_2} = \frac{10^6}{203.83} \ \bigg\| \ 100(24.53) = 4.906\text{ k}\Omega \| 2.453\text{ k}\Omega = 1.6353\text{ k}\Omega$$

so that
$$A_{v_1} = \frac{-(4.7\text{ k}\Omega \| 1.6353\text{ k}\Omega)}{(1\text{ k}\Omega + 0.0197\text{ k}\Omega)} = -\frac{1.213}{1.0197} \cong \mathbf{-1.19}$$

and
$$A_{v_T} = A_{v_1} \cdot A_{v_2} = (-1.19)(-203.83) \cong \mathbf{242.56}$$

(d) A_i: $\qquad |A_{i_T}| = |A_{v_T}| \left| \dfrac{Z_{i_1}}{Z_L} \right| = \dfrac{(242.56)(14.26\text{ k}\Omega)}{10\text{ k}\Omega} \cong \mathbf{345.89}$

(e) A_p: $\qquad |A_p| = |A_{i_T}| \cdot |A_{v_T}| = (345.89)(242.56) \cong \mathbf{83.9 \times 10^3}$

EXAMPLE 9.3 *FET RC-Coupled Amplifier. RC* coupling is not limited to transistor stages, as Fig. 9.7 indicates. Determine the total voltage gain.

Solution: Substituting the small-signal equivalent circuit results in the configuration of Fig. 9.8. Combining parallel elements and eliminating those having no effect on the desired overall voltage gain will result in Fig. 9.9.

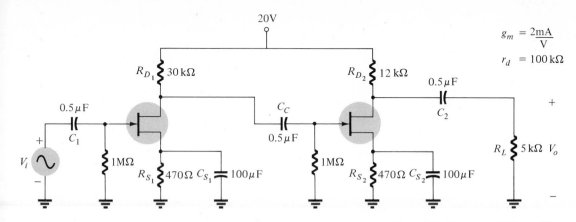

Figure 9.7 Two-stage FET amplifier.

$$g_m = \frac{2\text{mA}}{\text{V}}$$

$$r_d = 100\,\text{k}\Omega$$

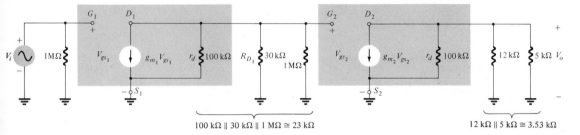

Figure 9.8 Network of Fig. 9.7 following the substitution of the small-signal ac equivalent circuits.

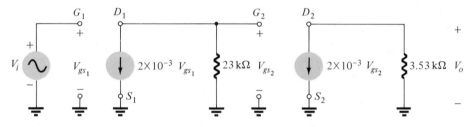

Figure 9.9 Network of Fig. 9.8 following the combination of parallel elements.

Obviously,

$$V_{gs_1} = V_i$$

and

$$V_{gs_2} = -(2 \times 10^{-3}\, V_{gs_1})(23\,\text{k}\Omega)$$

so that

$$V_{gs_2} = -46\, V_{gs_1}$$

The minus sign indicates that the polarity of the voltage across the 23-kΩ resistor due to the current source is the reverse of the defined polarities for V_{gs_2}.

In conclusion:

$$V_o = -(2 \times 10^{-3}\, V_{gs_2})(3.53\,\text{k}\Omega) = -7.06\, V_{gs_2}$$

so that

$$V_o = -7.06\, V_{gs_2} = -7.06(-46\, V_{gs_1}) = 324.8\, V_{gs_1} = 324.8\, V_i$$

and
$$A_v = \frac{V_o}{V_i} = 324.8$$

9.4 TRANSFORMER-COUPLED TRANSISTOR AMPLIFIERS

A two-stage transformer-coupled transistor amplifier is shown in Fig. 9.10. Note that step-down transformers are employed between stages while a step-up transformer is connected to the source V_i. The step-up transformer increases the signal level while the step-down transformer matches, as closely as possible, the loading of each stage to the output impedance of the preceding stage. This is done in an effort to be as close to maximum-power-transfer conditions as possible. The effect of this matching technique through the use of transformer coupling will be clearly demonstrated in the following analysis.

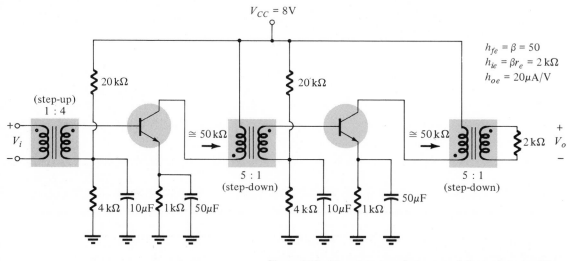

Figure 9.10 Two-stage transformer-coupled transistor amplifier.

Recall that a coupling capacitor was inserted to prevent any dc levels of one stage from affecting the bias conditions of another stage. The transformer provides this dc isolation very nicely.

The basic operation of this circuit is somewhat more efficient than the *RC*-coupled transistors due to the low dc resistance of the collector circuit of the transformer-coupled system. The primary resistance of the transformer is seldom more than a few ohms as compared to the large collector resistance R_C of the *RC*-coupled system. This lower dc resistance results in a lower dc power loss under operating conditions. The efficiency, as determined by the ratio of the ac power out to the dc power in, is therefore somewhat improved.

There are some decided disadvantages, however, to the transformer-coupled system. The most obvious is the increased size of such a system (due to the transformers) compared to *RC*-coupled stages. The second is a poorer frequency response due to the newly introduced reactive elements (inductance of coils and capacitance between

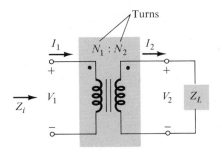

Turns

$N_1 : N_2$

I_1 I_2

$+$ $+$

Z_i V_1 V_2 Z_L

$-$ $-$

Figure 9.11 Basic transformer configuration.

turns). A third consideration, frequently an important one, is the increased cost of the transformer-coupled (as compared to the RC-coupled) system.

Before we consider the ac response of the system, the fundamental equations related to transformer action must be reviewed. For the configuration of Fig. 9.11,

$$\frac{V_1}{V_2} = \frac{N_1}{N_2} = a \qquad \text{(transformation ratio)} \tag{9.5}$$

$$\frac{I_1}{I_2} = \frac{N_2}{N_1} = \frac{1}{a} \tag{9.6}$$

and
$$Z_i = a^2 Z_L \tag{9.7}$$

which states, in words, that the input impedance of a transformer is equal to the turns ratio squared times the load impedance.

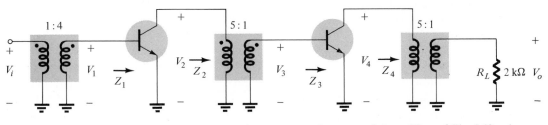

Figure 9.12 Cascaded transformer-coupled amplifiers of Fig. 9.10 redrawn to determine the small-signal ac response.

For the ac response, the circuit of Fig. 9.10 will appear as shown in Fig. 9.12. For maximum power transfer the impedances Z_2 and Z_4 should be equal to the output impedance of each transistor: $Z_o \cong 1/h_{oe} = 1/20 \ \mu S = 50 \ k\Omega$. This is one system where the effect of $1/h_{oe}$ must be considered. The hybrid parameters will therefore be employed in its solution. Applying $Z_i = a^2 Z_L$, $Z_4 = a^2 R_L = (5)^2 2 \ k\Omega = 50 \ k\Omega$. Z_2 is also 50 kΩ since the input resistance to each stage (Z_1 and Z_3) is $\cong h_{ie} = 2 \ k\Omega$. Frequency considerations may not always permit Z_2 or Z_4 to be equal to $1/h_{oe}$. For situations of this type Z_2 and Z_4 are usually made as close as possible to $1/h_{oe}$ in magnitude.

Further analysis of the circuit of Fig. 9.12 results in

$$V_1 = \frac{N_2}{N_1} V_i = 4 V_i$$

and

$$A_{v1} = \frac{-h_{fe} Z_L}{h_{ie}} = \frac{-h_{fe}(\cong 1/h_{oe} \| Z_2)}{h_{ie}} = \frac{-50(50 \text{ k}\Omega \| 50 \text{ k}\Omega)}{2 \text{ k}\Omega} = -625$$

so that

$$V_2 = -625 V_1 = -625(4 V_i) = -2500 V_i$$

but

$$V_3 = \frac{N_2}{N_1} V_2 = \frac{1}{5} V_2 = \frac{1}{5}(-2500 V_i) = -500 V_i$$

and

$$A_{v2} = \frac{-h_{fe} Z_L}{h_{ie}} = \frac{-(50)(25 \text{ k}\Omega)}{2 \text{ k}\Omega} = -625 = \frac{V_4}{V_3}$$

so

$$V_4 = -625 V_3 = -625(-500 V_i)$$
$$= 312.50 \times 10^3 V_i$$

with

$$V_L = \frac{1}{5} V_4 = \frac{1}{5}(312.50 \times 10^3 V_i)$$

and

$$A_{vT} = \frac{V_L}{V_i} = \mathbf{62.50 \times 10^3}$$

9.5 DIRECT-COUPLED TRANSISTOR AMPLIFIERS

The third type of coupling between stages to be introduced in this chapter is *direct coupling*. The circuit of Fig. 9.13 is an example of a two-stage direct-coupled transistor system. Coupling of this type is necessary for very low-frequency applications. For a configuration of this type the dc levels of one stage are obviously related to the dc levels of the other stages of the system. For this reason the biasing arrangement must be designed for the entire network rather than for each stage independently.

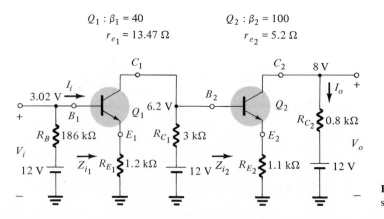

$Q_1 : \beta_1 = 40$ $r_{e_1} = 13.47 \ \Omega$
$Q_2 : \beta_2 = 100$ $r_{e_2} = 5.2 \ \Omega$

Figure 9.13 Direct-coupled transistor stages.

CH. 9 MULTISTAGE SYSTEMS AND FREQUENCY CONSIDERATIONS

Although three separate 12-V supplies are indicated, only one is required if the three terminals of higher (positive) potential for each supply are paralleled.

One of the biggest problems associated with direct-coupled networks is stability. Any variation in dc level in one stage is transmitted on an amplified basis to the other stages. The addition of the emitter resistor aids as a stabilizing element in each stage.

(dc) Bias Conditions

For an output voltage $V_{C_2} = 8$ V as indicated in Fig. 9.13,

$$I_{0.8k\Omega} = \frac{12 - 8}{0.8 \text{ k}\Omega} = 5 \text{ mA}$$

therefore $$I_{C_2} \cong I_{E_2} \cong 5 \text{ mA}$$

and $$V_{E_2} = (5 \text{ mA})(1.1 \text{ k}\Omega) = 5.5 \text{ V}$$

For $$V_{BE_2} = 0.7 \text{ V}$$

$$V_{B_2} = V_{C_1} = 5.5 + 0.7 = 6.2 \text{ V}$$

as indicated. Applying

$$I_C \cong \beta_2 I_B$$

$$I_{B_2} \cong \frac{I_{C_2}}{\beta_2} = \frac{5 \text{ mA}}{100} = 50 \text{ } \mu\text{A}$$

and $$I_{3k\Omega} = \frac{12 - 6.2}{3 \text{ k}\Omega} = \frac{5.8}{3 \text{ k}\Omega} = 1.93 \text{ mA}$$

and since $$I_{3k\Omega} \gg I_{B_2}$$

assume $$I_{C_1} \cong I_{3k\Omega} = 1.93 \text{ mA}$$

and $$I_{E_1} = 1.93 \text{ mA}$$

so $$V_{E_1} = (1.93 \text{ mA})(1.2 \text{ k}\Omega) = 2.32 \text{ V}$$

and $$V_{B_1} = V_{E_1} + V_{BE_1} = 2.32 + 0.7 = 3.02 \text{ V}$$

as indicated.

The above verification of the potential levels appearing in Fig. 9.13 demonstrates clearly the close tie-in required between bias levels of a direct-coupled amplifier.

Now for the ac response. The approach uses the approximate technique introduced earlier in this chapter.

The input impedance to each emitter-follower configuration is $\cong \beta R_E$. Therefore,

$$Z_{i_1} = \beta_1 R_{E_1} = 40(1.2 \text{ k}\Omega) = 48 \text{ k}\Omega$$

and $$Z_{i_2} \cong \beta_2 R_{E_2} = 100(1.1 \text{ k}\Omega) = 110 \text{ k}\Omega$$

$$A_{v_1} = \frac{-R_{L_1}}{R_{E_1}} = \frac{-R_{C_1} \| \beta_2 R_{E_2}}{R_{E_1}} = -\frac{3 \text{ k}\Omega \| 110 \text{ k}\Omega}{1.2 \text{ k}\Omega} \cong \frac{-3 \text{ k}\Omega}{1.2 \text{ k}\Omega} = -2.5$$

$$A_{v2} = \frac{-R_{L2}}{R_{E2}} = \frac{-R_{C2}}{R_{E2}} = \frac{-0.8 \text{ k}\Omega}{1.1 \text{ k}\Omega} = -0.7273$$

and

$$A_{v_T} = A_{v_1}A_{v_2} = (-2.5)(-0.7273) = \mathbf{1.818}$$

$$|A_i| = |A_v|\left|\frac{Z_{i_1}}{Z_L}\right| = \frac{(1.818)(48 \text{ k}\Omega)}{0.8 \text{ k}\Omega} = \mathbf{109.08}$$

with

$$|A_{p_T}| = |A_v| \cdot |A_i| = (1.818)(109.08) = \mathbf{198.3}$$

9.6 CASCODE AMPLIFIER

For high-frequency applications the CB configuration has the most desirable characteristics of the three configurations. However, it suffers from a very low input impedance ($Z_i \cong h_{ib} = r_e$). The *cascode* configuration in Fig. 9.14 is designed to improve the input impedance level for the CB configuration through the use of a typical CE network. The gain of the CE configuration is low to ensure that the input Miller capacitance level is a minimum (recall the discussion of Miller capacitance in Chapter 8 for the FET) for high-frequency applications.

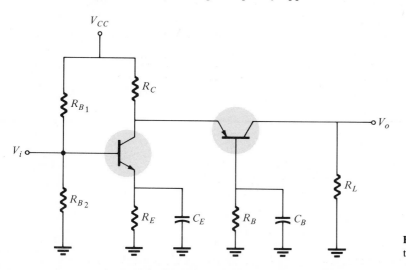

Figure 9.14 Cascode configuration.

A practical version of a cascode amplifier appears in Fig. 9.15. Note that the collector of the CE configuration is still tied directly to the emitter of the CB configuration.

For dc conditions:

$$I_{E_2} \cong I_{E_1} \quad \text{or} \quad I_{C_2} \cong I_{C_1}$$

or dividing each side by β since $\beta_1 = \beta_2 = \beta$

$$\frac{I_{C_2}}{\beta} \cong \frac{I_{C_1}}{\beta} \quad \text{or} \quad I_{B_2} \cong I_{B_1}$$

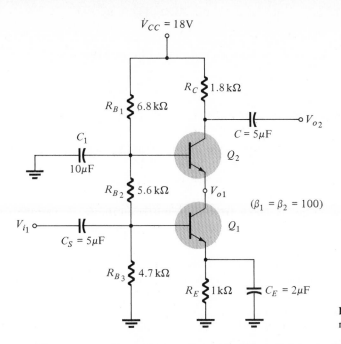

$V_{CC} = 18\text{V}$

R_C ⦚ $1.8\,\text{k}\Omega$

R_{B_1} ⦚ $6.8\,\text{k}\Omega$

C_1
$10\mu\text{F}$

R_{B_2} ⦚ $5.6\,\text{k}\Omega$

V_{o_1}

$(\beta_1 = \beta_2 = 100)$

V_{i_1}

$C_S = 5\mu\text{F}$

R_{B_3} ⦚ $4.7\,\text{k}\Omega$

R_E ⦚ $1\,\text{k}\Omega$ $C_E = 2\mu\text{F}$

$C = 5\mu\text{F}$ V_{o_2}

Q_2

Q_1

Figure 9.15 Practical cascode arrangement.

The current I_{B_1} will pass through βR_E of the parallel combination of R_{B_3} and βR_E. Since $\beta R_E = (100)(1\ \text{k}\Omega) = 100\ \text{k}\Omega$ and $R_{B_3} = 4.7\ \text{k}\Omega$, we shall assume that I_{B_1} is significantly smaller than $I_{4.7\text{k}\Omega}$ to permit ignoring its effect. Since this approximation is applied to I_{B_1}, it can also be applied to I_{B_2} (since $I_{B_2} \cong I_{B_1}$) and

$$V_{B_1} = \frac{R_{B_3}(V_{CC})}{R_{B_3} + R_{B_2} + R_{B_1}} = \frac{4.7\ \text{k}\Omega\ (18)}{4.7\ \text{k}\Omega + 5.6\ \text{k}\Omega + 6.8\ \text{k}\Omega} = \frac{84.6}{17.1}$$

$$= 4.95\ \text{V}$$

and

$$I_{E_1} = \frac{V_{E_1}}{R_E} = \frac{V_{B_3} - V_{BE}}{R_E} = \frac{4.95 - 0.7}{1\ \text{k}\Omega} = 4.25\ \text{mA}$$

with

$$r_{e_1} = \frac{26\ \text{mV}}{I_{E_1}} = \frac{26}{4.25} = 6.12\ \Omega$$

and since

$$I_{E_1} \cong I_{E_2},$$

$$r_{e_2} = 6.12\ \Omega$$

For ac conditions all capacitors assume their short-circuit equivalent and

$$A_{v_I} = \frac{V_{o_1}}{V_{i_1}} \cong \frac{-R_L}{r_{e_1}}$$

with $R_L = r_{e_2} = h_{ib_2}$ of Q_2, the input impedance of that connected stage and

$$A_{v_1} = \frac{-r_{e_2}}{r_{e_1}} \cong -1 \quad \text{(low as desired because of the Miller effect)}$$

with

$$A_{v_2} = \frac{R_L}{R_{e_2}} = \frac{R_C}{R_{e_2}} = \frac{1.2\ \text{k}\Omega}{6.12} \cong 196$$

and
$$A_{v_T} = \frac{V_{o_2}}{V_{i_1}} = A_{v_1} \cdot A_{v_2} = (-1)(+196) = -196$$

9.7 DARLINGTON COMPOUND CONFIGURATION

The Darlington circuit is a compound configuration that results in a set of improved amplifier characteristics. The configuration of Fig. 9.16 has a high input impedance with low output impedance and high current gain, all desirable characteristics for a current amplifier. We shall momentarily see, however, that the voltage gain will be less than one if the output is taken from the emitter terminal. A variation in the configuration can result in a trade-off between the output impedance and voltage gain.

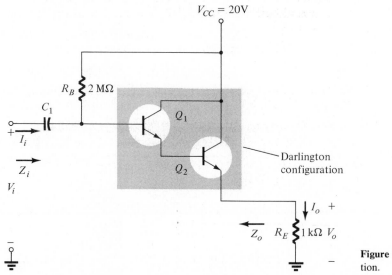

Figure 9.16 Darlington configuration.

The description of the biasing arrangement is similar to that of a single-stage emitter-follower configuration with current feedback (Chapter 4). Note for the Darlington configuration that the emitter current of the first transistor is the base current for the second active device.

In its small-signal ac form the circuit will appear as shown in Fig. 9.17.

For the second stage:

$$Z_{i_2} \cong h_{fe_2} R_E$$

and

$$A_{i_2} = \frac{I_o}{I_2} = \frac{I_{e_2}}{I_{b_2}} \cong h_{fe_2}$$

On a *good* approximate basis, these equations cannot be applied to the first stage. The "fly in the ointment" is the closeness with which Z_{i_2} compares with $1/h_{oe_1}$. You will recall that $1/h_{oe_1}$ could be eliminated in the majority of situations because

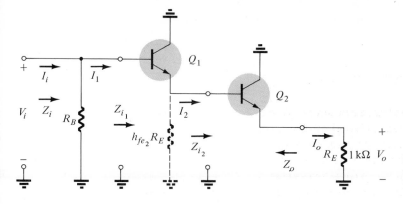

Figure 9.17 Darlington configuration of Fig. 9.16 redrawn to determine the small-signal ac response.

the load impedance $Z_L \ll 1/h_{oe_1}$. For the Darlington configuration the input impedance Z_{i_2} is close enough in magnitude to $1/h_{oe_1}$ to necessitate considering the effects of h_{oe_1}. In Chapter 7 it was found that for the single-stage grounded-emitter transistor amplifier where $1/h_{oe}$ was considered,

$$A_i \cong \frac{h_{fe}}{1 + h_{oe}Z_L}$$

Applying the above equation to this situation, $Z_L = Z_{i_2} \cong h_{fe_2}R_E$ and

$$A_{i_1} = \frac{I_2}{I_1} = \frac{I_{c_1}}{I_{b_1}} \cong \frac{h_{fe_1}}{1 + h_{oe_1}(h_{fe_2}R_E)}$$

with

$$\boxed{A_i = \frac{I_o}{I_1} = A_{i_1}A_{i_2} = \frac{h_{fe_1}h_{fe_2}}{1 + h_{oe_1}(h_{fe_2}R_E)}} \qquad (9.8)$$

For $h_{fe_1} = h_{fe_2}$ and $h_{oe_1} = h_{oe_2} = h_{oe}$

$$\boxed{A_i \cong \frac{h_{fe}^2}{1 + h_{oe}h_{fe}R_E}} \qquad (9.9)$$

For $h_{oe}h_{fe}R_E \leq 0.1$ a fairly good approximation (within 10%) is

$$\boxed{A_i \cong h_{fe}^2 = \beta^2} \qquad (9.10)$$

The current gain $A_{i_T} = I_o/I_i$, as defined by Fig. 9.17, can be determined through the use of the current-divider rule

$$I_1 = \frac{R_B I_i}{R_B + Z_{i_1}}$$

Since $Z_{i_2} \cong h_{fe_2}R_E$ is the "emitter resistor" of the first stage (note Fig. 9.17), the input impedance to the first stages is $Z_{i_1} \cong h_{fe_1}(Z_{i_2} \| 1/h_{oe_1})$ since $Z_{i_2} = h_{fe_2}R_{E_1}$, and $1/h_{oe_1}$ will appear in parallel in the small-signal equivalent circuit. The result is

$$Z_{i_1} \cong h_{fe_1}\left(h_{fe_2}R_E \left\| \frac{1}{h_{oe_1}}\right.\right) = \frac{h_{fe_1}h_{fe_2}R_E(1/h_{oe_1})}{h_{fe_2}R_E + 1/h_{oe_1}}$$

and
$$\boxed{Z_{i_1} = \frac{h_{fe_1}h_{fe_2}R_E}{h_{oe_1}h_{fe_2}R_E + 1}} \qquad (9.11)$$

which for $h_{fe_1} = h_{fe_2} = h_{fe}$ and $h_{oe_1} = h_{oe_2} = h_{oe}$

$$\boxed{Z_{i_1} \cong \frac{h_{fe}^2 R_E}{1 + h_{oe}h_{fe}R_E}} \qquad (9.12)$$

For $h_{oe}h_{fe}R_E \leq 0.1$

$$\boxed{Z_{i_1} \cong h_{fe}^2 R_E = \beta^2 R_E} \qquad (9.13)$$

For the following parameter values:

$$h_{fe_1} = h_{fe_2} = h_{fe} = 50$$
$$h_{ie_1} = 1 \text{ k}\Omega, \quad h_{ie_2} = 0.5 \text{ k}\Omega$$
$$h_{oe_1} = h_{oe_2} = h_{oe} = 20 \text{ }\mu\text{A/V}$$

$$A_i = \frac{I_o}{I_1} \cong \frac{h_{fe}^2}{1 + h_{oe}h_{fe}R_E} = \frac{(50)^2}{1 + (20 \times 10^{-6})(50)(1 \text{ k}\Omega)}$$

$$= \frac{2500}{1+1} = \mathbf{1250}$$

and
$$Z_{i_1} \cong \frac{h_{fe}^2 R_E}{1 + h_{oe}h_{fe}R_E} = \frac{(50)^2 1 \text{ k}\Omega}{2} = 1250 \text{ k}\Omega = \mathbf{1.25 \text{ M}\Omega}$$

so that for $R_B = 2 \text{ M}\Omega$

$$\frac{I_1}{I_i} = \frac{R_B}{R_B + Z_{i_1}} = \frac{2 \text{ M}\Omega}{2 \text{ M}\Omega + 1.25 \text{ M}\Omega} = \frac{2}{3.25} = 0.615$$

and
$$A_{i_T} = \frac{I_o}{I_i} = \left[\frac{I_o}{I_1}\right]\left[\frac{I_1}{I_i}\right] = A_i \times \frac{I_1}{I_i}$$

$$= (1250)(0.615) = \mathbf{770}$$

with
$$Z_i = 2 \text{ M}\Omega \| Z_{i_1} = 2 \text{ M}\Omega \| 1.25 \text{ M}\Omega = \mathbf{770 \text{ k}\Omega}$$

Too frequently, the current gain of a Darlington circuit is assumed to be simply $A_i^2 \cong h_{fe}^2$ without any regard to the output impedance $1/h_{oe}$. In this case $A_i \cong (h_{fe})^2 = 2500$. Certainly, 2500 versus 1250 is *not* a good approximation. The effect of h_{oe_1} must therefore be considered when the current gain of the first stage is determined.

The output impedance Z_o can be determined directly from the emitter-equivalent circuits, as follows.

For the first stage

$$Z_{o_1} \cong \frac{R_{s_1} + h_{ie_1}}{h_{fe_1}} \qquad (9.14)$$

$$= \frac{0 + 1\,\text{k}\Omega}{50} \cong \mathbf{20.0\ \Omega}$$

and

$$Z_{o_2} \cong \frac{(Z_{o_1} \,\|\, 1/h_{oe_1}) + h_{ie_2}}{h_{fe_2}} \qquad (9.15)$$

$$= \frac{(20.0 \,\|\, 50\,\text{k}\Omega) + 0.5\,\text{k}\Omega}{50} \cong \frac{20.0 + 0.5\,\text{k}\Omega}{50} = \frac{520}{50}$$

$$= \mathbf{10.40\ \Omega}$$

Note, as indicated in the introductory discussion, that the input impedance is high, output impedance very low, and current gain high. We shall now examine the voltage gain of the system. Applying Kirchhoff's voltage law to the circuit of Fig. 9.16:

$$V_o = V_i - V_{be_1} - V_{be_2}$$

That the output potential is the input *less* the base-to-emitter potential of each transistor clearly indicates that $V_o < V_i$. It is closer in magnitude to one than to zero. On an approximate basis it is given by

$$A_v \cong \frac{1}{1 + \dfrac{h_{ie_2}}{h_{fe_2} R_E}} \qquad (9.16)$$

as derived from the emitter equivalent circuit.

Substituting the numerical values of this general example:

$$A_v \cong \frac{1}{1 + \dfrac{0.5\,\text{k}\Omega}{50\,\text{k}\Omega}} = \frac{1}{1 + 0.01} = \mathbf{0.99}$$

The ratings and characteristics for a 10-A RCA *npn* Darlington power transistor are provided in Fig. 9.18. Some of the characteristics appear in Figs. 9.19 through 9.24. The data provided are for the complete device—the individual β values are not provided. Note in the characteristics that the level of V_{BE} is increased since it includes the drop across two transistors. Consider also that the minimum value of h_{fe} is 1000 at 1 kHz but drops to only 20 at 1 MHz. Frequency will obviously have a pronounced effect on its performance. Note in Fig. 9.19 that the collector current is in amperes and that a pulsed operation results in an increased level of current—the longer the pulse, the lower the permitted current. In Fig. 9.20 we note a drop in power rating starting with room temperature, and in Fig. 9.21 we find

2N6383 2N6384 2N6385

10-Ampere, N-P-N Darlington Power Transistors

40-60-80 Volts, 100 Watts
Gain of 1000 at 5 A

JEDEC TO-3

H 1570

TERMINAL CONNECTIONS

Pin 1 - Base
Pin 2 - Emitter
Case - Collector
Mounting Flange - Collector

Features:

■ Operates from IC without predriver
■ Low leakage at high temperature
■ High reverse second-breakdown capability

Applications:

■ Power switching ■ Audio amplifiers
■ Hammer drivers
■ Series and shunt regulators

The 2N6383, 2N6384, and 2N6385● are monolithic n-p-n silicon Darlington transistors designed for low- and medium-frequency power applications. The double epitaxial construction of these devices provides good forward and reverse second-breakdown capability; their high gain makes it possible for them to be driven directly from integrated circuits.

●Formerly RCA Dev. Nos. TA8349, TA8486, and TA8348.

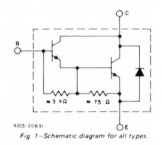

92CS-20691
Fig. 1—Schematic diagram for all types.

MAXIMUM RATINGS, *Absolute-Maximum Values:*

		2N6385	2N6384	2N6383	
* COLLECTOR-TO-BASE VOLTAGE	V_{CBO}	80	60	40	V
COLLECTOR-TO-EMITTER VOLTAGE:					
With external base-to-emitter resistance (R_{BE}) = 100Ω, sustaining	V_{CER}(sus)	80	60	40	V
With base open, sustaining	V_{CEO}(sus)	80	60	40	V
* With base reverse-biased V_{BE} = −1.5 V, R_{BB} = 100Ω	V_{CEX}	80	60	40	V
* EMITTER-TO-BASE VOLTAGE	V_{EBO}	5	5	5	V
COLLECTOR CURRENT:	I_C				
* Continuous		10	10	10	A
Peak		15	15	15	A
* CONTINUOUS BASE CURRENT	I_B	0.25	0.25	0.25	A
* TRANSISTOR DISSIPATION:	P_T				
At case temperatures up to 25°C		100	100	100	W
At case temperatures above 25°C		◄──── See Fig. 9.20 ────►			
* TEMPERATURE RANGE:					
Storage and Operating (Junction)		◄──── −65 to +200 ────►			°C
* PIN TEMPERATURE (During Soldering):					
At distances ≥ 1/32 in. 0.8 mm from seating plane for 10 s max.		◄──── 235 ────►			°C

*In accordance with JEDEC registration data format JS-6 RDF-2.

Figure 9.18   RCA NPN Darlington power transistors. (Courtesy RCA Solid State Division.)

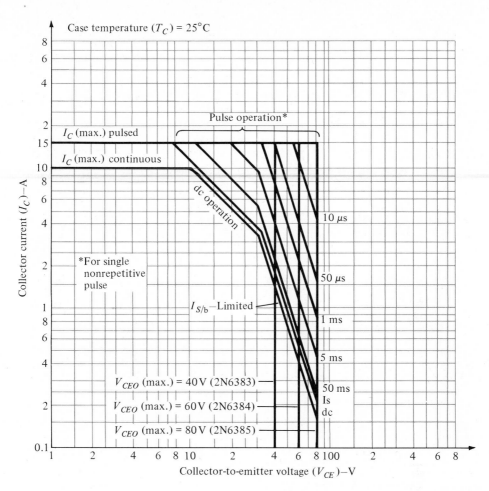

Figure 9.19 Maximum operating area for all types.

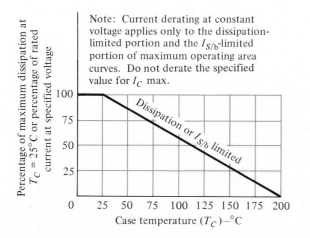

Note: Current derating at constant voltage applies only to the dissipation-limited portion and the $I_{S/b}$-limited portion of maximum operating area curves. Do not derate the specified value for I_C max.

Figure 9.20 Derating curve for all types.

341

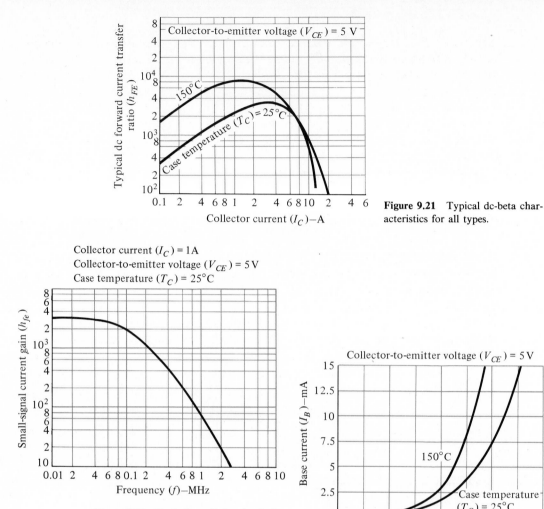

Figure 9.21 Typical dc-beta characteristics for all types.

Collector current (I_C) = 1A
Collector-to-emitter voltage (V_{CE}) = 5 V
Case temperature (T_C) = 25°C

Figure 9.22 Typical small-signal gain for all types.

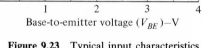

Figure 9.23 Typical input characteristics for all types.

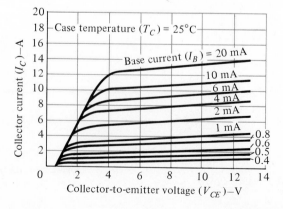

Figure 9.24 Typical output characteristics for all types.

that the dc β is very sensitive to collector current. The effect of frequency on the ac gain is more carefully defined in Fig. 9.22. It begins a severe drop at about 0.1 MHz or 100 kHz. Note in Fig. 9.23 the increased level of V_o (cut-in voltage) due to the two transistors and the base current in mA in Fig. 9.24.

In this case, the hybrid parameters and admittance parameters are provided. The admittance parameters, not presented in this text, simply represent the parameters of another equivalent network for the transistor. Note also the high value of input impedance for the pairs and the low value of the equivalent $1/h_{oe} = 9.3$ kΩ. The dB gain in power is defined by equations to be derived in the next section.

9.8 DECIBELS

The concept of the *decibel* (dB) and the associated calculations will become increasingly important in the remaining sections of this chapter. The background surrounding the term *decibel* has its origin in the old established fact that power and audio levels are related on a logarithmic basis. That is, an increase in power level, say 4 to 16 W, for discussion purposes, does not mean that the audio level will increase by a factor of $16/4 = 4$. It will increase by a factor of 2 as derived from the power of 4 in the following manner: $(4)^2 = 16$. For a change of 4 to 64 W the audio level will increase by a factor of 3 since $(4)^3 = 64$. In logarithmic form, the relationship can be written as

$$\log_4 64 = 3$$

In words, the equation states that the logarithm of 64 to the base 4 is 3. In general: $\log_b a = x$ relates the variables in the same manner as $b^x = a$.

Because of pressures for standardization, the *bel* (B) was defined by the following equation to relate power levels P_1 and P_2:

$$G = \log_{10} \frac{P_2}{P_1} \qquad \text{(bel)} \qquad (9.17)$$

Note that the common, or base 10, system was chosen to eliminate variability. Although the base is no longer the original power level, the equation will result in a basis for comparison of audio levels due to changes in power levels. The term *bel* was derived from the surname of Alexander Graham Bell.

It was found, however, that the bel was too large a unit of measurement for practical purposes, so the decibel (dB) was defined such that 10 decibels = 1 bel. Therefore,

and

$$G_{dB} = 10 \log_{10} \frac{P_2}{P_1} \qquad \text{(dB)} \qquad (9.18)$$

The terminal rating of electronic communication equipment (amplifiers, microphones, etc.) is commonly rated in decibels. Equation (9.18) indicates clearly, however, that the decibel rating is a measure of the difference in magnitude between *two* power levels. For a specified terminal (output) power (P_2) there must be a reference power level (P_1). The reference level is generally accepted to be 1 mW although on occasion the 6-mW standard of earlier years is applied. The resistance to be associated with the 1-mW power level is 600 Ω, chosen because it is the characteristic impedance of audio transmission lines. When the 1-mW level is employed as the reference level, the decibel symbol frequently appears as dBm. In equation form

$$G_{\text{dBm}} = 10 \log_{10} \frac{P_2}{1 \text{ mW}} \bigg|_{600\Omega} \qquad \text{(dBm)} \qquad (9.19)$$

There exists a second equation for decibels that is applied frequently. It can be best described through the circuit of Fig. 9.25a. For V_i equal to some value V_1, $P_1 = V_1^2/R_i$ where R_i is the input resistance of the system of Fig. 9.25a. If V_i should

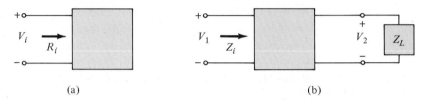

(a) (b)

Figure 9.25 Configurations employed in the discussion of Eq. 9.18.

be increased (or decreased) to some other level, V_2, then $P_2 = V_2^2/R_i$. If we substitute into Eq. (9.18) to determine the resulting difference in decibels between the power levels,

$$G_{\text{dB}} = 10 \log_{10} \frac{P_2}{P_1} = 10 \log_{10} \frac{V_2^2/R_i}{V_1^2/R_i} = 10 \log_{10} \left(\frac{V_2}{V_1}\right)^2$$

and

$$G_{\text{dB}} = 20 \log_{10} \frac{V_2}{V_1} \qquad \text{(dB)} \qquad (9.20)$$

Keep in mind, however, that this equation is only correct if the associated resistance for each applied voltage is the same. For the system of Fig. 9.25b, where output and input levels are being compared, $Z_i \neq Z_L$ and Eq. (9.20) will not result. Equation (9.18) should therefore be employed.

If $Z_i = Z_i \cos \theta_i$ and $Z_L = Z_L \cos \theta_L$ are substituted into Eq. (9.18) such that $P_L = \dfrac{V_L^2}{Z_L \cos \theta_L}$, etc., the following general equation will result

$$G_{\text{dB}} = 20 \log_{10} \frac{V_L}{V_i} + 10 \log_{10} \frac{Z_i}{Z_L} + 10 \log_{10} \frac{\cos \theta_i}{\cos \theta_L} \qquad (9.21)$$

For resistive elements, which are most commonly encountered, $\cos \theta_i = \cos \theta_L$, and the last term, $\log_{10}(1) = 0$. In addition, if $Z_i = Z_L$, the second term will also drop out, resulting in Eq. (9.20).

Frequently the effect of different impedances $(Z_i \neq Z_L)$ is ignored and Eq. (9.20) applied to simply establish a basis of comparison between levels—voltage or current. For situations of this type the decibel gain should more correctly be referred to as the *voltage or current gain in decibels* to differentiate it from the common usage of decibel as applied to power levels.

One of the advantages of the logarithmic relationship is the manner in which it can be applied to cascaded stages. For example, the overall voltage gain of a cascaded system is given by

$$A_{v_T} = A_{v_1} A_{v_2} A_{v_3} \cdots A_{v_n}$$

Applying the proper logarithmic relationship gives

$$G_v = 20 \log_{10} A_{v_T} = 20 \log_{10} A_{v_1} + 20 \log_{10} A_{v_2} \\ + 20 \log_{10} A_{v_2} + \cdots + 20 \log_{10} A_{v_n} \quad \text{(dB)} \quad (9.22)$$

In words, the equation states that the decibel gain of a cascaded system is simply the sum of the decibel gains of each stage, that is,

$$\boxed{G_v = G_{v_1} + G_{v_2} + G_{v_3} + \cdots + G_{v_n}} \quad (9.23)$$

The above equations can also be applied to current considerations. For $P_2 = I_2^2 R_o$ and $P_1 = I_1^2 R_o$,

$$\boxed{G_{\text{dB}} = 20 \log_{10} \frac{I_2}{I_1}} \quad \text{(dB)} \quad (9.24)$$

and

$$\boxed{G_i = G_{i_1} + G_{i_2} + G_{i_3} + \cdots + G_{i_n}} \quad (9.25)$$

Before considering a few examples, the fundamental operations associated with logarithmic functions will be considered. For many it will be simply a review. For some, an extended amount of time may be required to fully understand the material to follow.

Each equation introduced in this section employs the common or base 10 logarithmic system. As indicated in the introductory discussion, the logarithms of numbers that are powers of the chosen base are easily determined. For example,

$$\log_{10} \overbrace{10,000}^{a} = x \implies (10)^x = 10,000$$
$$\underset{b}{\uparrow}$$

and
$$x = 4$$

Similarly,

$$\log_{10} 1000 = \log_{10} (10)^3 = 3$$
$$\log_{10} 100 = \log_{10} (10)^2 = 2$$
$$\log_{10} 10 = \log_{10} (10)^1 = 1$$
$$\log_{10} 1 = \log_{10} (10)^0 = 0$$

For the logarithm of a number such as 24.8

$$\log_{10} 24.8 = x$$

or

$$10^x = 24.8$$

The unknown quantity x is obviously between 1 and 2, but a further determination would be purely a trial-and-error process if it were not for the logarithmic function. The procedure for determining the logarithm of a number requires that two components of the result be found separately. These two components are the *characteristic* and *mantissa*. The characteristic is simply the power of 10 associated with the number for which the logarithm is to be determined.

$$24.8 = 2.48 \times 10^1 \Longrightarrow 1 = \text{characteristic}$$

$$4860.0 = 4.860 \times 10^3 \Longrightarrow 3 = \text{characteristic}$$

The mantissa, or decimal portion of the logarithm, must be determined from a set of tables or a calculator.

$$\log_{10} 24.8 = 1.3945$$

$$\log_{10} 4860.0 = 3.6866$$

Of course most calculators provide the characteristic and mantissa directly.

There will be many occasions in which the antilogarithm of a number must be determined; that is, for the example above, determine 24.8 and 4860.0 from the logarithm of these numbers. The process is simply the reverse of that applied to determine the logarithm. For example, find the antilogarithm of 2.140.

$$2.140 \left\} \begin{array}{l} \text{characteristic} \Longrightarrow 10^2 \\ \text{mantissa} \quad\;\; \Longrightarrow 138 \end{array} \right\} \; 1.38 \times 10^2 = 138$$

For the calculator, keep in mind that

$$\log_{10} x = 2.140$$

is equivalent to

$$10^{2.140} = x$$

and 10^y is a common calculator function.

For ratios less than 1, the logarithm can be determined by simply inverting the ratio and introducing a negative sign. Most calculators will include the insertion of the negative sign.

$$\log_{10} \frac{1.6}{24} = -\log_{10} \frac{24}{1.6} = -\log_{10} 15 = -1.1761$$

$$\log_{10} 0.788 = -\log_{10} \frac{1}{0.788} = -\log_{10} 1.27 = -0.1038$$

For power ratios, a negative decibel rating simply indicates a reduction in power level as compared to the initial or input power.

EXAMPLE 9.4 Find the magnitude gain corresponding to a decibel gain of 100.

Solution: By Eq. (9.18)

$$G_{\text{dB}} = 10 \log_{10} \frac{P_2}{P_1} \; 100 \text{ dB} \Longrightarrow \log_{10} \frac{P_2}{P_1} = 10$$

so that

$$\frac{P_2}{P_1} = 10^{10} = \textbf{10,000,000,000}$$

This example clearly demonstrates the range of decibel values to be expected from practical devices. Certainly a future calculation giving a decibel result in the neighborhood of 100 should be questioned immediately. In fact, a decibel gain of 50 corresponds with a magnitude gain of 100,000, which is still very large.

EXAMPLE 9.5 The input power to a device is 10,000 W at a voltage of 1000 V. The output power is 500 W while the output impedance is 20 Ω.
(a) Find the power gain in decibels.
(b) Find the voltage gain in decibels.
(c) Explain why (a) and (b) agree or disagree.

Solution:

(a) $\quad G_{\text{dB}} = 10 \log_{10} \dfrac{P_2}{P_1} = 10 \log_{10} \dfrac{0.5 \times 10^3}{10 \times 10^3} = 10 \log_{10} \dfrac{1}{20} = -10 \log_{10} 20$

$\qquad = -10(1.301) = \textbf{-13.01 dB}$

(b) $\quad G_v = 20 \log_{10} \dfrac{V_o}{V_i} = 20 \log_{10} \dfrac{\sqrt{PR}}{1000} = 20 \log_{10} \dfrac{\sqrt{500 \times 20}}{1000}$

$\qquad = 20 \log_{10} \dfrac{100}{1000} = 20 \log_{10} \dfrac{1}{10} = -20 \log_{10} 10 = \textbf{-20 dB}$

(c) $\quad R_i = \dfrac{V^2}{P} = \dfrac{10^6}{10^4} = 10^2 \neq R_o = \textbf{20 } \Omega$

EXAMPLE 9.6 An amplifier rated at 40-W output is connected to a 10-Ω speaker. (a) Calculate the input power required for full power output if the power gain is 25 dB. (b) Calculate the input voltage for rated output if the amplifier voltage gain is 40 dB.

Solution:
(a) Eq. (9.18)

$$25 = 10 \log_{10} \frac{40}{P_i} \Longrightarrow P_i = \frac{40}{\text{antilog}\,(2.5)} = \frac{40}{3.16 \times 10^2}$$

$$= \frac{40}{316} \cong \textbf{126.5 mW}$$

(b) $\quad G_v = 20 \log_{10} \dfrac{V_o}{V_i} \Longrightarrow 40 = 20 \log_{10} \dfrac{V_o}{V_i}$

$$\frac{V_o}{V_i} = \text{antilog } 2 = 100$$

$$V_o = \sqrt{PR} = \sqrt{40 \times 10} = 20 \text{ V}$$

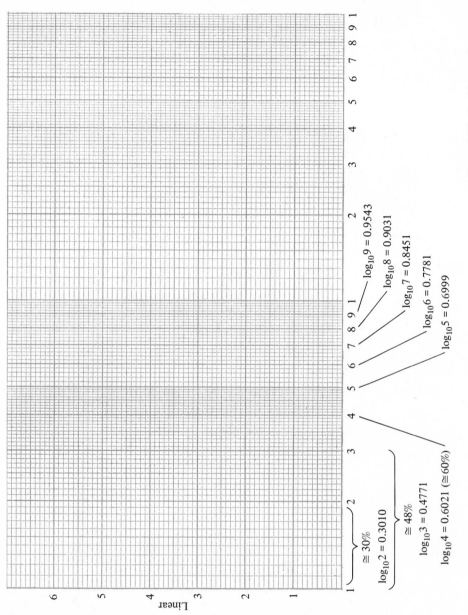

Figure 9.26 Semi-log graph paper.

348

$$V_i = \frac{V_o}{100} = \frac{20}{100} = 200 \text{ mV}$$

The use of log scales on a graph can significantly expand the range of variation of a particular variable—an effect to become obvious in the following section. Most plots provided, or graph paper available, is of the semi-log or double-log (log-log) variety. The term *semi* (meaning one-half) indicates that only one of the two scales is a log scale, whereas double-log indicates that both scales are log scales. A semi-log scale appears in Fig. 9.26. Note that the vertical scale is a linear scale with equal divisions. The source of the spacing between the lines of the log plot is shown on the graph. The log of 2 to the base 10 is approximately 0.3. The distance from $1(\log_{10}1 = 0)$ to 2 is therefore 30% of the span. Since $\log_{10}5 \cong 0.7$, it is marked off at a point 70% of the distance. Note that between any two digits the same compression of the lines appears as you progress from the left to right. It is important to note the resulting numerical value and the spacing since plots will typically only have the tic marks indicated in Fig. 9.27 due to a lack of space. You must realize from past experience that the longer bars for this figure have the numerical values of 0.3, 3, and 30 associated with them, whereas the next shorter bars have values of 0.5, 5, and 50, and the shortest bars 0.7, 7, and 70.

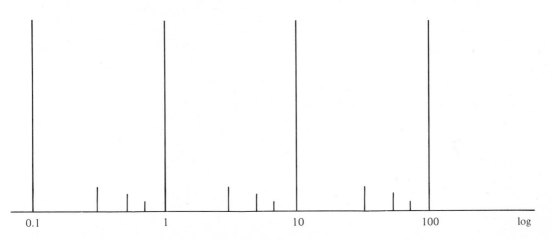

Figure 9.27 Identifying the numerical values of the tic marks on a log scale.

9.9 GENERAL FREQUENCY CONSIDERATIONS

The frequency of the applied signal can have a pronounced effect on the response of a single or multistage network. The analysis thus far has been for the mid-frequency spectrum. At low frequencies we shall find that the coupling and bypass capacitors can no longer be replaced by the short-circuit approximation because of the resulting change in reactance of these elements. The frequency-dependent parameters of the small-signal equivalent circuits and the stray capacitive elements associated with the active device and the network will limit the high-frequency response of the system. An increase in the number of stages of a cascaded system will limit both the high- and low-frequency response.

The magnitude gain of an *RC*-coupled, direct-coupled, and transformer-coupled amplifier system are provided in Fig. 9.28. Note that the horizontal scale is a logarithmic scale to permit a plot extending from the low- to the high-frequency regions. For each plot, a low-, high-, and mid-frequency region has been defined. In addition, the primary reasons for the drop in gain at low and high frequencies have also been indicated within the parentheses. For the *RC*-coupled amplifier the drop at low frequencies is due to the increasing reactance of C_C, C_S, or C_E, while its upper frequency

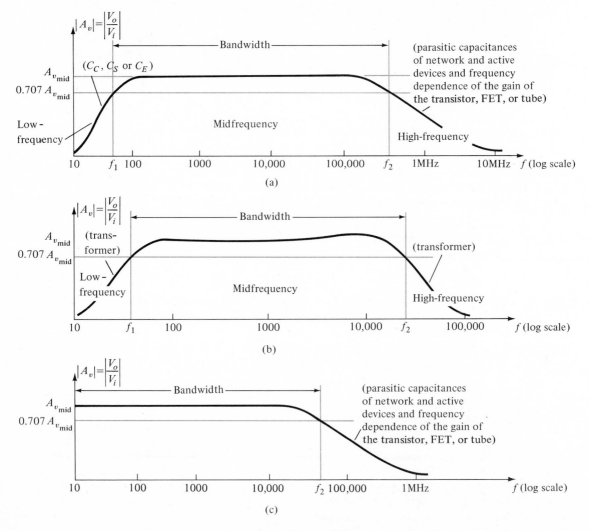

Figure 9.28 Gain vs. frequency for: (a) *RC*-coupled amplifiers; (b) transformer-coupled amplifiers; (c) direct-coupled amplifiers.

CH. 9 MULTISTAGE SYSTEMS AND FREQUENCY CONSIDERATIONS

limit is determined by either the parasitic capacitive elements of the network and active device or the frequency dependence of the gain of the active device. An explanation of the drop in gain for the transformer-coupled system requires a basic understanding of "transformer action" and the transformer equivalent circuit. For the moment let us say that it is simply due to the shorting effect (across the input terminals of the transformer) of a magnetizing inductive reactance at low frequencies ($X_L = 2\pi fL$). The gain must obviously be zero at $f = 0$ since at this point there is no longer a changing flux established through the core to induce a secondary or output voltage. As indicated in Fig. 9.28, the high-frequency response is controlled primarily by the stray capacitance between the turns of the primary and secondary windings. For the direct-coupled amplifier, there are no coupling or bypass capacitors to cause a drop in gain at low frequencies. As the figure indicates, it is a flat response to the upper cutoff frequency which is determined by either the parasitic capacitances of the circuit and active device or the frequency dependence of the gain of the active device.

For each system of Fig. 9.28 there is a band of frequencies in which the magnitude of the gain is either equal or relatively close to the mid-band value. To fix the frequency boundaries of relatively high gain, $0.707 A_{v_{\mathrm{mid}}}$ was chosen to be the gain cutoff level. The corresponding frequencies f_1 and f_2 are generally called the corner, cutoff, band, break, or half-power frequencies. The multiplier 0.707 was chosen because at this level the output power is half the mid-band power output, that is, at mid-frequencies,

$$ P_{o_{\mathrm{mid}}} = \frac{|V_o^2|}{R_o} = \frac{|A_{v_{\mathrm{mid}}} V_i|^2}{R_o} $$

and at the half-power frequencies,

$$ P_{o_{HPF}} = \frac{|0.707\,A_{v_{\mathrm{mid}}} V_i|^2}{R_o} = 0.5 \frac{|A_{v_{\mathrm{mid}}} V_i|^2}{R_o} $$

and

$$ \boxed{P_{o_{HPF}} = 0.5 P_{o_{\mathrm{mid}}}} \tag{9.26} $$

The bandwidth (or pass band) of each system is determined by f_1 and f_2, that is,

$$ \boxed{\text{bandwidth (BW)} = f_2 - f_1} \tag{9.27} $$

For applications of a communications nature (audio, video), a decibel plot of the voltage gain versus frequency is more useful than that appearing in Fig. 7.22. Before obtaining the logarithmic plot, however, the curve is generally normalized as shown in Fig. 9.29. In this figure the gain at each frequency is divided by the mid-band value. Obviously, the mid-band value is then 1 as indicated. At the half-power frequencies the resulting level is $0.707 = 1\sqrt{2}$. A decibel plot can now be obtained by applying Eq. (9.20) in the following manner:

$$ \boxed{\left| \frac{A_v}{A_{v_{\mathrm{mid}}}} \right|_{\mathrm{dB}} = 20 \log_{10} \left| \frac{A_v}{A_{v_{\mathrm{mid}}}} \right|} \tag{9.28} $$

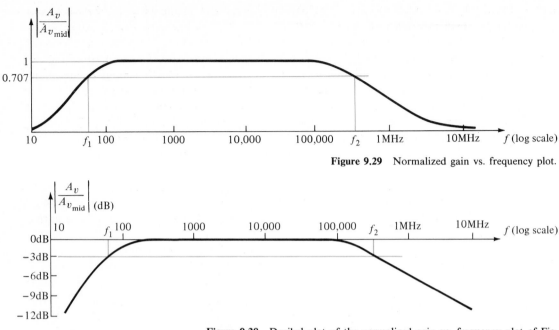

Figure 9.29 Normalized gain vs. frequency plot.

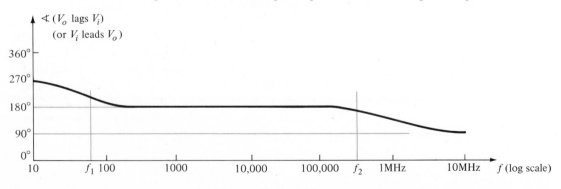

Figure 9.30 Decibel plot of the normalized gain vs. frequency plot of Fig. 9.29.

At mid-band frequencies, $20 \log_{10} 1 = 0$, and at the cutoff frequencies, $20 \log_{10} 1/\sqrt{2} = -3$ dB. Both values are clearly indicated in the resulting decibel plot of Fig. 9.30. The smaller the fractional ratio, the more negative the decibel level due to the inversion process discussed in Section 9.8.

For the greater part of the discussion to follow, a decibel plot will be made only for the low- or high-frequency regions. Keep Fig. 9.30 in mind, therefore, to permit a visualization of the broad system response.

It should be understood that an amplifier usually introduces an inversion between input and output signals. This fact must now be expanded to indicate that this is only the case in the mid-band region. At low frequencies there is a phase shift such that V_o lags V_i by an increased angle. At high frequencies the phase shift will drop below 180°. Fig. 9.31 is a standard phase plot for an RC-coupled amplifier.

Figure 9.31 Phase plot for an RC-coupled amplifier system (for each stage).

CH. 9 MULTISTAGE SYSTEMS AND FREQUENCY CONSIDERATIONS

9.10 SINGLE-STAGE TRANSISTOR AMPLIFIER— LOW-FREQUENCY CONSIDERATIONS

Before considering the basic BJT amplifier let us first examine the series RC network appearing in Fig. 9.32 and the effect the applied frequency will have on the gain $|A_v| = |V_o/V_i|$. At mid-frequencies and high frequencies the capacitive reactance $X_C = (1/2\pi fC)$ is sufficiently small compared to R to permit a short-circuit approximation and $V_o = V_i$. However, at low frequencies the capacitive reactance can become sufficiently large to cause a significant voltage drop across C. The result is a drop in V_o and the gain $A_v = (V_o/V_i)$. When conditions are such that $|X_C| = R$, then

$$|V_o| = \frac{R}{\sqrt{R^2 + X_C^2}}|V_i| = \frac{R}{\sqrt{R^2 + R^2}}|V_i| = \frac{R}{\sqrt{2R^2}}|V_i| = \frac{R}{\sqrt{2}R}|V_i| = \frac{1}{\sqrt{2}}|V_i|$$

and

$$|A_v| = \left|\frac{V_o}{V_i}\right| = \frac{1}{\sqrt{2}} = 0.707$$

with

$$20\log_{10}\frac{1}{\sqrt{2}} = -3 \text{ dB}$$

We can conclude, therefore, that if an RC configuration is encountered, the condition $|X_C| = R$ will result in a 3-dB drop in voltage from the mid-frequency value. A similar drop occurs for the current since at $X_C = R$,

$$|I| = \left|\frac{V_i}{Z}\right| = \frac{|V_i|}{\sqrt{R^2 + X_C^2}} = \frac{|V_i|}{\sqrt{2R^2}} = \frac{1}{\sqrt{2}}\frac{|V_i|}{R} = 0.707\frac{|V_i|}{R}$$

where $|V_i|/R$ is its mid-frequency value, if we replace X_C by its short-circuit equivalent.

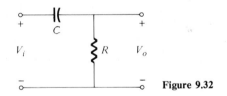

Figure 9.32

Using the condition $X_C = R$, we can determine the frequency at which the 3-dB voltage drop will occur. That is,

$$|X_C| = R$$

$$\frac{1}{2\pi fC} = R$$

and

$$\boxed{f_1 = \frac{1}{2\pi RC}}$$
(9.29)

If the gain equation is written in the following manner

$$A_v = \frac{V_o}{V_i} = \frac{R}{R - jX_C} = \frac{1}{1 - j\dfrac{X_C}{R}} = \frac{1}{1 - j\dfrac{1}{\omega CR}} = \frac{1}{1 - j\dfrac{1}{2\pi fCR}}$$

and using the defined frequency above,

$$A_v = \boxed{\frac{1}{1 - j(f_1/f)}}$$ (9.30)

In the magnitude and phase form:

$$A_v = \frac{V_o}{V_i} = \underbrace{\frac{1}{\sqrt{1 + (f_1/f)^2}}}_{\text{magnitude of } A_v} \underbrace{\underline{/\tan^{-1}(f_1/f)}}_{\substack{\text{phase } \measuredangle \text{ by which} \\ V_o \text{ leads } V_i}}$$

For the magnitude, when $f_1 = f$,

$$|A_v| = \frac{1}{\sqrt{1 + (1)^2}} = \frac{1}{\sqrt{2}} = 0.707 \implies -3 \text{ dB}$$

In the logarithmic form:

$$|A_v|_{\text{dB}} = 20 \log_{10} \frac{1}{\sqrt{1 + (f_1/f)^2}} = -20 \log_{10}[1 + (f_1/f)^2]^{1/2}$$

$$= -(\tfrac{1}{2})(20) \log_{10} [1 + (f_1/f)^2]$$

$$= -10 \log_{10}[1 + (f_1/f)^2]$$

For frequencies where $f_1 \gg f$ the above equation can be approximated by

$$= -10 \log_{10}(f_1/f)^2$$

and

$$\boxed{|A_v|_{\text{dB}} = -20 \log_{10} f_1/f} \quad {}_{f_1 \gg f}$$ (9.31)

Ignoring the condition $f_1 \gg f$ for a moment, a plot of Eq. (9.31) on a frequency log scale will yield some results of a useful nature for future decibel plots.

At $f = f_1$, or $\dfrac{f_1}{f} = 1$, $\left(\dfrac{f}{f_1} = 1\right)$, $-20 \log_{10} 1 = 0 \text{ dB}$

At $f = 0.5f_1$, or $\dfrac{f_1}{f} = 2$, $\left(\dfrac{f}{f_1} = 0.5\right)$, $-20 \log_{10} 2 = -6 \text{ dB}$

At $f = 0.25f_1$, or $\dfrac{f_1}{f} = 4$, $\left(\dfrac{f}{f_1} = 0.25\right)$, $-20 \log_{10} 4 = -12 \text{ dB}$

At $f = 0.1f_1$, or $\dfrac{f_1}{f} = 10$, $\left(\dfrac{f}{f_1} = 0.1\right)$, $-20 \log_{10} 10 = -20 \text{ dB}$

A plot of these points is indicated in Fig. 9.33 from $f/f_1 = 0.1$ to $f/f_1 = 1$. Note that this results in a straight line when plotted against a log scale. In the same figure a straight line is also drawn for the condition of 0 dB for $f \gg f_1$. As stated earlier, the straight-line segments (asymptotes) are only accurate for 0 dB when $f \gg f_1$, and the sloped line when $f_1 \gg f$. We know, however, that when $f =$

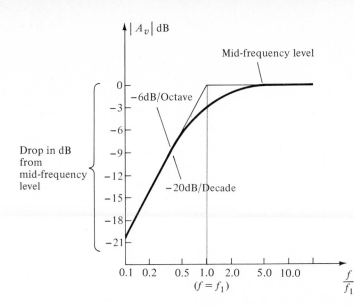

Figure 9.33 Bode plot for the low-frequency region.

f_1, there is a 3-dB drop from the mid-band level. Employing this information in association with the straight-line segments permits a fairly accurate plot of the frequency response as indicated in the same figure. The piecewise linear plot of the asymptotes and associated breakpoints is called a *Bode plot*.

The above calculations and the curve itself demonstrate clearly that a change in frequency by a factor of 2 (equivalent to 1 octave) results in a 6-dB change in the ratio. For a 10:1 change in frequency (equivalent to 1 decade) there is a 20-dB change in the ratio. In the future, therefore, a decibel plot can be easily obtained for a function having the format of Eq. (9.31). First simply find f_1 from the circuit parameters and then sketch two asymptotes—one along the 0-dB line and the other drawn through f_1 sloped at 6 dB/octave or 20 dB/decade. Then find the 3-dB point corresponding to f_1 and sketch the curve.

Let us now turn our attention to the basic BJT amplifier appearing in Fig. 9.34 with the capacitors that will affect the low-frequency response. We shall examine the effect of each independently.

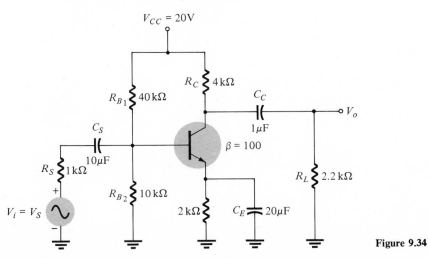

Figure 9.34

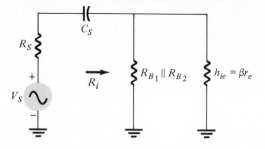

Figure 9.35 Localized ac equivalent for C_S.

The reduced equivalent circuit surrounding C_S appears in Fig. 9.35. When conditions are such that $R_T = |X_{C_S}|$ or $R_S + R_i = |X_{C_S}|$, the input current I_i is reduced by 3 dB and a lower break frequency is defined by

$$R_S + R_i = \frac{1}{2\pi f C_S}$$

or

$$\boxed{f_{L_S} = \frac{1}{2\pi(R_S + R_i)C_S}}$$

(9.32)

The network of Fig. 9.35 has the basic configuration of Fig. 9.36. As the applied frequency increases, the reactance of the capacitor will decrease, resulting in an increase in V' to a value equal to V when the capacitor can be replaced (for all practical purposes) by its short-circuit equivalent. As V' increases, the resulting gain (V_o/V_S) will increase until the mid-frequency level is attained.

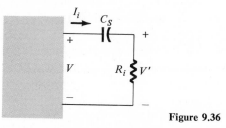

Figure 9.36

Although not described in detail here, analysis will show that a 3-dB drop in input current will result in a similar drop in overall voltage gain.

C_C

The resulting equivalent circuit appearing in Fig. 9.37 defines a lower break frequency by

$$R_C + R_L = X_{C_C}$$

or

$$R_C + R_L = \frac{1}{2\pi f C_C}$$

and

$$\boxed{f_{L_C} = \frac{1}{2\pi(R_C + R_L)C_C}}$$

(9.33)

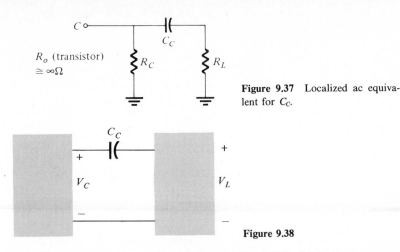

Figure 9.37 Localized ac equivalent for C_C.

Figure 9.38

The network of Fig. 9.37 has the basic configuration of Fig. 9.38. For reasons similar to the description applied to the network of Fig. 9.36 above, the voltage V_L will increase as the reactance of the capacitor decreases. The equivalent models of Figs. 9.36 and 9.38 are being presented primarily to ensure a clear understanding of the results obtained using the computer in the paragraphs to follow.

C_E

The equivalent circuit external to the capacitor C_E appears in Fig. 9.39. The condition $R_e = R_E \| [(R'_s/\beta) + r_e] = |X_{C_E}|$ defines the break frequency determined by the emitter bypass capacitor:

$$f_{L_E} = \frac{1}{2\pi R_e C_E} \tag{9.34}$$

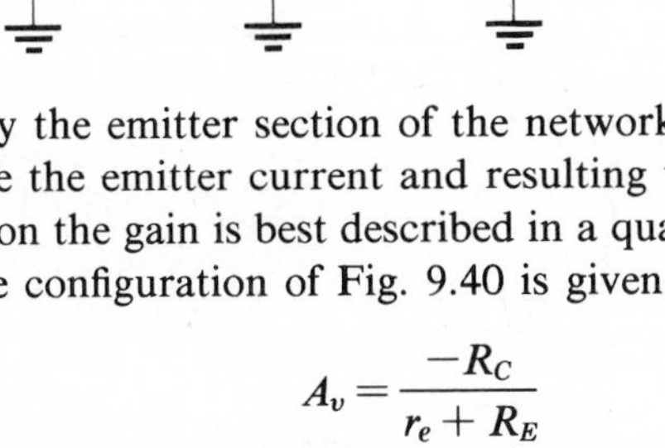

Figure 9.39 Localized ac equivalent for C_E.

At this frequency the emitter section of the network will have a sufficiently high impedance to reduce the emitter current and resulting voltage gain by 3 dB.

The effect of C_E on the gain is best described in a quantitative manner by recalling that the gain for the configuration of Fig. 9.40 is given by

$$A_v = \frac{-R_C}{r_e + R_E}$$

The maximum gain is obviously available when R_E is zero ohms. At low frequencies, with a bypass capacitor C_E, all of R_E appears in the above gain equation resulting

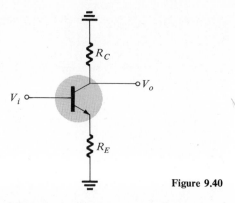

Figure 9.40

in the minimum gain. At higher frequencies, the reactance of the capacitor C_E will decrease reducing the parallel impedance of R_E and C_E. The result is an increase in gain until the mid-band gain as determined by R_C and r_e is obtained. At f_{L_E} the gain will be 3 dB below the mid-band value determined with R_E "shorted" out.

EXAMPLE 9.7   Determine the lower corner frequency for the network of Fig. 9.34 and sketch the Bode plot. Include an estimate of the actual gain-versus-frequency curve.

Solution: Determining r_e:
Dc conditions:

$$\beta R_E = (100)(2 \text{ k}\Omega) = 200 \text{ k}\Omega \gg 10 \text{ k}\Omega$$

βR_E can therefore be ignored on an approximate basis and

$$V_B \cong \frac{R_{B_2}(V_{CC})}{R_{B_2} + R_{B_1}} = \frac{10 \text{ k}\Omega(20)}{10 \text{ k}\Omega + 40 \text{ k}\Omega} = \frac{200}{50} = 4 \text{ V}$$

with

$$I_E = \frac{V_E}{R_E} = \frac{4 - 0.7}{2 \text{ k}\Omega} = \frac{3.3}{2 \text{ k}\Omega} = 1.65 \text{ mA}$$

so that

$$r_e = \frac{26 \text{ mV}}{1.65 \text{ mA}} \cong \mathbf{15.76} \ \mathbf{\Omega}$$

The following analysis will include computer printouts from the HP-85 desktop computer shown in Fig. 9.41. The software support for the system included a circuit analysis pac that permitted obtaining a plot of the ratio of the output to input voltage for the range of frequencies of interest. The mid-band gain will be required for comparison purposes:

$$A_v' = \frac{V_o}{V_b} = -\frac{R_L'}{r_e} = -\frac{R_C \| R_L}{r_e} = -\frac{(4 \text{ k}\Omega) \| (2.2 \text{ k}\Omega)}{15.76}$$

$$= \frac{-1419}{15.76} \cong -90$$

$$R_i = R_{B_1} \| R_{B_2} \| \beta r_e = 40 \text{ k}\Omega \| 10 \text{ k}\Omega \| 1.576 \text{ k}\Omega \cong 1.32 \text{ k}\Omega$$

Figure 9.41 HP-85 Scientific Personal Computer. (Courtesy Hewlett Packard Corporation.)

The network of Fig. 9.42 indicates that

$$V_b = \frac{R_i (V_i)}{R_i + R_S} \qquad \text{(voltage-divider rule)}$$

and

$$A_{v_{\text{mid}}} = \frac{V_o}{V_i} = \left[\frac{V_o}{V_b}\right]\left[\frac{V_b}{V_i}\right] = \left[-90\right]\left[\frac{R_i}{R_i + R_S}\right]$$

$$= \left[-90\right]\left[\frac{1.32\ \text{k}\Omega}{1.32\ \text{k}\Omega + 1.0\ \text{k}\Omega}\right] = \frac{-118.80}{2.32} = \boldsymbol{-51.21}$$

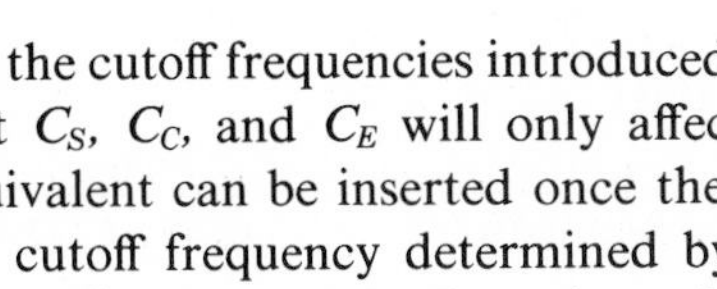

R_S

$+$

$V_S = V_i$ R_i V_b

$+$

$-$

$-$

Figure 9.42

Before determining the cutoff frequencies introduced independently by each capacitor, keep in mind that C_S, C_C, and C_E will only affect the low-frequency response. Their short-circuit equivalent can be inserted once the mid-band gain level has been achieved. The highest cutoff frequency determined by C_S, C_C, or C_E will have the greatest impact since it will determine when the gain can approach the mid-band level.

C_S:

$$R_i = R_{B_1} \| R_{B_2} \| \beta r_e = 40 \text{ k}\Omega \| 10 \text{ k}\Omega \| 1.576 \text{ k}\Omega \cong 1.32 \text{ k}\Omega$$

$$f_{L_S} = \frac{1}{2\pi (R_S + R_i) C_S} = \frac{1}{(6.28)(1 \text{ k}\Omega + 1.32 \text{ k}\Omega)(10 \times 10^{-6})}$$

$$= \frac{1}{(6.28)(2.32 \times 10^3)(10 \times 10^{-6})}$$

$$= \frac{1}{(6.28)(2.32)(10) \times 10^{-3}} = \frac{10^3}{145.7}$$

$$f_{L_S} \cong \textbf{6.86 Hz}$$

The resulting HP-85 computer plot for the network with just C_S appears in Fig. 9.43. Recall that the 3-dB point corresponds with a gain equal to 0.707 of the mid-band value, or 36.21, as indicated in Fig. 9.43.

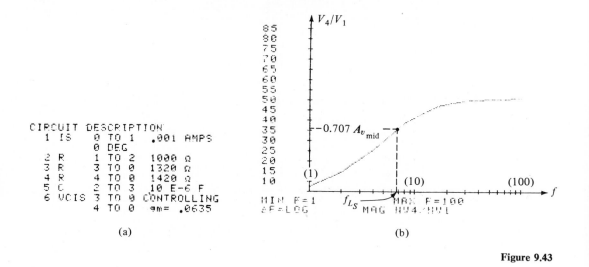

(a)   (b)

Figure 9.43

Dropping down to the log scale of Fig. 9.43, the cutoff frequency is as predicted ($\cong 6.86$ Hz).

C_C:

$$f_{L_C} = \frac{1}{2\pi (R_C + R_L) C_C}$$

$$= \frac{1}{(6.28)(4 \text{ k}\Omega + 2.2 \text{ k}\Omega)(1 \times 10^{-6})}$$

$$= \frac{1}{(6.28)(6.2 \times 10^3)(10^{-6})} = \frac{10^3}{(6.28)(6.2)} = \frac{10^3}{38.94}$$

$$\cong \textbf{25.68 Hz}$$

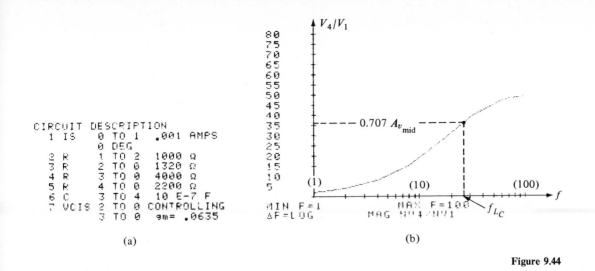

(a)

(b)

Figure 9.44

For C_C the computer plot appears in Fig. 9.44. Note again that the 0.707 level is achieved near the forecasted cutoff frequency of 25.68 Hz.

C_E:

$$R'_S = R_S \| R_{B_1} \| R_{B_2} = 1\,\text{k}\Omega \| 40\,\text{k}\Omega \| 10\,\text{k}\Omega \cong 1\,\text{k}\Omega$$

$$R_e = R_E \left\| \left(\frac{R'_S}{\beta} + r_e\right)\right. = 2\,\text{k}\Omega \left\| \left(\frac{1\,\text{k}\Omega}{100} + 15.76\right)\right. = 2\,\text{k}\Omega \| (10 + 15.76)$$

$$= 2\,\text{k}\Omega \| 25.76\,\Omega \cong 25.76\,\Omega$$

$$f_{L_E} = \frac{1}{2\pi R_e C_E} = \frac{1}{(6.28)(25.76)(20 \times 10^{-6})} = \frac{10^6}{3235.46} \cong 309.1\,\text{Hz}$$

For C_E the computer plot appears in Fig. 9.45. The 0.707 level is again attained near the forecasted cutoff frequency of 309.1 Hz.

Figure 9.45

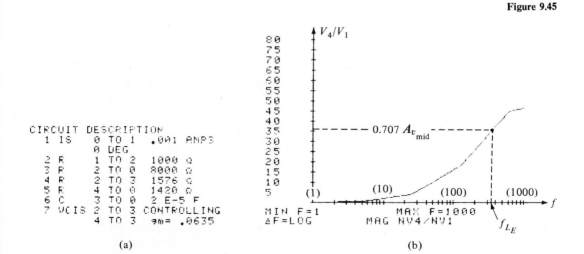

(a)

(b)

The fact that f_{L_E} is significantly higher than f_{L_S} or f_{L_C} suggests that it will be the predominant factor in determining the low-frequency response for the complete system. To test the accuracy of our hypothesis, the complete network was fed into the computer and the plot of Fig. 9.46 obtained. Note the very strong similarity with the plot of Fig. 9.44. For all practical purposes, C_C and C_S have only had an impact on the gain at frequencies below 100 Hz.

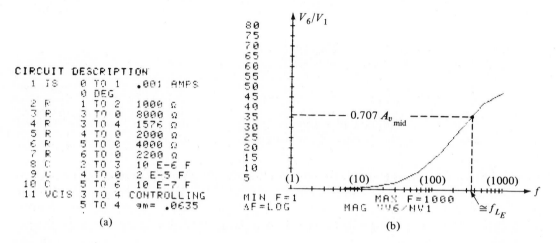

```
CIRCUIT DESCRIPTION
   1  IS    0 TO 1   .001 AMPS
            0 DEG
   2  R     1 TO 2   1000 Ω
   3  R     3 TO 0   8000 Ω
   4  R     3 TO 4   1576 Ω
   5  R     4 TO 0   2000 Ω
   6  R     5 TO 0   4000 Ω
   7  R     6 TO 0   2200 Ω
   8  C     2 TO 3   10 E-6 F
   9  C     4 TO 0   2 E-5 F
  10  C     5 TO 6   10 E-7 F
  11  VCIS  3 TO 4   CONTROLLING
            5 TO 4   gm= .0635
```

(a)

(b)

Figure 9.46

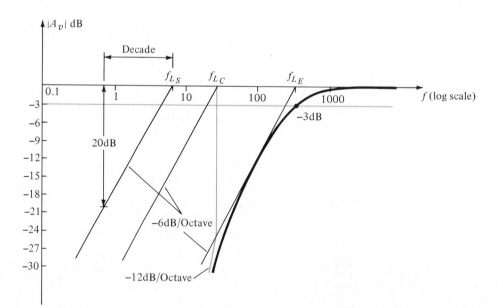

Figure 9.47 Low-frequency plot for the network of Example 9.7.

As described earlier, each cutoff (corner, break) frequency will define a −3 dB point and an asymptote as appearing in Fig. 9.47. The highest frequency as determined by C_E (in this example) will determine the resulting lower corner frequency (−3 dB point) for the network. Note that the lower frequency determined by C_C increased the slope of the asymptote originating at 309.1 Hz by −6 dB/octave. A rough sketch of the lower frequency region appears on the same plot.

For an unbypassed emitter resistor there will obviously be only two cutoff frequencies to be determined: that due to C_S and to C_C. The equation for f_{L_S} will be altered accordingly. There is an exercise on this very topic at the end of the chapter.

9.11 SINGLE-STAGE TRANSISTOR AMPLIFIER— HIGH-FREQUENCY CONSIDERATIONS

At the high-frequency end there are two factors that will define the −3 dB point: the network capacitance (parasitic and introduced) and the frequency dependence of $h_{fe}(\beta)$.

In the high-frequency region the RC network of concern has the configuration appearing in Fig. 9.48. At increasing frequencies the reactance X_C will decrease in magnitude, resulting in a shorting effect across the output and a decrease in gain. The derivation leading to the corner frequency for this RC configuration follows along similar lines to that encountered for the low-frequency region. The most significant difference is in the general form of A_v appearing below:

$$A_v = \frac{1}{1 + j(f/f_2)} \tag{9.35}$$

which results in an asymptotic plot such as shown in Fig. 9.49 that drops off at 6 dB/octave with increasing frequency.

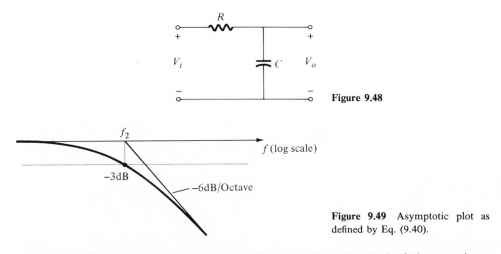

Figure 9.48

Figure 9.49 Asymptotic plot as defined by Eq. (9.40).

In Fig. 9.50 the various parasitic capacitances (C_{be}, C_{bc}, C_{ce}) of the transistor have been included with the wiring capacitance (C_{W_1}, C_{W_2}) introduced during construc-

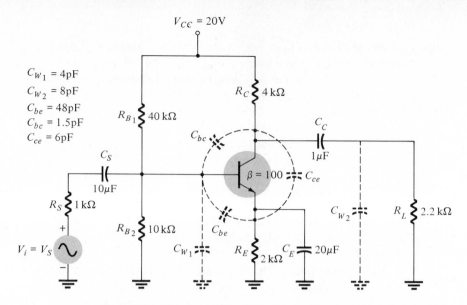

Figure 9.50 Network of Fig. 9.34 with the capacitors that affect the high-frequency response.

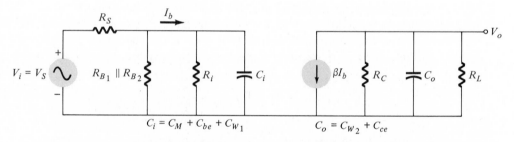

Figure 9.51 High-frequency model for the network of Fig. 9.50.

tion. The high-frequency equivalent model for the network of Fig. 9.50 appears in Fig. 9.51. Note the absence of the capacitors C_S, C_C, and C_E, which are all assumed to be in the short-circuit state at these frequencies. Keep in mind that the capacitors that determined the low-frequency cutoff point appear between the input and output terminals of the RC network and not across the output as indicated in Fig. 9.50. The capacitance C_i includes the Miller capacitance C_M as introduced for the FET in Chapter 8, the capacitance C_{be}, and the input wiring capacitance C_{W_1}. The capacitance C_o includes the output wiring capacitance and the collector capacitance.

Determining the Thévenin equivalent circuit of the input and output networks will result in the configurations of Fig. 9.52. For the input network the −3 dB frequency is defined by

$$f_{H_i} = \frac{1}{2\pi R_{Th_1} C_i} \qquad (9.36)$$

with $\qquad R_{Th_1} = R_S \| R_{B_1} \| R_{B_2} \| R_i$

and $\qquad C_i = C_{W_1} + C_{be} + C_M = C_{W_1} + C_{be} + (1 + |A_v|)C_{bc}$

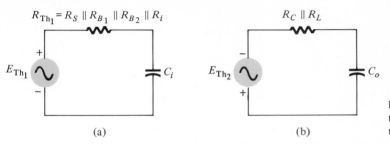

Figure 9.52 Thévenin circuits for the input and output networks of the network of Fig. 9.51.

At very high frequencies the effect of C_i is to reduce the total impedance of the parallel combination of R_{B_1}, R_{B_2}, R_i, and C_i in Fig. 9.51. The result is a reduced level of voltage across C_i as compared to that across R_S and a reduction in I_b. The net result is a reduced level of gain for the complete system.

For the condition $R_S = 0 \ \Omega$, $R_{Th_1} = 0 \ \Omega$ and f_{H_i} is infinite. In other words, with $R_S = 0 \ \Omega$, the applied voltage appears directly across C_i in the original network and an upper cutoff frequency for the input circuit is not defined.

Obviously, in the high-frequency range the voltage gain A_v will be a function of frequency due to the capacitive elements. It cannot be determined by a simple ratio of resistors as developed in Chapter 7. As a first approximation to the value of A_v, however, we shall use the mid-band value. As the frequency increases, the gain will obviously drop from the mid-band value due to the capacitive effects. If we therefore use the mid-band value, we have the maximum value of A_v and the maximum Miller capacitance. This will result in the maximum value for C_i and the minimum f_{H_i}. In essence, we have determined the lowest cutoff frequency due to C_i and defined a *worst-case* design situation. In other words, the actual cutoff due to C_i will always be higher than the level determined using the mid-band gain.

For the output network:

$$f_{H_o} = \frac{1}{2\pi R_{Th_2} C_o} \tag{9.37}$$

with
$$R_{Th_2} = R_C \| R_L$$

and
$$C_o = C_{W_2} + C_{Ce}$$

At very high frequencies the capacitive reactance of C_o will decrease and consequently reduce the total impedance of the output parallel branches of Fig. 9.51. The net result is that V_o will also decline toward zero as the reactance X_C becomes smaller.

For the network of Fig. 9.42:

$$R_{Th_1} = R_S \| R_{B_1} \| R_{B_2} \| R_i = 1 \ \text{k}\Omega \| 40 \ \text{k}\Omega \| 10 \ \text{k}\Omega \| 1.576 \ \text{k}\Omega \cong 0.575 \ \text{k}\Omega$$

$$C_i = C_{W1} + C_{be} + (1 + |A_v|)C_{bc}$$

$$= 4 \ \text{pF} + 48 \ \text{pF} + (1 + 90)(1.5 \ \text{pF})$$

$$= 188.5 \ \text{pF}$$

Note that A_v' as defined in Example 9.7 was employed since it provides the level of gain associated with the Miller capacitance.

$$f_{H_i} = \frac{1}{2\pi R_{Th_1} C_i} = \frac{1}{(6.28)(0.575 \times 10^3)(188.5 \times 10^{-12})}$$

$$= \frac{1000 \times 10^6}{680.7} \cong 1.47 \text{ MHz}$$

$$R_{Th_1} = R_C \,\|\, R_L = 4 \text{ k}\Omega \,\|\, 2.2 \text{ k}\Omega = 1.419 \text{ k}\Omega$$

$$C_o = C_{W2} + C_{ce} = 8 \text{ pF} + 6 \text{ pF} = 14 \text{ pF}$$

$$f_{H_o} = \frac{1}{2\pi R_{Th_2} C_o} = \frac{1}{(6.28)(1.419 \times 10^3)(14 \times 10^{-12})}$$

$$= \frac{1000 \times 10^6}{124.76} \cong 8.02 \text{ MHz}$$

The h_{fe} or β variation with frequency must now be considered to ensure that it does not determine the high cutoff frequency of the amplifier. The variation of h_{fe} with frequency will approach, with some degree of accuracy, the following relationship:

$$h_{fe} = \frac{h_{fe_{\text{mid}}}}{1 + jf/f_\beta} \tag{9.38}$$

The only undefined quantity, f_β, is determined by a set of parameters employed in the *hybrid π* or *Giacoletto* model frequently applied to best represent the transistor in the high-frequency region. It appears in Fig. 9.53. The various parameters warrant a moment of explanation. The resistance $r_{bb'}$ includes the base contact, base bulk, and base spreading resistance. The first is due to the actual connection to the base. The second includes the resistance from the external terminal to the active region of the transistors, while the last is the actual resistance within the active base region.

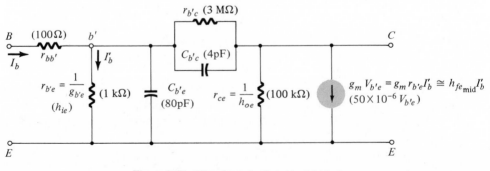

Figure 9.53 Giacoletto (or hybrid π) high-frequency transistor small-signal ac equivalent circuit.

The resistances $r_{b'e}$, r_{ce}, and $r_{b'c}$ are the resistances between the indicated terminals when the device is in the active region. The same is true for the capacitance $C_{b'c}$ and $C_{b'e}$, although the former is a transition capacitance while the latter is a diffusion capacitance. A more detailed explanation of the frequency dependence of each can be found in a number of readily available texts.

In terms of these parameters:

$$f_{\beta}\text{(sometimes appearing as } f_{h_{fe}}) = \frac{g_{b'e}}{2\pi(C_{b'e} + C_{b'c})}$$

(9.39)

or since the hybrid parameter h_{fe} is related to the hybrid parameter $g_{b'e}$ through $g_m = h_{fe_{mid}}g_{b'e}$

$$f_{\beta} = \frac{1}{h_{fe_{mid}}}\left[\frac{g_m}{2\pi(C_{b'e} + C_{b'c})}\right]$$

(9.40)

The basic format of Eq. (9.38) should suggest some similarities between it and the curves obtained for the low-frequency response. The most noticeable difference is the fact that f_{β} appears in the denominator while f_1 appears in the numerator of the frequency ratio. This particular difference will have the effect depicted in Fig. 9.54; the plot will drop off from the mid-band value rather than approach it with increase in frequency. The same figure has a plot of h_{fb} versus frequency. Note that it is almost constant for the frequency range. In general, the common-base configuration displays improved high-frequency characteristics over the common-emitter configuration. For this reason, common-base high-frequency parameters, rather than common-emitter parameters, are often specified for a transistor. The following equation permits a direct conversion for determining f_{β} if f_{α} and α are specified.

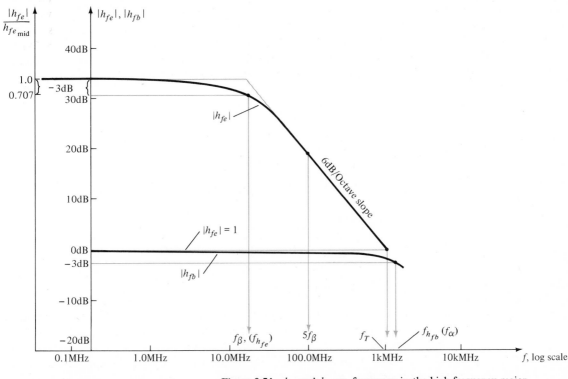

Figure 9.54 h_{fe} and h_{fb} vs. frequency in the high-frequency region.

$$\boxed{f_\beta = f_\alpha(1 - \alpha)} \tag{9.41}$$

A quantity called the *gain-bandwidth product* is defined for the transistor by the condition

$$\left| \frac{h_{fe_{\text{mid}}}}{1 + jf/f_\beta} \right| = 1$$

so that

$$|h_{fe}|_{\text{dB}} = 20 \log_{10} \left| \frac{h_{fe_{\text{mid}}}}{1 + jf/f_\beta} \right| = 20 \log_{10} 1 = 0 \text{ dB}$$

The frequency at which $|h_{fe}|_{\text{dB}} = 0$ dB is clearly indicated by f_T in Fig. 9.54. The magnitude of h_{fe} at the defined condition point $(f_T \gg f_\beta)$ is given by

$$\frac{h_{fe_{\text{mid}}}}{\sqrt{1 + (f_T/f_\beta)^2}} \cong \frac{h_{fe_{\text{mid}}}}{f_T/f_\beta} = 1$$

so that

$$\boxed{f_T \cong h_{fe_{\text{mid}}} \overbrace{f_\beta}^{(\cong \text{BW})} \quad \text{(Gain-bandwidth product)}} \tag{9.42}$$

or

$$f_T \cong \beta f_\beta$$

with

$$\boxed{f_\beta = \frac{f_T}{\beta}} \tag{9.43}$$

Substituting for f_β in Eq. (9.42) gives

$$f_T \cong h_{fe_{\text{mid}}} f_\beta = h_{fe} \left[\frac{1}{h_{fe}} \left(\frac{g_m}{2\pi(C_{b'e} + C_{b'c})} \right) \right]$$

$$\boxed{f_T \cong \frac{g_m}{2\pi(C_{b'e} + C_{b'c})}} \tag{9.44}$$

Ignoring the effect of $r_{bb'}$ as a first approximation, $C_{b'e} = C_{be}$ and $C_{b'c} = C_{bc}$. For the network of Fig. 9.50.

$$f_\beta = \frac{g_{b'e}}{2\pi(C_{b'e} + C_{b'c})}$$

$$= \frac{1 \times 10^{-3}}{(6.28)(49.5 \times 10^{-12})} = \frac{1000 \times 10^6}{310.86}$$

$$f_\beta = \textbf{3.217 MHz}$$

and

$$f_T \cong h_{fe} f_\beta$$

$$= 100 (3.217 \times 10^6)$$

$$= \textbf{321.7 MHz}$$

For the low-, mid-, and high-frequency regions a Bode plot for the network of

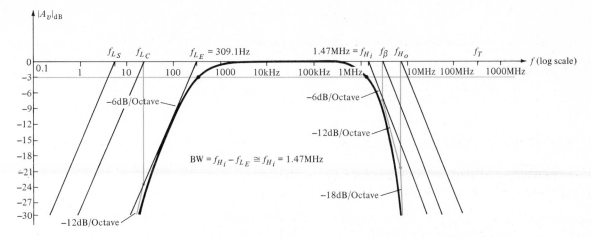

Figure 9.55 $|A_v|_{dB}$ versus frequency (log scale) for the network of Fig. 9.50.

Fig. 9.50 appears in Fig. 9.55. Note in the high- and low-frequency regions that a cutoff frequency defines a −6 dB/octave asymptote and the slope increases by −6 dB/octave (for the asymptote that will define the actual response) each time it passes a cutoff frequency. A curve approximating the actual response also appears in the figure. Note the −3 dB drop at the highest low-frequency cutoff frequency and at the lowest high-frequency cutoff frequency.

9.12 MULTISTAGE FREQUENCY EFFECTS

For a second transistor stage connected directly to the output of a first stage there will be a significant change in the overall frequency response. In the high-frequency region the output capacitance C_o must now include the wiring capacitance (C_{W_1}), parasitic capacitance (C_{be}), and Miller capacitance (C_M) of the following stage. Further, there will be additional low-frequency cutoff levels due to the second stage that will further reduce the overall gain of the system in this region. For each additional stage the upper cutoff frequency will be determined primarily by that stage having the lowest cutoff frequency. The low-frequency cutoff is primarily determined by that stage having the highest cutoff frequency. Obviously, therefore, one poorly designed stage can offset an otherwise well-designed cascaded system.

The effect of increasing the number of *identical* stages can be clearly demonstrated by considering the situations indicated in Fig. 9.56. In each case the upper and lower cutoff frequencies of each of the cascaded stages are identical. For a single stage the cutoff frequencies are f_1 and f_2 as indicated. For two identical stages in cascade the drop-off rate in the high- and low-frequency regions has increased to −12 dB/octave or −40 dB/decade. At f_1 and f_2, therefore, the decibel drop is now −6 dB rather than the defined band frequency gain level of −3 dB. The −3 dB point has shifted to f_1' and f_2' as indicated with a resulting drop in the bandwidth. A −18 dB/octave or −60 dB/decade slope will result for a three-stage system of identical stages with the indicated reduction in bandwidth (f_1'' and f_2'').

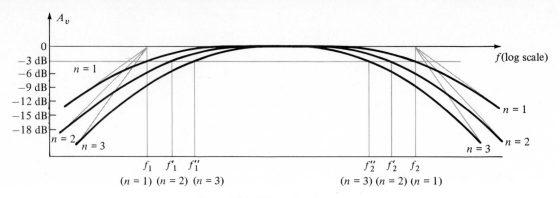

Figure 9.56 Effect of an increased number of stages on the cutoff frequencies and the bandwidth.

Assuming identical stages, an equation for each band frequency as a function of the number of stages *(n)* can be determined in the following manner:

For the low-frequency region,

$$A_{v_{\text{low,(overall)}}} = A_{v_{1\text{low}}}A_{v_{2\text{low}}}A_{v_{3\text{low}}} \cdot \cdot \cdot A_{v_{n\text{low}}}$$

but since each stage is identical, $A_{v_{1\text{low}}} = A_{v_{2\text{low}}} =$ etc.

and

$$A_{v_{\text{low,(overall)}}} = (A_{v_{1\text{low}}})^n$$

or

$$\frac{A_{v_{\text{low}}}}{A_{v_{\text{mid}}}}(\text{overall}) = \left(\frac{A_{v_{1\text{low}}}}{A_{v_{\text{mid}}}}\right)^n = \frac{1}{(1 + jf_1/f)^n}$$

Setting the magnitude of this result equal to $1/\sqrt{2}(-3$ dB level) results in

$$\frac{1}{[\sqrt{1 + (f_1/f_1')^2}]^n} = \frac{1}{\sqrt{2}}$$

or

$$\left\{\left[1 + \left(\frac{f_1}{f_1'}\right)^2\right]^{1/2}\right\}^n = \left\{\left[1 + \left(\frac{f_1}{f_1'}\right)^2\right]^n\right\}^{1/2} = \{2\}^{1/2}$$

so that

$$\left[1 + \left(\frac{f_1}{f_1'}\right)^2\right]^n = 2$$

and

$$1 + \left(\frac{f_1}{f_1'}\right)^2 = 2^{1/n}$$

with the result

$$\boxed{f_1' = \frac{f_1}{\sqrt{2^{1/n} - 1}}} \qquad (9.45)$$

In a similar manner, it can be shown that for the high-frequency region,

$$\boxed{f_2' = \sqrt{2^{1/n} - 1}\, f_2} \qquad (9.46)$$

Note the presence of the same factor $\sqrt{2^{1/n} - 1}$ in each equation. The magnitude of this factor for various values of *n* is listed below.

CH. 9 MULTISTAGE SYSTEMS AND FREQUENCY CONSIDERATIONS

n	$\sqrt{2^{1/n}-1}$
1	1
2	0.64
3	0.51
4	0.43
5	0.39

For $n = 2$, consider that the upper cutoff frequency $f_2' = 0.64f_2$ or 64% of the value obtained for a single stage, while $f_1' = (1/0.64)f_1 = 1.55f_1$. For $n = 3$, $f_2' = 0.51f_2$ or approximately $\frac{1}{2}$ the value of a single stage with $f_1' = (1/0.52)f_1 = 1.96f_1$ or approximately *twice* the single-stage value.

Consider the example of the past few sections, where $f_2 = 1.47$ MHz and $f_1 = 309.1$ Hz. For $n = 2$,

$$f_2' = 0.64f_2 = 0.64(1.47 \text{ MHz}) = 0.94 \text{ MHz}$$

and

$$f_1' = 1.56f_1 = 1.56(309.1) \cong 482.2 \text{ Hz}$$

The bandwidth is now 0.94 MHz $-$ 482.2 Hz $\cong f_2' \implies 64.4\%$ of its single-stage value, a drop of some significance.

For the *RC*-coupled transistor amplifier, if $f_2 = f_\beta$, or if they are close enough in magnitude for both to affect the upper 3-dB frequency, the number of stages must be increased by a factor of 2 when determining f_2' due to the increased number of factors $1/(1 + jf/f_x)$.

A decrease in bandwidth is not always associated with an increase in the number of stages if the mid-band gain can remain fixed independent of the number of stages. For instance, if a single-stage amplifier produces a gain of 100 with a bandwidth of 10,000 Hz, the resulting gain-bandwidth product is $10^2 \times 10^4 = 10^6$. For a two-stage system the same gain can be obtained by having two stages with a gain of 10 since $(10 \times 10 = 100)$. The bandwidth of each stage would then increase by a factor of 10 to 100,000 due to the lower gain requirement and fixed gain-bandwidth product of 10^6. Of course, the design must be such as to permit the increased bandwidth and establish the lower gain level.

This discussion of the effects of an increased number of stages on the frequency response was included here, rather than at the conclusion of this chapter, to add a note of completion to the analysis of cascaded transistor amplifier systems. The results, however, can be applied directly to the discussion of FET and vacuum-tube cascaded systems.

9.13 FREQUENCY RESPONSE OF CASCADED FET AMPLIFIERS

A representative cascaded system employing FET amplifiers appears in Fig. 9.57.

The equivalent circuit for the section a-a' indicated in Fig. 9.57 is presented in Fig. 9.58 for both the high- and low-frequency regions. It is assumed for the low-frequency response that the breakpoint frequency due to C_S is sufficiently less than that due to C_C, to result in C_C determining the lower-band frequency. For this reason

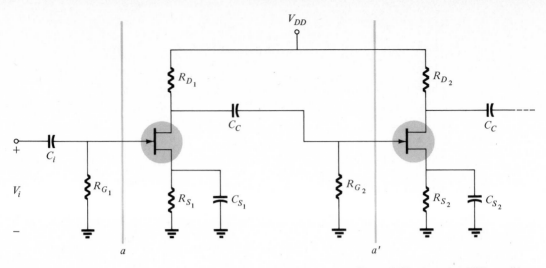

Figure 9.57 Cascaded FET amplifiers.

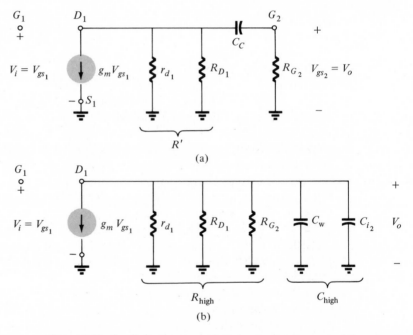

Figure 9.58 Small-signal ac equivalent circuit of section a-a' of the network of Fig. 9.57: (a) low frequency; (b) high frequency.

C_S does not appear in the low-frequency model. For future reference, the breakpoint frequency determined by C_S is given by

$$f_S = \frac{1 + R_S(1 + g_m r_d)/(r_d + R_D \parallel R_L)}{2\pi C_S R_S} \qquad (9.47)$$

It was demonstrated in this chapter and in Chapter 8 that the mid-band voltage gain is given by

$$A_{v\text{mid}} = \frac{V_o}{V_i} = -g_m R \qquad \text{where } R = r_{d_1} \| R_{D_1} \| R_{G_2}$$

The similarities between the circuits of Fig. 9.58 and those introduced in the last few sections on the transistor cascaded system are obvious. With this in mind, the steps leading to the following results should also be somewhat obvious.

$$\boxed{\frac{A_{v\text{low}}}{A_{v\text{mid}}} = \frac{1}{1 - jf_1/f}} \tag{9.48}$$

where

$$f_1 = \frac{1}{2\pi C_C (R' + R_{G_2})}$$

and

$$R' = r_{d_1} \| R_{D_1}$$

In addition,

$$\boxed{\frac{A_{v\text{high}}}{A_{v\text{mid}}} = \frac{1}{1 + jf/f_2}} \tag{9.49}$$

where

$$f_2 = \frac{1}{2\pi C_{\text{high}} R_{\text{high}}}$$

and

$$C_{\text{high}} = C_W + C_{i_2}$$

with

$$R_{\text{high}} = R = r_{d_1} \| R_{D_1} \| R_{G_2}$$

For FET amplifiers, the input capacitance of the succeeding stage, as derived in Chapter 8 is given by

$$\boxed{C_i = C_{gs} + C_{gd}(1 + |A_v|)} \tag{9.50}$$

The gain-bandwidth product is

$$[\text{gain}][\text{BW}] = [A_{v\text{mid}}][f_2 - f_1 \cong f_2] = [g_m R]\left[\frac{1}{2\pi R(C_W + C_i)}\right]$$

and

$$\boxed{\text{GBW} = \frac{g_m}{2\pi(C_W + C_i)} = f_T = A_{v\text{mid}} f_2} \tag{9.51}$$

PROBLEMS

§ 9.2

1. (a) Determine A_{i_T} and A_{v_T} for a cascaded system if the power gain is 12.8×10^3 and $Z_L = 4$ kΩ with the input impedance of the first stage $Z_{i_1} = 2$ kΩ.

   (b) If the system of part (a) is comprised of two identical stages, determine the voltage and current gain of each stage.

2. For the two-stage RC-coupled amplifier in Fig. 9.59:
   (a) Determine Z_i and Z_o.
   (b) Calculate the voltage gain $A_v = V_o/V_i$
   (c) Determine the current gain $A_i = I_o/I_i$.

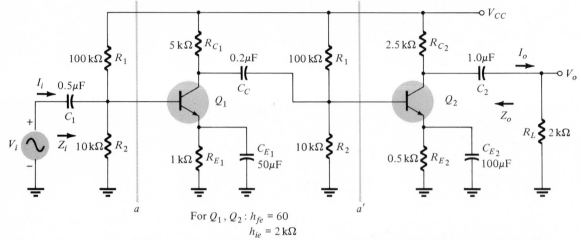

For Q_1, Q_2: $h_{fe} = 60$
$h_{ie} = 2\,k\Omega$

Figure 9.59 Two-stage RC-coupled amplifier.

3. Determine Z_i, Z_o, A_{v_T}, A_{i_T}, and A_{p_T} for the RC-coupled amplifier of Fig. 7.2 if $R_{B_1} = 56\,k\Omega$ and $R_{B_2} = 5.6\,k\Omega$ for each transistor and $R_{C_1} = 6.8\,k\Omega$, $R_{C_2} = 3.3\,k\Omega$, $R_{E_1} = R_{E_2} = 0.5\,k\Omega$, $R_L = 2.2\,k\Omega$. All capacitor values are the same. For each transistor $\beta = 120$. Since h_{ie} is not provided, r_e will have to be determined for each transistor. Use all appropriate approximations.

4. Repeat Problem 3 if both emitter capacitors C_E are removed.

5. Repeat Example 9.1 if C_E is removed.

6. Repeat Example 9.1 if C_E is removed and the output load Z_L is connected through a capacitor to the collector of Q_2.

7. Repeat Example 9.2 if R_{E_1} is removed and Z_L is reduced to a significantly lower level of 0.5 kΩ.

8. Design a two-stage RC-coupled amplifier to provide an overall gain of $\simeq 2000$. The circuit is to operate into a load of 10 kΩ; the signal is supplied from a perfect voltage source ($R_S = 0\ \Omega$). Show typical (commercially available) component values for each element and calculate the voltage gain of the resulting circuit as a check. The list of commercially available resistors can be found in any electronic products publication.

9. Repeat Example 9.3 if C_{S_1} and C_{S_2} are removed and $R_{S_1} = 2\,k\Omega$ with $R_{S_2} = 1\,k\Omega$.

§ 9.4

10. Calculate the impedance seen looking into the primary of a $5\!:\!1$ step-down transformer connected to a load of 20 Ω.

11. Calculate the necessary transformer turns ratio to match a 50-Ω load to a 20-kΩ source impedance.

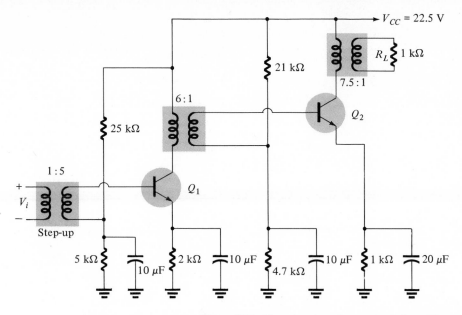

Figure 9.60 Two-stage transformer-coupled amplifier.

12. (a) Calculate the voltage gain (V_o/V_i) of the transformer-coupled amplifier of Fig. 9.60.
   (b) What is the voltage gain of the circuit of Fig. 9.65 if the load is reduced to 0.5 kΩ?

§ 9.5

13. Determine the new dc levels in Fig. 9.13 if the 12-V batteries are replaced by 16-V

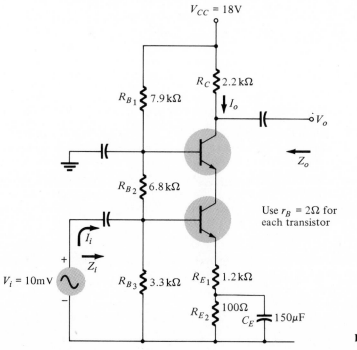

Figure 9.61

supplies. In addition, determine the new value of r_e for each transistor. How are the ac voltage and current gain affected? What are their new levels if affected by this change?

§ 9.6

14. Determine the following for the cascade amplifier of Fig. 9.61:
    (a) V_o.
    (b) Z_i, Z_o.
    (c) I_o, I_i, and A_i.
    (d) A_{p_T}.

15. Determine the following for the cascade amplifier of Fig. 9.62:
    (a) r_{e_1} and r_{e_2} for $\beta = 50$.
    (b) A_{v_T} and V_o if $V_i = 10$ mV.
    (c) Z_i and Z_o.

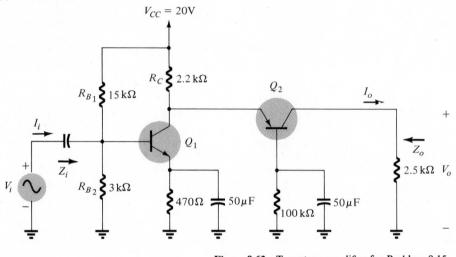

Figure 9.62 Two-stage amplifier for Problem 9.15.

§ 9.7

16. Determine A_i, Z_i, Z_o, and A_v for the Darlington configuration of Fig. 9.63.

17. Repeat Problem 16 if a collector resistor of 2.2 kΩ is added between the collector of Q_1 and V_{CC} and the output is taken off the collector of the Darlington configuration. I_o is the current through the added 2.2-kΩ resistor.

18. Repeat Problem 16 if R_E is changed to 150 Ω.

19. Repeat Problem 17 if R_E is changed to 150 Ω.

§ 9.8

20. Calculate the decibel power gain for the following:
    (a) $P_o = 100$ W, $P_i = 5$ W.
    (b) $P_o = 100$ mW, $P_i = 5$ mW.
    (c) $P_o = 100$ μW, $P_i = 20$ μW.

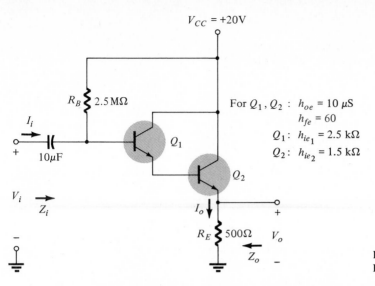

$V_{CC} = +20\text{V}$

R_B 2.5 MΩ

Q_1

I_i

10μF

V_i

Z_i

Q_2

For Q_1, Q_2: $h_{oe} = 10\ \mu S$

$h_{fe} = 60$

Q_1: $h_{ie_1} = 2.5\ k\Omega$

Q_2: $h_{ie_2} = 1.5\ k\Omega$

I_o

R_E 500Ω V_o

Z_o

Figure 9.63   Amplifier circuit for Problem 9.16.

21. Two voltage measurements made across the same resistance are $V_1 = 25$ V and $V_2 = 100$ V. Calculate the decibel power gain of the second reading over the first reading.

22. Input and output voltage measurements of $V_i = 10$ mV and $V_o = 25$ V are made. What is the voltage gain in decibels?

23. (a) The total decibel gain of a three-stage system is 120 dB. Determine the decibel gain of each stage if the second stage has twice the decibel gain of the first and the third has 2.7 times the decibel gain of the first.

(b) Determine the voltage gain of each stage.

§ 9.9–9.11

24. For the network of Fig. 9.64:

(a) Determine the low cutoff frequencies f_{L_S}, f_{L_C}, and f_{L_E}.

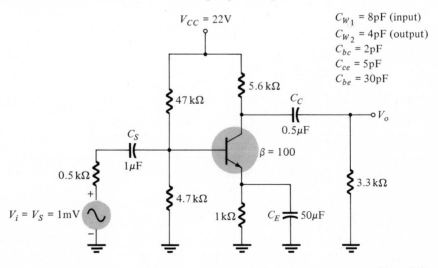

$V_{CC} = 22\text{V}$

$C_{W_1} = 8\text{pF (input)}$
$C_{W_2} = 4\text{pF (output)}$
$C_{bc} = 2\text{pF}$
$C_{ce} = 5\text{pF}$
$C_{be} = 30\text{pF}$

47 kΩ

5.6 kΩ

C_C

0.5μF

V_o

C_S

1μF

0.5 kΩ

β = 100

3.3 kΩ

4.7 kΩ

$V_i = V_S = 1\text{mV}$

1kΩ

C_E 50μF

Figure 9.64

(b) Calculate the mid-band voltage gain.
(c) Determine the high cutoff frequencies f_{H_i} and f_{H_o}.
(d) Sketch a rough plot of $A_v = V_o/V_i$ on a log plane.

25. Repeat Problem 25 if the capacitor C_E is removed.

26. Repeat Problem 25 if C_E is reduced to 1 μF.

§ 9.12

27. Calculate the overall voltage gain of four identical stages of an amplifier, each having a gain of 20.

28. Calculate the overall upper 3-dB frequency for a four-stage amplifier having an individual stage value of $f_2 = 2.5$ MHz.

29. A four-stage amplifier has a lower 3-dB frequency for an individual stage of $f_1 = 40$ Hz. What is the value of f_1 for this full amplifier?

30. (a) Determine the low cutoff frequencies for the two stage amplifier of Fig. 9.65.

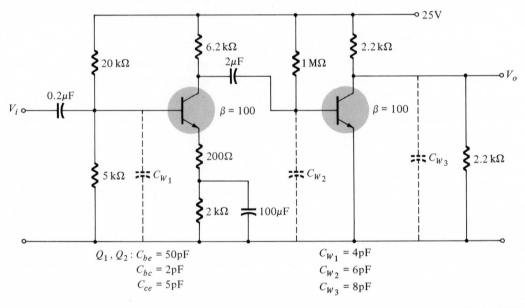

$Q_1, Q_2: C_{be} = 50\text{pF}$
$\qquad\quad C_{bc} = 2\text{pF}$
$\qquad\quad C_{ce} = 5\text{pF}$

$C_{W_1} = 4\text{pF}$
$C_{W_2} = 6\text{pF}$
$C_{W_3} = 8\text{pF}$

Figure 9.65

(b) Determine the high cutoff frequencies for the network of Fig. 9.65 ($f_\beta = 5$ MHz).
(c) Calculate the mid-band voltage gain and make a rough sketch of $A_v = V_o/V_i$ versus frequency (log plot).

§ 9.13

31. Calculate the mid-frequency gain for a stage of an FET amplifier as in Fig. 9.57 for circuit values: $g_m = 6000$ μmhos, $r_d = 50$ kΩ (Q_1 and Q_2); $R_{D_1} = R_{D_2} = 10$ kΩ; R_{G_1} $= R_{G_2} = 1$ MΩ; $C_S = 10$ μF, $R_S = 1$ kΩ, $C_C = 0.1$ μF, $C_{gs} = C_{gd} = 4$ pF, $C_W = 7$ pF.

32. For the circuit and values of Problem 31 calculate f_1 and f_2.

33. Calculate the gain-bandwidth product for the circuit and values of Problem 31.

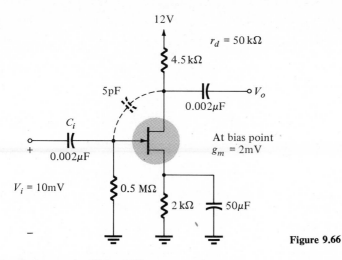

At bias point
$g_m = 2mV$

Figure 9.66

34. For the network of Fig. 9.66:
   (a) Calculate the low and high cutoff frequencies as determined by the provided levels of capacitance. Use $V_{CC} = 20$ V, $r_B = 0$ Ω.
   (b) What value of C_i will make the low cutoff frequency 10 Hz?
   (c) Determine V_o at the high cutoff frequency.

CHAPTER
10

Large-Signal Amplifiers

10.1 GENERAL

An amplifier system consists of a signal pickup transducer, followed by a small-signal amplifier, a large-signal amplifier, and an output transducer device. The input transducer signal is generally small and must be amplified sufficiently to be used to operate some output device. The factors of prime interest in small-signal amplifiers then are usually linearity and gain. Since the signal voltage and current from the input transducer is usually small, the amount of power-handling capacity and power efficiency are of slight concern. Voltage amplifiers provide a large enough voltage signal to the large-signal amplifier stages to operate such output devices as speakers and motors. A large-signal amplifier must operate efficiently and be capable of handling large amounts of power—typically, a few watts to hundreds of watts. This chapter concentrates on the amplifier stage used to handle large signals, typically a few volts to tens of volts. The amplifier factors of greatest concern are the power efficiency of the circuit, the maximum amount of power that the circuit is capable of handling, and impedance matching to the output device.

A class-A series-fed amplifier stage is considered first to show some of the limitations in using such a circuit connection. The single-ended transformer-coupled stage is then discussed to show one method of impedance matching between driver stage and load (output transducer). The push-pull connection, a very popular connection for low distortion and efficient coupling of the signal to a speaker or motor device, is discussed next. Finally, circuits using complementary transistors for push-pull operation without a transformer are presented.

10.2 SERIES-FED CLASS-A AMPLIFIER

The simple fixed-bias circuit connection can be used as a large-signal class-A amplifier as shown in Fig. 10.1. The only difference between this circuit and the small-signal version considered previously is that the signals handled by the large-signal circuit are in the range of volts and the transistor used is a power transistor capable of operating in the range of a few watts. As will be shown, this circuit is not the best to use for a large-signal amplifier.

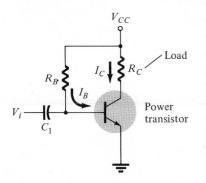

Figure 10.2a shows a typical class-A series-fed circuit with appropriate circuit values. The transistor-collector characteristic of Fig. 10.2b shows the load line and input and output signals. The input signal is an ac voltage of 0.5 V, peak amplitude. The circuit has an ac input resistance of 50 Ω, and the output ac signal is developed across the 20-Ω load resistor. For the bias value of base current shown, the transistor is operated at a collector-emitter voltage of 10 V and a collector current of about 500 mA.

Figure 10.1 Series-fed class-A large-signal amplifier.

The input ac current variation is calculated to be

$$I_i = \frac{V_i}{R_i} = \frac{0.5 \text{ V, peak}}{50 \text{ }\Omega} = 10 \text{ mA, peak}$$

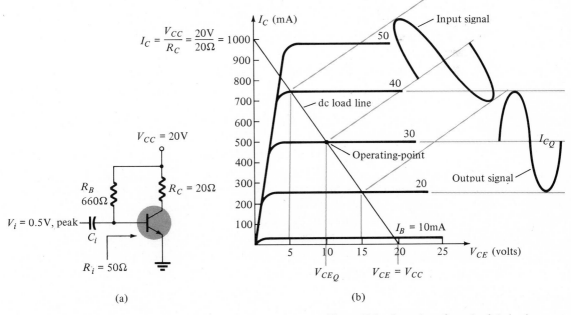

(a)

(b)

Figure 10.2 Operation of a series-fed circuit.

As shown on the transistor-collector characteristic (Fig. 10.2b) the corresponding current swing in the output (across the load resistor) is about 250 mA, peak. Corresponding to the indicated current swing is a voltage variation of 5 V, peak. From the present information we can calculate the current gain, voltage gain, and power gain of the circuit as follows:

$$A_v = \frac{V_o}{V_i} = \frac{5 \text{ V, peak}}{0.5 \text{ V, peak}} = 10$$

$$A_i = \frac{I_o}{I_1} = \frac{250 \text{ mA, peak}}{10 \text{ mA, peak}} = 25$$

$$A_p = \text{power gain} = A_v A_i = (10)(25) = 250$$

Up to now the power distribution in the amplifier has not been considered since the amount of power handled by small-signal circuits is small. In the large-signal circuit the amount of power in different parts of the amplifier is considerable and the efficiency of the amplifier circuit is of great interest. The dc battery is the source of power in a transistor amplifier. The dc power drawn from the battery is dissipated as heat lost in the load resistor and the transistor. The output power (ac signal developed across the load) is a part of the power taken from the dc battery. In fact, the operation of the circuit of Fig. 10.2a is to convert as much of the dc power drawn from the battery into ac power (output power) across the load (the 20-Ω resistor in this case). If the load were a speaker, the power delivered would result in audible sound.

In the present circuit we can calculate separately a number of different power terms as follows:

The average power taken from the dc battery is

$$\boxed{P_i(\text{dc}) = V_{CC} I_{C_Q}} \tag{10.1}$$

the product of the dc battery voltage times the *average* current drawn from the battery—the quiescent or average current for the nondistorted output signal, as shown in Fig. 10.2b.

The ac signal power developed across the load is

$$\boxed{P_o(\text{ac}) = I_C^2(\text{rms}) R_C} \tag{10.2}$$

the square of the root-mean square value of the ac current *through* the load resistor times the value of collector resistance.

The dc power dissipated as heat by the collector resistor is equal to the square of the average or dc current through the collector resistor times the value of the resistance.

$$\boxed{P_{RC}(\text{dc}) = I_{C_Q}^2 R_C} \tag{10.3}$$

The dc power dissipated by the transistor is

$$P_t = P_i - P_o - P_{RC} \qquad (10.4)$$

which is the input power minus that dissipated by the collector resistor (both ac and dc). This power is dissipated as heat with the transistor power rating indicating the maximum amount of power the particular device is capable of handling. (Transistor ratings and specifications are discussed in Section 10.7.)

The efficiency of the circuit in converting the dc power drawn from the battery into ac signal power across the load is calculated to be

$$\text{efficiency} = \eta = \frac{P_o(\text{ac})}{P_i(\text{dc})} \times 100\% \qquad (10.5)$$

The numerical calculations for the circuit of Fig. 10.2 are carried out in Example 10.1.

EXAMPLE 10.1   Calculate the various power terms for the circuit of Fig. 10.2a and transistor characteristic of Fig. 10.2b. Obtain the circuit power efficiency from these calculated values.

Solution:
(a) $P_i(\text{dc}) = V_{CC}I_{C_Q} = (20 \text{ V})(500 \text{ mA}) = \textbf{10 W}$

(b) $P_o(\text{ac}) = I_C^2(\text{rms}) R_C = \left(\dfrac{0.25}{\sqrt{2}}\right)^2 20 = \textbf{0.625 W}$

(c) $P_{RC}(\text{dc}) = I_{C_Q}^2 R_C = (0.5 \text{ A})^2(20 \text{ } \Omega) = \textbf{5 W}$

(d) $P_t(\text{dc}) = P_i - P_o - P_{RC} = 10 - 0.625 - 5 = \textbf{4.375 W}$

(e) $\eta = \dfrac{P_o}{P_i} \times 100 = \dfrac{0.625}{10} \times 100 = \textbf{6.25\%}$

The power efficiency of 6.25% is extremely poor and the circuit of Fig. 10.2 is therefore not a good amplifier for handling large amounts of power. Of the 10 W of power drawn from the battery, about half is dissipated in heat by the transistor and almost the same amount is wasted as heat in the load resistor, with only a very small amount of ac power being developed across the load. Although a 20-Ω speaker could be directly connected in the circuit as the load resistor, the operation of the circuit would be inefficient.

The maximum efficiency for the class-A series-fed amplifier can be seen to be 25% since the largest output power before distortion occurs is (see Fig. 10.2b)

$$V_o(p-p) = 20 \text{ V}, \qquad I_o(p-p) = 1000 \text{ mA} = 1 \text{ A}$$

which represents an output power of[1]

$$P_o = \frac{V_o(p-p)I_o(p-p)}{8} = \frac{(20 \text{ V})(1 \text{ A})}{8} = 2.5 \text{ W}$$

---

[1] $P_o(\text{ac}) = V_o(\text{rms})I_o(\text{rms}) = \dfrac{V_o(p)}{\sqrt{2}} \cdot \dfrac{I_o(p)}{\sqrt{2}} = \dfrac{V_o(p-p)/2}{\sqrt{2}} \cdot \dfrac{I_o(p-p)/2}{\sqrt{2}} = \dfrac{V_o(p-p)I_o(p-p)}{8}$

for an efficiency of

$$\eta = \frac{P_o(\text{ac})}{P_i(\text{dc})} \times 100 = \frac{2.5\text{ W}}{10\text{ W}} \times 100 = 25\%$$

For any output signal swing less than the maximum, the efficiency is less than the maximum value of 25%. Note that the input power remains 10 W in this example, regardless of the amount of ac voltage swing, the bias current remains the same whether an input signal is applied or not.

10.3 TRANSFORMER-COUPLED AUDIO POWER AMPLIFIER

A more reasonable class-A amplifier connection uses a transformer to couple the load to the amplifier stage as shown in Fig. 10.3a. This is a simple version of the circuit for the presentation of a few basic concepts. More practical circuit versions will be covered shortly. Figure 10.3b shows the output coupling transformer with voltage, current, and impedances indicated.

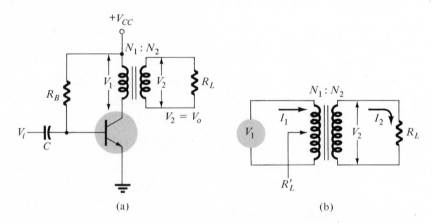

(a)   (b)

Figure 10.3 Transformer-coupled audio power amplifier.

Transformer Impedance Matching

The resistance seen looking into the primary of the transformer is related to the resistance connected across the secondary. The ratio of secondary resistance to primary resistance may be expressed as follows:

$$\frac{R_L' = V_1/I_1}{R_L = V_2/I_2} = \frac{R_L'}{R_L} = \frac{V_1}{I_1}\frac{I_2}{V_2} = \frac{V_1}{V_2}\frac{I_2}{I_1} = \frac{N_1}{N_2}\frac{N_1}{N_2} = \left(\frac{N_1}{N_2}\right)^2$$

where $V_1/V_2 = N_1/N_2$ and $I_2/I_1 = N_1/N_2$. Hence the ratio of the transformer input and output resistance varies directly as the *square* of the transformer turns ratio:

$$\boxed{\frac{R_L'}{R_L} = \left(\frac{N_1}{N_2}\right)^2 = a^2}$$

(10.6)

CH. 10   LARGE-SIGNAL AMPLIFIERS

and

$$R_L' = a^2 R_L = \left(\frac{N_1}{N_2}\right)^2 R_L$$

(10.7)

where
R_L = resistance of load connected across the transformer secondary;
R_L' = effective resistance seen looking into primary of transformer;
$a = N_1/N_2$ is the step-down turns ratio needed to make the load resistance appear as a larger effective resistance seen from the transformer primary.

EXAMPLE 10.2   Calculate the effective resistance (R_L') seen looking into the primary of a 15:1 transformer connected to an output load of 8 Ω.

Solution: Using Eq. (10.7), we get

$$R_L' = \left(\frac{N_1}{N_2}\right)^2 R_L = (15)^2 8 = 1800 \ \Omega = \mathbf{1.8 \ k\Omega}$$

EXAMPLE 10.3   What transformer turns ratio is required to match a 16-Ω speaker load to an amplifier so that the effective load resistance is 10 kΩ?

Solution: Using Eq. (8.6), we get

$$\frac{R_L'}{R_L} = \left(\frac{N_1}{N_2}\right)^2 = \frac{10,000}{16} = 625$$

$$\left(\frac{N_1}{N_2}\right) = \sqrt{625} = \mathbf{25:1}$$

Figure 10.4 shows a transistor-collector characteristic, load lines, input, and output signals for the circuit of Fig. 10.3.

dc Load Line

The transformer dc (winding) resistance determines the dc load line for the circuit. Typically, this dc resistance is small and is shown in Fig. 10.4a to be 0 Ω, providing a straight (vertical) load line. This is the ideal load line for the transformer. Practical transformer windings would provide a slight slope for the load line, but only the ideal case will be considered in this discussion. There is no dc voltage drop across the dc load resistance in the ideal case and the load line is drawn straight vertically from the voltage point, $V_{CEQ} = V_{CC}$.

Quiescent Operating Point

The operating point is obtained graphically as the point of intersection of the dc load line and the transistor base current curve. From the operating point the quiescent collector current I_{CQ} is read. The value of the base current is calculated separately from the circuit as considered in the dc bias calculations of Chapter 4.

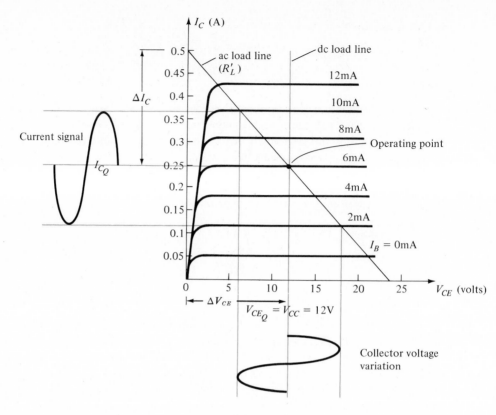

(a)

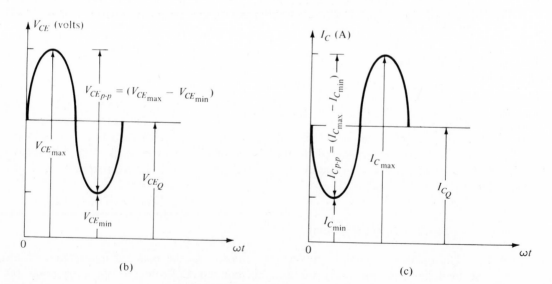

(b)   (c)

Figure 10.4 Graphical operation of transformer-coupled class-A audio power amplifier.

ac Load Line

In order to obtain the ac signal operation it is necessary to first calculate the ac load resistance seen looking into the primary side of the transformer and then to draw the ac load line on the transistor characteristic. The effective load resistance is calculated using Eq. (10.7) from the values of secondary load resistance and transformer turns ratio. Having obtained the value of R_L' the ac load line must be drawn so that it passes through the operating point and has a slope equal to $-1/R_L'$, the load-line slope being the negative reciprocal of the ac load resistance. Since the collector signal passes through the operating point when no signal is applied, the load line must pass through the operating point.

In order to simplify drawing a load line of slope $-1/R_L'$ through the operating point the following technique may be used:

If the ac signal were to vary from the quiescent level to 0 V, it would also vary from the quiescent current level, I_{C_Q}, by an amount

$$\Delta I_C = \frac{\Delta V_{CE}}{R_L'} \qquad (10.8)$$

Mark a point on the y-axis of the transistor characteristic ΔI_C units above the quiescent level and connect this point through the operating point to draw the ac load line desired (see Fig. 10.4a). Notice that the ac load line shows that the output signal swing can exceed the value of V_{CC}, the supply voltage. In fact, the voltage developed across the transformer primary can be large. One of the maximum operating values that will have to be checked carefully is that of $V_{CE_{max}}$, as specified by the transistor manufacturer, to see that the value of maximum voltage obtained after drawing the ac load line does not exceed the transistor rated maximum value. Assuming that maximum power and voltage ratings are not exceeded, the ac signal swings of current and voltage are obtained as shown in Fig. 10.4a and redrawn in detail in Fig. 10.4b and 10.4c.

Signal Swing and Output ac Power

From the signal variations shown in Figs. 10.4b and 10.4c the values of the peak-to-peak signal swings are

$$V_{\text{swing}} = V_{CE} \text{ (peak-to-peak)} = (V_{CE_{max}} - V_{CE_{min}}) \qquad (10.9)$$

$$I_{\text{swing}} = I_C \text{ (peak-to-peak)} = (I_{C_{max}} - I_{C_{min}}) \qquad (10.10)$$

The ac power developed across the transformer primary can be calculated to be

$$P_o(\text{ac}) = V_{CE} \text{ (rms) } I_C \text{ (rms)}$$

$$= \frac{V_{CE} \text{ (peak)}}{\sqrt{2}} \frac{I_C \text{ (peak)}}{\sqrt{2}}$$

$$= \frac{V_{CE} \text{ (peak-to-peak)}/2}{\sqrt{2}} \times \frac{I_C \text{ (peak-to-peak)}/2}{\sqrt{2}}$$

$$P_o(\text{ac}) = \frac{(V_{CE\text{max}} - V_{CE\text{min}})(I_{C\text{max}} - I_{C\text{min}})}{8} \qquad (10.11)$$

The ac power calculated is that developed across the primary of the transformer. Assuming a highly efficient transformer the power across the speaker is approximately equal to that calculated by Eq. (10.11). For our purposes an ideal transformer will be assumed so that the ac power calculated using Eq. (10.11) is also the ac power delivered to the load.

For the ideal transformer considered, the voltage across the secondary of the speaker can be calculated from

$$V_2 = V_L = \left(\frac{N_2}{N_1}\right)V_1 \qquad (10.12)$$

where the secondary voltage (V_2) equals the speaker or load voltage (V_L). The load voltage is related by the transformer turns ratio, N_2/N_1, to the voltage developed across the transformer primary (V_1). The voltage across the primary was previously labelled V_{CE} (rms), and for power considerations the rms values of voltage are usually used [unless otherwise stated, as in Eq. (10.11)].

The power across the load can be expressed as

$$P_L = \frac{V_L^2}{R_L} \qquad (10.13)$$

and equals the power calculated using Eq. (10.11). Thus the ac power can be calculated in a number of ways, including the following:

$$I_L(\text{rms}) = \frac{N_1}{N_2} I_C(\text{rms}) \qquad (10.14)$$

where I_L is the rms value of the current through the load resistor (or speaker resistance) and the load current is related to the rms value of the ac component of collector current by the transformer turns ratio.

The ac power is then calculated from

$$P_L = I_L^2 R_L \qquad (10.15)$$

EXAMPLE 10.4 The circuit of Fig. 10.5a shows a transformer-coupled class-A audio power amplifier driving an 8-Ω speaker. The coupling transformer has a 3:1 step-down turns ratio. If the circuit component values result in a dc base current of 6 mA and the input signal (V_i) results in a peak base current swing of 4 mA, calculate the following circuit values using the transistor characteristic of Fig. 10.5b: $V_{CE\text{max}}$, $V_{CE\text{min}}$, $I_{C\text{max}}$, $I_{C\text{min}}$, the rms values of load current and voltage, and the ac power developed across the load. Calculate the ac power using different equations as a check, that is, Eq. (10.11), Eq. (10.13), and Eq. (10.15).

Solution:
(a) The dc load line can be drawn vertically from the voltage point $V_{CE_Q} = V_{CC} = 10$ V (see Fig. 10.5c).
(b) For $I_B = 6$ mA the operating point on Fig. 10.5c is

$$V_{CE_Q} = 10 \text{ V} \qquad I_{C_Q} = 140 \text{ mA}$$

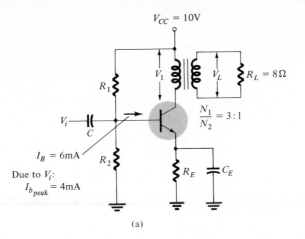

(a)

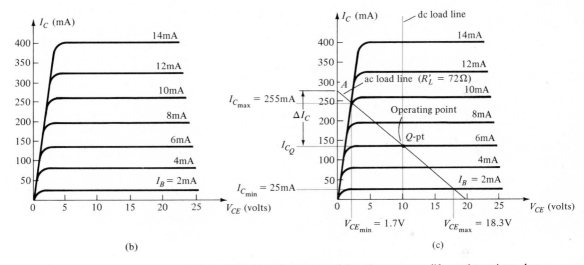

(b)

(c)

Figure 10.5 Transformer-coupled audio power amplifier and transistor characteristic for Example 10.4.

(c)   The effective ac resistance R_L' is [use Eq. (10.7)]

$$R_L' = \left(\frac{N_1}{N_2}\right)^2 R_L = (3)^2 8 = 72 \ \Omega.$$

(d)   Draw the ac load line as follows: Use Eq. (10.8) to calculate the current swing above the operating current

$$\Delta I_C = \frac{\Delta V_{CE}}{R_L'} = \frac{10 \ V}{72 \ \Omega} = 139 \ mA$$

Mark point A (Fig. 10.5c) $= I_{CE_Q} + \Delta I_C = 140 + 139 = 279$ mA along the *y*-axis. Connect point A to Q-point to draw ac load line.

(e)   For the given peak base current swing of 4 mA, the maximum and minimum values of collector current and voltage obtained from Fig. 10.5c are

$$V_{CE\text{min}} = 1.7 \ V \qquad I_{C\text{min}} = 25 \ mA$$
$$V_{CE\text{max}} = 18.3 \ V \qquad I_{C\text{max}} = 255 \ mA$$

(f)   Calculate the ac power across the transformer primary using Eq. (10.11)

$$P_o(ac) = \frac{(V_{CE\max} - V_{CE\min})(I_{C\max} - I_{C\min})}{8}$$

$$= \frac{(18.3 - 1.7)(255 - 25) \times 10^{-3}}{8} = \mathbf{0.477\ W}$$

(g)   Calculate the rms voltage across the primary

$$V_1\ (rms) = \frac{V_1\ (\text{peak-to-peak})}{2\sqrt{2}} = \frac{V_{CE\max} - V_{CE\min}}{2\sqrt{2}}$$

$$= \frac{16.6}{2.828} = 5.87\ V$$

(h)   The rms value of the load voltage is [using Eq. (10.12)]

$$V_L\ (rms) = \left(\frac{N_2}{N_1}\right) V_1\ (rms) = \left(\frac{1}{3}\right)(5.87) = 1.96\ V$$

(i)   Using Eq. (10.13) to calculate the ac power, we get

$$P_L\ (ac) = \frac{V_L^2}{R_L} = \frac{(1.96)^2}{8} = \mathbf{0.480\ W}$$

(j)   Using Eq. (10.14) to calculate the rms component of the load current, we get

$$I_L\ (rms) = \left(\frac{N_1}{N_2}\right)\frac{(I_{C\max} - I_{C\min})}{2\sqrt{2}} = (3)\frac{(230 \times 10^{-3})}{2.828}$$

$$= 244\ mA$$

(k)   Calculating the ac power using Eq. (10.15), we get

$$P_L(ac) = I_L^2 R_L = (244 \times 10^{-3})^2 8 = \mathbf{0.476\ W}$$

Power and Efficiency Calculations

So far we have considered calculating the ac power delivered to the load (the output ac power). We next consider the input power from the battery, power losses in the amplifier, and the overall power efficiency of the transformer-coupled class-A amplifier. The input dc power obtained from the battery is calculated from the values of dc battery voltage and average current from the battery as in Eq. (10.1).

$$P_i(dc) = V_{CC}I_{C_Q}$$

For the transformer-coupled amplifier, as shown in Fig. 10.3, the power dissipated by the transformer is small and will be ignored in the present calculations. Thus for the transformer-coupled amplifier the only power lost is that dissipated by the power transistor as calculated by the following equation

$$\boxed{P_Q = P_i - P_o} \qquad\qquad (10.16)$$

when P_Q is the power dissipated as heat. The equation seems simple but is significant in operating a power amplifier. The amount of power dissipated by the transistor (which then sets the transistor power rating) is the difference between the average dc input power from the battery (which is a constant for a fixed battery and operating point) and the output ac power drawn by the load. If the output power is zero, then the transistor must handle the maximum amount, that set by the battery voltage and bias current. If the load does draw some of the power, then the transistor has to handle that much less (for the moment). In other words, the transistor has to work hardest (dissipate the most power) when the load is disconnected from the amplifier circuit and the transistor dissipates least power when the load is drawing maximum power from the circuit. Obviously, the safest rating of the transistor used is the maximum set when the load is disconnected. Since normal operation with the load connected requires the transistor to dissipate less power, it is always preferable to keep the load connected as long as the amplifier unit is turned on.

EXAMPLE 10.5   Calculate the efficiency of the amplifier circuit of Example 10.4. Also calculate the power dissipated by the transistor.

Solution: Using Eq. (10.1) to calculate the input power, we get

$$P_i = V_{CC}I_{C_Q} = (10)(140 \times 10^{-3}) = 1.4 \text{ W}$$

Using Eq. (10.16), we see that the power dissipated by the transistor is

$$P_Q = P_i - P_o = 1.4 - 0.48 = \textbf{0.92 W}$$

$$\eta = \frac{P_o}{P_i} \times 100 = \frac{0.48}{1.4} \times 100 = \textbf{34.3\%}$$

Maximum Theoretical Efficiency

For a class-A amplifier the maximum theoretical efficiency for the series-fed circuit is 25% and for the transformer-coupled circuit it is 50%. From analysis of the operating range for a series-fed amplifier circuit the efficiency can be stated in the following form:

$$\eta = 25\left[\frac{V_{CE\text{max}} - V_{CE\text{min}}}{V_{CC}}\right]\% \tag{10.17}$$

In practical operation the efficiency is less than 25%. In the circuit of Example 10.1 the efficiency was only 6.25%, indicating a poorly operating series-fed circuit.

The efficiency of a transformer-coupled class-A amplifier can be expressed by

$$\eta = 50\left[\frac{V_{CE\text{max}} - V_{CE\text{min}}}{V_{CE\text{max}} + V_{CE\text{min}}}\right]\% \tag{10.18}$$

The larger the value of $V_{CE\text{max}}$ and the smaller the value of $V_{CE\text{min}}$ the closer the efficiency approaches the theoretical limit of 50%. In the circuit of Example 10.4 the value obtained was 34.3%. Well-designed circuits can approach the limit of 50% closely so that the circuit of Fig. 10.5a would be considered average in operation. The larger the amount of power handled by the amplifier the more critical the efficiency becomes. For a few watts of power a simpler, cheaper circuit with less than maximum efficiency is acceptable (and sometimes desirable). For power levels in the tens to hundreds of watts, efficiency as close as possible to the theoretical maximum would be desired.

The value of 50% considered as the maximum for transformer-coupled amplifiers is only for class-A operation. There are additional ways of operating (biasing) the amplifier to obtain even higher efficiency, as will now be considered.

10.4 CLASSES OF AMPLIFIER OPERATION AND DISTORTION

Operating Classes

The only class of amplifier operation so far considered has been class A. By definition class-A operation provides collector (output) current during the complete signal cycle (over a 360° interval). Figure 10.6a shows the output for class-A circuit

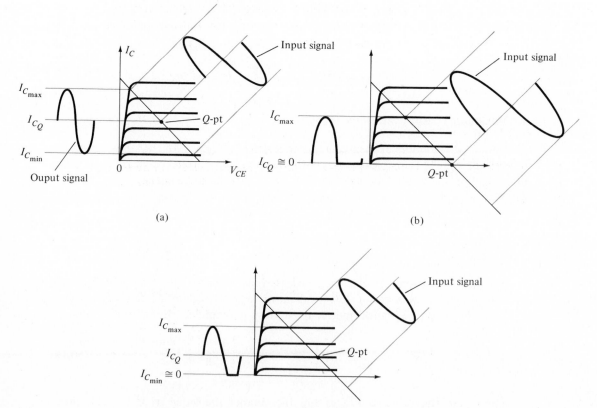

Figure 10.6 Various amplifier operating classes: (a) class A; (b) class B; (c) class AB.

operation. The bias level of current is I_{C_Q} and for the load line shown the output signal does not exceed values of $I_{C\max}$ or $I_{C\min}$, which would take the operation out of the linear region of device operation. Figure 10.6b shows *class-B* operation. The bias point is set at cutoff, the output current varying for only about 180° of the cycle, this being the definition of class-B operation. Note that the device is biased with no collector current and therefore no power dissipated by the transistor. Only when signal is applied does the transistor handle an average current which increases for larger input signals. Contrary to class-A operation, in which the worst condition occurs with no input signal and the least power is dissipated by the transistor for maximum input signal, the operation of a class-B circuit is to increase transistor dissipation for increased input signal. Since the average current in class-B operation is less than in class A, the amount of power dissipated by the transistor is less in class B.

A relation for the circuit efficiency of a class-B operating circuit is

$$\eta = 78.5 \left(1 - \frac{V_{CE\min}}{V_{CC}} \right) \% \tag{10.19}$$

which shows that the efficiency approaches the theoretical maximum value of 78.5% as the value of $V_{CE\min}$ approaches 0 V and the supply voltage used is made larger.

In between class-A and class-B operation is *class-AB* operation, shown in Fig. 10.6c. The collector current occurs for more than 180° of the signal cycle but less than 360°. The maximum operating efficiency of class AB is between that of class A and class B—that is, between 50% and 78.5%.

Operation with the output conducting for less than 180° is called *class-C* operation and is found in resonant or tuned amplifier circuits, as, for example, in radio or television. Operation on pulse-type signals is class D.

EXAMPLE 10.6 Determine the efficiency of transformer-coupled circuits indicated for the following operating conditions:
(a) Class-A operation with $V_{CE\max} = 30$ V and $V_{CE\min} = 3$ V.
(b) Class-B operation with $V_{CE\min} = 3$ V and $V_{CC} = 20$ V.

Solution:
(a) Using Eq. (10.18), we get

$$\eta = 50 \left[\frac{V_{CE\max} - V_{CE\min}}{V_{CE\max} + V_{CE\min}} \right] = 50 \left(\frac{30 - 3}{30 + 3} \right) = \mathbf{40.91\%}$$

(b) Using Eq. (10.19), we get

$$\eta = 78.5 \left(1 - \frac{V_{CE\min}}{V_{CC}} \right) = 78.5 \left(1 - \frac{3}{20} \right) = \mathbf{66.7\%}$$

Distortion

Output signal variations of less than 360° of the signal cycle are considered to have *distortion*. This means that the output signal is no longer just an amplified version of the input signal but in some ways is distorted or changed from that of the input. The poor quality of music coming from a radio or hi-fi system with the music or voice no longer sounding like that which was originally recorded or transmitted is a result of distortion. Distortion can come from a number of different places in any audio system.

Distortion can occur because the device characteristic is not linear: *nonlinear or amplitude distortion.* This can occur with all classes of operation. In addition, the circuit elements and the amplifying device can respond to the signal differently at various frequency ranges of operation: *frequency distortion.*

When amplitude distortion occurs, the output signal no longer represents the input signal exactly. One technique of accounting for this change in the output signal is the method of *Fourier analysis,* which provides a means for describing a periodic signal in terms of its fundamental frequency component and frequency components at integer multiples—components called *harmonic components* or *harmonics.* For example, a signal which is originally 1000 Hz could result, after distortion, in a frequency component at 1000 Hz, and harmonic components at 2 kHz (2×1000 Hz), at 3 kHz (3×1000 Hz), 4 kHz (4×1000 Hz), and so on. The original frequency of 1000 Hz is called the fundamental frequency and those at integer multiples are the harmonics—that at 2 kHz is the second harmonic, the component at 3 kHz is the third harmonic, and so on. The fundamental signal is considered the first harmonic. (No harmonics at fractional amounts of the fundamental frequency exist using this technique.)

An instrument such as a spectrum analyzer would allow measurement of the harmonics present in the signal by providing display of the fundamental component of the signal and a number of its harmonics on a CRT screen as in Fig. 10.7a. Similarly, a wave analyzer instrument allows more precise measurement of the harmonic components of a distorted signal by filtering out each of these components and providing a calibrated dial reading of these components, one at a time (see Fig. 10.7b).

In any case the technique of considering any distorted signal as containing a fundamental component and harmonic components is practical and useful. For a signal occurring in class AB or class B the distortion may be mainly even harmonics, of which the second harmonic component is greatest. Thus, although the distorted signal contains all harmonic components from second harmonic on up, the most important in terms of the amount of distortion for the classes of operation we will consider is the second harmonic.

A current output waveform is shown in Fig. 10.8 with the quiescent, minimum, and maximum signal levels, and the times they occur, marked on the waveform. The signal shown indicates that some distortion is present. An equation which approximately describes the distorted signal waveform is

$$i_C \cong I_{C_Q} + I_o + I_1 \cos \omega t + I_2 \cos 2\omega t \qquad (10.20)$$

The current waveform contains the original quiescent current I_{C_Q}, which occurs with zero input signal, an additional dc current I_o, due to the nonzero average of the distorted signal, the fundamental component of the distorted ac signal I_1, and a second harmonic component I_2, at twice the fundamental frequency. Although other harmonics are also present, only the second is considered here. Equating the resulting current from Eq. (10.20) at a few points in the cycle to that shown on the current

(a)

(b)

Figure 10.7 (a) Spectrum analyzer; (b) wave analyzer.

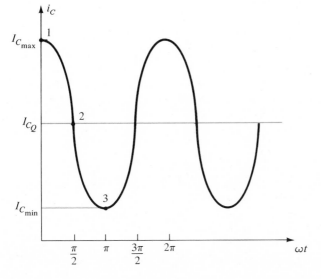

Figure 10.8 Waveform for obtaining second-harmonic distortion.

waveform provides the following three relations:

At $\omega t = 0$

$$i_C = I_{C\max} = I_{CQ} + I_o + I_1 \cos(0) + I_2 \cos(0)$$

$$I_{C\max} = I_{CQ} + I_o + I_1 + I_2 \tag{10.21}$$

At $\omega t = \pi/2$

$$i_C = I_{CQ} = I_{CQ} + I_o + I_1 \cos\left(\frac{\pi}{2}\right) + I_2 \cos\left(\frac{2\pi}{2}\right)$$

$$I_{CQ} = I_{CQ} + I_o - I_2 \tag{10.22}$$

At $\omega t = \pi$

$$i_C = I_{C\min} = I_{CQ} + I_o + I_1 \cos(\pi) + I_2 \cos(2\pi)$$

$$I_{C\min} = I_{CQ} + I_o - I_1 + I_2 \tag{10.23}$$

Solving Eq. (10.21), (10.22), and (10.23) simultaneously gives the following results

$$I_o = I_2 = \frac{I_{C\max} + I_{C\min} - 2 I_{CQ}}{4}$$

$$\tag{10.24}$$

$$I_1 = \frac{I_{C\max} - I_{C\min}}{2}$$

By definition the percent of the second harmonic distortion is given by

$$D_2 = \left| \frac{I_2}{I_1} \right| \times 100\% \tag{10.25}$$

The second harmonic distortion is the percent of the second harmonic component present in the output current waveform with respect to the amount of the fundamental component. Obviously, 0% distortion is the ideal condition of no distortion.

Using the results in Eq. (10.24) to express the second harmonic distortion defined by Eq. (10.25) gives

$$D_2 = \left| \frac{\frac{1}{2}(I_{C\max} + I_{C\min}) - I_{CQ}}{I_{C\max} - I_{C\min}} \right| \times 100\% \tag{10.26}$$

In a similar manner, the amount of second harmonic distortion can be related to the measured values of the distorted output voltage waveform

$$D_2 = \left| \frac{\frac{1}{2}(V_{CE\max} + V_{CE\min}) - V_{CEQ}}{V_{CE\max} - V_{CE\min}} \right| \times 100\% \tag{10.27}$$

EXAMPLE 10.7   An output waveform displayed on a scope provides the following measured values:

(a) $V_{CE\min} = 1$ V, $V_{CE\max} = 22$ V, $V_{CEQ} = 12$ V

(b) $V_{CE\min} = 4$ V, $V_{CE\max} = 20$ V, $V_{CEQ} = 12$ V

For each set of values calculate the amount of the second harmonic distortion.

Solution: Using Eq. (10.27), we get

(a) $D_2 = \left| \dfrac{\frac{1}{2}(22+1) - 12}{22-1} \right| \times 100 = \mathbf{2.38\%}$

(b) $D_2 = \left| \dfrac{\frac{1}{2}(20+4) - 12}{20+4} \right| \times 100 = \mathbf{0\%}$ (no distortion)

The method used to obtain the amount of second harmonic distortion was called the three-point method since it involved equating the assumed form of the output voltage to the measured voltage at three points in the signal cycle. Using an assumed output signal equation containing more harmonic terms, along with choosing more points in the waveform, results in obtaining relations for the magnitude of the harmonic components at higher harmonic frequencies. Using a five-point method provides the dc component, first harmonic (fundamental), second harmonic, third harmonic, and fourth harmonic components. The harmonic distortion for each of these components is then defined as

$$D_2 = \left| \frac{I_2}{I_1} \right|, \qquad D_3 = \left| \frac{I_3}{I_1} \right|, \qquad D_4 = \left| \frac{I_4}{I_1} \right| \qquad (10.28)$$

The total distortion may be defined, in general, using the individual distortion components

$$D = \sqrt{D_2{}^2 + D_3{}^2 + D_4{}^2 + \cdots} \qquad (10.29)$$

When distortion does occur, the output power calculated by the undistorted case is no longer correct. Equation (10.11), for example, is true *only* for the nondistorted case. When distortion is present, the output power due to the fundamental component of the distorted signal is

$$P_1 = \frac{I_1{}^2 R_C}{2} \qquad (10.30)$$

The total output power due to all the harmonic components of the distorted signal is

$$P = (I_1{}^2 + I_2{}^2 + I_3{}^2 + \cdots) \frac{R_C}{2} \qquad (10.31)$$

The total power can also be expressed in terms of the total distortion

$$P = (1 + D_2{}^2 + D_3{}^2 + \cdots) I_1{}^2 \frac{R_C}{2} = (1 + D^2) P_1 \qquad (10.32)$$

EXAMPLE 10.8   Using a five-point method to calculate harmonic components gives the following results: $D_2 = 0.1$, $D_3 = 0.02$, $D_4 = 0.01$, with $I_1 = 4$ A and $R_C = 8 \ \Omega$. Calculate the total distortion, fundamental power component, and total power.

Solution: Total distortion, using Eq. (10.29), gives

$$D = \sqrt{D_2{}^2 + D_3{}^2 + D_4{}^2} = \sqrt{(0.1)^2 + (0.02)^2 + (0.01)^2} \cong 0.1$$

Fundamental power, using Eq. (10.30), gives

$$P_1 = \frac{I_1{}^2 R_C}{2} = \frac{(4)^2 8}{2} = 64 \text{ W}$$

Total power, using Eq. (10.32), gives

$$P = (1 + D^2)P_1 = [1 + (0.1)^2]64 = (1.01)64 = 64.64 \text{ W}$$

(Total power mainly due to fundamental component even with 10% second harmonic distortion.)

GRAPHICAL DESCRIPTION OF HARMONIC
COMPONENTS OF DISTORTED SIGNAL

A demonstration of the use of harmonic components to represent a distorted signal is provided to help clarify the concept. As an example, a distorted waveform such as that resulting from class-B operation is shown in Fig. 10.9a. The signal is clipped on the negative half-cycle so that only the positive sinusoidal half-cycle provides an output signal.

Using Fourier analysis techniques, we can calculate a fundamental component of the distorted signal as shown in Fig. 10.9b. Figure 10.9b does not show the distorted waveform, only the fundamental component (which is a perfectly sinusoidal signal itself). Similarly, the second and third harmonic components can be obtained and are shown in Figs. 10.9c and 10.9d, respectively.

We now wish to check whether these components, each a purely undistorted sinusoidal signal, add up approximately to the original distorted signal. Figure 10.9e shows the resulting waveform when adding the fundamental and second harmonic components together. Note the flattening of the second half of the cycle. In Fig. 10.9f the third harmonic component is added to give a resulting waveform that comes fairly close to the original distorted signal. The addition of higher harmonic components of the right amplitude and correct phase will further alter the resulting waveform to approximate the original distorted signal. In a relatively simple manner we can observe that addition of a fundamental component and harmonic components can result in the original distorted waveform. In general, any periodic waveform can be represented by a fundamental component and harmonic components, each of varying amplitudes and at various phase angles.

The concept of harmonics is useful in both analyzing distorted (nonsinusoidal) waveforms and in providing a means of working with such signals. Since all the harmonic components are sinusoidal signals, we can separately consider the effect of each component on the circuit and obtain the total effect using superposition— adding together the voltages or currents being considered. Particular mention of harmonic content of a waveform will be made in the discussion of push-pull amplifiers, where it will be shown that the particular circuit connection eliminates the even harmonic components and leaves only the fundamental component and the odd harmonic components. Since the largest distortion component is the second harmonic, the elimination of that component will considerably reduce the total amount of distortion.

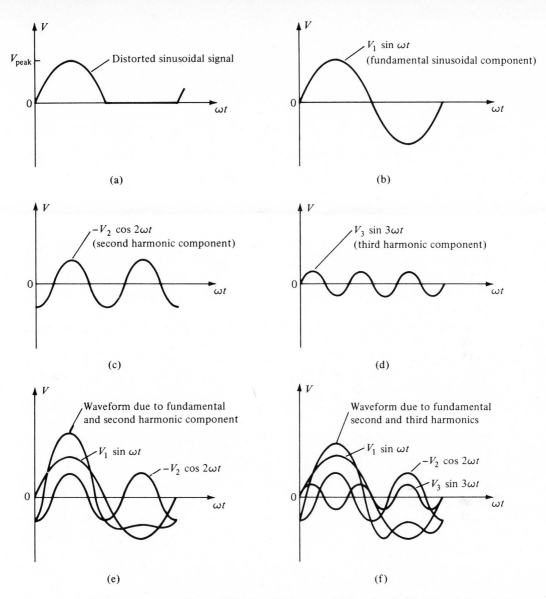

Figure 10.9 Graphical representation of a distorted signal through the use of harmonic components.

10.5 PUSH-PULL AMPLIFIER CIRCUIT

Previous discussion has shown that efficient operating conditions occur for class-AB or class-B operation. On the other hand, class-AB or class-B operation, as discussed so far, results in considerable distortion. To be forced to choose one or the other—more efficient operation or less distortion—is not the best of choices. Ideally, one wants to have the efficient operation of class-B but also the low distortion of class-A operation. This is not realistic, but a surprisingly low-distortion, high-efficiency operation can be obtained using the circuit connection known as *push-pull*. A typical

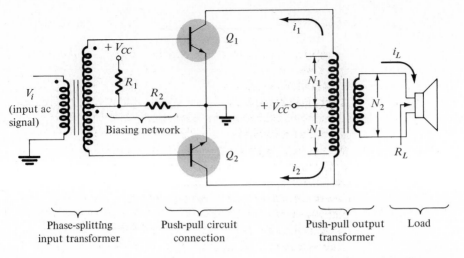

Figure 10.10   Push–pull circuit.

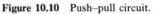

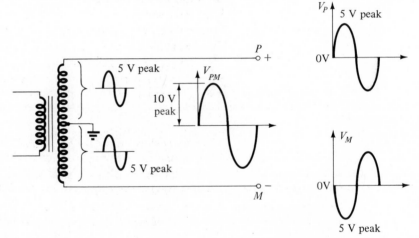

(a)

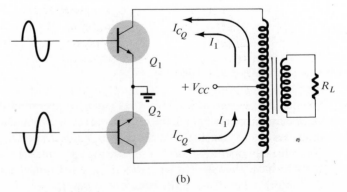

(b)

Figure 10.11   Details of push–pull operation.

transistor push-pull circuit connection is shown in Fig. 10.10. Similar circuits are also made using vacuum tubes or FETs.

The circuit of Fig. 10.10 uses an input transformer to produce opposite polarity signals to the two transistor inputs and an output transformer to drive the load in a push-pull mode of operation to be described. Figure 10.11a shows the input transformer with center tap at ground. As example, the voltage across the secondary from the plus (+) to the minus (−) terminal is 10 V, peak. As shown in Fig. 10.11a, the voltage across each half of the transformer is 5 V, peak, adding up to the total of 10 V, peak across the transformer. With the center tap of the transformer connected to ground (0-V potential) the signal observed from the plus terminal to ground is, as shown in Fig. 10.11a, in phase with the voltage across the full transformer. However, since the voltage measured across the bottom half of the transformer is also in phase with the total signal when measured from center to minus, it is opposite when measured from minus to center tap (or from minus to ground). The signals at the plus and minus terminals measured with respect to ground are therefore opposite in polarity as shown in Fig. 10.11a.

Having obtained opposite-polarity input to the two transistor units the push-pull nature of the circuit operation can be considered as shown in the partial circuit diagram of Fig. 10.11b. Consider first the dc bias current (I_{C_Q}) for each transformer. Figure 10.11b shows that the bias currents for each transistor are in opposite directions through the transformer winding. The magnetic flux set up by each of these currents results in opposite flux through the magnetic core so that the net flux, in the perfectly matched case, is zero. Thus the transformer need not handle a large flux due to the dc bias currents, resulting in a smaller size core, biased to operate near zero flux.

Considering only the positive half-cycle of operation, transistor Q_1 is driven further into conduction whereas transistor Q_2 is driven less into conduction. The varying component of current for each transformer is marked I_1 in Fig. 10.11b. Since Q_1 is driven further into conduction, the varying component of current is in the same direction as the dc bias current, resulting in a large total current. The varying component of the current in Q_2, however, is in an opposite direction to the bias current for that transistor, resulting in a net decrease in current for transistor Q_2. The overall operation of a net current through the transformer primary results in a half-cycle of voltage across the secondary and thus across the load.

During the second half-cycle of the input signal, the current through Q_1 decreases while that through Q_2 increases, resulting in an opposite polarity signal—the overall load signal varying over the full cycle of signal operation. We now need to consider how the push-pull connection with class-AB or class-B operation of the transistors will provide low-distortion output.

Figure 10.12a shows the output signal for each section of the push-pull circuit when operated in class B. Note that since the two transistors are identically biased, they distort the output waveform on the negative half-cycle for each signal—the second half-cycle for stage 1 and the first half-cycle for stage 2, as shown. Intuitively, one might say that the positive half-cycle due to stage 1 provides the output signal during the first half-cycle and that the positive half-cycle signal operates stage 2 during the second half-cycle. Since the two stages would cause current in opposite directions, the two half-cycles, as indicated in Fig. 10.12a, would result in opposite

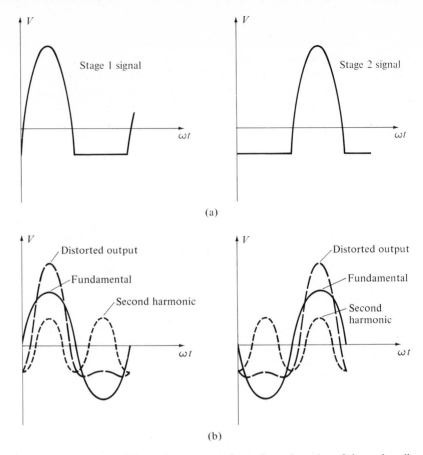

Figure 10.12 Distorted output waveforms for each section of the push–pull circuit operation.

half-cycles of signal in the transformer—thus providing a full or complete cycle of signal even with half-cycles of operation of each stage.

For a more detailed discussion of the circuit operation the approximate distorted output waveforms in each stage of the circuit are shown in Fig. 10.12b as being made up of a fundamental component and a second harmonic component. Although other harmonic components would be necessary to describe more fully the actual distorted waveforms of Fig. 10.12a, consideration of only the fundamental and second harmonic components will provide the necessary information here.

Notice in Fig. 10.12b that the fundamental components of each stage are opposite in polarity. For the push-pull circuit connection, opposite-polarity output signals will result in a net voltage in the secondary which is the sum of the two component signals. The second-harmonic components, however, are of the same polarity and therefore cancel, providing a net voltage in the secondary of 0 V due to these signal components. In other words, *the action of the push-pull connection would be to cancel out the second harmonic components of the signals applied to the primary of the output transformer,* and the same cancellation can be shown to result for all other even harmonic components. From our previous consideration of harmonics we can now appreciate that a distorted signal applied to the push-pull transformer connection, such as those shown in Fig. 10.12a, would result in cancellation of all even harmonic components so that the resulting output signal across the secondary can be considered to be made up of the fundamental component, third-harmonic component, and all odd-harmonic components of the distorted signal. Since the second harmonic is the largest component for the distorted signal of Fig. 10.12a, elimination of that component will result in an output signal having considerably less distortion. In calculating the distortion of the output signal we need to consider the amount of the third, fifth, and so on, harmonic components, which we expect will not be too large, so that the total distortion as calculated by Eq. (10.30) will not be much more than the small amount of the third harmonic distortion.

Of course, the above conclusion of even-harmonic cancellation is an ideal condition which will occur only if the circuit is perfectly balanced—transistors exactly matched, transformer center-tap connection perfect, and input signals exactly equal and opposite. This will never practically occur, but still the circuit can be operated class AB or class B for improved power efficiency with low distortion occurring (although more than for class-A operation).

To summarize, then, the advantages of the push-pull circuit connection are

1. The dc components of the collector currents oppose each other in the transformer resulting in no net flux due to the bias current and, hence, smaller size transformer cores can be used.
2. Class-AB or class-B operation is possible with small resulting distortion due to the cancellation of all the even harmonic components.
3. Ripple voltage in the voltage supply is cancelled out by the push-pull operation of the circuit.
4. High-efficiency operation is possible by using class-AB or class-B biasing.

To complete the picture it must be pointed out that the power supply used in class-B operation must have good voltage regulation since the amount of current drawn is about zero for no signal operation and rises in value for larger amounts of signal level. Since the current drawn from the supply ranges considerably from no load to full load, the supply voltage regulation must be good. Also, hum voltages (60-Hz pickup), which are picked up on the input lines of the circuit, are *not* eliminated by the push-pull since these are brought along with the input signals to the driver transformer and connected with the same polarity as the input to the push-pull circuit, thereby acting as proper input signals to drive the output device.

10.6 VARIOUS PUSH-PULL CIRCUITS INCLUDING TRANSFORMERLESS CIRCUITS

A number of other circuit arrangements for obtaining push-pull operation without transformers are possible. We shall consider a few of these and their various advantages and disadvantages. It is important to keep in mind the overall operation of the circuit in order to appreciate the different methods of obtaining the advantages of push-pull operation. For the push-pull circuit it is necessary to develop the output voltage

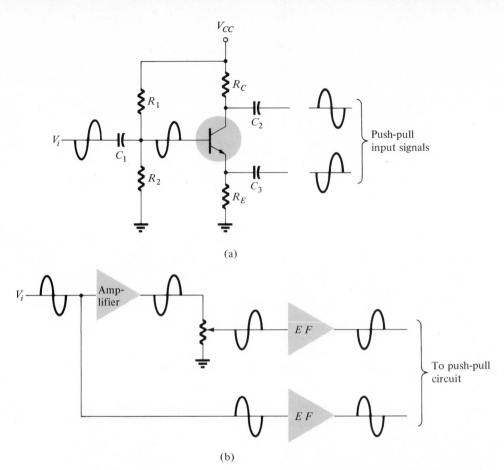

(a)

(b)

Figure 10.13 Phase-splitter circuits.

across the load in such a manner that two stages operating in class B will provide a full cycle of signal by conducting on alternate half-cycles.

Starting with an input signal obtained from a driver amplifier stage, it is necessary to operate the two-stage push-pull circuit on alternate half-cycles for class-B operation. The opposite-polarity input signals to the two stages of the push-pull circuit can be obtained in a number of ways. Figure 10.10 shows the use of an input transformer to provide the polarity inversion between the two push-pull input signals. Other circuits used to obtain opposite-polarity input signals are shown in Fig. 10.13. The input signal applied to the circuit of Fig. 10.13a appears of opposite polarity at the collector. The output from the emitter is the same polarity as the input so that the two output signals are of opposite polarity as shown in Fig. 10.13a. Values of R_C, R_E, and h_{fe} can be chosen to make the voltage gain for the collector output signal to be 1. The gain for the signal taken from the emitter is 1 (emitter-follower operation). Thus, the circuit results in opposite-polarity signals to drive the push-pull amplifier stage. The advantage of this driver connection is the saving on the use of a center-tapped transformer which is expensive and bulky and has a limited frequency operating range. A disadvantage is that the two signals do not come from similar impedance sources. The signal from the emitter provides a good driver connection since the source resistance viewed from the emitter is low. The collector circuit resistance, however, is relatively high and although the unloaded output signals are equal, they are different under load conditions. One possible improvement would be to add an additional emitter-follower stage to connect the output to the load since such stage would provide no additional voltage gain or polarity inversion but would drive the push-pull stage from a low-resistance source.

Another means of obtaining opposite-polarity signals to drive the push-pull stage is illustrated by the block diagram of Fig. 10.13b. The input signal is amplified and inverted by one amplifier stage and then attenuated for an overall gain of unity. The use of two emitter followers (possibly Darlington circuits) drives the push-pull stage from low-impedance sources.

Complementary-Symmetry Circuits

A number of circuits go beyond eliminating only the input polarity-inverting transformer from the circuit. These circuits also remove the output transformer so that the circuit is completely transformerless. A simple version of a transformerless push-pull amplifier circuit is shown in Fig. 10.14. Complementary transistors are used, that is, *npn* and *pnp* transistors are used instead of using two of the same type. The single input signal required is applied to both base inputs. However, since the transistors are of opposite type, they will conduct on opposite half-cycles of the input. During the positive half-cycle of the input signal, for example, the *pnp* transistor

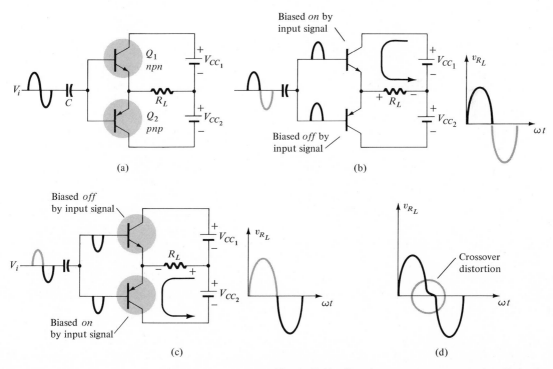

Figure 10.14 Complementary symmetry push–pull circuit.

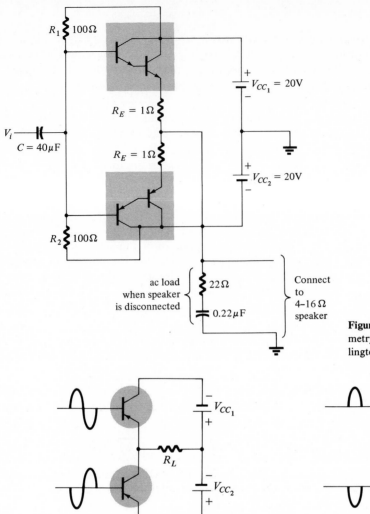

Figure 10.15 Complementary symmetry push–pull circuit using Darlington transistors.

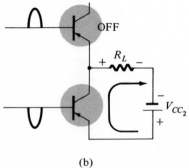

Figure 10.16 Transformerless push–pull circuit.

will be reverse biased by the positive half-cycle signal and will not conduct. The *npn* transistor will be biased into conduction by the positive half-cycle signal with a resulting half-cycle of output across the load resistor *(R_L)* as shown in Fig. 10.14b. During the negative half-cycle of input signal the *npn* transistor is biased off and the output half-cycle developed across the load is due to the operation of the *pnp* transistor at this time, as shown in Fig. 10.14c.

During a complete cycle of the input, a complete cycle of output signal is developed across the load. It should be obvious that one disadvantage of this circuit connection is the need for two supply voltages. Another, less obvious, but important, disadvantage with the complementary circuit as shown is the resulting *crossover* distortion in the output signal. Crossover distortion refers to the fact that during the signal crossover from positive to negative (or vice versa) there is some nonlinearity in the output signal as indicated in Fig. 10.14d. This results from the fact that for the simple circuit shown in Fig. 10.14a the operation of the circuit does not provide exact switching of one transistor *off* and the other *on* at the zero-voltage condition. Both may be off or partially conducting so that the output voltage is not exactly following the input and distortion occurs. This occurrence at the crossover point is of concern for the push-pull circuit of Fig. 10.10 as well, although not necessarily to the same degree. Bias of the transistors in class AB improves the operation by biasing the transistors so that each stays on for more than half of the cycle. For the circuit of Fig. 10.14a considerable effort must be made to reduce the crossover distortion. More practical circuit connections include additional biasing components in the base circuit to try to effect this improved operation.

Note that the load is driven as the output of an emitter-follower circuit so that the low resistance of the load is matched by low resistance from the driving source. Improved versions of the complementary circuit include the transistors, each connected in the Darlington arrangement, to provide even lower driver resistance than that with single transistors. The circuit of Fig. 10.15 shows a practical circuit connection using the Darlington connection of the transistors and additional emitter resistors for temperature bias stabilization.

Another transformerless push-pull circuit, shown in Fig. 10.16a, uses the same type of transistors rather than complementary transistors. However, the inputs applied to the circuit must be opposite in polarity so that some inverting circuit is required. Figures 10.16b and 10.16c show the operation of the transistors for alternate halves of the cycle and the resulting polarity of the signal across the load resistor.

Quasi-complementary Push-Pull Amplifier

The push-pull circuit form is achieved in the circuit of Fig. 10.17 by using complementary transistors (Q_1 and Q_2) before the power output transistors (Q_3 and Q_4) so that both power output transistors can be *npn* types. This is a practically preferable arrangement, for the *npn* power transistors are presently the best available. Notice that transistors Q_1 and Q_3 form a *Darlington connection* that provides output at a low-impedance level from the emitter. The connection of transistors Q_2 and Q_4 forms a *feedback pair,* which similarly provides a low-impedance drive to the load. Resistor R_2 can be adjusted to minimize crossover distortion. The single signal applied as input to the push-pull stage then results in a full cycle output to the load R_L, each half of the circuit operating class-B for efficient power operation. This *quasi-comple-mentary* push-pull power amplifier is presently a most popular circuit connection.

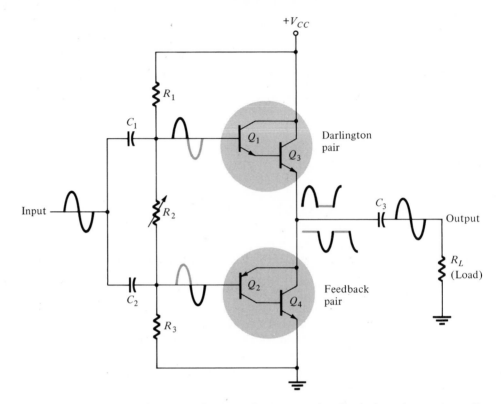

Figure 10.17 Quasi-complementary push–pull transformerless power amplifier.

Power and Efficiency Calculations in Class-B Amplifiers

The power and efficiency calculations of a variety of class-B power amplifiers should help in understanding how these circuits operate and provide some comparison between important circuit values.

INPUT DC POWER

The power provided to the speaker of a power amplifier circuit is drawn from the power supply (or power supplies) and is considered an input or dc power. The amount of this input power can be calculated from

$$P_i(\text{dc}) = V_{CC} I_{\text{dc}} \tag{10.33}$$

where I_{dc} is the average or dc current drawn from the power supply.

In class-B operation the current drawn from a single power supply is a full-wave rectified signal, while that drawn from a circuit having two power supplies is a half-wave rectified waveform from each supply. In either case the value of average power can be expressed as

$$I_{dc} = \frac{2}{\pi} I_{peak} \tag{10.34}$$

where I_{peak} is the peak value of the output current waveform.

OUTPUT AC POWER

The power delivered to the load (usually referred to as a resistance, R_L) can be calculated from any one of a number of equal relations.

$$P_o(ac) = \frac{V_L^2(p-p)}{8R_L} = \frac{V_L^2(p)}{2R_L} = \frac{V_L^2(rms)}{R_L} \tag{10.35}$$

POWER DISSIPATED BY OUTPUT TRANSISTORS

The power dissipated (as heat) by the output power transistors is the difference between the input power from the supplies and the output power delivered to the load.

$$P_{2Q} = P_i - P_o \tag{10.36}$$

where P_{2Q} is the power dissipated by the *two* output power transistors.

The circuit's power efficiency is then calculated as

$$\eta = \frac{P_o}{P_i} \times 100\% \tag{10.37}$$

Maximum-Power Considerations

For class-B operation the maximum output power to the load resulting when $V_L(p) = V_{CC}$ is

$$\text{maximum } P_o\,(ac) = \frac{V_{CC}^2}{2R_L} \tag{10.38}$$

A corresponding ac current signal developed through the load then varies to a peak value of

$$I_{peak} = \frac{V_{CC}}{R_L}$$

so that the average current from the power supply is

$$I_{dc} = \frac{2}{\pi} I_{peak} = \frac{2}{\pi} \cdot \frac{V_{CC}}{R_L} \tag{10.39}$$

The input power drawn by the circuit is then

$$\boxed{\text{maximum } P_i(\text{dc}) = V_{CC} I_{\text{dc}} = V_{CC}\left(\frac{2}{\pi}\cdot\frac{V_{CC}}{R_L}\right)} \qquad (10.40)$$

The maximum circuit efficiency for class-B operation is then

$$\text{maximum } \eta = \frac{P_o}{P_i}\,100 = \frac{\dfrac{V_{CC}^2}{2R_L}}{V_{CC}\left(\dfrac{2}{\pi}\dfrac{V_{CC}}{R_L}\right)}\,100 = \frac{\pi}{4}\cdot 100 = 78.54\% \qquad (10.41)$$

When the input signal results in less than the maximum output signal swing, the circuit efficiency is less than 78.5%. For class-B operation the maximum power dissipated by the output transistors *does not* occur at the maximum efficiency condition. The maximum power dissipated by the two output transistors occurs when the output voltage across the load is $0.636 V_{CC}$ [$= (2/\pi) V_{CC}$] and is

$$\boxed{\text{maximum } P_{2Q} = \frac{2}{\pi^2}\cdot\frac{V_{CC}^2}{R_L}} \qquad (10.42)$$

EXAMPLE 10.9 For the circuit of Fig. 10.18:

(a) Calculate the input and output power handled by the circuit and the power dissipated by *each* output transistor for an input of 12 V, rms.

(b) If the input signal is increased to provide the maximum undistorted output, calculate the values of maximum input and output power and the power dissipated by *each* output transistor for this condition.

(c) Calculate the maximum power that *each* output transistor will have to handle.

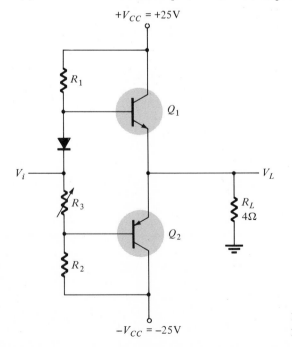

Figure 10.18 Class-B power amplifier for Example 10.9.

Solution:

(a) The peak input voltage is

$$V_i(p) = \sqrt{2}\ V_i(\text{rms}) = \sqrt{2}(12) = 16.97 \cong 17 \text{ V}$$

Since the resulting voltage across the load is ideally the same as the input signal (amplifier has, ideally, a voltage gain of unity),

$$V_L(p) = 17 \text{ V}$$

$$P_o(\text{ac}) = \frac{V_L^2(p)}{2R_L} = \frac{(17)^2}{2(4)} = \mathbf{36.125\ W}$$

$$I_L(p) = \frac{V_L(p)}{R_L} = \frac{17 \text{ V}}{4\ \Omega} = 4.25 \text{ A}$$

The dc current drawn from the two power supplies is then

$$I_{\text{dc}} = \frac{2}{\pi} I_L(p) = \frac{2(4.25 \text{ A})}{\pi} = 2.71 \text{ A}$$

so that the power supplied to the circuit is

$$P_i(\text{dc}) = V_{CC}\,I_{\text{dc}} = (25 \text{ V})(2.71 \text{ A}) = \mathbf{67.75\ W}$$

The circuit efficiency (for an input of $V_i = 12$ V, rms) is

$$\eta = \frac{P_o}{P_i} \cdot 100 = \frac{36.125 \text{ W}}{67.75 \text{ W}} \cdot 100 = 53.3\%$$

and the power dissipated by each output transistor is

$$P_Q = \frac{P_{2Q}}{2} = \frac{P_i - P_o}{2} = \frac{67.75 - 36.125 \text{ W}}{2} = \mathbf{15.8\ W}$$

(b) If the input signal is increased to $V_i = 25$ V, *peak* ($V_i = 17.68$ V, rms) so that $V_L(p) = V_{CC} = 25$ V, the calculations are

$$\text{maximum } P_o = \frac{V_{CC}^2}{2R_L} = \frac{(25)^2}{2(4)} = \mathbf{78.125\ W}$$

$$\text{maximum } P_i = \frac{2}{\pi}\frac{V_{CC}^2}{R_L} = \frac{2}{\pi}\frac{(25)^2}{4} = \mathbf{99.47\ W}$$

and

$$\eta = \frac{P_o}{P_i} \cdot 100 = \frac{78.125}{99.47} \cdot 100 = 78.54\%$$

(the maximum circuit efficiency).

At this maximum signal condition the power dissipated by *each* output transistor is

$$P_Q = \frac{P_{2Q}}{2} = \frac{P_i - P_o}{2} = \frac{99.47 - 78.125}{2} = \mathbf{10.67\ W}$$

(Note that this is less than the power dissipated with a smaller input signal.)[2]

(c) The maximum power dissipation required of the output transistors is

---

[2] The power dissipation of the output transistors actually increases until the condition of $V_i = (2/\pi)\ V_{CC}$ after which the power dissipation decreases to a value of $(2/\pi - 1/2) \cdot V_{CC}/R_L$ at maximum circuit efficiency.

$$\text{maximum } P_{2Q} = \frac{2}{\pi^2} \frac{V_{CC}^2}{R_L} = \frac{2}{\pi^2} \frac{(25)^2}{4} = 31.66 \text{ W}$$

so that

$$P_Q = \frac{P_{2Q}}{2} = \frac{31.66}{2} = \mathbf{15.83 \text{ W}}$$

10.7 POWER TRANSISTOR HEAT SINKING

While integrated circuits are used for small-signal and low-power applications, most high-power applications still require individual power transistors. Improvements in production techniques have provided higher power ratings in smaller-sized packaging cases, have increased the maximum transistor breakdown voltage, and have provided faster-switching power transistors.

Some discussion of power transistor rating was provided in Chapter 3. The maximum power handled by a particular device and the temperature of the transistor junctions are related since the power dissipated by the device causes an increase in temperature at the junctions of the device. Obviously, a 100-W transistor will provide more power capability than a 10-W transistor. On the other hand, proper heat sinking techniques will allow operation of a device closer to its maximum power rating.

We should note that of the two types of transistors—germanium and silicon—silicon transistors provide greater maximum temperature ratings. Typically, the maximum junction temperature of these types of power transistors is

germanium   100–110°C
silicon   150–200°C

For many applications the average power dissipated may be approximated by

$$P_D = V_{CE} I_C$$

This power dissipation, however, is only allowed up to some maximum temperature. Above this temperature the device power dissipation capacity must be reduced (or *derated*) so that at higher case temperatures the power-handling capacity is reduced—down to 0 W at the device maximum case temperature.

The greater the power handled by the transistor (dependent on power level set by the circuit) the higher the case temperature of the transistor. Actually, the limiting factor in power handled by a particular transistor is the temperature of the device collector junction. Power transistors are mounted in large metal cases to provide a large area from which the heat generated by the device may radiate. Even so, operating a transistor directly into air (mounting it on a plastic board, for example) severely limits the device power rating. If, instead (as is usual practice), the device is mounted on some form of *heat sink,* its power-handling capacity can approach the rated maximum value more closely. A few heat sinks are shown in Fig. 10.19. When the heat sink is used, the heat

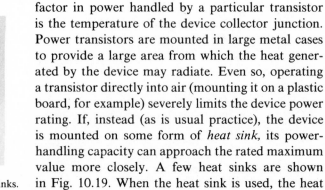

Figure 10.19   Typical power heat sinks.

produced by the transistor dissipating power has a larger area from which to radiate the heat into the air, thereby holding the case temperature to a much lower value than would result without the heat sink. Even with an infinite heat sink (which, of course, is not available), for which the case temperature is held at the *ambient* (air) temperature, the junction will be heated above the case temperature and a maximum power rating must be considered.

Since even a good heat sink cannot hold the transistor case temperature at ambient (which, by the way could be more than 25°C if the transistor circuit is in a confined area where other devices are also radiating a good deal of heat), it is necessary to derate the amount of *maximum power* allowed for a particular transistor as a function of increased case temperature.

Figure 10.20 shows typical power derating curves for silicon transistors. The curves show that the manufacturer will specify an upper temperature point (not necessarily 25°C) after which a linear derating takes place. For silicon the maximum power that should be handled by the device does not reduce to 0 W until a case temperature of 200°C (or 150° in some devices).

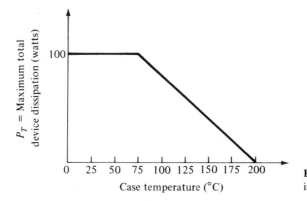

Figure 10.20 Typical power derating curve for silicon transistors.

It is not necessary to provide a derating curve since the same information could be given simply as a listed derating factor on the device specification sheet. For example, a derating factor for a germanium transistor may be stated as follows:

Derate linearly to 100°C case temperature at the rate of 1 watt per degree above 50°C.

For a 50-W transistor (rated below 50°C) this means that at 80°C the maximum power rating of the device must be reduced by

$$(80°C - 50°C)\left(1\,\frac{W}{°C}\right) = 30\text{ W}$$

so that the rated power dissipated should only be

$$50\text{ W} - 30\text{ W} = 20\text{ W}$$

at 80° case temperature. Stated mathematically,

$$P_D\,(\text{temp}_1) = P_D\,(\text{temp}_0) - (\text{Temp}_1 - \text{Temp}_0)\,(\text{Derating factor}) \qquad (10.43)$$

where the value of Temp$_0$ is the temperature at which derating should begin, the value of Temp$_1$ is the particular temperature of interest (above the value Temp$_0$), P_D (temp$_0$) and P_D (temp$_1$) are the maximum power dissipations at the temperatures specified, and the derating factor is the value given by the manufacturer in units of watts (or milliwatts) per degree of temperature.

EXAMPLE 10.10 Determine what maximum dissipation will be allowed for an 80-W silicon transistor (rated at 25°C) if derating is required above 25°C by a derating factor of 0.5 W/°C at a case temperature of 125°C.

Solution:

$$P_D(125°C) = P_D(25°C) - (125°C - 25°C)(0.5 \text{ W/°C})$$
$$= 80 \text{ W} - 100(0.5) = \textbf{30 W}$$

It is interesting to note what power rating results using a power transistor without a heat sink. For example, a silicon transistor rated at 100 W at (or below) 100°C is rated only 4 W at (or below) 25°C, free-air temperature. Thus, operated without a heat sink the device can handle a maximum of only 4 W at a room temperature of 25°C. Using a heat sink large enough to hold the case temperature to 100°C at 100 W allows operation at the maximum power rating.

Thermal Analogy of Power Transistor

Selection of a suitable heat sink requires a considerable amount of detail which is not appropriate to our present basic considerations of the power transistor. However, more detail about the thermal characteristics of the transistor and its relation to the power dissipation of the transistor may help provide a clearer understanding of power as limited by temperature. The following discussion should provide some background information.

A picture of how the junction temperature (T_J), case temperature (T_C), and am-

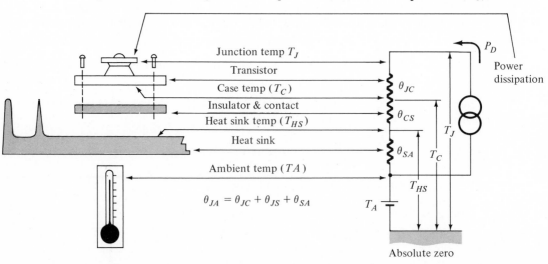

Figure 10.21 Thermal-to-electrical analogy.

bient (air) temperature (T_A) are related by the device heat-handling capacity—a temperature coefficient usually called *thermal resistance*—is presented in the thermal-electrical analogy shown in Fig. 10.21.

In providing a thermal-electrical analogy the term *thermal resistance* is used to describe heat effects by an electrical term. The terms in Fig. 10.21 are defined as follows:

θ_{JA} = total thermal resistance (junction to ambient)
θ_{JC} = transistor thermal resistance (junction to case)
θ_{CS} = insulator thermal resistance (case to heat sink)
θ_{SA} = heat sink thermal resistance (heat sink to ambient)

Using the electrical analogy for thermal resistances, we can write

$$\theta_{JA} = \theta_{JC} + \theta_{CS} + \theta_{SA} \qquad (10.44)$$

The analogy can also be used in applying Kirchhoff's law to obtain

$$T_J = P_D\theta_{JA} + T_A \qquad (10.45)$$

The last relation shows that the junction temperature "floats" on the ambient temperature and that the higher the ambient temperature the lower the allowed value of device power dissipation.

The thermal factor θ provides information about how much temperature drop (or rise) results for a given amount of power dissipation. For example, the value of θ_{JC} is usually about 0.5°C/W. This means that for a power dissipation of 50 W the difference in temperature between case temperature (as measured by a thermocouple) and the inside junction temperature is only

$$T_J - T_C = \theta_{JC}P_D = (0.5°C/W)(50 \text{ W}) = 25°C$$

Thus, if the heat sink can hold the case at, say 50°C, the junction is then only at 75°C. This is a relatively small temperature difference, especially at lower power-dissipation levels.

The value of thermal resistance from junction to free air (using no heat sink) is, typically,

$$\theta_{JA} = 40°C/W \qquad \text{(into free air)}$$

Note that in this case only 1 W of power dissipation results in a junction temperature 40°C greater than the ambient. A silicon transistor operating into an ambient temperature of 25°C could not dissipate 5 W without exceeding the junction temperature limit of 100°C.

$$T_J = T_A + \theta_{JA}P_D = 25°C + (40°C/W)(5 \text{ W}) = 225°C$$

A heat sink can now be seen to provide a low thermal resistance between case and air—much less than the 40°C/W value of the transistor case alone. Using a heat sink having

$$\theta_{SA} = 2°C/W$$

and, with an insulating thermal resistance (from case to heat sink) of

$$\theta_{CS} = 0.8°C/W$$

and, finally for the transistor,

$$\theta_{CJ} = 0.5°C/W$$

we can obtain

$$\theta_{JA} = \theta_{SA} + \theta_{CS} + \theta_{CJ}$$
$$= 2.0 + 0.8 + 0.5 = 3.3°C/W$$

So, with a heat sink, the thermal resistance between air and the junction is only 3.3°C/W as compared to, say 40°C/W for the transistor operating directly into free air. Using the value of θ_{JA} above for a transistor operated at, say, 2 W we calculate

$$(T_J - T_A) = \theta_{JA}P_D = (3.3°C/W)(2\ W) = 6.6°C$$

In other words, the use of a heat sink in this example provided only a 6.6°C increase in junction temperature as compared to an 80°C rise without a heat sink.

EXAMPLE 10.11 A silicon power transistor is operated with a heat sink ($\theta_{SA} = 1.5°C/W$). The transistor, rated at 150 W (25°C), has $\theta_{JC} = 0.5°C/W$ and the mounting insulation has $\theta_{CS} = 0.6°C/W$. What maximum power can be dissipated if the ambient temperature is 40°C and $T_{Jmax} = 200°C$?

Solution:

$$P_D = \frac{T_J - T_A}{\theta_{JC} + \theta_{CS} + \theta_{SA}} = \frac{200 - 40}{0.5 + 0.6 + 1.5} = \frac{160°C}{2.6°C/W} \cong \mathbf{61.5\ W}$$

PROBLEMS

§ 10.3

1. A class-A transformer-coupled amplifier uses a 25:1 transformer to drive a 4-Ω load. Calculate the effective ac load (seen by the transistor connected to the larger turns side of the transformer).

2. What turns ratio transformer is needed to couple to an 8-Ω load so that it appears as a 10-kΩ effective load?

3. Calculate the transformer turns ratio required to connect four parallel 16-Ω speakers so that they appear as an 8-kΩ effective load.

4. A transformer-coupled class-A amplifier drives a 16-Ω speaker through a $\sqrt{15}:1$ transformer. Using a power supply of 36 V (V_{CC}) the circuit delivers 2 W to the load. Calculate the following:
   (a) The ac power across the transformer primary.
   (b) The rms value of load voltage.
   (c) The rms value of primary voltage.

(d) The rms values of load and primary current.

5. Calculate the efficiency of the circuit of Problem 4 if the bias current is $I_{C_Q} = 150$ mA.

6. Draw the circuit diagram of a class-A transformer-coupled amplifier using an *npn* transistor.

§ 10.4

7. Calculate the efficiency of the following amplifier classes and voltages:
   (a) Class-A operation with $V_{CE_{max}} = 24$, and $V_{CE_{min}} = 2$ V.
   (b) Class-B operation with $V_{CE_{min}} = 4$ V, and $V_{CC} = 22$ V.

8. For the following voltage values measured on a scope calculate the amount of the second-harmonic distortion: $V_{CE_{max}} = 27$ V, $V_{CE_{min}} = 14$ V, $V_{CE_Q} = 20$ V.

§ 10.5

9. Draw the circuit diagram of a class-B *npn* push-pull power amplifier.

10. Sketch the input and output waveforms of the ac signal in the circuit of Problem 9.

11. Draw the circuit diagram of an *npn* push-pull power amplifier operated class AB. Show the driver stage before the push-pull stage.

§ 10.6

12. Sketch the circuit diagram of a quasi-complementary amplifier, showing voltage waveforms in circuit.

13. List any advantages of a transformerless circuit over a transformer-coupled circuit.

14. For a class-B power amplifier as in Fig. 10.17 using a 30-V power supply and $R_L = 8$ Ω, calculate (a) maximum P_o (ac), (b) maximum P_i (dc), (c) maximum %η, and (d) maximum power dissipated by both output transistors. Use $R_L = 8$ Ω.

15. If the input voltage to the power amplifier in Fig. 10.17 is 15 V, rms using a 30-V power supply and $R_L = 8$ Ω, calculate (a) P_o (ac), (b) P_i (dc), (c) %η, and (d) power dissipated by both output power transistors.

16. For power amplifier as in Fig. 10.18 using a 40-V power supply and signal input of 18 V, rms calculate (a) P_o (ac), (b) P_i (dc), (c) %η, and (d) power dissipated by both output power transistors, when load is $R_L = 8$ Ω.

§ 10.7

17. Determine the maximum dissipation allowed for a 100-W silicon transistor (rated at 25°C) for a derating factor of 0.6 W/°C, at a case temperature of 150°C.

18. A 160-W silicon power transistor operated with a heat sink ($\theta_{SA} = 1.5$°C/W) has $\theta_{JC} = 0.5$°C/W and mounting insulation of $\theta_{CS} = 0.8$°C/W. What maximum power can be handled by the transistor at an ambient temperature of 80°C? (Junction temperature should not exceed 200°C.)

19. What maximum power can a silicon transistor ($T_{J,max} = 200$°C) dissipate at 80°C into free air at room temperature?

COMPUTER PROBLEMS

Write BASIC programs to:

1. Calculate I_{dc} and P_i for a class-A series-fed circuit as in Fig. 10.1.

2. Calculate P_o for a class-A series-fed circuit as in Fig. 10.1.

3. Tabulate the efficiency of a class-A series-fed circuit as in Fig. 10.1 over a range of output peak voltages from $0.1 V_{CC}$ to V_{CC}.

4. Calculate the efficiency of a class-A series-fed circuit as in Fig. 10.1 over a range of values for R_C.

5. Calculate the P_i for a transformer-coupled class-A amplifier as in Fig. 10.3.

6. Calculate the transformer turns ratio needed to optimally connect a given load, R_L, into a class-A transformer-coupled amplifier as in Fig. 10.3.

7. Tabulate the efficiency for a class-A transformer-coupled amplifier as in Fig. 10.3 for supply voltage values from $0.1 V_{CC}$ to V_{CC}.

8. Tabulate the value of P_i for a class-B amplifier as in Fig. 10.18 for supply voltage from $0.1 V_{CC}$ to V_{CC}.

9. Tabulate P_o for a class-B power amplifier as in Fig. 10.18 for values of supply voltage varying from $0.1 V_{CC}$ to V_{CC}.

10. Tabulate the power dissipated by both output transistors over a range of supply voltages from $0.1 V_{CC}$ to V_{CC}.

CHAPTER 11

pnpn and Other Devices

11.1 INTRODUCTION

In this chapter a number of important devices not discussed in detail in earlier chapters will be introduced. The two-layer semiconductor diode has led the way to three-, four-, and even five-layer devices. A family of four-layer *pnpn* devices will first be considered (SCR, SCS, GTO, LASCR, Shockley diode, diac, and triac), followed by an increasingly important device—the UJT (unijunction transistor). Those four-layer devices with a control mechanism are commonly referred to as *thyristors*, although the term is most frequently applied to the SCR (the semiconductor equivalent of the *thyratron* gas-filled tube). The chapter will close with an introduction to the phototransistor, opto-isolators, and the PUT (programmable unijunction transistor).

pnpn DEVICES

11.2 SILICON CONTROLLED RECTIFIER (SCR)

Within the family of *pnpn* devices the silicon controlled rectifier (SCR) is unquestionably of the greatest interest today. It was first introduced in 1956 by Bell Telephone Laboratories. A few of the more common areas of application for SCRs include relay controls, time-delay circuits, regulated power suppliers, static switches, motor controls, choppers, inverters, cyclo-converters, battery chargers, protective circuits, heater controls, and phase controls.

419

In recent years, SCRs have been designed to *control* powers as high as 10 MW with individual ratings as high as 2000 A at 1800 V. Its frequency range of application has also been extended to about 50kHz, permitting some high-frequency applications such as induction heating and ultrasonic cleaning.

11.3 BASIC SILICON CONTROLLED RECTIFIER (SCR) OPERATION

As the terminology indicates, the SCR is a rectifier constructed of silicon material with a third terminal for control purposes. Silicon was chosen because of its high temperature and power capabilities. The basic operation of the SCR is different from the fundamental two-layer semiconductor diode in that a third terminal, called a *gate*, determines when the rectifier switches from the open-circuit to short-circuit state. It is not enough simply to forward-bias the anode-to-cathode region of the device. In the conduction region the dynamic resistance of the SCR is typically 0.01 to 0.1 Ω. The reverse resistance is typically 100 kΩ or more.

The graphic symbol for the SCR is shown in Fig. 11.1 with the corresponding connections to the four-layer semiconductor structure. As indicated in Fig. 11.1a, if forward conduction is to be established, the anode must be positive with respect to the cathode. This is not, however, a sufficient criterion for turning the device on. A pulse of sufficient magnitude must also be applied to the gate to establish a turn-on gate current, represented symbolically by I_{GT}.

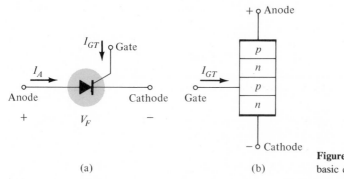

Figure 11.1 (a) SCR symbol; (b) basic construction.

A more detailed examination of the basic operation of an SCR is best effected by splitting the four-layer *pnpn* structure of Fig. 11.1b into two three-layer transistor structures as shown in Fig. 11.2a and then considering the resultant circuit of Fig. 11.2b.

Note that one transistor for Fig. 11.2 is an *npn* device while the other is a *pnp* transistor. For discussion purposes, the signal shown in Fig. 11.3a will be applied to the gate of the circuit of Fig. 11.2b. During the interval $0 \rightarrow t_1$, $V_{\text{gate}} = 0$ V, the circuit of Fig. 11.2b will appear as shown in Fig. 11.3b ($V_{\text{gate}} = 0$ V is equivalent to the gate terminal being grounded as shown in the figure). For $V_{BE_2} = V_{\text{gate}} = 0$ V, the base current $I_{B_2} = 0$ and I_{C_2} will be approximately I_{CO}. The base current of Q_1, $I_{B_1} = I_{C_2} = I_{CO}$, is too small to turn Q_1 on. Both transistors are therefore in the OFF state, resulting in a high impedance between the collector and emitter of

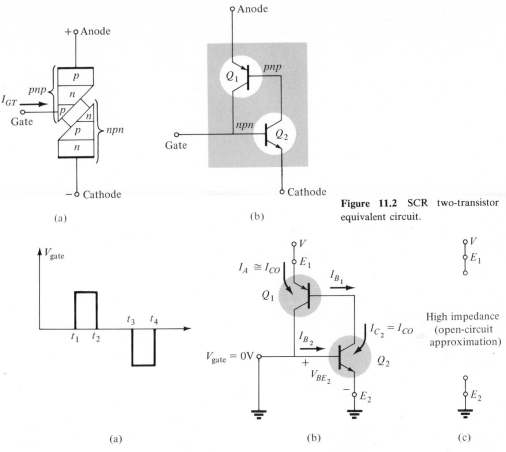

Figure 11.2 SCR two-transistor equivalent circuit.

(a)

(b)

Figure 11.3 OFF state of the SCR.

each transistor and the open-circuit representation for the controlled rectifier as shown in Fig. 11.3c.

At $t = t_1$ a pulse of V_G volts will appear at the SCR gate. The circuit conditions established with this input are shown in Fig. 11.4a. The potential V_G was chosen sufficiently large to turn Q_2 on $(V_{BE_2} = V_G)$. The collector current of Q_2 will then rise to a value sufficiently large to turn Q_1 on $(I_{B_1} = I_{C_2})$. As Q_1 turns on, I_{C_1} will increase, resulting in a corresponding increase in I_{B_2}. The increase in base current for Q_2 will result in a further increase in I_{C_2}. The net result is a regenerative increase in the collector current of each transistor. The resulting anode-to-cathode resistance $[R_{SCR} = V/(I_A - \text{large})]$ is then very small, resulting in the short-circuit representation for the SCR as indicated in Fig. 11.4b. The regenerative action described above results in SCRs having typical turn-on times of 0.1 to 1 μs. However, higher-power devices in the 100- to 400-A range may have 10- to 25-μs turn-on times.

In addition to gate triggering, SCRs can also be turned on by significantly raising the temperature of the device or raising the anode-to-cathode voltage to the breakover value shown on the characteristics of Fig. 11.7.

The next question of concern is: How long is the turn-off time and how is turn-off accomplished? An SCR *cannot* be turned off by simply removing the gate signal, and only a special few can be turned off by applying a negative pulse to the gate

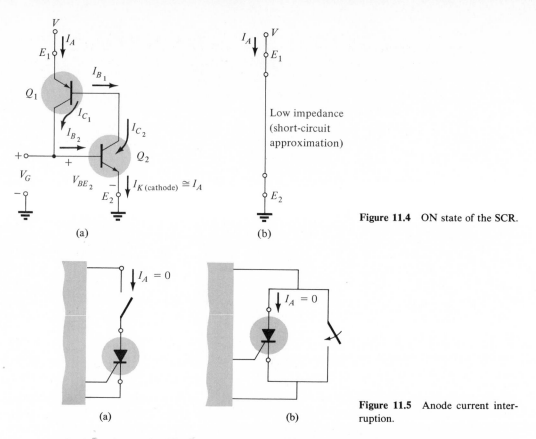

(a)   (b)

Figure 11.4 ON state of the SCR.

(a)   (b)

Figure 11.5 Anode current interruption.

terminal as shown in Fig. 11.3a at $t = t_3$. The two general methods for turning off an SCR are categorized as the *anode-current interruption* and the *forced-commutation technique*. The two possibilities for current interruption are shown in Fig. 11.5.

In Fig. 11.5a, I_A is zero when the switch is opened (series interruption) while in Fig. 11.5b the same condition is established when the switch is closed (shunt interruption). Forced commutation is the "forcing" of current through the SCR in the direction opposite to forward conduction. There are a wide variety of circuits for performing this function, a number of which can be found in the manuals of major manufacturers in this area. One of the more basic types is shown in Fig. 11.6. As indicated in the figure, the turn-off circuit consists of an *npn* transistor, a dc battery V_B, and a pulse

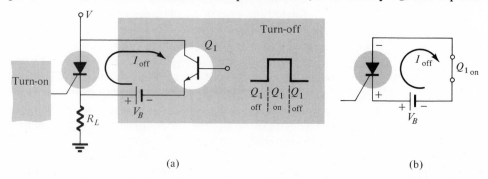

(a)   (b)

Figure 11.6 Forced-commutation technique.

generator. During SCR conduction the transistor is in the "off state"; that is, $I_B = 0$ and the collector-to-emitter impedance is very high (for all practical purposes an open circuit). This high impedance will isolate the turn-off circuitry from affecting the operation of the SCR. For turn-off conditions, a positive pulse is applied to the base of the transistor, turning it heavily on, resulting in a very low impedance from collector to emitter (short-circuit representation). The battery potential will then appear directly across the SCR as shown in Fig. 11.6b, forcing current through it in the reverse direction for turn-off. Turn-off times of SCRs are typically 5 to 30 μs.

11.4 SCR CHARACTERISTICS AND RATINGS

The characteristics of an SCR are provided in Fig. 11.7 for various values of gate current. The currents and voltages of usual interest are indicated on the characteristic. A brief description of each follows.

1. *Forward breakover voltage* $V_{(BR)F*}$ is that voltage above which the SCR enters the conduction region. The asterisk (*) is a letter to be added that is dependent on the condition of the gate terminal as follows:

$$O = \text{open-circuit from } G \text{ to } K$$

$$S = \text{short-circuit from } G \text{ to } K$$

$$R = \text{resistor from } G \text{ to } K$$

$$V = \text{fixed bias (voltage) from } G \text{ to } K$$

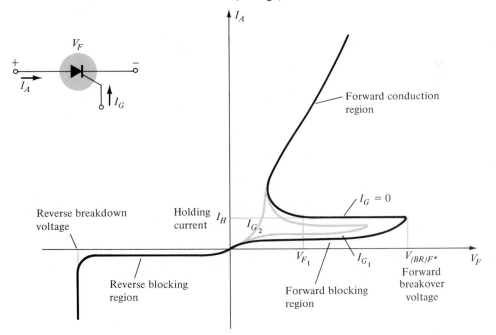

Figure 11.7 SCR characteristics.

2. *Holding current (I_H)* is that value of current below which the SCR switches from the conduction state to the forward blocking region under stated conditions.

3. *Forward and reverse blocking regions* are the regions corresponding to the open-circuit condition for the controlled rectifier which *block* the flow of charge (current) from anode to cathode.

4. *Reverse breakdown voltage* is equivalent to the Zener or avalanche region of the fundamental two-layer semiconductor diode.

It should be immediately obvious that the SCR characteristics of Fig. 11.7 are very similar to those of the basic two-layer semiconductor diode except for the horizontal offshoot before entering the conduction region. It is this horizontal jutting region that gives the gate control over the response of the SCR. For the characteristic having the solid line in Fig. 11.7 ($I_G = 0$), V_F must reach the largest required breakover voltage before the "collapsing" effect will result and the SCR can enter the conduction region corresponding to the *on* state. If the gate current is increased to I_{G_1}, as shown in the same figure, by applying a bias voltage to the gate terminal the value of V_F required for the conduction is considerably less. Note also that I_H drops with increase in I_G. If increased to I_{G_2} the SCR will fire at very low values of voltage and the characteristics begin to approach those of the basic *p-n* junction diode. Looking at the characteristics in a completely different sense, for a particular V_F voltage, say V_{F_1} (Fig. 11.7), if the gate current is increased from $I_G = 0$ to $I_G = I_{G_1}$, the SCR will fire.

The gate characteristics are provided in Fig. 11.8. The characteristics of Fig.

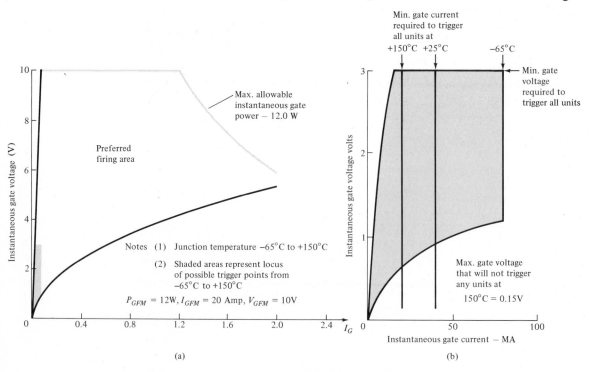

Figure 11.8 SCR gate characteristics (GE series—C38).

CH. 11 PNPN AND OTHER DEVICES

11.8b are an expanded version of the shaded region of Fig. 11.8a. In Fig. 11.8a the three gate ratings of greatest interest, P_{GFM}, I_{GFM}, and V_{GFM} are indicated. Each is included on the characteristics in the same manner employed for the transistor. Except for portions of the shaded region, any combination of gate current and voltage that falls within this region will fire any SCR in the series of components for which these characteristics are provided. Temperature will determine which sections of the shaded region must be avoided. At $-65°C$ the minimum current that will trigger the series of SCRs is 80 mA, while at $+150°C$ only 20 mA are required. The effect of temperature on the minimum gate voltage is usually not indicated on curves of this type since gate potentials of 3 V or more are usually obtained easily. As indicated on Fig. 11.8b, a minimum of 3 V is indicated for all units for the temperature range of interest.

Other parameters usually included on the specification sheet of an SCR are the turn-on time (t_{on}), turn-off time (t_{off}), junction temperature (T_J), and case temperature (T_C), all of which should by now be, to some extent, self-explanatory.

11.5 SCR CONSTRUCTION AND TERMINAL IDENTIFICATION

The basic construction of the four-layer pellet of an SCR is shown in Fig. 11.9a. The complete construction of a thermal-fatigue-free, high-current SCR is shown in

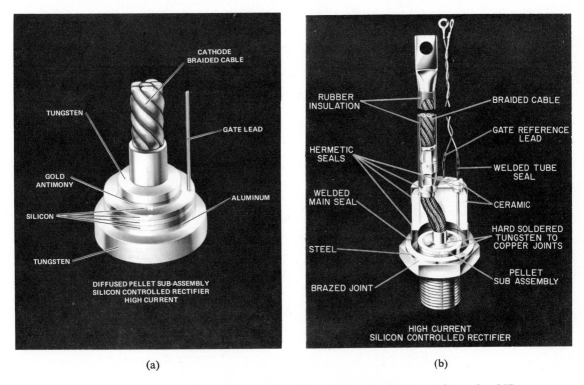

(a)   (b)

Figure 11.9 (a) Alloy-diffused SCR pellet; (b) thermal fatigue-free SCR construction. (Courtesy General Electric Company.)

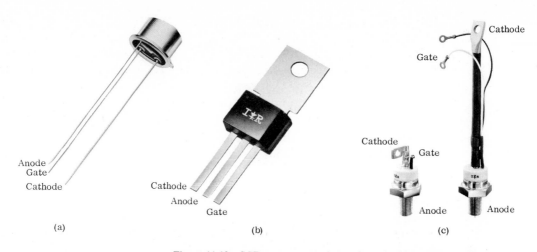

Figure 11.10 SCR case construction and terminal identification. [(a) courtesy General Electric Company; (b) and (c) courtesy International Rectifier Corporation, Inc.)]

Fig. 11.9b. Note the position of the gate, cathode, and anode terminals. The pedestal acts as a heat sink by transferring the heat developed to the chassis on which the SCR is mounted. The case construction and terminal identification of SCRs will vary with the application. Other case-construction techniques and the terminal identification of each are indicated in Fig. 11.10.

11.6 SCR APPLICATIONS

A few of the possible applications for the SCR are listed in the introduction to the SCR (Section 11.2). In this section we consider five: a static switch, phase-control system, battery charger, temperature controller, and single-source emergency-lighting system.

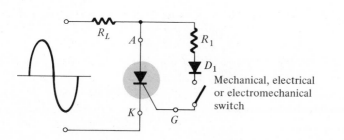

(a)

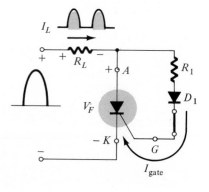

(b)

Figure 11.11 Half-wave series static switch.

A half-wave *series static switch* is shown in Fig. 11.11a. If the switch is closed as shown in Fig. 11.11b, a gate current will flow during the positive portion of the input signal, turning the SCR on. Resistor R_1 limits the magnitude of the gate current. When the SCR turns on, the anode-to cathode voltage (V_F) will drop to the conduction value, resulting in a greatly reduced gate current and very little loss in the gate circuitry. For the negative region of the input signal the SCR will turn off, since the anode is negative with respect to the cathode. The diode D_1 is included to prevent a reversal in gate current.

The waveforms for the resulting load current and voltage are shown in Fig. 11.11b. The result is a half-wave rectified signal through the load. If less than 180° conduction is desired, the switch can be closed at any phase displacement during the positive portion of the input signal. The switch can be electronic, electromagnetic, or mechanical, depending on the application.

A circuit capable of establishing a conduction angle between 90° and 180° is shown in Fig. 11.12a. The circuit is similar to that of Fig. 11.11a except for the addition of a variable resistor and the elimination of the switch. The combination

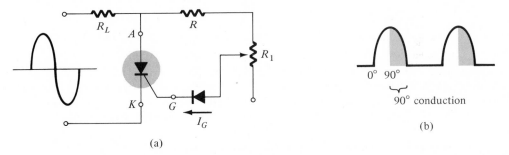

Figure 11.12   Half-wave variable-resistance phase control.

of the resistors R and R_1 will limit the gate current during the positive portion of the input signal. If R_1 is set to its maximum value, the gate current may never reach turn-on magnitude. As R_1 is decreased from the maximum the gate current will increase for the same input voltage. In this way, the required turn-on gate current can be established in any point between 0° and 90° as shown in Fig. 11.12b. If R_1 is low, the SCR will fire almost immediately, resulting in the same action as that obtained from the circuit of Fig. 11.11a (180° conduction). However, as indicated above, if R_1 is increased, a larger input voltage (positive) will be required to fire the SCR. As shown in Fig. 11.12b, the control cannot be extended past a 90° phase displacement since the input is its maximum at this point. If it fails to fire at this and lesser values of input voltage on the positive slope of the input, the same response must be expected from the negatively sloped portion of the signal waveform. The operation here is normally referred to in technical terms as *half-wave variable-resistance phase control*. It is an effective method of controlling the rms current and therefore power to the load.

A third popular application of the SCR is in a *battery-charging regulator*. The fundamental components of the circuit are shown in Fig. 11.13. You will note that the control circuit has been blocked off for discussion purposes.

As indicated in the figure, D_1 and D_2 establish a full-wave rectified signal across

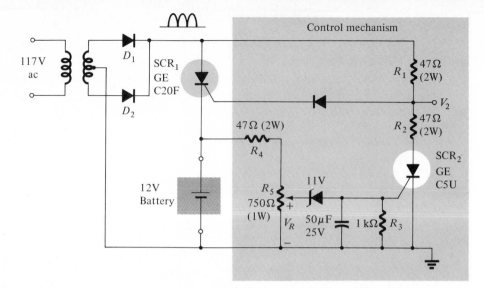

Figure 11.13 Battery-charging regulator.

SCR$_1$ and the 12-V battery to be charged. At low battery voltages SCR$_2$ is in the off state for reasons to be explained shortly. With SCR$_2$ open, the SCR$_1$ controlling circuit is exactly the same as the series static switch control discussed earlier in this section. When the full-wave rectified input is sufficiently large to produce the required turn-on gate current (controlled by R_1), SCR$_1$ will turn on and charging of the battery will commence. At the start of charging, the low battery voltage will result in a low voltage V_R as determined by the simple voltage-divider circuit. Voltage V_R is in turn too small to cause 11.0-V Zener conduction. In the off state, the Zener is effectively an open-circuit maintaining SCR$_2$ in the off state since the gate current is zero. The capacitor C_1 is included to prevent any voltage transients in the circuit from accidentally turning on SCR$_2$. Recall from your fundamental study of circuit analysis that the voltage cannot instantaneously change across a capacitor. In this way C_1 prevents transient effects from affecting the SCR.

As charging continues, the battery voltage rises to a point where V_R is sufficiently high to both turn on the 11.0-V Zener and fire SCR$_2$. Once SCR$_2$ has fired, the short-circuit representation for SCR$_2$ will result in a voltage-divider circuit determined by R_1 and R_2 that will maintain V_2 at a level too small to turn SCR$_1$ on. When this occurs, the battery is fully charged and the open-circuit state of SCR$_1$ will cut off the charging current. Thus, the regulator recharges the battery whenever the voltage drops and prevents overcharging when fully charged.

The schematic diagram of a 100-W heater control using an SCR appears in Fig. 11.14. It is designed such that the 100-W heater will turn on and off as determined by thermostats. Mercury-in-glass thermostats are very sensitive to temperature change. In fact, they can sense changes as small as 0.1°C. It is limited in application, however, in that it can only handle very low levels of current—below 1 mA. In this application, the SCR serves as a current amplifier in a load-switching element. It is not an amplifier in the sense that it magnifies the current level of the thermostat. Rather it is a device whose higher current level is controlled by the behavior of the thermostat.

It should be clear that the bridge network is connected to the ac supply through the 100-W heater. This will result in a full-wave rectified voltage across the SCR.

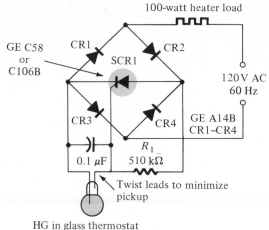

100-watt heater load

GE C58
or
C106B

CR1   SCR1   CR2

120 V AC
60 Hz

CR3   CR4   GE A14B
CR1–CR4

R_1
510 kΩ

0.1 μF

Twist leads to minimize
pickup

HG in glass thermostat
(such as vap. air div. 206-44
series; princo #T141, or
equivalent)

Figure 11.14 Temperature controller. (Courtesy General Electric Semiconductor Products Division.)

When the thermostat is open, the voltage across the capacitor will charge to a gate-firing potential through each pulse of the rectified signal. The charging time constant is determined by the *RC* product. This will trigger the SCR during each half-cycle of the input signal, permitting a flow of charge (current) to the heater. As the temperature rises, the conductive thermostat will short-circuit the capacitor, eliminating the possibility of the capacitor charging to the firing potential and triggering the SCR. The 510-kΩ resistor will then contribute to maintaining a very low current (less than 250 μA) through the thermostat.

The last application for the SCR to be described is shown in Fig. 11.15. It is a single-source emergency-lighting system that will maintain the charge on a 6-V battery to ensure its availability and also provide dc energy to a bulb if there is a power shortage.

A full-wave rectified signal will appear across the 6-V lamp due to diodes D_2 and D_1. The capacitor C_1 will charge to a voltage slightly less than a difference between the peak value of the full-wave rectified signal and the dc voltage across R_2 established by the 6-V battery. In any event, the cathode of SCR_1 is higher than

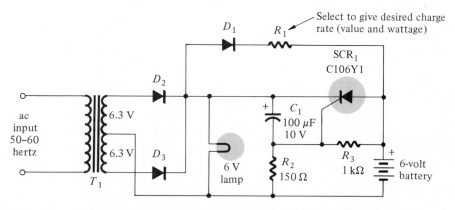

D_1 R_1 Select to give desired charge
rate (value and wattage)

SCR_1
C106Y1

D_2

ac
input
50–60
hertz

6.3 V

6.3 V D_3

T_1

6 V
lamp

C_1
100 μF
10 V

R_2
150 Ω

R_3
1 kΩ

6-volt
battery

Figure 11.15 Single-source emergency lighting system. (Courtesy General Electric Semiconductor Products Division.)

the anode and the gate-to-cathode voltage is negative, ensuring that the SCR is nonconducting. The battery is being charged through R_1 and D_1 at a rate determined by R_1. Charging will only take place when the anode of D_1 is more positive than its cathode. The dc level of the full-wave rectified signal will ensure that the bulb is lit when the power is on. If the power should fail, the capacitor C_1 will discharge through D_1, R_1, and R_3 until the cathode of SCR_1 is less positive than the anode. At the same time the junction of R_2 and R_3 will become positive and establish sufficient gate-to-cathode voltage to trigger the SCR. Once fired, the 6-V battery would discharge through the SCR_1 and energize the lamp and maintain its illumination.

Once power is restored the capacitor C_1 will recharge and reestablish the nonconducting state of SCR_1 as described above.

11.7 SILICON CONTROLLED SWITCH

The silicon controlled switch (SCS), like the silicon controlled rectifier, is a four-layer *pnpn* device. All four semiconductor layers of the SCS are available due to the addition of an anode gate, as shown in 11.16a. The graphic symbol and transistor equivalent circuit are shown in the same figure.

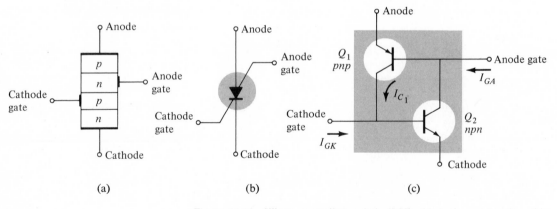

Figure 11.16 Silicon controlled switch (SCS): (a) basic construction; (b) graphic symbol; (c) equivalent transistor circuit.

The characteristics of the device are essentially the same as those for the SCR. The effect of an anode gate current is very similar to that demonstrated by the gate current in Fig. 11.7. The higher the anode gate current, the lower the required anode-to-cathode voltage to turn the device on.

The anode gate connection can be used to turn the device either on or off. To turn on the device, a negative pulse must be applied to the anode gate terminal, while a positive pulse is required to turn off the device. The need for the type of pulse indicated above can be demonstrated using the circuit of Fig. 11.16c. A negative pulse at the anode gate will forward-bias the base-to-emitter junction of Q_1, turning it on. The resulting heavy collector current I_{C_1} will turn on Q_2, resulting in a regenerative action and the on state for the SCS device. A positive pulse at the anode gate will reverse-bias the base-to-emitter junction of Q_1, turning it off, resulting in the open-circuit off state of the device. In general, the triggering (turn-on) anode gate current is larger in magnitude than the required cathode gate current. For one representative SCS device, the triggering anode gate current is 1.5 mA while the required cathode gate current is 1 μA. The required turn-on gate current at either terminal is affected by many factors. A few include the operating temperature, anode-to-cathode voltage, load placement, and type of cathode, gate-to-cathode or anode gate-to-anode connection (short-circuit, open-circuit, bias, load, etc.). Tables, graphs, and curves are normally available for each device to provide the type of information indicated above.

Three of the more fundamental types of turn-off circuits for the SCS are shown in Fig. 11.17. When a pulse is applied to the circuit of Fig. 11.17a, the transistor

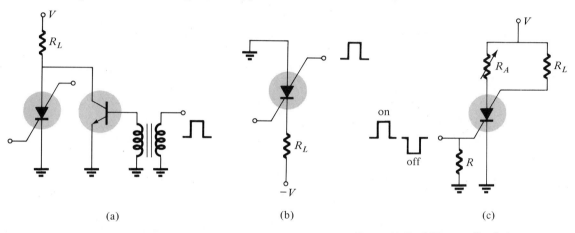

Figure 11.17 SCS turn-off techniques.

conducts heavily, resulting in a low-impedance ($\cong$ short-circuit) characteristic between collector and emitter. This low-impedance branch diverts anode current away from the SCS, dropping it below the holding value and consequently turning it off. Similarly, the positive pulse at the anode gate of Fig. 11.17b will turn the SCS off by the mechanism described earlier in this section. The circuit of Fig. 11.17c can be turned either off *or* on by a pulse of the proper magnitude at the cathode gate. The turn-off characteristic is possible only if the correct value of R_A is employed. It will control the amount of regenerative feedback, the magnitude of which is critical for this type of operation. Note the variety of positions in which the load resistor R_L can be placed. There are a number of other possibilities that can be found in any comprehensive semiconductor handbook or manual.

An advantage of the SCS over a corresponding SCR is the reduced turn-off time, typically within the range 1 to 10 μs for the SCS and 5 to 30 μs for the SCR.

Some of the remaining advantages of the SCS over an SCR include increased control and triggering sensitivity and a more predictable firing situation. At present, however, the SCS is limited to low power, current, and voltage ratings. Typical maximum anode currents range from 100 to 300 mA with dissipation (power) ratings of 100 to 500 mW.

A few of the more common areas of application include a wide variety of computer circuits (counters, registers, and timing circuits) pulse generators, voltage sensors, and oscillators. One simple application for an SCS as a voltage-sensing device is shown in Fig. 11.18. It is an alarm system with *n* inputs from various stations. Any single input will turn that particular SCS on, resulting in an energized alarm relay and light in the anode gate circuit to indicate the location of the input (disturbance).

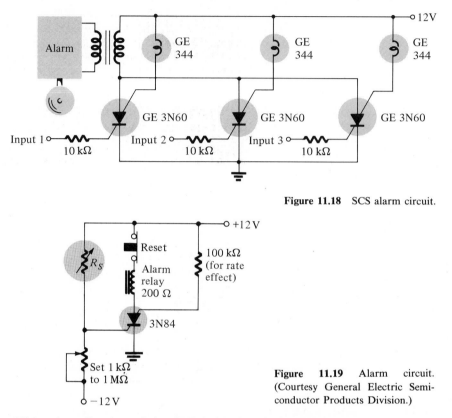

Figure 11.18 SCS alarm circuit.

Figure 11.19 Alarm circuit. (Courtesy General Electric Semiconductor Products Division.)

One additional application of the SCS is in the alarm circuit of Fig. 11.19. R_S represents a temperature-, light-, or radiation-sensitive resistor, that is, an element whose resistance will decrease with the application of any of the three energy sources listed above. The cathode gate potential is determined by the divider relationship established by R_S and the variable resistor. Note that the gate potential is at approximately 0 volts if R_S equals the value set by the variable resistor, since both resistors will have 12 V across them. However, if R_S decreases, the potential of the junction will increase until the SCS is forward-biased, causing the SCS to turn on and energize the alarm relay.

The 100-kΩ resistor is included to reduce the possibility of accidental triggering of the device through a phenomena known as *rate effect*. It is caused by the stray capacitance levels between gates. A high-frequency transient can establish sufficient base current to turn the SCS on accidentally. The device is reset by pressing the reset button, which in turn opens the conduction path of the SCS and reduces the anode current to zero.

CH. 11 PNPN AND OTHER DEVICES

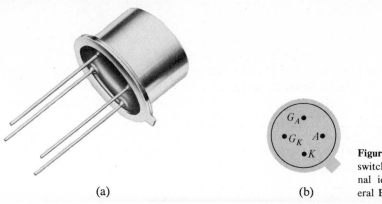

(a)

G_A•

•G_K A•

•K

(b)

Figure 11.20 Silicon controlled switch (SCS): (a) device; (b) terminal identification. (Courtesy General Electric Company.)

Sensitivity to resistors R_S that increase in resistance due to the application of any of the three energy sources described above can be accommodated by simply interchanging the location of R_S and the variable resistor.

The terminal identification of an SCS is shown in Fig. 11.20 with a packaged SCS.

11.8 GATE TURN-OFF SWITCH

The gate turn-off switch (GTO) is the third *pnpn* device to be introduced in this chapter. Like the SCR, however, it has only three external terminals, as indicated in Fig. 11.21a. Its graphic symbol is also shown in Fig. 11.21b. Although the graphic symbol is different from either the SCR or the SCS, the transistor equivalent is exactly the same and the characteristics are similar.

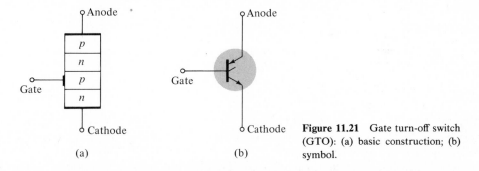

(a)

(b)

Figure 11.21 Gate turn-off switch (GTO): (a) basic construction; (b) symbol.

The most obvious advantage of the GTO over the SCR or SCS is the fact that it can be turned on *or* off by applying the proper pulse to the cathode gate (without the anode gate and associated circuitry required for the SCS). A consequence of this turn-off capability is an increase in the magnitude of the required gate current for triggering. For an SCR and GTO of similar maximum rms current ratings, the gate-triggering current of a particular SCR is 30 μA, while the triggering current of the GTO is 20 mA. The turn-off current of a GTO is slightly larger than the required triggering current. The maximum rms current and dissipation ratings of GTOs manufactured today are limited to about 3 A and 20 W, respectively.

A second very important characteristic of the GTO is improved switching characteristics. The turn-on time is similar to the SCR (typically 1 μs), but the turn-off time of about the *same* duration (1 μs) is much smaller than the typical turn-off time of an SCR (5 to 30 μs). The fact that the turn-off time is similar to the turn-on time rather than considerably larger permits the use of this device in high-speed applications.

A typical GTO and its terminal identification are shown in Fig. 11.22. The GTO gate input characteristics and turn-off circuits can be found in a comprehensive manual or specification sheet. The majority of the SCR turn-off circuits can also be used for GTOs.

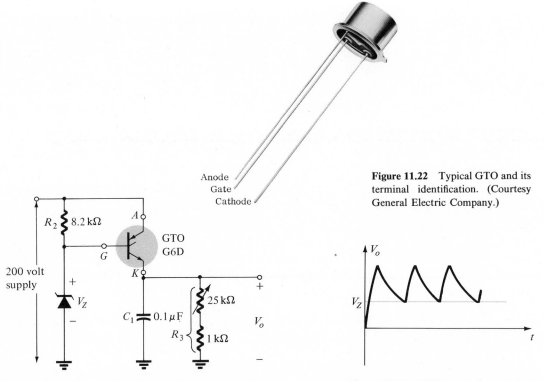

Figure 11.22 Typical GTO and its terminal identification. (Courtesy General Electric Company.)

Figure 11.23 GTO sawtooth generator.

Some of the areas of application for the GTO include counters, pulse generators, multivibrators, and voltage regulators. Figure 11.23 is an illustration of a simple sawtooth generator employing a GTO and a Zener diode.

When the supply is energized, the GTO will turn on, resulting in the short-circuit equivalent from anode to cathode. The capacitor C_1 will then begin to charge toward the supply voltage as shown in Fig. 11.23. As the voltage across the capacitor C_1 charges above the Zener potential, a reversal in gate-to-cathode voltage will result, establishing a reversal in gate current. Eventually, the negative gate current will be large enough to turn the GTO off. Once the GTO turns off, resulting in the open-circuit representation, the capacitor C_1 will discharge through the resistor R_3. The discharge time will be determined by the circuit time constant $\tau = R_3 C_1$. The proper choice of R_3 and C_1 will result in the sawtooth waveform of Fig. 11.23. Once the output potential V_o drops below V_Z, the GTO will turn on and the process will repeat.

11.9 LIGHT-ACTIVATED SCR

The next in the series of *pnpn* devices is the light-activated SCR (LASCR). As indicated by the terminology, it is an SCR whose state is controlled by the light falling upon a silicon semiconductor layer of the device. The basic construction of an LASCR is shown in Fig. 11.24a.

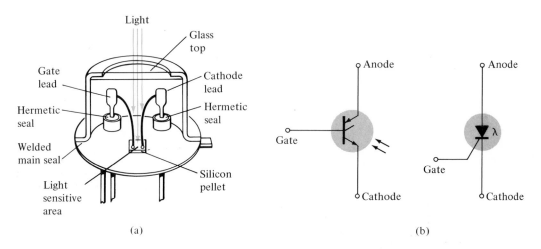

Figure 11.24 Light-activated SCR (LASCR): (a) basic construction; (b) symbols.

As indicated in Fig. 11.24a, a gate lead is also provided to permit triggering the device using typical SCR methods. Note also in the figure that the mounting surface for the silicon pellet is the anode connection for the device.

The graphic symbols most commonly employed for the LASCR are provided in Fig. 11.24b. The terminal identification and typical LASCRs are shown in Fig. 11.25a.

Some of the areas of application for the LASCR include optical light controls, relays, phase control, motor control, and a variety of computer applications. The maximum current (rms) and power (gate) ratings for LASCRs commercially available today are about 3 A and 0.1 W. The characteristics (light triggering) of a representative LASCR are provided in Fig. 11.25b. Note in this figure that an increase in junction temperature results in a reduction in light energy required to activate the device.

One interesting application of an LASCR is in the AND and OR circuits of Fig. 11.26. Only when light falls on LASCR$_1$ *and* LASCR$_2$ will the short-circuit representation for each be applicable and the supply voltage appear across the load. For the OR circuit, light energy applied to LASCR$_1$ *or* LASCR$_2$ will result in the supply voltage appearing across the load.

The LASCR is most sensitive to light when the gate terminal is open. Its sensitivity

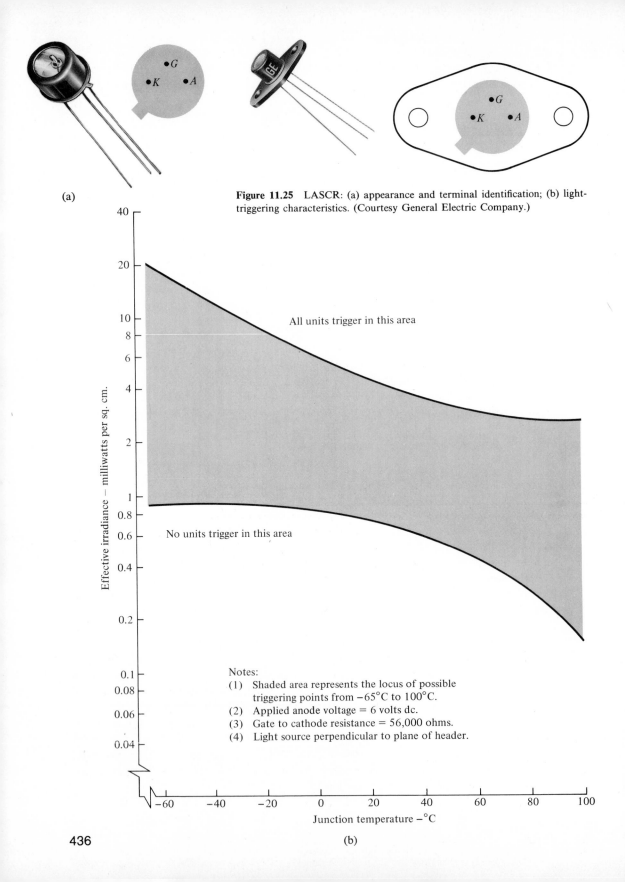

(a)

Figure 11.25 LASCR: (a) appearance and terminal identification; (b) light-triggering characteristics. (Courtesy General Electric Company.)

All units trigger in this area

No units trigger in this area

Notes:
(1) Shaded area represents the locus of possible triggering points from −65°C to 100°C.
(2) Applied anode voltage = 6 volts dc.
(3) Gate to cathode resistance = 56,000 ohms.
(4) Light source perpendicular to plane of header.

Effective irradiance – milliwatts per sq. cm.

Junction temperature –°C

(b)

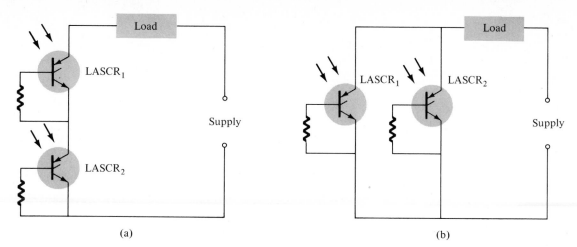

(a)

(b)

Figure 11.26 LASCR optoelectronic logic circuitry: (a) AND gate—input to LASCR$_1$ *and* LASCR$_2$ required for energization of the load; (b) OR gate—input to either LASCR$_1$ *or* LASCR$_2$ will energize the load.

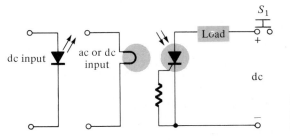

Figure 11.27 Latching relay. (Courtesy General Electric Semiconductor Products Division.)

can be reduced and controlled somewhat by the insertion of a gate resistor, as shown in Fig. 11.26.

A second application of the LASCR appears in Fig. 11.27. It is the semiconductor analog of an electromechanical relay. Note that it offers complete isolation between the input and switching element. The energizing current can be passed through a light-emitting diode or a lamp, as shown in the figure. The incident light will cause the LASCR to turn on and permit a flow of charge (current) through the load as established by the dc supply. The LASCR can be turned off using the reset switch S_1. This system offers the additional advantages over an electromechanical switch of long life, microsecond response, small size, and the elimination of contact bounce.

11.10 SHOCKLEY DIODE

The Shockley diode is a four-layer *pnpn* diode with only two external terminals, as shown in Fig. 11.28a with its graphic symbol. The characteristics (Fig. 11.28b) of the device are exactly the same as those encountered for the SCR with $I_G = 0$. As indicated by the characteristics, the device is in the off state (open-circuit representation) until the breakover voltage is reached, at which time avalanche conditions develop and the device turns on (short-circuit representation).

One common application of the Shockley diode is shown in Fig. 11.29, where it is employed as a trigger switch for an SCR. When the circuit is energized, the voltage

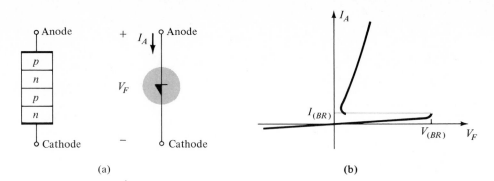

(a)

(b)

Figure 11.28 Shockley diode: (a) basic construction and symbol; (b) characteristics.

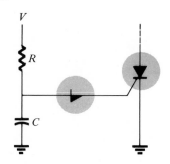

Figure 11.29 Shockley diode application—trigger switch for an SCR.

across the capacitor will begin to change toward the supply voltage. Eventually, the voltage across the capacitor will be sufficiently high to first turn on the Shockley diode and then the SCR.

11.11 DIAC

The diac is basically a two-terminal parallel-inverse combination of semiconductor layers that permits triggering in either direction. The characteristics of the device, presented in Fig. 11.30a, clearly demonstrate that there is a breakover voltage in either direction. This possibility of an on condition in either direction can be used to its fullest advantage in *ac* applications.

The basic arrangement of the semiconductor layers of the diac is shown in Fig. 11.30b, along with its graphic symbol. Note that neither terminal is referred to as the cathode. Instead, there is an anode 1 (or electrode 1) and an anode 2 (or electrode 2). When anode 1 is positive with respect to anode 2, the semiconductor layers of particular interest are $p_1 n_2 p_2$ and n_3. For anode 2 positive with respect to anode 1 the applicable layers are $p_2 n_2 p_1$ and n_1.

For the unit appearing in Fig. 11.30, the breakdown voltages are very close in magnitude but may vary from a minimum of 28 V to a maximum of 42 V. They are related by the following equation provided in the specification sheet:

$$V_{BR_1} = V_{BR_2} \pm 10\% \ V_{BR_2} \tag{11.1}$$

The current levels (I_{BR_1} and I_{BR_2}) are also very close in magnitude for each device.

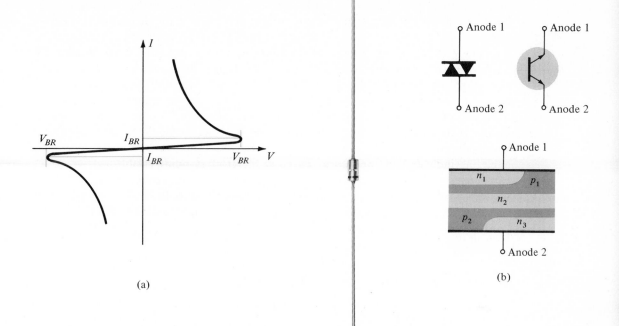

(a)

(b)

Figure 11.30 Diac: (a) characteristics; (b) symbols and basic construction. (Courtesy General Electric Company.)

For the unit of Fig. 11.30, both current levels are about 200 μA = 0.2 mA.

The use of the diac in a proximity detector appears in Fig. 11.31. Note the use of an SCR in series with the load and the programmable unijunction transistor (to be described shortly) connected directly to the sensing electrode.

As the human body approaches the sensing electrode, the capacitance between the electrode and ground will increase. The programmable UJT (PUT) is a device

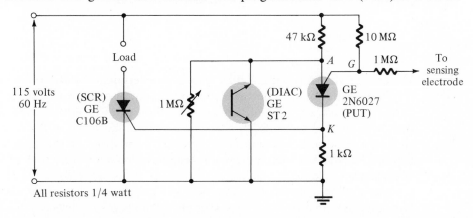

Figure 11.31 Proximity detector or touch switch. (Courtesy General Electric Semiconductor Products Division.)

that will fire (enter the short-circuit state) when the anode voltage (V_A) is at least 0.7 V (for silicon) greater than the gate voltage (V_G). Before the programmable device turns on, the system is essentially as shown in Fig. 11.32. As the input voltage rises, the diac voltage V_G will follow as shown in the figure until the firing potential is reached. It will then turn on and the diac voltage will drop substantially, as shown. Note that the diac is in essentially an open-circuit state until it fires. Before the capacitive element is introduced, the voltage V_G will be the same as the input. As indicated in the figure, since both V_A and V_G follow the input, V_A can never be greater than V_G by 0.7 V and turn on the device. However, as the capacitive element is introduced, the voltage V_G will begin to lag the input voltage by an increasing angle, as indicated in the figure. There is therefore a point established where V_A can exceed V_G by 0.7 V and cause the programmable device to fire. A heavy current is established through the PUT at this point, raising the voltage V_K and turning on the SCR. A heavy SCR current will then exist through the load, reacting to the presence of the approaching individual.

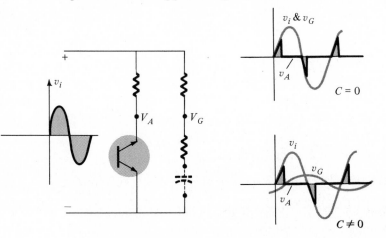

Figure 11.32 Effect of capacitive element on the behavior of the network of Fig. 11.31.

A second application of the diac appears in the next section (Fig. 11.34) as we consider an important power-control device: the triac.

11.12 TRIAC

The triac is fundamentally a diac with a gate terminal for controlling the turn-on conditions of the bilateral device in either direction. In other words, for either direction the gate current can control the action of the device in a manner very similar to that demonstrated for an SCR. The characteristics, however, of the triac in the first and third quadrants are somewhat different from those of the diac, as shown in Fig. 11.33c. Note the holding current in each direction not present in the characteristics of the diac.

The graphic symbol for the device and the distribution of the semiconductor layers are provided in Fig. 11.33 with photographs of the device. For each possible direction of conduction there is a combination of semiconductor layers whose state will be controlled by the signal applied to the gate terminal.

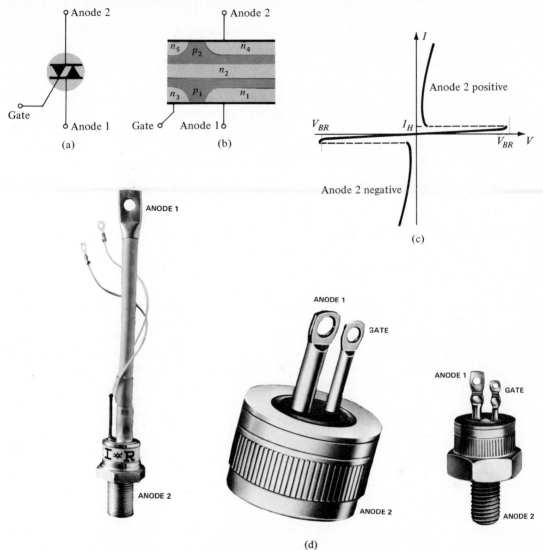

Figure 11.33 Triac: (a) symbol; (b) basic construction; (c) characteristics; (d) photographs.

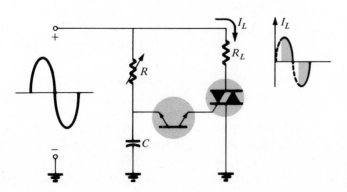

Figure 11.34 Triac application: phase (power) control.

One fundamental application of the triac is presented in Fig. 11.34. In this capacity, it is controlling the ac power to the load by switching on and off during the positive and negative regions of input sinusoidal signal. The action of this circuit during the positive portion of the input signal is very similar to that encountered for the Shockley diode in Fig. 11.29. The advantage of this configuration is that during the negative portion of the input signal the same type of response will result, since both the diac and triac can fire in the reverse direction. The resulting waveform for the current through the load is provided in Fig. 11.34. By varying the resistor R the conduction angle can be controlled. There are units available today that can handle in excess of 10-kW loads.

OTHER DEVICES

11.13 UNIJUNCTION TRANSISTOR

Recent interest in the unijunction transistor (UJT) has, like that for the SCR, been increasing at an exponential rate. Although first introduced in 1948, the device did not become commercially available until 1952. The low cost per unit, combined with the excellent characteristics of the device, have warranted its use in a wide variety of applications. A few include oscillators, trigger circuits, sawtooth generators, phase control, timing circuits, bistable networks, and voltage- or current-regulated supplies. The fact that this device is, in general, a low-power-absorbing device under normal operating conditions is a tremendous aid in the continual effort to design relatively efficient systems.

The UJT is a three-terminal device having the basic construction of Fig. 11.35.

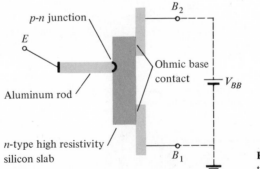

Figure 11.35 Unijunction transistor (UJT): basic construction.

A slab of lightly doped (increased resistance characteristic) n-type silicon material has two base contacts attached to both ends of one surface and an aluminum rod alloyed to the opposite surface. The p-n junction of the device is formed at the boundary of the aluminum rod and the n-type silicon slab. The single p-n junction accounts for the terminology unijunction. It was originally called a duo (double) base diode due to the presence of two base contacts. Note in Fig. 11.35 that the aluminum rod is alloyed to the silicon slab at a point closer to the base-2 contact than the base-1 contact and that the base-2 terminal is made positive with respect to the base-1 terminal by V_{BB} volts. The effect of each will become evident in the paragraphs to follow.

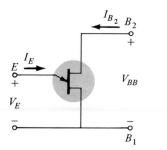

The symbol for the unijunction transistor is provided in Fig. 11.36. Note that the emitter leg is drawn at an angle to the vertical line representing the slab of *n*-type material. The arrowhead is pointing in the direction of conventional current (hole) flow when the device is in the forward-biased, active, or conducting state.

The circuit equivalent of the UJT is shown in Fig. 11.37. Note the relative simplicity of this equivalent circuit: two resistors (one fixed, one variable) and a single diode. The resistance R_{B_1} is shown as a variable resistor since its magnitude will vary with the current I_E. In fact, for a representative

Figure 11.36 Symbol and basic biasing arrangement for the unijunction transistor.

unijunction transistor, R_{B_1} may vary from 5 kΩ down to 50 Ω for a corresponding change of I_E from 0 to 50 μA. The interbase resistance R_{BB} is the resistance of the device between terminals B_1 and B_2 when $I_E = 0$. In equation form,

$$R_{BB} = (R_{B_1} + R_{B_2})|_{I_E=0} \tag{11.2}$$

(R_{BB} is typically within the range of 4 to 10 kΩ.) The position of the aluminum rod of Fig. 11.35 will determine the relative values of R_{B_1} and R_{B_2} with $I_E = 0$. The magnitude of $V_{R_{B1}}$ (with $I_E = 0$) is determined by the voltage-divider rule in the following manner:

$$V_{R_{B_1}} = \frac{R_{B_1} V_{BB}}{R_{B_1} + R_{B_2}} = \eta V_{BB} \Big|_{I_E=0} \tag{11.3}$$

The Greek letter η (eta) is called the *intrinsic stand-off* ratio of the device and is defined by

$$\eta = \frac{R_{B_1}}{R_{B_1} + R_{B_2}} \Big|_{I_E=0} \tag{11.4}$$

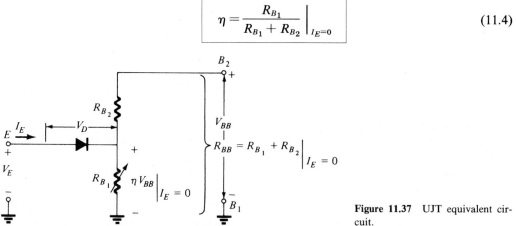

Figure 11.37 UJT equivalent circuit.

For applied emitter potentials (V_E) greater than $V_{R_{B1}} = \eta V_{BB}$ by the forward voltage drop of the diode, V_D (0.35 → 0.70 V) the diode will fire, assume the short-circuit representation (on an ideal basis), and I_E will begin to flow through R_{B_1}. In equation form the emitter firing potential is given by

$$V_P = \eta V_{BB} + V_D \qquad (11.5)$$

The characteristics of a representative unijunction transistor are shown for $V_{BB} = 10$ V in Fig. 11.38. Note that for emitter potentials to the left of the peak point, the magnitude of I_E is never greater than I_{EO} (measured in microamperes). The current I_{EO} corresponds very closely with the reverse leakage current I_{CO} of the conventional bipolar transistor. This region, as indicated in the figure, is called the cutoff region. Once conduction is established at $V_E = V_P$, the emitter potential V_E will drop with increase in I_E. This corresponds exactly with the decreasing resistance R_{B_1} for increasing current I_E, as discussed earlier. This device, therefore, has a *negative resistance* region which is stable enough to be used with a great deal of reliability in the areas of application listed earlier. Eventually, the valley point will be reached, and any further increase in I_E will place the device in the saturation region. In this region the characteristics approach that of the semiconductor diode in the equivalent circuit of Fig. 11.37.

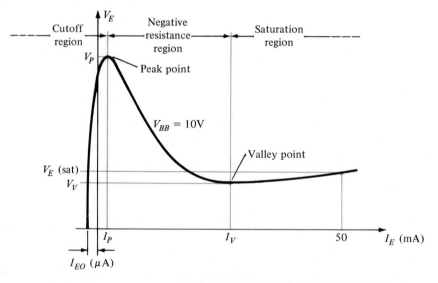

Figure 11.38 UJT static emitter-characteristic curve.

The decrease in resistance in the active region is due to the holes injected into the *n*-type slab from the aluminum *p*-type rod when conduction is established. The increased hole content in the *n*-type material will result in an increase in the number of free electrons in the slab, producing an increase in conductivity (G) and a corresponding drop in resistance ($R\downarrow = 1/G\uparrow$). Three other important parameters for the unijunction transistor are I_P, V_V, and I_V. Each is indicated on Fig. 11.38. They are all self-explanatory.

The emitter characteristics as they normally appear are provided in Fig. 11.39.

CH. 11 PNPN AND OTHER DEVICES

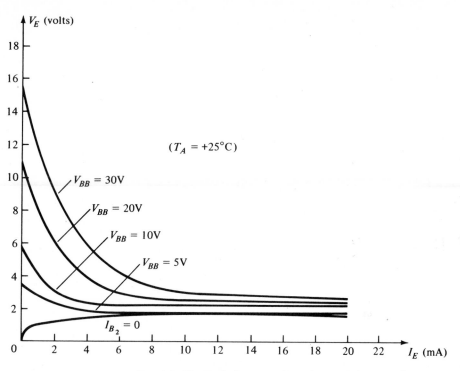

Figure 11.39 Typical static emitter-characteristic curves for a UJT.

Note that I_{EO} (μA) is not in evidence since the horizontal scale is in milliamperes. The intersection of each curve with the vertical axis is the corresponding value of V_P. For fixed values of η and V_D, the magnitude of V_P will vary as V_{BB}, that is,

$$V_p\uparrow = \eta V_{BB}\uparrow + V_D$$

$$\underset{\text{fixed}}{\underleftrightarrow{}}$$

A typical set of specifications for the UJT is provided in Fig. 11.40b. The discussion of the last few paragraphs should make each quantity readily recognizable. The terminal identification is provided in the same figure with a photograph of a representative UJT. Note that the base terminals are opposite each other while the emitter terminal is between the two. In addition, the base terminal to be tied to the higher potential is closer to the extension on the lip of the casing.

One rather common application of the UJT is in the triggering of other devices such as the SCR. The basic elements of such a triggering circuit are shown in Fig. 11.41. The resistor R_1 must be chosen to ensure that the load line determined by R_1 passes through the device characteristics to the right of the peak point but to the left of the valley point. If the load line fails to pass to the right of the peak point, the device cannot turn on. An equation for R_1 that will ensure a turn-on condition can be established if we consider the peak point at which $I_p = I_{R_1}$ and $V_E = V_p$. (The equality $I_p = I_{R_1}$ is valid since the charging current of the capacitor, at this instant, is zero; that is, the capacitor is at this particular instant changing from a charging to a discharging state.) Then $V - I_P R_1 = V_P$ or $(V - V_P)/I_P = R_1$.

absolute maximum ratings: (25°C)

Power Dissipation	300 mw
RMS Emitter Current	50 ma
Peak Emitter Current	2 amperes
Emitter Reverse Voltage	30 volts
Interbase Voltage	35 volts
Operating Temperature Range	−65°C to +125°C
Storage Temperature Range	−65°C to +150°C

electrical characteristics: (25°C)

		Min.	Typ.	Max.
Intrinsic Standoff Ratio				
($V_{BB} = 10V$)	η	0.56	0.65	0.75
Interbase Resistance				
($V_{BB} = 3V, I_E = 0$)	R_{BB}	4.7	7	9.1
Emitter Saturation Voltage				
($V_{BB} = 10V, I_E = 50$ ma)	$V_{E(\text{SAT})}$		2	
Emitter Reverse Current				
($V_{BB} = 30V, I_{B1} = 0$)	I_{EO}		0.05	12
Peak Point Emitter Current	I_P		0.4	5
($V_{BB} = 25V$)				
Valley Point Current				
($V_{BB} = 20V, R_{B2} = 100\Omega$)	I_V	4	6	

(a)

(b)

(c)

Figure 11.40 UJT: (a) appearance; (b) specification sheet; (c) terminal identification. (Courtesy General Electric Company.)

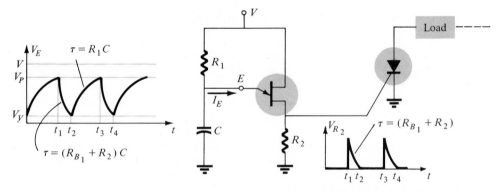

Figure 11.41 UJT triggering of an SCR.

To ensure firing,

$$\frac{V - V_P}{I_P} > R_1 \tag{11.6}$$

At the valley point $I_E = I_V$ and $V_E = V_V$, so that to ensure turning off,

$$\frac{V - V_V}{I_V} < R_1 \tag{11.7}$$

For the typical values of $V = 30$ V, $\eta = 0.5$, $V_V = 1$ V, $I_V = 10$ mA, $I_P = 10$ μA, and $R_{BB} = 5$ kΩ.

CH. 11 PNPN AND OTHER DEVICES

$$\frac{V - V_P}{I_P} = \frac{30 - [0.5(30) + 0.5]}{10 \times 10^{-6}} = \frac{14.5}{10 \times 10^{-6}} = 1.45 \text{ M}\Omega > R_1$$

and

$$\frac{V - V_V}{I_V} = \frac{30 - 1}{10 \times 10^{-3}} = 2.9 \text{ k}\Omega < R_1$$

Therefore, $1.45 \text{ M}\Omega > R_1 > 2.9 \text{ k}\Omega$.

The range for R_1 is therefore extensive. The resistor R_2 must be chosen small enough to ensure that the SCR is not turned on by the interbase current I_{BB} through R_2 when $I_E = 0$.

The capacitor C will determine, as we shall see, the time interval between triggering pulses and the time span of each pulse.

At the instant the dc supply voltage V is applied, the voltage V_E will charge toward V volts since the emitter circuit of the UJT is in the open-circuit state. The time constant of the charging circuit is $R_1 C$. When $V_E = V_P$, the UJT will enter the conduction state and the capacitor C will discharge through R_{B_1} and R_2 at a rate determined by the time constant $(R_{B_1} + R_2)C$. This time constant is much smaller than the former, resulting in the patterns of Fig. 7.46. Once V_E decays to V_V, the UJT will turn off and the charging phase will repeat itself. Since I_{R_2} and V_{R_2} are related by Ohm's law (linear relationship), the waveform for I_{R_2} appears the same as for V_{R_2}. The positive pulse of V_{R_2} is designed to be sufficiently large to turn the SCR on.

If we remove the SCR from Fig. 11.41, we have the basic construction of a *UJT relaxation oscillator*—that is, a source of a continually repeating waveform such as appears at V_E in Fig. 11.41. The frequency of the generated waveform is given approximately by

$$f \cong \frac{1}{R_1 C \log_e[1/(1 - \eta)]} \tag{11.8}$$

The conditions for oscillation are the same as described for the network of Fig. 11.41.

11.14 PHOTOTRANSISTORS

The fundamental behavior of photoelectric devices was introduced earlier with the description of the photodiode. This discussion will now be extended to include the phototransistor, which has a photosensitive collector-base *p-n* junction. The current induced by photoelectric effects is the base current of the transistor. If we assign the notion I_λ for the photoinduced base current, the resulting collector current, on an approximate basis, is

$$\boxed{I_c \cong h_{fe} I_\lambda} \tag{11.9}$$

A representative set of characteristics for a phototransistor is provided in Fig. 11.42 with the symbolic representation of the device. Note the similarities between

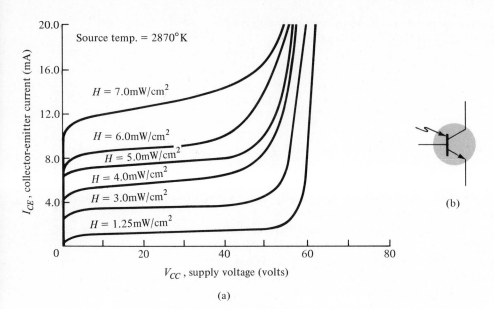

Source temp. = 2870°K

$H = 7.0 \text{mW/cm}^2$

$H = 6.0 \text{mW/cm}^2$

$H = 5.0 \text{mW/cm}^2$

$H = 4.0 \text{mW/cm}^2$

$H = 3.0 \text{mW/cm}^2$

$H = 1.25 \text{mW/cm}^2$

I_{CE}, collector-emitter current (mA)

V_{CC}, supply voltage (volts)

(a)

(b)

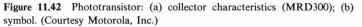

Figure 11.42 Phototransistor: (a) collector characteristics (MRD300); (b) symbol. (Courtesy Motorola, Inc.)

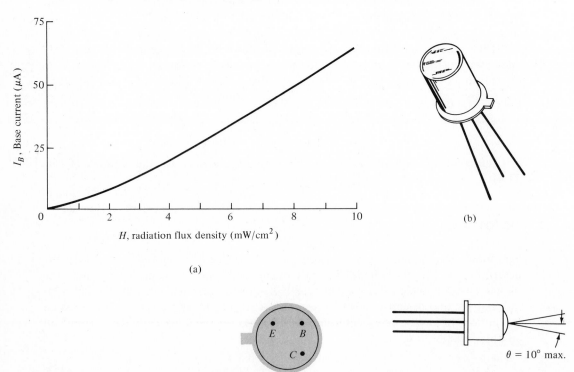

I_B, Base current (μA)

H, radiation flux density (mW/cm^2)

(a)

(b)

(c)

(d)

$\theta = 10°$ max.

E B C

Figure 11.43 Phototransistor: (a) base current vs. flux density; (b) device; (c) terminal identification; (d) angular alignment. (Courtesy Motorola, Inc.)

these curves and those of a typical bipolar transistor. As expected, an increase in light intensity corresponds with an increase in collector current. To develop a greater degree of familiarity with the light-intensity unit of measurement, milliwatts per square centimeter, a curve of base current versus flux density appears in Fig. 11.43a. Note the exponential increase in base current with increasing flux density. In the same figure a sketch of the phototransistor is provided with the terminal identification and the angular alignment.

Some of the areas of application for the phototransistor include punch-card readers, computer logic circuitry, lighting control (highways, etc.), level indication, relays, and counting systems.

A high-isolation AND gate is shown in Fig. 11.44 using three phototransistors and three LEDs (light-emitting diodes). The LEDs are semiconductor devices that emit light at an intensity determined by the forward current through the device. With the aid of discussions in Chapter 2, the circuit behavior should be relatively easy to understand. The terminology "high isolation" simply refers to the lack of an electrical connection between the input and output circuits.

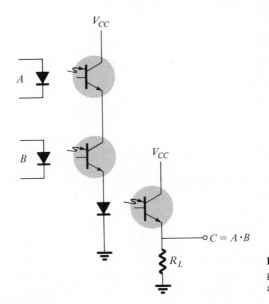

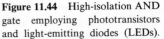

Figure 11.44 High-isolation AND gate employing phototransistors and light-emitting diodes (LEDs).

11.15 OPTO-ISOLATORS

The *opto-isolator* is a device that incorporates many of the characteristics described in the preceding section. It is simply a package that contains both an infrared LED and a photodetector such as a silicon diode, transistor Darlington pair, or SCR. The wavelength response of each device is tailored to be as identical as possible to permit the highest measure of coupling possible. In Fig. 11.45, two possible chip configurations are provided, with a photograph of each. There is a transparent insulating cap between each set of elements embedded in the structure (not visible) to permit the passage of light. They are designed with response times so small that they can be used to transmit data in the megahertz range.

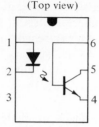

ISO-LIT 1

(Top view)

Pin No.	Function
1	anode
2	cathode
3	nc
4	emitter
5	collector
6	base

LED chip on Pin 2
PT chip on Pin 5

ISO-LIT Q1

(Top view)

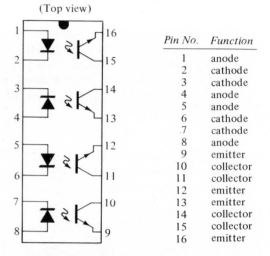

Pin No.	Function
1	anode
2	cathode
3	cathode
4	anode
5	anode
6	cathode
7	cathode
8	anode
9	emitter
10	collector
11	collector
12	emitter
13	emitter
14	collector
15	collector
16	emitter

Figure 11.45 Two Litronix opto-isolators. (Courtesy Litronix, Inc.)

The maximum ratings and electrical characteristics for the IL-1 model are provided in Fig. 11.46. Note that I_{CEO} is measured in nanoamperes and that the power dissipation of the LED and transistor are about the same.

The typical optoelectronic characteristic curves for each channel are provided in Figs. 11.47 to 11.51. Note the very pronounced effect of temperature on the output current at low temperatures but the fairly level response at or above room temperature (25°C). As mentioned earlier, the level of I_{CEO} is improving steadily with improved design and construction techniques (the lower the better). In Fig. 11.47 we do not reach 1 μA until the temperature rises above 75°C. The transfer characteristics of Fig. 11.48 compare the input LED current (which establishes the luminous flux) to the resulting collector current of the output transistor (whose base current is determined by the incident flux). In fact, Fig. 11.49 demonstrates that the V_{CE} voltage affects the resulting collector current only very slightly. It is interesting to note in Fig. 11.50 that the switching time of an opto-isolator decreases with increased current, while for many devices it is exactly the reverse. Consider that it is only 2 μs for a collector current of 6 mA and a load R_L of 100 Ω. The relative output versus temperature appears in Fig. 11.51.

The schematic representation for a transistor coupler appears in Fig. 11.46. The schematic representations for a photodiode, photo-Darlington, and photo-SCR opto-isolator appear in Fig. 11.52.

(a) Maximum Ratings

Gallium arsenide LED (each channel) IL-1
  Power dissipation @ 25°C   200 mW
  Derate linearly from 25°C   2.6 mW/°C
  Continuous forward current   150 mA
Detector silicon phototransistor (each channel) IL-1
  Power dissipation @ 25°C   200 mW
  Derate linearly from 25°C   2.6 mW/°C
  Collector-emitter breakdown voltage   30 V
  Emitter-collector breakdown voltage   7 V
  Collector-base breakdown voltage   70 V
Package IL-1
  Total package dissipation at 25°C ambient (LED plus detector)   250 mW
  Derate linearly from 25°C   3.3 mW/°C
  Storage temperature   −55°C to +150°C
  Operating temperature   −55°C to +100°C

(b) Electrical Characteristics per Channel (at 25°C Ambient)

Parameter	Min.	Typ.	Max.	Units	Test Conditions
Gallium arsenide LED					
Forward voltage		1.3	1.5	V	$I_F = 60$ mA
Reverse current		0.1	10	μA	$V_R = 3.0$ V
Capacitance		100		pF	$V_R = 0$
Phototransistor detector					
BV_{CEO}	30			V	$I_C = 1$ mA
I_{CEO}		5.0	50	nA	$V_{CE} = 10$ V, $I_F = 0$
Collector-emitter capacitance		2.0		pF	$V_{CE} = 0$
BV_{ECO}	7			V	$I_E = 100$ μA
Coupled characteristics					
dc current transfer ratio	0.2	0.35			$I_F = 10$ mA, $V_{CE} = 10$ V
Capacitance, input to output		0.5		pF	
Breakdown voltage	2500			V	DC
Resistance, input to output		100		GΩ	
V_{SAT}			0.5	V	$I_C = 1.6$ mA, $I_F = 16$ mA
Propagation delay					
$t_{D\ ON}$		6.0		μs	$R_L = 2.4$ kΩ, $V_{CE} = 5$ V
$t_{D\ OFF}$		25		μs	$I_F = 16$ mA

Figure 11.46   The Litronix IL-1 opto-isolator.

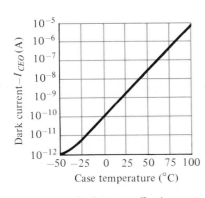

Figure 11.47   Dark current (I_{CEO}) vs. temperature.

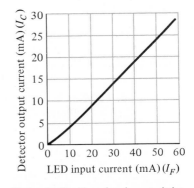

Figure 11.48   Transfer characteristics.

451

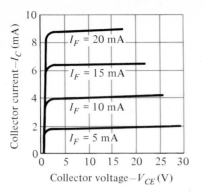

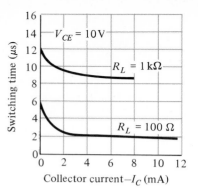

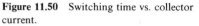

Figure 11.49 Detector output characteristics.

Figure 11.50 Switching time vs. collector current.

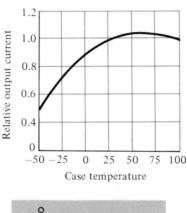

Figure 11.51 Relative output vs. temperature.

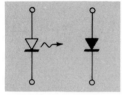

(a)

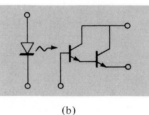

(b)

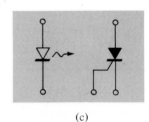

(c)

Figure 11.52 Opto-isolators: (a) photodiode; (b) photo-Darlington; (c) photo-SCR.

11.16 PROGRAMMABLE UNIJUNCTION TRANSISTOR

Although there is a similarity in name, the actual construction and mode of operation of the programmable unijunction transistor (PUT) is quite different from the unijunction transistor. The fact that the *I–V* characteristics and applications of each are similar prompted the choice of labels.

As indicated in Fig. 11.53, it is a four-layer *pnpn* device with a gate connected directly to the sandwiched *n*-type layer. The symbol for the device and the basic biasing arrangement appears in Fig. 11.54. As the symbol suggests, it is essentially an SCR with a control mechanism that permits a duplication of the characteristics of the typical SCR. The term "programmable" is applied because R_{BB}, η, and V_P

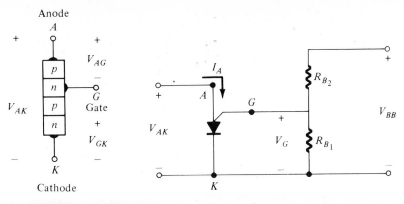

Anode
A

V_{AG}

p
n
p
n

V_{AK}

G
Gate

V_{GK}

K

Cathode

I_A

A

V_{AK}

G

V_G

R_{B_2}

R_{B_1}

V_{BB}

K

Figure 11.53 Programmable UJT (PUT).

Figure 11.54 Basic biasing arrangement for the PUT.

as defined for the UJT can be controlled through the resistors R_{B_1}, R_{B_2}, and the supply voltage V_{BB}. Note in Fig. 11.54 that through an application of the voltage-divider rule when $I_G = 0$:

$$V_G = \frac{R_{B_1}}{R_{B_1} + R_{B_2}} V_{BB} = \eta V_{BB} \qquad (11.10)$$

where

$$\eta = \frac{R_{B_1}}{R_{B_1} + R_{B_2}}$$

as defined for the UJT.

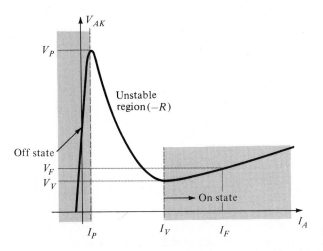

Figure 11.55 PUT characteristics.

The characteristics of the device appear in Fig. 11.55. As noted on the diagram, the off state (I low, V between 0 and V_P) and the on state ($I \geq I_V$, $V \geq V_V$) are separated by the unstable region as occurred for the UJT. That is, the device cannot stay in the unstable state—it will simply shift to either the off or on stable states.

The firing potential (V_P) or voltage necessary to "fire" the device is given by

$$V_P = \eta V_{BB} + V_D \tag{11.11}$$

as defined for the UJT. However, V_P represents the voltage drop V_{AG} in Fig. 11.53 (the forward voltage drop across the conducting diode). For silicon V_D is typically 0.7 V. Therefore,

$$V_P = \eta V_{BB} + V_D = \eta V_{BB} + V_{AG}$$

$$\boxed{V_P = \eta V_{BB} + 0.7}_{\text{silicon}} \tag{11.12}$$

We noted above, however, that $V_G = \eta V_{BB}$ with the result that

$$\boxed{V_P = V_G + 0.7}_{\text{silicon}} \tag{11.13}$$

Recall that for the UJT both R_{B_1} and R_{B_2} represent the bulk resistance and ohmic base contacts of the device—both inaccessible. In the development above, we note that R_{B_1} and R_{B_2} are external to the device permitting an adjustment of η and hence V_G above. In other words, a measure of control on the level of V_P required to turn on the device.

Although the characteristics of the PUT and UJT are similar, the peak and valley currents of the PUT are typically lower than those of a similarly rated UJT. In addition, the minimum operating voltage is also less for a PUT.

If we take a Thévenin equivalent of the network to the right of gate terminal in Fig. 11.54, the network of Fig. 11.56 will result. The resulting resistance R_S is important because it is often included in specification sheets since it affects the level of I_V.

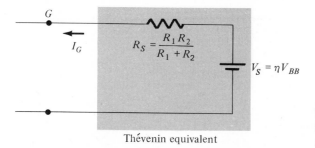

Thévenin equivalent

Figure 11.56 Thévenin equivalent for the network to the right of the gate terminal in Fig. 11.54.

The basic operation of the device can be reviewed through reference to Fig. 11.55. A device in the off state will not change state until the voltage V_P as defined by V_G and V_D is reached. The level of current until I_P is reached is very low, resulting in an open-circuit equivalent since $R = V(\text{high})/I(\text{low})$ will result in a high resistance level. When V_P is reached the device will switch through the unstable region to the on state, where the voltage is lower but the current higher, resulting in a terminal resistance $R = V(\text{low})/I(\text{high})$ which is quite small, representing short-circuit equivalent on an approximate basis. The device has therefore switched from essentially an open-circuit to a short-circuit state at a point determined by the choice of R_{B_1}, R_{B_2}, and V_{BB}. Once the device is in the on state, the removal of V_G will not turn the device off. The level of voltage V_{AK} must be dropped sufficiently to reduce the current below a holding level.

EXAMPLE 11.1 Determine R_{B_1} and V_{BB} for a silicon PUT if it is determined that η should be 0.8, $V_P = 10.3$ V, and $R_{B_2} = 5$ kΩ.

Solution: From Eq. (11.10):

$$\eta = \frac{R_{B_1}}{R_{B_1} + R_{B_2}} = 0.8$$

$$R_{B_1} = 0.8(R_{B_1} + R_{B_2})$$

$$0.2\,R_{B_1} = 0.8\,R_{B_2}$$

$$R_{B_1} = 4R_{B_2}$$

$$R_{B_1} = 4(5\text{ k}\Omega) = \textbf{20 k}\Omega$$

From Eq. (11.11):

$$V_P = \eta V_{BB} + V_D$$

$$10.3 = (0.8)(V_{BB}) + 0.7$$

$$9.6 = 0.8\,V_{BB}$$

$$V_{BB} = \textbf{12 V}$$

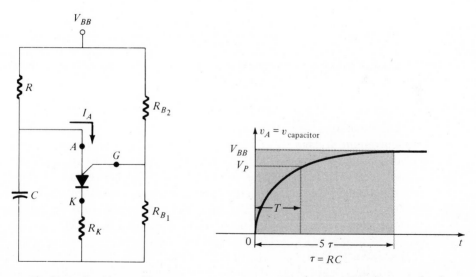

Figure 11.57 PUT relaxation oscillator.

Figure 11.58 Changing wave for the capacitor C of Fig. 11.57.

One popular application of the PUT is in the relaxation oscillator of Fig. 11.57. The instant the supply is connected, the capacitor will begin to charge toward 10 V, since there is no anode current at this point. The charging curve appears in Fig. 11.58.

The period T required to reach the firing potential V_P is given approximately by

$$T \cong RC \log_e \left(\frac{V_{BB}}{V_{BB} - V_P} \right) \qquad (11.14)$$

or

$$T \cong RC \log_e \left(1 + \frac{R_{B_1}}{R_{B_2}} \right) \qquad (11.15)$$

The instant the voltage across the capacitor equals V_P, the device will fire and a current I_P established through the PUT. If R is too large, the current I_P cannot be established and the device will not fire. At the point of transition,

$$I_P R = V_{BB} - V_P$$

and

$$R_{\max} = \frac{V_{BB} - V_P}{I_P} \qquad (11.16)$$

The subscript is included to indicate that any R greater than $R_{\max}$ will result in a current less than I_P. The level of R must also be such to ensure it is less than I_V if oscillations are to occur. In other words, we want the device to enter the unstable region and then return to the off state. From reasoning similar to that above:

$$R_{\min} = \frac{V_{BB} - V_V}{I_V} \qquad (11.17)$$

The discussion above requires that R be limited to the following for an oscillatory system:

$$R_{\min} < R < R_{\max}$$

The waveforms of v_A, v_G, and v_K appear in Fig. 11.59. Note that T determines the maximum voltage v_A can charge to. Once the device fires, the capacitor will rapidly discharge through the PUT and R_K, producing the drop shown. Of course, v_K will peak at the same time due to the brief but heavy current. The voltage v_G will rapidly drop down from V_G to a level just greater than 0 volts. When the capacitor voltage drops to a low level, the PUT will once again turn off and the charging cycle repeated. The effect on V_G and V_K is shown in Fig. 11.59.

EXAMPLE 11.2 If $V_{BB} = 12$ V, $R = 20$ kΩ, $C = 1$ μF, $R_K = 100$ Ω, $R_{B_1} = 10$ kΩ, $R_{B_2} = 5$ kΩ, $I_P = 100$ μA, $V_V = 1$ V, and $I_V = 5.5$ mA, determine:
(a) V_P.
(b) $R_{\max}$ and $R_{\min}$.
(c) T and frequency of oscillation.
(d) The waveforms of v_A, v_G, and v_K.

Solution:
(a) From Eq. (11.11):

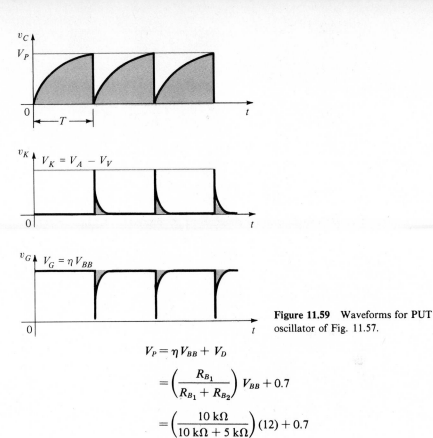

Figure 11.59 Waveforms for PUT oscillator of Fig. 11.57.

$$V_P = \eta V_{BB} + V_D$$

$$= \left(\frac{R_{B_1}}{R_{B_1} + R_{B_2}} \right) V_{BB} + 0.7$$

$$= \left(\frac{10 \text{ k}\Omega}{10 \text{ k}\Omega + 5 \text{ k}\Omega} \right) (12) + 0.7$$

$$= (0.67)(12) + 0.7 = \textbf{8.7 V}$$

(b)   From Eq. (11.16):

$$R_{\max} = \frac{V_{BB} - V_P}{I_P}$$

$$= \frac{12 - 8.7}{100 \times 10^{-6}} = \textbf{33 k}\Omega$$

From Eq. (11.17):

$$R_{\min} = \frac{V_{BB} - V_V}{I_V}$$

$$= \frac{12 - 1}{5.5 \times 10^{-3}} = \textbf{2 k}\Omega$$

$$R\colon \quad 2 \text{ k}\Omega < 20 \text{ k}\Omega < 33 \text{ k}\Omega$$

(c)   Eq. (11.14):

$$T = RC \log_e \left(\frac{V_{BB}}{V_{BB} - V_P} \right)$$

$$= (20 \times 10^3)(1 \times 10^{-6}) \log_e \left(\frac{12}{12 - 8.7} \right)$$

$$= 20 \times 10^{-3} \log_e (3.64)$$

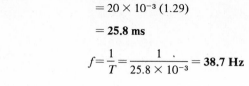

$$= 20 \times 10^{-3} (1.29)$$

$$= \textbf{25.8 ms}$$

$$f = \frac{1}{T} = \frac{1}{25.8 \times 10^{-3}} = \textbf{38.7 Hz}$$

(d)   as indicated in Fig. 11.60.

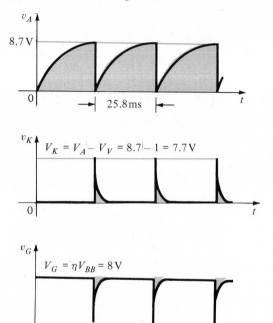

Figure 11.60   Waveforms for the oscillator of Example 11.1.

PROBLEMS

§ 11.3

1. Describe in your own words the basic behavior of the SCR using the two-transistor equivalent circuit.

2. Describe two techniques for turning an SCR off.

3. Consult a manufacturer's manual or specification sheet and obtain a turn-off network. If possible, describe the turn-off action of the design.

§ 11.4

4. (a)   At high levels of gate current the characteristics of an SCR approach those of what two-terminal device?
(b)   At a fixed anode-to-cathode voltage less than $V_{(BR)F*}$, what is the effect on the firing of the SCR as the gate current is reduced from its maximum value to the zero level?
(c)   At a fixed gate current greater than $I_G = 0$, what is the effect on the firing of the SCR as the gate voltage is reduced from $V_{(BR)F*}$?
(d)   For increasing levels of I_G, what is the effect on the holding current?

5. (a) In Fig. 11.8, will a gate current of 50 mA fire the device at room temperature (25°C)?

(b) Repeat part (a) for a gate current of 10 mA.

(c) Will a gate voltage of 2.6 V trigger the device at room temperature?

(d) Is $V_G = 6$ V, $I_G = 800$ mA a good choice for firing conditions? Would $V_G = 4$ V, $I_G = 1.6$ A be preferred? Explain.

§ 11.5

6. In Fig. 11.11b, why is there very little loss in potential across the SCR during conduction?

7. Fully explain why reduced values of R_1 in Fig. 11.12 will result in an increased angle of conduction.

8. Refer to the charging network of Fig. 11.13.

(a) Determine the dc level of the full-wave rectified signal if a 1:1 transformer were employed.

(b) If the battery in its uncharged state is sitting at 11 V, what is the anode-to-cathode voltage drop across SCR$_1$?

(c) What is the maximum possible value of V_R ($V_{GK} \cong 0.7$ V)?

(d) At the maximum value of part (c), what is the gate potential of SCR$_2$?

(e) Once SCR$_2$ has entered the short-circuit state, what is the level of V_2?

§ 11.7

9. Fully describe in your own words the behavior of the networks of Fig. 11.17.

§ 11.8

10. (a) In Fig. 11.23, if $V_Z = 50$ V, determine the maximum possible value the capacitor C_1 can charge to ($V_{GK} \cong 0.7$ V).

(b) Determine the approximate discharge time (5τ) for R$_3 = 20$ kΩ.

(c) Determine the internal resistance of the GTO if the rise time is one-half the decay period determined in part (b).

§ 11.9

11. (a) Using Fig. 11.25b, determine the minimum irradiance required to fire the device at room temperature (20°C).

(b) What percent reduction in irradiance is allowable if the junction temperature is increased from 0°C (32°F) to 100°C (212°F)?

§ 11.10

12. For the network of Fig. 11.29, if $V_{(BR)} = 6$ V, $V = 40$, $R = 10$ kΩ, $C = 0.2$ μF, and V_{GK} (firing potential) $= 3$ V, determine the time period between energizing the network and the turning on of the SCR.

§ 11.11

13. Using whatever reference you require, find an application of a diac and explain the network behavior.

14. If V_{BR_2} is 6.4 V, determine the range for V_{BR_1} using Eq. (11.1).

15. Repeat Problem 13 for the triac.

16. For the network of Fig. 11.41, in which $V = 40$ V, $\eta = 0.6$, $V_v = 1$ V, $I_v = 8$ mA, and $I_p = 10$ μA, determine the range of R_1 for the triggering network.

17. For a unijunction transistor with $V_{BB} = 20$ V, $\eta = 0.65$, $R_{B_1} = 2$ kΩ ($I_E = 0$), and $V_D = 0.7$ V, determine the following:
    (a) R_{B_2}.
    (b) R_{BB}.
    (c) $V_{R_{B_1}}$.
    (d) V_P.

18. For a phototransistor having the characteristics of Fig. 11.43, determine the photoinduced base current for a radiant flux density of 5 mW/cm². If $h_{fe} = 40$, find I_C.

19. Design a high-isolation OR-gate employing phototransistors and LEDs.

20. (a) Determine an average derating factor from the curve of Fig. 11.51 for the region defined by temperatures less than 25°C.
    (b) Is it fair to say that for temperature greater than room temperature (up to 100°C), the output current is somewhat unaffected by temperature?

21. (a) Determine, from Fig. 11.47, the average change in I_{CEO} per degree change in temperature for the range 25 to 50°C.
    (b) Can the results of part (a) be used to determine the level of I_{CEO} at 35°C? Test your theory.

22. Determine, from Fig 11.48, the ratio of LED input current to detector output current for an output current of 20 mA. Would you consider the device to be relatively efficient in its purpose?

23. (a) Sketch the maximum-power curve of $P_D = 200$ mW on the graph of Fig. 11.49. List any noteworthy conclusions.
    (b) Determine β_{dc} (defined by I_C/I_F) for the system at $V_{CE} = 15$ V, $I_F = 10$ mA.
    (c) Compare the results of part (b) with those obtained from Fig. 11.48 at $I_F = 8$ mA. Do they compare? Should they? Why?

24. (a) Referring to Fig. 11.50, determine the collector current above which the switching time does not change appreciably for $R_L = 1$ kΩ and $R_L = 100$ Ω.
    (b) At $I_C = 6$ mA, how does the ratio of switching times for $R_L = 1$ kΩ and $R_L = 100$ Ω compare to the ratio of resistance levels?

25. Determine η and V_G for a PUT with $V_{BB} = 20$ V and $R_{B_1} = 3$ R_{B_2}.

26. Using the data provided in Example 11.2 determine the impedance of the PUT at the firing and valley points. Are the approximate open- and short-circuit states verified?

27. Can Eq. (11.15) be derived exactly as shown from Eq. (11.14)? If not, what element is missing in Eq. (11.15)?

28. (a) Will the network of Example 11.2 oscillate if V_{BB} is changed to 10 V? What minimum value of V_{BB} is required (V_V a constant)?
(b) Referring to the same example, what value of R would place the network in the stable on state and remove the oscillatory response of the system?
(c) What value of R would make the network a 2-ms time-delay network? That is, provide a pulse v_k 2 ms after the supply is turned on and then stay in the on state.

CHAPTER 12

Integrated Circuits (ICs)

12.1 INTRODUCTION

During the past decade the characteristics and benefits associated with a relatively new type of electronic package have substantially altered the course of many areas of research and development. This product, called an *integrated circuit* (IC), has, through expanded usage and the various media of advertising, become a product whose basic function and purpose are now understood by the layperson. The most noticeable characteristic of an IC is its size. It is typically hundreds and even thousands of times smaller than a semiconductor structure built in the usual manner with discrete components. For example, the integrated circuit shown in Fig. 12.1 has 68,000 transistors, although it is only 246 × 281 mils (0.246″ × 0.281″). For comparison purposes, the integrated circuit appears alongside a discrete transistor package in the same figure. The MC6800 is a microprocessor unit that is the heart of a microcomputer manufactured by Motorola, Inc.

Integrated circuits are seldom, if ever, repaired; that is, if a single component within an IC should fail, the entire structure (complete circuit) is replaced—a more economical approach. There are three types of ICs commercially available on a large scale today. They include the monolithic, thin (or thick) film, and hybrid integrated circuits. Each will be introduced in this chapter.

12.2 RECENT DEVELOPMENTS

Although the sequence of steps leading to the manufacture of an integrated circuit has not changed substantially during the past decade, the manner in which each

462

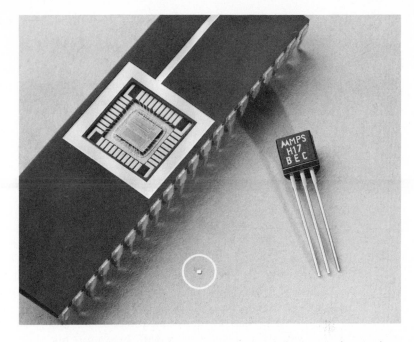

Figure 12.1 The MC6800 microprocessor and a discrete transistor package. The chip employed in the discrete device appears within the circled region. (Courtesy Motorola, Inc.)

step is performed has changed dramatically. In the early days the IC manufacturer designed, built, and maintained the equipment employed in the production cycle. Today, however, new industries have emerged that have assumed the responsibility of introducing the latest technological advances into the processing equipment. The result is that the manufacturer can concentrate on design, quality control, improved performance and reliability characteristics, and further miniaturization. The equipment available from the peripheral companies carries a high price tag (unit costs in excess of $\frac{1}{2}$ million are not unusual), however, and 24-hour operation is almost a necessity to ensure sound economic policy. In an effort to ensure continuous operation, the larger IC manufacturers have their own service staff rather than having to rely on immediate response from the equipment manufacturer.

The added responsibility associated with the cost of the equipment and its proper utilization requires personnel capable of sound judgment and reliability. During a recent visit to the Motorola facility in Phoenix, Arizona, both authors were quite impressed with those individuals we met at all levels. We left with the general impression that the field is in very capable hands and that further developments and improvements can be expected on a regular basis in the years to come.

Automation has become increasingly important in the production cycle. A great deal of microprocessor control introduced in the form of "cassette addressment" has significantly reduced the possibility of error owing to incorrect transfer of information to the processing unit. It also has a sensitivity to the process being performed that is unavailable through the human response curve. A complete set of instructions is placed on a magnetic tape cassette and identified for use with the preparation of a particular IC wafer. The increased level of automation also reduces the amount of "handling" required and thereby increases the yield factor.

If there is one area in which the industry appears to be most sensitive, it is the yield level. The average number of "good" individual dies resulting from an IC wafer is only in the range of 20–30%. The development of methods that would increase the average level to 50% or 60% would have a significant impact on the cost of IC components. Methods including the use of larger ingots (to 5" diameter), automation, and clean room environments have all contributed to increase the yield level—but, unfortunately, by only a few percent. The "clean room" environment as noted above has become increasingly important in recent years. In common use at present are the 0.3-micron filtration systems (Class 100 Type Rooms), which are classified as "cleaner" than required in hospital environments. In fact, most house-dust particles have diameters many times (4X and more) that of the filtration system. The white robe, boots, and hat ("bunny suits") as appearing in some of the photographs of this chapter have become part of the expected in the production area of such facilities. The control is so tight that the women working in many of these areas cannot wear makeup in order to eliminate introduction of foreign particles into the environment.

A natural requirement of an increased number of components in an individual die is an increase in the yield level due to the increase in investment level (to cover the additional processing steps and production time). The 64K byte memory unit has four times the memory of a 16K byte unit, but the added cost necessitates a higher level of yield compared to the 16K byte module, which can be discarded with significantly less concern about the investment involved.

The line widths of current manufacturing techniques extend from 10 to 15 microns (1 micron $= 10^{-6}$ m) for high-power devices to between 1 and 2 microns for RF devices. The impact of such efforts is best demonstrated by Fig. 12.1.

One element that plays a very important role in the manufacturing process is the high-resistivity water used for rinsing and cleaning operations. It is filtered at 0.2 microns, has an 18-MΩ resistivity level, and must be free of organic contamination. It is so "clean" that it will not support cultural growth.

Due to large investments, it is an absolute necessity that the product processing be tightly controlled through a strong management system. The computer is now playing a very important role in providing the data required for such continuous surveillance of the production cycle.

A number of improvements in the manufacturing process will be introduced in this chapter as each production step is described.

12.3 MONOLITHIC INTEGRATED CIRCUIT

The term *monolithic* is derived from a combination of the Greek words *monos,* meaning single, and *lithos,* meaning stone, which in combination result in the literal translation, single-stone, or more appropriately, single-solid structure. As this descriptive term implies, the monolithic IC is constructed within a *single* wafer of semiconductor material. The greater portion of the wafer will simpy act as a supporting structure for the very thin resulting IC. An overall view of the stages involved in the fabrication of monolithic ICs is provided in Fig. 12.2. The actual number of steps leading to a

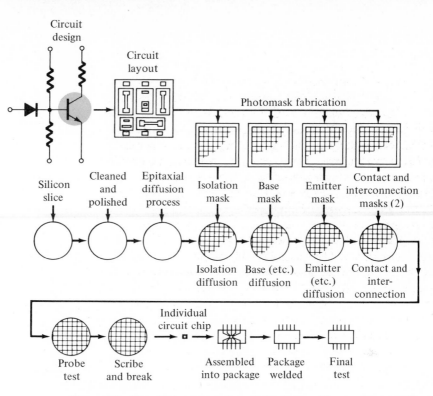

Figure 12.2 Monolithic integrated-circuit fabrication. (Courtesy Robert Hibberd.)

finished product is many times that appearing in Fig. 12.2. The figure does, however, point out the major production phases of forming a monolithic IC.

As indicated in the previous section, the processing equipment and not the steps have changed significantly in recent years. The initial preparation of the semiconductor wafer of Fig. 12.2 was discussed in Chapter 1 in association with the fabrication of diodes. As indicated in the figure, it is first necessary to design a circuit that will meet the specifications. The circuit must then be laid out in order to ensure optimum use of available space and a minimum of difficulty in performing the diffusion processes to follow. The appearance of the mask and its function in the sequence of stages indicated will be introduced in Section 12.5. For the moment, let it suffice to say that a mask has the appearance of a negative through which impurities may be diffused (through the light areas) into the silicon slice. The actual diffusion process for each phase is similar to that applied in the fabrication of diffused transistors in Chapter 3. The last mask of the series will control the placement of the interconnecting conducting pattern between the various elements. The wafer then goes through various testing procedures, is scribed and broken into individual chips, packaged and assembled as indicated. A processed silicon wafer appears in Fig. 12.3. The original wafer can be anywhere from $\frac{1}{2}$ to 5 in. in diameter. The size of each chip will, of course, determine the number of individual circuits resulting from a single wafer. The dimensions of each chip of the wafer in Fig. 12.3 are 50×50 mils. To point out the microminiature size of these chips, consider that 20 of them can be lined up along a 1-in. length. The average relative size of the elements of a monolithic IC appear

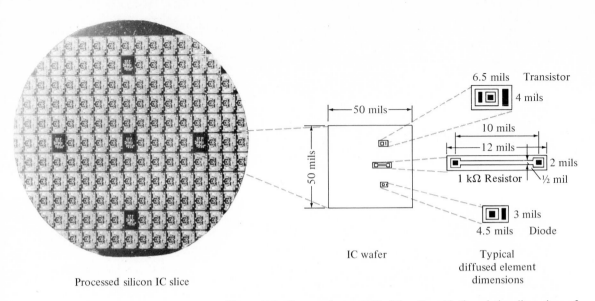

Processed silicon IC slice

IC wafer

Typical diffused element dimensions

Figure 12.3 Processed monolithic IC wafer with the relative dimensions of the various elements. (Courtesy Robert Hibberd.)

in Fig. 12.3. Note the large area required for the 1-kΩ resistor as compared to the other elements indicated. The next section will examine the basic construction of each of these elements.

A recent article indicated, by percentage, the relative costs of the various stages in the production of monolithic ICs as compared to discrete transistors. The resulting graphs appear in Fig. 12.4. The processing phase includes all stages leading up to the individual chips of Fig. 12.3. Note the high cost of packaging the integrated circuits and of testing the large-scale silicon integrated circuits (LSI). The cost of

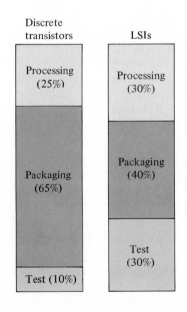

Figure 12.4 Cost breakdown for the manufacturing of discrete transistors and large-scale integrated circuits (LSIs). (Courtesy J. J. Suran, General Electric Co.)

packaging has resulted in an increase (wherever feasible) in the number of IC chips within a single package. This multichip hybrid type of integrated circuit will be considered in Section 12.8.

12.4 MONOLITHIC CIRCUIT ELEMENTS

The surface appearance of the transistor, diode, and resistor appear in Fig. 12.3. We shall now examine the basic construction of each in more detail.

Resistor

You will recall that the resistance of a material is determined by the resistivity, length, area, and temperature of the material. For the integrated circuit, each necessary element is present in the sheet of semiconductor material appearing in Fig. 12.5. As indicated in the figure, the semiconductor material can be either p- or n-type, although the p-type is most frequently employed.

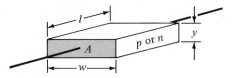

Figure 12.5 Parameters determining the resistance of a sheet of semiconductor material.

The resistance of any bulk material is determined by

$$R = \rho \frac{l}{A}$$

For $l = w$, resulting in a square sheet,

$$R = \frac{\rho l}{yw} = \frac{\rho l}{yl}$$

and

$$\boxed{R_S = \frac{\rho}{y}} \quad \text{(ohms)} \tag{12.1}$$

where ρ is in ohm-centimeters and y is in centimeters.

R_S is called the sheet resistance and has the units ohms per square. The equation clearly reveals that the sheet resistance is independent of the size of the square.

In general, where $l \neq w$,

$$\boxed{R = R_S \frac{l}{w}} \quad \text{(ohms)} \tag{12.2}$$

For the resistor appearing in Fig. 12.3, $w = \frac{1}{2}$ mil, $l = 10$ mils, and $R_S = 100\Omega/$ square:

$$R = R_S \frac{l}{w} = 100 \times \frac{10}{\frac{1}{2}} = 2 \text{ k}\Omega$$

A cross-sectional view of a monolithic resistor appears in Fig. 12.6 along with the surface appearence of two monolithic resistors. In Fig. 12.6a the sheet resistive material *(p)* is indicated with its aluminum terminal connections. The *n*-isolation region performs exactly that function indicated by its name; that is, it isolates the monolithic resistive elements from the other elements of the chip. Note in Fig. 12.6b the method employed to obtain a maximum *l* in a limited area. The resistors of Fig. 12.6 are called base-diffusion resistors since the *p*-material is diffused into the *p*-type substrate during the base-diffusion process indicated in Fig. 12.2.

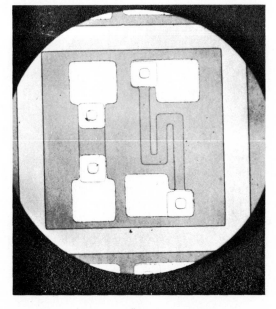

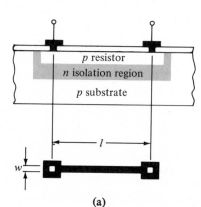

(a)

(b)

Figure 12.6 Monolithic resistors: (a) cross section and determining dimensions; (b) surface view of two monolithic resistances in a single die. (Courtesy Motorola, Inc.)

Capacitor

Monolithic capacitive elements are formed by making use of the transition capacitance of a reverse-biased *p-n* junction. At increasing reverse-bias potentials, there is an increasing distance at the junction between the *p*- and *n*-type impurities. The region between these oppositely doped layers is called the depletion region (see Chapter 3) due to the absence of "free" carriers. The necessary elements of a capacitive element are therefore present—the depletion region has insulating characteristics that separate the two oppositely charged layers. The transition capacitance is related to the width *(W)* of the depletion region, the area *(A)* of the junction, and the permittivity *(ε)* of the material within the depletion region by

$$C_T = \frac{\epsilon A}{W}$$

(12.3)

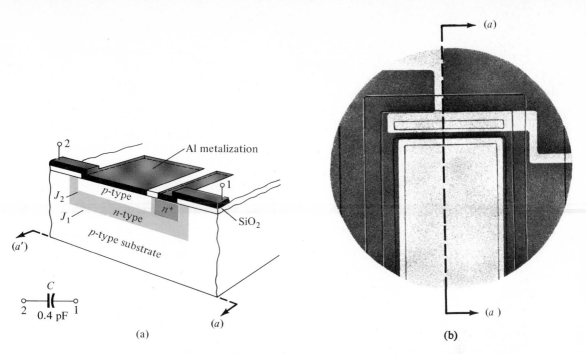

Figure 12.7 Monolithic capacitor: (a) cross section; (b) photograph. (Courtesy Motorola, Inc.)

The cross section and surface appearance of a monolithic capacitive element appear in Fig. 12.7. The reverse-biased junction of interest is J_2. The undesirable parasitic capacitance at junction J_1 is minimized through careful design. Due to the fact that aluminum is p-type impurity in silicon, a heavily doped n^+ region is diffused into the n-type region as shown to avoid the possibility of establishing an undesired p-n junction at the boundary between the aluminum contact and the n-type impurity region.

Inductor

Whenever possible, inductors are avoided in the design of integrated circuits. An effective technique for obtaining nominal values of inductances has so far not been devised for monolithic integrated circuits. In many instances, the need for inductive elements can be eliminated through the use of a technique known as RC synthesis. Thin (or thick) film or hybrid integrated circuits have an option open to them that cannot be employed in monolithic integrated circuits: the addition of discrete inductive elements to the surface of the structure. Even with this option, however, they are seldom employed due to their relatively bulky nature.

Transistors

The cross section of a monolithic transistor appears in Fig. 12.8a. Note again the presense of the n^+ region in the n-type epitaxial collector region. The vast majority of monolithic IC transistors are *npn* rather than *pnp* for reasons to be found in more advanced texts on the subject. Keep in mind when examining Fig. 12.8 that

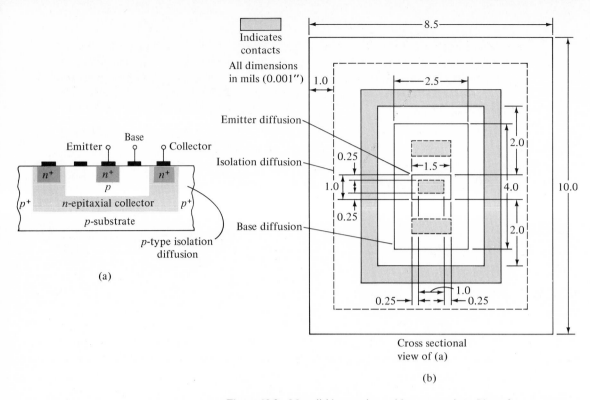

Figure 12.8 Monolithic transistor: (a) cross section; (b) surface appearance and dimension for a typical monolithic transistor. (Courtesy Motorola Monitor.)

the *p*-substrate is only a supporting and isolating structure forming no part of the active device itself. The base, emitter, and collector regions are formed during the corresponding diffusion processes of Fig. 12.2.

The top view of a typical monolithic transistor appears in Fig. 12.8b. Note that two base terminals are provided while the collector has the outer rectangular aluminum contact surface.

Diodes

The diodes of a monolithic integrated circuit are formed by first diffusing the required regions of a transistor and then masking the diode rather than transistor

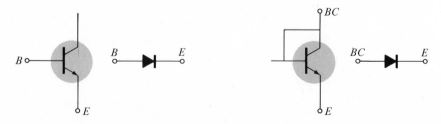

Figure 12.9 Transistor structure and two possible connections employed in the formation of monolithic diodes.

terminal connections. There is, however, more than one way of hooking up a transistor to perform a basic diode action. The two most common methods applied to monolithic integrated circuits appear in Fig. 12.9. The structure of a *BC-E* diode appears in Fig. 12.10. Note that the only difference between the cross-sectional view of Fig. 12.10 and that of the transistor of Fig. 12.8a is the position of the ohmic aluminum contacts.

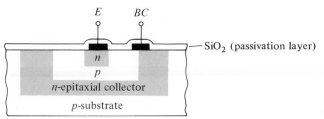

Figure 12.10 The cross-sectional view of a *BC-E* monolithic diode.

12.5 MASKS

The selective diffusion required in the formation of the various active and passive elements of an integrated circuit is accomplished through the use of masks such as that shown in Fig. 12.11. We shall find in the next section that the light areas are the only areas through which donor and acceptor impurities can pass. The dark areas will block the diffusion of impurities somewhat as a shade will prevent sunlight from changing the pigment of the skin. The next section will demonstrate the use of these masks in the formation of a computer logic circuit.

The sequence of steps leading to the final mask is controlled by the micron width of the smallest features on the wafer. For the 0.5–2-micron range, electron beam lithography must be employed, whereas for the 3–5-micron range, less sophisticated (and less expensive) methods can be employed.

At one time the making of a mask first required a large-scale stabilene drawing of all layers. The artwork was then transferred to a clear mylar coated with red

Figure 12.11 Mask. (Courtesy Motorola, Inc.)

plastic called *Rubylith*. Very precise cuts were made in the red material and sections peeled off to reveal the regions through which the diffusion of impurities could take place. The resulting pattern was then photographed and reduced by 500X (500 times the desired size for production) in a series of steps until the desired master (redicle) was obtained.

Today, the same stabilene drawing is taken to a CAD (Computer-Aided-Design) area called Calma (the manufacturer of the equipment employed) and is mounted on a digitizing board as shown in Fig. 12.12. The operator performs a function called *digitizing* through which the geometrical features from the stabilene drawing are transferred to the computer memory of the Calma system. The same procedure is repeated for each layer of the integrated circuit. Any section of the design can then be recalled and placed on the CRT of an editing station to modify the pattern, check continuity, etc., as shown in Fig. 12.13. One distinct advantage of the Calma system is that repetitive cells can be added from memory at the editing stage and not appear as part of the stabilene drawing. In addition, a library of cells could be established that could be introduced as required. The net result is that a skilled operator could design and develop a major portion of the complete reticle with a minimum of effort at the stabilene stage.

If required, a line drawing can then be generated using a Xynetics plotter as

Figure 12.12 Calma digitizing station. (Courtesy Motorola, Inc., and Calma, Inc.)

(a)

(b)

Figure 12.13 Circuit design/modification at a Calma (interactive computer graphics system) editing/tablet station. (Courtesy Motorola, Inc., and Calma, Inc.)

473

shown in Fig. 12.14. It is on-line with the Calma System and reacts directly to the descriptive data in storage. It is an amazing piece of equipment to watch in operation with its multicolor option and very quick precise movements. It can enlarge a region to 10,000X if required.

The photoreduction process can then be applied to reduce the pattern to 100X and then 10X. A second method for obtaining the 10X pattern in one step utilizes a *pattern generator*. The magnetic tape on which the pattern data are stored is loaded into the console, and a 10X pattern is obtained through computer control of the exposure shapes and placement. For a complex LSI, 200,000 to 300,000 flashes may be required over a period of 20–36 hours.

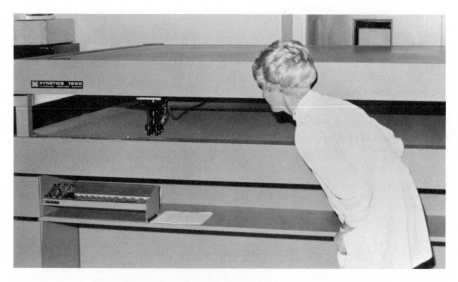

Figure 12.14 Xynetics 1200 Plotter (on line with the Calma system). (Courtesy Motorola, Inc., and Xynetics, Inc.)

A programmed step-and-repeat machine will then generate the required format (multiple images on the same mask) from the 10X pattern for the product group.

For LSI units, the time involved from initial design to mask availability may extend from a few months to 1 or 2 years.

12.6 MONOLITHIC INTEGRATED CIRCUIT—THE NAND GATE

This section is devoted to the sequence of production stages leading to a monolithic NAND-gate circuit (the operation of which is covered in Chapter 15). A detailed examination of each process would require many more pages that it is possible to include in this text. The description, however, should be sufficiently complete and informative to aid the reader in any future contact with this highly volatile area. The circuit to be prepared appears in Fig. 12.15a. The criteria of space allocation, placement of pin connection, and so on require that the elements be situated in the relative positions indicated in Fig. 12.15b. The regions to be isolated from one another

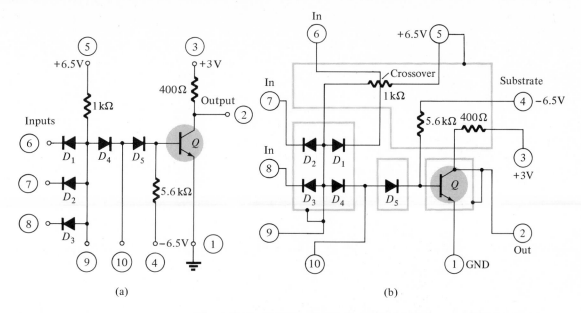

Figure 12.15 NAND gate: (a) circuit; (b) layout for monolithic fabrication.

appear within the solid heavy lines. A set of masks for the various diffusion processes must then be made up for the circuit as it appears in Fig. 12.15b.

Now we shall proceed slowly through the first diffusion process to demonstrate the natural sequence of steps that must be followed through each diffusion process indicated in Fig. 12.12.

p-Type Silicon Wafer Preparation

After being sliced from the grown ingot, a *p*-type silicon wafer is lapped, polished, and cleaned (Fig. 12.16a) to produce the structure of Fig. 12.16b. A chemical etch process is also applied to further smooth the surface and remove a layer of the wafer that may have been damaged during the lapping and polishing sequence.

n-Type Epitaxial Region

An *n*-type epitaxial region is then diffused into the *p*-type substrate as shown in Fig. 12.17. It is deposited in a way that will result in a single-crystal structure having the same crystal structure and orientation as the substrate but with a different conductivity level. It is in this thin epitaxial layer that the active and passive elements will be diffused. The *p*-type area is essentially a supporting structure that adds some thickness to the structure to increase its strength and permit easier handling.

The apparatus employed in the deposition process includes a long quartz tube surrounded by a radio-frequency (RF) induction coil. The wafers are placed on a rectangular graphite structure called a *boat* and are inserted into the chamber as indicated in Fig. 12.18. The RF coils then heat the chamber to above 1100°C, while the impurities are introduced in a gaseous state. The entire process can be carefully monitored and controlled to ensure the proper epitaxial growth.

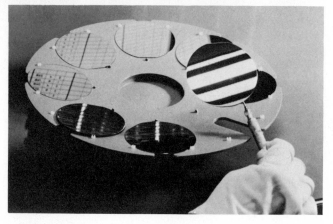

(a)

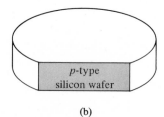

(b)

Figure 12.16 (a) Lapping and polishing stage of wafer preparation. (b) *p*-type silicon wafer. [(a) courtesy Motorola, Inc.; (b) courtesy Texas Instruments, Inc.]

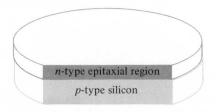

Figure 12.17 *p*-type silicon wafer after the *n*-type epitaxial diffusion process.

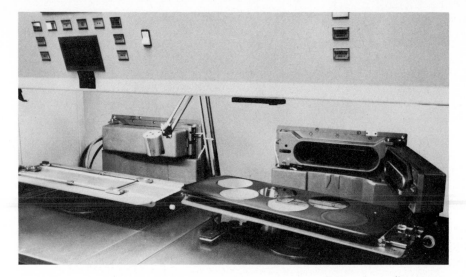

Figure 12.18 Deposition of the epitaxial layer. (Courtesy Motorola, Inc.)

Silicon Oxidation (SiO₂)

The resulting wafer is then subjected to an oxidation process resulting in a surface layer of SiO_2 (silicon dioxide) as shown in Fig. 12.19. This surface layer will prevent any impurities from entering the n-type epitaxial layer. However, selective etching of this layer will permit the diffusion of the proper impurity into designated areas of the n-type epitaxial region of the silicon wafer.

The apparatus employed in the oxidation process is similar to that used to establish the epitaxial layer in that the wafers are placed in a boat (now made of quartz) and inserted into a quartz tube. Typically, about 20 wafers are introduced at the same time. In this case, however, a resistance heater is wound around the tube to raise the temperature to about 1100°C. Oxygen is introduced in the wet or dry form until the desired SiO_2 layer is established. Recent developments include raising the atmospheric pressure in the vessel to permit a measurable reduction in the processing temperature. For every 1 atm (atmosphere) increase in pressure, there is a 30°C reduction in required temperature. At 10 atm, the temperature can be reduced 300°C. Silicon is reasonably well behaved at 800°C but an entirely different animal at 1100°C. At lower processing temperatures there is also an improved quality of oxide, a reduction in introduced stresses, and reduction or elimination of a number of device design limitations.

The time involved for the oxidation process can extend from a few hours to as long as 24 hours depending on the thickness of the oxide and the quality desired.

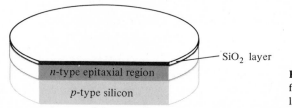

SiO₂ layer
n-type epitaxial region
p-type silicon

Figure 12.19 Wafer of Fig. 12.17 following the deposit of the SiO_2 layer.

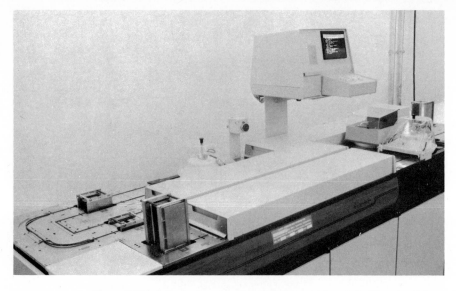

Photolithographic Process

The selective etching of the SiO$_2$ layer is accomplished through the use of a photolithographic process. A mask is first prepared on a glass plate as explained in Section 12.5. The first mask will determine those areas of the SiO$_2$ layer to be removed in preparation for the isolation diffusion process. The wafer is first coated with a thin layer of photosensitive material, commonly called *photoresist,* as demonstrated in Fig. 12.20.

The application of the photoresist is now entirely microprocessor controlled. A deck of wafers is deposited into the receiving trays appearing in the left-hand region

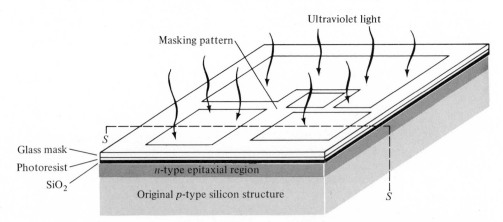

Figure 12.21   Photolithographic process: the application of ultraviolet light after the mask is properly set; the structure is only one of the 200, 400, or even 500 individual NAND gate circuits being formed on the wafer of Figs. 12.16 through 12.19.

of Fig. 12.20. The equipment automatically applies a high-pressure scrub, a dehydration process, the resist coating, and a soft bake; then it develops and hard bakes the wafers.

This new layer is then covered by the mask, and an ultraviolet light is applied that will expose those regions of the photosensitive material not covered by the masking pattern (Fig. 12.21). The resulting wafer is then subjected to a chemical solution that will remove the unexposed photosensitive material. A cross section of a chip (S-S) of Fig. 12.21 will then appear as indicated in Fig. 12.22. A second solution will then etch away the SiO_2 layer from any region not covered by the photoresist material (Fig. 12.23). The final step before the diffusion process is the removal, by solution, of the remaining photosensitive material. The structure will then appear as shown in Fig. 12.24.

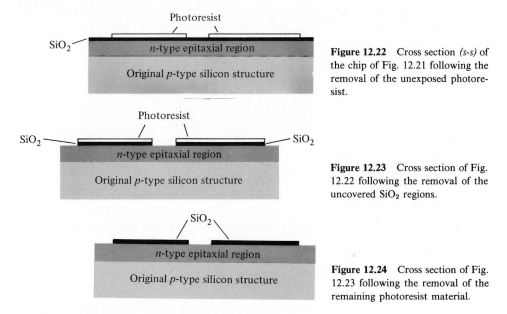

Figure 12.22 Cross section *(s-s)* of the chip of Fig. 12.21 following the removal of the unexposed photoresist.

Figure 12.23 Cross section of Fig. 12.22 following the removal of the uncovered SiO_2 regions.

Figure 12.24 Cross section of Fig. 12.23 following the removal of the remaining photoresist material.

Isolation Diffusion

The structure of Fig. 12.24 is then subjected to a *p*-type diffusion process resulting in the islands of *n*-type regions indicated in Fig. 12.25. The diffusion process ensures a heavily doped *p*-type region (indicated by p^+) between the *n*-type islands. The p^+ regions will result in improved *isolation* properties between the active and passive components to be formed in the *n*-type islands.

The apparatus employed includes a quartz boat and tube (to minimize the possibility of contaminating the processing environment) that is heated by a high-resistance wire wrapped around the tube. The diffusion operation normally occurs at a temperature neighboring on 1200°C. The system as appearing in Fig. 12.26 is totally microprocessor controlled (through cassette addressment). Three or four people can operate 16 furnaces; and the entire operation, from pulling the boats in and out of the furnaces to monitoring the temperature and dopant level, is computer controlled.

An alternative to the high temperature diffusion process is *ion implantation*. A

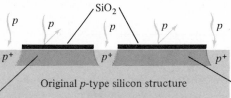

Figure 12.25 Cross section of Fig. 12.24 following the isolation diffusion process.

n-type epitaxial region

Original *p*-type silicon structure

n-type epitaxial region

(a)

(b)

Figure 12.26 Microprocessor controlled diffusion operation (with 5-in. wafers). (Courtesy Motorola, Inc.)

beam of dopant ions (about the size of a pencil) is directed toward a wafer at a very high velocity by an ion-accelerator gun. The ions will penetrate the medium to a level that can be controlled to within $\frac{1}{2}$ micron as compared to $2\frac{1}{2}$ microns using other methods. In addition to improved control, the processing temperature

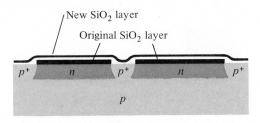

Figure 12.27 In preparation for the next diffusion process, the entire wafer is coated with an SiO$_2$ layer.

is lower, and a broader range of electrical parameters is now available. At present, the primary use of this method is in establishing bases. In time, with the proper modifications, emitter diffusions will also be a possibility.

In preparation for the next masking and diffusion process, the entire surface of the wafer is coated with an SiO$_2$ layer as indicated in Fig. 12.27.

Base and Emitter Diffusion Processes

The isolation diffusion process is followed by the base and emitter diffusion cycles. The sequence of steps in either case is the same as that encountered in the description of the isolation diffusion process. Although "base" and "emitter" refer specifically to the transistor structure, necessary parts (layers) of each element (resistor, capacitor, and diodes) will be formed during each diffusion process. The surface appearance of the NAND gate after the isolation base and emitter diffusion processes appears in Fig. 12.28. The mask employed in each process is also provided above each photograph.

Figure 12.28 The masks (a) employed in each diffusion process (b). The surface appearance of the monolithic NAND gate after the isolation, base, and emitter diffusion processes. (Courtesy Motorola Monitor.)

(a)

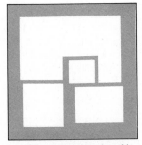
Isolation diffusion (mask)

Emitter diffusion (mask)

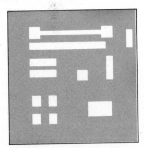
Base diffusion (mask)

(b)

Isolation diffusion

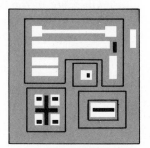
Emitter diffusion

Base diffusion

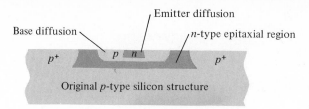

Base diffusion

Emitter diffusion

n-type epitaxial region

p^+ p n p^+

Original p-type silicon structure

Figure 12.29 Cross section of the transistor of Fig. 12.15 after the base and emitter diffusion cycles.

The cross section of the transistor of Fig. 12.15 will appear as shown in Fig. 12.29 after the base and emitter diffusion cycles.

Preohmic Etch

In preparation for a good ohmic contact, n^+ regions (see Section 12.4) are diffused into the structure as clearly indicated by the light areas of Fig. 12.30. Note the correspondence between the light areas and the mask pattern.

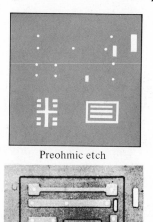

Preohmic etch

Preohmic etch

Figure 12.30 Surface appearance of the chip of Fig. 12.28 after the preohmic etch cycle. The mask employed is also included. (Courtesy Motorola Monitor.)

Metalization

A final masking pattern exposes those regions of each element to which a metallic contact must be made. The entire wafer is then coated with a thin layer of aluminum, gold, molybdenum, or tantalum (all high-conductivity, low-boiling-point metals) that after being properly etched will result in the desired interconnecting conduction pattern. A photograph of the completed metalization process appears in Fig. 12.31.

The two methods most commonly applied to establish the uniform layer of conducting material are *evaporation* and *sputtering*. In the former, the metal is either melted through the use of heating coils or bombarded with an electron-gun (E-gun) to result in the evaporation of the source metal. The metalization material is thereby sprayed over the wafers, which are held by clips in a drum or hemisphere structure such as is shown in Fig. 12.32.

Metalization

Figure 12.31 Completed metalization process. (Courtesy Motorola Monitor.)

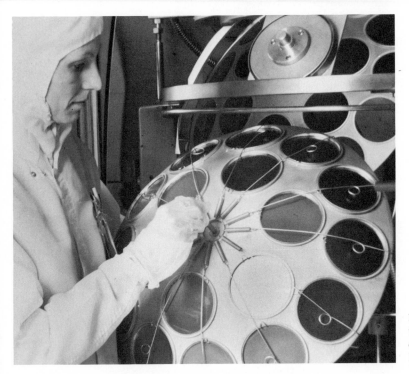

Figure 12.32 Deposition of the interconnect metal through metal evaporation. (Courtesy Motorola, Inc.)

A sputtering system such as that shown in Fig. 12.33 places the source metal (at a very high negative potential) opposite, but not touching, an anode plate at a positive potential. An inert gas, such as argon, introduced between the plates will release positive ions that will bombard the negative plate and release some of the surface metal from the source. The "free" metal will then be deposited on the wafers on the surface of the anode.

The sputtering technique is often preferred over the evaporation method because the coverage is less line-of-sight. There is therefore a more uniform layer of deposition over abrupt junctions. For the decade to come it would appear that the sputtering and evaporation methods will share the metalization role in the production cycle.

The complete structure with each element indicated appears in Fig. 12.34. Try to relate the interconnecting metallic pattern to the original circuit of Fig. 12.15a.

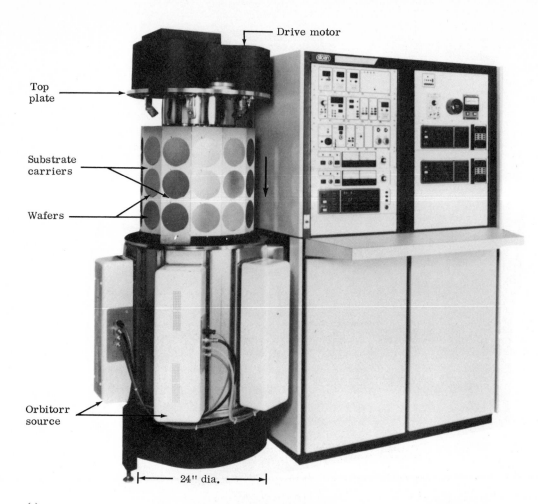

Drive motor

Top
plate

Substrate
carriers

Wafers

Orbitorr
source

←— 24" dia. —→

(a)

(b)

Figure 12.33 Deposition of the interconnect metal through the sputtering process. (Courtesy Sloan Technology Corporation.)

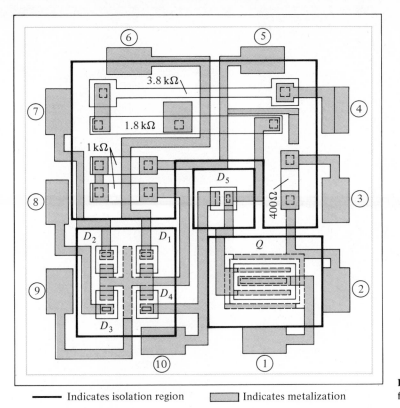
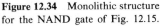

Figure 12.34 Monolithic structure for the NAND gate of Fig. 12.15.

—— Indicates isolation region   ▨ Indicates metalization

Passivation

An SiO$_2$ layer deposited on the surface of the entire structure will be an effective protection layer for water vapor and some contaminants. However, certain metal ions can migrate through the SiO$_2$ layer and disturb the device characteristics. In an effort to improve the passivation process, a layer (2000–5000 Å) of glass (plasma silicon nitride) is applied to further reduce the degradation problem.

Testing

Before breaking the wafer into the individual dies, an electrical test of each die is performed by the inspection system appearing in Fig. 12.35. The system automatically loads/unloads the wafers using carousels to further cut down on the degree of "handling." This process, as with most in the production cycle, is also microprocessor controlled. There is a probe card for each IC that will permit not only rejection but categorization of the type of faults (open, short, gain, etc.). The bad die are identified by a red dot automatically deposited by the inspection system.

Packaging

Once the metalization and testing processes are complete, the wafer must be broken down into its individual chips. This is accomplished through the scribing and breaking processes depicted in Fig. 12.36. Each individual chip may then be packaged in one of the forms shown in Fig. 12.37. The name of each is provided in the figure.

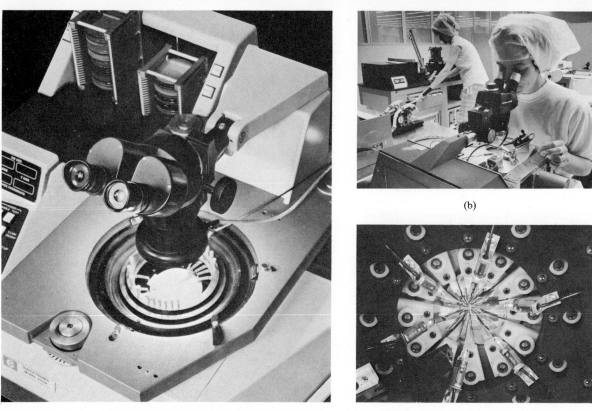

(a)

(b)

(c)

Figure 12.35 Electrical testing of the individual dies. [(a) courtesy Electroglas, Inc.; (b) courtesy Texas Instruments, Inc.; (c) courtesy Autonetics, North American Rockwell Corporation.]

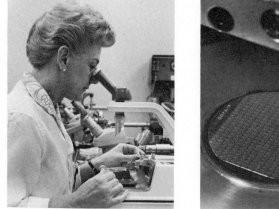

(a)

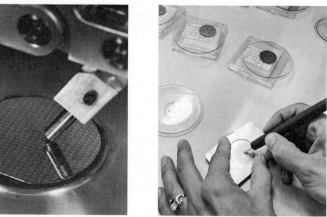

(b)

Figure 12.36 Scribing (a) and breaking (b) of the monolithic wafer into individual chips. (*Left,* courtesy Autonetics, North American Rockwell Corporation; *middle,* courtesy Texas Instruments, Inc.; *right,* courtesy Motorola, Inc.)

(a)

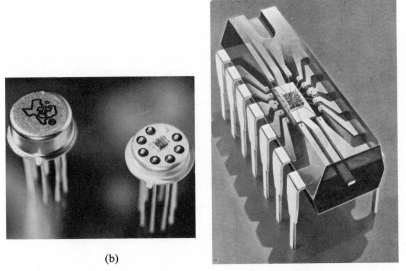

(b)

(c)

Figure 12.37 Monolithic packaging techniques: (a) flat package; (b) TO (top-hat)-type package; (c) dual in-line plastic package. (Courtesy Texas Instruments, Inc.)

12.7 THIN- AND THICK-FILM INTEGRATED CIRCUITS

The general characteristics, properties, and appearance of thin- and thick-film integrated circuits are similar although they both differ in many respects from the monolithic integrated circuit. They are not formed within a semiconductor wafer but *on* the surface of an insulating substrate such as glass or an appropriate ceramic material. In addition, *only* passive elements (resistors, capacitors) are formed through thin or thick film techniques on the insulating surface. The active elements (transistors, diodes) are added as *discrete* elements to the surface of the structure after the passive elements have been formed. The discrete active devices are frequently produced using the monolithic process.

The primary difference between the thin and thick film techniques is the process employed for forming the passive components and the metallic conduction pattern. The thin-film circuit employs an evaporation or cathode-sputtering technique; the thick film employs silk-screen techniques. Priorities do not permit a detailed description of these processes here.

In general, the passive components of film circuits can be formed with a broader range of values and reduced tolerances as compared to the monolithic IC. The use of discrete elements also increases the flexibility of design of film circuits although, obviously, the resulting circuit will be that much larger. The cost of film circuits with a larger number of elements is also, in general, considerably higher than that of monolithic integrated circuits.

12.8 HYBRID INTEGRATED CIRCUITS

The term *hybrid integrated circuit* is applied to the wide variety of multichip integrated circuits and also those formed by a combination of the film and monolithic IC techniques. The multichip integrated circuit employs either the monolithic or film technique to form the various components, or set of individual circuits, which are then interconnected on an insulating substrate and packaged in the same container. Integrated circuits of this type appear in Fig. 12.38. In a more sophisticated type of

Figure 12.38 Hybrid integrated circuits. (Courtesy Texas Instruments, Inc.)

hybrid integrated circuit the active devices are first formed within a semiconductor wafer which is subsequently covered with an insulating layer such as SiO_2. Film techniques are then employed to form the passive elements on the SiO_2 surface. Connections are made from the film to the monolithic structure through "windows" cut in the SiO_2 layer.

Linear ICs: Operational Amplifiers (Op-Amps)

13.1 BASIC DIFFERENTIAL AMPLIFIER

An amplifier is an electronic circuit containing BJT and FET devices, usually packaged in IC circuits, that provides voltage or current gain. It may also provide power gain, or allow impedance transformation. Since it is a basic part of practically every electronic application, the amplifier is an essential circuit. Amplifiers, as we have already discovered, may be classified in many ways. There are low-frequency amplifiers, audio amplifiers, ultrasonic amplifiers, radio-frequency (RF) amplifiers, wide-band amplifiers, video amplifiers, each type operating in a prescribed frequency range. We have considered small-signal and large-signal amplifiers and amplifiers that may be interconnected as either *RC*-coupled or transformer-coupled.

The *differential amplifier* is a special type of circuit that is used in a wide variety of applications. Let us first consider a number of basic properties of differential amplifiers. Figure 13.1 shows a block symbol of a differential amplifier unit. As shown, there are two separate input (1 and 2) and two separate output (3 and 4) terminals. We must first consider the relation between these terminals to obtain an understanding of how the differential amplifier may be applied. Notice that in Fig. 13.1 a ground connection is shown separately since both input or output terminals may be different from ground. Voltages may be applied to either or both input terminals and output voltages will appear at both output terminals. However, there are some very specific polarity relations between both input and both output terminals.

Figure 13.2 shows the block and circuit diagrams of a basic differential amplifier to be used in the following discussion. There are two inputs and two outputs shown in the block diagram. Inputs are applied essentially to each base of the two separate

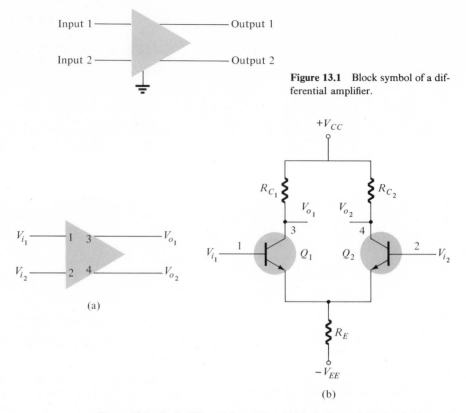

Figure 13.1 Block symbol of a differential amplifier.

(a)

(b)

Figure 13.2 Basic differential amplifier: (a) block diagram; (b) circuit diagram.

transistors. As shown, however, the transistor emitters are connected to a common-emitter resistor so that the two output terminals V_{o_1} and V_{o_2} are affected by either or both input signals. The outputs are taken from the collector terminals of each transistor. The input and output terminals are also numbered to facilitate reference. There are two supply voltages shown in the circuit diagram and it should be carefully noted that no ground terminal is indicated within the circuit although the opposite points of both positive- and negative-voltage supplies are understood to be connected to ground.

Single-Ended Input Differential Amplifier

Consider first the operation of the differential amplifier with a single input signal applied to terminal 1, with terminal 2 connected to ground (0 V). Figure 13.3 shows the block and circuit diagrams for input signal V_{i_1} at terminal 1 and output V_{o_1} at terminal 3. The block diagram shows a sinusoidal input and an amplified, inverted output. The circuit diagram shows the sinusoidal input applied to the base of a transistor with the amplified output at the collector inverted, as we would expect from past knowledge of a single-stage transistor amplifier.

With input 2 grounded it might seem that there is no output at terminal 4, but this is incorrect. The block diagram of Fig. 13.4 shows the operation of the differential

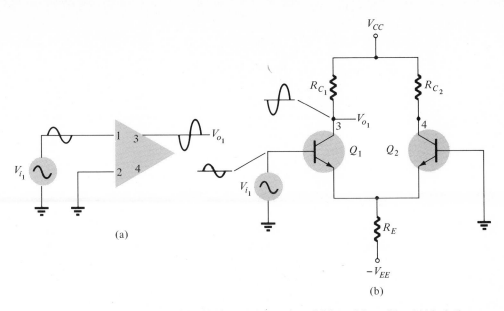

(a)

(b)

Figure 13.3 Single-ended operation of differential amplifier: (a) block diagram; (b) circuit diagram.

amplifier with the V_{o_2} output at terminal 4 resulting from an input V_{i_1} at terminal 1. The input at terminal 1, V_{i_1}, is a small sinusoidal voltage measured with respect to ground. Since the emitter resistor is connected in common with both emitters, a voltage due to V_{i_1} appears at the common-emitter point. This sinusoidal voltage, measured with respect to ground, is approximately one-half the magnitude and in phase with V_{i_1} because it results from emitter-follower action of the circuit.

The part of the circuit acting as an emitter follower is shown in Fig. 13.4c. An input applied to the base of Q_1 appears of same polarity and about one-half the

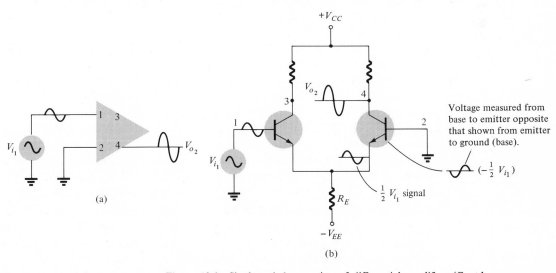

(a)

(b)

Figure 13.4 Single-ended operation of differential amplifier. (Contd. on p. 492.)

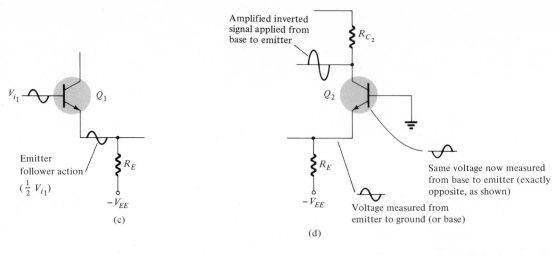

Emitter
follower action
($\frac{1}{2} V_{i_1}$)

R_E

$-V_{EE}$

(c)

Amplified inverted
signal applied from
base to emitter

R_{C_2}

Q_2

R_E

$-V_{EE}$

Voltage measured from
emitter to ground (or base)

Same voltage now measured
from base to emitter (exactly
opposite, as shown)

(d)

Figure 13.4 (continued)

magnitude at the emitter of Q_1 for the emitter-follower part of the circuit shown. Recall that for an emitter follower the gain is less than unity[1] (with no polarity reversal). This emitter signal is measured with respect to ground. Figure 13.4d shows the part of the circuit with the emitter voltage affecting the operation of transistor Q_2. The voltage at the emitter of Q_2 is the same as that of Q_1 (since the emitters are connected together) and appears from the emitter of Q_2 to ground or to the base of Q_2 (since that is connected to ground). If the voltage measured from the emitter to base of Q_2 is in phase with input V_{i_1} as shown, the voltage measured from base to emitter of Q_2 is the same signal with opposite polarity. Thus, by measuring from base to emitter of Q_2 a voltage of about one-half the magnitude of V_{i_1} is obtained, but the signal is opposite in polarity to that of V_{i_1}. The amplifier action of transistor Q_2 and load resistor R_{C_2} provides an output at the collector of Q_2 that is amplified and inverted from the signal developed across base to emitter of Q_2.

In summary, an input V_{i_1} is applied to input 1 and an amplified, in-phase signal V_{o_2} results at output terminal 4. Although the input at terminal 2 is grounded, an output still occurs at terminal 4. It should be clear that the internal common-emitter connection provides that the input at terminal 1 results in an output at terminal 4. In fact, we can now see that the input at terminal 1 causes output signals at both terminals 3 and 4. In addition, these outputs are opposite in polarity and of about the same magnitude. Finally, we should see (as in Fig. 13.5a) that the output at terminal 4 is the same polarity as the input at terminal 1, while the output at terminal 3 is opposite in polarity to the input at terminal 1. It follows from the previous discussion that an input applied to terminal 2 (with terminal 1 grounded) will result in output voltages as shown in Fig. 13.5b.

---

[1] Emitter-follower gain in this circuit connection is actually one-half since the load impedance equals the output impedance of the emitter follower.

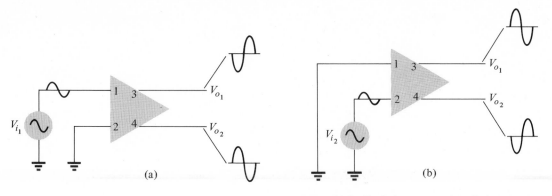

Figure 13.5 Single-input, opposite-polarity outputs.

Differential (Double-Ended) Input Operation

In addition to using only one input to operate the differential-amplifier circuitry, it is possible to apply signals to each input terminal, with opposite polarity outputs appearing at the two output terminals. The usual use of the *double-ended* or *differential* mode of input is when the two input signals are themselves opposite in polarity and about the same magnitude. Figure 13.6 shows such a situation.

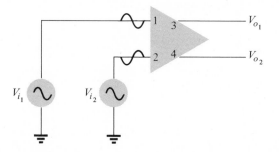

Figure 13.6 Operation with differential input signals.

We now must consider how each input affects the outputs and what the resulting output signal looks like. This can be done using the *superposition principle,* considering each input applied separately with the other at 0 V and summing the resulting output voltages at each terminal. Figures 13.7a and 13.7b show the result of each input acting alone and Fig. 13.7c shows the resulting overall operation. The input applied to terminal 1 results in an opposite-polarity amplified output at terminal 3 *and* a same-polarity amplified output at terminal 4. Assume that the inputs are about equal in magnitude and that the output magnitudes are about equal, of value V, for discussion purposes.

The input applied to terminal 2 results in an opposite-polarity amplified output at terminal 4 *and* a same polarity amplified output at terminal 3. The magnitudes of the outputs will both be V since the input magnitudes were assumed to be about

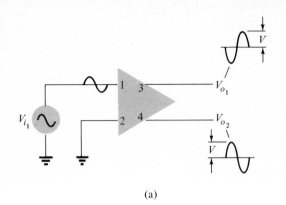

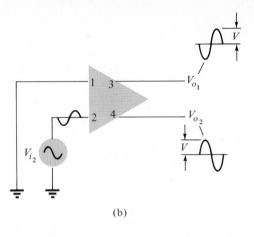

(a)

(b)

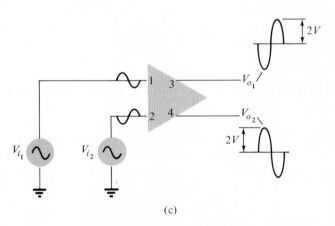

(c)

Figure 13.7 Differential operation of amplifier: (a) $V_{i_2} = 0$; (b) $V_{i_1} = 0$; (c) both inputs present.

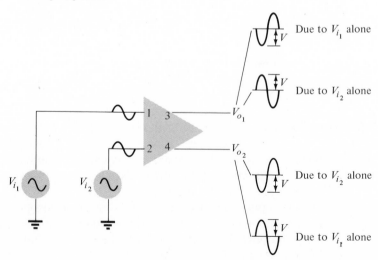

Figure 13.8 Operation with in-phase input signals.

the same. It is important to note that the outputs in each case are the same polarity at each output terminal. By superposition, the resulting signals at each output terminal are added, and we obtain the full operation of the circuit shown in Fig. 13.7c. The output at each terminal is twice that resulting from single-ended operation because the outputs due to each input have the same polarity. If the inputs applied were both the same polarity (or if the same input were applied to both input terminals), the resulting signals due to each input acting alone would be opposite in polarity at each output and the resulting output would *ideally* be about 0 V, as shown in Fig. 13.8.

To bring the operation as single- and double-ended differential-amplifier states into full perspective consider the connection of two differential amplifiers shown in Fig. 13.9. From the previous discussion, if the amplifiers had identical single-ended gains, then the outputs of stage 1 would be larger than the inputs by the amount of amplifier gain, while the outputs of stage 2 would be larger than the inputs to stage 2 by twice the amplifier gain. The initial signal, from an antenna of a radio, or a phonograph pickup cartridge, etc., is single-ended and is used as such. The second differential-amplifier stage, however, could be operated double-ended to obtain twice the stage gain. Either output of stage 2 (or both) could then be used as amplified signals to the next section of the system. Although differential operation requires about equal and opposite polarity signals, this is often available, especially after one single-ended stage of gain.

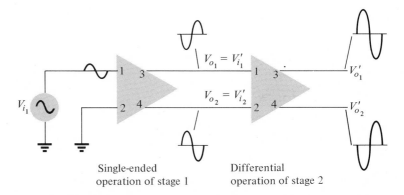

Figure 13.9 Single- and double-ended operation of differential-amplifier stages.

13.2 DIFFERENTIAL-AMPLIFIER CIRCUITS

Having considered some basic features of a differential amplifier, we shall now look into some details of differential-amplifier circuits. In particular, we shall consider the voltage gain of the stage and its input and output impedance. We will first present discrete-circuit versions of a difference amplifier to allow introducing circuit concepts. Then IC versions of various parts of the difference amplifier are presented, these being the typical circuits used in building IC op-amps. A basic circuit of a discrete differential amplifier is shown in Fig. 13.10. Input signals are shown as a voltage source with source resistance in the general case.

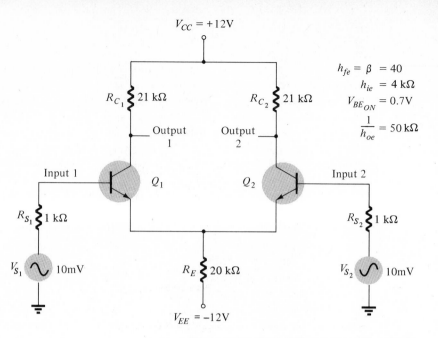

Figure 13.10 Basic differential-amplifier circuit.

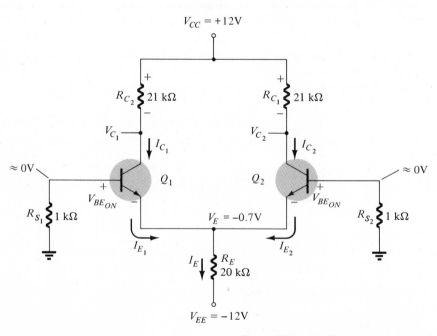

Figure 13.11 Dc bias action of circuit.

dc Bias Action of Circuit

Before considering the operation of the circuit as a voltage amplifier, let us see how the circuit is biased. Figure 13.11 shows the main dc voltage and current values of the circuit. Ac signal sources have been set to 0 V with only the source resistances

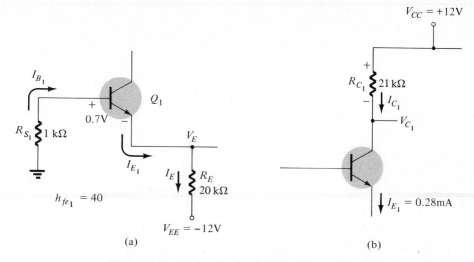

Figure 13.12 Partial circuits of differential amplifier: (a) input section; (b) output section.

present. The base-emitter of Q_1 is forward biased by the V_{EE} battery from ground through resistor R_{S_1}, through the base emitter, through resistor R_E to $-V_{EE}$ (see Fig. 13.12a). We would have to write a number of equations to solve for the dc voltages and currents. However, it is possible to use good approximations to make the calculations more direct. For example, the dc voltage drop across source resistor R_{S_1} will be small as the following calculation indicates (assuming a typical base current in the order of microamperes)

$$I_{B_1} R_{S_1} = (100\ \mu A)(1\ k\Omega) = 100\ mV = 0.1\ V$$

If the base current were only 10 μA, then the dc voltage drop across R_{S_1} would be 10 mV, which is negligible. On the other hand, a source resistance of 10 kΩ with base current of 100 μA would result in a voltage drop of 1 V, which is not negligible. For our purposes we shall assume the voltage drop to be small (which is often correct) and check later in the calculations to be sure we were able to make such an assumption.

If we assume that[2]

$$V_{B_1} \cong 0\ V$$

then the emitter voltage is

$$V_E = V_{E_1} = V_{B_1} - V_{BE_1} \tag{13.1}$$

$$= 0 - 0.7\ V = -0.7\ V$$

The current through resistor R_E is then

$$I_E = \frac{V_E - V_{EE}}{R_E} \tag{13.2}$$

$$= \frac{-0.7\ V - (-12\ V)}{20\ k\Omega} = \frac{11.3\ V}{20\ k\Omega} = 0.565\ mA$$

---

[2] This assumes that $(\beta + 1)\ 2R_2 \gg R_{S_1}$, usually a very good assumption.

The current through resistor R_E is made up of the emitter currents from each transistor. If the transistors are matched, the emitter current of each transistor is one-half the total current through R_E.

$$I_{E_1} = I_{E_2} = \frac{I_E}{2} \tag{13.3}$$

$$= \frac{0.565 \text{ mA}}{2} = 0.2825 \text{ mA} \cong 0.28 \text{ mA}$$

We can now check our assumption of V_{B_1} by calculating I_{B_1} as follows:

$$I_{B_1} = \frac{I_{E_1}}{1 + h_{fe_1}} \tag{13.4}$$

$$= \frac{0.28 \text{ mA}}{1 + 40} = 6.8 \ \mu\text{A}$$

$$V_{B1} = I_{B_1} R_{S_1} = 6.8 \ \mu\text{A} \times 1 \text{ k}\Omega = 6.8 \text{ mV}$$

which is negligible compared to the other voltage drops in the circuit. Figure 13.12b shows a partial circuit diagram of the output section of the circuit. The collector current is obtained from the calculation of emitter current.

$$I_{C_1} \cong I_{E_1} \tag{13.5}$$

$$= 0.28 \text{ mA}$$

and the collector voltage is

$$V_{C_1} = V_{CC} - I_{C_1} R_{C_1} \tag{13.6}$$
$$= 12 - (0.28 \text{ mA})(21 \text{ k}\Omega) = 6.12 \text{ V}$$

ac Operation of Differential-Amplifier Circuit

To consider the ac operation of the circuit all dc voltage supplies are set at zero and the transistors are replaced by small-signal ac equivalent circuits. Figure 13.13

Figure 13.13 Ac equivalent circuit of differential amplifier.

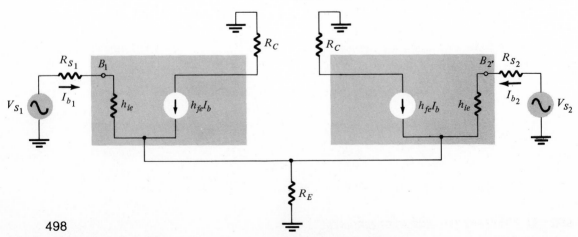

shows the resulting ac equivalent circuit, with the transistors replaced by hybrid equivalent circuits. The circuit obviously appears complex, and analyzing the total circuit would become involved. Again, we can break up the calculations by using some simplifying approximations so that smaller parts of the circuit can be analyzed separately. We can assume that

$$h_{ie_1} = h_{ie_2} = h_{ie}, \qquad h_{fe_1} = h_{fe_2} = h_{fe}, \; h_{oe_1} = h_{oe_2} \cong 0$$

and $\qquad R_{C_1} = R_{C_2} = R_C, \qquad R_{S_1} = R_{S_2} = R_S$

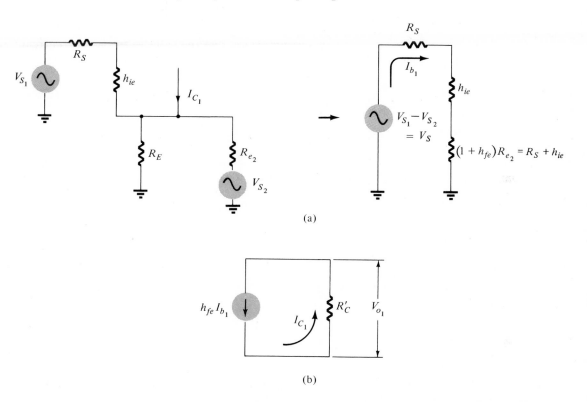

(a)

(b)

Figure 13.14 Partial ac equivalent circuit of differential amplifier.

INPUT AC SECTION

Figure 13.14a shows the partial ac equivalent circuit of the input for transistor Q_1. Looking into the emitter of transistor Q_2 a small ac equivalent resistance is present, equal in value to

$$R_{e_2} = \frac{R_S + h_{ie}}{1 + h_{fe}} \tag{13.7}$$

For the values of Fig. 13.10

$$R_{e_2} = \frac{1 \text{ k}\Omega + 4 \text{ k}\Omega}{1 + 40} = 122 \; \Omega$$

The parallel combination of resistors R_E and R_{E_2} gives an equivalent ac resistance of

$$R_E \parallel R_{e_2} = \frac{R_{e_2} R_E}{R_{e_2} + R_E} = \frac{122 \times 20{,}000}{122 + 20{,}000} \cong 121\ \Omega$$

Since the differential-amplifier circuit generally has an R_E of large value, we can make the approximate statement that if

$$R_E \gg R_{e_2}$$

the parallel combination is approximately R_{e_2} in value, as shown in Fig. 13.14a. Using the resulting ac equivalent circuit, we see that the value of the ac base current is calculated to be

$$I_{b_1} = \frac{V_{S_1} - V_{S_2}}{R_S + h_{ie} + (1 + h_{fe})R_{e_2}} = \frac{V_{S_1} - V_{S_2}}{2\,(R_S + h_{ie})} \tag{13.8a}$$

and defining $V_d \equiv V_{S_1} - V_{S_2}$ as the difference input voltage

$$I_{b_1} = \frac{V_d}{2\,(R_S + h_{ie})} \tag{13.8b}$$

OUTPUT AC SECTION

The output voltage can be written as

$$V_{o_1} = -I_{c_1} R_{C_1}$$

Using $I_{c_1} = h_{fe} I_{b_1}$ and I_{b_1} as expressed in Eq. (13.8b) results in

$$V_{o_1} = -\frac{h_{fe} R_{C_1}}{2\,(R_S + h_{ie})} V_d$$

The circuit ac difference gain is then

$$\boxed{A_{v_1} = \frac{V_{o_1}}{V_d} = -\frac{h_{fe} R_{C_1}}{2\,(R_S + h_{ie})} = -\frac{\beta R_{C_1}}{2\,(R_S + \beta r_e)}} \tag{13.9}$$

Using the values of the circuit of Fig. 13.10 results in

$$A_{v_1} \cong -\frac{40 \times 21\ k\Omega}{2(1\ k\Omega + 4\ k\Omega)} = -84$$

INPUT RESISTANCE

From the ac equivalent circuit of Fig. 13.14a the input resistance of the circuit seen from the source is

$$R_{i_1} = h_{ie} + (1 + h_{fe})R_{e_2} \tag{13.10a}$$

which can be expressed as

$$R_{i_1} = 2h_{ie} + R_S = 2\beta r_e + R_S \qquad (13.10b)$$

For the circuit of Fig. 13.10

$$R_{i_1} = 2(4\ k\Omega) + 1\ k\Omega = 9\ k\Omega$$

OUTPUT RESISTANCE

From the ac equivalent circuit of Fig. 13.14b the resulting approximate output resistance (assuming $h_{oe} \cong 0$) is

$$R_{o_1} = R_C \qquad (13.11)$$

which is 21 kΩ.

> **EXAMPLE 13.1** Calculate the input and output resistance of a differential-amplifier circuit as in Fig. 13.10 for the following circuit values: $R_{C_1} = R_{C_2} = 15\ k\Omega$, $R_E = 10\ k\Omega$, $h_{fe} = 60$, $h_{ie} = 2.5\ k\Omega$, and $R_{S_1} = R_{S_2} = 600\ \Omega$.
>
> **Solution:**
> $$\begin{aligned} R_{i_1} &= 2h_{ie} + R_S \\ &= 2(2.5\ k\Omega) + 0.6\ k\Omega \\ &= 5.6\ k\Omega \\ R_{o_1} &= R_C = 15\ k\Omega \end{aligned}$$

Differential-Amplifier Circuit with Constant-Current Source

One important thing to note in the previous circuit considerations was that with $R_{e_2} \ll R_E$, the value of R_E was very large and therefore negligible. In fact, the larger the value of R_E, the better certain desirable aspects of a differential-amplifier circuit. The main reason for R_E being very large is a circuit factor called *common-mode rejection*, which will be discussed in detail in Section 13.3.

However, dc bias calculations showed that the emitter (and thus the collector) current is determined partly by the value of R_E. For a fixed negative-voltage supply of, say, $V_{EE} = -20$ V a value of R_E of 10 kΩ would limit the emitter-resistor current to about

$$I_E \cong \frac{V_{EE}}{R_E} = \frac{20\ V}{10\ k\Omega} = 2\ mA$$

If a preferably larger value of $R_E = 100$ kΩ were used, the value of dc emitter-resistor current would then be

$$I_E \cong \frac{V_{EE}}{R_E} = \frac{20\ V}{100\ k\Omega} = 0.2\ mA = 200\ \mu A$$

and if a very large value of $R_E = 1$ MΩ were used,

$$I_E = \frac{20 \text{ V}}{1 \text{ M}\Omega} = 20 \text{ }\mu\text{A}$$

We see that as larger values of R_E are used the resulting dc emitter current becomes much too small for proper operation of the transistors since the emitter and collector current of each transistor is one-half the already very small emitter current.

One way to achieve high ac resistance while still allowing reasonable dc emitter currents is to use a constant-current source as shown in Fig. 13.15. The value of I_E could be set by the constant-current circuit to any desired value—1, 10, 20 mA, and so on. The ac resistance of a constant-current source is ideally infinite and practically from 100 kΩ to about 1 MΩ.

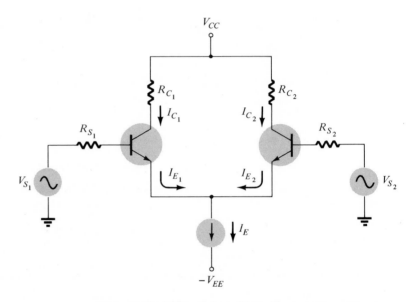

Figure 13.15 Differential amplifier with constant-current source.

A practical differential-amplifier circuit containing a constant-current source is shown in Fig. 13.16. To determine the dc currents and voltages let us first consider the details of the constant-current circuit.

DC OPERATION

The constant-current section of the difference amplifier is shown in Fig. 13.17a. No connection to the collector is shown since the collector current is determined by the value of emitter current set by the base-emitter section of the circuit. To a great degree the amount of collector current can be set at any desired value without regard to the circuit connected to the collector.

To simplify the calculation of I_E we note that if the resistance looking into the transistor base is much larger than R_2, the base voltage can be calculated by the voltage divider of R_1 and R_2. That is, if

$$(1 + h_{fe})R_E \gg R_2$$

CH. 13 LINEAR ICs: OPERATIONAL AMPLIFIERS (OP-AMPs)

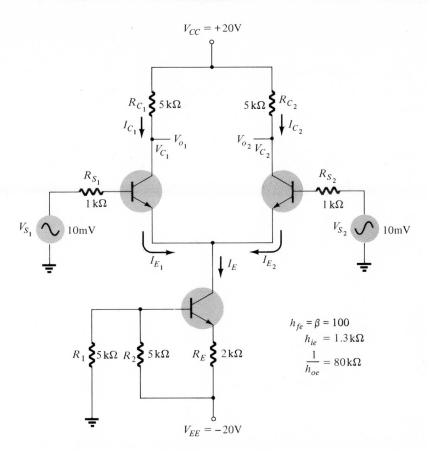

Figure 13.16 Practical differential-amplifier circuit with constant-current source.

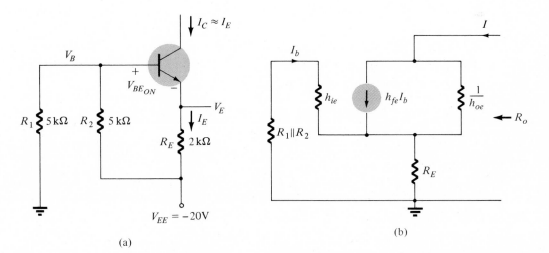

(a)

(b)

Figure 13.17 Constant-current circuit: (a) dc considerations; (b) ac considerations.

then
$$V_B = \frac{R_1}{R_1 + R_2}(-V_{EE}) \tag{13.12}$$

For the present circuit
$$(1 + 100)(2 \text{ k}\Omega) = 202 \text{ k}\Omega \gg 5 \text{ k}\Omega$$

so that
$$V_B = \frac{5 \text{ k}\Omega}{5 \text{ k}\Omega + 5 \text{ k}\Omega}(-20) = -10 \text{ V}$$

The emitter voltage is less than the base voltage by the voltage drop $V_{BE_{ON}}$
$$V_E = V_B - V_{BE_{ON}} \tag{13.13}$$
$$= -10 - (0.7) = -10.7 \text{ V}$$

The emitter current is then calculated to be
$$I_E = \frac{V_E - V_{EE}}{R_E} \tag{13.14}$$
$$= \frac{-10.7 - (-20) \text{ V}}{2 \text{ k}\Omega} = 4.65 \text{ mA}$$

Once I_E is determined, the remaining dc bias calculations are the same as those previously considered.

The emitter current of each transistor is then
$$I_{E_1} = I_{E_2} = \frac{I_E}{2} \tag{13.15}$$
$$= \frac{4.65}{2} \text{ mA} \cong 2.325 \text{ mA}$$

and
$$I_{C_1} = I_{C_2} \cong I_{E_1} = I_{E_2} \tag{13.16}$$
$$= 2.325 \text{ mA}$$

The collector voltage is, as before,
$$V_C = V_{CC} - I_C R_C$$
$$= 20 - (2.325 \text{ mA})(5 \text{ k}\Omega) = 8.375 \text{ V}$$

AC OPERATION

The ac action of the constant-current source is that of a very high resistance—ideally infinite. An ac equivalent of the constant-current circuit is shown in Fig. 13.17b. An expression for the circuit ac output impedance can be derived from the equivalent circuit

$$R_o \cong \frac{1}{h_{oe}}\left[1 + \frac{h_{fe}R_E}{R_E + h_{ie} + R_1 \| R_2}\right] \tag{13.18}$$

For the circuit of Fig. 13.16 the calculation of R_o is

$$R_o = 80 \text{ k}\Omega \left[1 + \frac{100(2 \text{ k}\Omega)}{2 \text{ k}\Omega + 1.3 \text{ k}\Omega + 2.5 \text{ k}\Omega} \right] = 2.84 \text{ M}\Omega$$

Thus, the circuit provides a high ac impedance of 2.84 MΩ while providing a dc bias current of 4.65 mA.

IMPROVED CONSTANT-CURRENT CIRCUIT

An improved version of the constant-current circuit is shown in Fig. 13.18. A Zener diode is used to help maintain the current constant. The Zener diode conducts when the reverse-bias voltage exceeds the Zener breakdown voltage, V_Z. The Zener diode will then conduct keeping the voltage across the diode fixed at V_Z, for a wide range of current values. In the circuit of Fig. 13.18, the emitter current is calculated to be

$$+V_Z - V_{BE_{ON}} - I_E R_E = 0$$

$$I_E = \frac{V_Z - V_{BE_{ON}}}{R_E} \qquad (13.19)$$

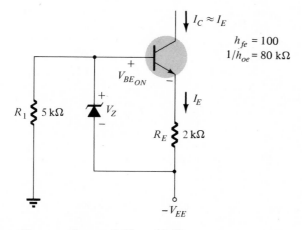

Figure 13.18 Constant-current circuit using Zener diode.

For a Zener voltage of $V_Z = 10$ V

$$I_E = \frac{V_Z - V_{BE}}{R_E} = \frac{10 - 0.7 \text{ V}}{2 \text{ k}\Omega} = 4.65 \text{ mA}$$

The ac output impedance for Fig. 13.18 can then be calculated as

$$R_o = \frac{1}{h_{oe}}(h_{fe} + 1)$$

which is $R_o = 80 \text{ k}\Omega(100 + 1) = 8.08 \text{ M}\Omega$

Another circuit used to provide a constant current is shown in Fig. 13.19. Dc bias calculations to obtain the emitter current, I_E, result in

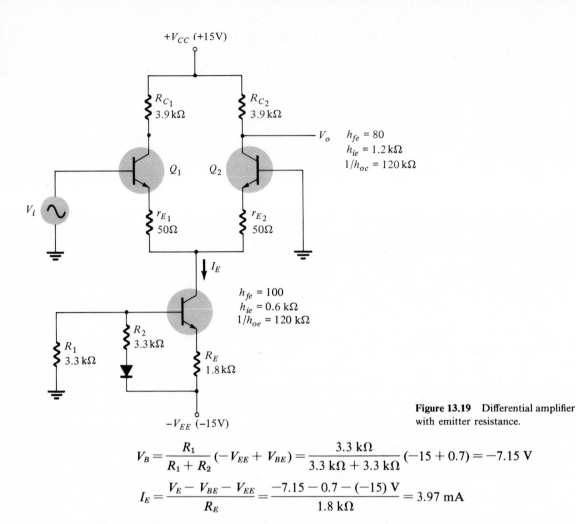

$h_{fe} = 80$
$h_{ie} = 1.2\,k\Omega$
$1/h_{oc} = 120\,k\Omega$

$h_{fe} = 100$
$h_{ie} = 0.6\,k\Omega$
$1/h_{oe} = 120\,k\Omega$

Figure 13.19 Differential amplifier with emitter resistance.

$$V_B = \frac{R_1}{R_1 + R_2}(-V_{EE} + V_{BE}) = \frac{3.3\,k\Omega}{3.3\,k\Omega + 3.3\,k\Omega}(-15 + 0.7) = -7.15\,V$$

$$I_E = \frac{V_E - V_{BE} - V_{EE}}{R_E} = \frac{-7.15 - 0.7 - (-15)\,V}{1.8\,k\Omega} = 3.97\,mA$$

The dc collector currents are then

$$I_C = \frac{I_E}{2} = \frac{3.97\,mA}{2} = 1.985\,mA$$

so that $V_C = V_{CC} - I_C R_C = 15\,V - 1.985\,mA\,(3.9\,k\Omega) = 7.259\,V$

The ac voltage gain is then (neglecting $1/h_{oe}$)

$$A_v = \frac{V_o}{V_i} = \frac{h_{fe}R_C}{2\,(h_{ie} + h_{fe}r_E)} = \frac{80(3.9\,k\Omega)}{2[1.2\,k\Omega + 80(50)]} = 30$$

where no polarity inversion takes place between the output taken from Q_2 with input applied to Q_1. Although the gain is reduced by the inclusion of resistors r_E in the circuit, the gain is more stable. In addition, the input impedance is increased because of R_E

$$R_i = 2\,(h_{ie} + h_{fe}r_E) = 2[0.6\,k\Omega + 100(50)] = 11.2\,k\Omega$$

The ac output impedance can be obtained from Eq. (13.18)

$$R_o = 120 \text{ k}\Omega \left(1 + \frac{100(1.8 \text{ k}\Omega)}{1.8 \text{ k}\Omega + 0.6 \text{ k}\Omega + 1.65 \text{ k}\Omega}\right) = 5.45 \text{ M}\Omega$$

IC Circuit Techniques

To construct the various parts of an op-amp on a single IC chip requires the use of a number of circuit techniques that achieve the desired function using mostly transistor components and keeping the number of resistors and their values low. The number of capacitors is also kept low as is their value. The following basic circuit parts are covered to aid in better understanding the overall IC circuit.

IC Voltage Sources (Bias Circuits):   When a reference or dc bias voltage is required in an integrated circuit, connections such as those of Fig. 13.20 can be used.

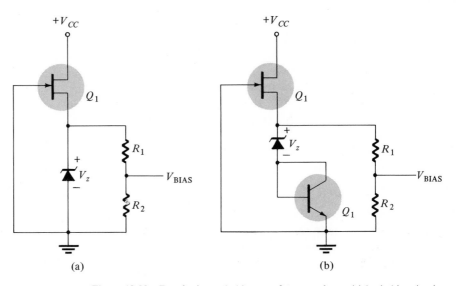

Figure 13.20   Developing a dc bias or reference voltage: (a) basic bias circuit; (b) temperature-compensated bias circuit.

Figure 13.20a shows a JFET used to provide a fixed bias current through a zener diode, which provides a fixed voltage. The resistor voltage divider then drops the zener diode voltage to a desired bias voltage value, V_{BIAS}; this bias voltage is available for use in other parts of the op-amp circuit. Figure 13.20b adds a bipolar transistor in series with the zener diode to provide temperature compensation to maintain the bias voltage over a range of temperature. The bipolar transistor base-emitter voltage drop varies with temperature opposite to that of the zener diode so that the bias voltage is maintained even as temperature varies. The tracking of the zener diode and bipolar transistor is quite good as both are formed near each other on the same IC chip.

IC Current Sources: The most popular IC circuit configuration is presently the current source. Figure 13.21 shows a few basic forms for constructing IC current sources. For low current values, the *current mirror* shown in Fig. 13.21a is quite

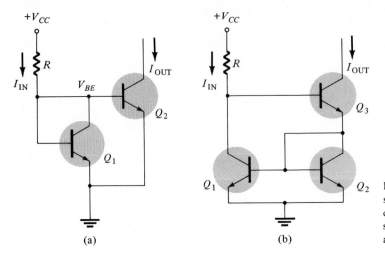

Figure 13.21 IC constant-current sources: (a) current mirror for low current values; (b) constant-current source with higher output impedance.

popular. Transistor Q_1 is a diode-connected transistor providing temperature compensation for transistor Q_2 in establishing a constant output current. If the two transistors are matched (typically the case when devices are formed on the same chip in very close proximity), then the output current will remain at the constant value set by supply V_{CC} and resistor R, regardless of the circuit connected to the current source, I_{OUT}. The circuit of Fig. 13.21b provides an additional transistor, Q_3, in series with the output to provide a much higher value of output impedance from the current source. The larger the value of the current-source impedance, the more ideal is the circuit's operation.

Differential Amplifier Stage with Constant-Current Load: To help appreciate how differential amplifier stages are built in ICs, Fig. 13.22a shows a basic stage with constant-current source, and Fig. 13.22b shows the load resistors replaced by *pnp* current-source loads for larger effective values of R_C and therefore larger stage voltage gain. When even larger gain is desired, the improved circuit of Fig. 13.22c can be used. The additional resistors and transistors provide larger effective load impedance and larger voltage gain for the single stage.

Level Shifting: To interface between input and output or couple stages together without the limitations imposed by using capacitor coupling, it is necessary to use a voltage-level-shifting circuit. Figure 13.23 shows a few typical circuits to provide a shift in dc level between input and output. In Fig. 13.23a, the output voltage follows the input voltage except for the lower dc voltage level of the output set by resistor R_1 and the current set through transistor Q_2. In the circuit of Fig. 13.23b, the voltage drop between input and output is essentially set by the zener diode voltage (and the value of transistor Q_1 base-emitter voltage drop).

Output Stage: After the inputs are amplified to the desired value of output voltage, an output stage is used to supply a signal capable of driving the load. Figure 13.24 shows a few output stage circuits. Figure 13.24a is simply a conventional emitter follower; Fig. 13.24b replaces R_E with a current source to provide large value of R_E using a small die area of the IC chip. The output circuit of Fig. 13.24c contains a driver with diode biasing that provides both current sourcing and sinking of the output. Figure 13.24d then shows a full-range output driver using transistor biasing.

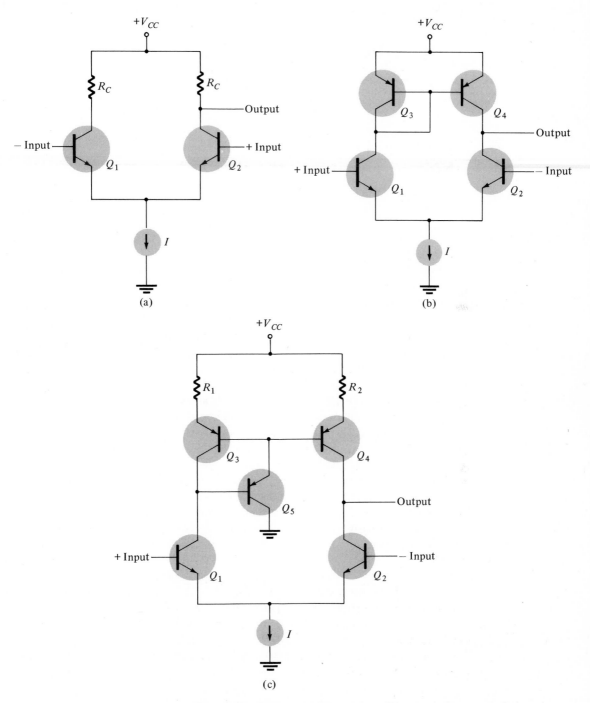

Figure 13.22 IC form of differential-amplifier stage using symmetrical constant-current sources.

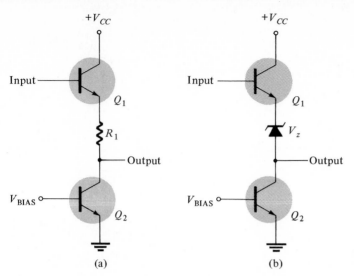

Figure 13.23 Level-shifting circuits: (a) resistive level shifting; (b) Zener-diode level shifting.

Finally, Fig. 13.24e shows the circuit of Fig. 13.24c modified by the addition of short-circuit protection for the output.

13.3 COMMON-MODE REJECTION

One of the more important features of a difference amplifier is its ability to cancel out or reject certain types of unwanted voltage signals. These unwanted signals are referred to as "noise" and can occur as voltages induced by stray magnetic fields in the ground or signal wires, as voltage variations in the voltage supply. What is important in this consideration is that these noise signals are not the signals that are desired to be amplified in the difference amplifier. Their distinguishing feature is that the noise signal appears *equally* at both inputs of the circuit.

We can say then that any unwanted (noise) signals that appear in polarity, or common to both input terminals, will be greatly rejected (cancelled out) at the output of the difference amplifier. The signal that is to be amplified appears at only one input or opposite in polarity at both inputs. What we wish to consider in this section is, if undesirable noise does occur, how much does the amplifier reject or cancel out this noise? A measure of this rejection of signals common to both inputs is called the amplifier's *common-mode rejection* and a numerical value is assigned, which is called the *common-mode rejection ratio* (CMRR).

Figure 13.25a shows an amplifier with two input signals. These signals can, in general, be considered to contain components that are exactly opposite in polarity *and* components that are the same polarity. For ideal operation we would want the differential amplifier to provide high gain for the opposite-polarity components of the signals and zero gain for the in-polarity components of the signals.

The voltage measured *from* terminal 1 *to* terminal 2 can be considered as a differential voltage

$$V_d = V_{i_1} - V_{i_2}$$

(13.20)

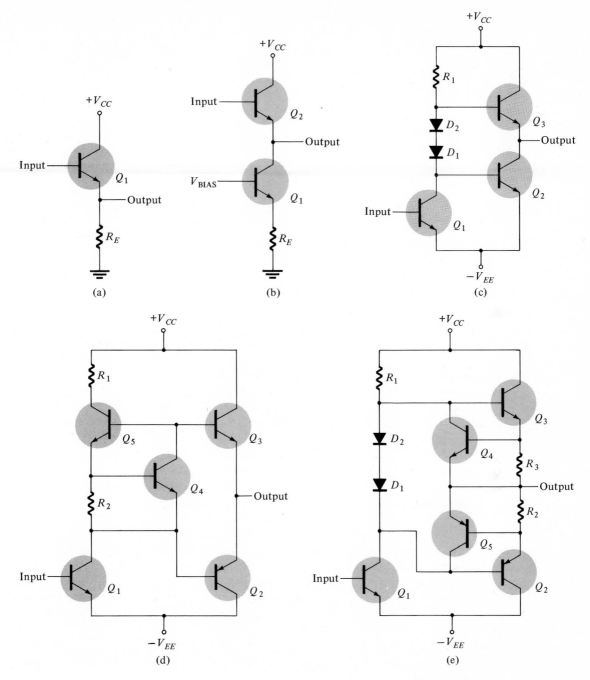

Figure 13.24 Output stages.

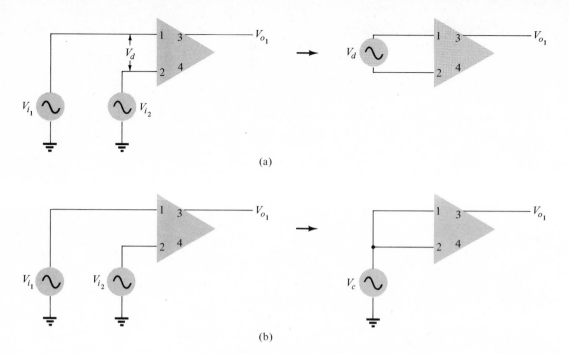

(a)

(b)

Figure 13.25 Differential and common-mode operation: (a) ideal differential-mode operation; (b) ideal common-mode operation.

If, as in the ideal case, $V_{i_1} = -V_{i_2}$, we note that

$$V_d = V_{i_1} - (-V_{i_1}) = 2V_{i_1} = -2V_{i_2}$$

In general, there may also exist common components to the input signals. We can define a common input as

$$V_c = \tfrac{1}{2}(V_{i_1} + V_{i_2}) \tag{13.21}$$

The ideal case (shown in Fig. 13.25b) is $V_{i_1} = V_{i_2}$, for which

$$V_c = \tfrac{1}{2}(V_{i_1} + V_{i_2}) = V_{i_1} = V_{i_2}$$

From Eqs. (13.20) and (13.21) we can obtain expressions for V_{i_1} and V_{i_2} based on V_c and V_d

$$V_{i_1} = V_c + \frac{V_d}{2} \tag{13.22a}$$

$$V_{i_2} = V_c - \frac{V_d}{2} \tag{13.22b}$$

The output voltages can then be expressed as

$$V_{o_1} = A_1 V_{i_1} + A_2 V_{i_2} \tag{13.23a}$$

$$V_{o_2} = A_2 V_{i_1} + A_1 V_{i_2} \tag{13.23b}$$

where A_1 = negative voltage gain from input terminal 1 to output terminal 3 (with input terminal 2 grounded), and

A_2 = positive voltage gain from input terminal 2 to output terminal 3 (with input terminal 1 grounded).

It is more important to consider the differential and common-mode operation of the amplifier since this allows determining the common-mode rejection of the circuit. This second way of considering the operation of the amplifier provides an output voltage as

$$V_{o_1} = A_d V_d + A_c V_c \tag{13.24a}$$

$$V_{o_2} = -A_d V_d + A_c V_c \tag{13.24b}$$

where A_d = differential-mode gain of the amplifier,

A_c = common-mode gain of the amplifier, and

V_d and V_c are defined in Eqs. (13.20) and (13.21), respectively.

Opposite-Polarity Inputs: If the inputs are equal and opposite in polarity, $V_{i_1} = V_s$ and $V_{i_2} = -V_s$, then from Eq. (13.20)

$$V_d = V_{i_1} - V_{i_2} = V_s - (-V_s) = 2V_s$$

and from Eq. (13.21)

$$V_c = \tfrac{1}{2}(V_{i_1} + V_{i_2}) = \tfrac{1}{2}[V_s + (-V_s)] = 0$$

so that in Eq. (13.24)

$$V_{o_1} = A_d V_d + A_c V_c = A_d(2V_s) + A_c(0)$$

$$V_{o_1} = 2A_d V_s$$

which shows that only differential-mode operation occurs (and that the overall gain is twice the value of A_d).

Same-Polarity Inputs: If the inputs are equal and the same polarity, $V_{i_1} = V_s = V_{i_2}$, then from Eq. (13.20)

$$V_d = V_{i_1} - V_{i_2} = V_s - V_s = 0$$

and from Eq. (13.21)

$$V_c = \tfrac{1}{2}(V_{i_1} + V_{i_2}) = \tfrac{1}{2}(V_s + V_s) = V_s$$

so that in Eq. (13.24)

$$\begin{aligned} V_{o_1} &= A_d V_d + A_c V_c = A_d(0) + A_c V_s \\ &= A_c V_s \end{aligned}$$

which shows that only common-mode operation occurs.

Common-Mode Rejection Ratio (CMRR)

The above calculations indicate how A_d and A_c can be measured in differential-amplifier circuits.

To measure A_d: Set $V_{i_1} = -V_{i_2} = V_s = 0.5$ V so that $V_d = 1$ V and $V_c = 0$ V. Under these conditions the output voltage is $A_d \times (1)$ so that the output voltage equals A_d.

To measure A_c: Set $V_{i_1} = V_{i_2} = V_S = 1$ V so that $V_d = 0$ V and $V_c = 1$ V. Then the output voltage measured equals A_c.

Having measured A_d and A_c for the amplifier we can now calculate a common-mode rejection ratio, which is defined as

$$\boxed{\text{CMRR} = \frac{A_d}{A_c}} \qquad (13.25a)$$

$$\boxed{\text{CMRR} = 20 \log \frac{A_d}{A_c}} \qquad (13.25b)$$

It should be clear that the desired operation will have A_d very large with A_c very small. That is, the signals of opposite polarity will appear greatly amplified at the output terminal, whereas the same-polarity signals will mostly cancel out so that the common-mode gain A_c is very small. Ideally, A_d is very large and A_c is zero so that the value of CMRR is infinite. The larger the value of CMRR the better the common-mode rejection of the circuit.

It is possible to obtain an expression for the output voltage as follows:

$$V_{o_1} = A_d V_d \left(1 + \frac{1}{\text{CMRR}} \frac{V_c}{V_d} \right) \qquad (13.26)$$

Even if both V_c and V_d components of voltage exist at the inputs, the value of $(1/\text{CMRR})(V_c/V_d)$ will be very small, for CMRR very large, and the output voltage will be approximately $A_d V_d$. In other words, the output will be almost completely due to the difference signal with the common-mode input signals rejected (or cancelled out). Some practical examples should help clarify these ideas.

EXAMPLE 13.2 Determine the output voltage of a differential amplifier for input voltages of $V_{i_1} = 150$ μV. The amplifier has a differential mode gain of $A_d = 1000$ and the value of CMRR is:
(a) 100
(b) 10^5

Solution:

$$V_d = V_{i_1} - V_{i_2} = (150 - 100)\ \mu\text{V} = 50\ \mu\text{V}$$

$$V_c = \tfrac{1}{2}(V_{i_1} + V_{i_2}) = \frac{(150 + 100)\ \mu\text{V}}{2} = 125\ \mu\text{V}$$

Note that the common signal is more than twice as large as the difference signal.

(a) $V_o = A_d V_d \left(1 + \frac{1}{\text{CMRR}} \frac{V_c}{V_d} \right)$

$$= A_d V_d \left(1 + \frac{1}{100} \times \frac{125}{50} \right)$$

$$= A_d V_d (1.025)$$

$$= (1000)(50 \ \mu V)(1.025) = \mathbf{51.25 \ mV}$$

The output is only 0.025 or 2.5% more than was the output due only to a difference signal of 50 μV.

The common-mode signal, even larger than the difference component, has been rejected so that only 1.25 mV appear as output.

(b) $\quad V_o = A_d V_d \left(1 + \dfrac{1}{10^5} \times \dfrac{125}{50} \right)$

$$= A_d V_d \ (1.000025)$$

$$\cong 100 \times 50 \ \mu V = \mathbf{50 \ mV}$$

Example 13.2 shows that the larger the value of CMRR the better the circuit rejects common-input signals. Thus, one of the important differential-amplifier factors to consider is the circuit's common-mode rejection ratio.

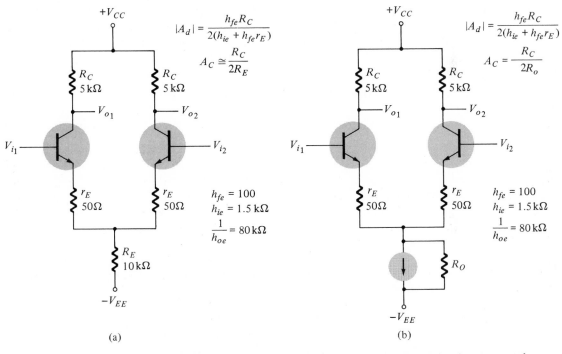

(a)

(b)

Figure 13.26 Differential amplifiers showing differential and common-mode gains.

As summarized in Fig. 13.26 the differential gain between any input and output terminal is

$$| A_d | = \frac{h_{fe} R_C}{2 h_{ie} + 2(h_{fe} + 1) r_E} \cong \frac{R_C}{2(r_e + r_E)} \tag{13.27}$$

the polarity relation between input and output depending on which set of terminals is used.

A common-mode gain can also be calculated as given by

$$A_C = \frac{R_C}{2R_E} \qquad (13.28)$$

using a circuit with emitter resistor, or

$$A_C = \frac{R_C}{2R_o} \qquad (13.29)$$

for a circuit with constant-current source having output resistance, R_o.

> **EXAMPLE 13.3** Calculate the differential and common-mode gains for the circuits of Fig. 13.26 and also the respective values of CMRR.
>
> **Solution:**
> (a) For Fig. 13.26a
>
> $$A_d = \frac{h_{fe}R_C}{2[h_{ie} + (h_{fe} + 1)r_E]} = \frac{100\,(5\text{ k}\Omega)}{2[1.5\text{ k}\Omega + 101\,(50\text{ }\Omega)]} = 38.17$$
>
> $$A_c = \frac{R_C}{2R_E} = \frac{5\text{ k}\Omega}{2(10\text{ k}\Omega)} = 0.25$$
>
> $$\text{CMRR} = \frac{A_d}{A_c} = \frac{38.17}{0.25} = 152.68\,(= \textbf{43.68 dB})$$
>
> (b) For the circuit of Fig. 13.26b, $A_d = 38.17$ [as in part (a)]. Using Eq. (13.18) to calculate R_o
>
> $$R_o = 80\text{ k}\Omega(101) = 8.08\text{ M}\Omega$$
>
> so that A_c is
>
> $$A_c = \frac{R_C}{2R_o} = \frac{5\text{ k}\Omega}{2(8.08\text{ M}\Omega)} = 3.09 \times 10^{-4}$$
>
> We then calculate
>
> $$\text{CMRR} = \frac{A_d}{A_c} = \frac{38.17}{3.09 \times 10^{-4}} = 1.24 \times 10^5\,(= \textbf{101.9 dB})$$

13.4 PRACTICAL OP-AMP CIRCUITS

Using multiple differential-amplifier stages in a single IC package results in an overall circuit called an *operational amplifier* or *op-amp*. Basic features of this circuit include extremely high voltage gain, high input resistance, and low output resistance. The differential-amplifier circuit considered earlier in this chapter is then a basic circuit used in building practical op-amp units. Integrated-circuit construction requires the use of the smallest sized components to form the many hundreds of components— mostly transistors—to build from one to four op-amps in a single IC chip. These circuits may be built using only BJT (bipolar), both bipolar and JFET (biFET), or

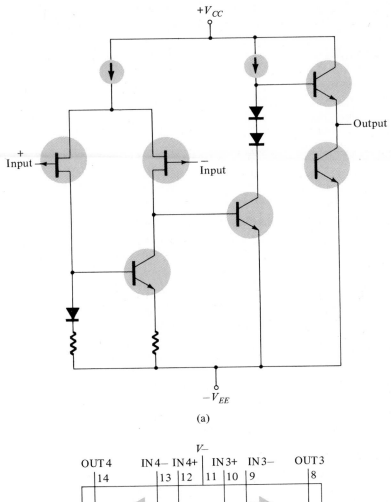

(a)

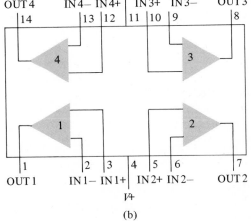

(b)

Figure 13.27 IC op-amp (347): (a) circuit diagram; (b) connection diagram.

bipolar and MOSFET (biMOS). At present, biFET op-amps are most popular with high resistance provided by the JFET input transistor, high gain using bipolar difference-amplifier circuits, and low output resistance using an emitter-follower output stage as shown in Fig. 13.24.

In a biFET op-amp, the JFET device is used in the input part of the circuit to obtain high input resistance. At present, mostly JFET transistors are used for input. For example, the schematic and connection diagram of a 347 op-amp are shown in Fig. 13.27. The 347 IC is a quad JFET input operational amplifier using biFET technology. The circuit of one op-amp stage is shown in Fig. 13.27a; the pin connection diagram showing the four op-amp units is detailed in Fig. 13.27b. Some of the manufacturer's listed unit specifications are:

R_{IN} Input resistance $= 10^{12} \, \Omega$

A_{VOL} Large-signal voltage gain $= 100 \, \text{V/mV} = 100{,}000 = 100 \, \text{dB}$

CMRR   Common-mode rejection ratio $= 100 \, \text{dB}$

GB Gain-bandwidth product $= 4 \, \text{MHz}$

SR   Slew rate $= 13 \, \text{V/}\mu\text{s}$

Notice how the circuit of Fig. 13.27a has many of the parts covered in Section 13.2. However, we need not dwell on the details of the circuit in order to use it. The external features that will be considered next are sufficient to allow using the amplifier.

Table 13.1 is part of the manufacturer's listing of electrical characteristics and definitions of terms. The following discussion will elaborate on a number of the more important characteristics and give some examples.

TABLE 13.1   **347 Op-Amp Electrical Characteristics at $T_A = 25°C$, $V_{CC} = +15$ V, $V_{EE} = -15$ V**

Characteristics	Symbols	Limits			Units
		Min.	Typ.	Max.	
Dynamic Characteristics					
Large-signal voltage gain	A_{VOL}	25	100		V/mV
Input resistance	R_{IN}		10^{12}		Ω
Common-mode rejection ratio	CMRR	70	100		dB
Output voltage swing	$V_o(p\text{-}p)$	± 12	± 13.5		V
Input common-mode voltage range	V_{CM}	± 11	-12		V
Gain-bandwidth product	GB		4		MHz
Static Characteristics					
Input offset voltage	V_{OS}		5	10	mV
Input offset current	I_{OS}		25	100	pA
Device dissipation	P_D			500	mW

13.5   DEFINITIONS OF OP-AMPS TERMS

Input Offset Voltage: The difference in the dc voltages that must be applied to the input terminals to obtain equal quiescent operating voltages (zero-output voltage) at the output terminals.

Input Offset Current: The difference in the currents at the two input terminals.

Quiescent Operating Voltage: The dc voltage at either output terminal, with respect to ground.

dc Device Dissipation: The total power drain of the device with no signal applied and no external load current.

Common-Mode Voltage Gain: The ratio of the signal voltages developed at either of the two output terminals to the common signal voltage applied to the two input terminals connected in parallel.

Differential Voltage Gain—Single-Ended Input–Output: The ratio of the change in output voltage at either output terminal with respect to ground, to difference in the input voltages.

Common-Mode Rejection Ratio: The ratio of the full differential voltage gain to the common-mode voltage gain.

Bandwidth at 3-dB Point (B):   The frequency at which the voltage gain of the device is 3 dB below the voltage gain at a specified lower frequency.

Maximum Output Voltage V_o(p-p):   The maximum peak-to-peak output voltage swing, measured with respect to ground, that can be achieved without clipping of the signal waveform.

Single-Ended Input Resistance (R_{in}):   The ratio of the change in input voltage to the change in input current measured at either input terminal with respect to ground.

Single-Ended Output Resistance (R_o):   The ratio of the change in output voltage to the change in output current measured at either output terminal with respect to ground.

Slew-Rate: Device parameter indicating how fast the output voltage changes with time.

13.6   DC ELECTRICAL PARAMETERS

Differential Voltage Gain—
Large-signal Voltage Gain, A_{VOL}

The typical value of 106 dB is the gain from one input terminal to either output terminal. This was considered the gain A_1 or A_2 in Sections 13.1–13.3. The manufacturer lists the gain in units of decibels (dB). The relation of decibels and the gain as numerical ratio of output voltage (V_o) to input voltage (V_i) is

$$A_{\text{dB}} = 20 \log |A_v| = 20 \log \left| \frac{V_o}{V_i} \right| \tag{13.30}$$

As an example, a gain of $A_v = 1000$ is the same as

$$A_{dB} = 20 \log 1000 = 20(3) = 60 \text{ dB}$$

and a gain of $A_v = 100,000$ is the same as

$$A_{dB} = 20 \log 100,000 = 20(5) = 100 \text{ dB}$$

A gain of 106 dB is then the same as a voltage gain above 100,000 and can be calculated exactly as

$$106 = 20 \log A_v$$

$$5.3 = \log A_v$$

$$A_v = \text{antilog } 5.3 \cong 2 \times 10^5 = 200,000$$

The large open-loop gain ($\cong 200,000$) and limited output voltage swing (± 13 V) requires that the input voltage for open-circuit operation be no greater than

$$V_d = V_o/A_{VOL} = \pm 13 \text{ V}/200,000 = \pm 65 \ \mu\text{V}$$

An op-amp with such large gain is typically operated closed loop as desired later in this chapter to allow operation with larger input voltages.

Single-Ended Input Resistance (R_{IN})

The input resistance is measured at either input terminal. A listed value of 10^{12} Ω indicates a high value. It should be recalled from Chapter 9 how important input resistance values are when interconnecting amplifier stages or driving the amplifier with a practical voltage source. If the input resistance is not much larger than the source resistance, loading will cause the input voltage to be less than that of the unloaded source signal, resulting in less output voltage.

Bipolar op-amps typically provide input resistances of around 1 MΩ, biFET op-amps are rated at 10^{12} Ω, and biMOS typically are 10^{15} Ω.

Output Resistance, R_o: The output resistance, typically 100 Ω, depends on the output stage used to drive the signal to a load. Output stages can provide signal swings good in only one direction of voltage swing, in both directions of voltage swing, and may be short-circuit protected depending on which output stage circuit is used (see Fig. 13.24).

Common-Mode Rejection Ratio (CMRR)

The common-mode rejection ratio defined as

$$\text{CMRR} = \left| \frac{A_d}{A_c} \right|$$

may also be calculated in decibel units as

$$\boxed{\text{CMRR} = 20 \log \left| \frac{A_d}{A_c} \right| \text{dB}}$$

A value of CMRR = 100 dB is a ratio of differential mode to common-mode gain of

$$\frac{A_d}{A_c} = \text{antilog}\,\frac{\text{CMRR(dB)}}{20} = \text{antilog}\left(\frac{100\ \text{dB}}{20}\right) = \text{antilog}\,5$$

$$= 10^5 = 100{,}000$$

The common or in-polarity signal is thus amplified by a gain of 100,000 less than the differential or opposite-polarity inputs.

The value of CMRR generally drops off at increasing frequency as shown by the plot of Fig. 13.28.

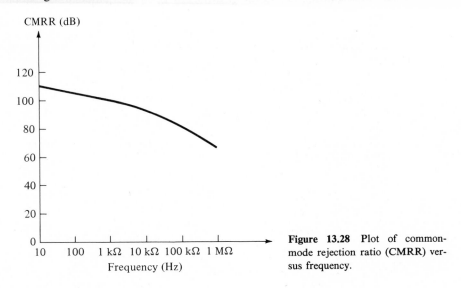

Figure 13.28 Plot of common-mode rejection ratio (CMRR) versus frequency.

OFFSET VOLTAGE, V_{OS}

The definition of input offset voltage, V_{OS}, can be stated as "that differential dc voltage required between inputs of an operational amplifier to force the output to

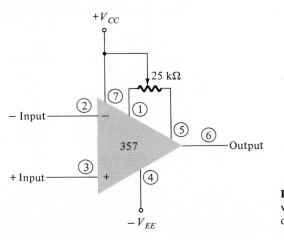

Figure 13.29 Operation of offset voltage (V_{OS}) adjustment using 357 op-amp.

0 V." Ideally the value of V_{OS} should be 0 V and practically the value of V_{OS} is a few millivolts. When the op-amp is used primarily for its large-signal operation, the small offset voltage is acceptable. When used in applications where a small output voltage represents some measured quantity as in a converter, meter, or measuring device, any nonzero output voltage can result in substantial error. In such circuit application, an op-amp with very low offset voltage is used, or one having input terminals allowing offset voltage adjustment is used.

The 357 op-amp, for example, has balance input terminals for adjusting the offset voltage as shown in Fig. 13.29. A 25-kΩ potentiometer is suggested by the manufacturer for connection between pins 1 and 5, the wiper arm being adjusted until the measured output voltage is adjusted to 0 V, when the inputs at pins 2 and 3 are grounded ($V_d = 0$ V).

INPUT BIAS CURRENT, I_{BIAS}

For the circuit inside the IC to operate properly, sufficient dc bias current must be provided as specified by the manufacturer's information. For BJT inputs, the current required is typically microamperes; for JFET input stages, the current required is some tens of picoamperes. Although the device value is typically rated at 25°C (room temperature), it increases greatly with temperature as shown in the plot of Fig. 13.30.

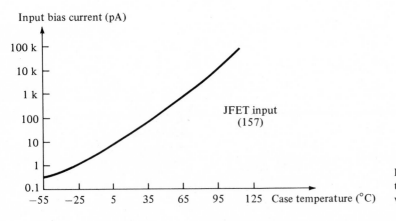

Figure 13.30 Plot showing variation of input bias current required versus device case temperature.

INPUT OFFSET CURRENT, I_{OS}

The small difference in bias currents at the inputs is magnified by the amplifier gain to provide an output offset voltage. Offset current for BJT input circuits is tens to hundreds of nanoamperes, whereas for JFET input stages the value is typically picoamperes.

DRIFT

Drift is the term describing the change in output voltage resulting from change in temperature. Even when the output voltage is adjusted to 0 V at room temperature, that value will change as temperature changes. Typically the offset-voltage drift,

$\Delta V_{OS}/\Delta T$, is in the range of 5–40 μV/°C. As a rule of thumb, the drift is 3.3 μV/°C for each mV of initial offset voltage. The drift resulting from the offset current is typically $\Delta I_{OS}/\Delta T = 0.01$ to 0.5 nA/°C.

Ac Electrical Parameters

BANDWIDTH, B

The op-amp unity-gain bandwidth specifies the upper frequency at which the gain drops to unity (gain = 1) owing to the capacitances resulting from manufacture of the circuit. Typical values of unity gain bandwidth or gain bandwidth product are greater than 1 MHz. Figure 13.31 shows a plot of open-loop voltage gain versus

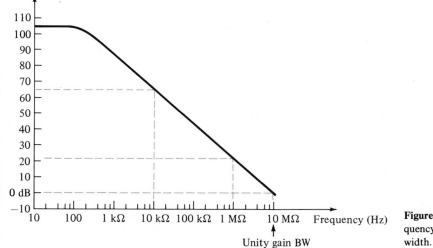

Figure 13.31 Gain versus frequency showing unity-gain bandwidth.

frequency for a 357 IC. Notice that below about 100 Hz, the gain stays constant at the rated dc open-loop gain value, dropping to unity gain (0 dB) at a frequency around 10 MHz.

An alternate parameter used to specify the device bandwidth is the rise time, t_r. The value of rise time and bandwidth are related by

$$B = 0.35/t_r \tag{13.31}$$

If the manufacturer lists the value of t_r at 0.3 μs, for example, the value of unity-gain bandwidth is then

$$B = 0.35/0.3 \ \mu s = 1.167 \ \text{MHz}$$

Various device spec sheets list unity-gain B, gain-bandwidth product, or rise time to specify the limited frequency range when using that op-amp. It should be clear that an op-amp having an open-loop gain rated at 100,000 (at dc) with gain-bandwidth of 1 MHz has a value of A_{VOL} much smaller at, say, 10 kHz. Using the plot of Fig. 13.31, the op-amp gain is greater than 100 dB at dc but decreases to about 65 dB at 10 kHz. At 1 MHz the op-amp gain is around 20 dB ($A_{\text{VOL}} = 10$) and obviously doesn't qualify as large gain for op-amp applications. In any case the larger value of bandwidth provides op-amp operation at higher frequency.

SLEW RATE, *SR*

Slew rate is a parameter indicating how fast the output voltage changes with time. Typical slew rate values are from 0.5 V/μs to 50 V/μs, with the larger values indicating the units that operate faster.

A comparison of the various device parameters for a number of ICs is listed in Table 13.2.

TABLE 13.2   Comparison of Op-Amp Parameters

Parameter		Bipolar	BiFET	Norton	Unit
Open-loop voltage gain	A_{VOL}	200	200	2.8	V/mV
Input resistance	R_{IN}	2	10^6	1	MΩ
Output resistance	R_{OUT}	75		8	Ω
Common-mode rejection ratio	CMRR	90	100		dB
Input offset voltage	V_{OS}	1	1		mV
Input bias current	I_{BIAS}	80 nA	30 pA	30 pA	nA or pA
Input offset current	I_{OS}	20 nA	3 pA		nA or pA
Drift	$\Delta V_{OS}/\Delta T$	15	3		μV/°C
Bandwidth	B	1	20	2.5	MHz
Slew rate	SR	0.5	50	0.5	V/μs

13.7   BASICS OF THE OPERATIONAL AMPLIFIER (OP-AMP)

An operational amplifier is a very high-gain differential amplifier that uses voltage feedback to provide a stabilized voltage gain. The basic amplifier used is essentially a difference amplifier having very high open-loop gain (no signal-feedback condition) as well as high input impedance and low output impedance. Typical uses of the operational amplifier are scale changing; analog computer operations, such as addition and integration; and a great variety of phase shift, oscillator, and instrumentation circuits.

Figure 13.32 shows an op-amp unit having two inputs and a single output. Recall how the two inputs affect an output in a difference amplifier. Here the inputs are marked with *plus* (+) and *minus* (−) to indicate noninverting and inverting inputs, respectively. A signal applied to the *plus* input will appear with the same polarity

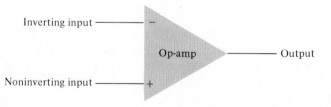

Figure 13.32   Basic op-amp.

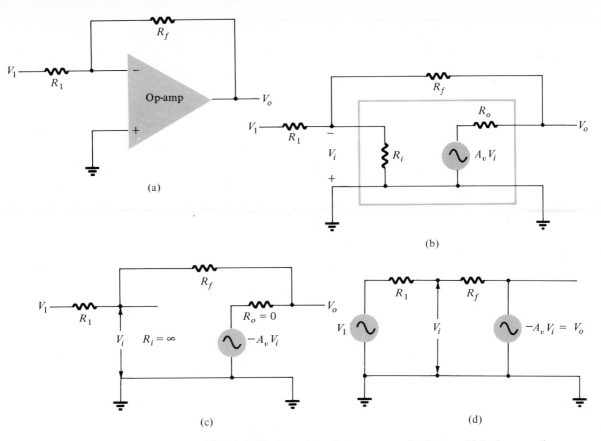

Figure 13.33 Operation of op-amp as scale changer: (a) basic connection (constant-gain multiplier); (b) effect of op-amp circuit; (c) ideal op-amp; (d) ideal equivalent circuit.

and amplified at the output, whereas an input applied to the *minus* (−) terminal will appear amplified but inverted at the output.

The basic circuit connection of an operational amplifier is shown in Fig. 13.33a. As shown the circuit operates as a scale changer or constant-gain multiplier. An input signal V_1 is applied through a resistor R_1 to the minus input terminal. The output voltage is fed back through resistor R_f to the same input terminal. The plus input terminal is connected to ground. We now wish to determine the overall gain of the circuit (V_o/V_1). To do this we must consider some more details of the op-amp unit.

Figure 13.33b shows the op-amp replaced by an equivalent circuit of input resistance R_i and output voltage source and resistance. An ideal op-amp as shown in Fig. 13.33c has infinite resistance ($R_i = \infty$), zero output resistance ($R_o = 0$), and very high voltage gain ($A_v \gg 1$). The connection for the ideal amplifier is shown redrawn in Fig. 13.33d.

Using superposition we can solve for the voltage V_i in terms of the components due to each of the sources. For source V_i only ($-A_v V_i$ set to zero)

$$V_{i_1} = \frac{R_f}{R_1 + R_f} V_1$$

For source $-A_v V_i$ only (V_1 set at zero)

$$V_{i_2} = \frac{R_1}{R_1 + R_f}(-A_v V_i)$$

The total voltage of V_i is then

$$V_i = V_{i_1} + V_{i_2} = \frac{R_f}{R_1 + R_f} V_1 + \frac{R_1}{R_1 + R_f}(-A_v V_i)$$

which can be solved for V_i as

$$V_i = \frac{R_f}{R_f + (1 + A_v)R_1} V_1 \qquad (13.32)$$

if $A_v \gg 1$ and $A_v R_1 \gg R_f$, as is usually true, then

$$V_i \cong \frac{R_f}{A_v R_1} V_1$$

Solving for V_o/V_1, we get

$$\frac{V_o}{V_1} = \frac{-A_v V_i}{V_1} = \frac{-A_v}{V_1}\left(\frac{R_f V_1}{A_v R_1}\right) = -\frac{R_f}{R_1}$$

$$\boxed{\frac{V_o}{V_1} = -\frac{R_f}{R_1}} \qquad (13.33)$$

The result shows that the ratio of overall output to input voltage is dependent only on the values of resistors R_1, R_f—provided that A_v is very large.

If $R_f = R_1$, the gain is

$$A_v = -\frac{R_1}{R_1} = -1$$

and the circuit provides a sign change with no magnitude change.

If $R_f = 2R_1$, then

$$A_v = \frac{-2R_1}{R_1} = -2$$

and the circuit provides a gain of 2 along with polarity inversion of the input signal.

If we select precise resistor values for R_f and R_1, we can obtain a wide range of gains, the gain being as accurate as the resistors used and only slightly affected by temperature and other circuit factors.

VIRTUAL GROUND

The output voltage is limited by the supply voltage of, typically, a few volts. Voltage gains as stated before are very high. If, for example, $V_o = -10$ V and $A_v = 10,000$, the input voltage is

$$V_i = -\frac{V_o}{A_v} = -\frac{(-10)}{10,000} = 1 \text{ mV}$$

If the circuit had an overall gain (V_o/V_1) of, say, 1, the value of V_1 would be 10 V. The value of V_i, compared to all other voltages, is then small and may be considered 0 V. Note that although $V_i \cong 0$ V, it is not exactly 0 V since the output is the value of V_i times the gain of the amplifier $(-A_v)$.

The fact that $V_i \cong 0$ V leads to the concept that at the input to the amplifier there exists a virtual short circuit or *virtual ground*. The concept of a virtual short implies that although the voltage is nearly 0 V, there is no current through the amplifier input to ground. Figure 13.34 depicts the virtual-ground concept. The heavy line is used to indicate that we may consider that a short exists with $V_i \cong 0$ V, but that this is a virtual short in that no current goes through the short to ground. Current is through resistor R_1 and through R_f as shown.

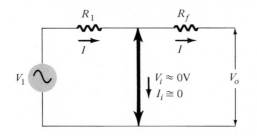

Figure 13.34 Virtual ground in an op-amp.

Using the virtual-ground concept we can write equations for the current I as follows:

$$I = \frac{V_1}{R_1} = -\frac{V_o}{R_f}$$

which can be solved for V_o/V_1

$$\frac{V_o}{V_1} = -\frac{R_f}{R_1}$$

The virtual-ground concept, which depended on A_v being very large, allowed simple solution of overall voltage gain. It should be understood that although the circuit of Fig. 13.34 is not a physical circuit, it does allow an easy means for determining the overall circuit gain.

13.8 OP-AMP CIRCUITS

Constant-Gain Multiplier

An inverting constant-multiplier circuit has already been considered but is repeated here to provide a fuller listing of basic op-amp circuits. Figure 13.35 shows an inverting constant-gain-multiplier circuit.

EXAMPLE 13.3   The circuit of Fig. 13.35 has $R_1 = 100$ kΩ and $R_f = 500$ kΩ. What is the output voltage for an input of $V_1 = -2$ V?

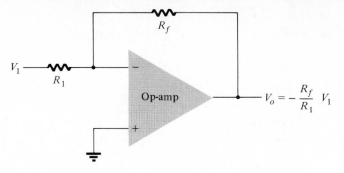

$$V_o = -\frac{R_f}{R_1} V_1$$

Figure 13.35 Inverting constant-gain multiplier.

Solution: Using Eq. (13.33) gives

$$V_o = -\frac{R_f}{R_1} V_1 = -\frac{500\ \text{k}\Omega}{100\ \text{k}\Omega}(-2) = \mathbf{+10\ V}$$

Noninverting Amplifier

The connection of Fig. 13.36 shows an op-amp circuit that works as a noninverting constant-gain multiplier. To determine the voltage gain of the circuit we can use the equivalent virtual-ground representation in Fig. 13.36b. Note that the voltage across R_1 is V_1, since $V_i \cong 0$ V. This must be equal to the voltage due to the output, V_o, through a voltage divider of R_1 and R_f so that

$$V_1 = \frac{R_1}{R_1 + R_f} V_o$$

and

$$\frac{V_o}{V_1} = \frac{R_1 + R_f}{R_1} = 1 + \frac{R_f}{R_1} \qquad (13.34)$$

EXAMPLE 13.4 Calculate the output voltage of a noninverting constant-gain multiplier (as in Fig. 13.36) for values of $V_1 = 2$ V, $R_f = 500$ kΩ, and $R_1 = 100$ kΩ.

Solution: Using Eq. (13.34), we get

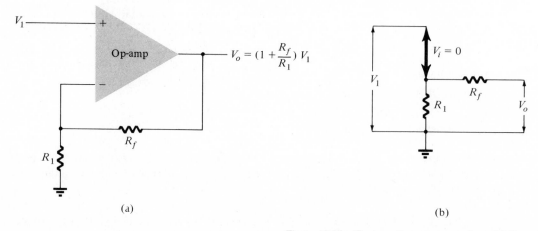

(a)

$$V_o = (1 + \frac{R_f}{R_1}) V_1$$

$$V_i = 0$$

(b)

Figure 13.36 Noninverting constant-gain multiplier.

CH. 13 LINEAR ICs: OPERATIONAL AMPLIFIERS (OP-AMPs)

$$V_o = \left(1 + \frac{R_f}{R_1}\right)V_1 = \left(1 + \frac{500 \text{ k}\Omega}{100 \text{ k}\Omega}\right)(2 \text{ V}) = 6(2) = +\mathbf{12 \text{ V}}$$

Unity Follower

The unity follower, as in Fig. 13.37, provides a gain of 1 with no polarity reversal. From the equivalent circuit with virtual ground it is clear that

$$\boxed{V_o = V_1}$$

(13.35)

and that the output is the same polarity and magnitude as the input. The circuit acts very much like an emitter follower except that the gain is very much closer to being exactly unity.

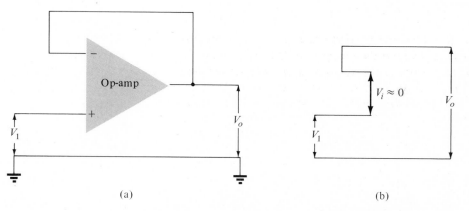

(a)   (b)

Figure 13.37 (a) Unity follower; (b) virtual-ground equivalent circuit.

Summing Amplifier

Probably the most useful of the op-amp circuits used in analog computers is the summing-amplifier circuit. Figure 13.38 shows a three-input summing circuit, which provides a means of algebraically summing (adding) three-input voltages, each multiplied by a constant-gain factor.

If the virtual equivalent circuit is used, the output voltage can be expressed in terms of inputs as

$$\boxed{V_o = -\left(\frac{R_f}{R_1}V_1 + \frac{R_f}{R_2}V_2 + \frac{R_f}{R_3}V_3\right)}$$

(13.36)

In other words, each input adds a voltage to the output as obtained for an inverting constant-gain circuit. If more inputs are used, they add additional components to the output.

EXAMPLE 13.5 What is the output voltage of an op-amp summing amplifier for the following sets of input voltages and resistors ($R_f = 1 \text{ M}\Omega$ in all cases)?

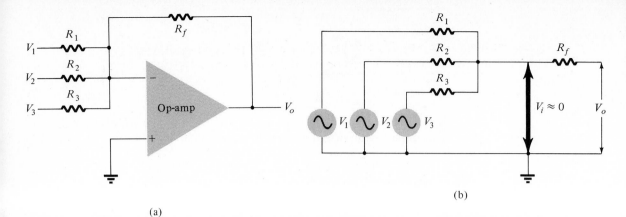

(a)

(b)

Figure 13.38 (a) Summing amplifier; (b) virtual-ground equivalent circuit.

(a) $V_1 = +1$ V, $V_2 = +2$ V, $V_2 = +3$ V,
$R_1 = 500$ kΩ, $R_2 = 1$ MΩ, $R_3 = 1$ MΩ

(b) $V_1 = -2$ V, $V_2 = +3$ V, $V_3 = +1$ V,
$R_1 = 200$ kΩ, $R_2 = 500$ kΩ, $R_3 = 1$ MΩ

Solution: Using Eq. (13.36), we get

(a) $V_o = -\left[\dfrac{1000 \text{ k}\Omega}{500 \text{ k}\Omega}(+1) + \dfrac{1000 \text{ k}\Omega}{1000 \text{ k}\Omega}(+2) + \dfrac{1000 \text{ k}\Omega}{1000 \text{ k}\Omega}(+3) \right]$

$= -[2(1) + 1(2) + 1(3)] = \mathbf{-7\ V}$

(b) $V_o = -\left[\dfrac{1000 \text{ k}\Omega}{200 \text{ k}\Omega}(-2) + \dfrac{1000 \text{ k}\Omega}{500 \text{ k}\Omega}(+3) + \dfrac{1 \text{ M}\Omega}{1 \text{ M}\Omega}(+1) \right]$

$= -[5(-2) + 2(+3) + 1(1)] = -(-10 + 6 + 1)$

$= \mathbf{+3\ V}$

Integrator

So far the input and feedback components have been resistors. If the feedback component used in a capacitor, as in Fig. 13.39, the resulting circuit is an integrator.

The virtual-ground equivalent circuit shows that an expression between input and output voltages can be derived from the current I, which flows from input to output. Recall that virtual ground means that we can consider the voltage at the

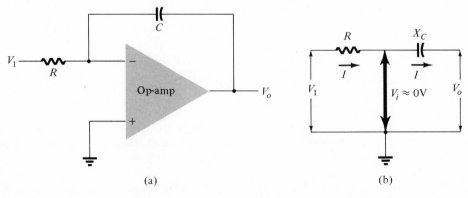

(a)

(b)

Figure 13.39 Integrator.

junction point of R and X_C to be ground (since $V_i \cong 0$ V) but that no current goes into ground at that point. The capacitive impedance can be expressed as

$$X_C = \frac{1}{j\omega C} = \frac{1}{sC}$$

where $s = j\omega$ is the Laplace notation.
Solving for V_0/V_1

$$I = \frac{V_1}{R} = -\frac{V_0}{X_C} = \frac{-V_o}{1/sC} = -sCV_0$$

$$\frac{V_0}{V_1} = \frac{-1}{sCR} \qquad\qquad (13.37a)$$

The last expression can be rewritten in the time domain as

$$v_0(t) = -\frac{1}{RC}\int v_1(t)\,dt \qquad\qquad (13.37b)$$

Equation (13.37b) shows that the output is the integral of the input, with an inversion and scale multiplier of $1/RC$. The ability to integrate a given signal provides the analog computer with the ability to solve differential equations and therefore allows setup of a wide variety of electrical circuit analogs of physical-system operations.

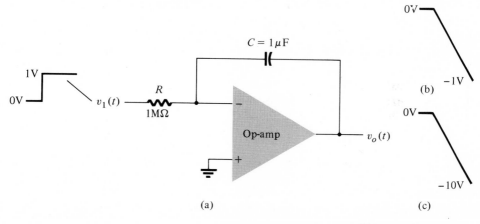

Figure 13.40 Operation of integrator with step input.

As an example, consider an input step voltage shown in Fig. 13.40a. The integral of the step voltage is a ramp or linearly changing voltage. The circuit scale factor of $-1/RC$ is

$$-\frac{1}{RC} = -\frac{1}{10^6 \times 10^{-6}} = -1$$

so that

$$v_0(t) = -\int v_i(t)\,dt$$

and the output is a negative ramp as shown in Fig. 13.40b.

If the scale factor is changed by making $R = 100$ kΩ, for example, then

$$-\frac{1}{RC} = -\frac{1}{10^5 \times 10^{-6}} = -10$$

and the output is
$$v_o(t) = -10 \int v_i(t)dt$$
which is shown in Fig. 13.40c.

More than one input may be applied to an integrator as shown in Fig. 13.41 with the resulting operation given by

$$v_o(t) = -\left[\frac{1}{R_1 C}\int v_1(t)dt + \frac{1}{R_2 C}\int v_2(t)dt + \frac{1}{R_3 C}\int v_3(t)dt\right] \qquad (13.38)$$

An example, showing a summing integrator as used in an analog computer, is given in Fig. 13.41. The actual circuit is shown with input resistors and feedback capacitor, whereas the analog-computer representation only indicates the scale factor for each input.

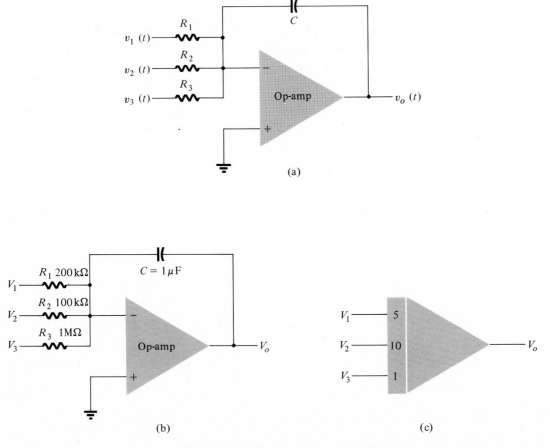

Figure 13.41 (a) Summing-integrator circuit; (b) op-amp; (c) analog-computer, integrator-circuit representation.

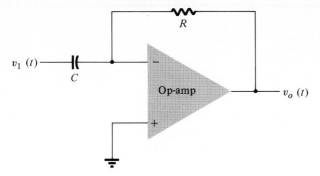

Figure 13.42 Differentiator circuit.

Differentiator

The differentiator circuit of Fig. 13.42 is not as useful a computer circuit as the integrator because of practical problems with noise. The resulting relation for the circuit is

$$v_o(t) = - RC\frac{dv_1(t)}{dt} \tag{13.39}$$

where the scale factor is $-RC$. Reference to any text on analog computers will show how differential equations are set up for solution using mainly summing and integrator circuits.

13.9 OP-AMP APPLICATIONS

To provide some indication of how useful op-amps can be in other than just analog-computer circuits (which is a very large op-amp application area) a few miscellaneous applications will be considered here, with additional oscillator applications in Chapter 17.

dc Millivoltmeter

Figure 13.43 shows a 741 op-amp used as the basic amplifier in a dc millivoltmeter. The amplifier provides a meter with high input impedance and scale factors dependent only on resistor value and accuracy. Notice that the meter reading represents millivolts of signal at the circuit input. An analysis of the op-amp circuit yields the circuit transfer function

$$\frac{I_0}{V_1} = \frac{R_f}{R_i}\left(\frac{1}{R_S}\right)$$
$$= \frac{100\text{ k}\Omega}{100\text{ k}\Omega} \times \frac{1}{10} = \frac{1\text{ mA}}{10\text{ mV}}$$

Thus, an input of 10 mV will result in a current through the meter of 1 mA. If the input is 5 mV, the current through the meter will be 0.5 mA, which is half-scale deflection.

Changing R_f to 200 kΩ, for example, would result in a circuit scale factor of

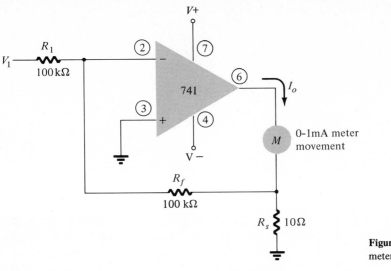

Figure 13.43 Op-amp dc millivolt-meter.

$$\frac{I_0}{V_1} = \frac{200 \text{ k}\Omega}{100 \text{ k}\Omega} \times \frac{1}{10} = \frac{1 \text{ mA}}{5 \text{ mV}}$$

showing that the meter now reads 5 mV, full scale. It should be kept in mind that building such a millivoltmeter requires purchasing an op-amp circuit, a few resistors, and a meter movement. The ability to obtain a completely operating, tested op-amp unit makes the overall meter unit easy to set up.

Constant-Current Source

Figure 13.44 shows op-amp circuits that provide constant current. The circuit of Fig. 13.44a provides an output current fixed at 1 mA by the three resistors. The voltage divider of R_2 and R_3 sets the input to the noninverting input to $+3$ V, resulting in $+2$ V across R_1. The output current is then fixed at a value of $(2 \text{ V})/R_1 = (2 \text{ V})/2 \text{ k}\Omega = 1$ mA. Low-current-source operation can be fixed over a range of values by selection of resistor R_1.

The circuit of Fig. 13.44b sets a fixed output sinking current (at 2 mA in this example). Setting the value of resistor R_1 fixes the output current at a desired current value.

Display Drivers

Figure 13.45 shows op-amp circuits that can be used to drive a lamp display or LED display. When the noninverting input to Fig. 13.45a goes above the inverting input, the output at terminal 1 goes to the positive saturation level (near $+5$ V in this example), and the lamp is driven on when transistor Q_1 goes on. As shown in the circuit, the output of the op-amp provides 30 mA current to the base of transistor Q_1, which then drives 600 mA through a suitably selected transistor (with $\beta > 20$) capable of handling that amount of current.

Figure 13.45b shows an op-amp circuit that can supply 20 mA to drive an LED display when the $+$ input goes positive compared to the $-$ input.

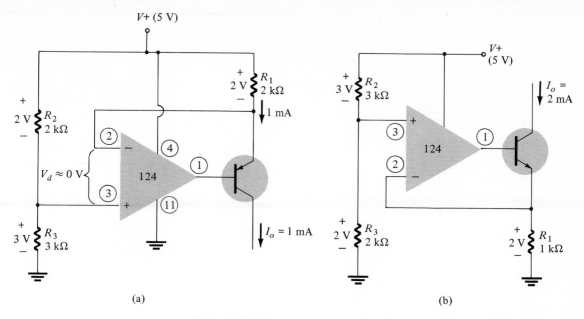

Figure 13.44 Constant-current circuits: (a) current source; (b) current sink.

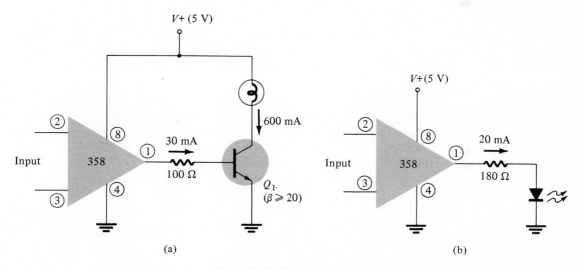

Figure 13.45 Display driver circuits: (a) lamp driver; (b) LED driver.

ac Millivoltmeter

As another example, an ac millivoltmeter circuit, is shown in Fig. 13.46. The resulting circuit transfer function is

$$\frac{I_0}{V_1} = \frac{R_f}{R_1}\left(\frac{1}{R_s}\right) = \frac{100 \text{ k}\Omega}{100 \text{ k}\Omega} \times \frac{1}{10} = \frac{1 \text{ mA}}{10 \text{ mV}}$$

which appears the same as the dc millivoltmeter, except that in this case it is for

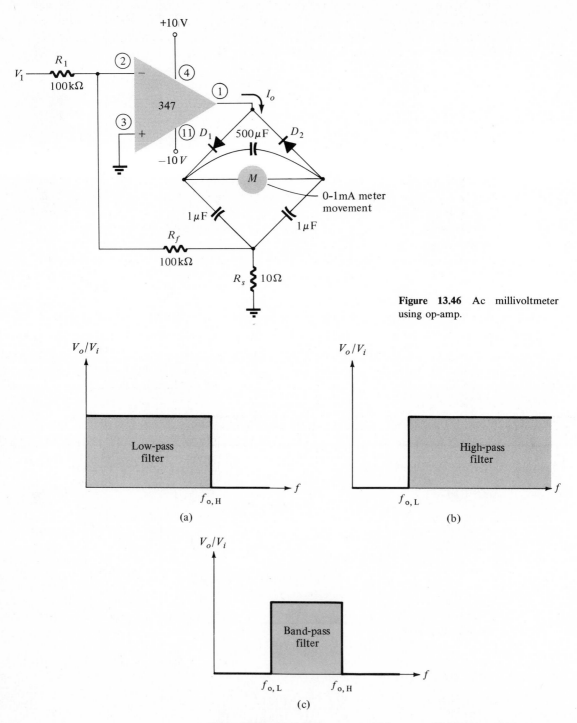

Figure 13.46 Ac millivoltmeter using op-amp.

Figure 13.47 Ideal filter response: (a) low-pass; (b) high-pass; (c) band-pass.

ac signals. The meter indication provides a full-scale deflection for an ac input voltage of 10 mV. An ac input signal of 10 mV will result in full-scale deflection, while an ac input of 5 mV will result in half-scale deflection, and the meter reading can be interpreted in millivolt units.

Active Filters

A popular application of op-amps is in building active filters. A filter circuit is built using passive components—resistors and capacitors. An active filter additionally uses an amplifier for voltage gain and signal isolation or buffering.

A filter that provides a constant output from dc up to a cutoff frequency, $f_{o,H}$ and then passes no signal is an ideal low-pass filter as described by the response plotted in Fig. 13.47a. A filter that passes signals only above a cutoff frequency is a high-pass filter, as idealized in Fig. 13.47b. When the filter circuit passes signals that are above one cutoff frequency and below a second cutoff frequency, it is a band-pass filter as idealized in Fig. 13.47c.

Low-Pass Filter: A first-order low-pass filter using a single resistor and capacitor as in Fig. 13.48a has a practical slope of 20 dB per decade, as shown in Fig. 13.48b

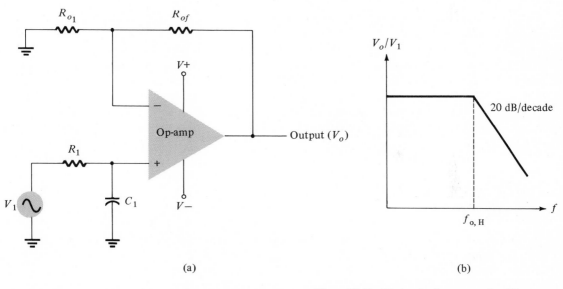

(a)   (b)

Figure 13.48 First-order low-pass active filter.

(rather than the ideal response of Fig. 13.47a). The voltage gain below the cutoff frequency is constant at

$$A_v = 1 + \frac{R_{of}}{R_{o_1}} \tag{13.40}$$

at a cutoff frequency of

$$f_{o,H} = \frac{1}{2\pi R_1 C_1} \tag{13.41}$$

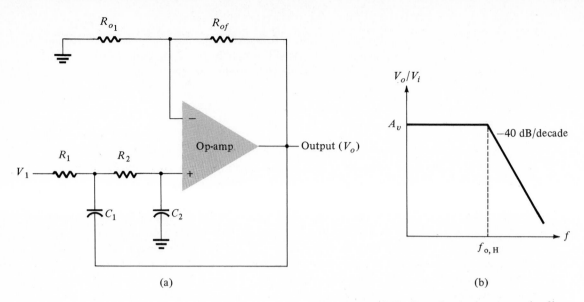

Figure 13.49 Second-order low-pass active filter.

Connecting two sections of filter as in Fig. 13.49 results in a second-order low-pass filter with cutoff at 40 dB per decade—closer to the ideal characteristic of Fig. 13.47a. The circuit voltage gain and cutoff frequency are the same for the second-order circuit as for the first-order filter circuit except that the filter response drops faster for a second-order filter circuit.

EXAMPLE 13.6 Calculate the cutoff frequency of a first-order low-pass filter for $R_1 = 1.2$ kΩ and $C_1 = 0.02$ μF.

Solution:

$$f_{o,H} = \frac{1}{2\pi R_1 C_1} = \frac{1}{2\pi (1.2 \times 10^3)(0.02 \times 10^{-6})} = \textbf{6.63 kHz}$$

High-Pass Active Filter: First- and second-order high-pass active filters can be built as shown in Fig. 13.50. The amplifier gain is calculated using Eq. (13.40), with cutoff frequency

$$f_{o,L} = \frac{1}{2\pi R_1 C_1} \tag{13.42}$$

(with second-order filter $R_1 = R_2$, and $C_1 = C_2$ results in the same cutoff frequency).

EXAMPLE 13.7 Calculate the gain and cutoff frequency of a second-order high-pass filter as in Fig. 13.50b for $R_1 = R_2 = 2.1$ kΩ, $C_1 = C_2 = 0.05$ μF, and $R_{o_1} = 10$ kΩ, $R_{of} = 50$ kΩ.

Solution:

$$A_v = 1 + \frac{R_{of}}{R_{o_1}} = 1 + \frac{50 \text{ k}\Omega}{10 \text{ k}\Omega} = 6$$

CH. 13 LINEAR ICs: OPERATIONAL AMPLIFIERS (OP-AMPs)

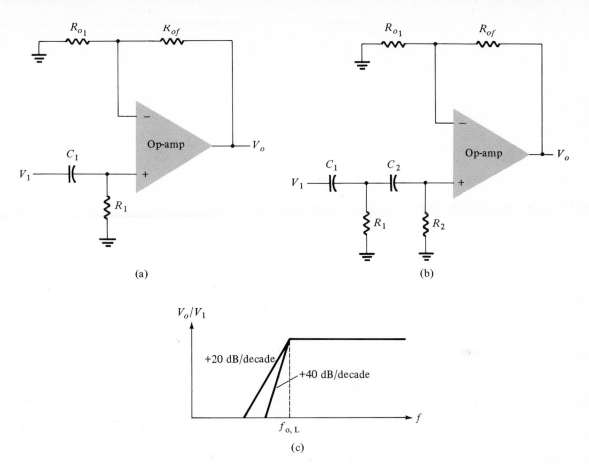

Figure 13.50 High pass filter: (a) first-order; (b) second-order; (c) response plot.

at cutoff frequency

$$f_{o,L} = \frac{1}{2\pi R_1 C_1} = \frac{1}{2\pi(2.1 \times 10^3)(0.05 \times 10^{-6})} = \textbf{1.5 kHz}$$

Band-Pass Filter: Figure 13.51 shows a band-pass filter using two stages, the first a high-pass filter and the second a low-pass, the combined operation being the desired band-pass response.

EXAMPLE 13.8 Calculate the cutoff frequencies of the band-pass filter circuit of Fig. 13.51 with $R_1 = R_2 = 10$ kΩ, $C_1 = 0.1$ μF, and $C_2 = 0.002$ μF.

Solution:

$$f_{o,L} = \frac{1}{2\pi R_1 C_1} = \frac{1}{2\pi(10 \times 10^3)(0.1 \times 10^{-6})} = \textbf{159.15 Hz}$$

$$f_{o,H} = \frac{1}{2\pi R_2 C_2} = \frac{1}{2\pi(10 \times 10^3)(0.002 \times 10^{-6})} = \textbf{7.96 kHz}$$

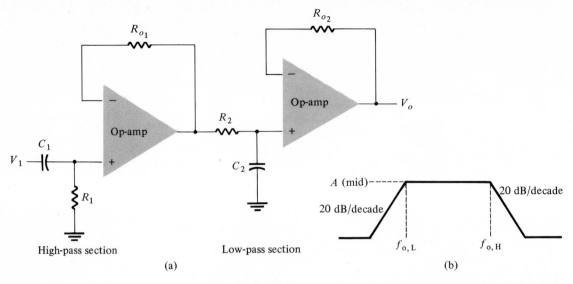

Figure 13.51 Band-pass active filter.

Voltage Buffer and Distributor

The circuit of Fig. 13.52 shows how a signal can be buffered and distributed to a number of outputs using op-amp circuits. The four op-amp circuits are all packaged in a single 347 quad op-amp IC unit. Supply voltages applied to pins 4 and 11 connect to all four op-amp circuits. The input stage, connected as a unity-gain amplifier provides output at pin 1, the same as the input signal. The other three stages are

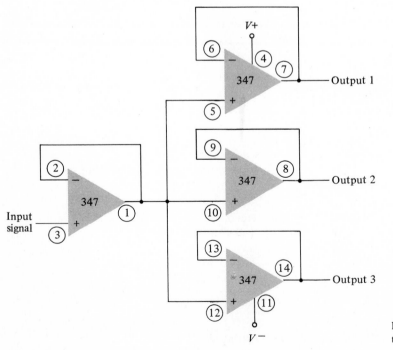

Figure 13.52 Use of buffer op-amp to provide signal distribution.

connected as unity-gain noninverting amplifiers and supply three identical signals to be distributed as three separate and isolated outputs.

PROBLEMS

§ 13.1

1. Draw the output waveforms for the input signal and differential amplifier of Fig. 13.53. (Refer to details of Fig. 13.2.)

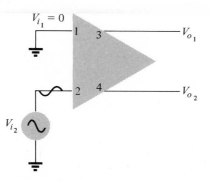

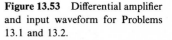

Figure 13.53 Differential amplifier and input waveform for Problems 13.1 and 13.2.

2. If the input signal V_{i_2} is an 8-mV, peak, 1000-Hz signal, sketch the output waveforms at V_{o_1} and V_{o_2} (the amplifier gain is 1000) in Fig. 13.53.

3. Draw the output waveform at V_{o_2} for the differential amplifier with input signal of Fig. 13.54.

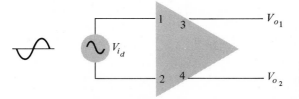

Figure 13.54 Differential amplifier and input for Problems 13.3 and 13.4.

4. Draw the output waveform at V_{o_1} from the circuit of Fig. 13.54 for an input of 5 mV, peak if the amplifier gain is 1000.

§ 13.2

5. Determine the dc collector voltages in the circuit of Fig. 13.55.

6. Calculate the ac voltage gain of the differential amplifier in Fig. 13.55.

7. Calculate the input and output resistance of the circuit in Fig. 13.55.

8. Calculate the value of the constant current, I_E, for the circuit of Fig. 13.56.

9. Calculate the dc collector voltage in the circuit of Fig. 13.56.

10. Calculate the constant current in a circuit such as that in Fig. 13.18 with values $V_Z = 6.8$ V, $R_E = 1.8$ kΩ, and $R_1 = 5$ kΩ.

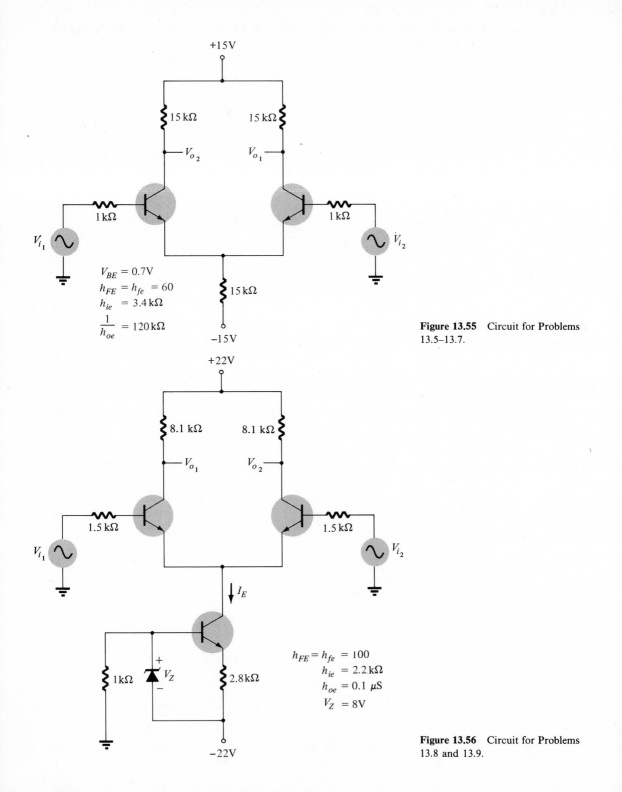

$V_{BE} = 0.7V$
$h_{FE} = h_{fe} = 60$
$h_{ie} = 3.4 \, k\Omega$
$\frac{1}{h_{oe}} = 120 \, k\Omega$

Figure 13.55 Circuit for Problems 13.5–13.7.

$h_{FE} = h_{fe} = 100$
$h_{ie} = 2.2 \, k\Omega$
$h_{oe} = 0.1 \, \mu S$
$V_Z = 8V$

Figure 13.56 Circuit for Problems 13.8 and 13.9.

11. Using the values of Problem 10, calculate the ac output resistance of circuit ($h_{fe} = 150$, $1/h_{oe} = 100$ kΩ, $h_{ie} = 1150$ Ω).

§ 13.3

12. An amplifier with differential voltage gain of 200 and common-mode rejection of 80 dB has what value common-mode gain?

13. What is the common-mode rejection ratio in dB of an amplifier having differential gain of 180 and common-mode gain of 0.5?

14. Calculate the differential- and common-mode gains for the circuit of Fig. 13.26a and the value of CMRR if $R_C = 4.3$ kΩ, $R_E = 11$ kΩ, and transistor $h_{fe} = 120$ and $h_{ie} = 1.5$ kΩ.

15. Calculate the differential- and common-mode gains and the value of CMRR for the circuit of Fig. 13.26b modified by $R_C = 4.3$ kΩ, $h_{fe} = 120$, $1/h_{oe} = 100$ kΩ, and $h_{ie} = 1.5$ kΩ.

16. Calculate the output voltage of a differential amplifier for inputs of $V_{i_1} = 0.5$ mV, $V_{i_2} = 0.45$ mV, $A_d = 450$, and CMRR $= 10^4$.

§ 13.4

17. The input resistance of a differential amplifier is measured using a 25-kΩ resistor in series with an input voltage of 5 V. What is the value of R_i if the voltage into the amplifier is 1.5 V?

18. A differential amplifier has single-ended gain of $A_1 = 120$. When determining A_c the circuit measurements are $V_i = 2$ V and $V_o = 20$ mV. Calculate the value of CMRR in dB.

§ 13.5

19. What is meant by "virtual ground"?

§ 13.6

20. Calculate the output voltage of a noninverting op-amp circuit (such as in Fig. 13.35) for values of $V_1 = 4$ V, $R_f = 250$ kΩ, and $R_1 = 50$ kΩ.

21. Calculate the output voltage of a three-input summing amplifier (as in Fig. 13.38) for the following values: $R_1 = 200$ kΩ, $R_2 = 250$ kΩ, $R_3 = 500$ kΩ, $R_f = 1$ MΩ, $V_1 = -2$ V, $V_2 = +2$ V, and $V_3 = 1$ V.

22. Determine the output voltage of the circuits of Fig. 13.57.

23. Repeat Problem 13.22 for the circuit of Fig. 13.58.

24. Draw the output waveform for the circuit and inputs of Fig. 13.59.

25. Repeat Problem 13.24 for Fig. 13.60.

§ 13.7

26. What is the constant current output of the circuit of Fig. 13.44a with $R_2 = 1.8$ kΩ, $R_3 = 3.3$ kΩ, and $R_1 = 1.8$ kΩ?

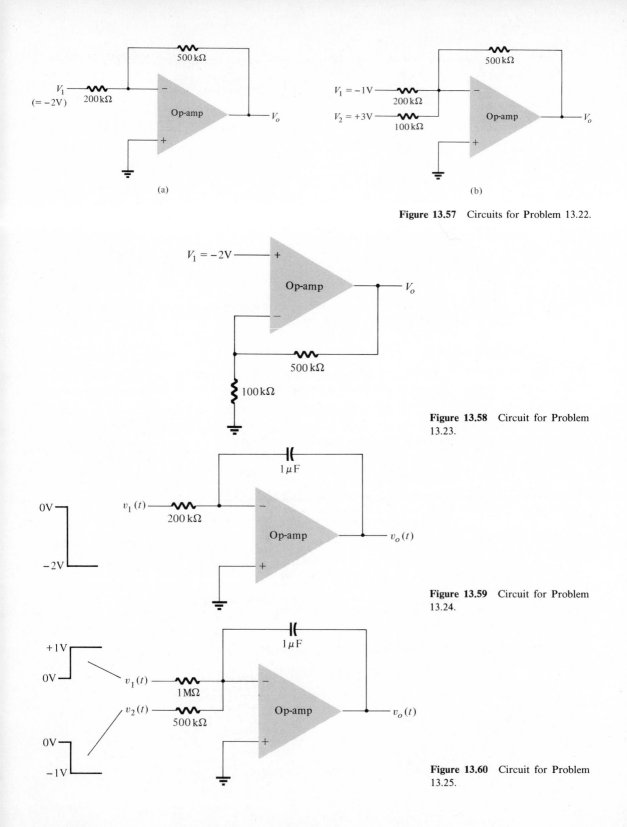

Figure 13.57 Circuits for Problem 13.22.

(a)

(b)

Figure 13.58 Circuit for Problem 13.23.

Figure 13.59 Circuit for Problem 13.24.

Figure 13.60 Circuit for Problem 13.25.

27. What is the value of output current, I_o, if $R_2 = 2.7$ kΩ, $R_3 = 1.8$ kΩ, and $R_1 = 1.2$ kΩ in the circuit of Fig. 13.44b?

28. Calculate the cutoff frequency of the filter circuit in Fig. 13.61a.

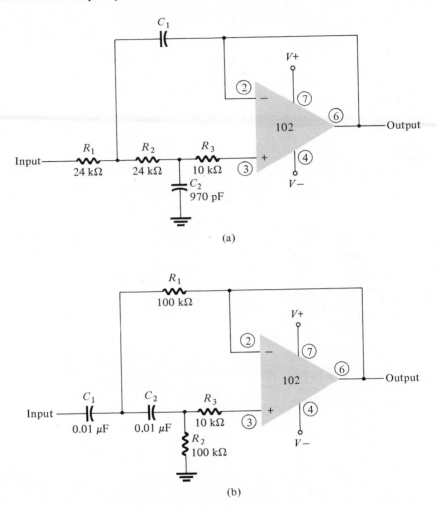

(a)

(b)

Figure 13.61 Active filter circuits for Problems 13.28 and 13.29.

29. Calculate the cutoff frequency of the filter circuit of Fig. 13.61b. Plot the ideal response curve of the filter.

14

Linear ICs:
Regulators
(Including Filters
and Power Supplies)

14.1 GENERAL

Voltage regulators are a popular group of linear ICs. A voltage regulator IC receives input of a fairly constant dc voltage and supplies as output a somewhat lower value of dc voltage, which the regulator maintains fixed or regulated over a wide range of load current, or input voltage, variation. Starting with an ac voltage supply, a steady dc voltage can be developed by rectifying the ac voltage, then filtering to a dc level, and finally regulating with an IC voltage-regulator circuit.

The present chapter introduces the operation of filter capacitors and the overall operation of converting an ac voltage into a dc voltage using transformer, rectifier, and filter. Refer back to Chapter 2 for coverage of diode rectifier circuits.

Voltage regulator ICs that provide a fixed output voltage are available in a range of output voltage values. Regulator ICs are selected to operate with either positive or negative voltages. Voltage regulators are also available to provide output at any set voltage over a range of values set by external resistor values.

14.2 GENERAL FILTER CONSIDERATIONS

A rectifier circuit is necessary to convert a signal having zero average value to one that has a nonzero average. However, the resulting pulsating dc signal is not pure dc or even a good representation of it. Of course, for a circuit such as a battery charger the pulsating nature of the signal is no great detriment as long as the dc level provided will result in charging of the battery. On the other hand, for voltage supply circuits for a tape recorder or radio the pulsating dc will result in a 60-(or 120-)Hz signal appearing in the output, thereby making the operation of the overall circuit poor. For these applications, as well as for many more, the output dc developed will have to be much "smoother" than that of the pulsating dc obtained directly from half-wave or full-wave rectifier circuits.

Filter Voltage Regulation and Ripple Voltage

Before going into the details of the filter circuit it would be appropriate to consider the usual method of rating the circuits so that we are able to compare a circuit's effectiveness as a filter. Figure 14.1 shows a typical filter output voltage, which will be used to define some of the signal factors. The filtered output voltage of Fig. 14.1 has a dc value and some ac variation *(ripple)*. Although a battery has essentially a constant or dc output voltage, the dc voltage derived from an ac source signal by rectifying and filtering will have some variation (ripple). The smaller the ac variation *with respect to* the dc level the better the filter circuit operation.

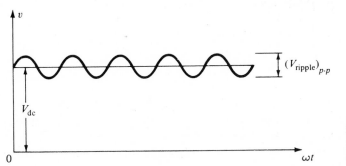

Figure 14.1 Filter voltage waveform showing dc and ripple voltages.

Consider measuring the output voltage of the filter circuit using a dc voltmeter and an ac (rms) voltmeter. The dc voltmeter will read only the average or dc level of the output voltage. The ac (rms) meter will read only the rms value of the ac component of the output voltage (assuming the signal is coupled to the meter through a capacitor to block out the dc level).

Definition: Ripple:

$$r = \text{ripple} = \frac{\text{ripple voltage (rms)}}{\text{dc voltage}} = \frac{V_r\,(\text{rms})}{V_{dc}} \times 100\% \qquad (14.1)$$

EXAMPLE 14.1 Using a dc and ac voltmeter to measure the output signal from a filter circuit, a dc voltage of 25 V and an ac ripple voltage of 1.5 V (rms) are obtained. Calculate the ripple of the filter output.

Solution:

$$r = \frac{V_r(\text{rms})}{V_{dc}} \times 100\% = \mathbf{6\%}$$

Another factor of importance in a voltage supply is the amount of change in the output dc voltage over the range of the circuit operation. The voltage provided at the output at no-load (no current drawn from the supply) is reduced when load current is drawn from the supply. How much this voltage changes with respect to either the loaded or unloaded voltage value is of considerable interest to anyone using the supply. This voltage change is described by a factor called *voltage regulation*.

Definition: Voltage Regulation:

$$\text{Voltage regulation} = \frac{\text{voltage at no-load} - \text{voltage at full-load}}{\text{voltage at full-load}}$$

$$V.R. = \frac{V_{NL} - V_{FL}}{V_{FL}} \times 100\% \qquad (14.2)$$

EXAMPLE 14.2 A dc voltage supply provides 60 V when the output is unloaded. When full-load current is drawn from the supply, the output voltage drops to 56 V. Calculate the value of voltage regulation.

Solution:

$$V.R. = \frac{V_{NL} - V_{FL}}{V_{FL}} = \frac{60 - 56}{56} \times 100\% = \textbf{7.1\%}$$

If the value of full-load voltage is the same as the no-load voltage, the *V.R.* calculated is 0%, which is the best to expect. This value means that the supply is a true voltage source for which the output voltage is independent of current drawn from the supply. The output voltage from most supplies decreases as the amount of current drawn from the voltage supply is increased. The smaller the voltage decreases, the smaller the percent of *V.R.* and the better the operation of the voltage supply circuit.

RIPPLE FACTOR OF RECTIFIED SIGNAL

Although the rectified voltage is not a filtered voltage, it nevertheless contains a dc component and a ripple component. We can calculate these values of dc voltage and ripple voltage (rms) and from them obtain the ripple factor for the half-wave and full-wave rectified voltages. The calculations will show that the full-wave rectified signal has less percent of ripple and is therefore a better rectified signal than the half-wave rectified signal, if lowest percent of ripple is desired. The percent of ripple is not always the most important concern. If circuit complexity or cost considerations are important (and the percent of ripple is secondary), then a half-wave rectifier may be satisfactory. Also, if the filtered output supplies only a small amount of current to the load and the filtering circuit is not critical, then a half-wave rectified signal may be acceptable. On the other hand, when the supply must have as low a ripple as possible, it is best to start with a full-wave rectified signal since it has a smaller ripple factor, as will now be shown.

For a half-wave rectified signal the output dc voltage is $V_{dc} = 0.318 V_m$. The rms value of the ac component of output signal can be calculated (see Appendix B), and is V_r (rms) $= 0.385 V_m$. Calculating the percent of ripple,

$$r = \frac{V_r \text{(rms)}}{V_{dc}}(100) = \frac{0.385 V_m}{0.318 V_m}(100) = 1.21(100) = 121\% \qquad \text{(half wave)} \qquad (14.3)$$

For the full-wave rectifier the value of V_{dc} is $V_{dc} = 0.636 V_m$. From the results obtained in Appendix B the ripple voltage of a full-wave rectified signal is V_r (rms) $= 0.308 V_m$. Calculating the percent of ripple,

$$r = \frac{V_r \text{(rms)}}{V_{dc}}(100) = \frac{0.308 V_m}{0.636 V_m}(100) = 48\% \qquad \text{(full-wave)} \qquad (14.4)$$

The amount of ripple factor of the full-wave rectified signal is about 2.5 times smaller than that of the half-wave rectified signal and provides a better filtered signal. Note that these values of ripple factor are absolute values and do not depend at all on the peak voltage. If the peak voltage is made larger, the dc value of the output increases but then so does the ripple voltage. The two increase in the same proportion so that the ripple factor stays the same.

14.3 SIMPLE-CAPACITOR FILTER

A popular filter circuit is the simple-capacitor filter circuit shown in Fig. 14.2. The capacitor is connected across the rectifier output and the dc output voltage is available across the capacitor. Figure 14.3a shows the rectifier output voltage of a full-wave rectifier circuit before the signal is filtered. Figure 14.3b shows the resulting waveform after the capacitor is connected across the rectifier output. As shown this filtered voltage has a dc level with some ripple voltage riding on it.

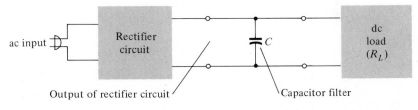

ac input

Rectifier circuit

C

dc load (R_L)

Output of rectifier circuit   Capacitor filter

Figure 14.2 Simple capacitor filter.

Figure 14.4 shows a full-wave rectifier and the output waveform obtained from the circuit when connected to an output load. If no load were connected to the filter, the output waveform would ideally be a constant dc level equal in value to the peak voltage *(V_m)* from the rectifier circuit. However, the purpose of obtaining a dc voltage is to provide this voltage for use by other electronic circuits, which then constitute a load on the voltage supply. Since there will always be some load on the filter, we must consider this practical case in our discussion. For the full-wave rectified signal indicated in Fig. 14.4b there are two intervals of time indicated. T_1 is the time during which a diode of the full-wave rectifier conducts and charges

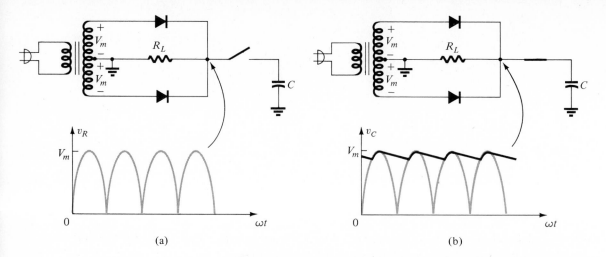

(a)

(b)

Figure 14.3 Capacitor filter operation: (a) full-wave rectifier voltage; (b) filtered output voltage.

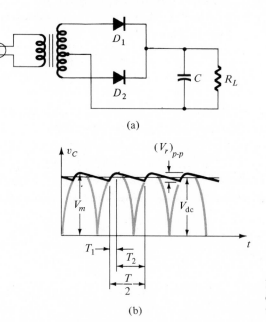

(a)

(b)

Figure 14.4 Capacitor filter: (a) capacitor filter circuit; (b) output voltage waveform.

the capacitor up to the peak rectifier output voltage (V_m). T_2 is the time during which the rectifier voltage drops below the peak voltage, and the capacitor discharges through the load.

If the capacitor were to discharge only slightly (due to a light load), the average voltage would be very close to the optimum value of V_m. The amount of ripple voltage would also be small for a light load. This shows that the capacitor filter circuit provides a large dc voltage with little ripple for light loads (and a smaller dc voltage with larger ripple for heavy loads). To appreciate these quantities better we must further examine the output waveform and determine some relations between the input signal to be rectified, the capacitor value, the resistor (load) value, the ripple factor, and the regulation of the circuit.

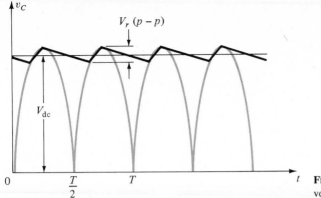

$V_r (p-p)$

V_{dc}

$0 \qquad \dfrac{T}{2} \qquad T \qquad\qquad\qquad\qquad t$

Figure 14.5 Approximate output voltage of capacitor filter circuit.

Figure 14.5 shows the output waveform approximated by straight line charge and discharge. This is reasonable since the analysis with the nonlinear charge and discharge that actually takes place is complex to analyze and because the results obtained will yield values that agree well with actual measurements made on circuits. The waveform of Fig. 14.5 shows the approximate output voltage waveform for a full-wave rectified signal. From an analysis of this voltage waveform the following relations can be obtained:

$$V_{dc} = V_m - \frac{V_r(p-p)}{2}$$ (half-wave   (14.5)

and

$$V_r(\text{rms}) = \frac{V_r(p-p)}{2\sqrt{3}}$$ full wave)   (14.6)

These relations, however, are only in terms of the waveform voltages and we must further relate them to the different components in the circuit. Since the form of the ripple waveform for half-wave is the same as for full-wave, Eqs. (14.5) and (14.6) apply to both rectifier-filter circuits.

Ripple Voltage, V_r (rms)

Appendix B provides the details for determining the value of the ripple voltage in terms of the other circuit parameters. The result obtained for V_r (rms) is the following:

$$V_r(\text{rms}) \cong \frac{I_{dc}}{4\sqrt{3}fC} \times \frac{V_{dc}}{V_m} \qquad \text{(full-wave)} \qquad (14.7a)$$

where f is the frequency of the sinusoidal ac power supply voltage (usually 60 Hz), I_{dc} is the average current drawn from the filter by the load, and C is the filter capacitor value.

Another simplifying approximation that can be made is to assume that when used typically for light loads[1] the value of V_{dc} is only slightly less than V_m so that

---

[1] Appendix B shows the relation of V_{dc} and V_m based on the amount of ripple. From Fig. B.3 we see that at ripple factors less than 6.5%, V_{de} is within 10% of V_m. We can therefore define a *light load* as one resulting in a ripple less than 6.5%.

$V_{dc} \cong V_m$, and the equation can be written as

$$V_r\,(\text{rms}) \cong \frac{I_{dc}}{4\sqrt{3}\,fC} \qquad \text{(full-wave, light load)} \qquad (14.7b)$$

Finally, we can include the typical value of line frequency ($f = 60$ Hz) and the other constants into the simpler equation

$$V_r\,(\text{rms}) = \frac{2.4\,I_{dc}}{C} = \frac{2.4\,V_{dc}}{R_L C} \qquad \text{(full-wave, light load)} \qquad (14.7c)$$

where I_{dc} is in milliamperes, C is in microfarads, and R_L is in kilohms.

> **EXAMPLE 14.3** Calculate the ripple voltage of a full-wave rectifier with a 100 μF filter capacitor connected to a load of 50 mA.
>
> **Solution:** Using Eq. (14.7c), we get
>
> $$V_r\,(\text{rms}) = \frac{2.4(50)}{100} = \textbf{1.2 V}$$

dc Voltage, V_{dc}

Using Eqs. (14.5), (14.6), and (14.7a), we see that the dc voltage of the filter is

$$V_{dc} = V_m - \frac{V_r\,(p-p)}{2} = V_m - \frac{I_{dc}}{4fC} \times \frac{V_{dc}}{V_m} \qquad \text{(full-wave)} \qquad (14.8a)$$

Again, using the simplifying assumption that V_{dc} is about the same as V_m for light loads, we get an approximate value of V_{dc} (which is less than V_m), of

$$V_{dc} = V_m - \frac{I_{dc}}{4fC} \qquad \text{(full-wave, light load)} \qquad (14.8b)$$

which can be written (using $f = 60$ Hz):

$$V_{dc} = V_m - \frac{4.17\,I_{dc}}{C} \qquad \text{(full-wave, light load)} \qquad (14.8c)$$

where V_m is the peak rectified voltage, in volts, I_{dc} is the load current in milliamperes, and C is the filter capacitor in microfarads.

> **EXAMPLE 14.4** If the peak rectified voltage for the filter circuit of Example 14.3 is 30 V, calculate the filter dc voltage.
>
> **Solution:** Using Eq. (14.8c), we get
>
> $$V_{dc} = V_m - \frac{4.17\,I_{dc}}{C} = 30 - \frac{4.17(50)}{100} = \textbf{27.9 V}$$

The value of dc voltage is less than the peak rectified voltage. Note, also, from

Eq. (14.13c), that the larger the value of average current drawn from the filter the less the value of output dc voltage, and the larger the value of the filter capacitor the closer the output dc voltage approaches the peak value of V_m.

Filter-Capacitor Ripple

Using the definition of ripple [Eq. (14.3)] and the equation for ripple voltage [Eq. (14.7c)], we obtain the expression for the ripple factor of a full-wave capacitor filter

$$r = \frac{V_r\,(\text{rms})}{V_{dc}} \times 100\% \cong \frac{2.4 I_{dc}}{C V_{dc}} \times 100\% \qquad \text{(full-wave, light load)} \qquad (14.9a)$$

Since V_{dc} and I_{dc} relate to the filter load R_L, we can also express the ripple as

$$r = \frac{2.4}{R_L C} \times 100\% \qquad \text{(full-wave, light load)} \qquad (14.9b)$$

where I_{dc} is in milliamperes, C is in microfarads, V_{dc} is in volts, and R_L is in kilohms.

This ripple factor is seen to vary directly with the load current (larger load current, larger ripple factor), and inversely with the capacitor size. This agrees with the previous discussion of the filter circuit operation.

EXAMPLE 14.5   A load current of 50 mA is drawn from a capacitor filter circuit ($C = 100\ \mu F$). If the peak rectified voltage is 30 V, calculate r.

Solution: Using the results of Examples 14.3 and 14.4 in Eq. (14.9a), we get

$$r = \frac{2.4 I_{dc}}{C V_{dc}} \times 100\% = \frac{2.4(50)}{(100)(27.9)} \times 100\% = \mathbf{4.3\%}$$

From the basic definition of r we could also calculate

$$r = \frac{V_r\,(\text{rms})}{V_{dc}} \times 100\% = \frac{1.2}{27.9} \times 100\% = 4.3\%$$

Diode Conduction Period and Peak Diode Current

From the previous discussion it should be clear that larger values of capacitance provide less ripple and higher average voltages, thereby providing better filter action. From this one may conclude that to improve the performance of a capacitor filter it is only necessary to increase the size of the filter capacitor. However, the capacitor also affects the peak current through the rectifying diode and, as will now be shown, the larger the value of capacitance used the larger the peak current through the rectifying diode.

Referring back to the operation of the rectifier and capacitor filter circuit, we see that there are two periods of operation to consider. After the capacitor is charged to the peak rectified voltage (see Fig. 14.4b), a period of diode nonconduction elapses

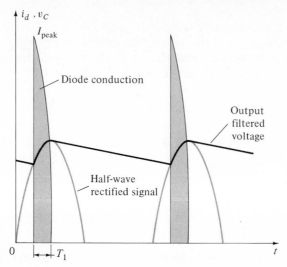

Figure 14.6 Diode conduction during charging part of cycle.

(time T_2) while the output voltage discharges through the load. After T_2 the input rectified voltage becomes greater than the capacitor voltage and for a time, T_1, the capacitor will charge back up to the peak rectified voltage. The average current supplied to the capacitor and load during this charge period must equal the average current drawn from the capacitor during the discharge period. Figure 14.6 shows the diode current waveform for half-wave rectifier operation. Notice that the diode conducts for only a short period of the cycle. In fact, it should be seen that the larger the capacitor, the less the amount of voltage decay and the shorter the interval during which charging takes place. In this shorter charging interval the diode will have to pass the same amount of *average current,* and can do so only by passing larger peak current. Figure 14.7 shows the output current and voltage waveforms for small and large capacitor values. The important factor to note is the increase in peak current through the diode for the larger values of capacitance. Since the average current drawn from the supply must equal the average of the current through the diode during the charging period, the following relation can be derived from Fig. 14.7:[2]

$$I_{dc} = \frac{T_1}{T} \times I_{peak} \qquad (14.10a)$$

from which we obtain

$$\boxed{I_{peak} = \frac{T}{T_1} \times I_{dc}} \qquad (14.10b)$$

where $T_1 =$ diode conduction time;
$T = 1/f = \frac{1}{60}$ for usual line 60-Hz voltage;
$I_{dc} =$ average current drawn from filter circuit; and
$I_{peak} =$ peak current through the conducting diode.

---

[2] Assuming a rectangular-shaped pulse of duration T_1, peak value, I_{peak}, and a period of T, the area under the pulse divided by the period gives the average value, I_{dc}:

$$\frac{1}{T} (I_{peak} T_1) = I_{dc}$$

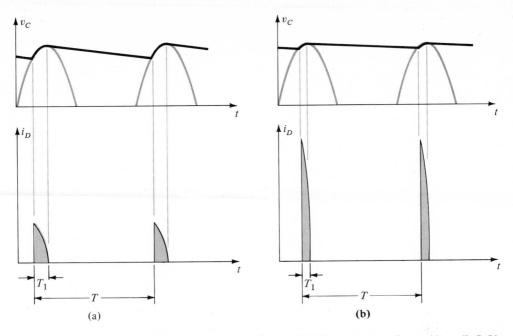

Figure 14.7 Output voltage and diode current waveforms: (a) small C; (b) large C.

14.4 *RC* FILTER

It is possible to further reduce the amount of ripple across a filter capacitor while reducing the dc voltage by using an additional *RC* filter section as shown in Fig. 14.8. The purpose of the added network is to pass as much of the dc component of the voltage developed across the first filter capacitor C_1 and to attenuate as much of the ac component of the ripple voltage developed across C_1 as possible. This action would reduce the amount of ripple in relation to the dc level, providing better filter operation than for the simple-capacitor filter. There is a price to pay for this improvement, as will be shown; this includes a lower dc output voltage due to the dc voltage drop across the resistor and the cost of the two additional components in the circuit.

Figure 14.9 shows the rectifier filter circuit for full-wave operation. Since the rectifier feeds directly into a capacitor, the peak currents through the diodes are many times the average current drawn from the supply. The voltage developed across capacitor C_1 is then further filtered by the resistor-capacitor section (R, C_2) providing

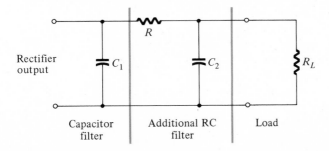

Figure 14.8 *RC* filter stage.

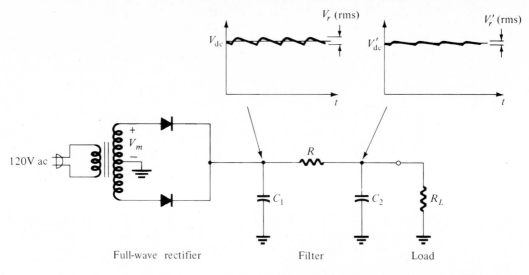

Figure 14.9 Full-wave rectifier and RC filter circuit.

an output voltage having less percent of ripple than that across C_1. The load, represented by resistor R_L, draws dc current through resistor R with an output dc voltage across the load being somewhat less than that across C_1 due to the voltage drop across R. This filter circuit, like the simple-capacitor filter circuit, provides best operation at light loads, with considerably poorer voltage regulation and higher percent of ripple at heavy loads.

The analysis of the resulting ac and dc voltages at the output of the filter from that obtained across capacitor C_1 can be carried out by using superposition. We can separately consider the RC circuit acting on the dc level of the voltage across C_1 and then the RC circuit action on the ac (ripple) portion of the signal developed across C_1. The resulting values can then be used to calculate the overall circuit voltage regulation and ripple.

dc Operation of *RC* Filter Section

Figure 14.10a shows the equivalent circuit to use when considering the dc voltage and current in the filter and load. The two filter capacitors are open circuit for dc

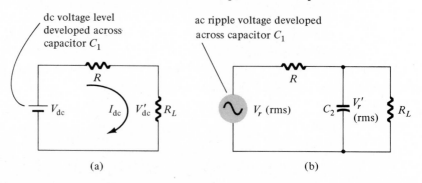

Figure 14.10 Dc and ac equivalent circuits of RC filter: (a) dc equivalent circuit; (b) ac equivalent circuit.

and are thus removed from consideration at this time. Calculation of the dc voltage across filter capacitor C_1 was essentially covered in Section 14.3 and the treatment of the additional RC filter stage will proceed from there. Knowing the dc voltage across the first filter capacitor (C_1), we can calculate the dc voltage at the output of the additional RC filter section. From Fig. 14.10a we see that the voltage, V_{dc}, across capacitor C_1 is attenuated by a resistor-divider network of R and R_L (the equivalent load resistance); the resulting dc voltage across the load being V'_{dc}:

$$V'_{dc} = \frac{R_L}{R + R_L} \times V_{dc} \qquad (14.11)$$

EXAMPLE **14.6**   The addition of an RC filter section with $R = 120 \ \Omega$, reduces the dc voltage across the initial filter capacitor from 60 V (V_{dc}). If the load resistance is 1 kΩ, calculate the value of the output dc voltage (V'_{dc}) from the filter circuit.

Solution: Using Eq. (14.1), we get

$$V'_{dc} = \frac{R_L}{R + R_L} \times V_{dc} = \frac{1000}{120 + 1000} \times 60 = \textbf{53.6 V}$$

In addition, we may calculate the drop across the filter resistor and the load current drawn:

$$V_R = V_{dc} - V'_{dc} = 60 - 53.6 = 6.4 \text{ V}$$

$$I_{dc} = \frac{V'_{dc}}{R_L} = \frac{53.6}{1 \times 10^3} = 53.6 \text{ mA}$$

ac Operation of *RC* Filter Section

Figure 14.10b shows the equivalent circuit for analyzing the ac operation of the filter circuit. The input to the filter stage from the first filter capacitor (C_1) is the ripple or ac signal part of the voltage across C_1, V_r (rms), which is approximated now as a sinusoidal signal. Both the RC filter stage components and the load resistance affect the ac signal at the output of the filter.

For a filter capacitor (C_2) value of 10 μF at a ripple voltage frequency (f) of 60 Hz, the ac impedance of the capacitor is[3]

$$X_C = \frac{1}{\omega C} = \frac{1}{2\pi f C} = \frac{1}{6.28(60)(10 \times 10^{-6})} = 0.265 \text{ k}\Omega$$

Referring to Fig. 2.22b, we see that this capacitive impedance is in parallel with the load resistance. For a load resistance of 2 kΩ, for example, the parallel combination of the two components would yield an impedance of magnitude:

$$Z = \frac{RX_C}{\sqrt{R^2 + X_C^2}} = \frac{2(0.265)}{\sqrt{2^2 + (0.265)^2}} = \frac{2}{2.02}(0.265) = 0.263 \text{ k}\Omega$$

This is close to the value of the capacitive impedance alone, as expected, since the

---

[3] X_C is understood here to represent only the *magnitude* of the capacitor's ac impedance.

capacitive impedance is much less than the load resistance and the parallel combination of the two would be smaller than the value of either. As a rule of thumb we can consider neglecting the loading by the load resistor on the capacitive impedance as long as the load resistance is at least five times as large as the capacitive impedance. Because of the limitation of light loads on the filter circuit the effective value of load resistance is usually large compared to the impedance of capacitors in the range of microfarads.

In the above discussion it was stated that the frequency of the ripple voltage was 60 Hz. Assuming that the line frequency was 60 Hz, the ripple frequency will also be 60 Hz for the ripple voltage from a half-wave rectifier. The ripple voltage from a full-wave rectifier, however, will be double since there are twice the number of half-cycles and the ripple frequency will then be 120 Hz. Referring to the relation for capacitive impedance $X_C = 1/\omega C$, we have value of $\omega = 377$ for 60 Hz and of $\omega = 754$ for 120 Hz. Using values of capacitance in μF, we can express the relation for capacitive impedance as

$$X_C = \frac{2.653}{C} \quad \text{(half-wave)} \tag{14.12a}$$

$$X_C = \frac{1.326}{C} \quad \text{(full-wave)} \tag{14.12b}$$

where C is in microfarads and X_C is in kilohms.[4]

> **EXAMPLE 14.7** Calculate the impedance of a 15-μF capacitor used in the filter section of a circuit using full-wave rectification.

> **Solution:**

$$X_C = \frac{1.326}{C} = \frac{1.326}{15} = 0.0884 \text{ k}\Omega = \mathbf{88.4 \ \Omega}$$

Using the simplified relation that the parallel combination of the load resistor and the capacitive impedance equals, approximately, the capacitive impedance, we can calculate the ac attenuation in the filter stage:

$$V_r'(\text{rms}) = \frac{X_C}{\sqrt{R^2 + X_C^2}} \, V_r(\text{rms}) \tag{14.13a}$$

The use of the square root of the sum of the squares in the denominator was necessary since the resistance and capacitive impedance must be added vectorially, not algebraically. If the value of the resistance is larger by a factor of 5 than that of the capacitive

---

[4] Equations (14,12a) and (14.12b) may also be expressed as

$$X_C = \frac{2653}{C} \quad \text{(half-wave)}$$

$$X_C = \frac{1326}{C} \quad \text{(full-wave)}$$

where C is in microfarads and X_C is in ohms.

impedance, then a simplification of the denominator may be made yielding the following result:

$$V'_r(\text{rms}) \cong \frac{X_C}{R} \cdot V_r(\text{rms}) \qquad (14.13b)$$

EXAMPLE 14.8 The output of a full-wave rectifier and capacitor filter is further filtered by an RC filter section (see Fig. 14.11). The component values of the *RC* section are $R = 500\ \Omega$ and $C = 10\ \mu\text{F}$. If the initial capacitor filter develops 150 V dc with a 15 V ac ripple voltage, calculate the resulting dc and ripple voltage across a 5-kΩ load.

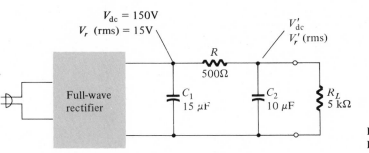

Figure 14.11 *RC* filter circuit for Example 14.8.

Solution:

dc calculations: Calculating the value of V'_{dc} from Eq. (14.11):

$$V'_{dc} = \frac{R_L}{R + R_L} \times V_{dc} = \frac{5000}{500 + 5000}(150) = \frac{5000}{5500}(150) = \textbf{136.4 V}$$

ac calculations: Calculating the value of the capacitive impedance first (for full-wave operation):

$$X_C = \frac{1.326}{C} = \frac{1.326}{10} = 0.133\ \text{k}\Omega = 133\ \Omega$$

Since this impedance is not quite 5 times smaller than that of the filter resistor ($R = 500\ \Omega$), we shall use Eq. (14.13a) for the calculation, and then repeat the calculation to show what the difference would have been using Eq. (14.13b) (since the components are almost 5 times different in size). Using Eq. (14.13a), we get

$$V'_r(\text{rms}) = \frac{X_C}{\sqrt{R^2 + X_C^2}}\, V_r(\text{rms}) = \frac{0.133}{\sqrt{(0.5)^2 + (0.133)^2}}(15)$$

$$= \frac{0.133}{0.517}(15) = \textbf{3.86 V}$$

Now using Eq. (14.13b), we get

$$V'_r(\text{rms}) = \frac{X_C}{R} \cdot V_r(\text{rms}) = \frac{0.133}{0.500}(15) = \textbf{3.99 V}$$

Comparing the results of 3.86 V and 3.99 V, use of Eq. (14.13b) would have yielded an answer within 3.5% of the more exact solution.

14.5 VOLTAGE-MULTIPLIER CIRCUITS

Voltage Doubler

A modification of the capacitor filter circuit allows building up a larger voltage than the peak rectified voltage (V_m). The use of this type of circuit allows keeping the transformer peak voltage rating low while stepping up the peak output voltage to two, three, four, or more times the peak rectified voltage.

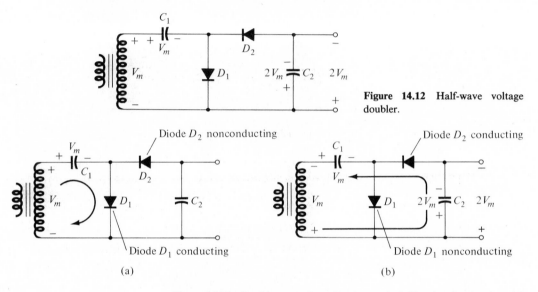

Figure 14.12 Half-wave voltage doubler.

Figure 14.13 Double operation, showing each half-cycle of operation: (a) positive half-cycle; (b) negative half-cycle.

Figure 14.12 shows a half-wave voltage doubler. During the positive-voltage half-cycle across the transformer, secondary diode D_1 conducts (and diode D_2 is cut off), charging capacitor C_1 up to the peak rectified voltage (V_m). Diode D_1 is ideally a short during this half-cycle and the input voltage charges capacitor C_1 to V_m with the polarity shown in Fig. 14.13a. During the negative half-cycle of the secondary voltage, diode D_1 is cut off and diode D_2 conducts charging capacitor C_2. Since diode D_2 acts as a short during the negative half-cycle (and diode D_1 is open), we can sum the voltages around the outside loop (see Fig. 14.13b):

$$-V_{C_2} + V_{C_1} + V_m = 0$$
$$-V_{C_2} + V_m + V_m = 0$$

from which

$$V_{C_2} = 2V_m$$

On the next positive half-cycle, diode D_2 is nonconducting and capacitor C_2 will discharge through the load. If no load is connected across capacitor C_2, both capacitors stay charged—C_1 to V_m and C_2 to $2V_m$. If, as would be expected, there is a load connected to the output of the voltage doubler, the voltage across capacitor C_2 drops during the positive half-cycle (at the input) and the capacitor is recharged up to $2V_m$ during the negative half-cycle.

The output waveform across capacitor C_2 is that of a half-wave signal filtered by a capacitor filter. The peak inverse voltage across each diode is $2V_m$.

Another doubler circuit is the full-wave doubler of Fig. 14.14. During the positive half-cycle of transformer secondary voltage (see Fig. 14.15a) diode D_1 conducts charging capacitor C_1 to a peak voltage V_m. Diode D_2 is nonconducting at this time.

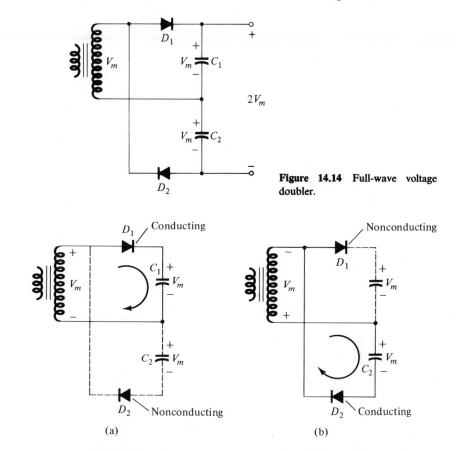

Figure 14.14 Full-wave voltage doubler.

Figure 14.15 Alternate half-cycles of operation for full-wave voltage doubler.

During the negative half-cycle (see Fig. 14.15b) diode D_2 conducts charging capacitor C_2 while diode D_1 is nonconducting. If no load current is drawn from the circuit, the voltage across capacitors C_1 and C_2 is $2V_m$. If load current is drawn from the circuit, the voltage across capacitors C_1 and C_2 is the same as that across a capacitor fed by a full-wave rectifier circuit. One difference is that the effective capacitance is that of C_1 and C_2 in series, which is less than the capacitance of either C_1 or C_2 alone. The lower capacitor value will provide poorer filtering action than the single-capacitor filter circuit.

The peak inverse voltage across each diode is $2V_m$ as it is for the filter capacitor circuit. In summary, the half-wave or full-wave voltage doubler circuits provide twice the peak voltage of the transformer secondary while requiring no center-tapped transformer and only $2V_m$ PIV rating for the diodes.

Voltage Tripler and Quadrupler

Figure 14.16 shows an extension of the half-wave voltage doubler, which develops three and four times the peak input voltage. It should be obvious from the pattern of the circuit connection how additional diodes and capacitors may be connected so that the output voltage may also be five, six, seven, etc., times the basic peak voltage (V_m).

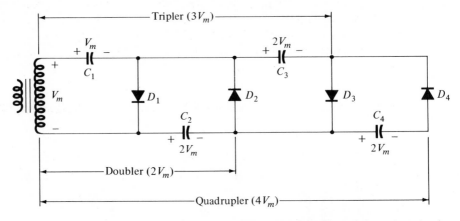

Figure 14.16 Voltage tripler and quadrupler.

In operation capacitor C_1 charges through diode D_1 to a peak voltage, V_m, during the positive half-cycle of the transformer secondary voltage. Capacitor C_2 charges to twice the peak voltage $2V_m$ developed by the sum of the voltages across capacitor C_1 and the transformer, during the negative half-cycle of the transformer secondary voltage.

During the positive half-cycle, diode D_3 conducts and the voltage across capacitor C_2 charges capacitor C_3 to the same $2V_m$ peak voltage. On the negative half-cycle, diodes D_2 and D_3 conduct with capacitor C_3, charging C_4 to $2V_m$.

The voltage across capacitor C_2 is $2V_m$, across C_1 and C_3 it is $3V_m$, and across C_2 and C_4 it is $4V_m$. If additional sections of diode and capacitor are used, each capacitor will be charged to $2V_m$. Measuring from the top of the transformer winding (Fig. 14.16) will provide odd multiples of V_m at the output, whereas measuring from the bottom of the transformer the output voltage will provide even multiples of the peak voltage, V_m.

The transformer rating is only V_m, maximum, and each diode in the circuit must be rated at $2V_m$ PIV. If the load is small and the capacitors have little leakage, extremely high dc voltages may be developed by this type of circuit, using many sections to step up the dc voltage.

14.6 DISCRETE VOLTAGE REGULATORS

Regulation Defined

A wide variety of circuit configurations are capable of performing voltage or current regulation. Only a few of the more commonly applied circuits will be considered in this text. Voltage and current regulation can best be defined through the use of the circuits of Figs. 14.17 and 14.18, respectively.

In Fig. 14.17a the no-load (open-circuit) terminal voltage of the supply is designated by V_{NL}. The corresponding load current is $I_{NL} = 0$. The full-load conditions are shown in Fig. 14.17b. The ideal situation would require that $V_L = V_{FL} = V_{NL}$ for every value of R_L between no-load and full-load conditions. In other words, the terminal voltage V_L would be unaffected by variations in R_L. Unfortunately, there is no supply available today, whether it be semiconductor or electromechanical (generator), that can provide a terminal voltage completely independent of the load applied. Voltage regulation is defined formally as in Eq. (14.2). Current regulation is defined by

$$\text{current regulation} = \frac{I_{NL} - I_{FL}}{I_{FL}} \times 100\% \qquad (14.14)$$

where I_{NL} and I_{FL} and the no-load full-load currents as defined by Fig. 14.18.

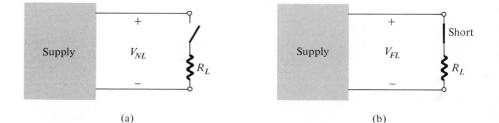

Figure 14.17 Voltage regulation: (a) no-load *(NL)* state; (b) full-load *(FL)* state.

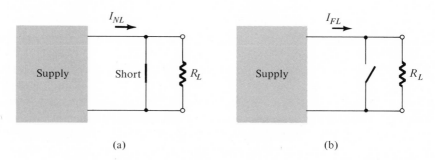

Figure 14.18 Current regulation: (a) no-load *(NL)* state; (b) full-load *(FL)* state.

Zener and Thermistor Voltage Regulators

There are, fundamentally, two basic configurations for establishing voltage or current regulation. Each is shown in Fig. 14.19. The usual terminology for each is indicated in the figure. You will find as you progress through some of the more typical circuit configurations that the more sophisticated regulators will apply the benefits to be derived from both series and shunt regulation in the same system. Before continuing, let us examine the unregulated supply of Fig. 14.20. It clearly demonstrates the need for voltage and/or current regulators. Recall from your experimental work that when your dc supply indicates 10 V, you want it to be that value for any load you apply to its terminals. At any except infinite ohms (open circuit) in Fig. 14.20, would this be the case? Obviously not. As R_L increases, the voltage

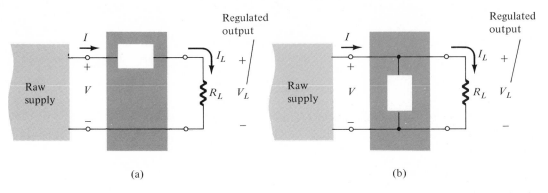

(a) (b)

Figure 14.19 Regulators: (a) series; (b) parallel (shunt).

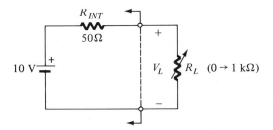

Figure 14.20 Circuit demonstrating the need for voltage and current regulators.

across R_L also increases, and V_L does not remain fixed. The function of the voltage regulator is to maintain V_L at 10 V for any R_L from 1 kΩ to ∞. Therefore, the voltage regulator appears between the unregulated supply and the load as shown in Fig. 14.21.

Our interest for the moment is in the voltage-regulating device. A simple shunt voltage regulating system is shown in Fig. 14.21. As indicated, it consists simply of a Zener diode and series resistance, R_S. Proper operation requires that the Zener diode be in the *on* state. The first requirement, therefore, is to find the minimum R_L (and corresponding I_L) to ensure that this condition is established. Before Zener conduction, the Zener diode is fundamentally an open circuit and the circuit of Fig. 14.21 can be replaced by that indicated in Fig. 14.22. As indicated in the figure, the load voltage will be determined by the voltage-divider rule (similar to the system

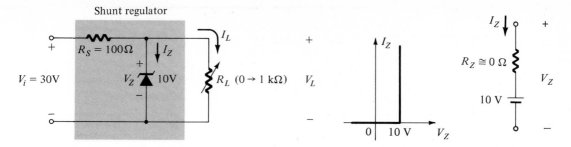

Figure 14.21 Zener diode shunt regulator.

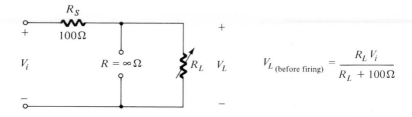

Figure 14.22 Regulator of Fig. 14.21 before firing.

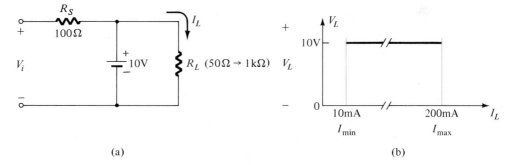

Figure 14.23 Zener diode shunt regulator: (a) after firing; (b) regulated output.

of Fig. 14.7). At Zener conduction, $V_L = V_Z = 10$ V. If these data are used, it is possible to find the minimum value of R_L that the Zener will tolerate in maintaining V_L constant. Applying the voltage-divider rule to the circuit of Fig. 14.23a results in

$$V_L = \frac{R_L V_i}{R_L + R_s}$$

Substituting values results in

$$10 = \frac{R_L \times 30}{R_L + 0.1 \text{ k}\Omega}$$

$$10R_L + 1 \text{ k}\Omega = 30R_L$$

$$20R_L = 1 \text{ k}\Omega$$

and

$$R_L = 50 \ \Omega$$

The *minimum* load R_L for this supply is, therefore, 50 Ω, corresponding to a *maximum* load current of

$$I_{max} = \frac{10\ V}{50\ \Omega} = 200\ mA$$

The maximum current is indicated on the plot of Fig. 14.23b. For any value of R_L from 50 Ω to 1 kΩ the Zener diode will be in the *on* state. An application of the voltage-divider rule to the circuit of Fig. 14.22 for any value of R_L less than 50 Ω will result in $V_L < V_Z$ and the diode will remain in the *off* state. At $R_L = 1$ kΩ, $I_L = (10/1\ k\Omega) = 10\ mA$ (indicated in Fig. 14.23b as I_{min}), and

$$I_{R_s} = \frac{30\ V - 10\ V}{0.1\ k\Omega} = 200\ mA$$

with

$$I_Z = I_{R_s} - I_L = 190\ mA$$

Our approximation, $R_Z \cong 0\ \Omega$, has resulted in the ideal characteristics of Fig. 14.23b between 10 and 200 mA. For this regulator the percent regulation would be determined between these two points of normal operation. The effect of R_Z on the regulation can be determined easily if we find the Thévenin equivalent circuit for the portion of the network of Fig. 14.24 at the left of points 1 and 2.

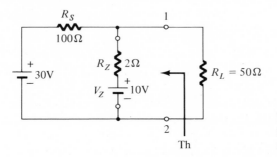

Figure 14.24 Determining the effect of R_z on the output of the Zener shunt regulator.

$$V_{Th} = 10 + \frac{2(30 - 10)}{102} = 10 + \frac{40}{102} \cong 10.4\ V$$

$$R_{Th} = 100 \parallel 2\ \Omega \cong 2\ \Omega$$

Substitute the Thévenin equivalent circuit (Fig. 14.25) and

$$V_L = \frac{50(10.4)}{52} = 10\ V$$

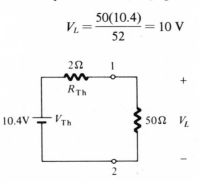

Figure 14.25 Thévenin equivalent circuit for the circuit of Fig. 14.24.

R_Z, therefore, has a negligible effect at minimum R_L (maximum I_L).

For $R_L = 1$ kΩ (minimum I_L):

$$V_L = \frac{1 \text{ k}\Omega(10.4)}{1 \text{ k}\Omega + 2 \text{ }\Omega} \cong 10.4 \text{ V} > 10 \text{ V}$$

and

$$V.R. = \frac{V_{1\text{k}\Omega} - V_{50\Omega}}{V_{50\Omega}} \times 100\% = \frac{10.4 - 10.0}{10.0} \times 100\%$$

$$= \frac{0.4}{10.0} \times 100\% = 4\%$$

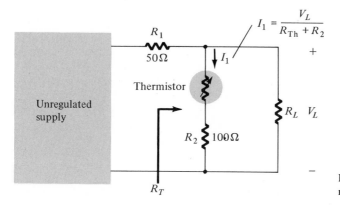

Figure 14.26 Thermistor shunt regulator.

A shunt regulator employing a thermistor appears in Fig. 14.26. Any tendency for V_L to decrease due to a change in load will result in a decrease in current through the thermistor. The temperature of the thermistor element will thereby decrease, resulting in an increase in its resistance. The resulting resistance $R_T = R_L \parallel (R_{\text{Th}} + 100 \text{ }\Omega)$ will increase somewhat and the load voltage $V_L = R_T V_i / (R_T + R_1)$ will tend to increase, offsetting the initial drop in V_L. The symbol $\uparrow$ is assigned to an increasing quantity and $\downarrow$ to a decreasing quantity in the following summary of the voltage-regulating action of this system (read from left to right):

$$V_L \downarrow, I_{\text{Th}} \downarrow, R_{\text{Th}} \uparrow, R_T \uparrow, V_L \uparrow$$
$$\underbrace{\qquad\text{balance}\qquad}$$

An increasing V_L will have the opposite effect on each element and quantity of the above summary.

Transistor Voltage Regulators

The characteristics of a voltage regulator can be markedly improved by using active devices such as the transistor. The simplest of the transistor-*series*-type voltage regulator appears in Fig. 14.27a. In this configuration the transistor behaves like a simple variable resistor whose resistance is determined by the operating conditions. The basic operation of the regulator is best described using the circuit of Fig. 14.27b, in which the transistor has been replaced by a variable resistor, R_T. For variations

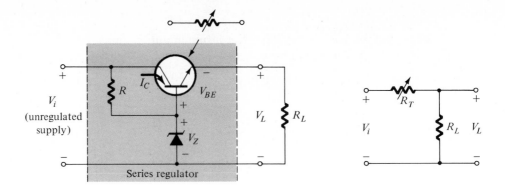

Figure 14.27 Transistor series voltage regulator.

in R_L, if V_L is to remain constant, the ratio of R_L to R_T must remain fixed. Applying the voltage-divider rule results in

$$[\text{fixed for constant } V_L(V_i = \text{constant})]$$

$$V_L = \overbrace{\frac{R_L}{R_L + R_T}}\, V_i$$

For

$$\frac{R_L}{R_T} = k_1 \qquad \text{or} \qquad R_L = k_1 R_T$$

$$\frac{R_L}{R_L + R_T} = \frac{k_1 R_T}{k_1 R_T + R_T} = \frac{k_1}{k_1 + 1} = k \,(\text{constant as required})$$

In summary, for a decreasing or increasing load *(R_L)*, *R_T* must change in the same manner and at the same rate to maintain the same voltage division.

Recall that the voltage regulation is determined by noting the variations in terminal voltage versus the load current demand. For this circuit an increasing current demand associated with a *decreasing R_L* will result in a tendency on the part of V_L to decrease in magnitude also. If, however, we apply Kirchhoff's voltage law around the output loop

$$V_{BE} = \overbrace{V_Z}^{(\text{fixed})} - V_L$$

A decrease in V_L (since V_Z is fixed in magnitude) will result in an increase in V_{BE}. This effect will, in turn, increase the level of conduction of the transistor, resulting in a *decrease* in its terminal (collector-to-emitter) resistance. This is, as described in the previous paragraphs of this section, the effect desired to maintain V_L at a fixed level.

A voltage regulator employing a transistor in the shunt configuration is provided in Fig. 14.28. Any tendency on the part of V_L to increase or decrease in magnitude will have the corresponding effect on V_{BE} since

$$V_{BE} = V_L - \overbrace{V_Z}^{(\text{fixed})}$$

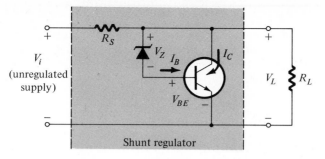

Figure 14.28 Transistor shunt voltage regulator.

For decreasing V_L, the current through the resistor R_S will decrease since the conduction level of the transistor has dropped $(V_{BE}\downarrow)$. The reduced drop in potential across R_S will offset any tendency on the part of V_L to decrease in magnitude. In sequential logic

$$V_L\downarrow, V_{BE}\downarrow, I_B\downarrow, I_C\downarrow, I_{R_S}\downarrow, V_{R_S}\downarrow, V_L\uparrow$$
$$\underbrace{\qquad\qquad\qquad}_{\text{balance}}$$

A similar discussion can be applied to increasing values of V_L.

A series voltage regulator employing a second transistor for control purposes can be found in Fig. 14.29. The base-to-emitter potential (V_{BE_2}) of the control transistor Q_2 is determined by the difference between V_1 and the reference voltage V_Z. The voltage level V_1 is sensitive to the changes in the terminal voltage V_L. Any tendency on the part of V_L to increase will result in an increase in V_1 and therefore in V_{BE_2} since $V_{BE_2} = V_1 - V_Z$. The difference in potential is amplified by the control transistor and carried to the variable series resistive element Q_1. An increase in V_{BE_2}, corresponding to an increase in I_{B_2} and I_{C_2} will result in a decreasing I_{B_1} (assuming I_{R_3} to be relatively constant or decreasing only slightly). The net result is a decrease in the conductivity of Q_1 corresponding to an increase in its terminal resistance and a stabilization of V_L. In sequential logic

$$V_L\uparrow, V_1\uparrow, V_{BE_2}\uparrow, I_{C_2}\uparrow, I_{B_1}\downarrow, R_{(Q1)}\uparrow, V_L\downarrow$$
$$\underbrace{\qquad\qquad\qquad}_{\text{balance}}$$

Again, a similar discussion can be applied to decreasing values of V_L.

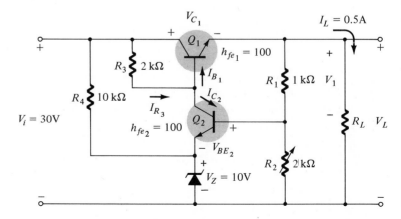

Figure 14.29   A series voltage regulator employing two transistors.

EXAMPLE 14.9 We shall now calculate various currents and voltages of the circuit of Fig. 14.29 for the input shown. Approximations, as introduced in previous chapters, will be the working tools of this analysis. Those of primary importance include $I_C \cong h_{fe}I_B$, $V_{BE} \cong 0$ V, and $I_C \cong I_E$.

Solution:
$$V_{R_4} = V_i - V_Z = 30 - 10 = \mathbf{20\ V}$$

and
$$I_{R_4} = \frac{20\ V}{10\ k\Omega} = \mathbf{2\ mA}$$

$$V_{R_2} \cong V_Z = \mathbf{10\ V} \qquad \text{since } V_{BE_2} \cong 0\ V$$

and
$$I_{R_2} = \frac{10\ V}{2\ k\Omega} = \mathbf{5\ mA}$$

Assuming
$$I_{B_2} \ll I_{R_1}, \quad I_{R_2}$$

then
$$I_{R_1} = I_{R_2} = \mathbf{5\ mA}$$

and
$$V_L = 5\ mA \times 3\ k\Omega = \mathbf{15\ V}$$

with
$$V_{R_3} = V_i - V_L (V_{BE_1} \cong 0\ V)$$
$$= 30 - 15 = \mathbf{15\ V}$$

and
$$I_{R_3} = \frac{15\ V}{2\ k\Omega} = \mathbf{7.5\ mA}$$

Similarly,
$$V_{C_1} = V_i - V_L = 30 - 15 = \mathbf{15\ V}$$

$$I_{E_1} \cong h_{fe}I_{B_1} = 100 I_{B_1}$$

and
$$I_{B_1} = \frac{I_{E_1}}{100} = \frac{(500 + 5)mA}{100} = \mathbf{5.05\ mA}$$

$$I_{C_2} = I_{R_3} - I_{B_1}$$
$$= (7.5 - 5.05)mA$$
$$= \mathbf{2.45\ mA}$$

$$I_{B_2} \cong \frac{I_{C_2}}{100} = \frac{2.45\ mA}{100} = \mathbf{24.5\ \mu A}$$

(Certainly, $I_{B_2}. \ll I_{R_2}.$ I_{R_2} as employed above is an excellent approximation.) and, finally,

$$I_Z = I_{R_4} + I_{C_2} = (2 + 2.45)mA = \mathbf{4.45\ mA}$$

The fact that $I_{B_2} \ll I_{R_1}$, I_{R_2} permits the use of the circuit of Fig. 14.30 to derive a rather useful equation for the circuit of Fig. 14.29. Applying the voltage-divider rule results in

$$V_Z = \frac{V_L R_2}{R_1 + R_2}$$

or since V_Z is fixed,

$$\boxed{V_L = V_Z\left(1 + \frac{R_1}{R_2}\right)} \tag{14.15}$$

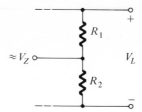

Figure 14.30 Circuit employed in the derivation of Eq. (14.15).

For the above case

$$V_L = 10(1 + \tfrac{1}{2}) = 15 \text{ V}$$

You will find in Fig. 14.29 that R_2 is a variable resistor. Variations in this resistance will control V_L as determined by Eq. (14.15). The maximum voltage available is obviously 30 V (for $V_i = 30$ V) since at this point $V_{C_1} = 0$ V (saturation). The minimum is 10 V attainable with either $R_1 = 0$ or $R_2 = \infty$.

Complete Power Supply (Voltage Regulated)

A power supply employing a voltage regulator similar to the one described in Fig. 14.29 appears in Fig. 14.31. A *Darlington* circuit has replaced the single series transistor of Fig. 14.27 in an effort to increase the sensitivity of the regulator to changes in V_L. In the circuit of Fig. 14.29 changes in I_{C_2} will be reflected in changes in I_{R_3} causing a reduction in the sensitivity of I_{B_1} to changes in V_L. To minimize this undesirable effect R_3 should be as large as possible while still permitting the necessary R_3 for proper circuit behavior. Greater efficiency is achieved by employing a current source in place of R_3. The current source has, ideally, infinite terminal resistance along with the capability to supply the necessary current. This portion of a power supply as indicated in Fig. 14.31 is sometimes referred to as a *preregulator*.

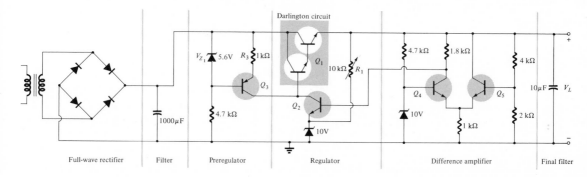

Figure 14.31 Complete voltage-regulated power supply.

For the circuit of Fig. 14.31

$$I_{\text{current source}} = I_{C_3} \cong \frac{V_{Z_1}}{R}$$

To improve further the sensitivity of the regulator to changes in V_L a *difference* amplifier has been introduced, the output of which is fed to the control transistor. The unregulated input is a full-wave rectified signal to be passed through a capacitive filter. The 10-μF capacitor at the output is to reduce the possibility of oscillations and further filter the supply voltage. The supply voltage V_L can be varied by changing R_1 while still maintaining regulation.

Current Regulator

The analysis of current regulators will be limited to a brief discussion of the circuit of Fig. 14.32. Recall from the introductory discussion that a current regulator is designed to maintain a fixed current through a load for variations in terminal voltage. A decrease in $I_L = I_C$ due to a drop in V_L would result in a decrease in $I_E \cong I_C$ and, in turn, a drop in V_{R_E}. The base-to-emitter potential is

$$V_{BE} = \overbrace{V_Z}^{\text{(fixed)}} - V_{R_E}$$

A decrease in V_{R_E} will result in an increase in V_{BE} and the conductivity of the transistor, maintaining I_L at a fixed level.

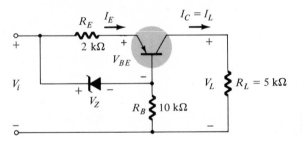

Figure 14.32 Series current regulator.

14.7 IC VOLTAGE REGULATORS

Voltage regulators comprise a class of widely used ICs. These units contain the circuitry for reference source, error amplifier, control device, and overload protection all in a single IC chip. Although the internal construction is somewhat different than that described for discrete voltage-regulator circuits, the external operation is much the same. We will examine operation using some of the popular 3-terminal fixed-voltage regulators (for both positive and negative voltages) and those allowing an adjustable output voltage.

A power supply can be built very simply using a transformer connected to the ac supply to step the voltage to a desired level, then rectifying with a half- or full-wave circuit, filtering the voltage using a simple capacitor filter, and finally regulating the dc voltage using an IC voltage regulator.

A basic category of voltage regulators includes those used with only positive voltages, those used with only negative voltages, and those further classified as having fixed or adjustable output voltages. These regulators can be selected for operation with load currents from hundreds of milliamperes to tens of amperes corresponding to power ratings from milliwatts to tens of watts. A representation of various types of voltage-regulator ICs is presented next. Various terms common to this area of electronics also will be introduced and defined.

Three-Terminal Voltage Regulators

Voltage regulators that provide a positive fixed regulated voltage over a range of load currents are schematically represented in Fig. 14.33. The fixed voltage regulator has an unregulated voltage, V_{IN}, applied to one terminal, delivers a regulated output voltage, V_o, from a second terminal, with the third terminal connected to ground.

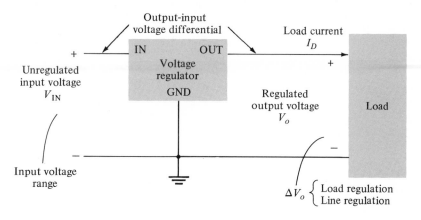

Figure 14.33 Block representation of 3-terminal voltage regulator.

For a particular IC unit, device specifications list a voltage range over which the input voltage can vary to maintain the regulated output voltage, V_o, over a range of load current, I_o. An output-input voltage differential must be maintained for the IC to operate, which means that the varying input voltage must always be kept large enough to maintain a voltage drop across the IC to permit proper operation of the internal circuit.

The device specifications also list the amount of output voltage change, V_o, resulting from changes in load current (load regulation) and also from changes in input voltage (line regulation).

A group of fixed-positive-voltage regulators is the series 78, which provide fixed voltages from 5 V to 24 V. Figure 14.34a shows how many of these regulators are connected. A rectified and filtered unregulated dc voltage is the input, V_{IN}, to pin 1 of the regulator IC. Capacitors connected from input or output to ground help to maintain the dc voltage and additionally to filter any high-frequency voltage variation. The output voltage from pin 2 is then available to connect to the load. Pin 3 is the IC circuit reference or ground. When selecting the desired fixed regulated output voltage, the two digits after the 78 prefix indicate the regulator output voltage. Table 14.1 lists some typical data.

Negative voltage regulator ICs are available in the 79 series (see Fig. 14.34b), which provide a series of ICs similar to the 78 series but operating on negative voltages, providing a regulated negative output voltage. Table 14.2 lists the 79XX series of fixed-negative-voltage regulators and their corresponding regulated voltages.

Voltage regulators are also available in circuit configurations that allow the user to set the output voltage to a desired regulated value. The LM317, for example,

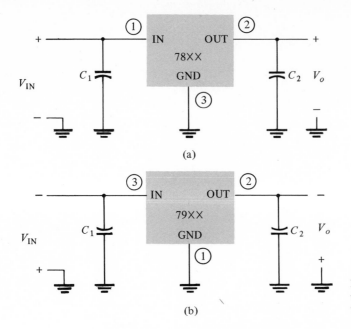

(a)

(b)

Figure 14.34 (a) Series 78XX positive-voltage regulator; (b) Series 79XX negative-voltage regulator.

TABLE 14.1 Positive Series 78XX Voltage Regulator ICs

IC Part Number	Regulated Positive Voltage	Minimum V_{IN}
7805	+5 V	7.3 V
7806	+6 V	8.35 V
7808	+8 V	10.5 V
7810	+10 V	12.5 V
7812	+12 V	14.6 V
7815	+15 V	17.7 V
7818	+18 V	21 V
7824	+24 V	27.1 V

TABLE 14.2 Fixed-Negative-Voltage Regulators in the 79XX Series

IC Part Number	Regulated Output Voltage	Minimum V_{IN}
7905	−5 V	−7.3 V
7906	−6 V	−8.4 V
7908	−8 V	−10.5 V
7909	−9 V	−11.5 V
7912	−12 V	−14.6 V
7915	−15 V	−17.7 V
7918	−18 V	−20.8 V
7924	−24 V	−27.1 V

can be operated with output voltage regulated at any setting over the range of voltage from 1.2 V to 37 V. Figure 14.35 shows a typical connection using the LM317 IC.

Selection of resistors R_1 and R_2 allow the setting of the output to any desired voltage over the adjustment range (1.2 V to 37 V). The output voltage desired can be calculated using

$$V_O = V_{REF}\left(1 + \frac{R_2}{R_1}\right) + I_{ADJ}R_2 \qquad (14.16)$$

with typical values of

$$V_{REF} = 1.25 \text{ V} \qquad \text{and} \qquad I_{ADJ} = 100 \ \mu\text{A}$$

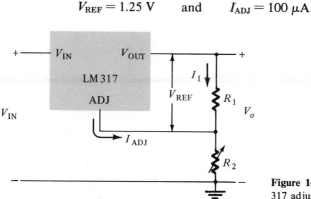

Figure 14.35 Connection of LM-317 adjustable-voltage regulator.

EXAMPLE 14.10 Determine the regulated output voltage using an LM317, as in Fig. 14.35, with $R_1 = 240 \ \Omega$ and $R_2 = 2.4 \ \text{k}\Omega$.

Solution: Using Eq. (14.16)

$$V_O = 1.25 \text{ V}\left(1 + \frac{2.4 \text{ k}\Omega}{240 \ \Omega}\right) + 100 \ \mu\text{A} \ (2.4 \text{ k}\Omega)$$

$$= 13.75 \text{ V} + 0.24 \text{ V} = \mathbf{13.99 \ V}$$

14.8 Practical Power Supplies

A practical power supply can be built to convert the 120-V supply voltage into a desired regulated dc voltage. The standard circuit includes a transformer to step the voltage to a desired ac level, a diode rectifier to half-wave or full-wave rectify the ac signal, and a capacitor filter to develop an unregulated dc voltage. The unregulated dc voltage is then connected as input to an IC voltage regulator, which provides the desired regulated output dc voltage. A few examples will show how a dc voltage supply can be built and how it operates.

EXAMPLE 14.11 Analyze the operation of the +12-V voltage supply shown in Fig. 14.36 operating into a load drawing 400 mA.

Solution: The transformer steps down the line voltage from 120 V, rms to a secondary voltage of 18 V, rms across each transformer half. This results in a peak voltage across the transformer of

$$V_m = \sqrt{2} \cdot V_{rms} = \sqrt{2} \cdot 18 \text{ V} = 25.456 \text{ V}$$

The ripple voltage [using Eq. (14.7c)] is then

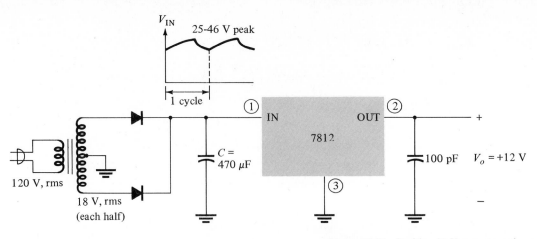

Figure 14.36 Positive 12-V power supply.

$$V_r(\text{rms}) = \frac{2.4 I_{dc}}{C} = \frac{2.4\,(400)}{470} = 2.043 \text{ V}$$

and the peak ripple voltage is then [using Eq. (14.6)]

$$V_r(\text{peak}) = \sqrt{3}\ V_r(\text{rms}) = \sqrt{3}\ (2.043 \text{ V}) = 3.539 \text{ V}$$

The dc level of the voltage across the 470-μF capacitor C is

$$V_{dc} = V_m - V_r(\text{peak}) = 25.456 \text{ V} - 3.539 \text{ V} = 21.917 \text{ V}$$

The ripple voltage across the filter capacitor when operating into a 400-mA load is then [Eq. (14.9a)]

$$r = \frac{2.4 I_{dc}}{C \cdot V_{dc}} \cdot 100\% = \frac{2.4(400)}{(470)(21.917)} \cdot 100\% = 9.319\%$$

The voltage across filter capacitor C has a ripple of about 9.3% and drops to a minimum voltage of

$$V_{\text{IN,min}} = V_m - 2 V_r(\text{peak}) = 25.456 \text{ V} - 2(3.539 \text{ V}) = 18.378 \text{ V}$$

Device specifications list V_{IN} as required to maintain line regulation at 14.6 V. The lowest voltage being maintained across the capacitor is somewhat greater at 18.378 V.

Lowering the value of the filter capacitor or increasing the load current will result in greater ripple voltage and lower minimum voltage across the capacitor. As long as this minimum voltage ramains above 14.6 V, the 7812 will maintain the output voltage regulated at +12 V.

Device specifications for the 7812 list the maximum voltage change as 60 mV. This means that the output voltage regulation will be less than

$$V.R. = \frac{60 \text{ mV}}{12 \text{ V}} \times 100\% = \mathbf{0.5\%}$$

EXAMPLE 14.12 Analyze the operation of the 5-V supply of Fig. 14.37 operating at a load current of: (a) 200 mA; and (b) 400 mA.

Solution: The specifications for the 7805 list an input of 7.3 V as the minimum allowable to maintain line regulation.

(a) At a load of $I_{dc} = 200$ mA, the ripple voltage is

$$V_r(\text{peak}) = \sqrt{3} \cdot V_r(\text{rms}) = \sqrt{3} \cdot \frac{(2.4)I_{dc}}{C} = \sqrt{3} \cdot \frac{2.4(200)}{(250)} = \textbf{3.326 V}$$

and the dc voltage across the 250-μF filter capacitor is

$$V_{dc} = V_m - V_r(\text{peak}) = 15 \text{ V} - 3.326 \text{ V} = \textbf{11.674 V}$$

The voltage across the filter capacitor will drop to a minimum value of

$$V_{\text{IN,min}} = V_m - 2V_r(\text{peak}) = 15 \text{ V} - 2(3.326 \text{ V}) = \textbf{8.348 V}$$

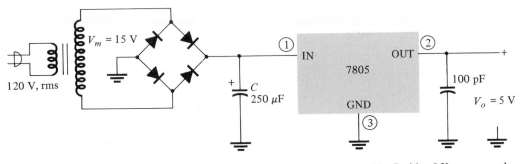

Figure 14.37   Positive 5-V power supply.

Since this is above the rated value of 7.3 V, the output will be maintained at the regulated +5 V.

(b) At a load of $I_{dc} = 400$ mA, the ripple voltage is

$$V_r(\text{peak}) = \sqrt{3} \cdot \frac{(2.4)(400)}{250} = \textbf{6.65 V}$$

around a dc voltage of

$$V_{dc} = 15 \text{ V} - 6.65 \text{ V} = \textbf{8.35 V}$$

which is above the rated 7.3-V lower level. However, the input swings around this dc level by 6.65 V, peak, dropping during part of the cycle to

$$V_{\text{IN,min}} = 15 \text{ V} - 2(6.65 \text{ V}) = \textbf{1.7 V}$$

which is well below the minimum allowed input voltage of 7.3 V. Therefore, the output is not maintained at the regulated +5-V level over the entire input cycle. Regulation is maintained for load currents below 200 mA, but not at or above 400 mA.

EXAMPLE 14.13   Determine the maximum value of load current at which regulation is maintained for the circuit of Fig. 14.37

Solution: To maintain $V_{\text{IN}} \geqslant 7.3$ V

$$V_r(\text{peak-peak}) \leq V_m - V_{\text{IN,min}} = 15 \text{ V} - 7.3 \text{ V} = 7.7 \text{ V}$$

so that

$$V_r(\text{rms}) = \frac{V_r(\text{peak-peak})/2}{\sqrt{3}} = \frac{7.7 \text{ V}/2}{\sqrt{3}} = 2.223 \text{ V}$$

We can then determine the value of I_{dc} (in mA)

$$I_{dc} = \frac{V_r(\text{rms})C}{2.4} = \frac{(2.223)(250)}{2.4} = \textbf{231.56 mA}$$

Any current above this value is too large for the circuit to maintain the regulator output at +5 V.

Using a positive adjustable voltage regulator IC it is possible to set the regulated output voltage to any desired voltage (within the device operating range).

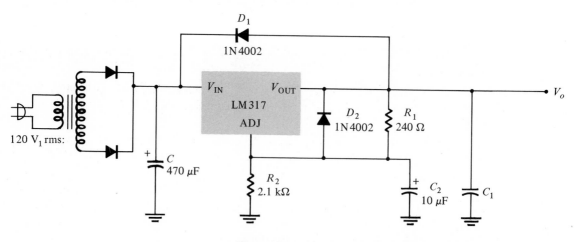

Figure 14.38   Positive adjustable-voltage regulator for Example 14.14.

EXAMPLE 14.14   Determine the regulated output voltage of the circuit in Fig. 14.38.

Solution: The output voltage is

$$V_o = 1.25 \text{ V}\left(1 + \frac{2.1 \text{ k}\Omega}{240 \text{ }\Omega}\right) + 100 \text{ }\mu\text{A}(1.8 \text{ k}\Omega) = 12.368 \text{ V}$$

A check of the filter-capacitor voltage shows that an input–output voltage differential of 2 V can be maintained up to at least 200 mA of load current.

PROBLEMS

§ 14.2

1. What is the ripple factor of a sinusoidal signal having a peak ripple of 2 V on an average of 50 V?

2. A filter circuit provides an output of 28 V unloaded and 25 V under full-load operation. Calculate the % voltage regulation.

3. A half-wave rectifier develops 20 V dc. What is the rms value of the ripple voltage?

4. What is the rms ripple voltage of a full-wave rectifier whose output voltage is 8 V dc?

5. A simple capacitor filter fed by a full-wave rectifier develops 14.5 V dc at 8.5% ripple factor. What is the output ripple voltage (rms)?

6. A full-wave rectified signal of 18 V peak is fed into a capacitor filter. What is the voltage regulation of the filter circuit if the output dc is 17 V dc at full load?

7. A full-wave rectified voltage of 18 V peak is connected to a 400-μF filter capacitor. What is the dc voltage across the capacitor at a load of 100 mA?

8. A full-wave rectifier operating from the 60-Hz ac supply produces a 20-V peak rectified voltage. If a 200-μF filter capacitor is used, calculate the ripple at a load of 120 mA.

9. A capacitor filter circuit ($C = 100$ μF) develops 12 V dc when connected to a load of 2.5 kΩ. Using a full-wave rectifier operating from the 60-Hz supply, calculate the ripple of the output voltage.

10. Calculate the size of the filter capacitor needed to obtain a filtered voltage with 15% ripple at a load of 150 mA. The full-wave rectified voltage is 24 V dc, and supply is 60 Hz.

11. A 500-μF filter capacitor provides a load current of 200 mA at 8% ripple. Calculate the peak rectified voltage obtained from the 60-Hz supply and the dc voltage across the filter capacitor.

12. Calculate the size of the filter capacitor needed to obtain a filtered voltage with 7% ripple at a load of 200 mA. The full-wave rectified voltage is 30 V dc, and supply is 60 Hz.

13. Calculate the percent ripple for the voltage developed across a 120-μF filter capacitor when providing a load current of 80 mA. The full-wave rectifier operating from the 60-Hz supply develops a peak rectified voltage of 25 V.

14. Calculate the amount of peak diode current through the rectifier diode of a full-wave rectifier feeding a capacitor filter when the average current drawn from the filter is 100 mA when diode conduction is for $\frac{1}{10}$ cycle.

§ **14.4**

15. An *RC* filter stage is added after a capacitor filter to reduce the percent of ripple to 2%. Calculate the ripple voltage at the output of the *RC* filter stage providing 80 V dc.

16. An *RC* filter stage ($R = 33$ Ω, $C = 120$ μF) is used to filter a signal of 24 V dc with 2 V (rms) operating from a full-wave rectifier. Calculate the percent ripple at the output of the *RC* section for a load of 100 mA. Calculate also the ripple of the filtered signal applied to the *RC* stage.

17. A simple-capacitor filter has an input of 40 V dc. If this voltage is fed through an *RC* filter section ($R = 50$ Ω, $C = 40$ μF), what is the load current for a load resistance of 500 Ω?

18. Calculate the ripple voltage (rms) at the output of an *RC* filter section that feeds a 1-kΩ load when the filter input is 50 V dc with 2.5 V rms ripple from a full-wave rectifier and capacitor filter. The *RC* filter section components are $R = 100$ Ω and $C = 100$ μF.

19. If the output no-load voltage for the circuit of Problem 18 is 60 V, calculate the percent of voltage regulation with the 1-kΩ load.

§ 14.5

20. Draw the circuit diagram of a voltage doubler. Indicate the value of the diode PIV rating in terms of the transformer peak voltage, V_m.

21. Draw a voltage tripler circuit. Indicate diode PIV ratings and voltage across each circuit capacitor. Include polarity.

22. Repeat Problem 21 for a voltage quadrupler.

§ 14.6

23. Determine the percent voltage regulation of a dc supply providing 100 V under no-load conditions and 95 V at full-load.

24. Determine the voltage regulation of the network of Fig. 14.39 if R_L is limited to a minimum value of 1 kΩ before exceeding the current limits of the supply.

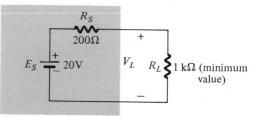

Figure 14.39

25. Determine the current regulation of the network of Fig. 14.40 if R_L is limited to a minimum value of 1 kΩ before exceeding the voltage limits of the supply.

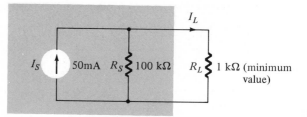

Figure 14.40

26. (a) For the Zener diode shunt regulator of Fig. 14.21, determine the minimum value of R_L to ensure that the Zener diode is in the "on" state if $V_i = 60$ V, $V_Z = 12.5$ V, and $R_S = 200$ Ω.
(b) For the conditions of part (a), determine the maximum load current I_L.
(c) Find the minimum value of I_L if $R_{L(max)}$ is 5 kΩ.
(d) If R_Z is 1.5 Ω, determine the voltage regulation for the range indicated above.

27. A Zener regulator uses a diode with $V_Z = 22$ V and a maximum dissipation of 2 W. If the applied voltage is 50 V, find the minimum value of the resistance R_S.

28. For the network of Fig. 14.27, if $R_L = 1$ kΩ, $R = 5$ kΩ, $V_Z = 10$ V, $V_{BE} = 0.7$ V, and $V_i = 20$ V, determine the following:
(a) The voltage V_L and the current I_L.

(b) The collector current I_C.

(c) The current through R.

(d) The supply current.

29. For the network of Fig. 14.28, if $R_L = 4$ kΩ, $R_S = 2$ kΩ, $V_Z = 10$ V, $V_{BE} = 0.7$ V, and $V_i = 20$ V, determine the following:

(a) The voltage V_L.

(b) The current I_L.

(c) The supply current through R_S.

(d) The Zener current if $\beta = 50$.

30. Determine the value of R_2 in the network of Fig. 14.29 to establish a load voltage of 20 V.

31. (a) Describe the complete behavior of the voltage-related supply of Fig. 14.31.

(b) Describe the resulting action to maintain a fixed V_L, if V_L begins to drop in level.

32. For the network of Fig. 14.32, calculate V_L if $V_{BE} = 0.7$ V, $V_Z = 10$ V, and $V_i = 20$ V. Resistor values remain the same.

§ 14.7

33. Determine the regulated output voltage using an LM317, as in Fig. 14.35, with $R_1 = 240$ Ω and $R_2 = 1.8$ kΩ.

34. What output voltage results in a circuit as in Fig. 14.35 with $R_1 = 240$ Ω and $R_2 = 3.3$ kΩ?

§ 14.8

35. Determine the % ripple across the filter capacitor of a voltage supply as in Fig. 14.36 operating into a load that draws 250 mA. Assume that $V_m = 25.5$ V.

36. If a voltage supply, as in Fig. 14.36, has a ripple of 12% at 20 V dc across capacitor C, to what lowest value does V_{IN} drop during the cycle? Assume that $V_m = 25.5$ V.

37. For a 5-V supply as in Fig. 14.37 with $C = 500$ μF and load $I_{dc} = 150$ mA, determine:

(a) V_{dc} across capacitor C.

(b) V_r(peak) across capacitor C.

38. A +5-V supply as in Fig. 14.37 with $C = 330$ μF and load of 300 mA has what value of $V_{IN,min}$? Will output be maintained at regulated +5-V level? (Assume that $V_m = 15$ V.)

Digital ICs

15.1 DIGITAL FUNDAMENTALS

Digital circuits are built using either bipolar or MOSFET transistors in integrated circuits. Basically, digital circuits are simple since they are operated either fully saturated *(on)* or in cutoff *(off),* usually represented by two distinct voltage levels. Our concern, then, is with the two states of a digital circuit (+5 V and 0 V, for example). Logically, these levels may be related to the binary conditions of 1 and 0. For example, +5 V = 1 and 0 V = 0 are possible relations that may be assigned to voltage levels and logic conditions. Digital circuits provide manipulation of the logical conditions representing the two different voltage levels of the circuit. A number of logic gates or circuits will be covered in this chapter. A logical AND gate, for example, provides specific output only when *all* inputs are present, whereas a logical OR gate provides output if *any* one input is present. An inverter circuit provides logical inversion—a 1 output for 0 input, or vice versa. The most popular configurations for a logic gate include an AND or OR gate, followed by an inverter, an AND-gate-inverter combination called a Not-AND or NAND gate, and an OR inverter called a Not-OR or NOR gate.

Also important in digital circuits is a class of multivibrator circuits. These circuits have two opposite output terminals (if one output is logical-1, the other is logical-0, or vice versa). The most useful of these is the bistable multivibrator or *flip-flop,* which can remain in either stable condition of output. The flip-flop can be used as a memory device—holding the state it was placed in after the initial signal operating the circuit has passed. It can also be used to build a binary counter or a data register for use in digital computers.

A second circuit, the monostable multivibrator, provides opposite voltage outputs but, as the name implies, it can remain stable in only one state. If the circuit is triggered to operate, it can go into the opposite state of output voltages but can remain in this state only for a fixed time interval. The monostable multivibrator, or *one-shot,* is used for pulse shaping, time delay, and other timing actions.

An astable multivibrator or *clock* circuit has no stable operating state and provides a changing output voltage (pulse train). The circuit is therefore an oscillator providing clock pulses to activate various computer operations.

A circuit that appears similar to the multivibrator class is the Schmitt trigger. This circuit operates from a slowly varying input signal and switches output voltage state when the input voltage goes above a preset voltage level or below a second voltage level.

Diode Logic Gates—AND, OR

A logic gate provides an output signal for a desired logical combination of input signals. An AND gate, for example, provides an output of "logical-1" only if all the inputs are present, that is, if each is at 1 (logical-1). This circuit can be compared to one using switches connected in series. Only if, as in Fig. 15.1 all switches are closed (at 1) does an output voltage appear (and drive the light indicator ON). Thus, a light ON (logical-1) appears if switch *A,* AND switch *B,* and AND switch *C* are closed. This can also be written as the Boolean or logical expression

$$\text{Light } On = A \cdot B \cdot C$$

(The dot indicates AND; read as A *AND* B *AND* C.)

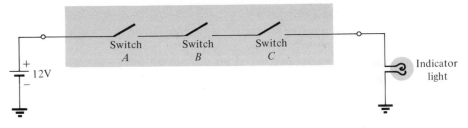

Figure 15.1   AND circuit using switches.

DIODE AND GATE

A more practical form of AND gate using diodes is shown in Fig. 15.2. Using positive-voltage levels of $+V$ for 1 and 0 V as 0, we see that the circuit provides a 1 output of $+V$ only if *all* inputs are $+V$ or 1. More specifically, any input at 0 V will hold the output at about 0 V, as shown below.

1. A zero-volt input at any (one or more) input terminal(s) will cause that diode to short the output to ground. In logic form this means that 0 at any input produces 0 output.
2. A 1 input at inputs 1 and 2, but a 0 input at 3, will produce a 0 output because diode 3 shorts the output to ground.

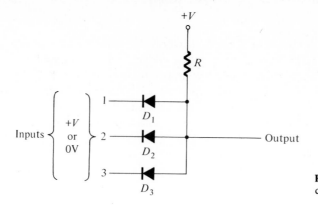

Figure 15.2 Diode AND-gate circuit.

3. Only when a 1 input is provided at inputs 1 AND 2 AND 3, with none of the diodes conducting, is the output 1.

Figure 15.3 shows a typical diode gate circuit and voltage truth table. Figure 15.4 shows the logic symbol and logic truth table for the circuit of Fig. 15.3 for positive logic operation (defined below).

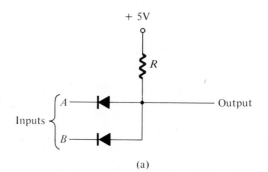

A	B	Output
0V	0V	0V
0V	+5V	0V
+5V	0V	0V
+5V	+5V	+5V

(a)   (b)

Figure 15.3 (a) Diode logic circuit; (b) voltage truth table.

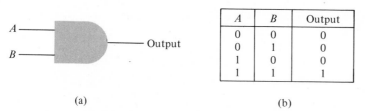

A	B	Output
0	0	0
0	1	0
1	0	0
1	1	1

(a)   (b)

Figure 15.4 (a) Diode AND-gate symbol; (b) logic truth table, for positive logic (+5 V = 1, 0 V = 0).

DIODE OR GATE

The circuit of Fig. 15.5 shows an OR-gate circuit connection using three switches connected in parallel. If either switch A OR B OR C (or any combination of these) is closed (logical-1), the indicator light will be turned ON. A practical version of such a circuit uses diodes as shown in Fig. 15.6. The circuit is a positive-logic diode OR gate, and for this example it uses the same $+V$ and 0-V levels as the AND-gate circuit previously considered.

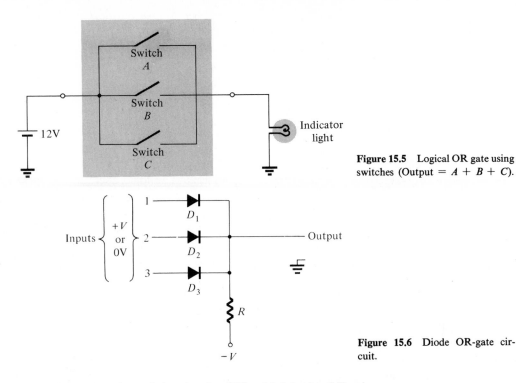

Figure 15.5 Logical OR gate using switches (Output = $A + B + C$).

Figure 15.6 Diode OR-gate circuit.

The operation of the circuit of Fig. 15.6 is the following:

1. A 1 $(+V)$ input at any (one or more) input terminal(s) will cause that diode to conduct, placing the output at the 1 level.

2. A 0 input at inputs 1 and 2, but a 1 input at 3, will produce a 1 output because diode 3 conducts, placing the output at $+V$, and thereby holding diodes 1 and 2 in cutoff.

3. Only when a 0 input is provided at all three inputs will the output be 0.

Figures 15.7 and 15.8 show a typical OR-gate circuit and logic symbol and the respective voltage and logic truth tables.

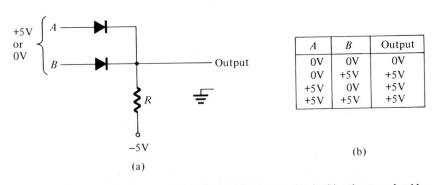

A	B	Output
0V	0V	0V
0V	+5V	+5V
+5V	0V	+5V
+5V	+5V	+5V

(b)

(a)

Figure 15.7 (a) OR-gate circuit; (b) voltage truth table.

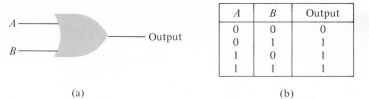

A	B	Output
0	0	0
0	1	1
1	0	1
1	1	1

(a)   (b)

Figure 15.8 (a) OR-gate logic symbol; (b) logic truth table.

POSITIVE LOGIC—NEGATIVE LOGIC

In the two circuits just considered the voltage operation of these circuits is described by the voltage truth tables of Figs. 15.3 and 15.7. The logic operation of these two circuits, however, is dependent on the definitions of logical-1 and logical-0. Positive-logic definitions were used so far, where **positive logic** meant that the *more positive* voltage was assigned as the logical-1 state. It is possible to use other logic definitions. Using the same two circuits and voltage levels of −V and 0 V provides **negative-logic** operation—with the definition of the *more negative* voltage *(−V)* as the logical-1 level (and 0 V as logical-0).

Transistor Inverter

A simple but important digital circuit is the inverter. Using voltage levels of 0 and +5 V, as an example, the inverter circuit will provide an output of 0 V for an input of +5 V and an output of +5 V for an input of 0 V. In logical terms the inverter, having single-input and single-output terminals, provides the opposite output—a 1 output for 0 input, or vice versa.

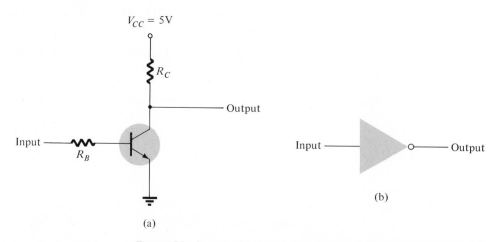

(a)

(b)

Figure 15.9 Inverter circuit and characteristics: (a) transistor inverter circuit; (b) inverter logic symbol.

Figure 15.9 shows the circuit diagram of an inverter. The transistor in this circuit is operated either in saturation *(on)* or in cutoff *(off)*. To operate the transistor in saturation there must be sufficient base current so that βI_B is greater than the collector saturation current. To hold the transistor in cutoff the base-emitter must be reverse-biased or not forward-biased.

15.2 INTEGRATED-CIRCUIT (IC) LOGIC DEVICES

Digital integrated circuits of various types find widespread use. Users of digital logic circuits depend on the IC units provided by the numerous manufacturers. At present there are a number of different types of popular logic circuits. Since no one circuit type has been universally accepted, it seems reasonable to consider some of the more popular circuit types to understand their operation and their relative advantages and disadvantages. It should be clear that each type provides the same basic logical function and that other more practical factors about the overall system generally dictate which particular logic circuit type is chosen.

Some of the important factors used in selecting a circuit type are (not necessarily in order of importance): (1) cost; (2) power dissipation; (3) speed of operation; and (4) noise immunity.

Transistor-Transistor Logic (TTL) Circuit

Transistor-transistor logic (T^2L) is one circuit form of logic gate that, although possible as discrete components, is appropriate for manufacture in integrated form. Figure 15.10 shows a basic form of the logic circuit. Notice that each input is made using an emitter, a single multiple emitter transistor providing the inputs.

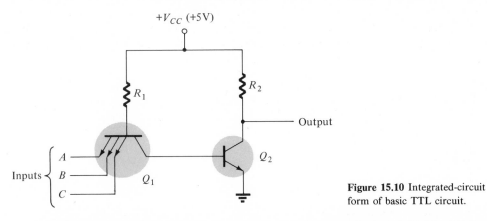

Figure 15.10 Integrated-circuit form of basic TTL circuit.

Figure 15.11 shows the circuit operation for output logical-0 (0 V). All inputs must either have high inputs (+5 V) or not be connected, so that transistor Q_1 is in reverse linear active mode. Transistor Q_2 is then driven *on* by a base current form $+V_{cc}$ through R_1, the base collector of Q_1 and forward-biased base emitter of Q_2. When Q_2 is on, the output voltage taken from the collector is V_{CEsat}, approximately 0 V.

Figure 15.11 shows one (or more) input of 0 V allowing Q_1 to be biased *on,* resulting in the collector voltage of Q_1 to be near 0 V, thereby holding Q_2 *off.* When Q_2 is *off,* the collector voltage is then +5 V. The circuit operates as a positive-logic NAND gate as summarized in Fig. 15.11c.

The T^2L unit is popular because it has fast speed, good noise immunity, and low cost. A practical T^2L logic unit is shown in Fig. 15.12a, the NAND logic symbol in Fig. 15.12b, and the package pin connections in Fig. 15.12c. Four T^2L NAND

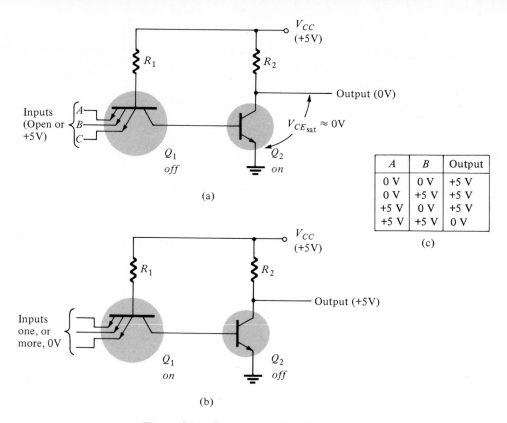

A	B	Output
0 V	0 V	+5 V
0 V	+5 V	+5 V
+5 V	0 V	+5 V
+5 V	+5 V	0 V

(c)

Figure 15.11 Operation of TTL (T²L) logic gate: (a) output transistor *on;* (b) output transistor *off.*

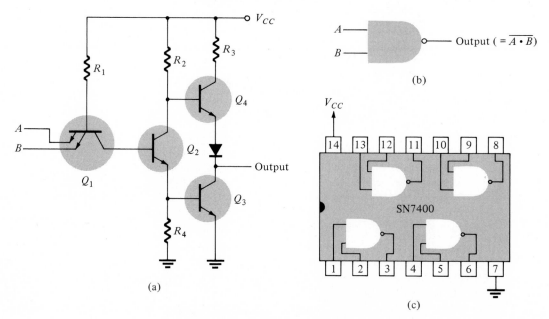

Figure 15.12 T²L logic unit: (a) circuit schematic; (b) logic symbol; (c) IC package of SN7400.

gates are contained in the single SN7400 IC package, the entire unit being a quad, 2-input positive NAND gate. The basic T²L gate can switch state in typically 10 ns at power dissipation of 10 mW.

CMOS Logic Circuits

Another popular IC logic circuit is made using MOSFET devices of both *p*-channel and *n*-channel transistors, the complementary-symmetry circuit being COS/MOS, COSMOS, or more simply CMOS logic. A basic CMOS inverter is shown in Fig. 15.13. A positive input voltage drives the *n*-channel FET *on* with output 0 V. An input of 0 V results in the *p*-channel FET *on* with the output a positive voltage.

A positive-logic NOR gate is made as shown in Fig. 15.14. When either input is positive, an *n*-channel FET is driven *on* with output then 0 V. When both inputs

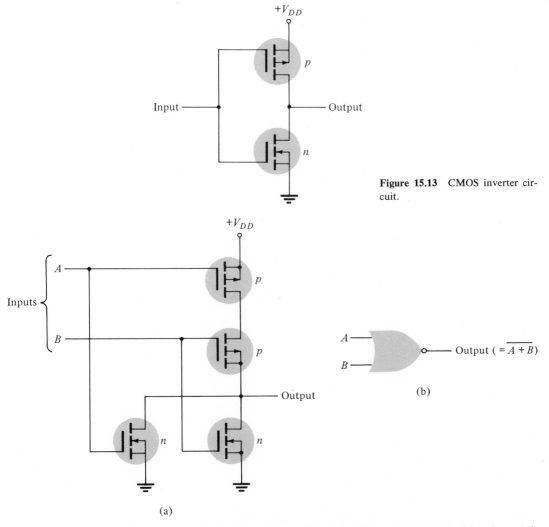

Figure 15.13 CMOS inverter circuit.

Figure 15.14 CMOS positive logic NOR gate: (a) circuit; (b) logic symbol.

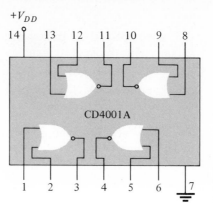

Figure 15.15 CMOS quad two-input positive NOR gate IC: CD4001A.

are 0 V, the two *p*-channel FETs are turned *on* with output then a positive voltage. The CD4001A, for example, is an IC package of four 2-input NOR gates as shown in Fig. 15.15.

CMOS logic units offer very low-power dissipation at relatively slow speed. Typical propagation delay is 100 ns, with 5-V supply at only microwatts of power dissipation, while providing noise immunity of over 1.5 V. A CMOS logic gate is at least five times smaller than the comparable T²L gate.

Integrated Injection Logic (I²L)

A relatively new IC logic circuit uses bipolar transistors in a circuit and physical arrangement called *integrated injection logic* (I²L) and has the better characteristics of T²L and CMOS units: The I²L unit has low-power dissipation, fast speed, and small size.

Complementary bipolar transitors are used, the *pnp* injector transistor serving as a current source, with a multiple collector *npn* transistor forming the basic gate. Figure 15.16 shows a dual-gate unit. I²L gates use microwatts of power at speeds in the tens of nanoseconds, and they require the least IC area of the gates considered.

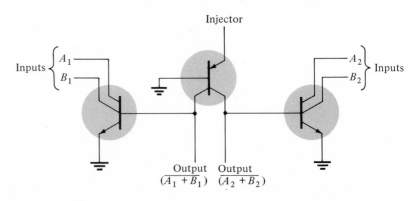

Figure 15.16 L²L dual NOR gate.

Emitter-Coupled Logic (ECL) Circuit

The technique of emitter-coupled logic (ECL) or current-mode logic (CML) circuits differs from those covered previously in one main respect. All previous circuits allowed the transistors to saturate. This means that an amount of charge is stored in the transistor base and collector regions, resulting in a time delay in turning off the transistor. Current-mode logic operates the transistor in a nonsaturated condition, thereby providing shorter propagation delays through the circuits.

An emitter-coupled logic circuit, such as that of Fig. 15.17, uses a transistor for each input of the circuit, which is desirable for IC manufacture. Whereas a direct-coupled logic circuit would connect the transistor emitters to ground (thereby allowing the base drive to be dependent on the input voltage), the present circuit uses an additional transistor stage providing a reference point for the common-emitter terminals. If more than the needed turn-on drive is applied, this will cause the emitter point to rise in voltage, maintaining the logic transistor in a nonsaturated mode of operation.

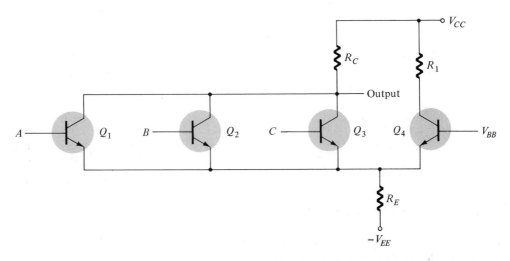

Figure 15.17 Emitter-coupled logic (ECL) circuit.

In summary, then, the nonsaturated operation of an emitter-coupled logic (ECL) circuit provides very fast switching speeds, simultaneous OR and NOR outputs from each circuit, high fan-in and fan-out capacity, and constant noise immunity of the power supply due to the relative constant-current demand of the logic circuit whether *off* or *on*. Disadvantages are the need for a bias driver to provide the reference supply voltage, the higher cost of ECL circuits, and the larger size of a circuit.

15.3 BISTABLE MULTIVIBRATOR CIRCUITS

Of equal importance to logic circuits in digital circuitry is the class of multivibrator circuits. There are three basic forms of the multivibrator—bistable, monostable, and astable; the most important of these, by far, is the bistable multivibrator or flip-flop. As an indication of the applications of the multivibrator circuits, consider the following:

Bistable *(flip-flop)*—storage stage, counter, shift register

Monostable *(one-shot)*—delay circuit, waveshaping, timing circuit
Astable *(clock)*—timing oscillator (square-wave)

In a logic system there will typically be a large number of flip-flop stages used as counters, storage registers, shift-registers, a few one-shot circuits in special timing or pulse-shaping uses, and a limited number of clock circuits (typically only one).

Characteristic of all three circuits is the availability of two outputs, where the outputs are logically inverse signals. One output is selected as the reference, this designation being indicated in a number of ways. The two outputs are sometimes marked as 0 and 1, *false* and *true,* or $\bar{A}$ and A, etc. The main point of the designation is to indicate that the outputs are logically opposite and to mark the output chosen as the reference output. Another means of indicating the state of the multivibrator circuit is the use of the designation of SET and RESET. When referring to the state of the circuit, the definitions of SET and RESET are the following:

SET: Q output is logical-1
$\bar{Q}$ output is logical-0
RESET: Q output is logical-0
$\bar{Q}$ output is logical-1

Bistable Multivibrator (Flip-Flop)

The flip-flop circuit, the most important of the multivibrator circuits, will be covered first. To provide some basic consideration of this circuit's operation a simple form of bistable circuit using two inverters is shown in Fig. 15.18. The inverters

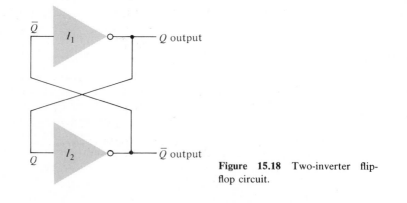

Figure 15.18 Two-inverter flip-flop circuit.

are essentially connected in series, with two output points indicated. The two outputs are labelled Q and $\bar{Q}$, respectively. If the Q output is a logical-1, then inverter I_2 will provide $\bar{Q}$ as a logical-0. Since $\bar{Q}$ is connected as input to inverter I_1, it will cause the output of that stage to be a logical-1, as assumed. Thus, the state of logical conditions, or the voltage they represent, forms a stable situation with Q output logical-1 and $\bar{Q}$ output logical-0. If some external means is used to cause the Q signal to change to logical-0, then, through inverter I_2, $\bar{Q}$ would change to logical-1. The $\bar{Q}$ input of logical-1 would then result in A being logical-0 as initially proposed. Thus, the circuit will also remain in a stable condition if the Q output is logical-0 and the $\bar{Q}$ output is logical-1. In effect, then, the circuit has two stable operating states that act as a memory of the last state it was placed into. Some external means is necessary, however, to cause the circuit to change state.

RS FLIP-FLOP

Figure 15.19 shows the use of NOR gates connected in series with additional inputs providing signals to cause the circuit to change state. The inputs are marked R for RESET and S for SET. Recall that a NOR gate provides logical-0 output if any of its inputs is logical-1. A logical-1 input to the S terminal will cause the output

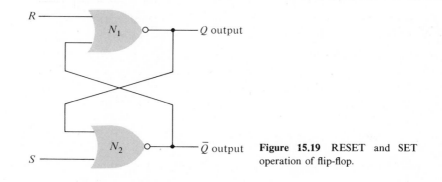

Figure 15.19 RESET and SET operation of flip-flop.

of N_2 to be logical-0. If it is assumed that $R = 0$ is connected to the R terminal at this time, the inputs to N_1 are both logical-0 with output of logical-1. The result of a SET input signal then is to cause the circuit to become SET, where the SET state was previously defined as Q output = logical-1 and $\bar{Q}$ output = logical-0. Similarly, the application of only a logical-1 to the R input will cause the Q output to become logical-0 and $\bar{Q}$ output logical-1, which is the RESET state of the circuit. It should be obvious that simultaneous application of logical-1 signals to both S and R inputs is ambiguous, forcing both outputs to the logical-0 condition. This would not be an accepted operation of this circuit in which the two output signals should be always logically opposite. If the R and S inputs are both logical-0, the circuit remains in whatever state it was last placed into.

T-TYPE FLIP-FLOP

One of the more common type of flip-flop circuits is the T-type or triggered flip-flop. This circuit is also called a *complementing* flip-flop, or *toggle* flip-flop, since its action is to change state every time an input is applied to the single T-input terminal. Figure 15.20a shows a block diagram of a T-type flip-flop. The waveforms in Fig. 15.20b show that the output changes stage, or toggles, every falling edge of the input signal. Notice that the output goes through a full cycle every two input cycles so that the output is at one-half the frequency of the input or that every two input pulses result in one full cycle of the output. The T-type flip-flop is the basic circuit used as a counter stage in binary counters.

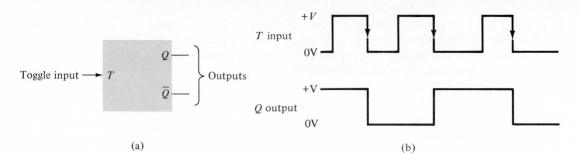

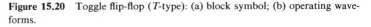

(a)

(b)

Figure 15.20 Toggle flip-flop (*T*-type): (a) block symbol; (b) operating waveforms.

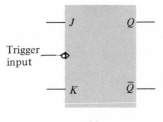

(a)

J	*K*	Circuit Action
0	0	Remains in same state
0	1	RESETS
1	0	SETS
1	1	Circuit toggles

(b)

Figure 15.21 Basics of *JK* flip-flop: (a) logic symbol; (b) truth table.

JK FLIP-FLOP

The *J* and *K* input terminals (Fig. 15.21a) are used to provide information or data inputs. When a trigger pulse is then applied, the circuit changes state corresponding to the inputs to the *J* and *K* terminals. A *JK* circuit is built as an integrated circuit form using TTL, CMOS, or ECL circuits. It is not at all important here to consider the circuit details, and in fact one only purchases a complete unit in IC form and has little to do with the details of circuit operation. The *JK* circuit is versatile and is presently the most popular version of the flip-flop circuit. A logic symbol of a *JK* flip-flop is shown in Fig. 15.21a. The *J* and *K* terminals shown are the data input terminals receiving the information of logical-1 or logical-0. These data inputs do not, however, change the state of the flip-flop circuit, which will remain in its present state until a trigger pulse is applied. Thus, for example, if the circuit were presently RESET and the input data were such as to result in the SET condition, the circuit would still maintain the RESET condition, even with the *J* and *K* input data signals applied. Only when the trigger pulse occurs are the input data used to determine the new state of the circuit—the SET state for the present example.

There are four possible combinations of the *J* and *K* inputs. To consider the complete operation of the circuit each of the possible conditions is listed in the truth

CH. 15 DIGITAL ICS

table of Fig. 15.21b. If both J and K inputs are logical-0, the circuit remains in the same state (no change takes place). If the J input is logical-0 and K input logical-1, the circuit ends up in the RESET state. If the K input is logical-0 and the J input logical-1, the occurrence of the trigger pulse will cause the circuit to be set SET. Finally, if both J and K inputs are logical-1, the action of the trigger pulse's becoming logical-1 is to toggle or complement the circuit. Thus, the circuit would toggle (change state) from whichever condition it happened to be in on application of the trigger pulse. In this last case, with J and K inputs both logical-1, the circuit would operate as a T flip-flop and could be used as such. When opposite data input signals are applied as J and K inputs, the trigger pulse will shift the data into the present flip-flop state, the stage then acting as a shift-register stage. Thus, the JK flip-flop can be used as a shift-register stage, a toggle state for counting operations, or generally as a control logic stage.

If positive logic is used (0 V = logical-0 and +V as logical-1, for example), the circuit triggers when the trigger pulse goes from the logical-0 to logical-1 condition—this being referred to as rising-edge triggering since the circuit is triggered at the time the voltage goes positive (from 0 to +V). When the circuit triggers on a voltage change from +V to 0 V, the triggering is referred to as falling-edge triggering. The manufacturer's information sheets should indicate the type of triggering required to operate the particular circuit so that it may be properly used.

Monostable Multivibrator (One-Shot)

The outputs of a monostable multivibrator circuit are stable in only one of the two states (SET and RESET). Figure 15.22a shows a logic block symbol of a one-

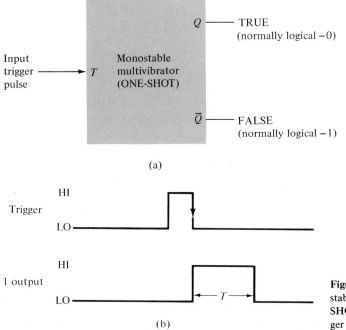

Figure 15.22 Operation of monostable multivibrator: (a) ONE-SHOT logic symbol; (b) input trigger and 1 output waveform.

shot circuit in the stable RESET state (Q output = logical-0, and $\overline{Q}$ output = logical-1). The input trigger signal is a pulse that operates the circuit in an edge-triggered manner. Figure 15.22b shows a typical input trigger pulse and corresponding output waveform (assuming triggering on the falling edge of the trigger pulse). The Q output is normally low (RESET state). When a negative-going voltage change triggers the circuit, the 1 output goes high (SET state), which is the unstable circuit state. It will remain high only for a fixed time interval, T, which is determined basically by a timing capacitor and resistor whose value may be selected externally. Thus, the output state remains in the SET state only for a preselected time T, after which the output returns to the RESET state, where it remains until another trigger pulse is applied. In the waveform of Fig. 15.22b the output of the one-shot can be viewed as a delayed pulse whose falling edge occurs at some set time T after the trigger pulse.

A popular TTL monostable multivibrator is the SN74121 shown in Fig. 15.23. The output provides a pulse of duration T,[1] the Q output going from its normally low state (0 V) to the high output state or SET state. Triggering can occur when either the A inputs are both grounded and the B input goes positive or when the B input is lifted high (or unconnected) and the A inputs go low. Example 15.1 will help show how the one-shot can be used.

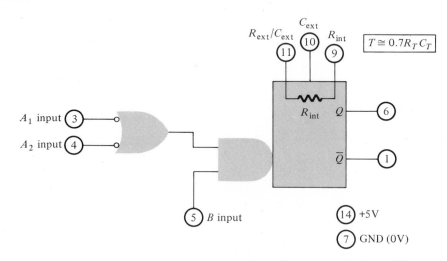

Figure 15.23 SN74121 monostable multivibrator.

EXAMPLE 15.1 Draw the Q output waveform for an SN74121 (see Fig. 15.24a) triggered by an input signal as shown in Fig. 15.24b.

Solution: The unit is triggered on a positive edge using the B input, and the time width of the output pulse from the normally low Q output is

$$T = 0.7\ R_T C_T = 0.7(5 \times 10^3)(1500 \times 10^{-12}) = \textbf{5.25}\ \boldsymbol{\mu}\textbf{s}$$

as shown in Fig. 15.24c.

---

[1] For the SN74121 the pulse width of the output pulse is $T \cong 0.7R_T C_T$ where R_T is the timing resistor, R_{ext} or R_{int} and C_T is the timing capacitor, C_{ext}.

CH. 15 DIGITAL ICS

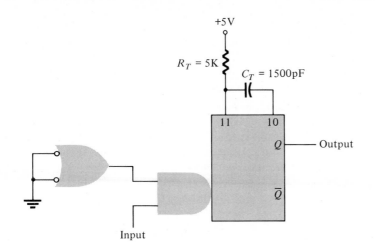

(a)

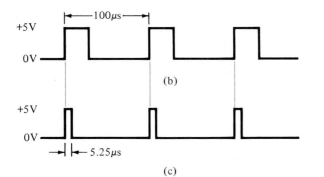

(b)

(c)

Figure 15.24 One-shot circuit and waveforms for Example 15.1.

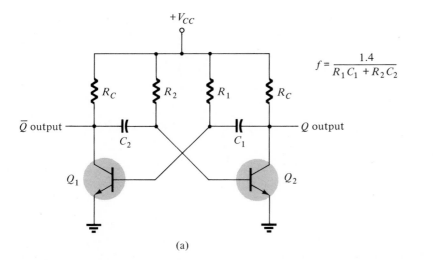

(a)

$$f = \frac{1.4}{R_1 C_1 + R_2 C_2}$$

(b)

Figure 15.25 (a) Astable multivibrator circuit; (b) logic symbol.

597

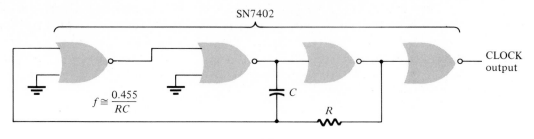

$$f \cong \frac{0.455}{RC}$$

CLOCK output

(a)

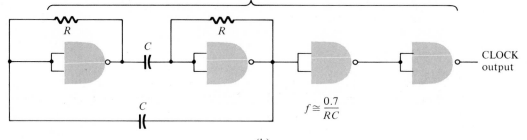

$$f \cong \frac{0.7}{RC}$$

CLOCK output

(b)

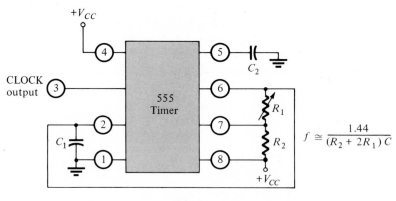

$$f \cong \frac{1.44}{(R_2 + 2R_1)\,C}$$

(c)

Figure 15.26 CLOCK circuits built with various IC units.

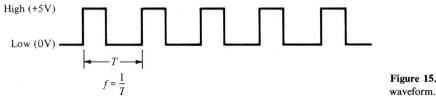

High (+5V)

Low (0V)

T

$$f = \frac{1}{T}$$

Figure 15.27 CLOCK output waveform.

Astable Multivibrator (Clock)

A third version of multivibrator has no stable operating state—it oscillates back and forth between RESET and SET states. The circuit provides a clock signal for use as a timing train of pulses to operate digital circuits. Figure 15.25 shows a circuit diagram of an astable multivibrator. Notice that both cross-coupling components are capacitors, thereby allowing no stable operating state. If the resistors and capacitors used are of equal value, the frequency of the clock is

$$f = \frac{1}{2T} = \frac{1}{2(0.7RC)} = \frac{1}{1.4RC} \cong \frac{0.7}{RC}$$

Integrated circuit units may be used to build the clock circuit. Figure 15.26 shows a few examples including the circuit parameters affecting the clock frequency. A typical clock output waveform is shown in Fig. 15.27.

Schmitt Trigger

A circuit that is somewhat like the multivibrator circuits considered is the Schmitt trigger circuit. Somewhat analogous to the one-shot, the Schmitt trigger is used for waveshaping purposes. Basically, the circuit has two opposite operating states as do all the multivibrator circuits. The trigger signal, however, is not typically a pulse waveform but a slowly varying ac voltage. The Schmitt trigger is level sensitive and switches the output state at two distinct triggering levels, one called a *lower-trigger level* (LTL) and the other an *upper-trigger level* (UTL). The circuit generally operates from a slowly varying input signal, such as a sinusoidal waveform, and provides a digital output—either the logical-0 or logical-1 voltage level.

The typical waveform of Fig. 15.28 shows a sinusoidal waveform input and squared

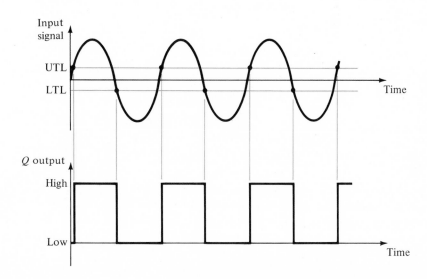

Figure 15.28 Typical waveforms for Schmitt trigger.

waveform output. Note that the output signal frequency is exactly that of the input signal, except that the output has a sharply shaped slope and remains at the low or high voltage level until it switches. One example of a Schmitt trigger application is converting a sinusoidal signal to one that is useful with digital circuits. Signals such as a 60-Hz line voltage or a slowly varying voltage obtained from a magnetic pickup are squared up for digital use. Another possibility is using the Schmitt trigger to provide a logic signal that indicates whenever the input goes above a threshold level (UTL).

15.4 DIGITAL IC UNITS

The basic digital IC units are packaged as small-scale integrated (SSI) units—logic gates, flip-flops, one-shots, and clocks. More functional arrangements, such as medium-scale integrated (MSI) units, provide counters, data registers, and decoders among the list of functional units. While the range of MSI units, mostly built as TTL, is quite extensive, a description of some typical units will provide an introduction of what is available.

Serial-Data Register

As an example, a 7491 (see Fig. 15.29) IC unit is an 8-bit serial-input/serial-output register containing eight flip-flop stages connected as a shift register plus some control gating. The data input is connected to the A, B terminals, which provide an ANDing of signals at those inputs. Usually, one of these inputs is not used, or used only for control, while the single serial data is connected as the other input.

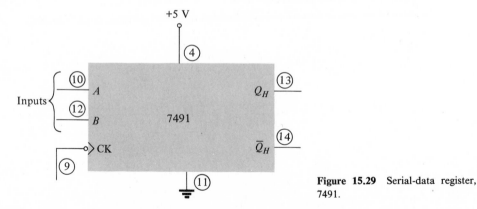

Figure 15.29 Serial-data register, 7491.

The serial data coming from such devices as a teletype, or phone line, are clocked into the eight-stage register of the 7491 by shift pulses applied to the clock (CK) input. Data stored in the eight register stages of the 7491 may also be shifted out at the Q_H (or $\overline{Q_H}$) data-output line when clocked by shift pulses provided at the clock input.

Parallel-Data Register

An example of an 8-bit parallel input/output data register is the 74199 (see Fig. 15.30). Eight input bits are provided as input and transfer into eight flip-flop stages in the 74199 when the SHIFT/LOAD lines goes low. This data remains stored in the IC unit with 8 bits appearing at the output terminals for transfer to other IC units. The register could be used, for example, to input an 8-bit character from a keyboard, the data being held in the 74199 IC until another keyboard character is entered, as indicated by a momentary low signal on the SHIFT/LOAD line. The IC can also be used as a teletype computer interface since it also operates as a shift register. Serial data from a teletype are applied at the JK input terminals and transferred into the eight register stages by shift pulses at the clock input (shifting can be prevented, if desired by a high-level signal at the clock-inhibit input). The data are then available in parallel as an 8-bit character to the computer. Another 74199 IC can be equally used to interface from the computer out to the teletype, with parallel transfer from the computer to the IC unit, followed by shifting out of the data using clock pulses (with data taken from the Q_H output).

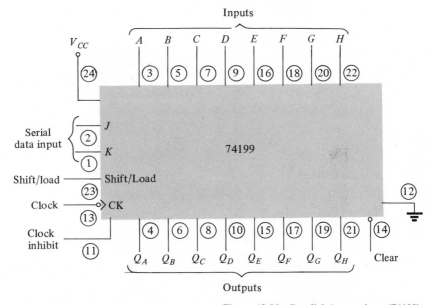

Figure 15.30   Parallel-data register (74199).

Counters

Another important application of digital MSI units is in digital counters of various count steps. The 7490 IC, for example, a popular decade counter is shown in Fig. 15.31a. The IC unit contains four flip-flop stages which can easily be connected to operate as a decade counter. The four output bits (Q_A, Q_B, Q_C, Q_D) advance from

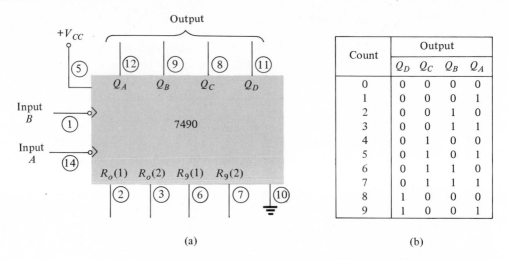

Figure 15.31 Decade counter, 7490: (a) IC unit; (b) BDC count sequence.

Count	Output			
	Q_D	Q_C	Q_B	Q_A
0	0	0	0	0
1	0	0	0	1
2	0	0	1	0
3	0	0	1	1
4	0	1	0	0
5	0	1	0	1
6	0	1	1	0
7	0	1	1	1
8	1	0	0	0
9	1	0	0	1

zero to nine as indicated in Fig. 15.31b. The 7490 counter can be part of a voltage or frequency converter, providing binary output to represent a decimal count from 0 to 9. [For decade-count operation, the reset inputs—$R_0(1)$, $R_0(2)$, $R_9(1)$, $R_9(2)$—must all be grounded and the Q_A output connected as the B input.] The counter advances from 0 to 9 and back to 0 on the following clock pulse (at input A). If two 7490 units are used, the count sequence obtained can go from 00 to 99, using each IC as a decade counter. In that case the Q_D output of the first decade counter is used as input A of the second-decade-counter stage.

Other count sequences are possible using any of the large variety of IC units.

Decoders

A most useful operation obtained from MSI units is decoding from one binary code into another. A popular decoder unit is the 7447 IC, which accepts a 4-bit binary-coded-decimal (see Fig. 15.31b) character as input and then outputs the signals to directly operate a seven-segment display as used in meters and clocks. Figure 15.32a shows the 7447 IC, while Fig. 15.32b shows the BCD-to-seven-segment code conversion. A logic-0 signal to a common-cathode LED will result in the LED segment going on. The 7447 provides logic-0 (0 V) outputs, which will result in the segment connected to that signal going on. Figure 15.33 shows a 7490 connected as a decade counter driving a 7447 connected to a seven-segment LED display. The clock input advances the 7490 through its count sequence from 0 to 9 (and back to 0). For each count step resulting as output from the 7490, the 7447 decodes into the seven-segment codes to display the character on an LED. Operating the unit of Fig. 15.33 results in the LED displaying 0 to 9 in decimal as the count proceeds in binary. This display could be one digit of a clock, a frequency counter, or a voltmeter.

The range and number of MSI units, although quite large, can be seen by the examples above to provide the small unit of a larger operating system. The particular units used depend on various features of the application.

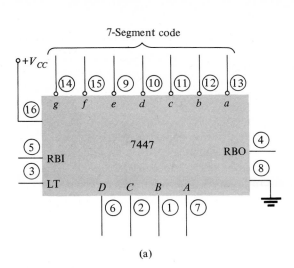

7-Segment code

+V_{CC}

(a)

D	C	B	A	a	b	c	d	e	f	g
0	0	0	0	0	0	0	0	0	0	1
0	0	0	1	1	0	0	1	1	1	1
0	0	1	0	0	0	1	0	0	1	0
0	0	1	1	0	0	0	0	1	1	0
0	1	0	0	1	0	0	1	1	0	0
0	1	0	1	0	1	0	0	1	0	0
0	1	1	0	1	1	0	0	0	0	0
0	1	1	1	0	0	0	1	1	1	1
1	0	0	0	0	0	0	0	0	0	0
1	0	0	1	0	0	0	1	1	0	0

1 = Open circut

0 = Low (0 V) level

(b)

Figure 15.32 Decoder IC, 7447.

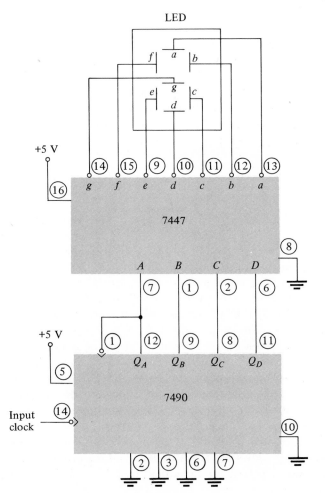

Figure 15.33 One-digit counter with display.

603

An important area of digital IC application is in semiconductor memory. Large-scale integrated (LSI) circuits provide memory cells arranged in various size groups to store information for use with computers, CRT terminals, and other digital devices. A typical IC unit contains many thousands of memory cells arranged in various ways. A 1K-bit (1024 memory cells) memory IC, for example, could be arranged as 1K memory groups of 1 bit each, or 256 groups (words) of 4 bits each, the first being referred to as a $1K \times 1$ (1K by 1), the second as 256×4. The designation of memory-cell organization provides information of the necessary IC signals to operate the unit. The number of words is a multiple of 2^n, $256 = 2^8$, $1K = 2^{10}$, $4K = 2^{12}$, as examples. The value n is the number of address bits needed to select each separate word. For 256 words it is necessary to use 8 bits to address each word of storage, the word size then being 1 bit or 4 bits or 8 bits. For a $1K \times 4$ memory the IC would require 10 address bits to select a word of memory, which is then 4 bits in size. Figure 15.34 shows a typical memory IC of 4K-bit capacity, organized as

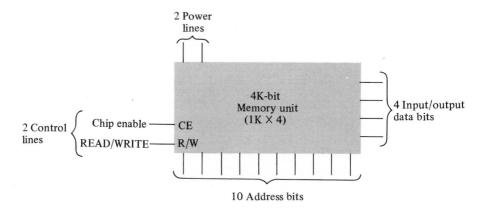

Figure 15.34 Typical 4K-bit memory IC.

$1K \times 4$. The IC can be housed in an 18-pin package with 4 pins providing 4 bits of data as input or output, 10 bits for selecting among the 1024 (2^{10}) memory addresses, 2 pins providing control signals, and 2 pins to supply the IC power (typically 5 V). The READ/WRITE control signal is used to specify whether the 4 data bits are input to the memory (READ/WRITE signal low) or output taken from the IC (READ/WRITE signal high). The chip enable (CE) is a control signal that can be used to select operation of each separate IC when a group of ICs are used to make a larger memory than is provided by only one IC.

Figure 15.35 shows an arrangement of two $1K \times 4$ ICs into $2K \times 4$ memory. There are 11 bits ($2^{11} = 2048$) necessary to select a location in a 2K memory. In this example the 11th bit is used to select among 1K located in one IC or 1K located in a second IC. The address bits are labeled to represent the positional value in address selection (i.e., A_0 is the 2^0 bit, A_9 is the 2^9 selection bit). In this example the chip-enable control line allows for the additional bit needed to extend the IC for use in a 2K memory unit. This chip-enable input can also be used with additional

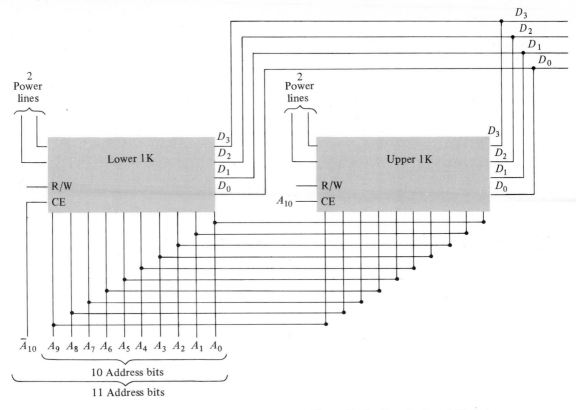

Figure 15.35 Organization of 2K × 4 memory.

logic gating or decoding to further extend the memory size above 2K using the basic 1K × 4 ICs. The data lines for both IC are connected in common with the address lines selecting which IC the data operate with for a specific address.

Memory Types

Semiconductor memory cells are built using bipolar or field-effect transistors, the latter being most popular at present. In either case the memory cell can be built to operate as static or dynamic. A static memory cell will hold a stored bit as long as power is maintained, while a dynamic memory cell could only hold the data bit a short time and must be *refreshed* or rewritten to maintain the data. The dynamic cell requires fewer transistor components than the static cell so that the greatest-capacity ICs use that cell type. The static ICs, however, handle data at a faster rate.

Memories that provide both reading and writing of information are RAM (random-access memory). IC capacities of 64K bits are presently available using dynamic MOSFET RAM cells. Eight such ICs could then be used to result in a 64K-byte memory.

When fixed data, as in programmed calculators, fixed conversion routines, or basic control or operating systems are used, ROM (read-only memory) ICs can be used. These ICs are organized similar to that of Fig. 15.34 except ROMs do not need a READ/WRITE control signal. If the data are programmed at the time of manufacture, the IC is called a ROM. If data are written into the IC after the IC is manufactured by the user, it is called a PROM (programmable ROM) and the stored data are also permanent. Of great popularity is the erasable programmable ROM (EPROM) made of MOSFET storage cells. Data are written into a cell as stored charge, trapped in the gate region of a MOSFET transistor.

While individual data bits cannot be altered, the total IC data can be erased using ultraviolet light, after which new information can be programmed into the IC memroy cells. The erasure is a slow, offline process, which is useful for occasional rewriting of "permanently" stored data.

PROBLEMS

§ 15.1

1. A positive-logic diode AND gate has voltage levels of 0 V and +5 V. Draw a circuit diagram and prepare voltage and logic truth tables for a two-input gate. (Assume ideal diodes.)

2. Draw the circuit diagram of a negative-logic AND gate for voltage levels of −5 V and 0 V. Assume ideal diodes and prepare voltage and logic truth tables for a two-input gate.

§ 15.2

3. Draw the circuit diagram of a four-input TTL NAND gate indicating voltage levels and logic definitions.

4. State two differences between CMOS and TTL operation.

5. Draw the circuit diagram of a three-input CMOS NOR gate.

6. How does an ECL-type gate differ from a TTL?

§ 15.3

7. Describe the operation of an RS flip-flop.

8. Draw the output waveform of a T flip-flop triggered by a 10-kHz clock.

9. Describe the operation of a JK flip-flop.

10. Draw the output waveform of an SN74121 circuit driven by an input clock signal of 100 kHz applied to the B input. The timing component values are $R = 10$ kΩ and $C = 100$ pF.

11. What is the frequency of an astable multivibrator circuit (as in Fig. 15.26a) having timing component values of $R = 2.7$ kΩ and $C = 750$ pF?

12. Draw the output voltage signal of a Schmitt trigger circuit for the input trigger signal of 5 V rms at 60 Hz. Circuit trigger levels are UTL $= +5$ V and LTL $= 0$ V.

§ 15.4

13. Describe the difference between a serial data register and a parallel data register.

14. What maximum count is possible using six count stages?

15. What maximum count is possible using two 7490 counter ICs?

§ 15.5

16. How many $1K \times 4$ ICs are needed to build a $4K \times 8$ memory?

17. What is the difference between a ROM and a RAM?

CHAPTER
16
Linear/Digital ICs

16.1 GENERAL

A group of IC units are available containing both linear and digital circuits. These include comparator circuits, digital/analog converter circuits, interfacing circuits, and timer circuits, among the more popular units. The comparator circuit receives input of a linear voltage, comparing it to a reference input voltage to determine which is greater. The output of the unit is a digital signal that indicates whether or not the input exceeded the reference. The IC unit thus accepts input of a linear voltage providing as output a digital voltage. Digital/analog converter circuits are used to convert a digital value to a proportional analog or linear voltage.

A wide variety of circuits are available to interface or interconnect different types of signals, both linear and digital. Some interface circuits operate by matching impedance levels, some operate by adjusting to various voltage levels, some operate from specified transducers, and some into prescribed loads. A variety of interface circuits deal with the conversion of different digital signal levels.

A timer circuit also contains both linear and digital parts. A flexible arrangement of both linear comparator circuits and digital circuits allows use of the timer in a variety of applications, including generation of pulse signals that are triggered by an input signal, and generation of a clock signal that operates at a frequency set by external resistor and capacitor. The 555 IC timer unit, one of the more popular circuits, is covered in this chapter.

16.2 COMPARATORS

A comparator circuit accepts input of linear voltages and provides a digital output that indicates when one input is less than or greater than the second. A basic comparator circuit can be represented as in Fig. 16.1a. The output is a digital signal that stays at a high level when the noninverting (+) input is greater than the inverting (−) input and switches to a lower voltage level when the noninverting input voltage goes below the inverting input reference voltage level. Figure 16.1b shows a typical connection with one input (the inverting input in this example) connected to a reference voltage, the other connected to the input signal voltage. As long as V_{IN} is less than the reference voltage level of +2 V, the output remains at a low-voltage level (near −10 V). When the input rises just above +2 V, the output quickly switches to a high-voltage level (near +10 V). Thus, the high output indicates that the input signal is greater than +2 V.

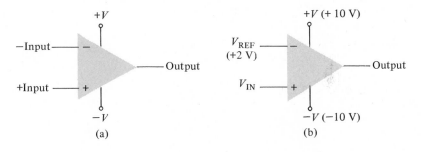

Figure 16.1 Comparator unit: (a) basic unit; (b) typical application.

Since the internal circuit used to build the comparator contains essentially an op-amp circuit with very high gain, we can first examine the operation of a comparator using a 741 op-amp, as shown in Fig. 16.2. With reference input (at pin 2) set to 0 V, a sinusoidal signal applied to the input terminal (at pin 3) will cause the output to switch between its two output states, as shown in Fig. 16.2. The input, V_i, going even a fraction of a millivolt above the 0 V reference level will be amplified by the very high gain (typically over 100,000) so that the output rises to its positive output saturation level and remains there while the input stays above $V_{ref} = 0$ V. When the input drops just below the reference 0-V level, the output is driven to its lower saturation level and stays there while the input remains below $V_{ref} = 0$ V. Figure 16.2b clearly shows that the input signal is linear while the output is digital.

In general, the reference level need not be 0 V, but any desired value, either positive or negative. Either op-amp (or comparator) input may be used as the reference input, the other then connected to the input signal.

Figure 16.3a shows a circuit operating with a positive voltage reference level, with output driving an indicator LED. The reference level is set at

$$V_{ref} = \frac{10 \text{ k}\Omega}{10 \text{ k}\Omega + 10 \text{ k}\Omega}(+12 \text{ V}) = +6 \text{ V}$$

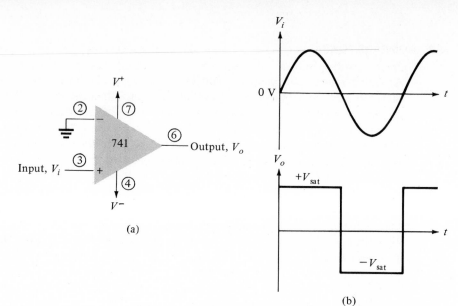

Figure 16.2   Operation of 741 op-amp as comparator.

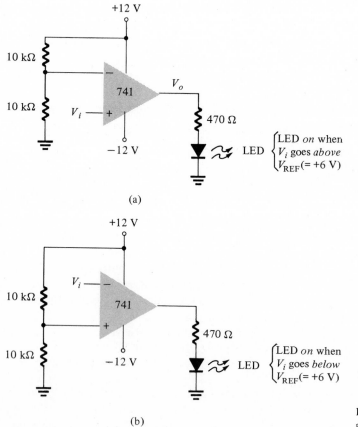

(a)

LED *on* when V_i goes *above* $V_{REF}(= +6 \text{ V})$

(b)

Figure 16.3   A 741 op-amp used as a comparator.

LED *on* when V_i goes *below* $V_{REF}(= +6 \text{ V})$

Since the reference voltage is connected to the inverting input, the output will switch to its positive saturation level when the input, V_i, goes more positive than the +6-V reference voltage level. The output voltage, V_o, then drives the LED on, as an indication that the input is more positive than the reference level.

As an alternative connection, the reference voltage could be connected to the noninverting input (see Fig. 16.3b). With this connection the input signal going below the reference level would cause the output to drive the LED on. The LED can thus be made to go on when the input signal goes either above or below the reference level, depending on which input is connected as the input signal and which as the reference.

While op-amp ICs can be used as comparator circuits, separate comparator ICs are available for use in such applications. Some improvements built into comparator IC units are faster switching between the two output levels, built-in noise immunity to prevent the output from oscillating when the input passes by the reference level, and outputs capable of directly driving a variety of loads. A few popular IC comparators are covered next to show how they are defined and how they may be used.

The 311 voltage comparator shown in Fig. 16.4 contains a comparator circuit that can operate as well from dual power supplies of ±15 V as from a single +5-V supply (as used for digital logic circuits). The output can provide voltage at one of two distinct levels or can be used to drive lamps or relays. Notice that the output is taken from a bipolar transistor to allow driving a variety of loads. The unit also has balance and strobe inputs, the strobe input allowing gating of the output. A few examples will show how the comparator unit can be used for common applications.

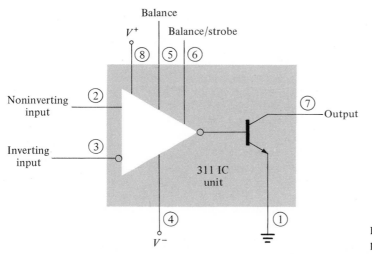

Figure 16.4 A 311 comparator (8-pin DIP unit).

A zero-crossing detector can be built using the 311 as shown in Fig. 16.5. The input going positive (above 0 V) drives the output transistor on, the output going low, to −10 V in this connection. The input going below 0 V will drive the output transistor off, the output going to +10 V.

The output is thus an indication of whether the input is above or below 0 V. When the input is any positive voltage (above 0 V), the output is a low level, while any negative input voltage will result in the output going to a high-voltage level.

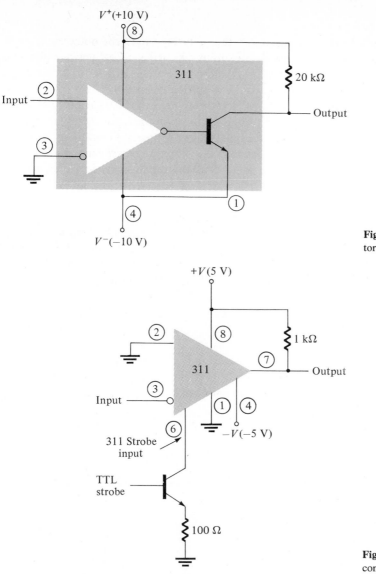

Figure 16.5 Zero-crossing detector using a 311 IC.

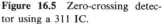

Figure 16.6 Operation of a 311 comparator with strobe input.

Figure 16.6 shows how a 311 comparator can be used with strobing. In this example the output will go high when the input goes above the reference level—but only if the TTL strobe input is off (or 0 V). If the TTL strobe input goes high, it drives the 311 strobe input at pin 6 low, causing the output to remain in the off state (with output high) regardless of the input signal. In effect, the output remains high unless strobed. If strobed, the output then acts normally, switching from high to low depending on the input signal level. In operation, the comparator output will respond to the input signal only during the time the strobe signal allows such operation.

Figure 16.7 shows the comparator output driving a relay. When the input goes below 0 V, driving the output low, the relay is activated, closing the normally open (N.O.) contacts at that time. These contacts can then be connected to operate a large variety of devices. For example, a buzzer or bell wired to the contacts can be driven on whenever the input voltage drops below 0 V. As long as a voltage is present at the input terminal, the buzzer will remain off.

Another popular comparator unit is packaged with four independent voltage comparator circuits in a single IC. The 339 is a quad comparator IC, all four comparator circuits connected to the external pins as shown in Fig. 16.8. Each comparator has inverting and noninverting input and an output. The supply voltage applied to a pair of pins connects to all four comparator circuits. Even if one wishes to use only some of the comparator circuits, all four are active, drawing power from the supply.

To see how these comparator circuits can be used, Fig. 16.9 shows one of the

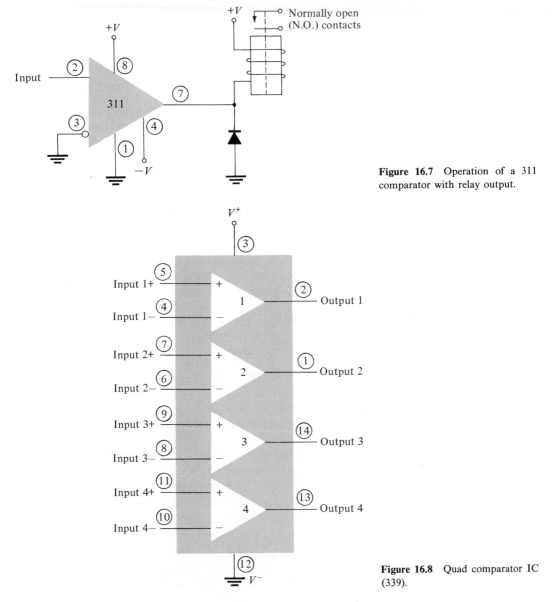

Figure 16.7 Operation of a 311 comparator with relay output.

Figure 16.8 Quad comparator IC (339).

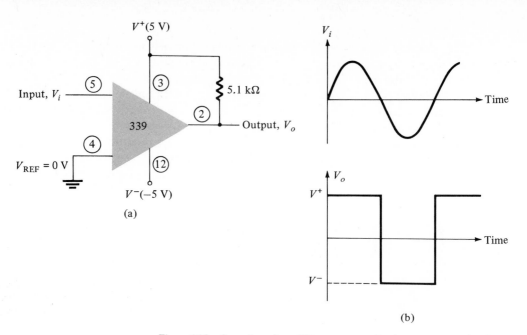

Figure 16.9 Operation of one 330 comparator circuit as a zero-crossing detector.

339 comparator circuits connected as a zero-crossing detector. Whenever the input signal goes above 0 V, the output switches to V^+. The output switches to V^- only when the input goes below 0 V.

A reference level of other than 0 V can also be used and either input terminal could be used as reference, the other input then being the signal input. The operation of one of the comparator circuits is described next.

The differential input (voltage difference across input terminals) going positive drives the output transistor off (open-circuit), while a negative differential input drives the output transistor on—the output then at the supply low level.

If the negative input is set at a reference level, V_{ref}, the positive input going above V_{ref} results in a positive differential input with output driven to the open-circuit state. When the noninverting input goes below V_{ref}, resulting in a negative differential input, the output will be driven to V^-.

If the positive input is set at the reference level, the inverting input going below V_{ref} results in the output open circuit, while the inverting input going above V_{ref} results in the output at V^-. This operation is summarized in Fig. 16.10.

Since the output of one of these comparator circuits is open-circuit collector, applications in which the outputs from more than one circuit can be wire-ORed are possible. Figure 16.11 shows two comparator circuits connected with common output, and also with common input. Comparator 1 has a +5-V reference voltage input connected to the noninverting input. The output will be driven low by comparator 1 when the input signal goes above +5 V. Comparator 2 has a reference voltage of +1 V connected to the inverting input. The output of comparator 2 will be driven low when the input signal goes below +1 V. In total the output will go low whenever the input is below +1 V or above +5 V, as shown in Fig. 16.11, the overall operation being that of a voltage window detector. The output high indicates that the input

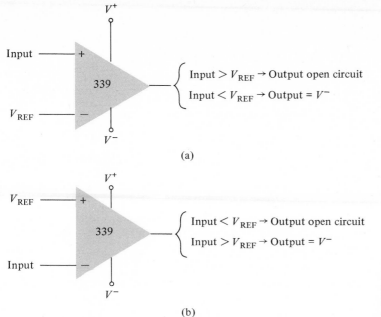

(a)

(b)

Figure 16.10 Operation of a 330 comparator circuit with reference input at: (a) minus input; (b) plus input.

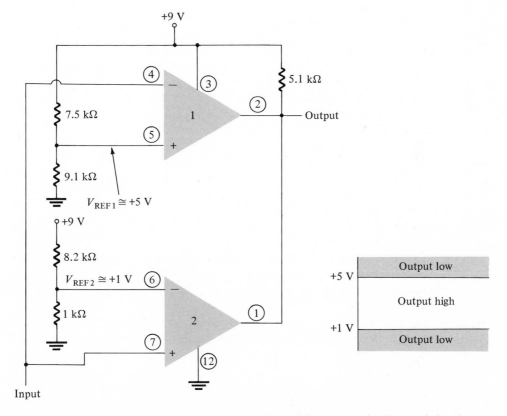

Figure 16.11 Operation of two 339 comparator circuits as a window detector.

is within a voltage window of +1 V to +5 V (these values being set by the reference voltage levels used).

16.3 DIGITAL/ANALOG CONVERTERS

Many of the voltage and current signals occurring in electronics are linear, in that they vary continuously over some range of values. In digital devices and computers the signals are digital, at one of two levels representing the binary values of one or zero.

If the signals to be used in some digital operations are linear (analog) voltages (e.g., dc voltages representing temperature or pressure, or position), a circuit must convert this analog voltage into a digital value—this conversion circuit being an analog-to-digital or A/D converter. When a computer has a digital value to be output as an analog voltage, a digital-to-analog or D/A converter circuit is used.

Digital-to-Analog (D/A) Conversion

Digital-to-analog conversion can be achieved using a number of different methods. One popular scheme uses a network of resistors, called a ladder network. A ladder network accepts inputs of binary values at, typically, 0 V or V_{ref}, and provides an output voltage proportional to the binary input value. Figure 16.12a shows a ladder network with four input voltages, representing 4 bits of digital data and a dc voltage output. The output voltage is proportional to the digital input value as given by the relation

$$V_o = \frac{D_0 \times 2^0 + D_1 \times 2^1 + D_2 \times 2^2 + D_3 \times 2^3}{2^4} \times V_{\text{ref}} \qquad (16.1)$$

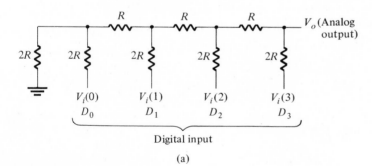

(a)

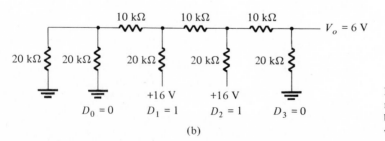

(b)

Figure 16.12 Four-stage ladder network used as D/A converter: (a) basic circuit; (b) circuit example with 0110 input.

In the example shown in Fig. 16.12b, the output voltage resulting should be

$$V_o = \frac{0 \times 1 + 1 \times 2 + 1 \times 4 + 0 \times 8}{16} \times 16 \text{ V} = 6 \text{ V}$$

Therefore, 0110_2 converts to 6 V.

Use superposition to verify that the resulting value of V_o is indeed 6 V. The function of the ladder network is to convert the 16 possible binary values from 0000 to 1111 into one of 16 voltage levels in steps of $V_{ref}/16$. Using more sections of ladder allows having more binary inputs and greater quantization for each step. For example, a 10-stage ladder network could extend the number of voltage steps or voltage resolution to $V_{ref}/2^{10}$ or $V_{ref}/1024$. A reference voltage of 10 V would then provide output voltage steps of 10 V/1024 or approximately 10 mV. More ladder stages provide greater voltage resolution, in general the voltage resolution for n ladder stages being

$$\frac{V_{ref}}{2^n} \tag{16.2}$$

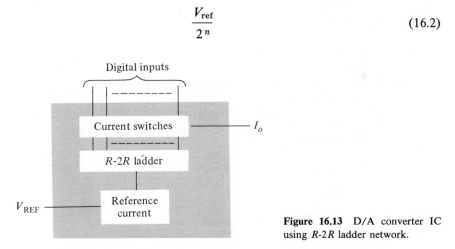

Figure 16.13 D/A converter IC using R-$2R$ ladder network.

A block diagram of the main components of a typical IC D/A converter is shown in Fig. 16.13. The ladder network, referred to in the diagram as an R-$2R$ ladder is sandwiched between the reference current supply and currents switches connected to each binary input with a resulting output current proportional to the input binary value. The binary inputs turn on selected legs of the ladder, the output current being a weighted summing of the reference current. Connecting the output across a resistor will produce an analog voltage, if desired.

Analog-to-Digital (A/D) Conversion

DUAL-SLOPE CONVERSION

A popular method of converting analog voltage into a digital value is the dual-slope method. Figure 16.14a shows a block diagram of the basic dual-slope converter.

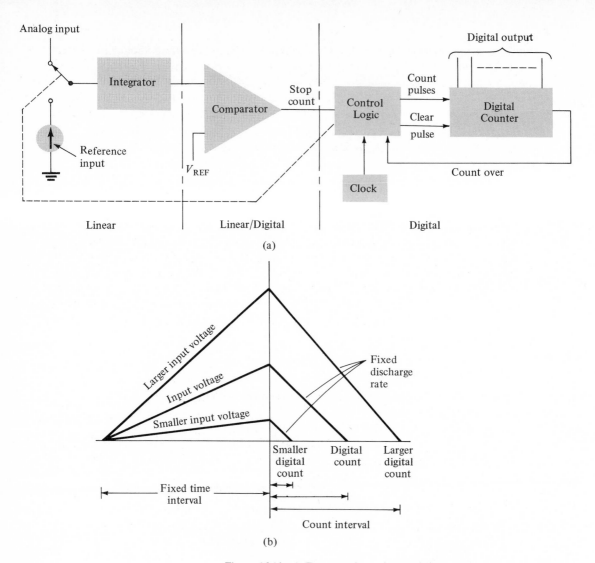

Figure 16.14 A/D conversion using dual-slope method: (a) logic diagram; (b) waveform.

The analog voltage to be converted is applied through an electronic switch to an integrator or ramp-generator circuit (essentially a constant current charging a capacitor to produce a linear-ramp voltage). The digital output is obtained from a counter operated during both positive and negative slope intervals of the integrator.

The conversion method proceeds as follows. For a fixed time interval (usually the full count range of the counter), the analog input voltage, connected to the integrator, raises the voltage in the comparator to some positive level. Figure 16.14b shows that at the end of the fixed time interval the voltage from the integrator is greater for larger input voltages. At the end of the fixed count interval, the count is set at zero and the electronic switch connects the integrator to a reference or fixed input. The integrator output (or capacitor input) then decreases at a fixed rate. The counter advances during this time. The integrator output voltage decreases at a fixed rate until it drops below the comparator reference voltage, at which time the control logic receives a signal (the comparator output) to stop the count. The digital value stored in the counter is then the digital output of the converter.

Using the same clock and integrator to perform the conversion during positive and negative slope intervals tends to compensate for clock frequency drift and integrator accuracy limitations. Setting the reference input value and clock rate can scale the counter output as desired. The counter can be binary or BCD or other digital form, if desired.

LADDER-NETWORK CONVERSION

Another popular method of analog-to-digital conversion uses the ladder network along with counter and comparator circuits (see Fig. 16.15). A digital counter advances from a zero count while a ladder network driven by the counter outputs a staircase voltage as shown in Fig. 16.15b, which increases one voltage increment for each count step. A comparator circuit, receiving both staircase voltage and analog input voltage, provides a signal to stop the count when the staircase voltage rises above the input voltage. The counter value at that time is the digital output.

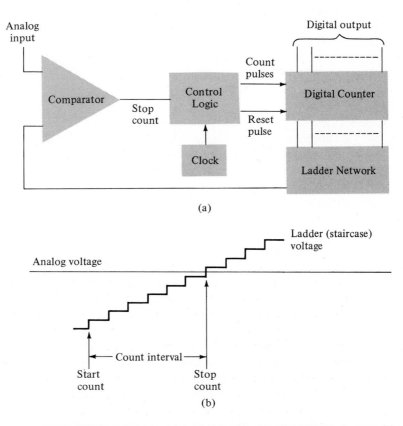

(a)

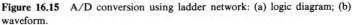

(b)

Figure 16.15 A/D conversion using ladder network: (a) logic diagram; (b) waveform.

The amount of voltage change stepped by the staircase signal depends on the reference voltage applied to the ladder network and on the number of count bits used. A 12-stage counter operating a 12-stage ladder network using a reference voltage of 10 V would step each count by a voltage of

$$V_{ref}/2^{12} = 10 \text{ V}/4096 = 2.4 \text{ mV}$$

This would result in a conversion resolution of 2.4 mV. The clock rate of the counter would affect the time required to carry out a conversion. A clock rate of 1 MHz operating a 12-stage counter would need a maximum conversion time of

$$4096 \times 1 \text{ } \mu s = 4096 \text{ } \mu s \cong 4.1 \text{ ms}$$

The minimum number of conversions that could be carried out each second would then be

$$\text{Number of conversions} = 1/4.1 \text{ ms} \cong 244 \text{ conversions/second}$$

Since on the average, with some conversions requiring little count time and others near maximum count time, a conversion time of $(4.1 \text{ ms})/2 = 2.05 \text{ ms}$ would be needed, the average number of conversions would be $2 \times 244 = 488$ conversions/second. A slower clock rate would result in fewer conversions per second. A converter using fewer count stages (and less conversion resolution) would carry out more conversions per second.

The conversion accuracy would depend on the accuracy of the voltage reference used in the ladder network and on the accuracy of the comparator.

16.4 INTERFACING

Connecting different types of circuits, different analog or digital units, and inputs or loads to other electronics all require some sort of interfacing. Interface circuits may be categorized as either driver or receiver units. A receiver essentially accepts inputs, providing high input impedance to minimize loading of the input signal. A driver circuit provides the output signal at voltage or current levels suitable to operate a number of loads, or to operate such devices as relays, displays, and power units. Furthermore, these inputs or outputs may require strobing, which provides the interface signal connection during specific time intervals as established by the strobe.

Figure 16.16a shows a dual line driver, each driver having input operation capable of accepting TTL signals and output which can operate TTL, DTL, or MOS devices. Often, the interface circuit is needed to receive signals from one type of circuit (TTL, DTL, ECL, MOS) and transmit the signal to another circuit type. The relation between the input and output signal may be such that the interface unit can be used as either noninverting unit or inverting unit. Interface circuits or all these various configurations are necessary and exist as IC units.

The circuit of Fig. 16.16b shows a dual line receiver having both inverting and noninverting inputs so that either operating condition can be selected. As example, connection of an input signal to the inverting input would result in an inverted output from the receiver unit. Connecting the input to the noninverting input would provide the same interfacing except that the output obtained would have the same polarity

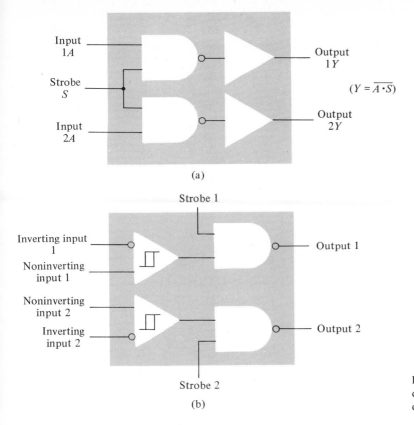

$(Y = \overline{A \cdot S})$

(a)

(b)

Figure 16.16 Interface units: (a) dual-line drivers (SN75150); (b) dual-line receivers (SN75152).

as the received signal. In the case of both driver and receiver circuits shown in Fig. 16.16, the outputs will be present only when the strobe signal is present—high level in the present circuits.

Another type of interfacing that is quite important occurs when connecting signals between various terminals of a digital system. Signals from such devices as a teletype, video terminal, card reader, or line printer are usually one of a number of signal forms. The EIA electronics industry's most popular standard is referred to as RS-232-C. Complete details of the expected signal conditions for this standard can be stated simply here as binary signals representing mark (logic-1) and space (logic-0) corresponding to the voltage levels of -12 V and $+12$ V, respectively. TTL circuits operate with signals defined as $+5$ V as mark and 0 V as space. Teletype units are sometimes wired to operate with current-loop signals, for which 20 mA represents a mark with no current representing a space. These different types of signals may occur as either input or output of a particular terminal, so that a variety of interface circuitry is necessary to convert from one signal type to one of the other. Some popular examples of interfaces will be described next.

Figure 16.17a shows the defined mark and space conditions for current loop, RS-232-C, and TTL signals.

RS-232-C-to-TTL Converter

If a unit having output defined by RS-232-C is to operate into another unit which operates with TTL-signal levels, the interface circuit of Fig. 16.17b could be used.

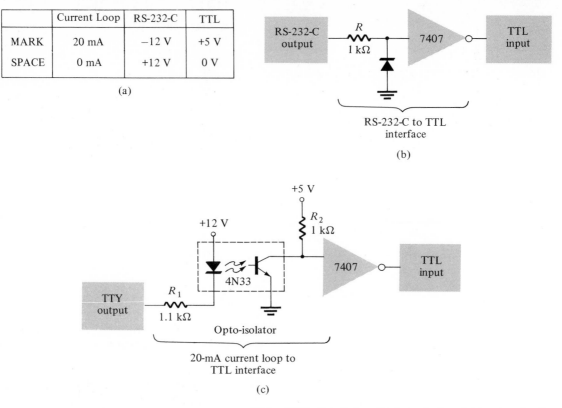

	Current Loop	RS-232-C	TTL
MARK	20 mA	−12 V	+5 V
SPACE	0 mA	+12 V	0 V

(a)

RS-232-C to TTL
interface

(b)

Opto-isolator

20-mA current loop to
TTL interface

(c)

Figure 16.17 Interfacing signal standards and converter circuits.

A mark output from the driver (−12 V) would get clipped by the diode so that the input to the inverter circuit is near 0 V, resulting in an output of +5 V or a TTL-level mark. A space output at +12 V would drive the inverter output low for a 0-V space (TTL).

Another example of interface is that between a current-loop input and TTL, as shown in Fig. 16.17c. An input mark results when 20 mA current is drawn through the output line of the Teletype (TTY). This current then goes through the diode element of an opto-isolator driving the output transistor on. The input to the inverter going low results in a +15-V signal to the TTL input, so that a mark from the Teletype results in a mark into the TTL input. A space from the Teletype current loop provides no current with opto-isolator transistor remaining off and inverter output then 0 V, which is a TTL space signal.

Other types of interface circuits can be considered, those of Fig. 16.17 being representative. Another means of interfacing digital signals is made using open-collector output and using tri-state buffer outputs. When a signal is output from a transistor collector (see Fig. 16.18) which is not connected to any other electronic components, the output is open-collector. This allows connecting a number of signals to the same signal wire or signal bus. Then any transistor going on provides a low output condition, while all transistors off provide a high output. An equally popular connection for tieing a number of digital signals to a common bus uses tri-state output as shown in Fig. 16.19b. The output can be either high level (near +5 V), low level (0 V), or

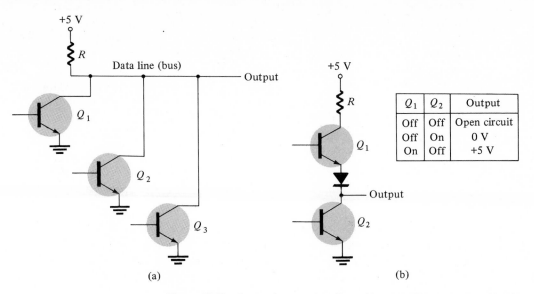

Q_1	Q_2	Output
Off	Off	Open circuit
Off	On	0 V
On	Off	+5 V

Figure 16.18 Connections to date lines: (a) open-collector output; (b) tri-state output.

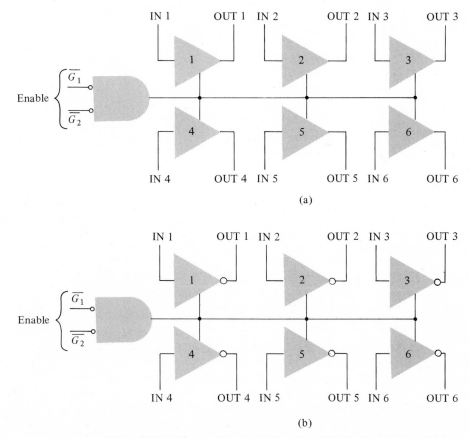

Figure 16.19 Tri-state ICs: (a) noninverting (74365); (b) inverting (74366).

open-circuit. With this circuit connection the various logic circuits connected to the common line must be gated so that only one circuit can operate the bus, the other outputs being open-circuit at that time.

The tri-state buffer circuits could be either noninverting output or inverting output. Figure 16.19 shows two popular IC units, each packaged to hold six identical gates. The enable inputs provide the strobe signal to disable the set of six gates or cause the output to be open-circuit. When the buffer gates are enabled, the output can go to either +5 V or 0 V, depending on the input signal.

16.5 TIMERS

Another popular analog/digital integrated circuit is the versatile 555 timer unit. The IC is made of a combination of linear comparators and digital flip-flop as described in Fig. 16.20. The entire circuit is usually housed in an eight-pin DIP package with pin numbers as specified in Fig. 16.20. A series connection of three resistors set the reference level inputs to the two comparators at $\frac{2}{3} V_{CC}$ and $\frac{1}{3} V_{CC}$, the outputs of these comparators setting or resetting the flip-flop unit. The flip-flop circuit output is then brought out through an output amplifier stage. The flip-flop circuit also operates a transistor inside the IC, the transistor collector usually being driven low to discharge a timing capacitor.

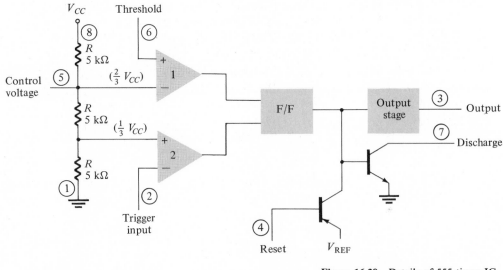

Figure 16.20 Details of 555 timer IC.

Astable

One popular application of the 555 timer IC is as an astable multivibrator or clock circuit. The following analysis of the operation of the 555 as an astable circuit will include details of the different parts of the unit and how the various inputs and outputs are utilized. Figure 16.21 shows an astable circuit using external resistor

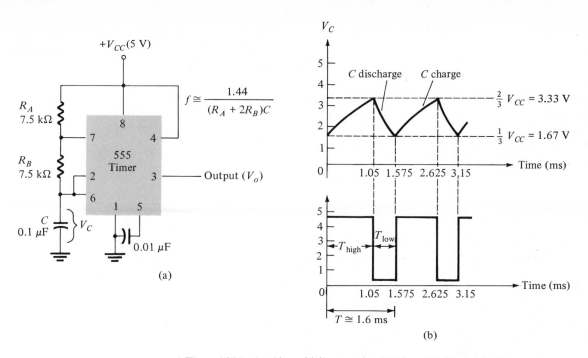

Figure 16.21 Astable multivibrator using 555 timer IC: (a) circuit; (b) waveforms.

and capacitor to set the timing interval of the output signal.

Capacitor C charges toward V_{CC} through external resistors R_A and R_B. Referring to Fig. 16.21, the capacitor voltage rises until it goes above $\frac{2}{3}V_{CC}$ ($=\frac{2}{3}(5\text{ V}) = 3.33$ V, in this example). This voltage is the threshold voltage at pin 6, which drives comparator 1 to trigger the flip-flop so that the output at pin 3 goes low. In addition, the discharge transistor is driven on, causing the output at pin 7 to discharge the capacitor through resistor R_B. The capacitor voltage then decreases until it drops below the trigger level ($V_{CC}/3 = 5$ V/3 = 1.67 V). The flip-flop is triggered so that the output goes back high and the discharge transistor is turned off, so that the capacitor can again charge through resistors R_A and R_B toward V_{CC}.

Figure 16.21b shows the capacitor and output waveforms resulting from the astable circuit connection. Calculation of the time intervals during which the output is high and low can be made using the relations:

$$\boxed{T_{\text{high}} \cong 0.7\ (R_A + R_B)C} \qquad (16.3)$$

$$= 0.7(7.5\text{ k}\Omega + 7.5\text{ k}\Omega)0.1\ \mu\text{F}$$

$$= 1.05\text{ ms}$$

$$\boxed{T_{\text{low}} \cong 0.7R_BC} \qquad (16.4)$$

$$= 0.7(7.5\text{ k}\Omega)0.1\ \mu\text{F} = 0.525\text{ ms}$$

The total period is then

$$\text{period} = T = T_{\text{high}} + T_{\text{low}} = (1.05 + 0.525)\text{ ms} = 1.575\text{ ms}$$

The frequency of the astable circuit is then calculated using[1]

$$f = \frac{1}{T} = \frac{1}{1.575 \text{ ms}} \cong 635 \text{ Hz}$$

The duty cycle of the output waveform can also be determined from[2]

$$\text{duty cycle} = \frac{T_{\text{low}}}{T} = \frac{0.525 \text{ ms}}{1.575 \text{ ms}} = 0.333 \ (= 33.3\%)$$

Monostable

The 555 timer can also be used as a one-shot or monostable multivibrator circuit. Figure 16.22 shows such connection. When the trigger input signal goes negative, it

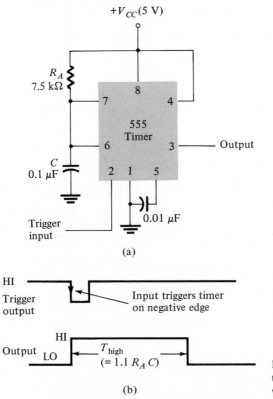

(a)

(b)

Figure 16.22 Operation of 555 timer as one-shot: (a) circuit; (b) waveforms.

---

[1] The period can be directly calculated from

$$T = 0.693 \ (R_A + 2R_B)C \cong 0.7(R_A + 2R_B)C$$

and frequency from

$$f \cong \frac{1.44}{(R_A + 2R_B)C}$$

[2] The duty cycle could also be calculated directly from

$$\text{duty cycle} = \frac{T_{\text{low}}}{T} \cdot 100\% = \frac{R_B}{R_A + 2R_B} \cdot 100\%$$

triggers the one-shot with output at pin 3 then going high for a time period

$$T_{high} = 1.1 R_A C$$ (16.5)

In the circuit of Fig. 16.22 this would be

$$T_{high} = 1.1(7.5 \text{ k}\Omega)(0.1 \text{ } \mu F) = 0.825 \text{ ms}$$

Referring back to Fig. 16.20, the negative edge of the trigger input causes comparator 2 to trigger the flip-flop with output at pin 3 going high. Capacitor C charges toward V_{CC} through resistor R_A. During the charge interval the output remains high. When the voltage across the capacitor reaches the threshold level of $\frac{2}{3} V_{CC}$ comparator 1 then triggers the flip-flop with output going low. The discharge transistor also goes low, causing the capacitor to remain at near 0 V until triggered again.

Figure 16.22 shows the input trigger signal and the resulting output waveform for the 555 timer IC operated as a one-shot. Time periods for this circuit can range from microseconds to many seconds, making this IC useful for a large range of applications.

16.6 VOLTAGE-CONTROLLED OSCILLATOR (VCO)

A voltage-controlled oscillator (VCO) is a circuit that provides an oscillating output signal (typically of square-wave or triangular-wave form) whose frequency can be adjusted over a range controlled by a dc voltage. An example of a VCO is the 566 IC unit, which contains circuitry to generate both square-wave and triangular-wave signals whose frequency is set by an external resistor and capacitor and then varied by an applied dc voltage. Figure 16.23a shows that the 566 contains current sources to charge and discharge an external capacitor, C_1, at a rate set by external resistor, R_1, and the modulating dc input voltage. A Schmitt Trigger circuit is used to switch the current sources between charging and discharging the capacitor, and the triangular voltage developed across the capacitor and square wave from the Schmitt Trigger are provided as outputs through buffer amplifiers.

Figure 16.23b shows the pin connection of the 566 unit and a summary of formula and value limitations. The oscillator can be programmed over a 10-to-1 frequency range by proper selection of an external resistor and capacitor, and then modulated over a 10-to-1 frequency range by a control voltage, V_C.

A free-running or center-operating frequency, f_o, can be calculated from

$$f_o = \frac{2}{R_1 C_1} \left[\frac{V^+ - V_c}{V^+} \right]$$ (16.6)

with the following practical circuit value restrictions:

R_1 should be within range $2 \text{ k}\Omega \leq R_1 \leq 20 \text{ k}\Omega$
V_c should be within range $\frac{3}{4} V^+ \leq V_c \leq V^+$
f_o should be below 1 MHz, and
V^+ should range between 10 V and 24 V.

Figure 16.24 shows an example in which the 566 function generator is used to

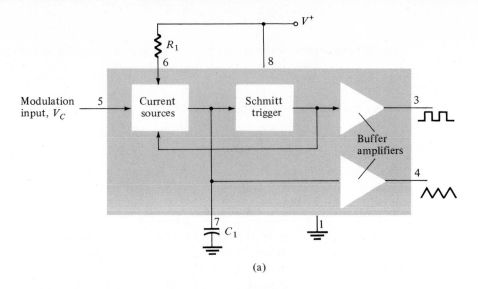

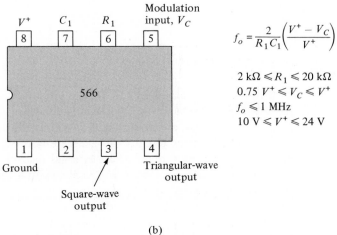

$$f_o = \frac{2}{R_1 C_1}\left(\frac{V^+ - V_C}{V^+}\right)$$

$2\ \text{k}\Omega \leqslant R_1 \leqslant 20\ \text{k}\Omega$
$0.75\ V^+ \leqslant V_C \leqslant V^+$
$f_o \leqslant 1\ \text{MHz}$
$10\ \text{V} \leqslant V^+ \leqslant 24\ \text{V}$

Figure 16.23 A 566 function generator: (a) block diagram; (b) pin configuration and summary of operating data.

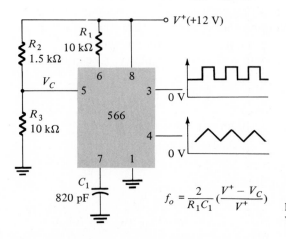

$$f_o = \frac{2}{R_1 C_1}(\frac{V^+ - V_C}{V^+})$$

Figure 16.24 Connection of 566 VCO unit.

provide both square-wave and triangular-wave signals at a fixed frequency set by R_1, C_1, and V_c. A resistor divider R_2 and R_3 sets the dc modulating voltage at a fixed value

$$V_c = \frac{R_3}{R_2 + R_3} V^+ = \frac{10 \text{ k}\Omega}{1.5 \text{ k}\Omega + 10 \text{ k}\Omega} 12 \text{ V} = +10.4 \text{ V}$$

(which falls properly in the voltage range $0.75 V^+ = 9$ V, and $V^+ = 12$ V). Using Eq. (16.6)

$$f_o = \frac{2}{(10 \times 10^3)(820 \times 10^{-12})} \left(\frac{12 - 10.4}{12}\right) = 32.5 \text{ kHz}$$

The circuit of Fig. 16.25 shows how the output square-wave frequency can be adjusted using the input voltage, V_c, to vary the signal frequency. Potentiometer R_3 allows varying V_c from about 9 V to near 12 V, over the full 10-to-1 frequency range. With the potentiometer wiper set at the top, the control voltage is

$$V_c = \frac{R_3 + R_4}{R_2 + R_3 + R_4} V^+ = \frac{5 \text{ k}\Omega + 18 \text{ k}\Omega}{510 \text{ }\Omega + 5 \text{ k}\Omega + 18 \text{ k}\Omega} (+12 \text{ V}) = 11.74 \text{ V}$$

resulting in a lower output frequency of

$$f_o = \frac{2}{(10 \times 10^3)(220 \times 10^{-12})} \left(\frac{12 - 11.74}{12}\right) = 19.7 \text{ kHz}$$

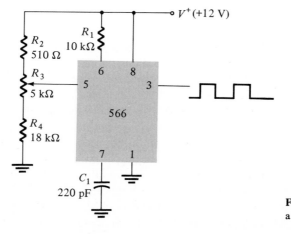

Figure 16.25 Connection of 566 as a VCO unit.

With the wiper arm of R_3 set at the bottom, the control voltage is

$$V_c = \frac{R_4}{R_2 + R_3 + R_4} V^+ = \frac{18 \text{ k}\Omega}{510 \text{ }\Omega + 5 \text{ k}\Omega + 18 \text{ k}\Omega} (+12 \text{ V}) = 9.19 \text{ V}$$

resulting in an upper output frequency of

$$f_o = \frac{2}{(10 \times 10^3)(220 \times 10^{-12})} \left(\frac{12 - 9.19}{12}\right) = 212.879 \text{ kHz}$$

The frequency of the output square wave can then be varied using potentiometer R_3 over a frequency range of at least 10 to 1.

Rather than varying a potentiometer setting to change the value of V_c, an input modulating voltage, V_{in}, can be applied as shown in Fig. 16.26. The voltage divider sets V_c at about 10.4 V. An input ac voltage of about 1.4 V, peak can drive V_c around the bias point between voltages of 9 V and 11.8 V, causing the output frequency to vary over about a 10-to-1 range. The input signal, V_{in}, thus frequency modulates the output voltage around the center frequency set by the bias value of $V_c = 10.4$ V ($f_o = 121.2$ kHz).

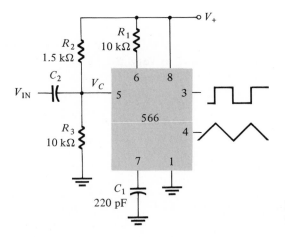

Figure 16.26 Operation of VCO with frequency modulating input.

16.7 PHASE-LOCKED LOOP (PLL)

A phase-locked loop (PLL) is an electronic circuit that consists of a phase detector, a low-pass filter, and a voltage-controlled oscillator connected as shown in Fig. 16.27. Common applications of a PLL include: (1) frequency synthesizers that provide multiples of a reference signal frequency (for example, the carrier frequency for the multiple channels of a citizens band (CB) unit or marine-radio-band unit can be generated using a single-crystal-controlled frequency and its multiples generated using a PLL); (2) FM demodulation networks for fm operation with excellent linearity between the input signal frequency and the PLL output voltage; (3) demodulation of the two data transmission or carrier frequencies in digital-data transmission used in frequency-shift keying (FSK) operation; and (4) a wide variety of areas including modems, telemetry receivers and transmitters, tone decoders, AM detectors, and tracking filters.

An input signal, V_i, and that from a VCO, V_o, are compared by a phase comparator (refer to Fig. 16.27) providing an output voltage, V_e, that represents the phase difference between the two signals. This voltage is then fed to a low-pass filter that provides an output voltage (amplified if necessary) that can be taken as the output voltage from the PLL and is used internally as the voltage to modulate the VCO's frequency. The closed-loop operation of the circuit is to maintain the VCO frequency locked to that of the input signal frequency.

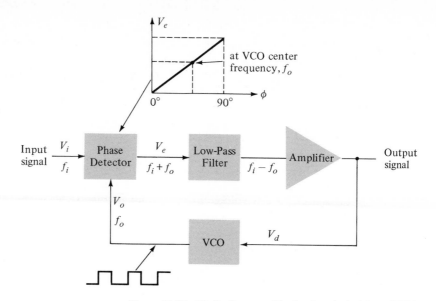

Figure 16.27 Block diagram of basic phase-locked loop (PLL).

Basic PLL Operation

The basic operation of a PLL circuit can be explained using the circuit of Fig. 16.27 as reference. We will first consider the operation of the various circuits in the phase-locked loop when the loop is operating in lock (input signal frequency and VCO frequency are the same). When the input signal frequency is the same as that from the VCO to the comparator, the voltage, V_d, taken as output is the value needed to hold the VCO in lock with the input signal. The VCO then provides output of a fixed-amplitude square-wave signal at the frequency of the input. Best operation is obtained if the VCO center frequency, f_o, is set with the dc bias voltage midway in its linear operating range. The amplifier allows this adjustment in dc voltage from that obtained as output of the filter circuit. When the loop is in lock, the two signals to the comparator are of the same frequency although not necessarily in phase. A fixed phase difference between the two signals to the comparator results in a fixed dc voltage to the VCO. Changes in the input signal frequency then result in change in the dc voltage to the VCO. Within a capture-and-lock frequency range, the dc voltage will drive the VCO frequency to match that of the input.

While the loop is trying to achieve lock, the output of the phase comparator contains frequency components at the sum and difference of the signals compared. A low-pass filter passes only the lower-frequency component of the signal so that the loop can obtain lock between input and VCO signals.

Owing to the limited operating range of the VCO and the feedback connection of the PLL circuit, there are two important frequency bands specified for a PLL. The *capture range* of a PLL is the frequency range centered about the VCO free-running frequency, f_o, over which the loop can *acquire* lock with the input signal. Once the PLL has achieved capture, it can maintain lock with the input signal over a somewhat wider frequency range called the *lock range*.

Applications

The PLL can be used in a wide variety of applications, including: (1) frequency demodulation; (2) frequency synthesis; and (3) FSK decoders. Examples of each of these follow.

FREQUENCY DEMODULATION

FM demodulation or detection can be directly achieved using the PLL circuit. If the PLL center frequency is selected or designed at the FM carrier frequency, the filtered or output voltage in the circuit of Fig. 16.27 is the desired demodulated voltage, varying in value proportional to the variation of the signal frequency. The PLL circuit thus operates as a complete intermediate-frequency (IF) strip, limiter, and demodulator as used in fm receivers.

One popular PLL unit is the 565, shown in Fig. 16.28a. The 565 contains a phase detector, amplifier, and voltage-controlled oscillator, which are only partially connected internally. An external resistor and capacitor, R_1 and C_1, are used to set the free-running or center frequency of the VCO. Another external capacitor, C_2, is used to set the low-pass filter pass band, and the VCO output must be connected back as input to the phase detector to close the PLL loop. The 565 typically uses two power supplies, V^+ and V^-.

Figure 16.28b shows the 565 PLL connected to work as an FM demodulator. Resistor R_1 and capacitor C_1 set the free-running frequency, f_o,

$$f_o = \frac{0.3}{R_1 C_1} \tag{16.7}$$

$$= \frac{0.3}{(10 \times 10^3)(220 \times 10^{-12})} = 136.36 \text{ kHz}$$

with limitation $2 \text{ k}\Omega \leq R_1 \leq 20 \text{ k}\Omega$. The lock range is then

$$f_L = \pm \frac{8 f_o}{V} \tag{16.8}$$

$$= \pm \frac{8(136.36 \times 10^3)}{6} = \pm 181.8 \text{ kHz}$$

and capture range is

$$f_C = \pm \frac{1}{2\pi} \sqrt{\frac{2\pi f_L}{(3.6 \times 10^3) C_2}} \tag{16.9}$$

$$= \pm \frac{1}{2\pi} \sqrt{\frac{2\pi (181.8 \times 10^3)}{(3.6 \times 10^3)(330 \times 10^{-12})}} = 156.1 \text{ kHz}$$

The signal at pin 4 is a 136.36 kHz square wave. An input within the lock range of 181.8 kHz will result in the output voltage at pin 7 varying around its dc voltage level set with input signal at f_o. Figure 16.28c shows the output at pin 7 as a function of the input signal frequency. The dc voltage at pin 7 is linearly related to the input signal frequency within the frequency range $f_L = 181.8$ kHz around the center fre-

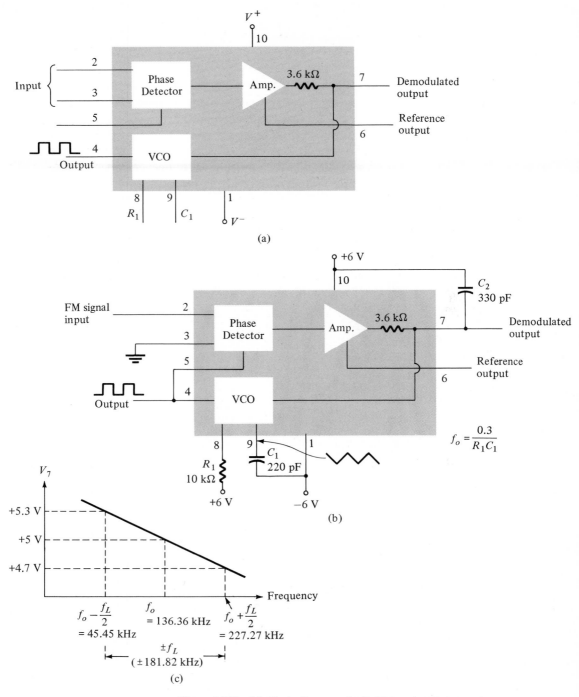

Figure 16.28 (a) Block diagram of 565 PLL unit; (b) connection as FM demodulator; and (c) output voltage-frequency relation.

quency 136.36 kHz. The output voltage is the demodulated signal that varies with frequency within the operating range specified.

FREQUENCY SYNTHESIS

A frequency synthesizer can be built around a PLL as shown in Fig. 16.29. A frequency divider is inserted between the VCO output and the phase comparator so that the loop signal to the comparator is at frequency f_o while the VCO output is Nf_o. This output is a multiple of the input frequency as long as the loop is in lock. The input signal can be crystal stabilized at f_1 with the resulting VCO output at Nf_1 if the loop is set up to lock at the fundamental frequency (when $f_o = f_1$). Figure 16.29b shows an example using a 565 PLL as frequency multiplier and 7490 as

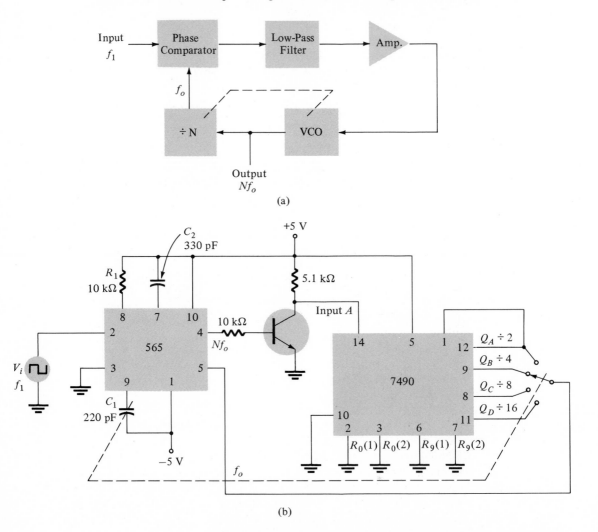

(a)

(b)

Figure 16.29 Frequency synthesizer: (a) block diagram; (b) implementation using 565 PLL unit.

divider. The input V_i at frequency f_1 is compared to the input (frequency f_o) at pin 5. An output at Nf_o ($4f_o$ in the present example) is connected through an inverter circuit to provide an input at pin 14 of the 7490, which varies between 0 V and +5 V. Using the output at pin 9, which is divided by four from that at the input to the 7490, the signal at pin 4 of the PLL is four times the input frequency as long as the loop remains in lock. Since the VCO can only vary over a limited range from its center frequency, it may be necessary to change the VCO frequency whenever the divider value is changed. As long as the PLL circuit is in lock, the VCO output frequency will be exactly N times the input frequency. It is only necessary to readjust f_o to be within the capture-and-lock range, the closed loop then resulting in the VCO output becomes exactly Nf_1 at lock.

FSK DECODERS

An FSK (frequency-shift keyed) signal decoder can be built as shown in Fig. 16.30. The decoder receives a signal at one of two distinct carrier frequencies, 1270 Hz or 1070 Hz, representing the RS-232-C logic levels of mark (−5 V) or space (+14 V), respectively. As the signal appears at the input, the loop locks to the input frequency and tracks it between the two possible frequencies with a corresponding dc shift at the output.

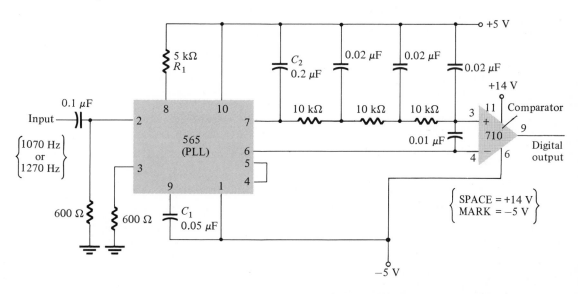

Figure 16.30 Connection of 565 as FSK decoder.

The RC ladder filter (three sections of $C = 0.02$ μF and $R = 10$ kΩ) is used to remove the sum frequency component. The free-running frequency is adjusted with R_1 so that the dc voltage level at the output (pin 7) is the same as that at pin 6. Then, an input at frequency 1070 Hz will drive the decoder output voltage to a more positive voltage level, driving the digital output to the high level (space or +14 V). An input at 1270 Hz will correspondingly drive the 565 dc output less positive with the digital output, which then drops to the low level (mark or −5 V).

PROBLEMS

§ 16.2

1. Draw the diagram of a 741 op-amp operated from ±15-V supplies with input to minus input terminal and plus input terminal connected to a +5-V reference voltage. Include terminal pin connections.

2. Sketch the voltage waveforms for the circuit of Problem 1, with input of 10 V rms.

3. Draw the connection diagram of a 311 op-amp showing 10 V rms input applied to pin 3 and ground to pin 2.

4. Using the ±12-V supply and output connected to positive supply through a 10-kΩ resistor. Sketch the input and resulting output waveforms.

5. Describe the operation of the circuit of Fig. 16.7 with input of 10 V rms applied.

6. Draw the circuit diagram of a zero-crossing detector using a 339 comparator stage with ±12-V supplies.

7. For the circuit of Problem 6, sketch the voltage waveforms with 10-V rms input applied to the minus input and plus input grounded.

8. Describe the operation of a window detector circuit as in Fig. 16.11 for resistor values of 7.5 kΩ and 8.2 kΩ changed to 6.2 kΩ.

§ 16.3

9. Sketch a three-input ladder network using 15-kΩ and 30-kΩ resistor values.

10. For a reference voltage of 16 V, calculate the output of the network in Problem 9 with an input of 110.

11. What voltage resolution is possible using a 12-stage ladder network with a 10-V reference voltage?

12. Describe what occurs during the fixed time interval and during the count interval of dual-slope conversion.

13. How many count steps occur using a 12-stage digital counter as the output of an A/D converter?

14. What is the maximum count interval resulting using a 12-stage counter operated at a clock rate of 2 MHz?

§ 16.4

15. Describe the signal conditions for current-loop and RS-232-C interfaces.

16. What is meant by a data bus?

17. What is the difference between open-collector and tri-state outputs?

§ 16.5

18. Sketch a 555 timer connected as an astable multivibrator for operation at 100 kHz. Determine the value of capacitor C needed if $R_A = R_B = 7.5$ kΩ.

19. Draw a 555 timer used as a one-shot using $R_A = 7.5$ kΩ for a time period of 25 μs. Determine the value of capacitor C needed.

20. Sketch the input and output waveforms for a one-shot as in Problem 19 for input triggered by a 10-kHz clock.

§ 16.6

21. Calculate the center frequency of a VCO using a 566 as in Fig. 16.24 for $R_1 = 4.7$ kΩ, $R_2 = 1.8$ kΩ, $R_3 = 11$ kΩ, and $C_1 = 0.001$ μF.

22. What frequency range results in the circuit of Fig. 16.25 for $C_1 = 0.001$ μF?

23. Determine the capacitor needed in the circuit of Fig. 16.24 to obtain a 100-kHz output.

§ 16.7

24. Calculate the VCO free-running frequency for the circuit of Fig. 16.28b with $R_1 = 4.7$ kΩ and $C_1 = 0.001$ μF.

25. What value capacitor, C_1, is required in the circuit of Fig. 16.28b to obtain a center frequency of 100 kHz?

26. What is the lock range of the PLL circuit in Fig. 16.28b for $R_1 = 4.7$ kΩ and $C_1 = 0.001$ μF?

CHAPTER 17

Feedback Amplifiers and Oscillator Circuits

17.1 FEEDBACK CONCEPTS

Feedback was mentioned when considering dc bias stabilization in Chapters 4 and 13. Amplifier gain was sacrificed in the circuit design for improvement in dc bias stability. We might say that a trade-off of gain for stability was made in the circuit design. Such trade-off is typical of engineering design compromises. If negative voltage feedback is used, for example, a circuit can be designed to couple some of the output voltage back to the input, reducing the overall voltage gain of the circuit. For this loss of gain, however, it is possible to obtain higher input impedance, lower output impedance, more stable amplifier gain, or higher cutoff frequency operation.

If the feedback signal is connected in order to aid or add to the input signal applied, however, *positive* feedback occurs, which could drive the circuit into operation as an oscillator.

Voltage-Feedback Connection

As an example of voltage feedback the circuit of Fig. 17.1 shows an FET amplifier with negative voltage feedback. Resistor R_f and capacitor C_f (used here to block dc bias voltage) form the feedback path. Because of the amplifier inversion any signal at the output is opposite in polarity to the signal at the input. The output signal fed back will then be opposite in polarity and is thus a negative feedback signal. The net result of the feedback action will be to decrease the overall voltage gain to a lower amount dependent on the original gain of the amplifier without feedback and on the amount of the feedback.

638

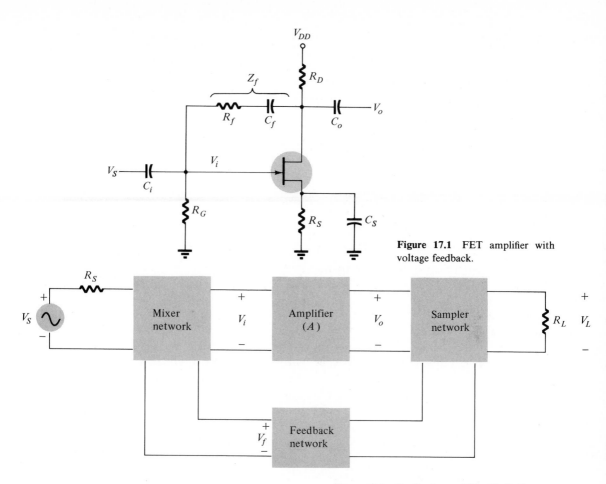

Figure 17.1 FET amplifier with voltage feedback.

Figure 17.2 Feedback amplifier, block diagram.

A general block diagram of a feedback circuit is shown in Fig. 17.2. The input signal (V_s) and feedback signal (V_f) are *mixed* or combined to form the single signal (V_i) which is then amplified by the amplifier section of the circuit. The amplifier output then goes into a sample circuit which feeds part of the amplified signal to the load and part of the signal to the feedback network.

In the circuit of Fig. 17.1 the components C_i and R_g form the mixer network combining input and feedback signals. No sampling network is used in this circuit because the output and signal to the feedback network are the same. The feedback network is comprised of resistor R_f and capacitor C_f.

A simpler version of the feedback amplifier of Figs. 17.1 and 17.2 is that of Fig. 17.3. The mixer is shown as a circle with two inputs that are opposite in polarity as indicated by the plus and minus input signs. The basic amplifier gain is A and the gain (attenuation, normally) of the feedback network is given as β (beta). Usually, a part of the output signal is coupled back to the input in order to oppose the applied input signal V_s. In return for this gain reduction a number of improvements can be obtained, such as the following:

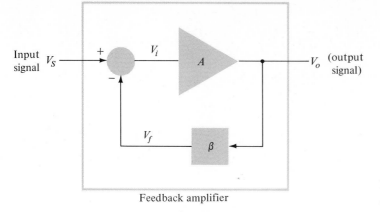

Feedback amplifier

Figure 17.3 Simpler block diagram of feedback amplifier.

1. Higher input impedance.
2. Better stabilized voltage gain.
3. Improved frequency response.
4. More linear operation.
5. Lower output impedance.
6. Reduced noise.

These improvements all occur with a voltage-series type of feedback. In addition to the voltage feedback just discussed, there is also current feedback, for which the list of changes with and without feedback is somewhat different than those listed above. At present we wish to obtain some concept of what feedback is all about. Only the voltage-series feedback connection of Figs. 17.2 or 17.3 will be used for the present. We shall demonstrate mathematically how some of the listed improvements are obtained and shall provide a means of numerically specifying the result of using feedback. Some relations between the amplifier operation with and without feedback will now be considered.

Voltage Gain with Feedback

In the feedback circuit of Fig. 17.3 the gain without feedback is A and the feedback factor is β. It is assumed that signal transmission goes only from input (V_i) to output (V_o) for the amplifier state and only from output (V_o) to feedback input (V_f) for the feedback network.

The input voltage to the basic amplifier is the difference between signal and feedback voltage

$$V_i = V_s - V_f \tag{17.1}$$

where the feedback voltage is a portion of the output voltage

$$V_f = \beta V_o \tag{17.2}$$

the proportionality factor being β. The gain of the basic amplifier is simply

$$A = \frac{V_o}{V_i}$$

so that the output voltage is given by

$$V_o = AV_i \qquad (17.3)$$

With feedback employed, the overall gain of the circuitry represented by Fig. 17.3 is

$$A_f = \frac{V_o}{V_s} \qquad (17.4)$$

We can solve for this factor using Eqs. (17.1), (17.2), and (17.3) as follows:

$$\frac{V_o}{A} = V_i = V_s - V_f = V_s - \beta V_o$$

$$V_o + \beta A V_o = A V_s$$

$$\boxed{A_f = \frac{V_o}{V_s} = \frac{A}{1 + \beta A}} \qquad (17.5)$$

Thus, Eq. (17.5) shows that the gain with feedback depends on the basic amplifier gain and the amount of the feedback factor.

If, for example, the quantities β and A are $\beta = 1/10$, $A = 90$, then the gain *without* feedback is 90 and the gain *with* feedback is

$$A_f = \frac{A}{1 + \beta A} = \frac{90}{1 + (1/10)(90)} = \frac{90}{10} = 9$$

the gain being reduced by a factor of 10. This is *negative feedback.*

We can also show that the gain *with* feedback is more stable than that without feedback. A change in amplifier gain from 90 to 100 due to component-value changes with temperature represents a change of 11.1%. With feedback the resulting gain is

$$A_f = \frac{100}{1 + (1/10)(100)} = 9.1$$

which represents a change of only 1.11%, an improvement by a factor of nearly 10.

If the amplifier gain and feedback values are $A = 90$, $\beta = -1/100$, the gain with feedback is then

$$A_f = \frac{90}{1 + (-1/100)(90)} = \frac{90}{1 - 0.9} = \frac{90}{0.1} = 900$$

The gain has been increased by a factor of 10. This is *positive feedback* and is the principle by which a feedback amplifier can be made into an oscillator circuit. Detailed discussion of the oscillator is deferred to Section 17.6.

Thus, as a general statement: if $|A_f| < |A|$, feedback is negative; if $|A_f| > |A|$, feedback is positive.

For a negative-feedback amplifier we see that $|1 + \beta A| > 1$. Typically, the value of $|\beta A| \gg 1$ so that

$$A_f = \frac{A}{1 + \beta A} \cong \frac{A}{\beta A} = \frac{1}{\beta} \qquad (17.6)$$

In other words, the feedback gain is dependent mainly on the factor β for the case of negative feedback with $|\beta A| \gg 1$. Whereas the amplifier gain A is dependent on temperature, device parameters, and so on, and may vary considerably, the reduced gain with negative feedback can be very stable, typically depending on a resistor feedback network. Since resistors can be selected precisely and with small change in resistive value due to temperature, highly precise and stable gain with negative feedback is possible.

EXAMPLE 17.1 Calculate the gain of a negative-feedback amplifier circuit having $A = 1000$ and $\beta = 1/10$.

Solution: Since $\beta A = (1/10)(1000) = 100 \gg 1$, the gain with feedback is

$$A_f \cong \frac{1}{\beta} = \frac{1}{0.1} = 10$$

In addition to the β factor setting a precise gain value, we are also interested in how stable the feedback amplifier is compared to an amplifier without feedback. Differentiating Eq. (17.5) leads to

$$\frac{dA_f}{A_f} = \frac{1}{|1 + \beta A|} \frac{dA}{A} \qquad (17.7a)$$

$$\frac{dA_f}{A_f} \cong \frac{1}{\beta A} \frac{dA}{A} \qquad \text{for } \beta A \gg 1 \qquad (17.7b)$$

This shows that the change in gain *(dA)* is reduced by the factor βA when feedback is employed.

EXAMPLE 17.2 If the amplifier in Example 17.1 has a gain change of 20% due to temperature, calculate the change in gain of the feedback amplifier.

Solution: Using Eq. (17.7b), we get

$$\frac{dA_f}{A_f} \cong \frac{1}{\beta A} \frac{dA}{A} = \frac{1}{0.1(1000)}(20\%) = 0.2\%$$

The improvement is 100 times. Thus, while the amplifier gain changes from $A = 1000$ by 20%, the feedback gain changes from $A_f = 100$ by only 0.2%.

17.2 FEEDBACK CONNECTION TYPES

There are four basic ways of connecting the feedback signal. Both *voltage* and *current* can be fed back to the input either in *series* or *parallel*. Specifically, there can be

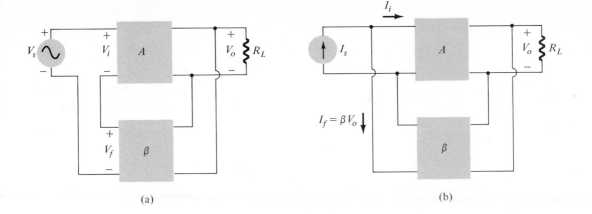

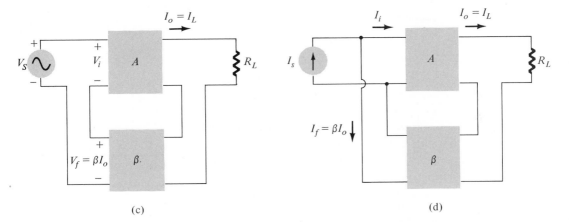

Figure 17.4 Feedback amplifier connection types: (a) voltage-series feedback; (b) voltage-shunt feedback; (c) current-series feedback; (d) current-shunt feedback.

1. Voltage-series feedback (Fig. 17.4a).
2. Voltage-shunt feedback (Fig. 17.4b).
3. Current-series feedback (Fig. 17.4c).
4. Current-shunt feedback (Fig. 17.4d).

In the above listing *voltage* refers to connecting the output voltage as input to the feedback network; *current* refers to tapping off some output current through the feedback network. *Series* refers to connecting the feedback signal is series with the input signal voltage; *shunt* refers to connecting the feedback signal in shunt (parallel) with an input current source.

Series feedback connections tend to *increase* the input resistance while shunt feedback connections tend to *decrease* the input resistance. Voltage feedback tends to *decrease* the output impedance while current feedback tends to *increase* the output impedance. Typically, higher input and lower output impedances are desired for most cascade amplifiers. Both of these are provided using the voltage-series feedback connection. We shall therefore concentrate first on this amplifier connection for practical feedback circuits.

Voltage-Series Feedback Amplifier

A block diagram of a voltage-series feedback amplifier is shown in Fig. 17.5. A practical voltage source having voltage V_s and source resistance R_S is shown in *series* with the feedback signal, V_f, the resulting input signal to the amplifier stage being V_i. The output voltage of the amplifier stage (with gain A_v) is V_o, which is developed directly across load resistor R_L. A parallel *voltage* pickup of the output voltage is connected to the feedback network.

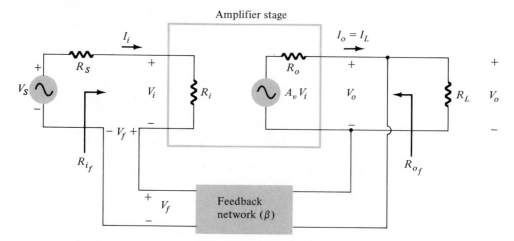

Figure 17.5   Practical voltage-series feedback amplifier.

If $V_f = 0$ (no feedback), the overall voltage gain A, including loading effects of R_S and R_i, is defined as

$$A_v = \frac{V_o}{V_s}\Bigg|_{R_L = \infty \text{ (output open circuit)}}$$

where

$$V_o = A_v V_i - I_o R_o$$

$$= A_v \underbrace{\frac{R_i}{R_i + R_S}}_{A} V_s - I_o R_o$$

$$V_o = A V_s - I_o R_o \qquad \text{(without feedback)}$$

(if $R_S = 0$, then $A = A_v$).

Including feedback connection ($V_f \neq 0$) the voltage gain is

$$A_f = \frac{V_o}{V_s}\Bigg|_{R_L = \infty}$$

where

$$V_o = A V_i - I_o R_o$$

$$= A(V_s - V_f) - I_o R_o$$

Now,

$$A(A_S - V_f) = A V_s - A V_f = A(\beta V_o)$$

$$= A(V_s - \beta V_o)$$

CH. 17   FEEDBACK AMPLIFIERS AND OSCILLATOR CIRCUITS

We have then

$$V_o = AV_S - A\beta V_o - I_o R_o$$

$$V_o(1 + A\beta) = AV_S - I_o R_o$$

$$V_o = \left(\frac{A}{1 + \beta A}\right) V_S - I_o \frac{R_o}{1 + \beta A}$$

$$V_o = A_f V_S - I_o R_{of}$$

with

$$A_f = \frac{A}{1 + \beta A} \qquad (17.8)$$

$$R_{of} = \frac{R_o}{1 + \beta A} \qquad (17.9)$$

The last two equations show that the gain without feedback is reduced by the factor $(1 + \beta A)$ with feedback connected. In addition, the output resistance is seen to be reduced from R_o (without feedback) by the factor $(1 + \beta A)$ with feedback. The larger the factor $(1 + \beta A)$, the lower the output resistance.

The input resistance with feedback is

$$R_{if} = \frac{V_S}{I_i} - R_S$$

which can be shown to be

$$R_{if} = R_i(1 + \beta A) \qquad (17.10)$$

This time the factor $(1 + \beta A)$ makes the feedback input resistance larger than for the amplifier without feedback. In summary, then, a voltage-series feedback amplifier circuit can improve the operation of a nonfeedback amplifier circuit (having gain A, input resistance R_i, and output resistance R_o) as follows:

1. Stabilized voltage gain $A_f = A/(1 + \beta A)$.
2. Higher input resistance $R_{if} = R_i(1 + \beta A)$.
3. Lower output impedance $R_{of} = R_o/(1 + \beta A)$.

> **EXAMPLE 17.3** Calculate the gain and input and output impedance of a voltage-series feedback amplifier if the amplifier without feedback has $A = 100$, $R_i = 2$ kΩ, $R_o = 40$ kΩ, and the amount of feedback is $\beta = 1/10$.
>
> **Solution:** For a feedback amplifier using voltage-series feedback we can use Eqs. (17.8)–(17.10)
>
> $$A_f = \frac{A}{1 + \beta A} = \frac{100}{1 + (1/10)(100)} = \frac{100}{11} = \mathbf{9.1}$$
>
> $$R_{if} = R_i(1 + \beta A) = 2 \text{ k}\Omega \ (11) = \mathbf{22 \text{ k}\Omega}$$

$$R_{of} = \frac{R_o}{1 + \beta A} = \frac{40 \text{ k}\Omega}{11} = \textbf{3.6 k}\Omega$$

REDUCTION IN FREQUENCY DISTORTION

Recall that Eq. (17.6) shows that for a negative-feedback amplifier having $\beta A \gg 1$ the gain with feedback is $A_f \cong 1/\beta$. It follows from this that if the feedback network is purely resistive, the gain with feedback is not dependent on frequency even though the basic amplifier gain is frequency dependent. Practically, the frequency distortion arising because of varying amplifier gain with frequency is considerably reduced in a negative-voltage feedback amplifier circuit.

REDUCTION IN NOISE AND NONLINEAR DISTORTION

Signal feedback connected to oppose the input signal as in a negative-feedback amplifier tends to hold down the amount of noise signal (such as power-supply hum) and nonlinear distortion. The factor $(1 + \beta A)$ reduces both input noise and resulting nonlinear distortion for considerable improvement. However, it should be noted that there is a reduction in overall gain (the price required for the improvement in circuit performance). If additional stages are used to bring the overall gain up to the level without feedback, it should be noted that the extra stage(s) might introduce as much noise back into the system as that reduced by the feedback amplifier. This problem can be somewhat alleviated by readjusting the gain of the feedback-amplifier circuit to obtain higher gain while also providing reduced noise signal.

EFFECT OF NEGATIVE FEEDBACK ON GAIN AND BANDWIDTH

In Eq. (17.6) the overall gain with negative feedback is shown to be

$$A_f = \frac{A}{(1 + \beta A)} \cong \frac{A}{\beta A} = \frac{1}{\beta} \qquad \text{for } \beta A \gg 1$$

As long as $\beta A \gg 1$ the overall gain is approximately $1/\beta$. We should realize that for a practical amplifier (for single low- and high-frequency breakpoints) the open-loop gain drops off at high frequencies due to the active device and circuit capacitances. Gain may also drop off at low frequencies for capacitively coupled amplifier stages. Once the open-loop gain A drops low enough and the factor βA is no longer much larger than 1, the conclusion of Eq. (17.6) that $A_f \cong 1/\beta$ no longer holds true.

Figure 17.6 shows that the amplifier with negative feedback has more bandwidth (B) than the amplifier without feedback. The feedback amplifier has a higher upper 3-dB frequency and smaller lower 3-dB frequency.

It is interesting to note that the use of feedback, while resulting in a lowering of voltage gain, has provided an increase in B and in the upper 3-dB frequency, particularly. In fact, the product of gain and frequency remains the same so that the gain-bandwidth product of the basic amplifier is the same value for the feedback amplifier. However, since the feedback amplifier has lower gain, the net operation was to *trade* gain for bandwidth (we use bandwidth for the upper 3-dB frequency since typically $f_2 \gg f_1$).

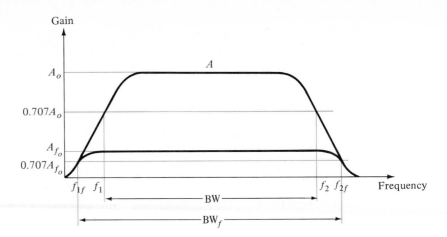

Figure 17.6 Effect of negative feedback on gain and bandwidth.

17.3 PRACTICAL VOLTAGE-SERIES NEGATIVE-FEEDBACK AMPLIFIER CIRCUITS

Transistor Stage

Figure 17.7a shows a transistor amplifier circuit with the output taken from the emitter terminal. Figure 17.7b shows an approximate small-signal equivalent circuit. The feedback signal is shown connected to the input in series with input voltage (V_s).

VOLTAGE GAIN

At first glance it might not appear that any feedback exists in this simple one-stage amplifier circuit. However, the ac equivalent circuit of Fig. 17.7b shows that

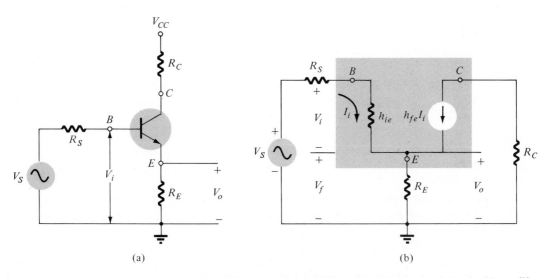

Figure 17.7 Transistor amplifier with voltage-series feedback: (a) amplifier circuit; (b) equivalent ac circuit.

the input voltage (V_i) is the difference of the source voltage (V_S) and feedback voltage (V_f)

$$V_i = V_S - V_f$$

The gain of the amplifier without feedback (ground side of V_S connected to emitter instead of ground) is

$$A = \frac{h_{fe}R_E}{R_S + h_{ie}} \tag{17.11}$$

The feedback voltage is equal in magnitude to the output voltage so that we have $\beta = +1$. Using Eq. (17.6), we calculate the gain with feedback to be

$$A_f = \frac{A}{1 + \beta A} = \frac{h_{fe}R_E/(R_S + h_{ie})}{1 + (1)[h_{fe}R_E/(R_S + h_{ie})]}$$

$$\boxed{A_f = \frac{h_{fe}R_E}{R_S + h_{ie} + h_{fe}R_E}} \tag{17.12}$$

INPUT RESISTANCE

Without feedback we note that

$$R_i = h_{ie}$$

With feedback and neglecting R_S ($R_S = 0$) we find, using Eq. (17.10),

$$R_{if} = R_i(1 + \beta A) = h_{ie}\left(1 + \frac{h_{fe}R_E}{h_{ie}}\right)$$

$$\boxed{R_{if} = h_{ie} + h_{fe}R_E} \tag{17.13}$$

OUTPUT RESISTANCE

Without feedback the equivalent circuit used has an output resistance of infinity. Had $1/h_{oe}$ been included in the transistor equivalent circuit, a practical value of around 50 kΩ would be present. Including feedback and using Eq. (17.9), we get

$$\boxed{R_{of} = \frac{R_o}{1 + \beta A} \cong \frac{R_S + h_{ie}}{h_{fe}}} \tag{17.14}$$

A numerical example using the above equations should show that negative feedback in a voltage-series connection reduces the gain while increasing input resistance and lowering output resistance.

EXAMPLE 17.4 For the circuit of Fig. 17.7 and the following circuit values, calculate the voltage gain and input and output resistances, with and without feedback: $R_E = 1.5$ kΩ, $R_S = 1$ kΩ, $h_{ie} = 2$ kΩ, $h_{fe} = 50$, $h_{oe} = 12.5$ μS.

Solution: Without feedback

$$A = \frac{h_{fe}R_E}{R_S + h_{ie}} = \frac{50(1.5\ k\Omega)}{1\ k\Omega + 2\ k\Omega} = 25$$

$$R_i = h_{ie} = 2\ k\Omega$$

$$R_o = \frac{1}{h_{oe}} = 80\ k\Omega$$

With feedback

$$A_f = \frac{h_{fe}R_E}{R_S + h_{ie} + h_{fe}R_E} = \frac{50(1.5\ k\Omega)}{1\ k\Omega + 2\ k\Omega + 50(1.5\ k\Omega)} = \frac{75}{78} = 0.96$$

$$R_{if} = h_{ie} + h_{fe}R_E = 2\ k\Omega + 50(1.5\ k\Omega) = 77\ k\Omega$$

$$R_{of} = \frac{R_S + h_{ie}}{h_{fe}} = \frac{1\ k\Omega + 2\ k\Omega}{50} = 60\ \Omega$$

FET Stage

Figure 17.8 shows a single-stage RC-coupled FET amplifier with negative feedback. A part of the output signal (V_o) is picked off by a feedback network made up of resistors R_1 and R_2. The feedback voltage V_f is connected in series with the source signal V_S, and their difference is the input signal V_i (measured from gate to drain).

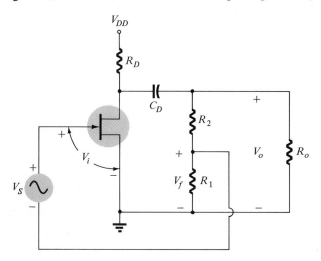

Figure 17.8 FET amplifier stage with voltage-series feedback.

Without feedback the amplifier gain is

$$A = -g_m R_L \tag{17.15}$$

where R_L is the parallel combination of resistors R_D, R_o, and a series equivalent of R_1 and R_2.

The feedback network provides a feedback factor of

$$\beta = \frac{R_1}{R_1 + R_2} \tag{17.16}$$

Using the above values of A and β in Eq. (17.6), we find the gain with negative feedback to be

$$A_f = \frac{A}{1 + \beta A} = \frac{g_m R_L}{1 + [R_1 R_L/(R_1 + R_2)]g_m} \qquad (17.17a)$$

If $\beta A \gg 1$, we have

$$\boxed{A_f \cong \frac{1}{\beta} = \frac{R_1 + R_2}{R_1}} \qquad (17.17b)$$

EXAMPLE 17.5 Calculate the gain without and with feedback for the FET amplifier circuit of Fig. 17.8 and the following circuit values: $R_1 = 20$ kΩ, $R_2 = 80$ kΩ, $R_o = 10$ kΩ, $R_D = 10$ kΩ, and $g_m = 4000$ μS.

Solution:

$$R_L \cong \frac{R_o R_D}{R_o + R_D} = \frac{10 \text{ k}\Omega \, (10 \text{ k}\Omega)}{10 \text{ k}\Omega + 10 \text{ k}\Omega} = 5 \text{ k}\Omega$$

(neglecting 100 kΩ resistance of R_1 and R_2 in series)

$$A = g_m R_L = (4000 \times 10^{-6})(5 \text{ k}\Omega) = \mathbf{20}$$

The feedback factor is

$$\beta = \frac{R_1}{R_1 + R_2} = \frac{20}{20 + 80} = 0.2$$

The gain with feedback is

$$A_f = \frac{A}{1 + \beta A} = \frac{20}{1 + 0.2(20)} = \frac{20}{5} = \mathbf{4}$$

17.4 OTHER PRACTICAL FEEDBACK CIRCUIT CONNECTIONS

Two-Stage Voltage-Series Feedback

A popular means of incorporating negative feedback to stabilize the gain of an amplifier is signal feedback in a multistage circuit. As an example, Fig. 17.9 shows two cascaded stages with voltage-series feedback. Cascaded amplifier stages with transistors Q_1 and Q_2 provide an overall gain A. A feedback network of resistors R_1 and R_2 is coupled by a capacitor (to block dc) between output and input while allowing feedback of the ac output signal. The feedback signal is taken from the collector of the second stage and is connected to the emitter (neglecting the bypassed 3.6-kΩ resistor) of the first amplifier stage. Negative feedback results from this connection since the in-phase output signal of the second stage collector connected through the feedback network to the emitter opposes the input signal between the base emitter of the first stage.

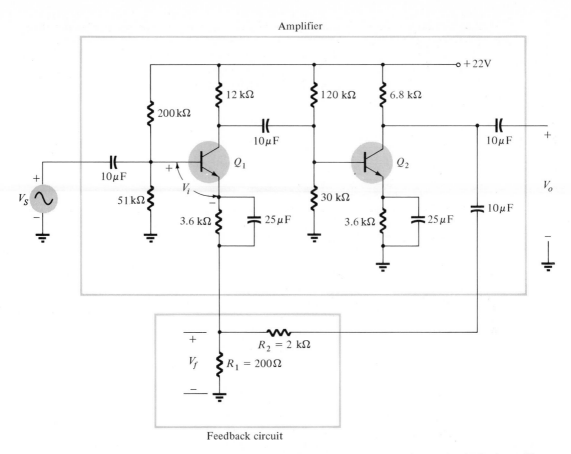

Figure 17.9 Cascaded voltage-series feedback amplifier.

Calculations of gain, input, and output resistance with and without feedback require no new theory or equations. The techniques for cascaded stages developed in Chapter 9 and the basic equations [Eqs. (17.8)–(17.10)] for voltage-series feedback are used in the following example to show how a circuit such as in Fig. 17.9 can be analyzed.

EXAMPLE 17.6   Calculate A, R_i, and R_o for the cascaded amplifier of Fig. 17.9, omitting feedback, and then A_f, R_{if}, and R_{of} with the feedback connection considered. Use transistor parameters $h_{fe} = 65$, $h_{ie} = 1.8$ kΩ, and $1/h_{oe} = \infty$.

Solution: *Without the feedback network:* Looking into the base transistor Q_1, we see that the resistance is $R_{i_1} = h_{ie} = 1.8$ kΩ. The emitter 3.6-kΩ resistor is neglected here for ac calculations due to the bypass capacitor. Since the parallel combination of the 51-kΩ and 200-kΩ bias resistors is in parallel with the input (as seen by the source signal), the overall amplifier input impedance is calculated to be

$$R_i = 1.8 \text{ k}\Omega \,\|\, 51 \text{ k}\Omega \,\|\, 200 \text{ k}\Omega \cong \mathbf{1.7 \text{ k}\Omega}$$

The output impedance looking back into stage 2 is approximately

$$R_o = 6.8 \text{ k}\Omega \,\|\, 2.2 \text{ k}\Omega \cong \mathbf{1.7 \text{ k}\Omega}$$

where $1/h_{oe}$ is neglected as being much larger than 6.8 kΩ, the bypassed emitter resistor is neglected, and the capacitive ac impedance is neglected—these being valid assumptions in the amplifier mid-frequency range, and feedback resistors (2-kΩ and 0.2-kΩ) provide an effective resistance in parallel with the output.

The gain of each stage can be obtained (including the loading of the second stage on the first) as follows: Effective load resistances of each stage are

$$R_{L_1} = 12 \text{ k}\Omega \parallel 120 \text{ k}\Omega \parallel 30 \text{ k}\Omega \parallel 1.8 \text{ k}\Omega \cong 1.5 \text{ k}\Omega$$

$$R_{L_2} = 6.8 \text{ k}\Omega \parallel 2.2 \text{ k}\Omega \cong 1.7 \text{ k}\Omega$$

(where 2.2 kΩ is an output load resistance of the 2-kΩ and 0.2-kΩ resistors connected in series to ground).

The voltage gains of each stage (magnitude only) are then

$$A_{v_1} \cong \frac{h_{fe}R_{L_1}}{h_{ie}} = \frac{65(1.5)}{1.8} \cong 54.2$$

$$A_{v_2} \cong \frac{h_{fe}R_{L_2}}{h_{ie}} = \frac{65(1.7)}{1.8} \cong 61.4$$

The overall gain of the cascaded amplifier, neglecting feedback, is then

$$A = A_{v_1}A_{v_2} = 54.2(61.4) \cong \mathbf{3327.9}$$

We can calculate the feedback factor to be

$$\beta = \frac{R_1}{R_1 + R_2} = \frac{0.2}{0.2 + 2} = \frac{1}{11} = 0.091$$

With the feedback network: Using Eqs. (17.8)–(17.10)

$$R_{of} = \frac{R_o}{1 + \beta A} = \frac{1.7 \text{ k}\Omega}{1 + 0.091(3327.9)} = \frac{1.7 \text{ k}\Omega}{303.8} = \mathbf{5.6 \ \Omega}$$

$$R_{if} = R_i(1 + \beta A) = 1.7 \text{ k}\Omega(303.8) = \mathbf{516.5 \text{ k}\Omega}$$

$$A_f = \frac{A}{1 + \beta A} = \frac{3327.9}{303.8} \cong \mathbf{11}$$

Current-Series Feedback Amplifier

So far we have considered only a feedback connection that samples the output voltage and feeds a portion of that voltage back to the input in series opposition with the source signal. Another feedback technique is to sample the output current (I_o) and return a proportional voltage in series with the input. While stabilizing the amplifier gain, the current-series feedback connection increases *both* input and output resistance.

Figure 17.10 shows a simple version of an amplifier with current-series negative feedback. If the input current is negligible as in a tube or FET amplifier, then the current through resistor R is the output current I_o. The voltage developed across R is a feedback voltage connected in series with the source signal. We can consider the amplifier as providing an output current dependent on the input source voltage—the amplifier then acting as a transconductance amplifier. Feedback acts to stabilize the transconductance so that the load current depends on the signal voltage and

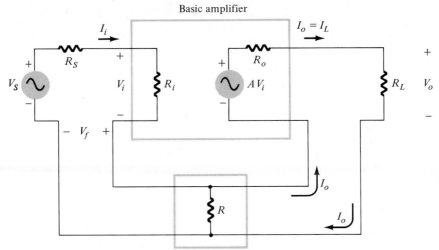

Figure 17.10 Amplifier with current-series negative-feedback connection.

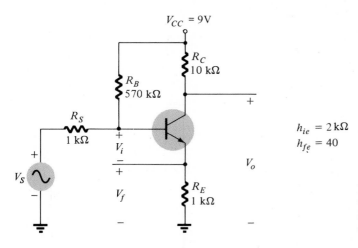

$h_{ie} = 2\,\text{k}\Omega$

$h_{fe} = 40$

Figure 17.11 Transistor amplifier with unbypassed emitter resistor (R_E) for current-series negative feedback.

resistance R only and is stabilized in regard to any other circuit changes.

Figure 17.11 shows a single transistor amplifier stage. Since the emitter of this stage has an unbypassed emitter, it effectively has current-series feedback. The current through resistor R_E results in a feedback voltage that opposes the source signal applied so that the output voltage V_o is reduced. To remove the current-series feedback the emitter resistor must be either removed or bypassed by a capacitor (as is usually done). An example will show how the amplifier operation is affected by current-series feedback due to resistor R_E.

EXAMPLE 17.7 For the circuit of Fig. 17.11 calculate the gain, input resistance, and output resistance without feedback (bypassed emitter resistor, $R_E = 0$) and with feedback (R_E present), and $1/h_{oe} = \infty$.

Solution: With R_E removed (bypassed)

$$A = \frac{-h_{fe}R_C}{h_{ie}} = \frac{-40(10 \text{ k}\Omega)}{2 \text{ k}\Omega} = -200$$

$$R_i = h_{ie} = 2 \text{ k}\Omega$$

$$R_o = R_C = 10 \text{ k}\Omega$$

With R_E in the circuit

$$A_f = \frac{-h_{fe}R_C}{h_{ie} + h_{fe}R_E} = \frac{-40(10 \text{ k}\Omega)}{2 \text{ k}\Omega + 40(1 \text{ k}\Omega)} = \frac{-400}{42} = -9.5$$

$$R_{if} = h_{fe}R_E + h_{ie} = 40(1 \text{ k}\Omega) + 2 \text{ k}\Omega = 42 \text{ k}\Omega$$

$$R_{of} \cong R_C = 10 \text{ k}\Omega$$

Example 17.7 shows that current-series feedback

1. Reduces amplifier gain.
2. Increases input resistance.

Voltage-Shunt Feedback

Negative feedback can be obtained by coupling a portion of the output voltage in parallel (shunt) with the input signal. Figure 17.12 shows a typical voltage-shunt feedback connection. A portion of the output voltage (which is opposite in polarity to the input voltage) is connected to the base through resistor R_f. A voltage-shunt connection stabilizes amplifier overall gain while decreasing both input and output resistances.

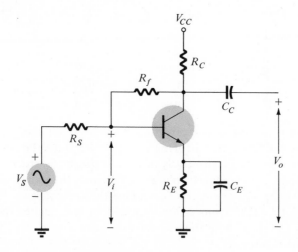

Figure 17.12 Voltage-shunt negative-feedback amplifier.

Current-Shunt Feedback

A fourth feedback connection samples the output current and develops a feedback voltage in shunt with the input signal. A practical circuit version is the two-stage amplifier of Fig. 17.13. The unbypassed emitter resistor of stage 2 provides current

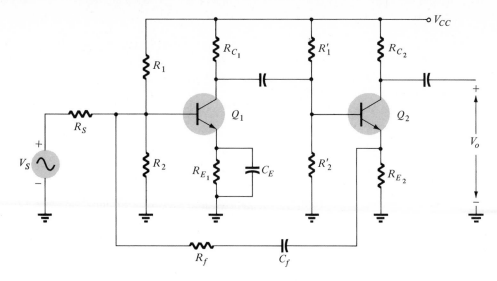

Figure 17.13 Amplifier with current-shunt negative-feedback connection.

sensing. The feedback signal is connected in shunt with the first stage input through a feedback network.

Checking the feedback signal polarity for an input to the base of stage 1, we see that the output of stage 1 is opposite in polarity. The input to the base of stage 2 and the voltage across emitter R_{E_1} is then opposite in polarity to the input to stage 1 so that negative feedback is achieved. A current-shunt feedback circuit typically increases output resistance and decreases input resistance while holding the gain with feedback constant.

The operation of the four types of feedback connections is summarized in Table 17.1. All types provide stabilized but reduced gain, increased bandwidth, and decreased nonlinear distortion.

TABLE 17.1   Effect of Feedback Connection Type on Input and Output Resistance

	Voltage-Series	Current-Series	Voltage-Shunt	Current-Shunt
R_{if}	increased	increased	decreased	decreased
R_{of}	decreased	increased	decreased	increased

17.5 FEEDBACK-AMPLIFIER STABILITY— PHASE AND FREQUENCY CONSIDERATIONS

So far we have considered the operation of a feedback amplifier in which the feedback signal was *opposite* to the input signal—negative feedback. In any practical circuit this condition occurs only for some mid-frequency range of operation. We know that an amplifier gain will change with frequency, dropping off at higher frequencies from the mid-frequency value. In addition, the phase shift of an amplifier will also change with frequency.

If, as the frequency increases, the phase shift changes then some of the feedback signal *adds* to the input signal. It is then possible for the amplifier to break into oscillations due to positive feedback. If the amplifier oscillates at some low or high frequency, it is no longer useful as an amplifier. Proper feedback-amplifier design requires that the circuit be stable at *all* frequencies, not merely those in the range of interest. Otherwise, a transient disturbance could cause a seemingly stable amplifier to suddenly start oscillating.

Nyquist Criterion

In judging the stability of a feedback amplifier, as a function of frequency, the factor of loop gain A_f, amplifier gain A, and feedback attenuation β as functions of frequency, can be used. One of the most popular techniques used to investigate stability is the Nyquist method. A Nyquist diagram is used to plot gain and phase shift as a function of frequency on a complex plane. The Nyquist plot, in effect, combines the two Bode plots of gain versus frequency and phase-shift versus frequency on a single plot. A Nyquist plot is used to quickly show whether an amplifier is stable for all frequencies and how stable the amplifier is relative to some gain or phase-shift criteria.

As a start, consider the *complex plane* shown in Fig. 17.14. A few points of various gain (βA) values are shown at a few different phase-shift angles. By using the positive real axis as reference $(0°)$ a magnitude of $\beta A = 2$ is shown at a phase shift of $0°$ at point 1. Additionally, a magnitude of $\beta A = 3$ at a phase shift of $-135°$ is shown at point 2 and a magnitude/phase of $\beta A = 1$ at $180°$ is shown at

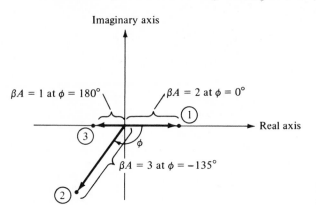

Figure 17.14 Complex plane showing typical gain-phase points.

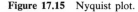

Figure 17.15 Nyquist plot.

point 3. Thus, points on this plot can represent *both* gain magnitude of βA and phase shift. If the points representing gain and phase shift for an amplifier circuit are plotted at increasing frequency, then a Nyquist plot is obtained as shown by the plot in Fig. 17.15. At the origin the gain is 0 at a frequency of 0 (for RC-type coupling). At increasing frequency points f_1, f_2, and f_3 the phase shift increased as did the magnitude of βA. At a representative frequency f_4 the value of A is the vector length from the origin to point f_4 and the phase shift is the angle ϕ. At a frequency f_5 the phase shift is 180°. At higher frequencies the gain is shown to decrease back to 0.

The Nyquist criteria for stability can be stated as follows:

The amplifier is unstable if the Nyquist curve plotted encloses (encircles) the -1 point, and it is stable otherwise.

An example of the Nyquist criteria is demonstrated by the curves in Fig. 17.16. The Nyquist plot in Fig. 17.16a is stable since it does not encircle the -1 point, whereas that shown in Fig. 17.16b is unstable since the curve does encircle the -1 point. Keep in mind that encircling the -1 point means that at a phase shift of 180° the loop gain (βA) is greater than 1; therefore, the feedback signal is in phase with the input and large enough to result in a larger input signal than that applied, with the result that oscillation occurs.

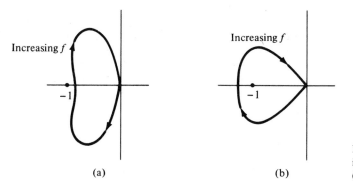

Figure 17.16 Nyquist plots showing stability conditions: (a) stable; (b) unstable.

Gain and Phase Margins

From the Nyquist criterion we know that a feedback amplifier is stable if the loop gain (βA) is less than unity (0 dB) when its phase angle is 180°. We can additionally determine some margins of stability to indicate how close to instability the unit is. That is, if the gain (βA) is less than unity but, say, 0.95 in value, this would not be as relatively stable as another amplifier having, say, $(\beta A) = 0.7$ (both measured at 180°). Of course, amplifiers with loop gains 0.95 and 0.7 are both stable, but one is closer to instability, if the loop gain increases, than the other. We can define the following terms:

Gain margin (GM) is defined as the value of βA in decibels at the frequency at which the phase angle is 180°. Thus, 0 dB, equal to a value of $\beta A = 1$, is on the border of stability and any negative decibel value is stable. The more negative the decibel gain value the more stable the feedback circuit. The GM may be evaluated in decibels from the curve of Fig. 17.17.

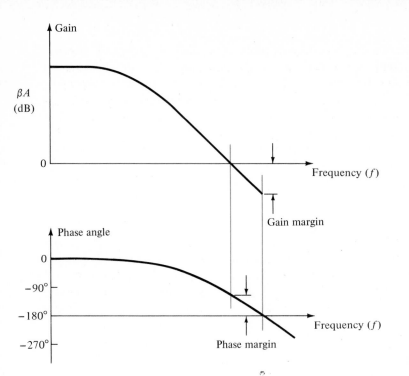

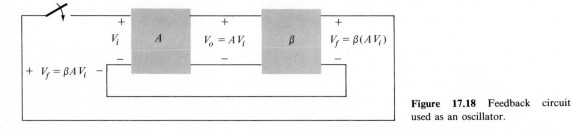

Figure 17.17 Bode plots showing gain and phase margins.

Phase margin (PM) is defined as the angle of 180° minus the magnitude of the angle at which the value βA is unity, 0 dB. The PM may also be evaluated directly from the curve of Fig. 17.17.

An example of these two amplifier factors is shown on the Bode plots of Fig. 17.17. Instability occurs, therefore, with a positive GM and PM greater than 180°.

17.6 OPERATION OF FEEDBACK CIRCUIT AS AN OSCILLATOR

The use of positive feedback which results in a feedback amplifier having closed-loop gain A_f greater than 1 and satisfies the phase conditions will result in operation as an oscillator circuit. An oscillator circuit then provides a constantly varying output signal. If the output signal varies sinusoidally, the circuit is referred to as a *sinusoidal oscillator*. If the output voltage rises quickly to one voltage level and later drops quickly to another voltage level, the circuit is generally referred to as a *pulse* or *square-wave oscillator*.

To understand how a feedback circuit performs as an oscillator consider the feedback circuit of Fig. 17.18. When the switch at the amplifier input is open, no oscillation

Figure 17.18 Feedback circuit used as an oscillator.

occurs. Consider that we have a *fictitious* voltage at the amplifier input *(Vi)*. This results in an output voltage $V_o = AV_i$ after the base amplifier stage and in a voltage $V_f = \beta(AV_i)$ after the feedback stage. Thus, we have a feedback voltage $V_f = \beta AV_i$, where βA is referred to as the *loop gain*. If the circuits of the base amplifier and feedback network provide βA of a correct magnitude and phase, V_f can be made equal to V_i. Then, when the switch is closed and fictitious voltage V_i is removed, the circuit will continue operating since the feedback voltage is sufficient to drive the amplifier and feedback circuits resulting in a proper input voltage to sustain the loop operation. The output waveform will still exist after the switch is closed if the condition

$$\beta A = 1 \qquad (17.18)$$

is met. This is known at the *Barkhausen criterion* for oscillation.

In reality, no input signal is needed to start the oscillator going. Only the condition $\beta A = 1$ must be satisfied for self-sustained oscillations to result. In practice βA is made greater than 1, and the system is started oscillating by amplifying noise voltage which is always present. Saturation factors in the practical circuit provide an "average" value of βA of 1. The resulting waveforms are never exactly sinusoidal. However, the closer the value βA is to exactly 1 the more nearly sinusoidal is the waveform. Figure 17.19 shows how the noise signal results in a build-up of a steady-state oscillation condition.

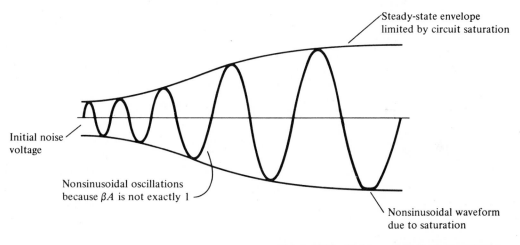

Figure 17.19 Build-up of steady-state oscillations.

Another way of seeing how the feedback circuit provides operation as an oscillator is obtained by noting the denominator in the basic feedback equation, (17.6) $A_f = A/(1 + \beta A)$. When $\beta A = -1$ or magnitude 1 at a phase angle of 180°, the denominator becomes 0 and the gain with feedback, A_f, becomes infinite. Thus, an infinitesimal signal (noise voltage) can provide a measurable output voltage, and the circuit acts as an oscillator even without an input signal.

The remainder of this chapter is devoted to various oscillator circuits that use a variety of components. Practical considerations are included so that workable circuits in each of the various cases are discussed.

17.7 PHASE-SHIFT OSCILLATOR

An example of an oscillator circuit that follows the basic development of a feedback circuit is the *phase-shift oscillator*. An idealized version of this circuit is shown in Fig. 17.20. Recall that the requirements for oscillation are that the loop gain, βA, is greater than unity *and* that the phase shift around the feedback network is 180° (providing positive feedback). In the present idealization we are considering the feedback network to be driven by a perfect source (zero source impedance) and the output of the feedback network is connected into a perfect load (infinite load impedance). The idealized case will allow development of the theory behind the operation of the phase-shift oscillator. Practical circuit versions will then be considered.

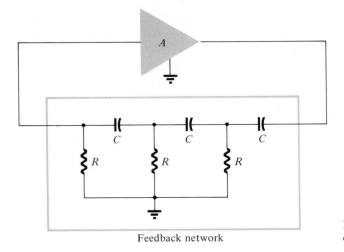

Figure 17.20 Idealized phase-shift oscillator.

Feedback network

Concentrating our attention on the phase-shift network we are interested in the attenuation of the network at the frequency at which the phase shift is exactly 180°. Using classical network analysis, we find that

$$f = \frac{1}{2\pi RC\sqrt{6}} \qquad (17.19a)$$

$$\beta = \frac{1}{29} \qquad (17.19b)$$

and the phase shift is 180°.

For the loop gain βA to be greater than unity the gain of the amplifier stage must be greater than $1/\beta$ or 29

$$A > 29 \qquad (17.19c)$$

When considering the operation of the feedback network one might naively select the values of R and C to provide (at a specific frequency) 60°-phase shift per section

CH. 17 FEEDBACK AMPLIFIERS AND OSCILLATOR CIRCUITS

for three sections, resulting in 180°-phase shift as desired. This, however, is not the case, since each section of the RC in the feedback network loads down the previous one. The net result that the *total* phase shift be 180° is all that is important. The frequency given by Eq. (17.19a) is that at which the *total* phase shift is 180°. If one measured the phase shift per RC section, each section would not provide the same phase shift (although the overall phase shift is 180°). If it were desired to obtain exactly 60°-phase shift for each of three stages, then emitter-follower stages would be needed after each RC section to prevent each from being loaded from the following circuit.

FET Phase-Shift Oscillator

A practical version of a phase-shift oscillator circuit is shown in Fig. 17.21a. The circuit is drawn to show clearly the amplifier and feedback network. The amplifier stage is self-biased with a capacitor bypassed source resistor R_S and a drain bias resistor R_D. The FET device parameters of interest are g_m and r_d. From FET amplifier theory the amplifier gain magnitude is calculated from

$$A = g_m R_L \tag{17.20}$$

where R_L in this case is the parallel resistance of R_D and r_d

$$R_L = \frac{R_D r_d}{R_D + r_d} \tag{17.21}$$

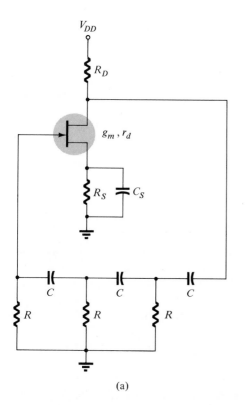

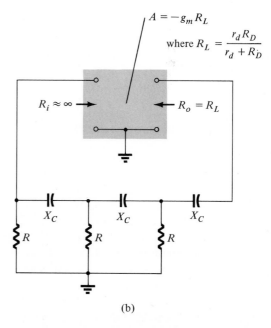

(a) (b)

Figure 17.21 FET phase-shift oscillator circuit.

We shall assume as a very good approximation that the input impedance of the FET amplifier stage is infinite (see Fig. 17.21b). This assumption is valid as long as the oscillator operating frequency is low enough so that FET capacitive impedances can be neglected. The output impedance of the amplifier stage given by R_L should also be small compared to the impedance seen looking into the feedback network so that no attenuation due to loading occurs. In practice, these considerations are not always negligible, and the amplifier stage gain is then selected somewhat larger than the needed factor of 29 to assure oscillator action.

EXAMPLE 17.8 It is desired to design a phase-shift oscillator (as in Fig. 17.21a) using an FET having $g_m = 5000 \ \mu S$, $r_d = 40 \ k\Omega$, and feedback circuit value of $R = 10 \ k\Omega$. Select the value of C for oscillator operation at 1 kHz and R_D for $A > 29$ to ensure oscillator action.

Solution: Equation (12.19a) is used to solve for the capacitor value. Since $f = 1/2\pi RC\sqrt{6}$, we can solve for C

$$C = \frac{1}{2\pi Rf\sqrt{6}} = \frac{1}{(6.28)(10 \times 10^3)(10^3)(2.45)} = \textbf{6.5 nF}$$

Using Eq. (17.20), we solve for R_L to provide a gain of, say, $A = 40$ (this allows for some loading between R_L and the feedback network input impedance)

$$A = g_m R_L$$

$$R_L = \frac{A}{g_m} = \frac{40}{5000 \times 10^{-6}} = \textbf{8 k}\Omega$$

Using Eq. (17.21), we solve for R_D

$$R_L = \frac{R_D r_d}{R_D + r_d}$$

$$8 \ k\Omega = \frac{R_D (40 \ k\Omega)}{R_D + 40 \ k\Omega}$$

$$R_D = \textbf{10 k}\Omega$$

Transistor Phase-Shift Oscillator

If a transistor is used as the active element of the amplifier stage, the output of the feedback network is loaded appreciably by the relatively low input resistance (h_{ie}) of the transistor. Of course, an emitter follower input stage followed by a common-emitter amplifier stage could be used. If a single transistor stage is desired, however, the use of voltage-shunt feedback (as shown in Fig. 17.22a) is more suitable. In this connection, the feedback signal is coupled through the feedback resistor R_S in *series* with the amplifier stage input resistance (R_i).

An ac equivalent circuit is shown in Fig. 17.22b. The figure shows that the input resistance R_i in series with feedback resistor R', is the parallel combination of resistors R_1, R_2, and h_{ie}. Also, the effective resistance for the third leg of the feedback network, the series combination of resistors R' and R_i, is made the same value as the resistance of the other two resistors of the feedback network to make calculations simpler.

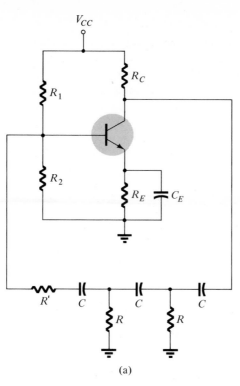

(a)

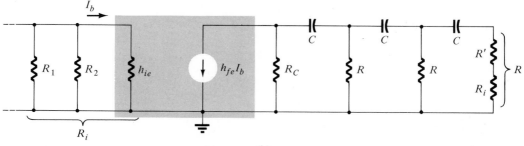

(b)

Figure 17.22 Transistor phase-shift oscillator: (a) transistor circuit; (b) ac equivalent circuit.

We shall assume that the transistor output impedance $1/h_{oe}$ is much larger than R_C.

Analysis of the ac circuit provides the following equation for the resulting oscillator frequency:

$$f = \left(\frac{1}{2\pi RC}\right)\frac{1}{\sqrt{6 + 4(R_C/R)}} \qquad (17.22)$$

For the loop gain to be greater than unity, the requirement on the current gain of the transistor is found to be

$$h_{fe} > 23 + 29\frac{R_C}{R} + 4\frac{R}{R_C} \qquad (17.23)$$

A practical example will demonstrate the use of the above information in designing an oscillator circuit.

EXAMPLE 17.9 Select the value of capacitor C and transistor gain h_{fe} to provide an oscillator frequency of $f = 2$ kHz. Circuit values are $h_{ie} = 2$ kΩ, $R_1 = 20$ kΩ, $R_2 = 80$ kΩ, $R_C = 10$ kΩ, and $R = 8$ kΩ.

Solution: Using Eq. (17.22), we can determine the required value of C

$$f = \left(\frac{1}{2\pi RC}\right)\frac{1}{\sqrt{6 + 4R_C/R}}$$

$$2 \times 10^3 = \left[\frac{1}{6.28(8 \times 10^3)C}\right]\frac{1}{\sqrt{6 + 4(1.25)}}$$

$$C = \left[\frac{1}{6.28(8 \times 10^3)(2 \times 10^3)}\right]\frac{1}{3.32} = \frac{10^{-6}}{332} = \textbf{3 nF}$$

Calculating R_i as the parallel resistance of R_1, R_2, and h_{ie} gives

$$R_i = 20 \text{ k}\Omega \parallel 80 \text{ k}\Omega \parallel 2 \text{ k}\Omega \cong 1.8 \text{ k}\Omega$$

For
$$R' + R_i = R = 8 \text{ k}\Omega$$

$$R' = R - R_i = (8 - 1.8) \text{ k}\Omega = 6.2 \text{ k}\Omega$$

To determine the value of h_{fe} necessary [using Eq. (17.23)]

$$h_{fe} > 23 + 29\frac{R}{R_C} + 4\frac{R_C}{R} = 23 + 29\left(\frac{8}{10}\right) + 4\left(\frac{10}{8}\right)$$

$$= 23 + 23.2 + 5 = \textbf{51.2}$$

A transistor with $h_{fe} > 51.2$ will provide sufficient loop gain for the circuit to operate as an oscillator. Practically, a transistor with at least $h_{fe} > 60$ would be selected.

Phase-shift oscillators are suited to operating frequencies in the range of a few hertz to a few hundred kilohertz. Other oscillator configurations (typically the tuned circuits) are more suitable to frequency in the megahertz range. To adjust the frequency of a phase-shift oscillator it is necessary to operate the three feedback capacitors as a single ganged component so that all three capacitor values are kept the same. A phase-shift oscillator is operated class-A to keep distortion low while providing a sinusoidal output waveform taken from the output of the amplifier stage. It should be clear that the output signal should not directly drive any low-impedance circuit that will load down the output. This would decrease the output voltage and thereby drop the loop gain below the necessary value for oscillator action. Feeding the output to a high-impedance stage, such as an FET amplifier stage or emitter-follower transistor stage, provides negligible loading of this oscillator circuit.

17.8   THE *LC*-TUNED OSCILLATOR CIRCUIT

An oscillator circuit can be made using a transformer for the feedback network. In addition, the inductance of the transformer and a parallel capacitor can be used to

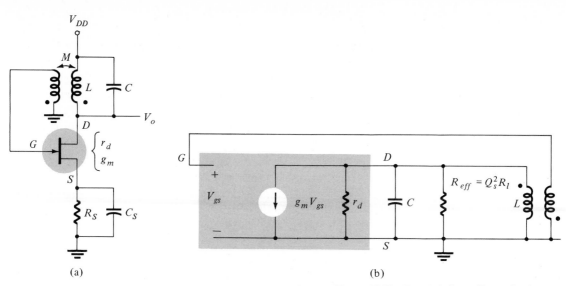

(a)   (b)

Figure 17.23   Tuned-drain oscillator circuit.

tune the circuit to the desired oscillator frequency. Figure 17.23a shows an FET amplifier with positive feedback provided by the transformer and tuning by an *LC* circuit. The circuit is described as a tuned-drain, untuned-gate oscillator.

An ac equivalent circuit is shown in Fig. 17.23b. The input resistance to the gate is assumed very high and is shown as an open circuit. The FET ac equivalent circuit is shown as a current source $(g_m V_{gs})$ in parallel with an output resistance r_d. The transformer can be represented as an inductance L in the primary side and a mutual coupling factor M. The series resistance of the transformer (representing its losses) can be accounted for by an effective resistance in parallel with the primary shown as $R_{\text{eff}} = Q_s^2 R_s$ in the circuit. The factor Q_s is the series Q of the transformer, defined as

$$Q_s = \frac{\omega_o L}{R_s} \tag{17.24}$$

Analysis of the ac equivalent circuit provides the oscillator frequency

$$f_o = \frac{1}{2\pi\sqrt{LC}} \tag{17.25}$$

and FET g_m

$$g_m = \frac{1}{R}\frac{L}{M} \tag{17.26}$$

where $$R = r_d \parallel Q_s^2 R_s \tag{17.27}$$

A summary of the results provided in the above equations follows:

1. The oscillator frequency is determined by the *LC*-resonant ("tank") circuit. This relationship is sometimes modified slightly due to circuit nonlinearities and unac-

counted for resistances, capacitances, and so on. However, it is good as a first-order approximation.

2. The minimum required FET g_m is dependent on the transformer effective resistance, the FET output resistance, the coil inductance and mutual coupling—in other words, on the parameters of the transformer chosen and the value of the FET output resistance. The value of g_m should be larger than this minimum value for oscillator action to take place.

3. The transformer primary and secondary windings must be connected in proper polarity sense to result in positive feedback. For this to occur the transformer should be connected to provide opposite relative polarity, which, added to the polarity inversion of the FET amplifier stage, results in overall positive feedback.

4. Loading of the circuit provided by either a lower value of r_d, lower effective transformer resistance, or external loading due to a connection of the oscillator to another circuit results in the value of the resulting resistance R being lower. This then requires a larger value of g_m to provide sufficient loop gain for oscillator action.

Another example of a tuned-circuit oscillator is the tuned-collector circuit shown in Fig. 17.24. The primary side of the transformer forms a tuned-tank circuit to set the oscillator frequency. The transformer is connected to provide positive feedback and the amplifier provides sufficient gain for oscillator action to take place.

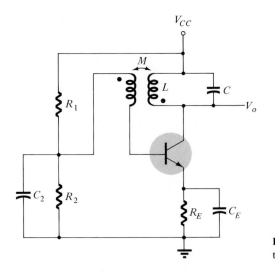

Figure 17.24 Bipolar transistor-tuned LC oscillator circuit.

Resistors R_1, R_2, and R_E are used to dc bias the transistor. Capacitors C_E and C_2 act to bypass resistors R_E and R_2, respectively, so that they have no effect on the ac operation of the circuit. Notice that although the low-resistance secondary winding of the transformer provides the dc bias voltage set by R_1 and R_2 to be connected to the base, the secondary essentially provides an ac feedback voltage in shunt with the transistor base emitter since the junction point of R_1 and R_2 is at ac ground (due to bypass capacitor C_2).

EXAMPLE 17.10   For the oscillator circuit of Fig. 17.23 and the following circuit values calculate the circuit frequency of oscillation and the minimum gain (g_m) of the FET unit: $r_d = 40$ kΩ, $L = 4$ mH, $M = 0.1$ mH, $R_s = 50$ Ω, and $C = 1$ nF.

Solution: The resonant frequency of the oscillator circuit is calculated using Eq. (17.25)

$$f_o = \frac{1}{2\pi\sqrt{LC}} = \frac{1}{6.28\sqrt{(4\times 10^{-3})(0.001\times 10^{-6})}} \cong \mathbf{80\ kHz}$$

$$\omega_o = 2\pi f_o = 6.28(80\ \text{kHz}) \cong 500\times 10^3\ \text{rad/s}$$

The coil Q is then

$$Q_s = \frac{\omega_o L}{R_s} = \frac{(500\times 10^3)(4\times 10^{-3})}{50} = 40$$

The effective coil resistance is

$$R_{\text{eff}} = Q_s^2 R_s = (40)^2(50) = 80\ \text{k}\Omega$$

Using Eq. (17.27) to calculate R gives

$$R = \frac{r_d R_{\text{eff}}}{r_d + R_{\text{eff}}} = \frac{40\ \text{k}\Omega\ (80\ \text{k}\Omega)}{40\ \text{k}\Omega + 80\ \text{k}\Omega} = 26.7\ \text{k}\Omega$$

The minimum value of g_m can now be calculated using Eq. (17.26)

$$g_m = \frac{1}{R}\frac{L}{M} = \frac{1}{26.7\times 10^3}\frac{4\times 10^{-3}}{0.1\times 10^{-3}} = 1.5\times 10^{-3} \cong \mathbf{1.5\ mS}$$

The FET selected should have a value of g_m greater than 1.5 mS.

17.9   TUNED-INPUT, TUNED-OUTPUT OSCILLATOR CIRCUITS

A variety of circuits shown in Fig. 17.25 provide tuning in both the input and output sections of the circuit. Analysis of the circuit of Fig. 17.25 reveals that the following types of oscillators are obtained when the reactance elements are as designated.

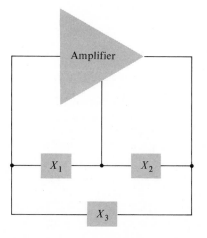

Figure 17.25   Basic configuration of resonant circuit oscillator.

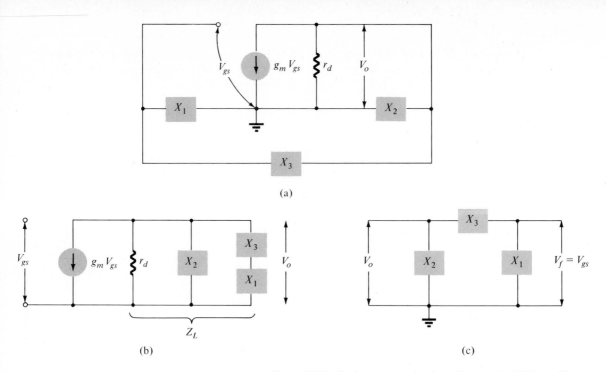

(a)

(b)   (c)

Figure 17.26 Basic resonant circuit oscillator using FET amplifier.

| | | Reactance Elements | |
Oscillator Type	X_1	X_2	X_3
1. Colpitts oscillator	C	C	L
2. Hartley oscillator	L	L	C
3. Tuned input, tuned output	LC	LC	—

The circuit form of the oscillator using an FET amplifier is shown in Fig. 17.26a. For operation as an oscillator the Barkhausen criterion is

$$\beta A = 1$$

The amplifier gain A is given simply by

$$A = -g_m Z_L \tag{17.28}$$

where Z_L is a parallel combination of impedances

$$Z_L = r_d \parallel X_2 \parallel (X_1 + X_3) \tag{17.29}$$

The circuit is redrawn in Fig. 17.26b so that the gain network is clearly shown. Figure 17.26b shows the parallel components that form an equivalent ac impedance, Z_L, as given in Eq. (17.29).

Figure 17.26c shows the feedback section of the circuit. Notice that the output voltage (V_o) is developed across X_2 and that the resulting feedback voltage (V_f) across X_1 is the input voltage to the amplifier (V_g). The feedback factor β is given by

$$\beta = \frac{X_1}{X_1 + X_3} \tag{17.30}$$

Plugging Eq. (17.28) for A and Eq. (17.30) for β into the basic equation (17.18) and using Eq. (17.29) for Z_L provide a means of determining the necessary device gain (g_m).

$$A = \frac{-g_m r_d X_2 (X_1 + X_3)}{r_d (X_1 + X_2 + X_3) + X_2 (X_1 + X_3)} \tag{17.31}$$

The oscillator frequency is obtained from

$$(X_1 + X_2 + X_3) = 0 \tag{17.32}$$

When X_1 and X_2 are capacitors and X_3 is an inductor, the circuit is called a *Colpitts oscillator*. When X_1 and X_2 are inductors and X_3 is a capacitor, the circuit is called a *Hartley oscillator*.

17.10 COLPITTS OSCILLATOR

FET Colpitts Oscillator

A practical version of an FET Colpitts oscillator is shown in Fig. 17.27a. The circuit is basically the same form as shown in Fig. 17.26 with the addition of the

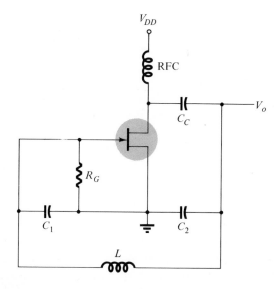

Figure 17.27 FET Colpitts oscillator.

components needed for dc bias of the FET amplifier. The oscillator frequency can be found to be

$$\boxed{f_o = \frac{1}{2\pi\sqrt{LC_{eq}}}} \tag{17.33}$$

where

$$C_{eq} = \frac{C_1 C_2}{C_1 + C_2}$$

Transistor Colpitts Oscillator

A transistor Colpitts oscillator circuit can be made as shown in Fig. 17.28. The circuit frequency of oscillation is given by Eq. (17.33).

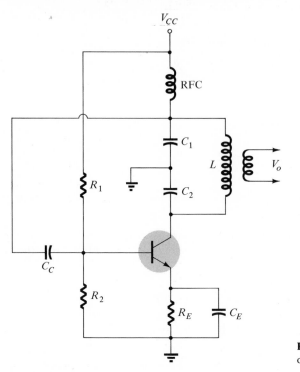

Figure 17.28 Transistor Colpitts oscillator.

17.11 HARTLEY OSCILLATOR

If the elements in the basic resonant circuit of Fig. 17.26 are X_1 and X_2 (inductors), and X_3 (capacitor), the circuit is a Hartley oscillator.

FET Oscillator

An FET Hartley oscillator circuit is shown in Fig. 17.29. The circuit is drawn so that the feedback network conforms to the form shown in the basic resonant circuit (Fig. 17.26). Note, however, that inductors L_1 and L_2 have a mutual coupling, M, which must be taken into account in determining the equivalent inductance for the resonant tank circuit.

AC CIRCUIT ANALYSIS

We can obtain an ac equivalent circuit for Fig. 17.29 by replacing capacitors C_C and C_G by shorts, RFC coil by an open circuit, and the FET by its equivalent circuit

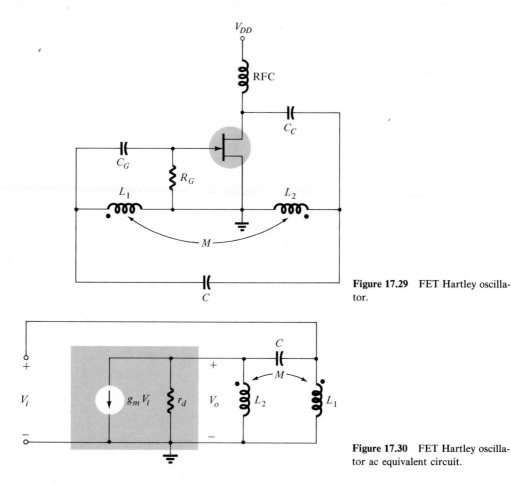

Figure 17.29 FET Hartley oscillator.

Figure 17.30 FET Hartley oscillator ac equivalent circuit.

as shown in Fig. 17.30. The tank circuit of mutually linked inductors L_1 and L_2 and capacitor C can be shown equivalent to a tank circuit of capacitor C in parallel with an equivalent inductance

$$L_{eq} = L_1 + L_2 + 2\,M \tag{17.34}$$

The circuit frequency of oscillation is then given approximately by

$$f_o = \frac{1}{2\pi\sqrt{L_{eq}C}} \tag{17.35}$$

with L_{eq} given in Eq. (17.34).

Transistor Hartley Oscillator

Figure 17.31 shows a transistor Hartley oscillator circuit. The circuit operates at a frequency given by Eq. (17.35).

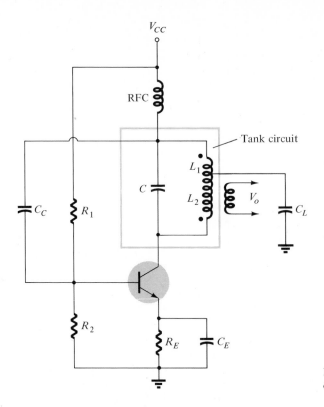

Figure 17.31 Transistor Hartley oscillator circuit.

17.12 CRYSTAL OSCILLATOR

A crystal oscillator is basically a tuned-circuit oscillator using a piezo-electric crystal as a resonant tank circuit. The crystal (usually quartz) has a greater stability in holding constant at whatever frequency the crystal is originally cut to operate. Crystal oscillators are used whenever great stability is required, for example, in communication transmitters and receivers.

Characteristics of a Quartz Crystal

A quartz crystal (one of a number of crystal types) exhibits the property that when mechanical stress is applied across the faces of the crystal, a difference of potential develops across opposite faces of the crystal. This property of a crystal is called the *piezoelectric effect*. Similarly, a voltage applied across one set of faces of the crystal causes mechanical distortion in the crystal shape.

When alternating voltage is applied to a crystal, mechanical vibrations are set up—these vibrations having a natural resonant frequency dependent on the crystal. Although the crystal has electromechanical resonance, we can represent the crystal action by an equivalent electrical resonant circuit as shown in Fig. 17.32. The inductor L and capacitor C represent electrical equivalents of crystal mass and compliance while resistance R is an electrical equivalent of the crystal structure's internal friction. The shunt capacitance C_M represents the capacitance due to mechanical mounting

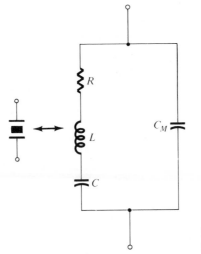

Figure 17.32 Electrical equivalent circuit of a crystal.

of the crystal. Because the crystal losses, represented by R, are small, the equivalent crystal Q (quality factor) is high—typically 20,000. Values of Q up to almost 10^6 can be achieved by using crystals.

The crystal as represented by the equivalent electrical circuit of Fig. 17.32 can have two resonant frequencies. One resonant condition occurs when the reactances of the series RLC leg are equal (and opposite). For this condition the *series-resonant* impedance is very low (equal to R). The other resonant condition occurs at a higher frequency when the reactance of the series-resonant leg equals the reactance of capacitor C_M. This is a parallel resonance or antiresonance condition of the crystal. At this frequency the crystal offers a very high impedance to the external circuit. The impedance versus frequency of the crystal is shown in Fig. 17.33. In order to use the crystal properly it must be connected in a circuit so that its low impedance in the series-resonant operating mode or high impedance in the antiresonant operating mode is selected.

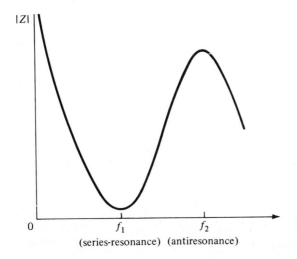

Figure 17.33 Crystal impedance vs. frequency.

Series-Resonant Circuits

To excite a crystal for operation in the series-resonant mode it may be connected as a series element in a feedback path. At the series-resonant frequency of the crystal its impedance is smallest and the amount of (positive) feedback is largest. A typical transistor circuit is shown in Figure 17.34. Resistors R_1, R_2, and R_E provide a voltage-divider stabilized dc bias circuit. Capacitor C_E provides ac bypass of the emitter resistor and the RFC coil provides for dc bias while decoupling any ac signal on the power lines from affecting the output signal. The voltage feedback from collector to base is a maximum when the crystal impedance is minimum (in series-resonant mode). The coupling capacitor C_C has negligible impedance at the circuit operating frequency but blocks any dc between collector and base.

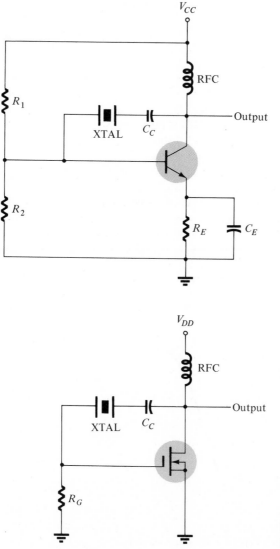

Figure 17.34 Crystal controlled oscillator using crystal in series-feedback path.

Figure 17.35 FET Pierce crystal controlled oscillator circuit.

The resulting circuit frequency of oscillation is set, then, by the series-resonant frequency of the crystal. Changes in supply voltage, transistor device parameters, and so on, have no effect on the circuit operating frequency which is held stabilized by the crystal. The circuit frequency stability is set by the crystal frequency stability, which is good.

The circuits shown in Figs. 17.34 and 17.35 are generally called Pierce crystal-controlled oscillators.

Another transistor circuit is shown in Fig. 17.36. The circuit provides tuning by an LC tank circuit in the collector and tuning by a series-resonant excited crystal connected as feedback from a capacitive voltage divider. The LC circuit is adjusted near the desired operating crystal frequency, but the exact circuit frequency is set by the crystal and stabilized by the crystal.

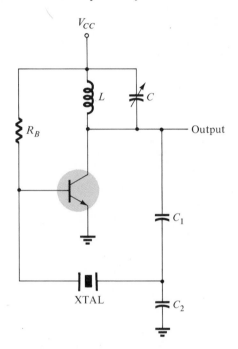

Figure 17.36 Transistor crystal os-cillator.

Parallel-Resonant Circuits

Since the parallel-resonant impedance of a crystal is a maximum value, it is con-nected in shunt. At the parallel-resonant operating frequency a crystal appears as an inductive reactance of largest value. Figure 17.37 shows a crystal connected as the inductor element in a modified Colpitts circuit. The basic dc bias circuit should be evident. Maximum voltage is developed across the crystal at its parallel-resonant frequency. The voltage is coupled to the emitter by a capacitor voltage divider—capacitors C_1 and C_2.

A *Miller* crystal-controlled oscillator circuit is shown in Fig. 17.38. A tuned LC circuit in the drain section is adjusted near the crystal parallel-resonant frequency. The maximum gate-source signal occurs at the crystal antiresonant frequency control-ling the circuit operating frequency.

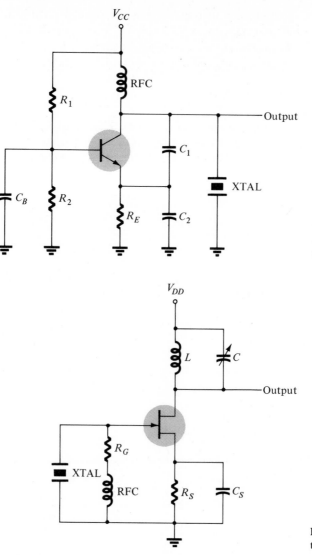

Figure 17.37 Crystal controlled oscillator operating in parallel-resonant triode.

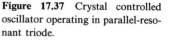

Figure 17.38 Miller crystal controlled oscillator.

17.13 OP-AMP OSCILLATOR CIRCUITS

Phase-Shift Oscillator

As IC circuits have become more popular they have been adapted to operate in oscillator circuits. One need buy only an op-amp to obtain an amplifier circuit of stabilized gain setting and incorporate some means of signal feedback to produce an oscillator circuit. For example, a phase-shift oscillator is shown in Fig. 17.39. The output of the op-amp is fed to a three-stage RC network which provides the needed 180° of phase shift (at an attenuation factor of 1/29). If the op-amp provides gain (set by resistors R_i and R_f) of greater than 29, a loop gain greater than unity

CH. 17 FEEDBACK AMPLIFIERS AND OSCILLATOR CIRCUITS

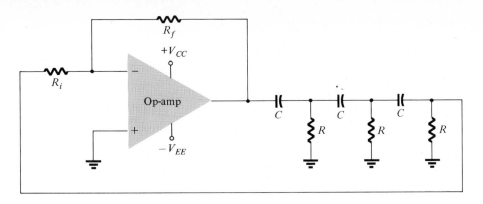

Figure 17.39 Phase-shift oscillator using op-amp.

results and the circuit acts as an oscillator [oscillator frequency is given by Eq. (17.19a)].

Colpitts Oscillator

An op-amp Colpitts oscillator circuit is shown in Fig. 17.40. Again, the op-amp provides the basic amplification needed while the oscillator frequency is set by an *LC* feedback network of a Colpitt configuration. The oscillator frequency is given by Eq. (17.33).

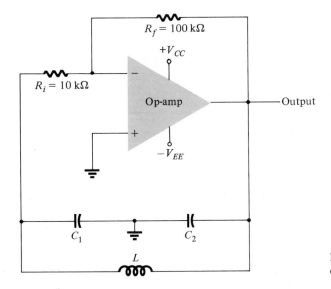

Figure 17.40 Op-amp Colpitts os-cillator.

Crystal Oscillator

An op-amp can be used in a crystal oscillator as shown in Fig. 17.41. The crystal is connected in the series-resonant path and operates at the crystal series-resonant frequency. The present circuit has a high gain so that an output square-wave signal results as shown in the figure. A pair of Zener diodes is shown at the output to provide output amplitude at exactly the Zener voltage (V_Z).

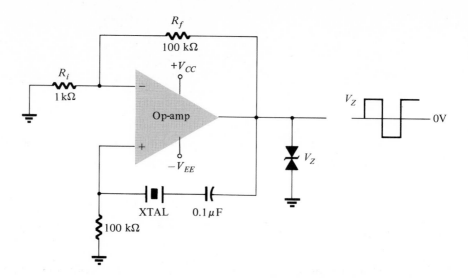

Figure 17.41 Crystal oscillator using op-amp.

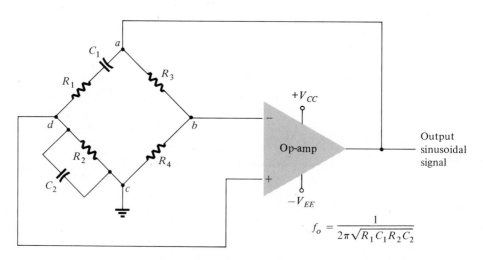

$$f_o = \frac{1}{2\pi\sqrt{R_1 C_1 R_2 C_2}}$$

Figure 17.42 Wien-bridge oscillator circuit using op-amp amplifier.

Wien-Bridge Oscillator

A practical oscillator circuit uses an op-amp and RC bridge circuit, with the oscillator frequency set by the R and C components. Figure 17.42 shows a basic version of a Wien bridge oscillator circuit. Note the basic bridge connection. Resistors R_1, R_2 and capacitors C_1, C_2 form the frequency-adjustment elements, while resistors R_3 and R_4 form part of the feedback path. The op-amp output is connected as the bridge input at points a and c. The bridge circuit output at points b and d is the input to the op-amp.

Neglecting loading effects of the op-amp input and output impedances, the analysis of the bridge circuit results in

$$\boxed{\frac{R_3}{R_4} = \frac{R_1}{R_2} + \frac{C_2}{C_1}} \qquad (17.36a)$$

and

$$\boxed{\omega_o = \frac{1}{\sqrt{R_1 C_1 R_2 C_2}}} \qquad (17.36b)$$

If, in particular, the values are $R_1 = R_2 = R$ and $C_1 = C_2 = C$, the resulting oscillator frequency is

$$\omega_o = \frac{1}{RC} \qquad (17.37a)$$

and

$$\frac{R_3}{R_4} = 2 \qquad (17.37b)$$

Thus, a ratio of R_3 to R_4 greater than 2 will provide sufficient loop gain for the circuit to oscillate at the frequency calculated using Eq. (17.37a).

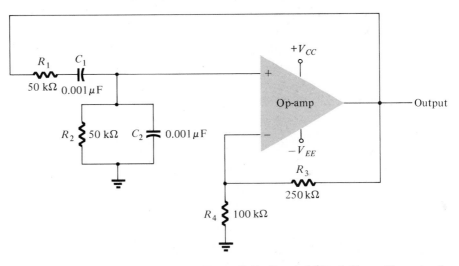

Figure 17.43 Practical Wien-bridge oscillator circuit.

A practical circuit design is shown in Fig. 17.43. Although the circuit is shown in somewhat different form from Fig. 17.42, you should compare the two to satisfy yourself that they are indeed identical in form.

For the circuit values given we calculate

$$f_o = \frac{1}{2\pi RC} = \frac{1}{6.28(50 \times 10^3)(0.001 \times 10^{-6})} = 3.18 \text{ kHz}$$

EXAMPLE 17.11 Design the RC elements of a Wien-bridge oscillator as in Fig. 17.43 for operation at $f_o = 10$ kHz.

Solution: Using equal values of R and C we can select $R = 100$ kΩ and calculate the required value of C using Eq. (17.37a):

$$f_o = \frac{1}{2\pi RC}$$

$$C = \frac{1}{2\pi f_o R} = \frac{1}{6.28(10 \times 10^3)(100 \times 10^3)} = \frac{10^{-9}}{6.28} = \mathbf{159 \ pF}$$

We can use $R_3 = 250 \ k\Omega$ and $R_4 = 100 \ k\Omega$ to provide a ratio R_3/R_4 greater than 2 for oscillation to take place.

17.14 UNIJUNCTION OSCILLATOR

A particular device, the unijunction transistor (discussed in Chapter 11), can be used in a single-stage oscillator circuit to provide a pulse signal suitable for digital-circuit applications. The unijunction transistor can be used in what is called a *relaxation oscillator* as shown by the basic circuit of Fig. 17.44. Resistor R_T and capacitor

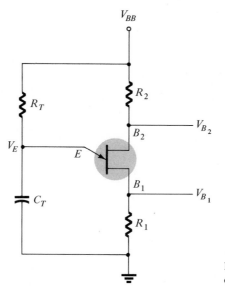

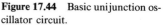

Figure 17.44 Basic unijunction oscillator circuit.

C_T are the timing components that set the circuit oscillating rate. The oscillating frequency may be calculated using Eq. (17.38) which includes the unijunction transistor *intrinsic stand-off ratio* η, as a factor (in addition to R_T and C_T) in the oscillator operating frequency.

$$f_o \cong \frac{1}{R_T C_T \ln \left[1/(1-\eta) \right]} \tag{17.38}$$

Typically, a unijunction transistor has a stand-off ratio from 0.4 to 0.6. Using a value of $\eta = 0.5$, we get

$$f_o \cong \frac{1}{R_T C_T \ln [1/(1 - 0.5)]} = \frac{1}{R_T C_T \ln 2} = \frac{1.44}{R_T C_T}$$

$$\cong \frac{1.5}{R_T C_T} \tag{17.39}$$

Capacitor C_T is charged through resistor R_T toward supply voltage V_{BB}. As long as the capacitor voltage V_E is below a stand-off voltage (V_P) set by the voltage across $B_1 - B_2$, and the transistor stand-off ratio η is given by Eq. (17.40),

$$V_P = \eta V_{B_1} V_{B_2} - V_D \tag{17.40}$$

and the unijunction emitter lead appears as an open circuit. When the emitter voltage across capacitor C_T exceeds this value (V_P), the unijunction circuit fires, discharging the capacitor, after which a new charge cycle begins. When the unijunction fires, a voltage rise is developed across R_1 and voltage drop is developed across R_2 as shown in Fig. 17.45. The signal at the emitter is a sawtooth voltage waveform, that at base 1 is a positive-going pulse, and that at base 2 is a negative-going pulse.

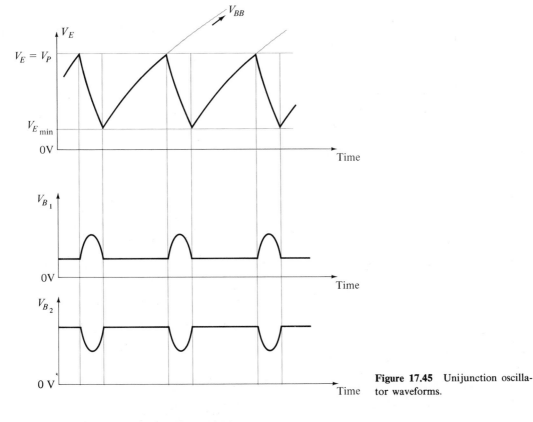

Figure 17.45 Unijunction oscillator waveforms.

Use of Nomograph to Design Circuit

For the basic circuit of Fig. 17.44 the relation between R_T, C_T, and the oscillator frequency (at a fixed value of η) can be given as a nomograph (see Fig. 17.46) instead

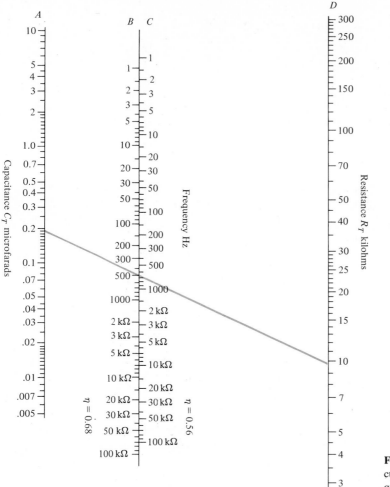

Figure 17.46 Nomograph to calculate unijunctions oscillator frequency.

of as an equation [Eq. (17.38)]. The line shown on the nomograph shows how to obtain the operating frequency for values of $R_T = 10$ kΩ and $C_T = 0.2$ μF. The straight line connecting these points intersects the frequency axis at about $f_o = 650$ Hz for $\eta = 0.56$, or $f_o = 480$ for $\eta = 0.68$. A transistor with $\eta = 0.6$, for example, would operate at a frequency between 480 and 650 Hz.

> **EXAMPLE 17.12** Using the nomograph determine the value of R_T for $C_T = 0.01$ μF and $\eta = 0.56$ for operation at $f_o = 10$ kHz.
>
> **Solution:** By connecting a straight line between $C_T = 0.01$ μF and $f_o = 10$ kHz (on $\eta = 0.56$ scale) points, the intersection with the R_T axis is read as
>
> $$R_T = 12 \text{ k}\Omega$$
>
> [As an exercise, check the result using Eq. (17.38).]

A few circuit variations of the unijunction oscillator are provided in Fig. 17.47.

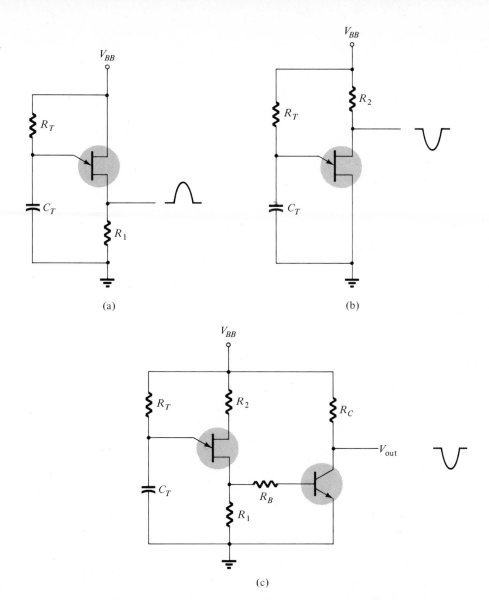

Figure 17.47 Some unijunction oscillator circuit configurations.

PROBLEMS

§ 17.1

1. Calculate the gain of a negative-feedback amplifier having $A = 2000$, $\beta = 1/10$.

2. If the gain of an amplifier changes from a value of 1000 by 10%, calculate the gain change if the amplifier is used in a feedback circuit having $\beta = 1/20$.

§ 17.2

3. Calculate the gain, input, and output impedances of a voltage-series feedback amplifier having $A = 300$, $R_i = 1.5$ kΩ, $R_o = 50$ kΩ, and $\beta = 1/15$.

4. Calculate the voltage gain, input, and output impedance with and without feedback for a circuit as in Fig. 17.7 for circuit values $R_e = 2$ kΩ, $R_S = 600$ Ω, $h_{ie} = 1.5$ kΩ, and $h_{fe} = 100$.

5. Calculate the gain with and without feedback for an FET amplifier as in Fig. 17.8 for circuit values $R_1 = 200$ kΩ, $R_2 = 800$ Ω, $R_o = 40$ kΩ, $R_D = 8$ kΩ, and $g_m = 5000$ μS.

§ 17.4

6. Calculate A, R_i, and R_o with and without feedback for a circuit as in Fig. 17.19 for transistor parameters $h_{fe} = 80$, $h_{ie} = 2.2$ kΩ.

7. For a circuit as in Fig. 17.11 and the following circuit values, calculate the circuit gain and the input and output impedance with and without feedback: $R_B = 600$ kΩ, $R_E = 1.2$ kΩ, $R_C = 12$ kΩ, $h_{ie} = 2$ kΩ, and $h_{fe} = 75$. Use $V_{CC} = 16$ V.

§ 17.7

8. An FET phase-shift oscillator having $g_m = 6000$ μS, $r_d = 36$ kΩ, and feedback resistor $R = 12$ kΩ is to operate at 2.5 kHz. Select R_D and C for specified oscillator operation.

9. Select values of capacitor C and transistor gain h_{fe} to provide operation of a transistor phase-shift oscillator at 5 kHz for circuit values $R_1 = 24$ kΩ, $R_2 = 75$ kΩ, $R_C = 18$ kΩ, $R = 6$ kΩ, and $h_{ie} = 2$ kΩ.

§ 17.8

10. Calculate the minimum gain (g_m) of the FET in the oscillator circuit of Fig. 17.23 and the circuit frequency of oscillation for circuit values $r_d = 50$ kΩ, $L = 5$ mH, $M = 0.2$ mH, $R_S = 60$ Ω, and $C = 0.002$ μF.

11. For an FET Colpitts oscillator as in Fig. 17.27a and the following circuit values determine the circuit oscillation frequency: $C_1 = 750$ pF, $C_2 = 2500$ pF, $L = 40$ μH, $R_g = 750$ kΩ, $L_{RFC} = 0.2$ mH, $C_C = 2000$ pF.

12. For the transistor Colpitts oscillator of Fig. 17.28 and the following circuit values calculate the oscillation frequency: $L = 100$ μH, $L_{RFC} = 0.5$ mH, $C_1 = 0.005$ μF, $C_2 = 0.01$ μF, $C_C = 10$ μF.

§ 17.11

13. Calculate the oscillator frequency for an FET Hartley oscillator as in Fig. 17.29 for the following circuit values: $C = 250$ pF, $L_1 = 1.5$ mH, $L_2 = 1.5$ mH, $M = 0.5$ mH.

14. Calculate the oscillation frequency for the transistor Hartley circuit of Fig. 17.31 and the following circuit values: $L_{RFC} = 0.5$ mH, $L_1 = 750$ μH, $L_2 = 750$ μH, $M = 150$ μH, $C = 150$ pF.

§ 17.12

15. Draw circuit diagrams of (a) a series-operated crystal oscillator, and (b) a shunt-excited crystal oscillator.

16. Design the RC elements of a Wien-bridge oscillator circuit (as in Fig. 17.42) for operation at $f_o = 2$ kHz.

17. Design a unijunction oscillator circuit for operation (a) at 1 kHz and (b) at 150 kHz.

CHAPTER 18

Cathode-Ray Oscilloscope (CRO)

18.1 GENERAL

One of the basic functions of electronic circuits is the generation and manipulation of electronic waveshapes. These electronic signals may represent audio information, computer data, television pictures, timing information (as used in radar work), and so on. The common meters used in electrical work—the dc or ac voltmeter—(VOM or DDM) measure either dc, peak, or rms. These measurements are correct only for nondistorted sinusoidal signals or they measure true rms for a particular signal with no indication of how the signal varies with time. Obviously, when signal processing is being done, these overall measurements are essentially meaningless. What is necessary is to "see" what is going on in the circuit, hopefully in the small fractions of time it takes for the signal waveshape to change. The cathode-ray oscilloscope (CRO) provides just this type of operation—visual presentation of the signal waveshape, allowing the technician or engineer to look at different points in the electronic circuit to see those changes taking place. In addition, the CRO may be calibrated and used to measure both voltage and time variations so that information is available on how much voltage is present, how much voltage changed, and how long it took to make the change (or a portion of the change).

Consider a radar circuit in which a pulse originates in an electronic circuit and is radiated by an antenna toward a distant object. At some later time a reflected pulse is received. The time it took to reach the object and return is measured on a CRO to provide an indication of distance. It is necessary when developing and setting up such circuits to be able to display the pulse sent out—its amplitude, the sharpness of the pulse, etc.—and then visually compare it with the return signal after reception and processing of the signal. The return signal can be checked on the CRO for sharpness and voltage amplitude. The time difference between the pulse transmitted and that received may be read on the calibrated time scale of the CRO.

A second example is the use of a CRO to view signals throughout the circuit of a television (TV) receiver (or transmitter). The video signals (information signals for TV) contain complex waveforms, which provide black-white intensity, synchronization information for the picture, and audio information. To properly test and adjust these circuits requires comparing their operating waveforms with expected waveforms. Such adjustments could not be made without the use of the CRO. (It is interesting to note that due to the cathode-ray tube (CRT), the TV can be its own test scope, showing where the circuit is operating well or poorly.)

Without the scope, electronic work would virtually be impossible—it would be like groping in a dark room. Unless one is able to "see" the circuit waveforms, there is no way to correct errors, understand mistakes in design, or make adjustments. As the voltmeter, ammeter, and power meter are basic tools of the power engineer or electrician, the CRO is the basic tool of the electronic engineer and technician.

There is a wide range of CROs available, some suited for general work in many areas of electronics, others for work only in a specific area. A CRO may be designed to operate from 100 Hz up to 500 kHz, or from dc up to 50 MHz; it may allow viewing signals to within a time span of 1 microsecond, or down to a few nanoseconds (10^{-9} s); it may provide one waveform or a number of waveforms simultaneously on the face of the CRO. When a number of waveforms are shown simultaneously, voltages at different places in the circuit can be viewed at the same time. All these

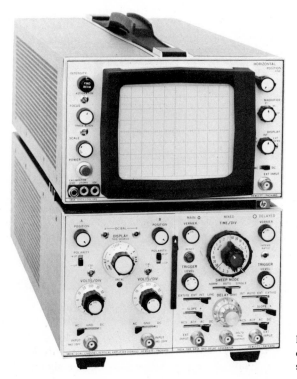

Figure 18.1 Typical cathode-ray oscilloscope (CRO): HP 180A scope.

CRO features provide flexibility and enable one to use a CRO that is well suited for the job at hand. Another CRO feature is its ability to hold the display for either a very short duration of time or for a long duration. In fact, CROs, called *storage scopes,* provide storage of a display for many hours so that an original signal (which appeared long before) may still be analyzed or compared with another signal at a later time. Figure 18.1 shows two representative scopes.

18.2 CATHODE-RAY TUBE—THEORY AND CONSTRUCTION

The cathode-ray tube (CRT) is the "heart" of the CRO. It provides the visual display that makes the instrument so useful. The tube contains the following four basic parts:

1. An *electron gun* to produce a stream of electrons.
2. *Focusing and accelerating* elements to produce a well-defined *beam* of electrons.
3. *Horizontal and vertical deflecting plates* to control the path of the beam.
4. An *evacuated glass envelope* with a *phosphorescent screen* that glows visibly when struck by the electron beam.

Figure 18.2 shows an overall view of a complete electrostatic-deflection CRT. We shall first briefly consider its operation. A *cathode (K)* containing an oxide coating is heated indirectly by a filament resulting in the release of electrons from the cathode surface. The preaccelerating and accelerating anodes are at a sufficiently high positive potential to attract the generated electrons in their direction. The control gird *(G)* (at a slightly negative potential) provides for control of the number of electrons passing on into the tube. The electrons are then focused into a tight beam and accelerated to higher velocity by the focusing and accelerating anodes. The parts discussed so far comprise the *electron gun* of the CRT.

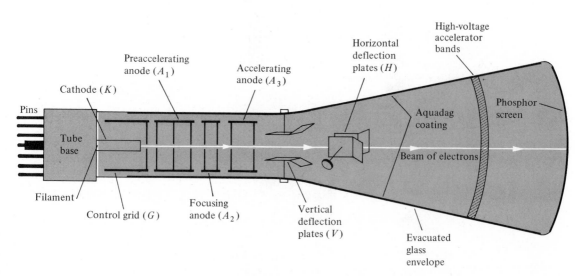

Figure 18.2 Cathode-ray tube: basic construction.

The high-velocity, well-defined electron beam then passes through two sets of deflection plates. The first set of plates is oriented to deflect the electron beam *vertically,* up or down. The *direction* of the vertical deflection is determined by the voltage *polarity* applied to the deflecting plates. The amount of deflection is set by the applied-voltage *magnitude.* The beam is also deflected horizontally (to the left or right) by a voltage applied to the horizontal deflecting plates. The deflected beam is further accelerated by very high voltages applied to the tube and it finally strikes a *phosphorescent* material on the inside face of the tube. The phosphor glows when struck by the energetic electrons—the visible glow seen at the front of the tube by the person using the scope.

The CRT is a self-contained unit with leads brought out through a base to pins, as in any electron tube. Various types of CRTs are manufactured for a variety of applications. The CRT of Fig. 18.2 is basic and allows discussion of the essential elements of the device. We can now consider each part of the tube in more detail and then the external CRO controls, which provide the useful operation of the tube.

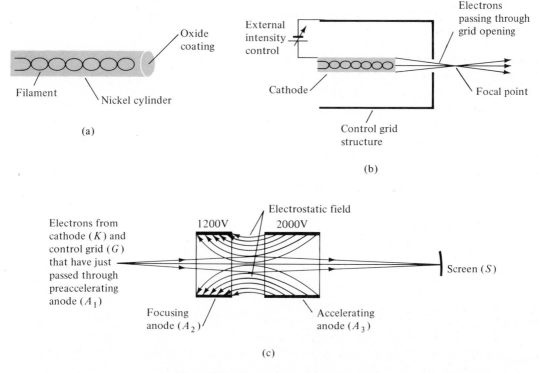

Figure 18.3 Components of the CRT electron gun: (a) cathode *(K)* and filament; (b) control grid *(G);* (c) focusing (A_1) and accelerating anodes (A_2).

Electron Gun

The cathode (Fig. 18.3a) is cylindrical, made of nickel, and capped on the end with an oxide coating. The oxide coating, typically made of oxides of barium and strontium, is applied to the cap of the cathode facing the screen direction. A filament, made of a tungsten or tungsten alloy, indirectly (not in direct contact) heats the cathode when a current is established through the filament. The filament acts like the heating elements of a device such as a toaster or rotisserie. Electrons will be liberated from the oxide coating on the cathode surface by the heating effect of the filament. These liberated electrons will travel in the general direction of the screen but at various angles and with various velocities.

Focusing and Accelerating Elements

To provide some focusing of the electrons a control grid (Fig. 18.3b) with a small opening in the direction of the screen is placed after the cathode. In addition, a biasing voltage is applied to the control grid to control the flow of electrons through the small opening of the grid structure. If the grid voltage is made negative (with respect to the cathode), there will be a reduction in the number of negatively charged electrons passing through the grid opening. With a large enough negative voltage all the electrons leaving the cathode due to the heating effect of the filament will be prevented from passing through the grid aperture. The grid therefore permits adjustment of the number of electrons generated by the electron gun with the ability to completely cut off the electron flow to the screen, if desired.

Those electrons that pass through the control grid structure are given an initial acceleration toward the screen of the tube by a preaccelerator anode (A_1) having a positive potential (100–200 V) with respect to the cathode. The focusing anode (A_2) and accelerating anode (A_3) operate to focus the electrons into a tight beam and also to further accelerate or speed up the electrons coming from the cathode. Fig. 18.3c shows the electric field set up by typical voltage applied to anodes A_2 and A_3 and their effect on the electron flow. The force on the electrons due to the electric field set up by these anodes will result in a well-shaped beam having a focal point at the screen of the tube. The person using the scope can control the focusing of the beam through an external FOCUS control.

The difference in potential between anodes A_2 and A_3 will set up an electrostatic field as shown in Fig. 18.3c. The effect of these electric field force lines will be to focus the electron beam as shown. The deflection will be such as to push those electrons diverging from the center back in toward the center of the beam. Once past the focusing anodes, the electrons move past the deflecting plates to the screen in a tight beam.

Deflecting Plates (Vertical and Horizontal)

Figure 18.4 shows the electron beam passing through a pair of plates. If the voltage of the upper plate is more positive than the voltage of the lower plate, the electron beam will be attracted upward. Reversing the plate polarity would cause the beam to be deflected downward. The voltage applied externally (of the CRT) to the deflection plates shown results in the signal's being deflected in a vertical manner; hence, the designation *vertical-deflection plates*. This voltage may be a steady (dc) voltage or a varying one. As a point of interest, the home TV tube is similar to that discussed so far except for a magnetic deflection system.

An expression relating the factors in calculating the amount of deflection is (see Fig. 18.4)

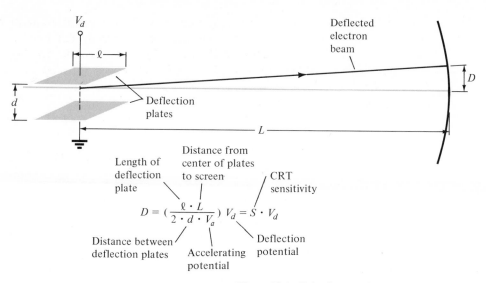

Figure 18.4 Tube factors affecting beam deflection.

$$D = \frac{lLV_d}{2dV_a} \qquad (18.1)$$

where the distance terms are shown in Fig. 18.4 and the voltages are V_d, the voltage applied to the deflection plates and V_a, the accelerating voltage of the tube (typically thousands of volts).

Electron Acceleration and Phosphor Screen Action

After the beam is acted upon by the deflection plates, it passes down the tube in a straight path (although at some deflection angle due to the deflecting voltages). To ensure that sufficiently energetic electrons strike the phosphor screen the electron beam may be further accelerated by intensifier bands as indicated in Fig. 18.5.

The high-energy electron beam strikes the phosphor material, causing it to glow. Because of the high energy of the striking electrons a secondary emission of electrons from the phosphor screen will occur. These electrons would build up a layer of negative charge, which would deteriorate the tube operation. A layer of material called *Aquadag* is coated along the side of the tube, however, and the electrons emitted by secondary action are picked up by the coating and are returned to the cathode.

If the glow seen were to be present only as long as the beam strikes the screen, the light given off would be termed fluorescent (as the lights in a home or plant). The phosphor screen, however, will continue to glow even after the beam is turned off. The length of time the glow continues can vary from a few milliseconds to a few seconds depending on the type of phosphor material used. CRO tubes used to observe very-high-frequency signals have short *persistence* (short amount of glow time after beam is removed). Other scopes used for observing only very-low-frequency signals have persistence times of a few seconds. There is also a special type of CRT used in memory-type scopes that has an effective persistence in the range of hours.

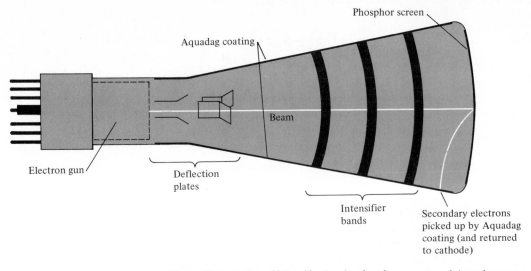

Figure 18.5 Action of intensifier bands, phosphor screen, and Aquadag coating.

18.3 CRO—DEFLECTION AND SWEEP OPERATION

We shall now consider the operation of the CRO for various inputs to the vertical and horizontal deflection plates. Although the actual CRO inputs are not *directly* connected to the deflection plates, but are coupled through attenuator networks and amplifiers, we shall still refer to such signals (applied to the vertical and horizontal inputs or vertical and horizontal channels of the CRO) as the deflection signals. Obviously, the positioning controls and the setting of the attenuators for each channel of the scope will affect the resulting signal actually applied to the deflection plates.

With 0 V connected to the vertical input terminals the electron beam may be positioned to the vertical center of the screen. If 0 V are also applied to the horizontal input, the beam is then at the center of the CRT face and remains a stationary dot on the face of the CRT. Note that the vertical and horizontal positioning controls

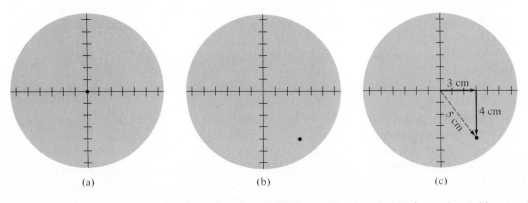

Figure 18.6 Dot on CRT screen due to stationary electron beam: (a) centered dot due to stationary electron beam; (b) off-center stationary dot; (c) stationary dot showing deflection components.

allow movement of this dot anywhere on the face of the CRT so that the zero input voltages cause the beam to strike the center of the screen *only* if the positioning controls are also zeroed. Unless otherwise noted in the following discussion, assume that the beam is properly centered and that any deflection off center is due to the input signals to the vertical and horizontal channels of the scope.

Zero input signals result in a dot on the screen (Fig. 18.6a). Any dc voltage applied to either input will result in shifting the dot on the screen with the resulting picture still only a dot. Figure 18.6b shows the resulting display due to a negative dc voltage applied at the vertical input and a positive dc voltage at the horizontal input. The negative voltage to the vertical plates deflects the beam downward by an amount proportional to the voltage magnitude and the voltage on the horizontal plates deflects the beam to the right, the resulting dot appearing in the lower-right sector of the screen as shown. The position of the resultant dot can be considered the vector sum of the two deflection voltages. A positive horizontal deflection of 3 cm and a negative vertical deflection of 4 cm, for example, would result in the beam spot as shown in Fig. 18.6c, at a distance of 5 cm from the screen center.

To view a signal on the CRT face (note that only a dot was visible even though a steady dc voltage was applied) it is necessary to deflect the beam across the CRT with a horizontal sweep signal so that the variations of the vertical signal can be observed. Figure 18.7 shows the resulting straight-line display for the positive dc

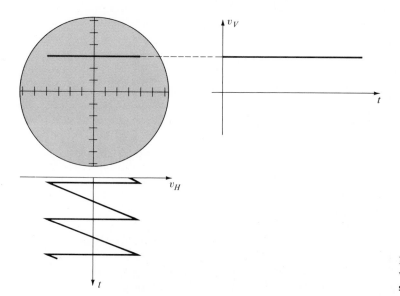

Figure 18.7 Scope display for dc vertical signal and linear horizontal-sweep signal.

voltage applied to the vertical input using a linear (sawtooth) sweep signal on the horizontal channel. With the electron beam held at a constant vertical distance, the horizontal voltage, going from negative to zero to positive voltage, causes the beam to move from the left side of the tube to the center and over to the right side. The resulting display is a straight line above the vertical center and a dc voltage now properly appears as a steady-voltage line. The sweep signal is indicated to be a continuous waveform and not just a single sweep. This is necessary if a long-term display is to be obtained. A single sweep across the tube would quickly fade out. By repeating the sweep, the display is generated over and over again and if enough sweeps are generated per second, the display always appears to be present. If the sweep rate is slowed down (the time scale controls of the scope allow this), the actual travel of the beam across the face of the tube can be observed.

Applying only a sinusoidal signal to the vertical input (no horizontal signal) results in a vertical straight line as shown in Fig. 18.8. If the sweep speed (frequency of the sinusoidal signal in this case) is reduced, it will be possible to see the electron beam moving up and down along the straight-line path.

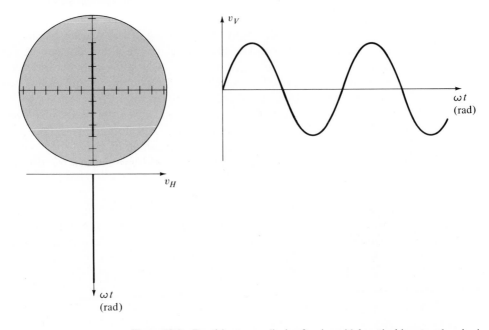

Figure 18.8 Resulting scope display for sinusoidal vertical input and no horizontal input.

Use of Linear Sawtooth Sweep To Display Vertical Input

A signal applied only to the vertical input will cause the electron beam to be deflected only up and down. If the vertical input is a dc voltage, the result will be a dot on the screen displaced from the screen center. If a varying voltage is applied to the vertical input, the beam will move up and down resulting in either the dot moving up and down or the appearance of a line, depending on how fast the input signal repeats a cycle and the persistence of the screen phosphor. Similar action results for an input to only the horizontal channel with resulting displacement to the right or left or a horizontal line.

To obtain a display that shows the form of the input signal applied to the vertical channel it is necessary to apply a linear sawtooth sweep signal to the horizontal channel as well. This is the normal operation of the CRO. The sawtooth sweep signal is provided as part of the scope circuitry with adjustment of the sweep rate

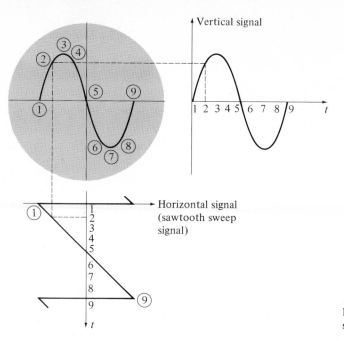

Figure 18.9 Use of linear sweep signal to view input signal.

provided as an external control. To understand the operation of the CRO we must consider how the electron beam is deflected as a result of signals applied to both the vertical and horizontal input at the same time.

To understand how the two deflecting voltages result in the CRT display we shall determine the path the beam takes by observing where it is at a few points during a single sweep of the electron beam (see Fig. 18.9). The voltage applied to the horizontal deflection plates is the sawtooth voltage. For convenience, the horizontal signal time axis is shown with increasing time in the downward vertical direction. The amount of the sweep voltage is shown horizontally with 0 V at the center of the screen display, positive voltage to the right of center, and negative voltage to the left of center. If the beam deflection is considered due to the sweep signal alone, the voltage starting off as negative deflects the beam to the left side of the screen. As the sweep voltage gets less negative the beam will move toward the center. The sweep voltage will pass through zero and go to some positive value at which time the beam will be deflected to the right side of the screen. (Note that the vertical deflection is not being considered for the moment.) When the sweep voltage reaches the largest positive voltage shown (point 9), it very quickly drops back to a negative value and the beam moves back to the starting point on the left of the screen.

The action of the sweep voltage, then, is to move the electron beam across the screen (from left to right) at a constant rate. When we now consider the additional deflection that occurs because of the vertical input signal, we shall obtain the actual display shown in Fig. 18.9. It should be clearly understood that the vertical signal applied alone (without horizontal sweep) results in a straight vertical line on the screen. The linear (horizontal) sawtooth sweep voltage must also be applied to cause the beam to move across the CRT so that a display of the input signal is obtained.

For the example shown in Fig. 18.9, the time for a complete cycle of the vertical signal and for a cycle of the sawtooth sweep signal are the same. When this is true, the display is a single cycle of the input signal. The horizontal sweep speed, however,

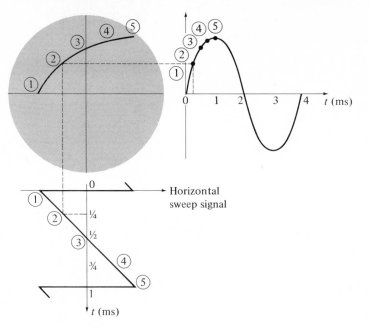

Figure 18.10 Time base at higher frequency than input signal—magnification of displayed signal.

is adjustable and need not be exactly the same as the input signal to be viewed. Figure 18.10 shows the resulting display if the sweep speed is faster than the input signal. In the example shown the time for 1 cycle of the input signal is 4 ms, whereas the time for 1 cycle of the sawtooth signal is only 1 ms. In this case the beam is moved across the tube in 1 ms during which time the vertical input is deflected from zero up to the maximum positive point as shown in the figure. Only a part of the input signal is shown in this case. If the time base of the sawtooth sweep is adjusted even shorter, the amount of the signal displayed will be an even shorter portion. This is an effective *magnification* of the signal, since the full screen width can now display a smaller portion of the input signal. For pulse-type computer signals such magnification is extremely important. It may also be necessary to view more than one full cycle of the input signal. In this case the sweep speed is made slower so that it takes longer for the sweep beam to be deflected once across the screen, thereby allowing a number of cycles of the input signal to be displayed. Figure 18.11 shows the resulting display when the sweep signal takes 16 ms and the input signal only 4 ms for one full cycle. In this case the horizontal sweep speed is 4 times slower and the display shown is 4 cycles of the input signal.

EXAMPLE 18.1   A sinusoidal waveform at a frequency of 4 kHz is applied to the vertical input of a scope. The horizontal sweep speed is set so that a full cycle takes 0.5 ms. Describe the resulting display for one sweep of the beam. (b) Repeat part (a) for a sweep frequency of 8 kHz.

Solution:
(a)   At a frequency of 4 kHz the time for one full cycle of the input signal is

$$T = \frac{1}{f} = \frac{1}{4 \times 10^3} = 0.25 \times 10^{-3}\ \text{s} = 0.25\ \text{ms}$$

During the sweep time of 0.5 ms **the input signal will go through 2 full cycles.**

Ch. 18   CATHODE-RAY OSCILLOSCOPE (CRO)

(b) At a sweep speed of 8 kHz the time for 1 cycle of sweep is

$$T = \frac{1}{f} = \frac{1}{8 \times 10^3} = 0.125 \text{ ms}$$

The sweep period is only $\frac{1}{2}$ the period of the 4-kHz signal. Therefore, only $\frac{1}{2}$ **cycle of the input signal will appear.**

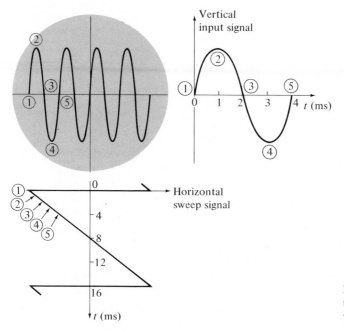

Figure 18.11 Time base slower than input signal—many cycles viewed in one sweep.

18.4 SYNCHRONIZATION AND TRIGGERING

Synchronization

The CRO display can be adjusted by setting the sweep speed to display either a number of cycles, 1 cycle, or part of a cycle. This is a very valuable feature of the CRO and helps make it the useful instrument it is. However, in discussing the sweep of the beam for a single cycle, we have only considered the case in which the input signal and sweep-signal frequencies are integer (whole number) related. More generally the horizontal-sweep-frequency setting is not the same as or even proportional to the frequency of the imput signal. When this occurs, the display is not synchronized and either appears to drift or is not recognizable.

Figure 18.12 shows the resulting display for a number of cycles of the sweep signal. Each time the horizontal sawtooth voltage goes through a linear sweep cycle (from maximum negative to zero to maximum positive voltage), the electron beam is caused to move once horizontally across the face of the tube. The sawtooth voltage then drops very quickly back to the negative starting voltage and the electron beam is suddenly caused to move back to the left side of the screen. In most CROs the electron beam is *blanked* during this *retrace* of the beam so that no line is shown

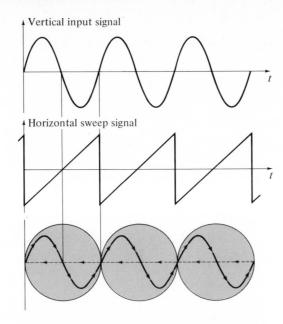

Figure 18.12 Steady scope display—input and sweep signals synchronized.

on the screen. After the very short retrace time the beam begins another sweep across the tube. If the input voltage is not the same each time a new sweep begins, the same display will not be seen each time. To see a steady display it is necessary that the input signal repeat exactly the same pattern for each sweep of the beam. In Fig. 18.13 the sweep signal frequency is too low and the CRO display will have an apparent "drift" to the left. Actually, a different display is formed each sweep of the beam as shown in Fig. 18.13. When viewed on the CROs such a display continuously drifts to the left. Observe carefully that each sweep of the beam starts at a different point in the cycle of the input signal and that a different display is formed. Adjustment of the sweep speed to a faster sweep time synchronizes or brings the display to a standstill—assuming that neither the sweep-generator frequency nor

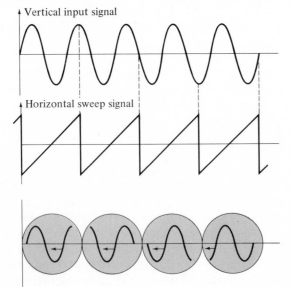

Figure 18.13 Sweep frequency too *low*—apparent drift to left.

the input-signal frequency changes. But if the sweep frequency is set too high, less than 1 cycle of the input signal will be viewed by each sweep of the beam, and the display will have an apparent drift to the right.

Triggering

The usual method of synchronizing the input signal uses a portion of the input signal to *trigger* the sweep generator so that the rate of the sweep signal is locked or synchronized to the input signal. This is easily done in most CROs using the INTERNAL sync. Fig. 18.14 shows that portion of the control panel of a CRO indicating the trigger and sync inputs and controls. We shall refer to these during the following discussion.

Figure 18.14   Scope sync and trigger controls.

When INTERNAL sync is used, a portion of the vertical input signal is taken from some point in the vertical amplifier circuit and fed as the trigger input signal to the synchronizing circuit section of the horizontal sweep generator. When triggered sweep is used, the start of a horizontal linear sweep voltage does not begin immediately after the end of the retrace time (as previously considered) but only when the triggering signal occurs. Thus, the sweep occurs not at a repetitive rate set by the cycle time of the sawtooth signal but by the cycle time of the *triggering* signal. Using the same triggering signal as that viewed on the vertical input achieves the desired synchronization without any necessary sweep-speed adjustment. Figure 18.15 shows the operation of a few cycles of the sweep signal and the triggering of the sweep generator. Since the trigger signal shown in Fig. 18.15 occurs at the beginning of each sinusoidal cycle, the display starts only when the input voltage, at the beginning of a positive-going cycle, triggers the sweep generator, and a steady picture is thereby obtained.

The exact triggering point at which the sweep begins can be adjusted using front-panel controls (see Fig. 18.14). Setting the trigger control to *INT+*, for example, means that the triggering signal to the sweep circuits will be obtained when the vertical input to the CRO has positive slope (voltage getting more positive with time). When the position marked *INT−* is used, the sweep will be started during the time the input signal is going more negative (having negative slope). Figure 18.16 shows the display of the same sinusoidal signal for INT+ and INT− trigger settings, respectively. In addition to these two control settings, the triggering *level* may also be adjusted. Setting the level to zero results in the display's starting when the input signal level crosses 0 V. This level can be adjusted so that the sweep is triggered to

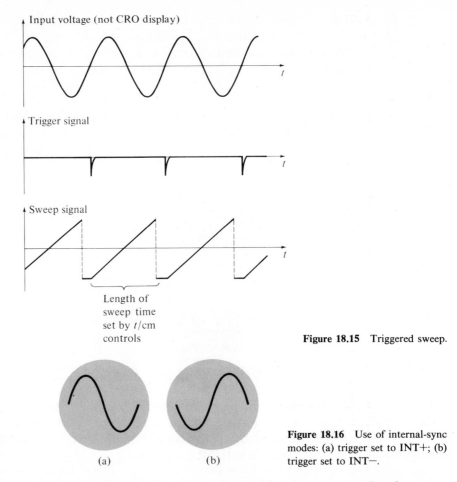

Figure 18.15 Triggered sweep.

Figure 18.16 Use of internal-sync modes: (a) trigger set to INT+; (b) trigger set to INT−.

start at *any* point of voltage during either the positive-slope or negative-slope part of the cycle. Using both trigger level and trigger slope on plus (+) or minus (−) allows a wide range of trigger-time adjustment for synchronization of a given waveform.

The controls shown in Fig. 18.14 also allow two other sync modes of operation. The *LINE* sync provides triggering at the line frequency rate (60 Hz) for measurements of signals derived from the main power line. An equally important sync mode is the *EXT* (external) mode of synchronization. A completely separate signal from that applied to the vertical input can be applied to the input terminals marked EXT. This external input signal is then used to trigger the CRO. In many applications a signal taken from a point in the circuit being tested is used as the external sync signal. Thus, any measurements made using the CRO at any other point in the same circuit will be synced by a signal having the same frequency rate. More important, these other signals may be out of phase with the sync signal and this relative phase difference will be both displayed and *measurable* using the CRO. The use of the scope both to display and allow measurement of phase difference (or more generally of time displacement) between signals is important and will be covered in detail in Section 18.4.

Synchronization of the CRO deflection is obtained by triggering the start of a new sweep cycle using the signal to be displayed (or some other signal derived from it) as the trigger signal. Figure 18.17 shows a simple block diagram of how the trigger signal and sweep signal are related. Either a part of the vertical input signal to be viewed (INT) or a separate external input signal (EXT) is connected as the trigger input signal. The *coupling* to the trigger circuit (ac or dc), the *slope* of the signal selected (+ or −), and the *level* at which the trigger signal is set to operate are all adjustable using the CRO external controls (see Fig. 18.18). The trigger signal so determined is then used to trigger the gate generator circuit. This causes the gate circuit to begin a timing operation providing an output pulse, which begins when the trigger pulse is received and ends after the amount of time for a single sweep of the beam, which is determined by the external time-base settings (e.g., 1

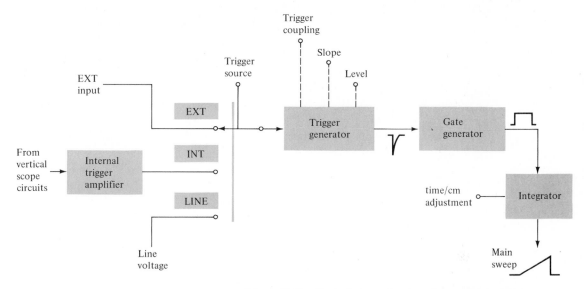

Figure 18.17 Block diagram showing trigger operation of scope.

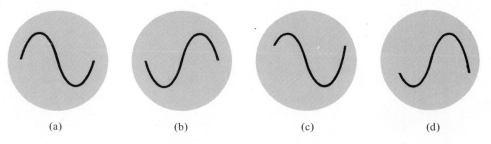

Figure 18.18 Triggering at various points of signal level (*note:* sign starts at same point in cycle each sweep and is therefore synchronized): (a) positive-going zero level; (b) negative-going zero level; (c) positive-voltage trigger level; (d) negative-voltage trigger level.

ms, 10 μs, etc.). A sweep voltage used to drive the horizontal deflection plates is derived from the pulse gate generator signal using an integrator circuit.

MULTITRACE OPERATION

There are CROs available that through *dual beam* CRTs or the use of external electronic circuitry have the ability to display two signals during the same time interval. The dual-beam CRO employs a CRT that has a second electron gun to generate the additional electron beam. The *dual-trace* CRO employs electronic circuitry to switch a single electron beam between two signals at a rate that will result in two clear displays on the face of the tube. Two dual-trace features are the ALTERNATE and the CHOPPED modes of display.

When the ALTERNATE mode of electronic switching is used, the two input signals to be viewed (connected to channels *A* and *B* of the scope to differentiate between the two vertical input signals) are connected to the vertical channel of the scope for alternate cycles of the sweep. Thus, on 1 cycle of the sweep the input to channel *A* is connected to the vertical section of the scope and drives the vertical-deflection plates. After 1 cycle of sweep is completed, the electronic switching circuits (see Fig. 18.19) connect the channel *B* input to the vertical section of the scope and on the next sweep of the electron beam the *B*-channel signal is displayed on the screen. If (as is usually the case) the sweep times are fast enough, the alternate sweep of the beam will trace out a second display *before* the first display has disappeared. Thus, two displays will appear to be present at the same time. Figure 18.20a

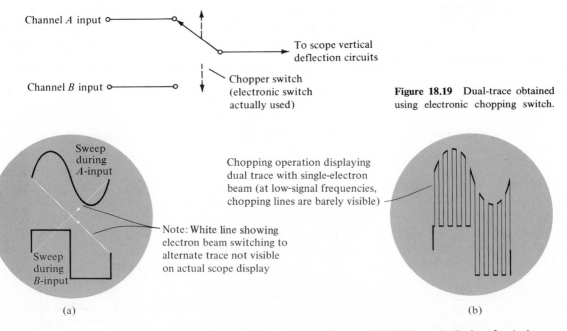

Channel *A* input

To scope vertical deflection circuits

Channel *B* input

Chopper switch (electronic switch actually used)

Figure 18.19 Dual-trace obtained using electronic chopping switch.

Sweep during *A*-input

Chopping operation displaying dual trace with single-electron beam (at low-signal frequencies, chopping lines are barely visible)

Note: White line showing electron beam switching to alternate trace not visible on actual scope display

Sweep during *B*-input

(a)

(b)

Figure 18.20 ALTERNATE and CHOPPED mode displays for dual-trace operations: (a) ALTERNATE mode for dual-trace using single electron beam; (b) CHOPPED mode for dual-trace using single electron beam.

shows the resulting display for a sine wave and square wave applied as inputs to the CRO when using the ALTERNATE mode of presentation.

When the signals to be viewed are of low frequency, a CHOPPED mode of display is used. The CHOPPED mode of electronic switching switches the vertical amplifier input signal from the input connected at channel *A* to the signal at channel *B* and back again, repetitively—many times—for a single sweep of the beam. Thus, the display obtained is actually comprised of small pieces of each signal with enough of these small pieces to provide the illusion of two steady display signals. Figure 18.20b shows the resulting display of two signals using the CHOPPED mode of electronic switching.

An important point to keep in mind in regard to the two modes of display is the triggering operation for each mode. In CHOPPED operation a single trigger signal starts each sweep, which then provides two displays on the screen. Whatever phase displacement exists between the two signals will be displayed *exactly,* since the two inputs are being shown, in time, as they occur. The ALTERNATE mode of display, however, may provide some problem. For example, the INTERNAL mode of trigger is used with ALTERNATE sweep and, say, the trigger setting is the zero voltage level with positive slope. When the channel *A* signal crosses 0 V, with positive slope, a trigger signal starts the sweep. The ALTERNATE sweep will be when the channel *B* signal crosses 0 V with positive slope—*whenever* that occurs. The display will thus be that of two in-phase signals and the true phase displacement between the signals is *not* shown. To get around this difficulty, EXTERNAL sync may be used so that each sweep is started by the same input signal and any phase offset from *that* reference signal will be observed. Using one of the inputs as the EXTERNAL sync-signal reference provides a display of the two signals in proper phase relation. Some newer multitrace CROs provide additional triggering for ALTERNATE mode display using the same input from, say, channel *B* as the trigger for initiating the sweep of both input signals. The ALTERNATE *B*-trigger mode, for example, uses the *B* input as the trigger for *each* sweep (for both channel-*A* and channel-*B* display). Any phase shift displayed between the two signals will then be properly shown.

18.5 MEASUREMENTS USING CALIBRATED CRO SCALES

The CRT face has two display axes—vertical and horizontal. In the normal operation of the CRO the input signal to be observed is applied to the vertical input (either single-channel or dual-channel) and the horizontal sweep is obtained using internal sweep circuitry (see Fig. 18.21). If the vertical amplifiers and attenuators are set to the calibrated settings, then the *amplitude* of the vertical (input) signal (or any part of it) can be accurately *measured*.

The term *calibrated* implies that the scope has been checked to ensure that each scale setting properly reflects the actual indication on the scope. For example, if the 1-V/cm setting is chosen, then the scope will, in fact, ensure a 1-cm movement of the display for each volt applied.

If the calibrated horizontal sweep scales of the CRO are used, the amount of time for 1 cycle of the input signal (the *period*) may be measured and used to calculate the signal frequency. In addition, the amount of time between two sinusoidal signals

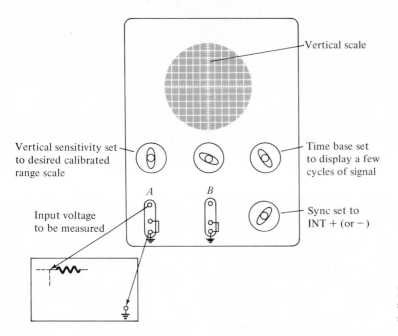

Figure 18.21 Connections to measure voltage amplitude using the scope.

crossing 0 V can be read and used to calculate the phase shift between the two signals. It is also possible to use the horizontal scales to measure the amount of time that separates the two signals being observed. Measurements of this type provide important information in pulse and digital circuitry.

Amplitude Measurements

The vertical scale of the scope is generally calibrated in units of volts per centimeter (V/cm). Figure 18.22 shows the typical CRT screen of a CRO with the vertical scale marked off in centimeters (cm). (The centimeter is a full box as indicated and the scale is sometimes referred to as volts per box.) Each centimeter or box is further subdivided into five parts so that each minor division mark represents 0.2 cm.

The scope calibrated scale can be used to measure voltage amplitude. For instance, the peak amplitude of the sinusoidal waveform shown in Fig. 18.22 can be read off the CRT screen as 3.6 cm. If the scale selected by the vertical sensitivity dial setting

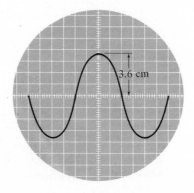

Figure 18.22 Scope scale for amplitude measurement: (a) scope scales; (b) measurement of the peak amplitude of a sinusoidal waveform.

is 1 V/cm, this reading would represent 3.6 cm × 1 V/cm = 3.6 V. If the vertical sensitivity dial setting were 0.1 V/cm, then the voltage amplitude would be 3.6 cm × 0.1 V/cm = 0.36 V.

The vertical and horizontal *position* of the displayed waveform may be adjusted without affecting the amplitude value. *It is important, however, that the vernier part of the sensitivity dial be set to the calibrated* (CAL) *position.* For the measurement considered above, the peak value of the waveform was measured with respect to the center line of the CRT screen. For this to correspond to the center of the waveform the vertical position of the beam would have had to be centered previously. When centering is *not* desired, the measurement of peak-to-peak amplitude is a more reasonable and accurate choice. The pulse-type waveform shown in Fig. 18.23a has a peak-to-peak amplitude of 4.6 cm so that a dial setting of 1 V/cm would indicate a voltage amplitude (peak-to-peak) of 4.6 cm × 1 V/cm = 4.6 V.

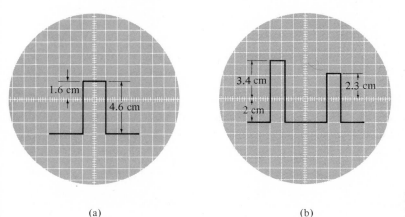

(a)

(b)

Figure 18.23 Measurement of pulse-type waveform amplitudes.

Figure 18.23b shows another pulse-type waveform. To find the amplitude (voltage) difference between the peaks of the two pulses shown the measurement is

$$\text{1st peak} = 3.4 \text{ cm}$$

$$\text{2nd peak} = 2.3 \text{ cm}$$

$$\text{pulse amplitude difference} = 3.4 - 2.3 = 1.1 \text{ cm}$$

If the scale setting were 10 V/cm, this would correspond to a voltage difference of 1.1 cm × 10 V/cm = 11 V.

> **EXAMPLE 18.2** A sinusoidal waveform is observed on a scope as having a peak-to-peak amplitude of 6.4 cm. If the CRO vertical sensitivity setting is 5 V/cm, calculate the peak-to-peak and the rms values of the voltage.
>
> **Solution:** The peak-to-peak amplitude is
>
> $$6.4 \text{ cm} \times 5 \text{ V/cm} = 32 \text{ V} = V_{p\text{-}p}$$
>
> $$V_p = \frac{V_{p\text{-}p}}{2} = \frac{32 \text{ V}}{2} = 16 \text{ V}$$

The rms value of voltage is

$$V_{rms} = 0.707 V_p = 0.707(16) = \textbf{11.31 V}$$

EXAMPLE 18.3 A pulse-type waveform is measured as having a peak amplitude of 15 V. If the vertical dial setting was 10 V/cm, how many centimeters of signal amplitude were observed (peak-to-peak)?

Solution: The number of centimeters for a peak reading of 15 V is

$$15 \text{ V} \times \frac{1 \text{ cm}}{10 \text{ V}} = 1.5 \text{ cm}$$

The peak-to-peak amplitude is 2(1.5 cm) = **3 cm.**

Most vertical-sensitivity dials indicate separate positions for ac and dc readings. (On many other quality CROs there is one set of sensitivity scales with a separate switch to change from ac to dc operation.) The difference between these two modes of measurement is simple but important in using the scope properly. The *dc input* results in the displayed waveform showing the dc level of the signal being measured. If, for example, the signal to be measured is a 2-V, peak-to-peak sinusoidal voltage riding on a 3-V dc level as shown in Fig. 18.24a (assuming the scope position controls were previously centered), the display indicates the presence of dc. We are then able to measure not only the ac variation of the signal, but also the exact dc levels at all parts of the signal, as shown in Fig. 18.24a.

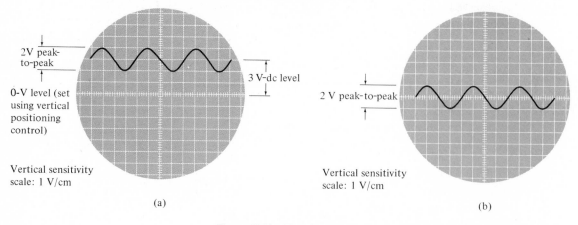

(a)   (b)

Figure 18.24 Use of dc and ac input modes: (a) dc input mode; (b) ac input mode.

When only the ac variation is of interest, then the ac scale setting may be used. When the ac input setting is used, the same input signal shown in Fig. 18.24a is displayed in Fig. 18.24b. Notice that the dc level has been removed and only the ac variation is shown. Essentially, the difference between the two is that the ac input scale position couples the signal through a capacitor to eliminate the dc level of the input signal and provide only the ac variation for measurement.

A practical example of the advantage of the ac over the dc setting is in the measurement of a signal having, say, a dc level of 80 V and an ac variation around this level of only 4 V. Imagine the scope display when using a dc scale setting of 20 V/cm. Consider how small the ac variation is compared to the large dc level of 80 V. (If a smaller scale setting of, say, 2 V/cm were used, the ac signal would not be visible at all since a deflection for 80 V would be well off the face of the screen.) Using an ac input scale setting of, say, 1 V/cm would provide a display in which only the ac part of the signal would be visible and be expanded to a reasonable viewing and measuring amplitude.

Although the ac scales are good for observing the ac part of a signal having a dc level, the dc scale setting is still important when dc levels must be measured. One additional feature is often found in a good scope—a zero voltage or GROUND position, which connects the input of the vertical amplifier to 0 V without requiring the input signal connection to be removed.

USING CALIBRATED SWEEP FOR TIME MEASUREMENTS

The horizontal sweep signal can be adjusted in calibrated steps from a few seconds to microseconds of time per centimeter. If the sweep-time selector were set at 1 ms/cm, each box (or centimeter) on the screen would correspond to a time of 1 ms. Thus, the CRO allows display of waveforms of all shapes and permits measurements of time so that all aspects of the signal observed can be measured accurately.

With the dual-trace feature of the CRO two different waveforms can be observed simultaneously and any time differences between parts of the respective signals may be observed and measured. For example, the waveforms of Fig. 18.25 show two pulses viewed at once (using, say, external sync to preserve their proper time displacement). If we use the calibrated horizontal sweep scales, we can measure, for example, the time for each pulse and the time between the two pulses. From the waveforms shown the pulse widths are calculated as

$$1.2 \text{ cm} \times 20 \text{ } \mu\text{s/cm} = 24 \text{ } \mu\text{s}$$

$$2.3 \text{ cm} \times 20 \text{ } \mu\text{s/cm} = 46 \text{ } \mu\text{s}$$

The time from the start of the first pulse until the start of the second is calculated as

$$2.1 \text{ cm} \times 20 \text{ } \mu\text{s/cm} = 42 \text{ } \mu\text{s}$$

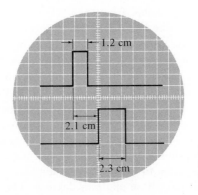

Figure 18.25 Pulse waveforms showing time measurement using calibrated sweep scales.

Using the scope in this way provides time period or interval information about the two waveforms and their relation to each other.

> **EXAMPLE 18.4** Two pulse-type signals are observed on a CRO using a sweep-scale setting of 5 ms/cm. If the pulse widths are measured as 3.5 and 4.2 cm, respectively, and the distance between the start of each pusle is 2.8 cm, calculate the time measurements for the pulse widths and delay between pulses.
>
> **Solution:** Pulse width 1 = 3.5 cm × 5 ms/cm = **17.5 ms**
> Pulse width 2 = 4.2 cm × 5 ms/cm = **21.0 ms**
> Delay time = 2.8 cm × 5 ms/cm = **14.0 ms**

Frequency Measurements Using Calibrated Scope Scales

It is also possible to use the calibrated time scales of the CRO to calculate the *frequency* of the observed signals. This requires using the calibrated horizontal sweep scale to measure the time for 1 cycle of the observed signal and then calculating the signal frequency using the relation

$$f = 1/T \tag{18.2}$$

where f is the signal frequency and T is the period or the time for one full cycle of the signal.

> **EXAMPLE 18.5** A square-wave signal is observed on the scope to have 1 cycle measured as 8 cm at a scale setting of 20 μs/cm. Calculate the signal frequency.
>
> **Solution:** Calculating, first, the period for 1 cycle of the observed signal:
>
> $$T = 8 \text{ cm} \times 20 \text{ } \mu s/cm = 160 \text{ } \mu s$$
>
> The frequency is then calculated to be
>
> $$f = \frac{1}{T} = \frac{1}{160 \text{ } \mu s} = \textbf{6.25 kHz}$$

Phase-Shift Measurements Using Calibrated Scope Scales

The calibrated time scales can also be used to calculate phase shift between two sinusoidal signals (of the same frequency, of course). If a dual-trace or dual-beam CRO is used to display the two sinusoidal signals simultaneously so that one signal is used for the EXT sync input, the two waveforms will appear in proper time perspective and the CRO can be used to measure the amount of time between the start of 1 cycle of each of the waveforms. This amount of time can then be used to calculate the phase angle between the two signals. Figure 18.26 shows two sinusoidal signals having a phase shift of theta (θ) degrees. We can measure distance in centimeters on the CRO scale and use these readings to obtain θ as follows. The value of the phase angle is related to the degrees in one full cycle of the sinusoidal signal. We can set up a simple relation between these values by equating the number of centimeters or boxes for one full cycle to 360° and the number of centimeters or boxes for the phase shift to the desired phase angle in degrees. The relation is

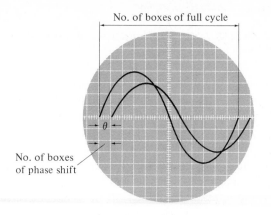

No. of boxes of full cycle

No. of boxes
of phase shift

θ

$$\theta = \frac{\text{No. of boxes of phase shift}}{\text{No. of boxes of full cycle}} \times 360°$$

Figure 18.26 Phase-shift measurement using horizontal scope scale.

$$\frac{\text{No. of boxes for one full cycle}}{360°} = \frac{\text{No. of boxes of phase shift}}{\theta} \qquad (18.3)$$

where θ is the phase angle (phase shift) in degrees. Using Eq. (18.3), we can calculate the phase angle from

$$\boxed{\theta = \frac{\text{No. of boxes of phase shift}}{\text{No. of boxes for full cycle}} \times 360°} \qquad (18.4)$$

Note that the calculation does not involve the actual calibrated time base setting and, in fact, the observed waveform can be varied using the horizontal amplifier vernier adjustment to obtain as many boxes for one full cycle as desired. This will not affect the actual phase shift calculated since the proportionality between the phase shift and one full cycle is preserved at any gain or scale setting used. Adjusting the time base so that, say, 12 boxes (or centimeters) correspond to one full cycle would rescale the measurement so that each box is 30° of phase angle (360°/12). Then, reading 2 boxes of phase shift would quickly convert to 60° phase shift by multiplying the boxes of phase shift by the 30°/box scale factor. If the reading obtained were 1.5 boxes, the phase shift would be 1.5 boxes × 30°/box = 45° phase shift. The relation for calculating phase angle is then

$$\boxed{\theta = \text{scale factor} \times \text{phase distance measured}} \qquad (18.5)$$
$$\text{(in boxes or centimeters)}$$

EXAMPLE **18.6** In measuring phase shift between two sinusoidal signals on a CRO the scale setting is adjusted so that one full cycle is 8 boxes. Calculate the scale factor for this adjustment and the amount of phase shift for a reading of 0.75 box.

Solution: Setting one full cycle to 8 boxes results in a scale factor of 360°/8 box = 45°/box.

$$\theta = \text{scale factor} \times \text{phase distance measured}$$

$$= \frac{45°}{\text{box}} \times 0.75 \text{ box} = \textbf{33.75°}$$

18.6 USE OF LISSAJOUS FIGURES FOR PHASE AND FREQUENCY MEASUREMENTS

Another method for measuring either phase shift between two sinusoidal signals or the frequency of an unknown sinusoidal signal is the use of Lissajous figures. The technique can be applied to a single-channel CRO and does not require the fine calibration scales previously considered. Basically, the two signals under study (to determine the phase shift between the two signals) are connected as vertical and horizontal inputs to the CRO. The usual (internal) horizontal sweep signal is *not* used at this time. A pattern or Lissajous figure is developed on the CRT and is used to determine the amount of the phase shift or frequency of the unknown signal.

Lissajous-figure techniques for measurement are more popular for low-quality CROs and for single-trace scopes (where two inputs cannot be compared at one time). Although not as popular a measurement technique as those considered in Section 18.5, the use of Lissajous figures is still interesting and sometimes helpful.

Use of Lissajous Figures To Calculate Phase Shift

Lissajous figures are obtained on the scope by applying the two sinusoidal inputs to be compared to the vertical and horizontal channels of the oscilloscope. The value of the phase shift is then calculated from measured values taken from the resulting Lissajous pattern.

Figure 18.27 shows a circuit arrangement with two different sinusoidal signals connected to the vertical and horizontal inputs of the CRO, respectively. The resulting pattern on the CRT face is a straight line, a circle, or an ellipse. To understand the relationship between the resulting figure and the applied inputs we shall investigate a variety of signals having different phase angles and determine the resulting Lissajous pattern.

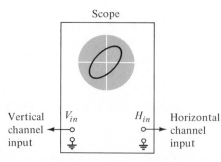

Figure 18.27 Signal-input connection for Lissajous-figure measurement of phase angle.

The procedure for measuring phase difference between two sinusoidal signals is simply to apply the two signals to the vertical and horizontal inputs and then make two measurements from the resulting display. The usual horizontal linear sweep is not used at all for this procedure. Note that the two signals must be the same frequency (otherwise the parameter phase angle is meaningless).

Figure 18.28 shows the resulting waveform for two inputs having a phase shift between 0° and 90°. The resulting pattern is an ellipse (at 45° if the two amplitudes are the same). The angle at which the ellipse is generated is of no importance for

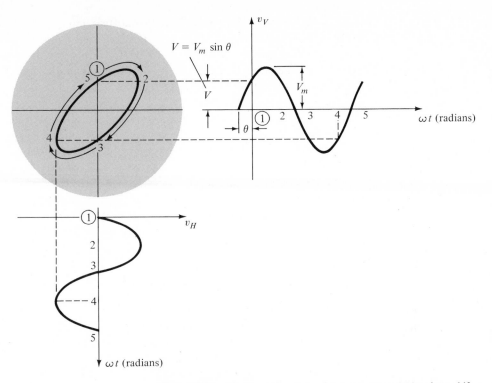

Figure 18.28 Lissajous figures for θ between 0° and 90° phase shift.

the phase angle calculation. Noting that the vertical signal amplitude at time 1 is $V = V_m \sin \theta$, we can calculate the angle θ from

$$\theta = \sin^{-1}(V/V_m) \qquad (18.6)$$

The values of V and V_m can be easily obtained from the ellipse by measuring the distance (amplitude) of the signal from the center line to where it crosses the center vertical axis, and V_m as the distance from the vertical center line to the top of the ellipse. Using these values in the above relation, we can calculate the phase angle θ.

Since the measurements of V and V_m will be used as a ratio, the actual size or values are not important—only their ratio. This being the case, the actual scale settings

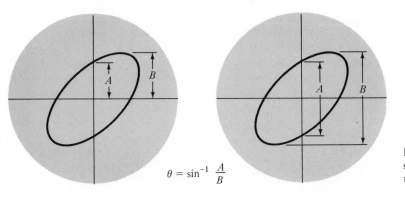

$$\theta = \sin^{-1} \frac{A}{B}$$

Figure 18.29 Calculation of phase shift from Lissajous figure for θ between 0° and 90°.

of the input signals are unimportant and the CRO adjustments may be used to get an ellipse on the CRT face of about maximum possible size for greatest accuracy. Once the deflection controls are properly centered by adjusting the beam to the center of the tube with no input signals, the two measurements marked A and B in Fig. 18.29 are read and the angle θ is calculated from

$$\theta = \sin^{-1}\left(\frac{A}{B}\right) \tag{18.7}$$

If, for example, the distance B is set to 10 boxes (for whatever scale setting and fine adjustments are necessary), the number of boxes of A can be read and the ratio of A/B obtained. The actual voltages are not important, only their ratio as obtained by the number of scale divisions or boxes on the CRT face.

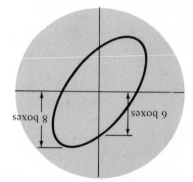

Figure 18.30 Lissajous figure for Example 18.7.

EXAMPLE 18.7 Calculate the phase shift θ for the Lissajous figure in Fig. 18.30.

Solution: $$\theta = \sin^{-1}\frac{A}{B} = \sin^{-1}\frac{6}{8} = \sin^{-1} 0.75 = \mathbf{48.59°}$$

If the two signals are out of phase by exactly 90°, the resulting waveform is a circle. This result also follows from the above calculation, since for a circle the measured values A and B are equal and the value of θ calculated is $\theta = \sin^{-1}(1) = 90°$. The relation for calculating θ would also show that for the straight line the measured value of $A = 0$ gives $0/B = 0$, and $\theta = \sin^{-1}(0) = 0°$. Thus, in summary,

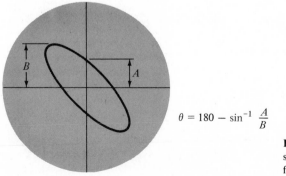

$$\theta = 180 - \sin^{-1}\frac{A}{B}$$

Figure 18.31 Calculation of phase shift from Lissajous figure for θ from 90° to 180°.

Ch. 18 CATHODE-RAY OSCILLOSCOPE (CRO)

the values A and B can be measured as indicated above and used to calculate the phase angle within the range of 0 and 90°.

For the phase angles of 90–180° the ellipse has a negative slope, as in Fig. 18.31, and the angle calculated by the above method must be subtracted from 180° to obtain the phase shift. Phase angles above 180° result in Lissajous figures such as those below 180°, and they cannot be directly distinguished. One technique for determining if the measured angle is less or more than 180° is to add an extra (slight) phase shift to the signal being measured. If the phase angle measured increases, the angle was less than 180°. If it decreases, the angle was greater than 180°, and the correct angle is then calculated by adding 180° for the angle computed with negative-sloped ellipse. A comprehensive summary, which clearly shows the required methods to compute the phase angle, is shown in Fig. 18.32.

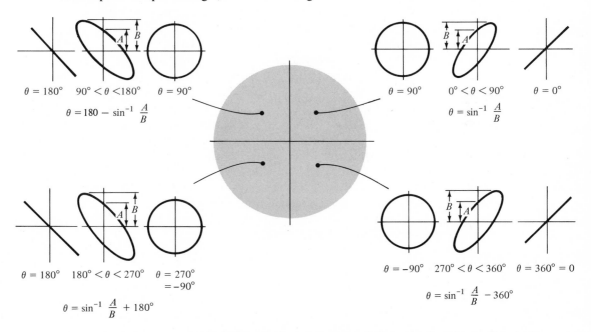

$\theta = 180°$ $90° < \theta < 180°$ $\theta = 90°$

$$\theta = 180 - \sin^{-1} \frac{A}{B}$$

$\theta = 90°$ $0° < \theta < 90°$ $\theta = 0°$

$$\theta = \sin^{-1} \frac{A}{B}$$

$\theta = 180°$ $180° < \theta < 270°$ $\theta = 270°$
$= -90°$

$$\theta = \sin^{-1} \frac{A}{B} + 180°$$

$\theta = -90°$ $270° < \theta < 360°$ $\theta = 360° = 0$

$$\theta = \sin^{-1} \frac{A}{B} - 360°$$

Figure 18.32 Lissajous phase-angle calculation in all quadrants. *Note:* Additional test adding phase-angle shift to signal to be measured is necessary to determine whether observed figure is for upper two quadrants or lower two quadrants. If added phase shift causes positive-sloped ellipse to become larger or negative-sloped ellipse to become smaller, upper quadrants are indicated; otherwise, lower quadrants.

EXAMPLE 18.8   Calculate the phase angle (between 0° and 180°) for the following Lissajous figures shown in Fig. 18.33.

Solution:
(a) $\theta = \sin^{-1} (A/B) = \sin^{-1} (3/5) = \mathbf{36.87°}$
(b) $\theta = \sin^{-1} (A/B) = \sin^{-1} (6/6) = \mathbf{90°}$ (could have been determined by inspection—circle indicates phase shift of 90°)
(c) $\theta = 180° - \sin^{-1} (A/B) = 180° - \sin^{-1} (4/5) = 180° - 53.13° = \mathbf{126.87°}$
(d)   By inspection, $\theta = \mathbf{180°}$

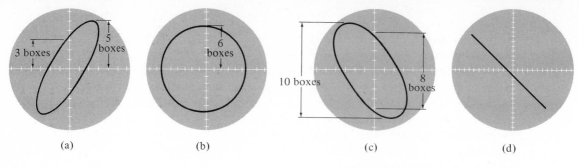

Figure 18.33 Lissajous figures for Example 18.8.

18.7 SPECIAL CRO FEATURES

The CRO is becoming increasingly more sophisticated and specialized in use. Whereas CROs were originally general in range and usage, modern CROs can be geared specifically to the field or area of interest providing those measurements of importance to a particular area of electronics. One important feature of many CROs is the use of *plug-ins* as shown in Fig. 18.34. Rather than manufacture a single integral unit with certain features of interest to a particular area or range of operation, the CRO is manufactured with only the power supply and deflection circuitry as integral to the unit. Plug-in units are available for both the vertical and horizontal sections of the CRO. These plug-in units may be selected to have features such as the following: single input ranging from 0.005 V/cm to 20 V/cm scale sensitivity; dual-trace capability with two vertical inputs and selection of channel *A* only, channel *B* only, CHOPPED, and ALTERNATE modes of display; differential input for two separate signals; dc-50-MHz frequency range; and dc input from 100 μV to 20 V. Time base plug-ins might provide single time base from 1 μs to 1 s/cm selection; two time bases allowing delayed and mixed sweep modes of operation; time base with sampling operation allowing viewing down to a few nanoseconds, etc.

A useful CRO feature uses two time bases to provide a selection of a small part of the signal viewed allowing expanded presentation of only that selected part of the signal. Figure 18.35 shows a digital-type signal containing a number of separate pulses. A delayed type of sweep presentation would be necessary if it were desired

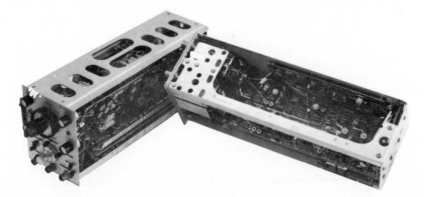

Figure 18.34 Plug-in scope attachments.

to view, say, just the third pulse in Fig. 18.35. If the usual single time base were expanded by changing the sweep generator rate so that a shorter period of time were viewed on the screen, this would only magnify the signal shown in Fig. 18.35 from the left (start of the sweep) out; that is, the display would be expanded for the part of the signal starting at the beginning of the sweep, but the part of the signal over to the right would be out of the screen area for this more detailed sweep setting. The delay sweep feature to be discussed allows selecting a part of the displayed presentation and then changing the display to show only a selected portion at whatever magnification is desired.

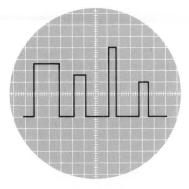

Figure 18.35 Digital-type pulse viewed on scope.

The main time base is referred to as the A time base and is the horizontal sweep, which provides the picture as shown in Fig. 18.35. An additional time base sweep generator is also provided (called B time base) when delayed sweep operation is available. A basic description of this delayed sweep operation is shown in Fig. 18.36. The pulse-type signal shown in Fig. 18.36a indicates a selected part of that signal by the more intensified display. This part of the signal is then shown in Fig. 18.36b in a more detailed (more magnified) presentation. The use of a delayed sweep time base allowed this selection. The detailed presentation in the block diagram of Fig. 18.36c shows how the circuitry is connected to accomplish the above. Note in Fig. 18.36c that the main and delayed sweep circuitry are two approximately identical units. Their connection in the overall CRO operation is what differentiates them. If the front panel control is set to main sweep, only the main sweep generator is activated and the input trigger signal is connected to the main sweep trigger circuit, resulting in the main sweep waveform being used as the horizontal deflection signal. When delayed sweep operation is set up, an additional sweep generator is activated—this is the delayed sweep circuit. For operational ease, the CRO is set up so that the sweep of the delayed generator can be added to that of the main generator to provide the intensified display shown in Fig. 18.36a. In effect, then, a potentiometer adjustment is provided on the front panel controls to set the sweep comparator of Fig. 18.36c so that at some selected time in the main sweep the delayed sweep circuitry will be activated (triggered). The amount of time that the delayed sweep is generated is independently set by the delayed sweep time base controls on the scope front panel. This time setting is always less than that of the main sweep. Thus, an intensified presentation is obtained with some adjustment for when the intensified part of the display begins and for how long it lasts.

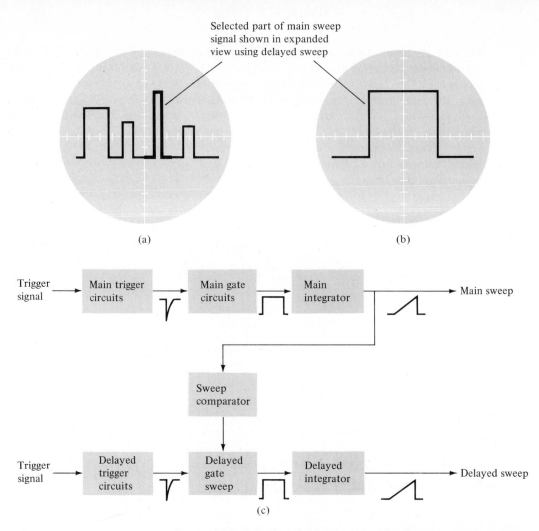

Figure 18.36 Operation of delayed sweep: (a) main sweep presentation; (b) delayed sweep presentation; (c) delayed sweep operation—block diagram.

The CRO mode of operation can now be changed from main time base to delayed time base operation. When this is done, the displayed signal is then only the portion previously shown as the intensified section. What has happened is that the main sweep no longer drives the horizontal deflection circuitry—but it still is used to trigger the delayed sweep as before. Thus, the delayed sweep does not start until the delayed time interval previously seen on the screen as the start of the intensified display. When this time in the signal operation occurs, the delayed sweep now drives the horizontal deflection circuitry and a display (at the delayed generator sweep rate) is provided. This display, as shown in Fig. 18.36b, shows only the previously intensified part of the signal and shows it at the expanded display setting of the delayed sweep generator.

Figure 18.37 shows the sweep waveforms for the main and delayed circuitry to show the resulting operation as described. As shown in Fig. 18.37a, the amount of the vertical input signal displayed is set by the main sweep horizontal time base. The delayed sweep time base is adjusted for some faster sweep rate providing less sweep time for a cycle, the start of the delayed sweep being set at some amount of time after the start of the main sweep. With the controls set to main sweep, the 3 pulses shown in Fig. 18.37a are displayed. Figure 18.37b shows the intensified main sweep and the intensity signal, which controls the signal intensity seen on the scope screen. (This intensity signal drives the control grid to vary the number of electrons in the electron beam.) The intensity is kept at a normally lower level and set to a higher level during the delay time interval. The display still shows the 4 pulses selected by the main sweep, but a particular part of the sweep as set by the delayed sweep circuitry is now intensified.

Finally, when the controls are switched to delayed sweep operation, the horizontal sweep is taken from the delayed sweep generator and the previously intensified presentation is now the complete screen display. It must be kept in mind that even when only the delayed sweep is operating the screen display, the signal seen is tied to that originally displayed by the main sweep and the delayed time is also based on the start of the main sweep.

To summarize, the use of the delayed sweep additional to the main sweep allows selection of a part of any displayed signal with complete and flexible control in displaying only the selected part of the signal at whatever shorter time display interval desired.

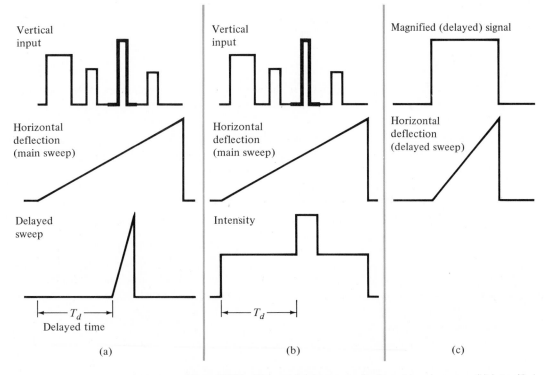

Figure 18.37 Main and delay sweep displays: (a) main sweep; (b) intensified main sweep; (c) delayed sweep.

Another special feature found in CROs having the delayed sweep operation is *mixed sweep*. This is nothing more than a mixing of the main sweep and delayed sweep signals at one time on the screen. The adjustment of the delayed trigger time sets the point at which the sweep changes from the main sweep rate to the delayed sweep rate.

PROBLEMS

§ 18.2

1. A scope CRT has a rated deflection factor of 50 V/in. Determine the deflection on the CRT screen for plate deflection voltages of: (a) +40 V, (b) −75 V.

2. What is the vertical deflection sensitivity of a CRT that has a screen deflection of 10 mm when a voltage of 50 V is applied to the vertical deflection plates?

3. A plate deflection voltage of 80 V results in a screen deflection of 2 cm. What is the tube deflection factor and how much deflection would result if 120 V were applied to the deflection plates?

4. A CRT having 0.3 mm/V sensitivity is used. How much deflection, in inches, results from a plate deflection voltage of 150 V?

5. A CRT with an accelerating potential of 10,000 V has a deflection sensitivity of 0.45 mm/V. If an applied voltage causes a deflection of 6 cm, how much deflection would result with an accelerating potential of 15,000 V? What is the applied voltage?

§ 18.3

6. A 1-kHz sinusoidal signal is fed to the vertical input of an oscilloscope. Draw the scope presentation for the following horizontal time base sweep frequencies (assume sweep triggered on positive-going slope at 0-V level): (a) 1 kHz, (b) 2kHz, (c) 500 Hz.

7. (a) A 50-kHz square wave is fed into the vertical input of a CRO. If the horizontal sweep speed is set to 2 μs/cm, draw the CRO display for a field of 10 cm on the CRT.
   (b) Repeat for a sweep speed of 4 μs/cm.
   (c) Repeat for a sweep speed of 1 μs/cm.

8. What is the CRO horizontal sweep frequency if 4 cycles of a 10-kHz signal are viewed?

§ 18.5

9. A sinusoidal signal is observed on a CRO as having a peak amplitude of 4.1 cm. If the scope vertical gain setting is 0.5 V/cm, calculate the peak and rms values of the input voltage.

10. Draw the CRO display for a 4-V, rms sinusoidal waveform for a vertical scale of 2 V/cm. Indicate vertical axis and scale markings clearly.

11. A square-wave signal measured on a CRO has a peak amplitude of 650 mV. If the CRO scale setting was 200 mV/cm, how many centimeters of signal amplitude were observed (peak-to-peak)?

12. A pulse-type signal is observed on a CRO to have a width of 6.4 cm. Calculate the pulse width time for the following sweep scale settings: (a) 5 ms/cm, (b) 100 μs/cm, (c) 2 μs/cm.

13. Two pulse-type signals are observed on a CRO at a scale setting of 20 μs/cm. If the pulse widths are measured as 1.8 and 3.2 cm, respectively, and both start at the same time, calculate the time width of each pulse and the time delay between the end of the pulses.

14. A sinusoidal signal observed on a CRO repeats 1 cycle in 6.3 cm. If the scale setting was 5 μs/cm, calculate the signal frequency.

15. Calculate the signal frequency of a square-wave signal having a width for 1 half-cycle of 10.5 cm at a scale setting of 10 μs/cm.

16. A 400-Hz signal is observed on a CRO. How many centimeters of the horizontal scale are required to observe 3 cycles of the signal if the scale setting is 1 ms/cm?

17. For the CRO display of Fig. 18.38, calculate the following: (a) peak-to-peak voltage (V_{p-p}) and V_{rms}, (b) time for one complete cycle (T), (c) frequency of waveform signal (f).

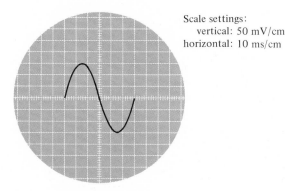

Scale settings:
vertical: 50 mV/cm
horizontal: 10 ms/cm

Figure 18.38 Waveform for Problem 17.

18. For the CRO display of Fig. 18.39 calculate the following: (a) V_{p-p}, (b) time for 2 cycles (T), (c) pulse repetition rate (f).

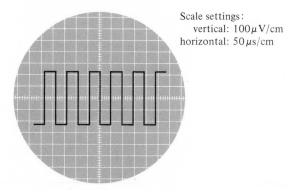

Scale settings:
vertical: 100 μV/cm
horizontal: 50 μs/cm

Figure 18.39 Waveform for Problem 18.

19. The CRO scale is adjusted so that a sinusoidal signal takes 6 cm for one full cycle. Calculate the scale factor for this adjustment and the amount of phase shift for a reading of 1.5 cm.

20. A full cycle is set to 8 cm. The phase displacement is measured as 0.5 cm. Calculate the following: (a) the scale factor and phase shift in degrees, (b) the number of centimeters observed for a phase shift of 40°.

§ 18.6

21. Draw the Lissajous figures for the following phase angles: (a) $\theta = 180°$, (b) $\theta + -90°$, (c) $\theta = 45°$, (d) $\theta = 135°$

22. Calculate the phase angle for the Lissajous figures in Fig. 18.40. (Assume that phase shift is less than 180°.)

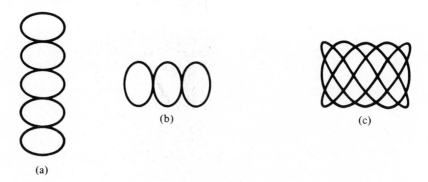

(a)

(b)

(c)

Figure 18.40 Lissajous figures for Problem 22.

A

Hybrid Parameters – Conversion Equations (Exact and Approximate)

A.1 EXACT

Common-Emitter Configuration

$$h_{ie} = \frac{h_{ib}}{(1 + h_{fb})(1 - h_{rb}) + h_{ob}\,h_{ib}} = h_{ic}$$

$$h_{re} = \frac{h_{ib}\,h_{ob} - h_{rb}(1 + h_{fb})}{(1 + h_{fb})(1 - h_{rb}) + h_{ob}\,h_{ib}} = 1 - h_{rc}$$

$$h_{fe} = \frac{-h_{fb}(1 - h_{rb}) - h_{ob}\,h_{ib}}{(1 + h_{fb})(1 - h_{rb}) + h_{ob}\,h_{ib}} = -(1 + h_{fc})$$

$$h_{oe} = \frac{h_{ob}}{(1 + h_{fb})(1 - h_{rb}) + h_{ob}\,h_{ib}} = h_{oc}$$

Common-Base Configuration

$$h_{ib} = \frac{h_{ie}}{(1 + h_{fe})(1 - h_{re}) + h_{ie}\,h_{oe}} = \frac{h_{ic}}{h_{ic}\,h_{oc} - h_{fc}\,h_{rc}}$$

$$h_{rb} = \frac{h_{ie}\,h_{oe} - h_{re}(1 + h_{fe})}{(1 + h_{fe})(1 - h_{re}) + h_{ie}\,h_{oe}} = \frac{h_{fc}(1 - h_{rc}) + h_{ic}\,h_{oc}}{h_{ic}\,h_{oc} - h_{fc}\,h_{rc}}$$

$$h_{fb} = \frac{-h_{fe}(1 - h_{re}) - h_{ie}\,h_{oe}}{(1 + h_{fe})(1 - h_{re}) + h_{ie}\,h_{oe}} = \frac{h_{rc}(1 + h_{fc}) - h_{ic}\,h_{oc}}{h_{ic}\,h_{oc} - h_{fc}\,h_{rc}}$$

$$h_{ob} = \frac{h_{oe}}{(1 + h_{fe})(1 - h_{re}) + h_{ie}\,h_{oe}} = \frac{h_{oc}}{h_{ic}\,h_{oc} - h_{fc}\,h_{rc}}$$

Common-Collector Configuration

$$h_{ic} = \frac{h_{ib}}{(1 + h_{fb})(1 - h_{rb}) + h_{ob}\,h_{ib}} = h_{ie}$$

$$h_{rc} = \frac{1 + h_{fb}}{(1 + h_{fb})(1 - h_{rb}) + h_{ob}\,h_{ib}} = 1 - h_{re}$$

$$h_{fc} = \frac{h_{rb} - 1}{(1 + h_{fb})(1 - h_{rb}) + h_{ob}\,h_{ib}} = -(1 + h_{fe})$$

$$h_{oc} = \frac{h_{ob}}{(1 + h_{fb})(1 - h_{rb}) + h_{ob}\,h_{ib}} = h_{oe}$$

A.2   APPROXIMATE

Common-Emitter Configuration

$$h_{ie} \cong \frac{h_{ib}}{1 + h_{fb}} \qquad \cong \beta\,r_e$$

$$h_{re} \cong \frac{h_{ib}\,h_{ob}}{1 + h_{fb}} - h_{rb}$$

$$h_{fe} \cong \frac{-h_{fb}}{1 + h_{fb}} \qquad \cong \beta$$

$$h_{oe} \cong \frac{h_{ob}}{1 + h_{fb}}$$

Common-Base Configuration

$$h_{ib} \cong \frac{h_{ie}}{1 + h_{fe}} \qquad \cong \frac{-h_{ic}}{h_{fc}} \qquad \cong r_e$$

$$h_{rb} \cong \frac{h_{ie}\,h_{oe}}{1 + h_{fe}} - h_{re} \cong h_{rc} - 1 - \frac{h_{ic}\,h_{oc}}{h_{fc}}$$

$$h_{fb} \cong \frac{-h_{fe}}{1 + h_{fe}} \qquad \cong \frac{-(1 + h_{fc})}{h_{fc}} \qquad \cong -\alpha$$

$$h_{ob} \cong \frac{h_{oe}}{1 + h_{fe}} \qquad \cong \frac{-h_{oc}}{h_{fc}}$$

Common-Collector Configuration

$$h_{ic} \cong \frac{h_{ib}}{1 + h_{fb}} \quad \cong \beta r_e$$

$$h_{rc} \cong 1$$

$$h_{fc} \cong \frac{-1}{1 + h_{fb}} \quad \cong -\beta$$

$$h_{oc} \cong \frac{h_{ob}}{1 + h_{fb}}$$

B

Ripple Factor
and
Voltage Calculations

B.1 RIPPLE FACTOR OF RECTIFIER

The ripple factor of a voltage is defined by

$$r \equiv \frac{\text{rms value of ac component of signal}}{\text{average value of signal}}$$

which can be expressed as

$$r = \frac{V_r\,(\text{rms})}{V_{\text{dc}}}$$

Since the ac voltage component of a signal containing a dc level is

$$v_{\text{ac}} = v - V_{\text{dc}}$$

the rms value of the ac component is

$$V_r\,(\text{rms}) = \left[\frac{1}{2\pi} \int_0^{2\pi} v_{\text{ac}}^2\, d\theta \right]^{1/2} = \left[\frac{1}{2\pi} \int_0^{2\pi} (v - V_{\text{dc}})^2\, d\theta \right]^{1/2}$$

$$= \left[\frac{1}{2\pi} \int_0^{2\pi} (v^2 - 2v\,V_{\text{dc}} + V_{\text{dc}}^2)\, d\theta \right]^{1/2}$$

$$= [V^2\,(\text{rms}) - 2V_{\text{dc}}^2 + V_{\text{dc}}^2]^{1/2} = [V^2\,(\text{rms}) - V_{\text{dc}}^2]^{1/2}$$

where $V(\text{rms})$ is the rms value of the total voltage.

For the half-wave rectified signal

$$V_r \text{ (rms)} = [V^2 \text{ (rms)} - V_{dc}^2]^{1/2}$$

$$= \left[\left(\frac{V_m}{2}\right)^2 - \left(\frac{V_m}{\pi}\right)^2\right]^{1/2}$$

$$= V_m\left[\left(\frac{1}{2}\right)^2 - \left(\frac{1}{\pi}\right)^2\right]^{1/2}$$

$$\boxed{V_r \text{ (rms)} = 0.385\,V_m \qquad \text{(half-wave)}} \qquad \text{(B.1)}$$

For the full-wave rectified signal

$$V_r \text{ (rms)} = [V^2 \text{ (rms)} - V_{dc}^2]^{1/2}$$

$$= \left[\left(\frac{V_m}{\sqrt{2}}\right)^2 - \left(\frac{2V_m}{\pi}\right)^2\right]^{1/2}$$

$$= V_m\left[\frac{1}{2} - \frac{4}{\pi^2}\right]^{1/2}$$

$$\boxed{V_r \text{ (rms)} = 0.308\,V_m \qquad \text{(full-wave)}} \qquad \text{(B.2)}$$

B.2 RIPPLE VOLTAGE OF CAPACITOR FILTER

Assuming a triangular ripple waveform approximation as shown in Fig. B.1, we can write (see Fig. B.2)

$$V_{dc} = V_m - \frac{V_r\,(p\text{--}p)}{2} \qquad \text{(B.3)}$$

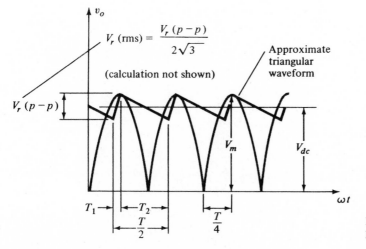

Figure B.1 Approximate triangular ripple voltage for capacitor filter.

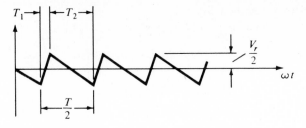

Figure B.2   Ripple voltage.

During capacitor-discharge the voltage change across C is

$$V_r\,(p\text{--}p) = \frac{I_{dc}\,T_2}{C} \tag{B.4}$$

From the triangular waveform in Fig. B.1

$$V_r\,(\text{rms}) = \frac{V_r\,(p\text{--}p)}{2\sqrt{3}} \tag{B.5}$$

(obtained by calculations, not shown).

Using the waveform details of Fig. B.1 results in

$$\frac{V_r\,(p\text{--}p)}{T_1} = \frac{V_m}{T/4}$$

$$T_1 = \frac{V_r\,(p\text{--}p)(T/4)}{V_m}$$

Also,

$$T_2 = \frac{T}{2} - T_1 = \frac{T}{2} - \frac{V_r\,(p\text{--}p)(T/4)}{V_m} = \frac{2\,T V_m - V_r\,(p\text{--}p)T}{4\,V_m}$$

$$T_2 = \frac{2\,V_m - V_r\,(p\text{--}p)}{V_m}\frac{T}{4} \tag{B.6}$$

Since Eq. (B.3) can be written as

$$V_{dc} = \frac{2\,V_m - V_r\,(p\text{--}p)}{2}$$

we can combine the last equation with B.6

$$T_2 = \frac{V_{dc}}{V_m}\frac{T}{2}$$

which, inserted into Eq. (B.4), gives

$$V_r\,(p\text{--}p) = \frac{I_{dc}}{C}\left(\frac{V_{dc}}{V_m}\frac{T}{2}\right)$$

$$T = \frac{1}{f}$$

RIPPLE FACTOR AND VOLTAGE CALCULATIONS

$$V_r\,(p-p) = \frac{I_{\text{dc}}}{2\,fC}\frac{V_{\text{dc}}}{V_m} \tag{B.7}$$

Combining Eqs. (B.5) and (B.7), we solve for V_r (rms)

$$V_r\,(\text{rms}) = \frac{V_r\,(p-p)}{2\sqrt{3}} = \frac{I_{\text{dc}}}{4\sqrt{3}\,fC}\frac{V_{\text{dc}}}{V_m} \tag{B.8}$$

B.3 RELATION OF V_{dc} AND V_m TO RIPPLE, r

The dc voltage developed across a filter capacitor from a transformer providing a peak voltage, V_m, can be related to the ripple as follows:

$$r = \frac{V_r\,(\text{rms})}{V_{\text{dc}}} = \frac{V_r\,(p-p)}{2\sqrt{3}\,V_{\text{dc}}}$$

$$V_{\text{dc}} = \frac{V_r\,(p-p)}{2\sqrt{3}\,r} = \frac{V_r\,(p-p)/2}{\sqrt{3}\,r} = \frac{V_r\,(p)}{\sqrt{3}\,r} = \frac{V_m - V_{\text{dc}}}{\sqrt{3}\,r}$$

$$V_m - V_{\text{dc}} = \sqrt{3}\,r V_{\text{dc}}$$

$$V_m = (1 + \sqrt{3}\,r)\,V_{\text{dc}}$$

$$\boxed{\frac{V_m}{V_{\text{dc}}} = 1 + \sqrt{3}\,r} \tag{B.9}$$

The relation of Eq. (B.9) applies to both half- and full-wave rectifier-capacitor filter circuits and is plotted in Fig. B.3. As example, at a ripple of 5% the dc voltage is $V_{\text{dc}} = 0.92\,V_m$, or within 10% of the peak voltage, where as at 20% ripple the dc voltage drops to only $0.74\,V_m$ which is more than 25% less than the peak value. Note that V_{dc} is within 10% of V_m for ripple less than 6.5%. This amount of ripple represents the borderline of the light-load condition.

B.4 RELATION OF V_r (RMS) AND V_m TO RIPPLE, r

We can also obtain a relation between $V_r(\text{rms})$, V_m, and the amount of ripple for both half-wave and full-wave rectifier-capacitor filter circuits as follows:

$$\frac{V_r\,(p-p)}{2} = V_m - V_{\text{dc}}$$

$$\frac{V_r\,(p-p)/2}{V_m} = \frac{V_m - V_{\text{dc}}}{V_m} = 1 - \frac{V_{\text{dc}}}{V_m}$$

$$\frac{\sqrt{3}\,V_r\,(\text{rms})}{V_m} = 1 - \frac{V_{\text{dc}}}{V_m}$$

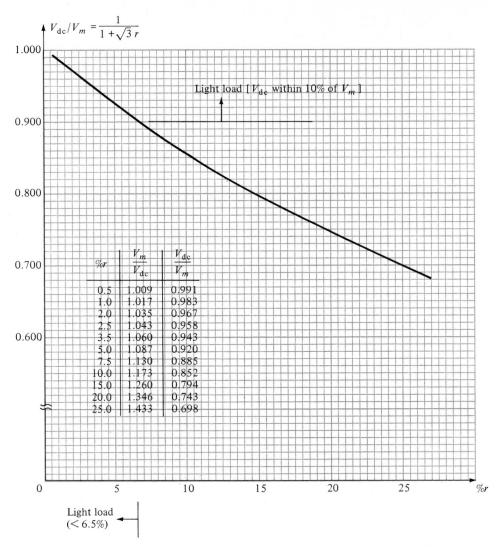

The figure shows a plot with the y-axis labeled $V_{dc}/V_m = \dfrac{1}{1+\sqrt{3}\,r}$ and the x-axis labeled $\%r$.

Annotations on the plot: "Light load [V_{dc} within 10% of V_m]"

Embedded table within the figure:

$\%r$	$\dfrac{V_m}{V_{dc}}$	$\dfrac{V_{dc}}{V_m}$
0.5	1.009	0.991
1.0	1.017	0.983
2.0	1.035	0.967
2.5	1.043	0.958
3.5	1.060	0.943
5.0	1.087	0.920
7.5	1.130	0.885
10.0	1.173	0.852
15.0	1.260	0.794
20.0	1.346	0.743
25.0	1.433	0.698

Light load (< 6.5%)

Figure B.3 Plot of (V_{dc}/V_m) as a function of $\%r$.

Using Eq. (B.9), we get

$$\frac{\sqrt{3}\,V_r\,(\text{rms})}{V_m} = 1 - \frac{1}{1+\sqrt{3}\,r}$$

$$\frac{V_r\,(\text{rms})}{V_m} = \frac{1}{\sqrt{3}}\left(1 - \frac{1}{1+\sqrt{3}r}\right) = \frac{1}{\sqrt{3}}\left(\frac{1+\sqrt{3}r-1}{1+\sqrt{3}r}\right)$$

$$\boxed{\frac{V_r\,(\text{rms})}{V_m} = \frac{r}{1+\sqrt{3}r}} \qquad (\text{B.10})$$

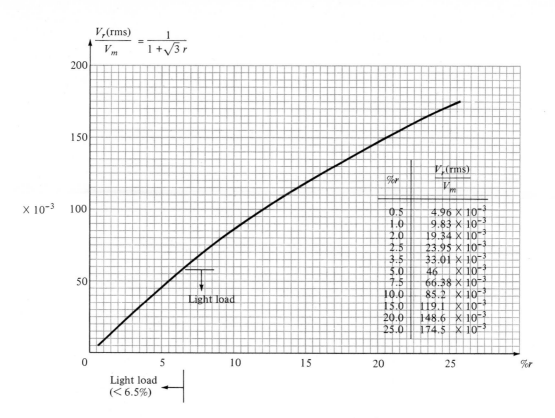

Figure B.4   Plot of $V_r(\text{rms})/V_m$ as a function of $\%r$.

Equation (B.10) is plotted in Fig. B.4.

Since V_{dc} is within 10% of V_m for ripple $\leq 6.5\%$,

$$\frac{V_r\,(\text{rms})}{V_m} \cong \frac{V_r\,(\text{rms})}{V_{\text{dc}}} = r \qquad \text{light load}$$

and we can use $V_r\,(\text{rms})/V_m = r$ for ripple $\leq 6.5\%$.

B.5   RELATION BETWEEN CONDUCTION ANGLE, % RIPPLE, AND $I_{\text{peak}}/I_{\text{dc}}$ FOR RECTIFIER-CAPACITOR FILTER CIRCUITS

In Fig. B.1 we can determine the angle at which the diode starts to conduct, θ, as follows: Since

$$v = V_m \sin \theta = V_m - V_r\,(p\text{--}p) \quad \text{at} \quad \theta = \theta_1$$

$$\theta_1 = \sin^{-1}\left[1 - \frac{V_r\,(p\text{--}p)}{V_m}\right]$$

Using Eq. (B.10) and V_r (rms) $= V_r (p\text{–}p)/2\sqrt{3}$ gives

$$\frac{V_r (p\text{–}p)}{V_m} = \frac{2\sqrt{3}\, V_r \,(\text{rms})}{V_m}$$

so that

$$1 - \frac{V_r (p\text{–}p)}{V_m} = 1 - \frac{2\sqrt{3}\, V_r \,(\text{rms})}{V_m} = 1 - 2\sqrt{3}\left(\frac{r}{1+\sqrt{3}r}\right)$$

$$= \frac{1 - \sqrt{3}r}{1 + \sqrt{3}r}$$

and

$$\boxed{\theta_1 = \sin^{-1}\frac{1 - \sqrt{3}r}{1 + \sqrt{3}r}} \qquad (B.11)$$

where θ_1 is the angle at which conduction starts.

When the current becomes zero after charging the parallel impedances R_L and C, we can determine that

$$\theta_2 = \pi - \tan^{-1} \omega R_L C$$

An expression for $\omega R_L C$ can be obtained as follows:

$$r = \frac{V_r\,(\text{rms})}{V_{dc}} = \frac{\dfrac{I_{dc}}{4\sqrt{3}fC}}{\dfrac{V_{dc}}{V_m}} \cdot \frac{V_{dc}}{V_m} = \frac{V_{dc}/R_L}{4\sqrt{3}fC} \cdot \frac{1}{V_m}$$

$$= \frac{V_{dc}/V_m}{4\sqrt{3}fCR_L} = \frac{2\pi\left(\dfrac{1}{1+\sqrt{3}r}\right)}{4\sqrt{3}\omega CR_L}$$

so that

$$\omega R_L C = \frac{2\pi}{4\sqrt{3}(1+\sqrt{3}r)r} = \frac{0.907}{r(1+\sqrt{3}r)}$$

Thus, conduction stops at an angle

$$\boxed{\theta_2 = \pi - \tan^{-1}\frac{0.907}{(1+\sqrt{3}r)r}} \qquad (B.12)$$

From Eq. (14.10b) we can write

$$\frac{I_{peak}}{I_{dc}} = \frac{I_p}{I_{dc}} = \frac{T}{T_1} = \frac{180°}{\theta} \qquad \text{full-wave} \qquad (B.13a)$$

$$= \frac{360°}{\theta} \qquad \text{full-wave}$$

A plot of I_p/I_{dc} as a function of ripple is provided in Fig. B.5 for both half- and full-wave operation.

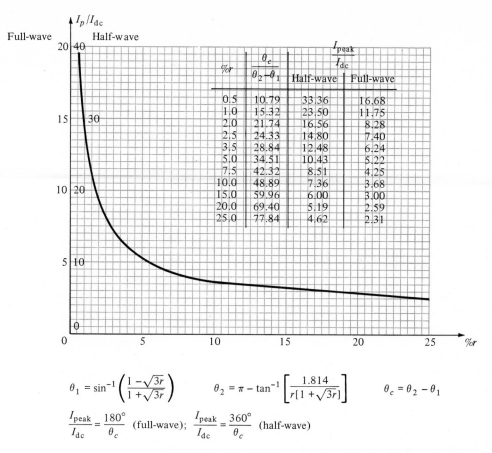

%r	$\dfrac{\theta_c}{\theta_2 - \theta_1}$	$\dfrac{I_{peak}}{I_{dc}}$ Half-wave	Full-wave
0.5	10.79	33.36	16.68
1.0	15.32	23.50	11.75
2.0	21.74	16.56	8.28
2.5	24.33	14.80	7.40
3.5	28.84	12.48	6.24
5.0	34.51	10.43	5.22
7.5	42.32	8.51	4.25
10.0	48.89	7.36	3.68
15.0	59.96	6.00	3.00
20.0	69.40	5.19	2.59
25.0	77.84	4.62	2.31

$$\theta_1 = \sin^{-1}\left(\frac{1 - \sqrt{3r}}{1 + \sqrt{3r}}\right) \qquad \theta_2 = \pi - \tan^{-1}\left[\frac{1.814}{r[1 + \sqrt{3r}]}\right] \qquad \theta_c = \theta_2 - \theta_1$$

$$\frac{I_{peak}}{I_{dc}} = \frac{180°}{\theta_c} \ \text{(full-wave)}; \quad \frac{I_{peak}}{I_{dc}} = \frac{360°}{\theta_c} \ \text{(half-wave)}$$

Figure B.5 Plot of I_p/I_{dc} versus %r, half- and full-wave operation.

APPENDIX C

Charts and Tables

TABLE C.1 Greek Alphabet and Common Designations

Name	Capital	Lowercase	Used to Designate
alpha	A	α	Angles, area, coefficients
beta	B	β	Angles, flux density, coefficients
gamma	Γ	γ	Conductivity, specific gravity
delta	Δ	δ	Variation, density
epsilon	E	ϵ	Base of natural logarithms
zeta	Z	ζ	Impedance, coefficients, coordinates
eta	H	η	Hysteresis coefficient, efficiency
theta	Θ	θ	Temperature, phase angle
iota	I	ι	
kappa	K	κ	Dielectric constant, susceptibility
lambda	Λ	λ	Wave length
mu	M	μ	Micro, amplification factor, permeability
nu	N	ν	Reluctivity
xi	Ξ	ξ	
omicron	O	o	
pi	Π	π	Ratio of circumference to diameter $= 3.1416$
rho	P	ρ	Resistivity
sigma	Σ	σ	Sign of summation
tau	T	τ	Time constant, time phase displacement
upsilon	Υ	υ	
phi	Φ	ϕ	Magnetic flux, angles
chi	X	χ	
psi	Ψ	ψ	Dielectric flux, phase difference
omega	Ω	ω	Capital: ohms; lower case: angular velocity

Logarithms

Formulas: $\log ab = \log a + \log b$

$$\log \frac{a}{b} = \log a - \log b$$

$$\log a^n = n \log a$$

TABLE C.2 Common Logarithms

No.	0	1	2	3	4	5	6	7	8	9
0	....	0000	3010	4771	6021	6990	7782	8451	9031	9542
1	0000	0414	0792	1139	1461	1761	2041	2304	2553	2788
2	3010	3222	3424	3617	3802	3979	4150	4314	4472	4624
3	4771	4914	5051	5185	5315	5441	5563	5682	5798	5911
4	6021	6128	6232	6335	6435	6532	6628	6721	6812	6902
5	6990	7076	7160	7243	7324	7404	7482	7559	7634	7709
6	7782	7853	7924	7993	8062	8129	8195	8261	8325	8388
7	8451	8513	8573	8633	8692	8751	8808	8865	8921	8976
8	9031	9085	9138	9191	9243	9294	9345	9395	9445	9494
9	9542	9590	9638	9685	9731	9777	9823	9868	9912	9956
10	0000	0043	0086	0128	0170	0212	0253	0294	0334	0374
11	0414	0453	0492	0531	0569	0607	0645	0682	0719	0755
12	0792	0828	0864	0899	0934	0969	1004	1038	1072	1106
13	1139	1173	1206	1239	1271	1303	1335	1367	1399	1430
14	1461	1492	1523	1553	1584	1614	1644	1673	1703	1732
15	1761	1790	1818	1847	1875	1903	1931	1959	1987	2014
16	2041	2068	2095	2122	2148	2175	2201	2227	2253	2279
17	2304	2330	2355	2380	2405	2430	2455	2480	2504	2529
18	2553	2577	2601	2625	2648	2672	2695	2718	2742	2765
19	2788	2810	2833	2856	2878	2900	2923	2945	2967	2989
20	3010	3032	3054	3075	3096	3118	3139	3160	3181	3201
21	3222	3243	3263	3284	3304	3324	3345	3365	3385	3404
22	3424	3444	3464	3483	3502	3522	3541	3560	3579	3598
23	3617	3636	3655	3674	3692	3711	3729	3747	3766	3784
24	3802	3820	3838	3856	3874	3892	3909	3927	3945	3962
25	3979	3997	4014	4031	4048	4065	4082	4099	4416	4133
26	4150	4166	4183	4200	4216	4232	4249	4265	4281	4298
27	4314	4330	4346	4362	4378	4393	4409	4425	4440	4456
28	4472	4487	4502	4518	4533	4548	4564	4579	4594	4609
29	4624	4639	4654	4669	4683	4698	4713	4728	4742	4757
30	4771	4786	4800	4814	4829	4843	4857	4871	4886	4900
31	4914	4928	4942	4955	4969	4983	4997	5011	5024	5038
32	5051	5065	5079	5092	5105	5119	5132	5145	5159	5172
33	5185	5198	5211	5224	5237	5250	5263	5276	5289	5302
34	5315	5328	5340	5353	5366	5378	5391	5403	5416	5428
35	5441	5453	5465	5478	5490	5502	5514	5527	5539	5551
36	5563	5575	5587	5599	5611	5623	5635	5647	5658	5670
37	5682	5694	5705	5717	5729	5740	5752	5763	5775	5786
38	5798	5809	5821	5832	5843	5855	5866	5877	5888	5899
39	5911	5922	5933	5944	5955	5966	5977	5988	5999	6010
No.	0	1	2	3	4	5	6	7	8	9

TABLE C.2 (Continued)

No.	0	1	2	3	4	5	6	7	8	9
40	6021	6031	6042	6053	6064	6075	6085	6096	6107	6117
41	6128	6138	6149	6160	6170	6180	6191	6201	6212	6222
42	6232	6243	6253	6263	6274	6284	6294	6304	6314	6325
43	6335	6345	6355	6365	6375	6385	6395	6405	6415	6425
44	6435	6444	6454	6464	6474	6494	6493	6503	6513	6522
45	6532	6542	6551	6561	6571	6580	6590	6599	6609	6618
46	6628	6637	6646	6656	6665	6675	6684	6693	6702	6712
47	6721	6730	6739	6749	6758	6767	6776	6785	6794	6803
48	6812	6821	6830	6839	6848	6857	6866	6875	6884	6893
49	6902	6911	6920	6928	6937	6946	6955	6964	6972	6981
50	6990	6998	7007	7016	7024	7033	7042	7050	7059	7067
51	7076	7084	7093	7101	7110	7118	7126	7135	7143	7152
52	7160	7168	7177	7185	7193	7202	7210	7218	7226	7235
53	7243	7251	7259	7267	7275	7284	7292	7300	7308	7316
54	7324	7332	7340	7348	7356	7364	7372	7380	7388	7396
55	7404	7412	7419	7427	7435	7443	7451	7459	7466	7474
56	7482	7490	7497	7505	7513	7520	7528	7536	7543	7551
57	7559	7566	7574	7582	7589	7597	7604	7612	7619	7627
58	7634	7642	7649	7657	7664	7672	7679	7686	7694	7701
59	7709	7716	7723	7731	7738	7745	7752	7760	7767	7774
60	7782	7789	7796	7803	7810	7818	7825	7832	7839	7846
61	7853	7860	7868	7875	7882	7889	7895	7903	7910	7917
62	7924	7931	7938	7945	7952	7959	7966	7973	7980	7987
63	7993	8000	8007	8014	8021	8028	8035	8041	8048	8055
64	8062	8069	8075	8082	8089	8096	8102	8109	8116	8122
65	8129	8136	8142	8149	8156	8162	8169	8176	8182	8189
66	8195	8202	8209	8215	8222	8228	8235	8241	8248	8254
67	8261	8267	8274	8280	8287	8293	8299	8306	8312	8319
68	8325	8331	8338	8344	8351	8357	8363	8370	8376	8382
69	8388	8395	8401	8407	8414	8420	8426	8432	8439	8445
70	8451	8457	8463	8470	8476	8482	8488	8494	8500	8506
71	8513	8519	8525	8531	8537	8543	8549	8555	8561	8567
72	8573	8579	8585	8591	8597	8603	8609	8615	8621	8627
73	8633	8639	8645	8651	8657	8663	8669	8675	8681	8686
74	8692	8698	8704	8710	8716	8722	8727	8733	8739	8745
75	8751	8756	8762	8768	8774	8779	8785	8791	8797	8802
76	8808	8814	8820	8825	8831	8837	8842	8848	8854	8859
77	8865	8871	8876	8882	8887	8893	8899	8904	8910	8915
78	8921	8927	8932	8938	8943	8949	8954	8960	8965	8971
79	8976	8982	8987	8993	8998	9004	9009	9015	9020	9025
80	9031	9036	9042	9047	9053	9058	9063	9069	9074	9079
81	9085	9090	9096	9101	9106	9112	9117	9122	9128	9133
82	9138	9143	9149	9154	9159	9165	9170	9175	9180	9186
83	9191	9196	9201	9206	9212	9217	9222	9227	9232	9238
84	9243	9248	9253	9258	9263	9269	9274	9279	9284	9289
No.	0	1	2	3	4	5	6	7	8	9

TABLE C.2 (Continued)

No.	0	1	2	3	4	5	6	7	8	9
85	9294	9299	9304	9309	9315	9320	9235	9330	9335	9340
86	9345	9350	9355	9360	9365	9370	9375	9380	9385	9390
87	9395	9400	9405	9410	9415	9420	9425	9430	9435	9440
88	9445	9450	9455	9460	9465	9469	9474	9479	9484	9489
89	9494	9499	9504	9509	9513	9518	9523	9528	9533	9538
90	9542	9547	9552	9557	9562	9566	9571	9576	9581	9586
91	9590	9595	9600	9605	9609	9614	9619	9624	9628	9633
92	9638	9643	9647	9652	9657	9661	9666	9671	9675	9680
93	9685	9689	9694	9699	9703	9708	9713	9717	9722	9727
94	9731	9736	9741	9745	9750	9754	9759	9763	9768	9773
95	9777	9782	9786	9791	9795	9800	9805	9809	9814	9818
96	9823	9827	9832	9836	9841	9845	9850	9854	9859	9863
97	9868	9872	9877	9881	9886	9890	9894	9899	9903	9908
98	9912	9917	9921	9926	9930	9934	9939	9943	9948	9952
99	9956	9961	9965	9969	9974	9978	9983	9987	9991	9996
100	0000	0004	0009	0013	0017	0022	0026	0030	0035	0039
No.	0	1	2	3	4	5	6	7	8	9

APPENDIX D

Computer Programs

The purpose of this appendix is to introduce the use of the computer in studying or solving electronic problems. Programs written in the BASIC language are used in various chapters to carry out the multistep solution of part or all of a given circuit. Since the BASIC language is available to the majority of small computer units and on larger computer systems, it is expected that a computer capable of running programs described in this text is readily available. Even the latest portable hand-held computer (HHC) units can run multistep BASIC programs similar to those described in this appendix. Since engineering study is so heavily involved with computer-aided solution, it is essential that the students begin to rely on the use of computers to solve problems as early in their studies as possible.

Although computer solution to problems in almost all chapters of this text could be provided, only a selection from Chapters 4, 6, 7, and 10 are provided. A listing of selected programs referenced to a particular chapter is provided with results using examples from that chapter for comparison.

```
10 PRINT
20 PRINT "THIS PROGRAM CALCULATES THE DC BIAS FOR "
30 PRINT "A VOLTAGE-DIVIDER BIAS CIRCUIT"
40 PRINT "AND FOR AN EMITTER-STABILIZED BIAS CIRCUIT"
50 PRINT "USE RB2= 1E30 FOR RB2=INFINITY)"
60 PRINT
70 PRINT "INPUT THE FOLLOWING CIRCUIT VALUES:"
80 INPUT "RB1=";R1:INPUT "RB2=";R2
90 INPUT "RE=";RE
100 INPUT "RC=";RC
110 INPUT "VCC=";SV
120 INPUT "AND TRANSISTOR BETA=";B
130 PRINT
140 PRINT "THE RESULTING DC BIAS CURRENTS ARE:"
150 VT=R2*SV/(R1+R2)
```

Figure D.1 Program (a) and run (b) of dc bias (Chapter 4).

```
160 RT=(R2/(R1+R2))*R1
170 IB=(VT-0.7)/(RT+B*RE)
180 IC=B*IB
190 IE=(B+1)*IB
200 PRINT "IB=";IB,"IC=";IC;" IE =";IE
210 PRINT
220 PRINT "AND THE DC VOLTAGES ARE:"
230 VE=IE*RE
240 VB = VE+0.7
250 VC = SV - IC*RC
260 PRINT "VB= ";VB;" VE= ";VE;" VC= ";VC
270 PRINT "and VCE= ";VC-VE
280 END
```

(a)

```
RUN

THIS PROGRAM CALCULATES THE DC BIAS FOR
A VOLTAGE-DIVIDER BIAS CIRCUIT
AND FOR AN EMITTER-STABILIZED BIAS CIRCUIT
USE RB2= 1E30 FOR RB2=INFINITY)

INPUT THE FOLLOWING CIRCUIT VALUES:
RB1=? 40E3
RB2=? 4E3                               } data input
RE=? 1.5E3
RC=? 10E3
VCC=? 22
AND TRANSISTOR BETA=? 140

THE RESULTING DC BIAS CURRENTS ARE:
IB= 6.09E-06   IC= 8.52E-04  IE = 8.58E-04    } computed
                                                results
AND THE DC VOLTAGES ARE:
VB=  1.99  VE=  1.29  VC= 13.48
and VCE=  12.19
```

(b)

```
10 DIGITS 6,2
20 PRINT "THIS IS A GENERAL PROGRAM TO CALCULATE"
30 PRINT "THE DC BIAS FOR A VARIETY OF JFET CIRCUIT CONFIGURATIONS"
40 PRINT
50 PRINT"ENTER THE FOLLOWING CIRCUIT VALUES"
60 INPUT"RD = ";RD
70 INPUT"RS1(UNBYPASSED SOURCE RESISTANCE)=";S1:INPUT"RS2=";S2:RS=S1+S2
80 INPUT "GATE RESISTANCE TO GROUND, RG2=";G2
90 INPUT"GATE RESISTANCE TO SUPPLY (USE 1E30 IF OPEN), RG1=";G1
100 INPUT "GATE SUPPLY VOLTAGE, VGG=";G6
110 INPUT "SUPPLY VOLTAGE, VDD= ";SV
120 PRINT "NOW FOR THE JFET DEVICE PARAMETERS"
130 INPUT "GATE-SOURCE OFF VOLTAGE, VGS(OFF) = ";VP
140 INPUT "SATURATION CURRENT, IDSS = ";ID
150 PRINT
160 PRINT "RESULTS OF BIAS CALCULATIONS ARE:"
170 VG=G6+(G2/(G1+G2))*SV
180 IF RS=0 THEN VQ=VG:GOTO 260
190 A=1:B=VP^2/(ID*RS) - 2*VP
200 C=VP^2*(1-VG/(ID*RS))
210 D=B^2-4*A*C
220 IF D>=0 THEN 230 ELSE PRINT "NO REAL SOLUTION":STOP
230 V1=(-B+SQR(B^2-4*A*C))/(2*A)
240 V2=(-B-SQR(B^2-4*A*C))/(2*A)
250 IF V1<0 THEN VQ=V1 ELSE VQ=V2
260 PRINT"BIAS GATE-SOURCE VOLTAGE IS, VGSQ = ";VQ
270 IQ=ID*(1-VQ/VP)^2
280 VS=IQ*RS
290 PRINT"B'1S DRAIN-SOURCE CURRENT,IDQ = ";IQ
300 VD=SV-IQ*RD
310 PRINT "DRAIN VOLTAGE TO GROUND IS ";VD
320 PRINT "SOURCE VOLTAGE TO GROUND IS ";VS
330 PRINT "AND DRAIN-SOURCE VOLTAGE IS ";VD-VS
340 END
```

(a)

RUN   (see Example 6.3, p. 212)

THIS IS A GENERAL PROGRAM TO CALCULATE
THE DC BIAS FOR A VARIETY OF JFET CIRCUIT CONFIGURATIONS

ENTER THE FOLLOWING CIRCUIT VALUES
RD = ? 4.1E3
RS1(UNBYPASSED SOURCE RESISTANCE)=? 0
RS2=? 2.2E3
GATE RESISTANCE TO GROUND, RG2=? 1.8E6
GATE RESISTANCE TO SUPPLY (USE 1E30 IF OPEN), RG1=? 1E30
GATE SUPPLY VOLTAGE, VGG=? 0
SUPPLY VOLTAGE, VDD= ? 16
NOW FOR THE JFET DEVICE PARAMETERS
GATE-SOURCE OFF VOLTAGE, VGS(OFF) = ? -6
SATURATION CURRENT, IDSS = ? 5E-3

RESULTS OF BIAS CALCULATIONS ARE:
BIAS GATE-SOURCE VOLTAGE IS, VGSQ = -2.91
BIAS DRAIN-SOURCE CURRENT,IDQ = 1.32E-03
DRAIN VOLTAGE TO GROUND IS 10.57
SOURCE VOLTAGE TO GROUND IS 2.91
AND DRAIN-SOURCE VOLTAGE IS 7.66

READY

RUN   (see Example 6.4, pp. 213-14)

THIS IS A GENERAL PROGRAM TO CALCULATE
THE DC BIAS FOR A VARIETY OF JFET CIRCUIT CONFIGURATIONS

ENTER THE FOLLOWING CIRCUIT VALUES
RD = ? 2.5E3
RS1(UNBYPASSED SOURCE RESISTANCE)=? 0
RS2=? 1.5E3
GATE RESISTANCE TO GROUND, RG2=? 280E3
GATE RESISTANCE TO SUPPLY (USE 1E30 IF OPEN), RG1=? 2E6
GATE SUPPLY VOLTAGE, VGG=? 0
SUPPLY VOLTAGE, VDD= ? 16
NOW FOR THE JFET DEVICE PARAMETERS
GATE-SOURCE OFF VOLTAGE, VGS(OFF) = ? -4
SATURATION CURRENT, IDSS = ? 8E-3

RESULTS OF BIAS CALCULATIONS ARE:
BIAS GATE-SOURCE VOLTAGE IS, VGSQ = -1.77
BIAS DRAIN-SOURCE CURRENT,IDQ = 2.49E-03
DRAIN VOLTAGE TO GROUND IS 9.78
SOURCE VOLTAGE TO GROUND IS 3.73
AND DRAIN-SOURCE VOLTAGE IS 6.04

(b)

Figure D.2(b) Runs of JFET dc bias (Chapter 6).

Figure D.3 Program (a) and run (b) for BJT ac analysis (Chapter 7).

(a)

```
 10 PRINT "THIS PROGRAM DOES AC CALCULATIONS USING THE COMPLETE "
 20 PRINT "HYBRID EQUIVALENT CIRCUIT MODEL"
 30 PRINT
 40 PRINT"ENTER THE FOLLOWING CIRCUIT COMPONENT VALUES:"
 50 INPUT "RB1=";R1
 60 INPUT "RB2= (USE 1E30 IF INFINITE) ";R2
 70 INPUT "RC=";RC
 80 INPUT "RE= (UNBYPASSED VALUE ONLY) ";RE
 90 INPUT "LOAD RESISTANCE, RL= (USE 1E30 IF OUTPUT OPEN) ";RL
100 PRINT
110 PRINT "NOW ENTER VALUES FOR TRANSISTOR HYBRID EQUIVALENT CIRCUIT:"
120 INPUT "hie=";HI
130 INPUT "hfe=";HF
140 INPUT "hoe=";HO
150 INPUT "hre=";HR
```

```
160 PRINT
170 REM AC CIRCUIT CALCULATIONS FOLLOW
180 R3=R1*(R2/(R1+R2))
190 RP=RC*(RL/(RC+RL))
200 D=1+HO*RP
210 Z1=HI-HF*HR*RP/D
220 AI=(R3/(R3+Z1))*(HF/D)
230 AV=-HF*RP/(HI+(HI*HO-HF*HR)*RP)
240 ZI=R3*(Z1/(R3+Z1))
250 Y2=HO-HF*HR/(HI+RS)
260 IF Y2=0 THEN Z2=1E30 ELSE Z2=1/Y2
270 ZO=RC*(Z2/(RC+Z2))
280 AP=ABS(AV*AI)
290 PRINT "RESULTS OF AC CIRCUIT ANALYSIS ARE:"
300 PRINT "CURRENT GAIN, Ai= ";AI
310 PRINT "VOLTAGE GAIN, Av= ";AV
320 PRINT "INPUT IMPEDANCE, Zi= ";ZI;" ohms"
330 PRINT "OUTPUT IMPEDANCE, Zo= ";ZO;" ohms"
340 PRINT "POWER GAIN, Ap= ";AP
350 END
```

(b)

```
THIS PROGRAM DOES AC CALCULATIONS USING THE COMPLETE
HYBRID EQUIVALENT CIRCUIT MODEL

ENTER THE FOLLOWING CIRCUIT COMPONENT VALUES:
RB1=? 50E3
RB2= (USE 1E30 IF INFINITE) ? 5E3
RC=? 5E3
RE= (UNBYPASSED VALUE ONLY) ? 0
LOAD RESISTANCE, RL= (USE 1E30 IF OUTPUT OPEN) ? 1E30

NOW ENTER VALUES FOR TRANSISTOR HYBRID EQUIVALENT CIRCUIT:
hie=? 1.5E3
hfe=? 80
hoe=? 20E-6
hre=? 3E-4

RESULTS OF AC CIRCUIT ANALYSIS ARE:
CURRENT GAIN, Ai= 55.69
VOLTAGE GAIN, Av= -261.44
INPUT IMPEDANCE, Zi= 1065.01  ohms
OUTPUT IMPEDANCE, Zo= 4901.96  ohms
POWER GAIN, Ap= 1.46E+04


READY

RUN

THIS PROGRAM DOES AC CALCULATIONS USING THE COMPLETE
HYBRID EQUIVALENT CIRCUIT MODEL

ENTER THE FOLLOWING CIRCUIT COMPONENT VALUES:
RB1=? 50E3
RB2= (USE 1E30 IF INFINITE) ? 5E3
RC=? 5E3
RE= (UNBYPASSED VALUE ONLY) ? 0
LOAD RESISTANCE, RL= (USE 1E30 IF OUTPUT OPEN) ? 1E30

NOW ENTER VALUES FOR TRANSISTOR HYBRID EQUIVALENT CIRCUIT:
hie=? 1.5E3
hfe=? 80
hoe=? 0
hre=? 0

RESULTS OF AC CIRCUIT ANALYSIS ARE:
CURRENT GAIN, Ai= 60.15
VOLTAGE GAIN, Av= -266.67
INPUT IMPEDANCE, Zi= 1127.82  ohms
OUTPUT IMPEDANCE, Zo= 5000  ohms
POWER GAIN, Ap= 1.6E+04
```

```
10 PRINT"PROGRAM TO CALCULATE DC BIAS AND THEN AC ANALYSIS"
20 PRINT"ENTER THE FOLLOWING CIRCUIT VALUES"
30 INPUT"RD = ";RD
40 INPUT"RS1(UNBYPASSED SOURCE RESISTANCE)=";S1:INPUT"RS2=";S2:RS=S1+S2
50 INPUT "GATE RESISTANCE TO GROUND, RG2=";G2
60 INPUT"GATE RESISTANCE TO SUPPLY (USE 1E30 IF OPEN), RG1=";G1
70 INPUT "GATE SUPPLY VOLTAGE, VGG=";GG
80 INPUT "SUPPLY VOLTAGE, VDD= ";SV
90 PRINT "NOW FOR THE JFET DEVICE PARAMETERS"
100 INPUT "GATE-SOURCE OFF VOLTAGE, VGS(OFF) = ";VP
110 INPUT "SATURATION CURRENT, IDSS = ";ID
120 PRINT
130 PRINT "RESULTS OF BIAS CALCULATIONS ARE:"
140 VG=GG+(G2/(G1+G2))*SV
150 IF RS=0 THEN VQ=VG:GOTO 230
160 A=1:B=VP^2/(ID*RS) - 2*VP
170 C=VP^2*(1-VG/(ID*RS))
180 D=B^2-4*A*C
190 IF D>=0 THEN 200 ELSE PRINT "NO REAL SOLUTION":STOP
200 V1=(-B+SQR(B^2-4*A*C))/(2*A)
210 V2=(-B-SQR(B^2-4*A*C))/(2*A)
220 IF V1<0 THEN VQ=V1 ELSE VQ=V2
230 PRINT"BIAS GATE-SOURCE VOLTAGE IS, VGSQ = ";VQ
240 IQ=ID*(1-VQ/VP)^2
250 VS=IQ*RS
260 PRINT"BIAS DRAIN-SOURCE CURRENT,IDQ = ";IQ
270 VD=SV-IQ*RD
280 PRINT "DRAIN VOLTAGE TO GROUND IS ";VD
290 PRINT "SOURCE VOLTAGE TO GROUND IS ";VS
300 PRINT "AND DRAIN-SOURCE VOLTAGE IS ";VD-VS
310 PRINT
320 INPUT "SOURCE SIGNAL VOLTAGE =";SG
330 INPUT "AND SOURCE GENERATOR RESISTANCE =";RG
340 INPUT "LOAD RESISTANCE (USE 1E30 IF OPEN) =";RL
350 G0=2*ID/ABS(VP)
360 GM=G0*(1-VQ/VP)
370 RM=1/GM
380 PRINT "JFET RESISTANCE IS ";RM
390 AV=-(RD*RL/(RD+RL))/(RM+S1)
400 RI=G1*G2/(G1+G2)
410 RO=RD
420 AI=ABS(AV)*RI/RL
430 VI=RI*SG/(RG+RI)
440 VO=AV*VI
450 PRINT "RESULTS OF AC ANALYSIS:"
460 PRINT "VOLTAGE GAIN,AV=";AV
470 PRINT "INPUT RESISTANCE, Ri=";RI
480 PRINT "OUTPUT RESISTANCE, Ro=";RO
490 PRINT "CURRENT GAIN, Ai=";AI
500 PRINT "AND THE OUTPUT VOLTAGE IS ";VO
510 END
```

Figure D.4 Program for JFET analysis.

Answers to Selected Odd-Numbered Problems

CHAPTER 1

7. 6.4×10^{-19}C. **29.** 80 mV. **31.** 25.14 mA. **35.** (a) 0 V: 3.2 pF, 0.25 V: 9 pF; (b) 23.20 pF/V. **37.** 0.2 V: 5.3 kΩ, -20 V: 442.5 kΩ. **39.** $-75°$C: 1.7 V, 0.1 μA; 25°C: 1.3 V, 0.4 μA; 100°C: 0.98 V, 1.0 μA; 200°C: 0.64 V, 2.4 μA. **41.** 5.0 W. **43.** 1.10°C/W. **45.** 200 mW (forward), 10 μW (reverse). **47.** 50°C: 416.75 mW. **51.** $t_t = t_s = t_{rr/2} = 1.5$ μs. **59.** reduced power levels and increased saturation currents. **61.** 240 mA, 1.4 V. **63.** $V_Q \cong 0.85$ V, $I_Q \cong$ 5.4 mA, $P_D = 4.59$ mW. **65.** 170 Ω. **67.** 149.12 Ω. **69.** 7.2 Ω. **71.** 160 Ω. **73.** $V_R \cong 8.79$ V, $I_D \cong 18.9$ mA, $V_D \cong 1.251$ V, $R_{dc} = 29.17$ Ω. **75.** $V_{R_{max}} = $ 4.3 V, $v_{d_{max}}$ (piecewise-linear model) $= 1.261$ V, $v_{d_{max}}$ (approximate) $= 0.7$ V. **77.** $v_{d(p-p)} = 0.71$ mV, $v_{R(p-p)} = 25.29$ mV. **79.** For all positive values of v_i, $v_o = v_i/2$; for all negative values of v_i, $v_o = v_i$. **81.** For all positive values of v_i, $v_o = v_i$; for all negative values of v_i, $v_o = 0$ V. **83.** For all positive values of $v_i > 5$ V, $v_o = 5$ V; for values of $v_i < 5$ V, $v_o = v_i$. **85.** clamped to 0 and -32 V. **87.** clamped to $+22$ V and -10 V. **91.** 18 V. **95.** $V_{peak} \cong 170$ V, turns ratio: 4.5, diode PIV $= 37.74$ V. **97.** 188.68 V. **101.** 78.62 V.

CHAPTER 2

3. 0.053%/°C. **5.** 13 Ω. **7.** curve: 175°C, $D_F = 3.33$ mW/°C: 195.12°C. **9.** $v_o = v_i$ for $v_i = 10$ sin wt and $v_i = 20$ sin wt. **11.** 100 Ω. **13.** (a) r_d (hot-carrier) $\cong r_d$ ($p-n$ junction) $< r_d$ (point-contact). **15.** 47.5°C, larger temperature

coefficients occur at lower current levels. **17.** $0 \to 2$ V: 33%, $8 \to 10$ V: 5.4%.
19. (a) 27 pF; (b) -8 V: 2 pF/V; -2 V: 9.25 pF/V. **21.** 6.38. **23.** lower levels
of V_r. **27.** 1 MHz: 31.85 kΩ, 100 MHz: 0.3185 kΩ, low-level magnitudes of X_L.
29. 3.97×10^{-9}J, 2.48 eV. **31.** 330 μA. **33.** 42.5 V. **39.** yellow.
41. (a) $\cong 0.77$. **45.** 2.3 V. **47.** (a) 40 mA; (b) 60 mA. **55.** lower levels.
59. 20 kΩ. **61.** 90 Ω.

CHAPTER 3

9. 7.92 mA. **11.** 25. **13.** (a) 4.95 mA; (b) 3 mA; (c) 800 mV.
17. (a) 114.3; (b) 0.99; (c) 300 μA; (d) 2.62 μA. **23.** 0.972.
27. (a) 3.3 mA; (b) 28 V; (c) 25 μA. **29.** (a) case: 17.14 mW/°C;
free-air: 4.57 mW/°C. **31.** (a) 7.5 nA; (b) 1.5 μA; (c) 0.267 nA/°C.

CHAPTER 4

1. $I_C = 1.094$ mA, $V_{CB} = 4.733$ V. **3.** $I_C = 2.49$ mA, $V_{CE} = 3.771$ V. **5.** $I_C = 10.782$ mA, $V_{CE} = 4.523$ V. **7.** $I_C = 1.397$ mA, $V_{CE} = 13.453$ V. **9.** $I_C = 1.395$ mA, $V_C = 2.909$ V. **11.** $I_C = 1.724$ mA, $V_{CE} = 9.161$ V. **13.** $V_E = 10.913$ V. **17.** $R_C = 6$ kΩ, $R_B = 852$ kΩ. **19.** $R_E = 300$ Ω, $R_C = 1.2$ kΩ, $R_B = 657.5$ kΩ. **21.** $R_E = 360$ Ω, $R_C = 1.64$ kΩ, $R_{B_1} = 29$ kΩ, $R_{B_2} = 4.68$ kΩ.
23. $V_B = 3.269$ V. **25.** $V_{CE} = -12.78$ V.

CHAPTER 5

1. (a) $I_D = 3.556$ mA; (b) $I_D = 8$ mA; (c) $I_D = 0.889$ mA. **3.** $V_{GS} = 0.628$ V.
5. $V_{GS(OFF)} = -5$ V. **7.** $V_{GS} = -1.5$ V. **9.** $I_{DSS} = 6.75$ mA. **11.** $g_{mo} = 3.579$ mS. **13.** $g_m = 4.041$ mS. **15.** $V_{GS(OFF)} = -3.988$ V. **17.** $g_m = 2.608$ mS.
19. $g_m = 4.381$ mS. **21.** $g_m = 5.25$ mS. **23.** $I_D = 0.675$ mA. **25.** $g_m = 1.92$ mS.
27. $g_m = 2.683$ mS.

CHAPTER 6

1. $I_D = 6.125$ mA. **3.** $V_{DSQ} = 9.583$ V. **5.** $V_S = 2.3$ V. **7.** $R_S = 800$ Ω.
9. $I_D = 2.32$ mA. **11.** $I_{DQ} = 1.7$ mA. **13.** $R_D = 5.73$ kΩ. **15.** $V_{DSQ} = 7.36$ V.
17. $V_{GSQ} = -2.61$ V. **19.** $I_D = 1.32$ mA. **21.** $V_{GS} = -2.43$ V. **23.** $I_{DQ} = 4.69$ mA. **25.** $R_D = 1050$ Ω. **27.** $V_{DSQ} = 7.8$ V. **29.** $I_{DQ} = 6.02$ mA.
31. $V_{DSQ} = 6.1$ V.

CHAPTER 7

1. (a) $h_{11} = 2$ Ω, $h_{12} = \frac{2}{3}$, $h_{21} = -\frac{2}{3}$, $h_{22} = \frac{4}{9}$ S. **3.** $h_{fe} = 110$, $h_{oe} = 40$ μS.
5. greatest $= h_{re}$, least $= h_{fe}$. **7.** (a) 55.56; (b) -558.67; (c) 397.78 Ω;
(d) 3.91 kΩ; (e) 3.10×10^4. **9.** (a) 79.78; (b) -276.26; (c) 1.126 kΩ;
(d) 3.87 kΩ; (e) 22.04×10^3. **11.** (a) 60; (b) -571.43; (c) 420; (d) 4 kΩ;
(e) 3.43×10^4. **13.** (a) 85.1; (b) -278.9; (c) 1.19 kΩ; (d) 3.9 kΩ;

(e) 23.7×10^3.   **15.** (a) $A_i = -25$, $A_v = -23.4$;   (b) $Z_i \cong 2.35$ kΩ, $Z_o \cong 1.1$ kΩ;
(c) -16.38.   **17.** $Z_i = 123.23$ kΩ, $Z_o \cong 5.6$ kΩ, $A_v = -4.67$.   **19.** $Z_i = 62.95$ kΩ,
$Z_o \cong 18.7$ Ω, $A_v = 0.994$, $A_i = 19$.   **21.** $I_B = 75.3$ μA, $I_E = 4.518$ mA, $r_e = 7$ Ω,
$\beta = 60$.   **23.** $V_B = 2.938$ V, $I_E = 1.865$ mA, $r_e = 15.04$ Ω, $\beta = 200$.   **25.** $V_B =$
5.71 V, $I_E = 1.52$ mA, $r_e = 19.01$ Ω, $\beta = 100$.   **27.** $r_e = 26$ Ω, $A_v = 14.43$, $Z_o =$
12 kΩ, $A_i = -0.25$, $Z_i = 207.88$.   **31.** $A_v = -3.09$, $Z_i = 49.5$ kΩ, $A_i = -22.5$.

CHAPTER 8

1. $A_v = -4.392$.   **3.** $A_v = -5.512$.   **5.** $A_v = -6.961$.   **7.** $V_o = -514.5$ mV,
peak.   **9.** $A_v = -3.372$, $R_i = 1$ MΩ, $R_o = 1.8$ kΩ.   **11.** $A_v = -4.075$.   **13.** $A_v =$
-7.615.   **15.** $V_i = 60.508$ mV, peak.   **17.** $A_v = 0.817$.   **19.** $V_o = 320$ mV,
peak.   **21.** $V_{o1} = -0.179$ V.   **23.** $V_{o2} = 9.857$ mV.   **25.** $A_v = -3.959$.   **27.** $R_i =$
1.5 MΩ, $R_o = 497.748$ Ω.   **29.** $A_v = 4.865$, $R_i = 330$ Ω, $R_o = 1.8$ kΩ.   **31.** $R_D =$
5 kΩ, $R_S = 1.8$ kΩ.   **33.** $R_i = 86.2$ MΩ.

CHAPTER 9

1. (a) $A_{i_T} = 80$, $A_{v_T} = 160$;   (b) $|A_i| = 8.94$, $|A_v| = 12.65$.   **3.** $r_e = 13.33$ Ω, $Z_i =$
1.217 kΩ, $Z_o \cong 3.3$ kΩ, $A_{v_T} = 7662.6$, $A_{i_T} = 4238.81$, $A_{p_T} = 32.5 \times 10^6$.   **5.** Q_1:
$r_e = 11.61$ Ω; Q_2: $r_e = 4.51$ Ω, $Z_i = 13.85$ kΩ, $Z_o = 24.51$ Ω, $A_{v_T} = -2$, $A_{i_T} =$
27.7.   **7.** $r_{e_1} = 14.86$ Ω, $r_{e_2} = 24.53$ Ω, $Z_i \cong 1.19$ kΩ, $Z_o \cong 10$ kΩ, $A_{v_T} = 16.4$,
$A_{i_T} = 39.03$, $A_{p_T} = 640.09$.   **9.** 26.142.   **11.** 20.   **13.** $V_{E_2} = 11$ V, $V_{E_1} = 1.716$ V,
$r_{e_1} = 18.18$ Ω, $r_{e_2} = 2.6$ Ω, on an approximate basis ac gain unaffected.
15. (a) $r_{e_1} = 5.28$ Ω, $r_{e_2} = 6.13$ Ω;   (b) $A_{v_T} = 473.08$, $V_o = 4.73$ V;   (c) $Z_i =$
0.239 kΩ, $Z_o \cong 2.5$ kΩ.   **17.** $Z_i = 0.89$ MΩ, $Z_o \cong 2.2$ kΩ, $A_{i_T} = 1783.38$, $A_{v_T} =$
4.4.   **19.** $Z_i = 0.44$ MΩ, $Z_o \cong 2.2$ kΩ, $A_{i_T} = 2952$, $A_{v_T} = 14.76$.   **21.** 12.
23. (a) $dB_1 = 21.05$, $dB_2 = 42.1$, $dB_3 = 56.835$;   (b) $A_{v_1} = 11.285$, $A_{v_2} = 127.35$,
$A_{v_3} = 691.83$.   **25.** (a) $r_e = 20$ Ω, $f_{L_S} = 85.61$ Hz, $f_{L_C} = 35.78$ Hz;   (b) -2.08;
(c) $f_{H_i} = 9.75$ MHz, $f_{H_o} = 8.52$ MHz.   **27.** 160×10^3.   **29.** 92.17 Hz.   **31.** $\cong -50$.
33. 4.445×10^6.

CHAPTER 10

1. $R_L' = 2.5$ kΩ.   **3.** $N = 44.72/1$.   **5.** $\eta = 37.04\%$.   **7.** (a) $\eta = 35.8\%$;
(b) $\eta = 64.2\%$.   **15.** (a) $P_o = 28.125$ W;   (b) $P_i = 50.643$ W;   (c) $\eta =$
55.536%;   (d) $P_{2Q} = 22.518$ W.   **17.** P_D $(125°C) = 25$ W.   **19.** $P_{D,\text{max}} = 3$ W.

CHAPTER 11

5. (a) yes, for $T = 25°C$, $I_G > 40$ mA;   (b) no;   (c) no, minimum of 3 V;
(d) (6 V, 800 mA): excellent, (4 V, 1.6 A): no.   **11.** (a) $\cong 0.7$ mW/cm^2;
(b) 81.25%.   **17.** typical: $V_o \cong 1$ V, maximum: $V_o \cong 1.75$ V.   **19.** 3 cycles.
21. (b) 0.6 A/°C;   (c) 17.8 A vs. 20 A (graph), 11% difference.

23. (a) 1.08 kΩ; (b) 3.08 kΩ; (c) 13 V; (d) 13.7 V. **27.** (a) 1.14 nA/°C;
(b) no. **29.** (b) 0.4. **31.** $\eta = 0.75$, $V_G = 15$ V. **33.** no, V_D.

CHAPTER 13

5. $V_C = 7.845$ V. **7.** $R_i = 7.8$ kΩ, $R_o = 15$ kΩ. **9.** $V_C = 11.438$ V. **11.** $R_o = 3.496$ MΩ. **13.** CMRR = 51.125 dB. **15.** CMRR = 113.7 dB. **17.** $R_i = 10.714$ kΩ. **21.** $V_o = 0$ V. **23.** $V_o = -12$ V. **25.** $v_o(t) = t$. **27.** $I_o = 1.667$ mA.

CHAPTER 14

1. $r = 0.028$. **3.** $V_r(\text{rms}) = 22.4$ V. **5.** $V_r(\text{rms}) = 1.233$ V. **7.** $V_{dc} = 16.958$ V.
9. $r = 0.96\%$. **11.** $V_{dc} = 12$ V, $V_r(\text{peak}) = 1.663$ V. **13.** $V_{dc} = 22.22$ V, $r = 7.201\%$. **15.** $V_r = 1.6$ V. **17.** $I_L = 73$ mA. **19.** V.R. = 32.01%. **23.** V.R. = 5.263%. **25.** C.R. = 1%. **27.** $R_{S,\text{min}} = 307.692$ Ω. **29.** (a) $V_L = 10.7$ V;
(b) $I_L = 2.675$ mA; (c) $I_{R_s} = 4.65$ mA; (d) $I_Z = 44.61$ μA, $I_C = 2.23$ mA.
33. $V_0 = 10.805$ V. **35.** $r = 5.485\%$. **37.** (a) $V_{dc} = 13.749$ V;
(b) $V_r(\text{peak}) = 1.247$ V.

CHAPTER 15

11. $f = 224.691$ kHz.

CHAPTER 16

11. 2.44 mV resolution. **13.** 2^{12} counts = 4096 count steps. **15.** MARK = 20 mA = -12 V; SPACE = 0 mA = $+12$ V. **19.** $C = 3030$ pF.

CHAPTER 17

1. $A_f = 9.95$. **3.** $A_f = 14.286$, $R_{if} = 31.5$ kΩ, $R_{of} = 2.381$ kΩ. **5.** $A = 33.335$,
$\beta = 0.996$, $A_f = 0.975$. **7.** R_e bypassed: $A = 450$, $R_i = 2$ kΩ, $R_o = 12$ kΩ, R_e in
circuit: $A_f = 0.998$, $R_{if} = 902$ kΩ, $R_{of} = 12$ kΩ. **9.** $C = 1251$ pF, $h_{fe} \geq 44.67$.
11. $f_o = 1.048$ MHz. **13.** $f_o = 159.155$ kHz. **17.** (a) $C_T = 0.15$ μF; (b) $C_T = 1000$ pF.

CHAPTER 18

1. (a) 08 in.; (b) -1.5 in. **3.** 3 cm. **5.** 4 cm, 133.3 V. **7.** (a) $T = 20$ μs/cycle,
1 full cycle; (b) 2 full cycles; (c) $\frac{1}{2}$ cycle. **9.** $V_{\text{peak}} = 2.05$ V, $V_{\text{rms}} = 1.45$ V.
11. 6.5 cm. **13.** 1.8 cm: 36 μs; 3.2 cm: 64 μs, 28 μs. **15.** 4.76 kHz.
17. (a) $V_{p-p} = 300$ mV, $V_{\text{rms}} = 106.05$ V; (b) 60 ms; (c) 15.6 Hz.
19. scale factor = 60°/cm, 90°.

Index

Oscillator (*cont.*)
Barkhausen criterion, 659
Colpitts, 669
crystal, 672
feedback, 658
Hartley, 670
LC-tuned, 664
op-amp, 676
phase shift, 660
relaxation, 446–447, 455–458
tuned-input tuned-output, 667
uni junction, 680
Wien bridge, 678

P

Parallel data registers, 601
Passivation, 485
Pattern generator, 474
Peak diode current, 553
Peak inverse voltage, 22, 92
Persistence, 691–692
Phase control, 427, 441–442
Phase locked loop (PLL), 630
Phase margin, 657
Phase shift, 708–714
Phase-shift oscillator:
    FET, 661
    op-amp, 676
    transistor, 662
Phosphorescence, 688–692
Photoconductive cell, 99–101
Photodiode, 95–98
Photolithographic process, 478–479
Photons, 95–98, 101, 103, 111–112
Photoresist, 478–479
Phototransistors, 447–479
Photovoltaic voltage, 112
pnp transistor, 124–138
pnpn device, 419–442, 452–458
Point-contact transistor, 145
Power:
    diodes, 92
    heat sinking, 412
    large-signal amplifiers, 380
    maximum, 409
    maximum theoretical efficiency, 391
Power supply, 571, 575
Preohmic etching, 482
Programmable UJT, 439–440, 452–458
PROM, 606
Proximity switch, 439–440
Push-pull operation, 399
PUT, 439–440, 452–458

Q

Quasi-complementary amplifier, 407
Quiescent point (*Q*-point), transistor, 180

R

Rate effect, 432
RC-coupled amplifiers, 323–330, 350
Rectification, 15–16, 59–65, 92, 428–429
Reference voltage, 78–82
Regulator:
    thermistor, 564
    transistor, 567
    Zener, 564
Relaxation oscillator, 446–447, 455–458
Resistivity, 2
Resistor, monolithic, 467–468
Reverse breakdown voltage, 424
Reverse recovery time, 24, 85
Reverse saturation current, 18–21, 25, 97, 126, 130, 144–145
Ripple, 547
Rubylith, 471–474

S

Saturation region (transistor), 130, 133, 162, 444
Sawtooth sweep (CRO), 434–435, 694–703
Schmitt trigger, 599
Schottky barrier diode, 82–87, 114
SCR, 419–430
SCS, 430–433
Selenium, 96
Semiconductor diode, 1–6
    acceptor atoms, 8–9
    alloy, 35
    average ac resistance, 47–48
    bulk resistance, 2, 20, 46
    characteristics, 19–22
    contact resistance, 20, 46
    covalent bonding, 3–9
    Czochralski technique, 11–12
    dc conditions, 41–44
    depletion region, 16–18, 23–24
    derating factor, 29–30
    diffusion, 35–36
    diffusion capacitance, 23–24
    donor atoms, 7
    doping, 3, 6, 17
    dynamic resistance, 44–47
    epitaxial growth, 36–37
    equivalent circuit, 48–53
    extrinsic materials, 6–9